Theory and Practice of Pile Foundations

Theory and Practice of Pile Foundations

Wei Dong Guo

CRC Press
Taylor & Francis Group
Boca Raton London New York

CRC Press is an imprint of the
Taylor & Francis Group, an **informa** business

CRC Press
Taylor & Francis Group
6000 Broken Sound Parkway NW, Suite 300
Boca Raton, FL 33487-2742

First issued in paperback 2019

ISBN-13: 978-0-415-80933-7 (hbk)
ISBN-13: 978-0-367-86677-8 (pbk)

Library of Congress Cataloging-in-Publication Data

Guo, Wei Dong.
Theory and practice of pile foundations / Wei Dong Guo.
p. cm.
Includes bibliographical references and index.
ISBN 978-0-415-80933-7 (hardback)
1. Piling (Civil engineering) I. Title.

TA780.G86 2012
624.1'54--dc23 2012032208

Visit the Taylor & Francis Web site at
http://www.taylorandfrancis.com

and the CRC Press Web site at
http://www.crcpress.com

Contents

List of Figures *xv*
Preface *xxxiii*
Author *xxxvii*
List of Symbols *xxxix*

1 Strength and stiffness from in situ tests 1

1.1 *Standard penetration tests (SPT)* 1
1.1.1 *Modification of raw SPT values* 1
1.1.1.1 *Method A* 2
1.1.1.2 *Method B* 3
1.1.2 *Relative density* 3
1.1.3 *Undrained soil strength vs. SPT* N 5
1.1.4 *Friction angle vs. SPT* N, D_r, *and* I_p 6
1.1.5 *Parameters affecting strength* 8
1.2 *Cone penetration tests* 10
1.2.1 *Undrained shear strength* 11
1.2.2 *SPT blow counts using* q_c 11
1.3 *Soil stiffness* 11
1.4 *Stiffness and strength of rock* 13
1.4.1 *Strength of rock* 13
1.4.2 *Shear modulus of rock* 15

2 Capacity of vertically loaded piles 19

2.1 *Introduction* 19
2.2 *Capacity of single piles* 19
2.2.1 *Total stress approach: Piles in clay* 19
2.2.1.1 *α Method* ($\tau_s = \alpha s_u$ *and* q_b) 19
2.2.1.2 *λ Method: Offshore piles* 22

2.2.2 *Effective stress approach* 23
2.2.2.1 *β Method for clay* ($\tau_s = \beta\sigma'_{vs}$) 23
2.2.2.2 *β Method for piles in sand* ($\tau_s = \beta\sigma'_{vs}$) 23
2.2.2.3 *Base resistance* q_b ($= N_q\sigma'_{vb}$) 27
2.2.3 *Empirical methods* 29
2.2.4 *Comments* 30
2.2.5 *Capacity from loading tests* 32
2.3 *Capacity of single piles in rock* 34
2.4 *Negative skin friction* 35
2.5 *Capacity of pile groups* 38
2.5.1 *Piles in clay* 39
2.5.2 *Spacing* 39
2.5.3 *Group interaction (free-standing groups)* 40
2.5.4 *Group capacity and block failure* 41
2.5.4.1 *Free-standing groups* 41
2.5.4.2 *Capped pile groups versus free-standing pile groups* 42
2.5.5 *Comments on group capacity* 44
2.5.6 *Weak underlying layer* 44

3 Mechanism and models for pile–soil interaction 47

3.1 *Concentric cylinder model (CCM)* 47
3.1.1 *Shaft and base models* 47
3.1.2 *Calibration against numerical solutions* 49
3.1.2.1 *Base load transfer factor* 51
3.1.2.2 *Shaft load transfer factor* 52
3.1.2.3 *Accuracy of load transfer approach* 55
3.2 *Nonlinear concentric cylinder model* 59
3.2.1 *Nonlinear load transfer model* 59
3.2.2 *Nonlinear load transfer analysis* 64
3.2.2.1 *Shaft stress-strain nonlinearity effect* 64
3.2.2.2 *Base stress-strain nonlinearity effect* 65
3.3 *Time-dependent CCM* 65
3.3.1 *Nonlinear visco-elastic stress-strain model* 67
3.3.2 *Shaft displacement estimation* 69
3.3.2.1 *Visco-elastic shaft model* 69
3.3.2.2 *Nonlinear creep displacement* 72
3.3.2.3 *Shaft model versus model loading tests* 74
3.3.3 *Base pile–soil interaction model* 77
3.3.4 *GASPILE for vertically loaded piles* 78

3.3.5 Visco-elastic model for reconsolidation 78
3.4 Torque-rotation transfer model 78
3.4.1 Nonhomogeneous soil profile 79
3.4.2 Nonlinear stress-strain response 79
3.4.3 Shaft torque-rotation response 80
3.5 Coupled elastic model for lateral piles 81
3.5.1 Nonaxisymmetric displacement and stress field 82
3.5.2 Short and long piles and load transfer factor 83
3.5.3 Subgrade modulus 87
3.5.4 Modulus k *for rigid piles in sand 90*
3.6 Elastic-plastic model for lateral piles 93
3.6.1 Features of laterally loaded rigid piles 94
3.6.1.1 Critical states 94
3.6.1.2 Loading capacity 96
3.6.2 Generic net limiting force profiles (LFP) (plastic state) 98

4 Vertically loaded single piles 105

4.1 Introduction 105
4.2 Load transfer models 106
4.2.1 Expressions of nonhomogeneity 106
4.2.2 Load transfer models 106
4.3 Overall pile–soil interaction 108
4.3.1 Elastic solution 109
4.3.2 Verification of the elastic theory 110
4.3.3 Elastic-plastic solution 113
4.4 Paramatric study 119
4.4.1 Pile-head stiffness and settlement ratio (Guo and Randolph 1997a) 119
4.4.2 Comparison with existing solutions (Guo and Randolph 1998) 119
4.4.3 Effect of soil profile below pile base (Guo and Randolph 1998) 122
4.5 Load settlement 122
4.5.1 Homogeneous case (Guo and Randolph 1997a) 125
4.5.2 Nonhomogeneous case (Guo and Randolph 1997a) 126
4.6 Settlement influence factor 127
4.7 Summary 131
4.8 Capacity for strain-softening soil 132

4.8.1 *Elastic solution* 132
4.8.2 *Plastic solution* 134
4.8.3 *Load and settlement response* 135
4.9 *Capacity and cyclic amplitude* 142

5 Time-dependent response of vertically loaded piles 147

5.1 *Visco-elastic load transfer behavior* 147
5.1.1 *Model and solutions* 148
5.1.1.1 *Time-dependent load transfer model* 148
5.1.1.2 *Closed-form solutions* 149
5.1.1.3 *Validation* 151
5.1.2 *Effect of loading rate on pile response* 152
5.1.3 *Applications* 152
5.1.4 *Summary* 156
5.2 *Visco-elastic consolidation* 158
5.2.1 *Governing equation for reconsolidations* 159
5.2.1.1 *Visco-elastic stress-strain model* 159
5.2.1.2 *Volumetric stress-strain relation of soil skeleton* 160
5.2.1.3 *Flow of pore water and continuity of volume strain rate* 162
5.2.1.4 *Comments and diffusion equation* 163
5.2.1.5 *Boundary conditions* 163
5.2.2 *General solution to the governing equation* 163
5.2.3 *Consolidation for logarithmic variation of* u_o 165
5.2.4 *Shaft capacity* 168
5.2.5 *Visco-elastic behavior* 169
5.2.6 *Case study* 171
5.2.6.1 *Comments on the current predictions* 175
5.2.7 *Summary* 175

6 Settlement of pile groups 177

6.1 *Introduction* 177
6.2 *Empirical methods* 178
6.3 *Shallow footing analogy* 178
6.4 *Numerical methods* 180
6.4.1 *Boundary element (integral) approach* 180
6.4.2 *Infinite layer approach* 181
6.4.3 *Nonlinear elastic analysis* 182

6.4.4 *Influence of nonhomogeneity* 182
6.4.5 *Analysis based on interaction factors and superposition principle* 183
6.5 *Boundary element approach: GASGROUP* 184
6.5.1 *Response of a pile in a group* 184
6.5.1.1 *Load transfer for a pile* 184
6.5.1.2 *Pile-head stiffness* 185
6.5.1.3 *Interaction factor* 186
6.5.1.4 *Pile group analysis* 187
6.5.2 *Methods of analysis* 187
6.5.3 *Case studies* 192
6.6 *Comments and conclusions* 199

7 Elastic solutions for laterally loaded piles 201

7.1 *Introduction* 201
7.2 *Overall pile response* 202
7.2.1 *Nonaxisymmetric displacement and stress field* 202
7.2.2 *Solutions for laterally loaded piles underpinned by* k *and* N_p 205
7.2.3 *Pile response under various boundary conditions* 208
7.2.4 *Load transfer factor* γ_b 209
7.2.5 *Modulus of subgrade reaction and fictitious tension* 211
7.3 *Validation* 212
7.4 *Parametric study* 212
7.4.1 *Critical pile length* 212
7.4.2 *Short and long piles* 213
7.4.3 *Maximum bending moment and the critical depth* 213
7.4.3.1 *Free-head piles* 213
7.4.3.2 *Fixed-head piles* 216
7.4.4 *Effect of various head and base conditions* 216
7.4.5 *Moment-induced pile response* 218
7.4.6 *Rotation of pile-head* 218
7.5 *Subgrade modulus and design charts* 218
7.6 *Pile group response* 220
7.6.1 *Interaction factor* 220
7.7 *Conclusion* 227

8 Laterally loaded rigid piles 229

8.1 Introduction 229
8.2 Elastic plastic solutions 230
8.2.1 Features of laterally loaded rigid piles 230
8.2.2 Solutions for pre-tip yield state (Gibson p_u*, either* k*)* 232
8.2.2.1 $\bar{H}$, $\bar{u}_g$, $\bar{\omega}$, *and* $\bar{z}_r$ *for Gibson* p_u *and Gibson* k 232
8.2.2.2 $\bar{H}$, $\bar{u}_g$, $\bar{\omega}$, *and* $\bar{z}_r$ *for Gibson* p_u *and constant* k 236
8.2.3 Solutions for pre-tip yield state (constant p_u *and constant* k*)* 237
8.2.4 Solutions for post-tip yield state (Gibson p_u*, either* k*)* 238
8.2.4.1 $\bar{H}$, $\bar{u}_g$, *and* $\bar{z}_r$ *for Gibson* p_u *and Gibson* k 238
8.2.4.2 $\bar{H}$, $\bar{u}_g$, *and* $\bar{z}_r$ *for Gibson* p_u *and constant* k 239
8.2.5 Solutions for post-tip yield state (constant p_u *and constant* k*)* 241
8.2.6 $\bar{u}_g$, ω, *and* p *profiles (Gibson* p_u*, tip-yield state)* 241
8.2.7 $\bar{u}_g$, ω, *and* p *profiles (constant* p_u*, tip-yield state)* 244
8.2.8 Yield at rotation point (YRP, either p_u*)* 245
8.2.9 Maximum bending moment and its depth (Gibson p_u*)* 245
8.2.9.1 Pre-tip yield ($z_o < \bar{z}_*$*) and tip-yield (*$z_o = \bar{z}_*$*) states* 245
8.2.9.2 Yield at rotation point (Gibson p_u*)* 248
8.2.10 Maximum bending moment and its depth (constant p_u*)* 248
8.2.10.1 Pre-tip yield ($z_o < \bar{z}_*$*) and tip-yield (*$z_o = \bar{z}_*$*) states* 248
8.2.10.2 Yield at rotation point (constant p_u*)* 249
8.2.11 Calculation of nonlinear response 250
8.3 Capacity and lateral-moment loading loci 253
8.3.1 Lateral load-moment loci at tip-yield and YRP state 253
8.3.2 Ultimate lateral load H_o *against existing solutions* 255
8.3.3 Lateral–moment loading locus 257

8.3.3.1 *Impact of p_u profile (YRP state) on $\bar{M}_o$ and $\bar{M}_m$* 257
8.3.3.2 *Elastic, tip-yield, and YRP loci for constant p_u* 261
8.3.3.3 *Impact of size and base (pile-tip) resistance* 264
8.4 *Comparison with existing solutions* 266
8.5 *Illustrative examples* 268
8.6 *Summary* 275

9 Laterally loaded free-head piles 277

9.1 *Introduction* 277
9.2 *Solutions for pile–soil system* 279
9.2.1 *Elastic-plastic solutions* 280
9.2.1.1 *Highlights for elastic-plastic response profiles* 280
9.2.1.2 *Critical pile response* 284
9.2.2 *Some extreme cases* 286
9.2.3 *Numerical calculation and back-estimation of LFP* 292
9.3 *Slip depth versus nonlinear response* 296
9.4 *Calculations for typical piles* 296
9.4.1 *Input parameters and use of GASLFP* 296
9.5 *Comments on use of current solutions* 308
9.5.1 *32 Piles in clay* 308
9.5.2 *20 Piles in sand* 313
9.5.3 *Justification of assumptions* 322
9.6 *Response of piles under cyclic loading* 325
9.6.1 *Comparison of* p-y(w) *curves* 325
9.6.2 *Difference in predicted pile response* 327
9.6.3 *Static and cyclic response of piles in calcareous sand* 328
9.7 *Response of free-head groups* 334
9.7.1 *Prediction of response of pile groups (GASLGROUP)* 335
9.8 *Summary* 339

10 Structural nonlinearity and response of rock-socket piles 341

10.1 *Introduction* 341

10.2 *Solutions for laterally loaded shafts* 343
10.2.1 *Effect of loading eccentricity on shaft response* 343
10.3 *Nonlinear structural behavior of shafts* 346
10.3.1 *Cracking moment* M_{cr} *and effective flexural rigidity* E_cI_c 346
10.3.2 M_{ult} *and* I_{cr} *for rectangular and circular cross-sections* 347
10.3.3 *Procedure for analyzing nonlinear shafts* 350
10.3.4 *Modeling structure nonlinearity* 350
10.4 *Nonlinear piles in sand/clay* 352
10.4.1 *Taiwan tests: Cases SN1 and SN2* 352
10.4.2 *Hong Kong tests: Cases SN3 and SN4* 356
10.4.3 *Japan tests: CN1 and CN2* 359
10.5 *Rock-socketed shafts* 361
10.5.1 *Comments on nonlinear piles and rock-socketed shafts* 371
10.6 *Conclusion* 372

11 Laterally loaded pile groups 375

11.1 *Introduction* 375
11.2 *Overall solutions for a single pile* 376
11.3 *Nonlinear response of single piles and pile groups* 379
11.3.1 *Single piles* 379
11.3.2 *Group piles* 381
11.4 *Examples* 386
11.5 *Conclusions* 401

12 Design of passive piles 405

12.1 *Introduction* 405
12.1.1 *Flexible piles* 406
12.1.2 *Rigid piles* 407
12.1.3 *Modes of interaction* 408
12.2 *Mechanism for passive pile–soil interaction* 409
12.2.1 *Load transfer model* 409
12.2.2 *Development of on-pile force* p *profile* 410
12.2.3 *Deformation features* 412
12.3 *Elastic-plastic (EP) solutions* 414
12.3.1 *Normal sliding (upper rigid–lower flexible)* 414
12.3.2 *Plastic (sliding layer)–elastic-plastic (stable layer) (P-EP) solution* 415

12.3.3 *EP solutions for stable layer* 417
12.4 p_u*-based solutions (rigid piles)* 421
12.5 *E-E, EP-EP solutions (deep sliding–flexible piles)* 430
12.5.1 *EP-EP solutions (deep sliding)* 430
12.5.2 *Elastic (sliding layer)–elastic (stable layer) (E-E) solution* 430
12.6 *Design charts* 433
12.7 *Case study* 435
12.7.1 *Summary of example study* 444
12.8 *Conclusion* 444

13 Physical modeling on passive piles 447

13.1 *Introduction* 447
13.2 *Apparatus and test procedures* 448
13.2.1 *Salient features of shear tests* 448
13.2.2 *Test program* 450
13.2.3 *Test procedure* 451
13.2.4 *Determining pile response* 455
13.2.5 *Impact of loading distance on test results* 455
13.3 *Test results* 456
13.3.1 *Driving resistance and lateral force on frames* 456
13.3.2 *Response of* M_{max}, y_o, ω *versus* w_i (w_f) 460
13.3.3 M_{max} *raises (T block)* 465
13.4 *Simple solutions* 467
13.4.1 *Theoretical relationship between* M_{max} *and* T_{max} 467
13.4.2 *Measured* M_{max} *and* T_{max} *and restraining moment* M_{oi} 468
13.4.3 *Equivalent elastic solutions for passive piles* 470
13.4.4 *Group interaction factors* F_m, F_k, *and* p_m 472
13.4.5 *Soil movement profile versus bending moments* 473
13.4.6 *Prediction of* T_{maxi} *and* M_{maxi} 474
13.4.6.1 *Soil movement profile versus bending moments* 474
13.4.7 *Examples of calculations of* M_{max} 475
13.4.8 *Calibration against in situ test piles* 478
13.5 *Conclusion* 482
Acknowledgment 484

References 485
Index 509

List of Figures

Figure 1.1 SPT N correction factor for overburden pressure. 3

Figure 1.2 Effect of overburden pressure. 4

Figure 1.3 Plasticity versus ratio of undrained shear strength over N_{60}. 5

Figure 1.4 Undrained shear strength versus N_{60} (clean sand). 7

Figure 1.5 SPT blow counts versus angle of internal friction. 7

Figure 1.6 Variations of friction angle with plasticity index data. 9

Figure 1.7 Shear resistance, ϕ'_{max}, ϕ'_{cv}, relative density D_r, and mean effective stress p'. 9

Figure 1.8 Results of regular and plane-strain triaxial tests. 10

Figure 1.9 Cone penetration resistance q_c over standard penetration resistance N_{60} versus particle size D_{50} of sand. 12

Figure 1.10 Moduli deduced from shafts in rock and piles in clay (q_u, E in MPa). 13

Figure 1.11 (a) SPT versus moisture content (density). (b) Moisture content (density) versus strength. 16

Figure 1.12 (a) Moisture content (density) versus strength. (b) Modulus versus uniaxial compressive strength. 17

Figure 2.1 Variation of α with normalized undrained shear strength s_u/σ_v'. 21

Figure 2.2 Variation of α as deduced from λ method. 22

Figure 2.3 α versus the normalized uniaxial compressive strength q_u; p_a = 100 kPa. 26

Figure 2.4 Variation of β with depth. 26

Figure 2.5 $K\tan\delta$ proposed by (a) Meyerhof (1976), (b) Vesić (1967). 27

Figure 2.6 N_q. 29

Figure 2.7 Variation of ultimate shaft friction with SPT blow counts. (a) China. (b) Southeast Asia. 31

Figure 2.8 Variation of ultimate base pressure with SPT blow counts. 32

Figure 2.9 Effect of soil stiffness on pile-head load versus settlement relationship. 33

Figure 2.10 Schematic of a single pile with negative skin friction (NSF). 36

Figure 2.11 Schematic of a pile in a group with negative skin friction (NSF). 38

Figure 2.12 Schematic modes of pile failure. 39

Figure 2.13 Comparison of η~s/d relationships for free-standing and piled groups. 42

Figure 2.14 Group interaction factor K_g for piles in (a) clay and (b) sand. 45

Figure 3.1 Schematic analysis of a vertically loaded pile. 48

Figure 3.2 Effect of back-estimation procedures on pile response ($L/r_o = 40$, $\nu_s = 0.4$, $H/L = 4$). (a) ζ with depth. (b) Shear stress distribution. (c) Head-stiffness ratio. (d) Load with depth. (e) Displacement with depth. 51

Figure 3.3 $1/\omega$ versus (a) slenderness ratio, L/d; (b) Poisson's ratio, υ_s; (c) layer thickness ratio, H/L. 53

Figure 3.4 Load transfer factor versus slenderness ratio ($H/L = 4$, $\nu_s = 0.4$). (a) $\lambda = 300$. (b) $\lambda = 1{,}000$. (c) $\lambda = 10{,}000$. 54

Figure 3.5 Load transfer factor versus Poisson's ratio relationship ($H/L = 4$, $L/r_o = 40$). (a) $\lambda = 300$. (b) $\lambda = 1{,}000$. (c) $\lambda = 10{,}000$. 56

Figure 3.6 Load transfer factor versus H/L ratio ($L/r_o = 40$, $\nu_s = 0.4$). (a) $\lambda = 300$. (b) $\lambda = 1{,}000$. (c) $\lambda = 10{,}000$. 58

Figure 3.7 Load transfer factor versus relative stiffness ($\nu_s = 0.4$, $H/L = 4$). (a) $L/r_o = 20$. (b) $L/r_o = 40$. (c) $L/r_o = 60$. 60

Figure 3.8 Effect of soil-layer thickness on load transfer parameters A and ζ ($L/r_o = 40$, $\nu_s = 0.4$, $\lambda = 1000$). (a) Parameter A. (b) Parameter ζ. 62

Figure 3.9 Variation of the load transfer factor due to using unit base factor ω and realistic value. (a) $L/r_o = 10$. (b) $L/r_o = 80$. 62

Figure 3.10 Comparison of pile behavior between the nonlinear (NL) and simplified linear (SL) analyses ($L/r_o = 100$). (a) NL and SL load transfer curves. (b) Load distribution. (c) Displacement distribution. (d) Load and settlement. 63

Figure 3.11 Comparison of pile-head load settlement relationship among the nonlinear and simple linear ($\psi = 0.5$) GASPILE analyses and the CF solution ($L/r_o = 100$). 65

Figure 3.12 Stress and displacement fields underpinned load transfer analysis. (a) Cylindrical coordinate system with stresses. (b) Vertical (P) loading. (c) Torsional (T) loading. (d) Lateral (H) loading. 66

Figure 3.13 Creep model and two kinds of loading adopted in this analysis. (a) Visco-elastic model. (b) One-step loading. (c) Ramp loading. 67

Figure 3.14 Normalized local stress displacement relationships for (a) 1-step and ramp loading, and (b) ramp loading. 72

Figure 3.15 (a) Variation of nonlinear measure with stress level. (b) Modification factor for step loading. (c) Modification factor for ramp loading. 73

Figure 3.16 (a) Predicted shaft creep displacement versus test results (Edil and Mochtar 1988). (b) Evaluation of creep parameters from measured time settlement relationship (Ramalho Ortigão and Randolph 1983). 76

Figure 3.17 Modeling radial consolidation around a driven pile. (a) Model domain. (b) Visco-elastic model. 79

Figure 3.18 Torsional versus axial loading: (a) Variation of shear modulus away from pile axis. (b) Local load transfer behavior. 80

Figure 3.19 Schematic of a lateral pile–soil system. (a) Single pile. (b) Pile element. 82

Figure 3.20 Schematic of pile-head and pile-base conditions (elastic analysis). (a) FreHCP (free-head, clamped pile). (b) FixHCP (fixed-head, clamped pile). (c) FreHFP (free-head, floating pile). (d) FixHFP (fixed-head, floating pile). 84

Figure 3.21 Load transfer factor γ_b versus E_p/G^* (clamped pile, free-head). 85

Figure 3.22 (a) Normalized modulus of subgrade reaction. (b) Normalized fictitious tension. 86

Figure 3.23 (a) Modulus of subgrade reaction. (b) Fictitious tension for various slenderness ratios. 89

Figure 3.24 Subgrade modulus for laterally loaded free-head (a) rigid and (b) flexible piles. 90

Figure 3.25 Comparison between the predicted and the measured (Prasad and Chari 1999) radial pressure, σ_r, on a rigid pile surface. 91

Figure 3.26 Schematic analysis for a rigid pile (a) pile–soil system (b) load transfer model ($p_u = A_r dz$ and $p = k_o dzu$). 92

Figure 3.27 Coupled load transfer analysis for a laterally loaded free-head pile. (a) The problem addressed. (b) Coupled load transfer model. (c) Load transfer (p-$y(w)$) curve. 96

Figure 3.28 Schematic profiles of limiting force, on-pile force, and pile deformation. (a_i) Tip-yield state. (b_i) Post-tip yield state. (c_i) Impossible yield at rotation point (YRP) (i = 1, 2 for p, p_u profiles and displacement profiles, respectively). 97

Figure 3.29 Schematic limiting force and deflection profiles. (a) A fixed-head pile. (b) Model of single pile. (c) Piles in a group. (d) LFPs. (e) Pile deflection and w_p profiles. (f) p-y curve for a single pile and piles in a group. 99

Figure 3.30 Determination of p-multiplier p_m. (a) Equation 3.69 versus p_m derived from current study. (b) Reduction in p_m with number of piles in a group. 104

Figure 4.1 (a) Typical pile–soil system addressed. (b) α_g and equivalent n_e. 107

Figure 4.2 Effect of the α_g on pile-head stiffness for a L/r_o of (a) 20, (b) 40, and (c) 100. 111

Figure 4.3 Features of elastic-plastic solutions for a vertically loaded single pile. (a) Slip depth. (b) τ_o~w curves along the pile. 113

Figure 4.4 Effect of the α_g on pile response (L/r_o = 100, λ = 1000, $n = \theta = 0.5$, A_g/A_v = 350, α_g = as shown). (a) P_t–w_t, (b) Load profiles. (c) Displacement profiles. 115

Figure 4.5 Comparison of pile-head stiffness among FLAC, SA (A = 2.5), and CF (A = 2) analyses for a L/r_o of (a) 20, (b) 40, (c) 60, and (d) 80. 120

Figure 4.6 Comparison between the ratios of head settlement over base settlement by FLAC analysis and the CF solution for a L/r_o of (a) 20, (b) 40, (c) 60, and (d) 80. 121

Figure 4.7 Pile-head stiffness versus slenderness ratio relationship. (a) Homogeneous soil (n = 0). (b) Gibson soil (n = 1). 123

Figure 4.8 Pile-head stiffness versus (a_1–c_1) the ratio of H/L relationship (ν_s = 0.4), (b_2–c_2) Poisson's ratio relationship (H/L = 4). 124

Figure 4.9 Comparison between various analyses of single pile load-settlement behavior. 125

Figure 4.10 Effect of slip development on pile-head response ($n = \theta$). 126

Figure 4.11 (a) Comparison among different predictions of load settlement curves (measured data from Gurtowski and Wu 1984) (Guo and Randolph 1997a); (b) Determining shaft friction and base resistance. 128

Figure 4.12 Comparison between the CF and the nonlinear GASPILE analyses of pile (a) displacement distribution and (b) load distribution. 129

Figure 4.13 Comparison among different predictions of the load distribution. 129

Figure 4.14 Comparison of the settlement influence factor (n = 1.0) by various approaches. 130

Figure 4.15 Comparison of settlement influence factor from different approaches for a L/r_o of: (a) 20, (b) 50, (c) 100, and (d) 200. 131

Figure 4.16 Schematic pile–soil system. (a) Typical pile and soil properties. (b) Strain-softening load transfer models. (c) Softening $\xi\tau_f$ ($\xi_c\tau_f$) with depth. 133

Figure 4.17 Capacity ratio versus slip length. (a) n = 0. (b) n = 1.0; or normalized displacement. (c) n = 0. (d) n = 1.0. 136

Figure 4.18 Development of load ratio as slip develops. 140

Figure 4.19 n_{max} versus π_v relationship (ξ as shown and $R_b = 0$). 141

Figure 4.20 μ_{max} versus π_v relationship (ξ as shown and $R_b = 0$). 141

Figure 4.21 Critical stiffness for identifying initiation of plastic response at the base prior to that occurring at ground level. 142

Figure 4.22 Effect of yield stress level (ξ_c) on the safe cyclic amplitude (n_c) ($\xi = 0.5$, $n = 0$, $R_b = 0$, one-way cyclic loading). 144

Figure 4.23 Effect of strain-softening factor on ultimate capacity ratio (n_{max}) ($R_b = 0$). 144

Figure 5.1 Comparison of the numerical and closed form approaches. (a) The settlement influence factor. (b) The ratio of pile head and base load. 151

Figure 5.2 Comparison of the settlement influence factor. 152

Figure 5.3 Comparison between closed-form solutions (Guo 2000b) and GASPILE analyses for different values of creep parameter G_1/G_2. 153

Figure 5.4 Loading time t_c/T versus settlement influence factor. (a) Different loading. (b) Influence of relative ratio of t_c/T. 153

Figure 5.5 Loading time t_c/T versus ratio of P_b/P_t. (a) Comparison among three different loading cases. (b) Influence of relative t_c/T. 154

Figure 5.6 Comparison between the measured and predicted load and settlement relationship. (a) Initial settlement. (b) Total settlement for the tests (33 days) (Konrad and Roy 1987). 155

Figure 5.7 Creep response of pile B1. (a) Load settlement. (b) Load distribution. (c) Creep settlement (measured from Bergdahl and Hult 1981). 157

Figure 5.8 Influence of creep parameters on the excess pore pressure. (a) Various values of *N*. (b) Typical ratios of $G_{\gamma 1}/G_{\gamma 2}$. 167

Figure 5.9 Variations of times T_{50} and T_{90} with the ratio $u_o(r_o)/s_u$. 168

Figure 5.10 Comparison between the calculated and measured (Seed and Reese 1955) load-settlement curves at different time intervals after driving. 172

Figure 5.11 Normalized measured time-dependent properties (Seed and Reese 1955) versus normalized predicted pore pressure. (a) Elastic analysis. (b) Visco-elastic analysis. 173

Figure 5.12 Normalized measured time-dependent properties (Konrad and Roy 1987) versus normalized predicted pore pressure. (a) Elastic analysis. (b) Visco-elastic analysis. 174

Figure 5.13 Comparison of load-settlement relationships predicted by GASPILE/closed-form solution with the measured (Konrad and Roy 1987): (a) Visco-elastic-plastic. (b) Elastic-plastic and visco-elastic-plastic solutions. 174

Figure 6.1 Load transmitting area for (a) single pile, (b) pile group. 179

Figure 6.2 Use of an imaginary footing to compute the settlement of pile. 179

Figure 6.3 Model for two piles in layered soil. 181

Figure 6.4 α~s/d relationships against (a) slenderness ratio, (b) pile–soil relative stiffness. 183

Figure 6.5 Pile group in nonhomogeneous soil. n_g = total number of piles in a group; R_s = settlement ratio; $w_g = R_s w_t$ (rigid cap); $P_t = P_g/n_g$ (flexible cap); $G = A_g(\alpha_g + z)^n$ (see Chapter 4, this book). 185

Figure 6.6 Group stiffness for 2×2 pile groups. (a_i) $H/L = \infty$. (b_i) $H/L = 3$. (c_i) $H/L = 1.5$ ($i = 1$, $s/d = 2.5$ and $i = 2$, $s/d = 5$) (lines: Guo and Randolph 1999; dots: Butterfield and Douglas 1981). 188

Figure 6.7 Group stiffness for 3×3 pile groups (a_i) $H/L = \infty$. (b_i) $H/L = 3$. (c_i) $H/L = 1.5$ ($i = 1$, $s/d = 2.5$ and $i = 2$, $s/d = 5$) (lines: Guo and Randolph 1999; dots: Butterfield and Douglas 1981). 189

Figure 6.8 Group stiffness for 4×4 pile groups (a_i) $H/L = \infty$. (b_i) $H/L = 3$. (c_i) $H/L = 1.5$ ($i = 1$, $s/d = 2.5$ and $i = 2$, $s/d = 5$). (lines: Guo and Randolph 1999; dots: Butterfield and Douglas 1981). 190

Figure 6.9 (a) Group stiffness factors. (b) Group settlement factors. 191

Figure 7.1 Schematic of (a) single pile, (b) a pile element, (c) free-head, clamped pile, (d) fixed-head, clamped pile, (e) free-head, floating pile, and (f) fixed-head, floating pile. 203

Figure 7.2 Soil response due to variation of Poisson's ratio at $z = 0$ (FreHCP(H)), $E_p/G^* = 44737$ ($\nu_s = 0$), 47695 (0.33). Rigid disc using intact model: $H = 10$ kN, $r_o = 0.22$ cm, and a maximum influence radius of $20r_o$). (a) Radial deformation. (b) Radial stress. (c) Circumferential stress. (d) Shear stress. 205

Figure 7.3 Comparison of pile response due to Poisson's ratio ($L/r_o = 50$). (a) Deflection. (b) Maximum bending moment. 212

Figure 7.4 Depth of maximum bending moment and critical pile length. 214

Figure 7.5 Single (free-head) pile response due to variation in pile–soil relative stiffness. (a) Pile-head deformation. (b) Normalized maximum bending moment. 215

Figure 7.6 Single (fixed-head) pile response due to variation in pile–soil relative stiffness. (a) Deformation. (b) Maximum bending moment. 217

Figure 7.7 Pile-head (free-head) (a) displacement due to moment loading, (b) rotation due to moment or lateral loading. 219

Figure 7.8 Impact of pile-head constraints on the single pile deflection due to (a) lateral load H, and (b) moment M_o. 221

Figure 7.9 Impact of pile-head constraints on the single pile bending moment. (a) Free-head. (b) Fixed-head. 222

Figure 7.10 Normalized pile-head rotation angle owing to (a) lateral load, H, and (b) moment M_o. 223

Figure 7.11 Pile-head constraints on normalized pile-head deflections: (a) clamped piles (CP) and (b) floating piles (FP). 224

Figure 7.12 Variations of the interaction factors ($l/r_o = 50$) (except where specified: $\nu_s = 0.33$, free-head, $E_p/G^* = 47695$): (a) effect of θ, (b) effect of spacing, (c) head conditions, and (d) other factors. 225

Figure 7.13 Radial stress distributions for free- and fixed-head (clamped) pile. For intact model: $\nu_s = 0.33$, $H = 10$ kN, $r_o = 0.22$ cm, $E_p/G^* = 47695$, $R = 20r_o$ (free-head), $30r_o$ (fixed-head). (a) Radial deformation.

(b) Radial stress. (c) Circumferential stress. (d) Shear stress. 226

Figure 8.1 Schematic analysis for a rigid pile. (a) Pile–soil system. (b) Load transfer model. (c) Gibson p_u (LFP) profile. (d) Pile displacement features. 230

Figure 8.2 Schematic limiting force profile, on-pile force profile, and pile deformation. (a_i) Tip-yield state, (b_i) Post-tip yield state, (c_i) Impossible yield at rotation point (YRP) (piles in clay) (i = 1 p profiles, 2 deflection profiles). 231

Figure 8.3 Predicted versus measured (see Meyerhof and Sastray 1985; Prasad and Chari 1999) normalized on-pile force profiles upon the tip-yield state and YRP: (a) Gibson p_u and constant k. (b) Gibson p_u and Gibson k. (c) Constant p_u and constant k. 235

Figure 8.4 Response of piles at tip-yield state. (a) z_o/l. (b) $u_g k_o l^{m-n}/A_r$. (c) $\omega k_o l^{1+m-n}/A_r$. (d) z_m/l. (e) Normalized ratio of u, ω. (f) Ratios of normalized moment and its depth. $n = 0, 1$ for constant p_u and Gibson p_u; $k = k_o z^m$, $m = 0$ and 1 for constant and Gibson k, respectively. 243

Figure 8.5 Normalized applied $\bar{M}_o$ (= $\bar{H}\,\bar{e}$) and maximum $\bar{M}_m$ (a), (b) Gibson p_u (= $A_r d z^n$, n = 1), (c) Constant p_u (= $N_g s_u d z^n$, n = 0). 247

Figure 8.6 Normalized response of (a_i) pile-head load $\bar{H}$ and groundline displacement $\bar{u}_g$. (b_i) $\bar{H}$ and rotation $\bar{\omega}$. (c_i) $\bar{H}$ and maximum bending moment $\bar{M}_{max}$ (subscript i = 1 for Gibson p_u and constant or Gibson k, and i = 2 for constant k and Gibson or constant p_u). Gibson p_u (= $A_r d z^n$, n = 1), and constant p_u (= $N_g s_u d z^n$, n = 0). 252

Figure 8.7 Normalized profiles of $\bar{H}(z)$ and $\bar{M}(z)$ for typical e/l at tip-yield and YRP states (constant k): (a)–(b) constant p_u; (c)–(d) Gibson p_u. 254

Figure 8.8 Predicted versus measured normalized pile capacity at critical states. (a) Comparison with measured data. (b) Comparison with Broms (1964b). 256

Figure 8.9 Tip yield states (with uniform p_u to l) and Broms' solutions (with uniform p_u between depths 1.5d and l). (a) Normalized $\bar{H}_o$. (b) $\bar{H}_o$ versus normalized moment ($\bar{M}_{max}$). 258

Figure 8.10 Yield at rotation point and Broms' solutions with $p_u = 3K_p\gamma_s' dz$. (a) Normalized $\bar{H}_o$. (b) $\bar{H}_o$ versus normalized moment ($\bar{M}_{max}$). 259

Figure 8.11 Tip-yield states and Broms' solutions. (a) Normalized stiffness versus normalized moment ($\bar{M}_{max}$). (b) $\bar{H}_o$ versus normalized moment ($\bar{M}_{max}$). 260

Figure 8.12 Normalized load $\bar{H}_o$-moment $\bar{M}_o$ or $\bar{M}_m$ at YRP state (constant p_u). 261

Figure 8.13 Impact of p_u profile on normalized load $\bar{H}_o$-moment $\bar{M}_o$ or $\bar{M}_m$ at YRP state. 262

Figure 8.14 Normalized $\bar{H}_o$, $\bar{M}_o$, and $\bar{M}_m$ at states of onset of plasticity, tip-yield, and YRP. (a) Constant p_u and k. (b) Gibson p_u. 263

Figure 8.15 Enlarged upper bound of $\bar{H}_o$~$\bar{M}_o$ for footings against rigid piles. (a) Constant p_u and k. (b) Non-uniform soil. 265

Figure 8.16 Comparison among the current predictions, the measured data, and FEA results (Laman et al. 1999). (a) M_o versus rotation angle ω (Test 3). (b) H versus mudline displacement u_g. (c) M_o versus rotation angle ω (effect of k profiles, tests 1 and 2). 267

Figure 8.17 Current predictions of model pile (Prasad and Chari 1999) data: (a) Pile-head load H and mudline displacement u_g. (b) H and maximum bending moment M_{max}. (c) M_{max} and its depth z_m. (d) Local shear force~displacement relationships at five typical depths. (e) Bending moment profiles. (f) Shear force profiles. 269

Figure 9.1 Comparison between the current predictions and FEA results (Yang and Jeremić 2002) for a pile in three layers. (a_i) p_u. (b_i) H-w_g and w_g-M_{max}. (c_i) Depth of M_{max} (i = 1, clay-sand-clay layers, and i = 2, sand-clay-sand layers). 295

Figure 9.2 Normalized (a_i) lateral load~groundline deflection, (b_i) lateral load~maximum bending moment, (c_i) moment and its depth ($e = 0$) ($i = 1$ for $\alpha_0\lambda = 0$, $n = 0$, 0.5, and 1.0, and $i = 2$ for $\alpha_0\lambda = 0$ and 0.2). 297

Figure 9.3 Calculated and measured (Kishida and Nakai 1977) response of piles A and C. (a_i) p_u. (b_i) H-w_g and H-M_{max}. (c_i) Depth of M_{max} (i = 1, 2 for pile A and pile C). 301

Figure 9.4 Calculated and measured response of: (a_i) p_u, (b_i) H-w_g, and H-M_{max}, and (c_i) bending moment profile [i = 1, 2 for Manor test (Reese et al. 1975), and Sabine (Matlock 1970)]. 305

Figure 9.5 LFPs for piles in clay. (a) Thirty-two piles with max x_p/d <10. (b) Seventeen piles with 6.5 < max x_p/d < 10. (c) Fifteen piles with max $x_p/d \le 6.5$. Note that Matlock's LFP is good for piles with superscript "M," but it underestimates the p_u with "U" and overestimates the p_u for the remaining 22 piles. 314

Figure 9.6 Normalized load, deflection (clay): measured versus predicted (n = 0.7). 315

Figure 9.7 Normalized load, maximum M_{max} (clay): measured versus predicted (n = 0.7). 315

Figure 9.8 LFPs for piles in sand. (a) Eighteen piles with max x_p/d < 10. (b) Nine piles with max $x_p/d \le 3$. (c) Nine piles with max $x_p/d > 3$. (Note that Reese's LFP is good for piles with superscript R, but it underestimates the p_u for the remaining piles.) 321

Figure 9.9 Normalized load, deflection (sand): Measured versus predicted (n = 1.7). 323

Figure 9.10 Normalized load, maximum M_{max} (sand): Measured versus predicted (n = 1.7). 323

Figure 9.11 (a) p-$y(w)$ curves at $x = 2d$ (d = 2.08 m). (b) Stress versus axial strain of Kingfish B sand. 326

Figure 9.12 Comparison of pile responses predicted using p-$y(w)$ curves proposed by Wesselink et al. (1988) and Guo (2006): (a) Pile-head deflection and maximum bending moment. (b) Deflection and bending moment profiles. (c) Soil reaction and p_u profile. 327

Figure 9.13 Predicted versus measured pile responses for Kingfish B. (a_1) H-w_g. (a_2 and a_3) H-w_t. (b_i) M profile. (c_i) LFPs (i = 1, and 2 onshore test A, and test B, i = 3 centrifuge test). 330

Figure 9.14 Predicted versus measured in situ pile test, North Rankin B. (a) H-w_t. (b) LFP. 333

Figure 9.15 Predicted versus measured pile responses for Bombay High model tests. (a) H-w_t. (b) LFPs. 333

Figure 9.16 Modeling of free-head piles and pile groups. (a) A free-head pile. (b) LFPs. (c) *p-y* curves for a single pile and piles in a group. (d) A free-head group. 334

Figure 9.17 Predicted versus measured (Rollins et al. 2006a) response: (a) Single pile. (b) 3×3 group (static, 5.65*d*). 336

Figure 9.18 Predicted versus measured (Rollins et al. 2006a) bending moment profiles (3×3, w_g = 64 mm). (a) Row 1. (b) Row 2. (c) Row 3. 337

Figure 9.19 Predicted versus measured (Rollins et al. 2006a) response of 3×4 group. (a) Static at 4.4*d*, (b) Cyclic at 5.65*d*. 338

Figure 9.20 Predicted versus measured (Rollins et al. 2006a) bending moment (3×4, w_g = 25 mm). 338

Figure 10.1 Calculation flow chart for the closed-form solutions (e.g., GASLFP) (Guo 2001; 2006). 344

Figure 10.2 Normalized bending moment, load, deflection for rigid and flexible shafts. (a_1–c_1) $n = 1.0$. (a_2–c_2) $n = 1.0$ and 2.3. 345

Figure 10.3 Simplified rectangular stress block for ultimate bending moment calculation. (a) Circular section. (b) Rectangular section. (c) Strain. (d) Stress. 346

Figure 10.4 Effect of concrete cracking on pile response. (a) LFPs. (b) Deflection ($w(x)$) profile. (c) Bending moment ($M(x)$) profile. (d) Slope (θ) profile. (e) Shear force ($Q(x)$) profile. (f) Soil reaction (p) profile. 352

Figure 10.5 Comparison of E_pI_p/EI~M_{max} curves for the shafts. 353

Figure 10.6 Comparison between measured (Huang et al. 2001) and predicted response of Pile B7 (SN1). (a) LFPs. (b) H-w_g and H~M_{max} curves. 355

Figure 10.7 Comparison between measured (Huang et al. 2001) and predicted response of pile P7 (SN2). (a) LFPs. (b) H-w_g and H~M_{max} curves. 357

Figure 10.8 Comparison between measured (Ng et al. 2001) and predicted response of Hong Kong pile (SN3). (a) LFPs. (b) H-w_g and H~M_{max} curves. 358

Figure 10.9 Comparison between measured (Zhang 2003) and predicted response of DB1 pile (SN4). (a) LFPs. (b) H-w_g and H~M_{max} curves. 359

Figure 10.10 Comparison between calculated and measured (Nakai and Kishida 1982) response of Pile D (CN1). (a) LFPs. (b) H-w_g and H~M_{max} curves. 360

Figure 10.11 Comparison between calculated and measured (Nakai and Kishida 1982) response of pile E. (a) LFPs. (b) H-w_g curves. (c) M profiles for H = 147.2 kN, 294.3 kN, 441.5 kN, and 588.6kN. 361

Figure 10.12 Normalized load, deflection: measured versus predicted (n = 0.7 and 1.7). 365

Figure 10.13 Normalized load, deflection: measured versus predicted (n = 2.3). (a) Normal scale. (b) Logarithmic scale. 366

Figure 10.14 Analysis of shaft in limestone (Islamorada test). (a) Measured and computed deflection. (b) Maximum bending moment. (c) Normalized LFPs. (d) E_cI_e. 368

Figure 10.15 Analysis of shaft in sandstone (San Francisco test). (a) Schematic drawing of shaft B. (b) Measured and computed deflection. (c) Maximum bending moment and E_cI_e. (d) Normalized LFPs. 370

Figure 10.16 Analysis of shaft in a Pomeroy-Mason test. (a) Measured and computed deflection. (b) Bending moment and E_eI_e. (c) Bending moment profiles (n = 0.7). 371

Figure 11.1 Schematic limiting force and deflection profiles. (a) Single pile. (b) Piles in a group. (c) LFP. (d) Pile deflection and w_p profiles. (e) p-y(w) curve for a single pile and piles in a group. 377

Figure 11.2 Schematic profiles of on-pile force and bending moment. (a) Fixed-head pile. (b) Profile of force per unit length. (c) Depth of second largest moment x_{smax} ($>x_p$). (d) Depth of x_{smax} ($<x_p$). 379

Figure 11.3 Calculation flow chart for current solutions (e.g., GASLGROUP). 381

Figure 11.4 Normalized pile-head load versus (a_1, a_2) displacement, (b_1, b_2) versus maximum bending moment. 382

Figure 11.5 Normalized load, deflection (clay): measured versus predicted ($n = 0.7$) with (a) fixed-head solution, or (b) measured data. 383

Figure 11.6 Normalized load, maximum M_{max} (clay): measured versus predicted ($n = 0.7$) with (a) fixed-head solution, or (b) measured data. 384

Figure 11.7 Normalized load, deflection (sand): measured versus predicted ($n = 1.7$) with (a) fixed-head solution, or (b) measured data. 385

Figure 11.8 Normalized load, maximum M_{max} (sand): measured versus predicted ($n = 1.7$). 386

Figure 11.9 Three typical pile-head constraints investigated and H2 treatment. 387

Figure 11.10 The current solutions compared to the model tests and FEM^{3D} results (Wakai et al. 1999) (Example 11.1). (a) Load. (b) Bending moment. 389

Figure 11.11 The current solutions compared to the group pile tests, and FEM^{3D} results (Wakai et al. 1999) (Example 11.1). 390

Figure 11.12 Predicted $H_g(H)$-w_g versus measured $H_g(H)$-w_t response of (a) the single pile and (b) pile group (Wakai et al. 1999) (Example 11.2). 392

Figure 11.13 (a) Group layout. (b) P_u for single pile. (c) P_u for a pile in 5-pile group. (d) P_u for a pile in 10-pile group. (Matlock et al. 1980) (Example 11.3). 394

Figure 11.14 Predicted versus measured (Matlock et al. 1980) (Example 11.3). (a) and (c) Load and deflection. (b) and (d) Maximum bending moment. 395

Figure 11.15 Predicted versus measured (Matlock et al. 1980) (Example 11.3) moment profiles of (a) single pile, and (b) a pile in 10-pile groups (Example 11.3). 397

Figure 11.16 Predicted H_g-w_g versus measured (Gandhi and Selvam 1997) H_g-w_t response of various groups (Example 11.4, three rows). (a) Load – displacement. (b) x_p for $s = 4d$. (c) x_p for $s = 6d$. 400

Figure 12.1 An equivalent load model for a passive pile: (a) the problem, (b) the imaginary pile, (c) equivalent load

H_2 and eccentricity e_{o2}, (d) normal sliding, and (e) deep sliding (Guo in press). 406

Figure 12.2 A passive pile modeled as a fictitious active pile. (a) Deflection profile. (b) H and on-pile force profile. 407

Figure 12.3 Limiting force and on-pile force profiles for a passive pile. (a) Measured p profiles for triangular soil movements. (b) Measured p profiles for arc and uniform soil movements. (c) p_u-based predictions. 411

Figure 12.4 Comparison between the current p_u-based solutions and the measured data (Guo and Qin 2010): (a) u_g~ω. (b) u_g~M_m. 413

Figure 12.5 Predicted (using $\theta_o = -\theta_{g2}$) versus measured (Esu and D'Elia 1974) responses (Case 12.I). (a) p_u profile. (b) Pile deflection. (c) Bending moment. (d) Shear force. 419

Figure 12.6 Normalized response based on p_u and fictitious force. (a) Normalized force and moment. (b) Normalized deflection and moment. 423

Figure 12.7 Current predictions versus measured data (AS50-0 and AS50-294). (a) M_m during overall sliding. (b) Pile rotation angle ω. (c) Mudline deflection u_g. (d) M_m for local interaction. 425

Figure 12.8 Current prediction (p_u-based solutions) versus measured data (AS50-0) at typical "soil" movements. (a) p profiles. (b) Shear force profiles. (c) Pile deflection profiles. (d) Bending moment profiles. 429

Figure 12.9 Normalized pile response in stable layer owing to soil movement w_s ($n_2 = 1$, $n_1 = 0$, $e_{o2} = 0$, $\xi = 0$). (a) Soil movement. (b) Load. (c) Slope. (d) Maximum moment. 434

Figure 12.10 Predicted versus measured pile responses: (a) moment and (b) deflection for Cases 12.2, 12.3, 12.4, and 12.5. 436

Figure 12.11 Predicted (using $\theta_o = -\theta_{g2}$) versus measured normalized responses (Cases 12.6 and 12.7). (a) Bending moment (b) Shear force. (c) Deflection. 437

Figure 12.12 Predicted versus measured pile response during excavation (Leung et al. 2000) (Case 12.8): (a) deflection and (b) bending moment at 2.5–4.5m. 438

Figure 12.13 Current predictions versus measured data (Smethurst and Powrie 2007). (a) w_s~M_m and u_g~H. (b) w_s-ω and u_g-ω. (c) Soil movement w_s and pile deflection u(z). (d) Bending moment profile. (e) p profiles (day 42). (f) p profiles (day 1,345). 440

Figure 12.14 Predicted versus measured pile responses for all cases. (a) Normalized load versus moment. (b) Normalized rotation angle versus moment. 445

Figure 13.1 Schematic of shear apparatus for modeling pile groups. (a) Plan view. (b) Testing setup. (c) An instrumented model pile. 449

Figure 13.2 Impact of distance of piles from loading side s_b on normalized M_{max} and T_{max}. 456

Figure 13.3 Jack-in resistance measured during pile installation. 457

Figure 13.4 Total applied force on frames against frame movements. 458

Figure 13.5 Variation of maximum shear force versus lateral load on loading block. 458

Figure 13.6 Development of pile deflection y_o and rotation ω with w_f. (a) Triangular block. (b) Uniform block. 461

Figure 13.7 Maximum response profiles of single piles under triangular movement. (a_i) Bending moment. (b_i) Shear force (i = 1, final sliding depth = 200 mm, and i = 2, final sliding depth 350 mm). 462

Figure 13.8 Evolution of maximum response of single piles (final sliding depth = 200 mm). (a) Triangular movement. (b) Uniform movement. 463

Figure 13.9 Evolution of M_{max} and T_{max} with w_c for two piles in a row (w/o axial load). (a) Triangular movement. (b) Uniform movement. 464

Figure 13.10 Evolution of maximum response of piles (final sliding depth = 350 mm). 466

Figure 13.11 Variation of maximum bending moment versus sliding depth ratio. 466

Figure 13.12 Maximum shear force versus maximum bending moment. (a) Triangular movement. (b) Uniform movement. 469

Figure 13.13 Soil reaction deduced from uniform tests (SD = 400 mm). 471

Figure 13.14 Soil reaction. (a) Triangular block. (b) Uniform block. 473

Figure 13.15 Calculated versus measured ratios of $M_{max}/(T_{max}L)$. 481

Figure 13.16 Measured M_{max} and T_{max} (Frank and Pouget 2008) versus predicted values. 482

Preface

Over the last 20 years or so, I have witnessed and grown to appreciate the difficulty undergraduate students sometimes have in grasping geotechnical courses. Geotechnical courses have many new concepts and indigestible expressions to initially be memorized. The expressions, which generally originate from advanced elastic mechanics, plastic mechanics, and numerical approaches, are technical and not easily accessible to students. These courses may leave the students largely with fragments of specifications. This naturally leads the students to have less confidence in their ability to learn geotechnical engineering. To resolve the problem, our research resolution has been to minimize the number of methods and input parameters for each method and to resolve as many problems as possible. We believe this will allow easier access to learning, practice, and integration into further research.

Piles, as a popular foundation type, are frequently used to transfer superstructure load into subsoil and stiff-bearing layer and to transfer impact of surcharge owing to soil movement and/or lateral force into underlying layers. They are installed to cater for vertical, lateral, and/or torsional loading to certain specified capacity and deformation criteria without compromising structural integrity. They are conventionally made of steel, concrete, timber, and synthetic materials (Fleming et al. 2009) and have been used since the Neolithic Age (6,000~7,000 years ago) (Shi et al. 2006). Their use is rather extensive and diverse. For instance, in 2006, ~50 million piles were installed in China, ranging (in sequence of high to low proportion) from steel pipe piles, bored cast-in-place piles, hand-dug piles, precast reinforced concrete piles, pre-stressed concrete pipe piles, and driven cast-in-place piles to squeezed branch piles. These piles, although associated with various degrees of soil displacement during installation, are generally designed using the same methods but with different parameters. It is critical that any method of design should allow reliable design parameters to be gained in a cost-effective manner. More parameters often mean more difficulty in warranting a verified and expeditious design.

We attempt to devise design methods that require fewer parameters but resolve more problems. This has yielded a systematic approach to model pile response in the context of load transfer models. This is summarized in this book of 13 chapters. Chapter 1 presents an overview of estimating soil shear modulus and strength using the conventional standard penetration tests and cone penetration tests. Chapter 2 provides a succinct summary of typical methods for estimating bearing capacity (including negative skin friction) of single piles and pile groups. Chapter 3 recaptures pile–soil interaction models under vertical, lateral, or torsional loading. Chapters 4 and 5 model the response of vertically loaded piles under static and cyclic loading and time-dependent behavior, respectively. The model is developed to estimate settlement of large pile groups in Chapter 6. A variational approach is employed to deduce an elastic model of lateral piles in Chapter 7, incorporating typical base and head constraints. Plastic yield between pile and soil (p_{u}-based model) is subsequently introduced to the elastic model to capture a nonlinear response of rigid (Chapter 8) and flexible piles (Chapter 9) under static or cyclic loading. Plastic yield (hinge) of pile itself is further incorporated into the model in Chapter 10. The p_{u}-based model is further developed to mimic a nonlinear response of laterally loaded pile groups (see Chapter 11) and to design passive piles in Chapter 12. The mechanism about passive piles is revealed in Chapter 13 using 1-g model tests. Overall, Chapter 3 provides a time-dependent load transfer model that captures pile–soil interaction under vertical loading in Chapters 4 through 6. Chapters 3 and 7 provide a framework of elastic modeling (e.g., the modulus of subgrade reaction) and p_u-based modeling for lateral piles, which underpins nonlinear simulation in Chapters 8 through 12 and simple solutions in Chapter 13 dealing with lateral loading. The p_u-based model essentially employs a limiting force profile and slip depth to capture evolution of nonlinear response. A model for torsional loading is presented in Chapter 3, but the prediction of the piles is not discussed in the book.

The input parameters for various solutions in Chapters 3 through 13 were gained extensively through study on measured data of 10 pile foundations (vertical loading), ~70 laterally loaded single piles (static and/or cyclic loading), ~20 laterally loaded, structurally nonlinear piles, ~30 laterally loaded pile groups, ~10 passive (slope stabilizing) piles, and ~20 rigid passive piles. The contents of Chapters 3 through 9 were taught to postgraduates successfully for five years and are updated herein. This book is aimed to facilitate prediction of nonlinear response of piles in an effective manner, an area of study that has largely been neglected.

I would like to express my gratitude to Professor Mark Randolph for his assistance over many years. My former students helped immensely in conducting model tests and case studies, including Dr. Bitang Zhu,

Dr. Enghow Ghee, and Dr. Hongyu Qin. Last but not least, my wife, Helen (Xiaochun) Tang, has always been supportive to this endeavor, otherwise this would not be possible.

Author

Wei Dong Guo earned his BEng in civil engineering from Hohai University, MEng in geotechnical engineering from Xi'an University of Architecture and Technology, China, and a PhD from the University of Western Australia. He has studied and worked at four world-renowned geotechnical groups in Hohai University (China), the University of Western Australia, the National University of Singapore, and Monash University, as well as the private sector.

Dr. Guo was awarded a talent grant from China and three prestigious research fellowships—a Singapore National Science Postdoctoral Fellowship, a Monash Logan Research Fellowship, and an Australian Research Council Postdoctoral Fellowship—and a significant medal from the Institute of Civil Engineering of the United Kingdom. He has developed one boundary element program (GASGROUP), one finite element program, and a few spreadsheet programs. As a sole or leading author, he has published a number of rigorous models and theories for sophisticated response of piles in high-impact international journals.

Dr. Guo is currently an associate professor at the University of Wollongong (a node of the Australian Research Council Centre of Excellence for Geotechnical Science and Engineering) after teaching nearly a decade (2002–2011) at Griffith University in Australia.

List of Symbols

The following symbols are used in this book:

Roman

A	=	a coefficient for estimating shaft load transfer factor
$A(t)$	=	time-dependent part of the shaft creep model
A_b, A_s	=	area of pile base and shaft, respectively
A_c	=	a parameter for the creep function of $J(t)$
A_g	=	constant for soil shear modulus distribution
A_h	=	a coefficient for estimating "A," accounting for the effect of H/L
A_L, A_{Li}	=	coefficient for the LFP; A_L for the ith layer $[FL^{-1-ni}]$
A_n, B_n	=	coefficients for predicting excess pore pressure
A_{oh}	=	the value of A_h at a ratio of $H/L = 4$
A_p	=	cross-sectional area of an equivalent solid cylinder pile
A_r	=	coefficient for the LFP $[FL^{-3}]$
A_s	=	$2(B_i + L_i)L$, perimeter area of pile group
A_v	=	a constant for shaft limit stress distribution
B	=	a coefficient for estimating shaft load transfer factor
$B(z)$	=	sub-functions reflecting base effect due to lateral load
B_c	=	a parameter for the creep function of $J(t)$
B_c	=	width of block
c, c'	=	cohesion $[FL^{-2}]$
C, C_y	=	$z_r u^*/u_g$ (constant k), used for post-tip yield state, and C at the tip-yield state
$C(z)$	=	subfunction reflecting base effect due to moment
C_A, C_p	=	factors for relative density D_r

C_B, C_R, C_s	=	borehole diameter correction, rod length correction, and sample correction
C_N	=	a modification factor for overburden stress
C_{OCR}	=	overconsolidation correction factor
c_{ub}	=	undrained cohesion at pile-base level [FL^{-2}]
c_v	=	coefficient of soil consolidation
$C_v(z)$	=	a function for assessing pile stiffness at a depth of z, under vertical loading
C_{vb}	=	limiting value of the function for the ratio of base and head loads as z approaches zero
C_{vo}	=	limiting value of the function, $C_v(z)$, as z approaches zero
C_λ	=	a coefficient for estimating "A," accounting for the effect of λ
D	=	outside diameter of a cylindrical pile [L]
D_{50}	=	maximum size of the smallest 50% of the sample [L]
d_{max}	=	depth of maximum bending moment in the shaft [L]
d_o	=	outside diameter of a pipe pile [L]
d_r	=	reference width, 0.3 m
D_r	=	relative density of sand
E	=	Young's modulus of soil or rock mass [FL^{-2}]
E_c	=	modulus of elasticity of concrete [FL^{-2}]
E_cI_p	=	EI, initial flexural rigidity of the shaft [FL^2]
E_m	=	hammer efficiency
E_m	=	deformation modulus of an isotropic rock mass [FL^2]
E_p	=	Young's modulus of an equivalent solid cylinder pile [FL^{-2}]
E_r	=	deformation modulus of intact rock mass [FL^2]
e, e_{oi}	=	eccentricity or free-length from the loading location to the mudline; or $e = M_o/H_t$; or distance from point O to incorporate dragging effect [L]
e_m, e_p, e_w	=	eccentricity of the location above the ground level for measuring the M_{max}, applying the $P(P_g)$, and measuring the w_t
$F(t)$	=	the creep compliance derived from the visco-elastic model
$FixH$	=	fixed-head, allowing translation but not rotation at head level

FreH	=	free head, allowing translation and rotation at head level
F_m	=	pile-pile interaction factor (passive pile tests)
f_c'	=	characteristic value of compressive strength of the concrete [FL^{-2}]
f_c''	=	design value of concrete compressive strength [FL^{-2}]
f_r	=	$M_{cr}y_r/I_g$, modulus of rupture [FL^{-2}]
f_r	=	friction ratio computed using the point and sleeve friction (side friction), in percent
f_y	=	yield strength of reinforcement [FL^{-2}]
G, $\tilde{G}$	=	(initial) elastic shear modulus, average of the G [FL^{-2}]
G^*	=	soil shear modulus, $G^* = (1 + 0.75\nu_s)G$ [FL^{-2}]
G_b, G_{bj}	=	shear modulus at just beneath pile base level; or initial G_b for spring j (= 1, 2) [FL^{-2}]
$G_b(t)$	=	time-dependent initial shear modulus at just beneath pile base level [FL^{-2}]
G_i, G_i^*	=	G, G_i^* for ith layer [FL^{-2}]
G_j	=	the instantaneous and delayed initial shear modulus for spring j (j = 1, 3) [FL^{-2}]
G_L	=	(initial) shaft soil shear modulus at just above the pile base level [FL^{-2}]
G_m	=	shear modulus of an isotropic rock mass [FL^{-2}]
$G_{\gamma j}$, $G_{1\%}$	=	shear modulus at a strain level of γ_j for spring j (j = 1, 2) within the visco-elastic model, or shear modulus at a shear strain of 1% [FL^{-2}]
g_s	=	$p_u/(q_u)^{1/n}$, a factor featuring the impact of rock roughness on the resistance p_u
GSI	=	geology strength index
h	=	distance above pile-tip level
h_d	=	the pile penetration into dense sand
H	=	the depth to the underlying rigid layer (Chapters 4 and 6) [L], or horizontal load applied on pile-head (Chapter 7) [F]
H, H_{max}	=	lateral load exerted on a single pile, and maximum imposed H [F]
$H(z)$	=	sub-function, due to lateral load (Chapter 7); lateral force induced in a pile at a depth of z [F]

H_{av}, H_g	=	average load per pile in a group, and total load imposed on a group [F]
H_e	=	lateral load applied when the slip depth is just initiated at mudline [F]
H_o	=	H at a defined (tip-yield or YRP) state [F]
H_2	=	lateral load applied at a distance of "e_{o2}" above point O, and $H_1 = -H_2$ [L]
$\bar{H}$	=	$H\lambda^{n+1}/A_L$, normalized pile-head load
$\bar{H}_i(\bar{z})$	=	normalized shear force induced in a pile at a normalized depth of $\bar{z}$ for $i = 1$ $(0 < z \le z_o)$, 2 $(z_o < z \le z_1)$, and 3 $(z_1 < z \le l)$, respectively
i	=	subscripts 1 and 2 denoting the upper sliding and lower stable layer, respectively
I	=	settlement influence factor for single piles subjected to vertical loading
I_{cr}	=	moment of inertia of cracked section [L^4]
I_e	=	effective moment of inertia of the shaft after cracking [L^4]
I_g	=	moment of inertia of a gross section about centroidal axis neglecting reinforcement [L^4]
I_g	=	$w_g dE_L/P_g$, settlement influence factor for a pile group
I_m, I_{m-1}	=	modified Bessel functions of the first kind of non-integer order, m and $m - 1$ respectively
I_{p}	=	moment of inertia of an equivalent solid cylinder pile [L^4]
I_p	=	the plasticity index
$I(z)$	=	sub-function due to moment
J	=	empirical factor lying between 0.5 and 3 for estimating N_g
J_i	=	Bessel functions of the first kinds and of order i $(i = 0, 1)$
$J(t)$	=	a creep function defined as ζ_c/G_1
k	=	permeability of soil (Chapter 5)
k, k_o	=	modulus of subgrade reaction [FL^{-3}], $k = k_o z^m$, m = 0 and 1 for constant and Gibson k, respectively; and k_o, a parameter [FL^{-3-m}]
k_i, k_j	=	parameters for estimating load transfer factor, $i = 1$, 3 (lateral loading), or a factor representing soil nonlinearity of elastic spring j (vertical loading)

k_i	=	modulus of subgrade reaction for *i*th layer (passive piles)
k_r	=	a constant for concrete rupture
k_s	=	a factor representing pile–soil relative stiffness
k_{sg}	=	k_s for a pile in a two-pile group
k_ϕ	=	pile rotational (constraining) stiffness about the head
K	=	average coefficient of earth pressure on pile shaft with minimum and maximum values of K_{min} and K_{max}
K_a	=	$\tan^2 (45° - \phi'/2)$, the coefficient of active earth pressure
K_g	=	0.6~1.5, group interaction factor, with higher values for dense cohesionless or stiff cohesive soils, otherwise for loose or soft soils
$K_i(\gamma)$	=	modified Bessel function of second kind of *i*-th order
K_m, K_{m-1}	=	modified Bessel functions of the second kind of non-integer order m and order $m - 1$, respectively
K_p	=	$\tan^2 (45° + \phi'/2)$, the coefficient of passive earth pressure
$L(l)$	=	embedded pile length
L_c	=	length of block
L_c	=	critical embedded pile length beyond which the pile is classified as infinitely long
LFP	=	net limiting force profile per unit length [FL^{-1}]
LR, MR, TR	=	leading row, middle row, and trailing row
L_1	=	the depth of transition from elastic to plastic phase, the slip part length of a pile under vertical loading
L_2	=	length of the elastic part of a pile under a given load
L_m	=	sliding depth during a passive soil test [L]
L_n	=	depth of neutral plane
m	=	1/(2 + n), or number of rows of piles
m_2	=	ratio of shear moduli, $G_{\gamma 1}/G_{\gamma 2}$
M, $M(x)$	=	moment induced on a pile element, or M at a depth of "x" [FL]
$M_A(x)$, $M_B(z)$	=	moment induced in a pile element, at depth x and z, respectively [FL]
M_B	=	moment induced at pile-base level [FL]
M_{cr}	=	cracking moment [FL]

M_{max}, M_m = maximum bending moment within a pile [FL]
M_n = nominal or calculated ultimate moment [FL]
M_o = moment applied on pile-head or at the mudline level [FL]
M_{oi} = $H_i e_{oi}$, bending moment about point O [FL]
M_t = moment applied at the shaft at the groundline [FL]
$\bar{M}(x)$ = $M(x)\lambda^{n+2}/A_L$, normalized bending moment at depth x
$\bar{M}_i(\bar{z})$ = bending moment induced in a pile at a normalized depth of $\bar{z}$ for $i = 1$ ($0 < z \leq z_o$), 2($z_o < z \leq z_1$), and 3 ($z_1 < z \leq l$), respectively
$\bar{M}_{max}$ = $M_{max}\lambda^{2+n}/A_L$, normalized M_{max}
n = number of piles enclosed by the square (NSF)
n = number of piles in each row
n = power of the shear modulus distribution, nonhomogeneity factor, or power for the LFP
n_c = the safe cyclic load amplitude
n_c^{FreH}, n_s^{FixH} = power of the LFP for free-head and fixed-head respectively. The subscripts *s* and *c* refer to *sand* and *clay*, respectively
n_e = equivalent nonhomogeneity factor
n_g = number of piles in the pile group
n_{max} = maximum ratio of pile-head load and the ultimate shaft load
n_p = ratio of pile-head load and the ultimate shaft load
N = visco-elastic time factor, or blow count of the Standard Penetration Test
N' = corrected blow counts of SPT
N_{60} = blow counts of SPT at a standard rod energy ratio of 60%
N_c, N_q = bearing capacity factors
N_{co} = lateral capacity factor–correlated soil, undrained strength, with the limiting pile–soil pressure at mudline
N_g = gradient-correlated soil, undrained strength, with the limiting pile–soil pressure
N_g^{FreH}, N_g^{FixH} = gradient-correlated soil strength to the p_u for free- and fixed-head piles, respectively
N_k = cone factor (a constant for each soil) ranging from 5 to 75

N_p	=	a factor for limiting force per unit length (plastic zone)
N_p	=	fictitious tension for a strecthed membrane tied together the springs around the pile shaft (elastic zone)
$\tilde{N}_c$	=	equivalent lateral capacity factor correlated average soil undrained strength with average net limiting force per unit length by $\tilde{p}_u = \tilde{s}_u \tilde{N}_c d$
p	=	surcharge loading on ground surface
p, p_u	=	force per unit length, and limiting value of the p [FL^{-1}]
p'	=	mean effective stress [FL^{-2}]
P	=	vertical load on a pile head under passive tests [F]
p_a	=	atmospheric pressure, ≈ 100 kPa
$P_b(P_{fb})$	=	load of pile base (ultimate P_b) [F]
P_{BL}	=	a total failure load of the group [F]
P_{cap}	=	the bearing capacity of the pile cap on the bearing stratum
P_e	=	axial load at the depth of transition (L_1) from elastic to plastic phase [F]
$P^{ex}_{cap}, P^{in}_{cap}$	=	components of capacity deduced from areas A^{ex}_c (lying outside the block) and A^{in}_c (inside block)
$P_f(P_u)$	=	ultimate pile bearing load [F]
P_{fs}	=	ultimate shaft load of a pile [F]
P_{ij}	=	pile number ij in a group (i = 1–3, j = 1–2)
p_m	=	p-multipliers used to reduce stiffness, and limiting force for individual piles in a group
P_{ns}	=	total downdrag load [F]
P_t	=	vertical load acting on pile head [F]
P_{ug}	=	ultimate capacity of the pile group
$\tilde{p}_u$	=	average limiting force per unit length over the pile embedment [FL^{-1}]
$p(\bar{x}), Q(\bar{x})$	=	net force per unit length, and shear force at the normalized depth $\bar{x}$
$P(z)$	=	axial force of pile body at a depth of z [F]
q'	=	stress on the top of the weaker layer [FL^{-2}]
q_b, q'_b	=	unit base resistance, net q_b [FL^{-2}]
$q_{b\max}$	=	maximum end-bearing capacity [FL^{-2}]
q_c	=	cone resistance q_c of a CPT test

q_u	=	uniaxial compressive strength of the weaker material (rock or concrete) [FL^{-2}]
q_{ui}	=	uniaxial compressive strength of intact rock [FL^{-2}]
Q, Q_B	=	shear force induced on a pile cross-section, and the Q at pile-base level [F]
$Q_A(x)$, $Q_B(x)$	=	shear force induced on a pile cross-section [F]
Q_g	=	the perimeter of the pile group (equivalent large pile)
r	=	radial distance from pile axis [L]
r^*	=	the radius at which the excess pore pressure, by the time it reaches here, is small and can be ignored [L]
r_m	=	radius of zone of shaft shear influence [L]
r_{mg}	=	a radius of influence of pile group [L]
r_o	=	radius of a cylindrical pile [L]
R	=	the radius beyond which the excess pore pressure is initially zero [L]
$R(\tilde{R})$	=	ratio of N_g of a single pile in a group over that of the single pile (average of R)
R_A, R_B, R_C, R_D, R_E	=	subratings for q_{ui}, RQD, spacing of discontinuities, conditions of discontinuities, and groundwater, respectively
R_b	=	ratio of settlement between that for pile and soil caused by P_b, base settlement ratio
R_{fb}, R_{fj}	=	a hyperbolic curve-fitting constant for pile base load settlement curve, or for the elastic element j within the creep models
RMR	=	rock mass rating
RQD	=	rock quality designation
R_{max}	=	roughness of the concrete
R_s	=	group settlement ratio
s	=	argument of the Laplace transform (Chapter 5), or pile center to center spacing (Chapter 7)
s_b	=	a loading distance between center of a pile (group) and loading block
s_g	=	an integral factor to cater for all sorts of influence
s_u, $\tilde{s}_u$	=	undrained shear strength, an average su over the pile embedment [FL^{-2}]
s_{uL}	=	undrained shear strength at pile tip or footing base level [FL^{-2}]
t (t^*)	=	time elapsed

t	=	wall thickness of a pipe pile or thickness of concrete cover [L]
t_{90}	=	time for $(u_o - u)/u_o = 0.9$
T	=	loading time
T	=	relaxation time, η/G_2, or rate of consolidation
T_{50}, T_{90}	=	time factor, T, for 50% and 90% degree of "consolidation," respectively
$T_2(T_3)$	=	relaxation time, $\eta_{\gamma 2}/G_{\gamma 2}$ $(\eta_{\gamma 3}/G_{\gamma 3})$
T_{max}	=	sliding force on a rigid pile
$T_n(t)$	=	the time for the solution of the reconsolidation theory, also written as $T(t)$
T_R	=	total resistance over the maximum slip depth x_p obtained under P_{max} [F]
T_{ult}	=	ultimate T_{max}
u	=	excess pore water pressure [FL^{-2}] or radial displacement [L]
u, u_g	=	lateral displacement, and u at mudline level [L]
u^*	=	local threshold u* above which pile–soil relative slip is initiated [L]
$u(z)$	=	axial pile displacement at a depth of z [L]
U_k	=	energy parameter for "y = 1" per unit pile length
U_m	=	energy parameter for "$d\phi/dr$ = 1" per unit radial length
U_N	=	energy parameter for "dy/dz = 1" per unit pile length
U_n	=	energy parameter for "ϕ = 1" per unit radial length
u_o	=	initial pore water pressure [FL^{-2}]
$u_o(r)$	=	initial excess pore water pressure at radius r [FL^{-2}]
u_v	=	vertical displacement along depth [L]
v	=	circumferential displacement [L]
V_i	=	cylinder function of i-th order
w, w_c	=	local shaft deformation, and creep part of w [L]
w(or y), $w(x)$, $w(z)$	=	lateral deflection of a pile, w in the plastic, and w in elastic zone, respectively [L]
$w(\bar{x})$, $w'(\bar{x})$	=	deflection [L] and rotation at the normalized depth $\bar{x}$
$w(z)$	=	deformation of pile body at a depth of z for a given time [L]
$w(z)$	=	pile body displacement at depth z, or simply written as w [L]

w_A, w_B	=	lateral deflection in the upper plastic zone and lower elastic zone, respectively [L] (Chapter 9)
w_B	=	pile base displacement at the base level [L] (Chapter 7)
w_A^{IV}, w_A''', w_A'', w_A'	=	fourth, third, second, and first derivatives, respectively, of deflection w with respect to the depth x
w_B^{IV}, w_B''', w_B'', w_B'	=	fourth, third, second, and first derivatives, respectively, of deflection w with respect to the depth x
w_b, w_t	=	settlement of pile base and head, respectively [L]
w_e	=	settlement due to elastic compression of pile [L]
w_e^c	=	a reduced limiting shaft displacement deduced from the limiting shaft stress, τ_f^c
w_f	=	frame (soil) movement during a shear test [L]
w_g, w_g'	=	lateral pile deflection [L], and rotation angle (in radian) at mudline, respectively, or point O for passive piles
$\overline{w}_g$	=	$w_g k\lambda^n/A_L$, normalized mudline deflection
w_{gi}	=	a lateral pile deflection at point O (sliding level) [L]
w_i	=	settlement or deformation of the i-th pile in a group of n_g piles [L]
w_i	=	initial frame movement during which pile response is negligible [L]
w_p	=	$P_tL/(E_pA_p)$
w_p	=	p_u/k, lateral deflection at the slip depth of x_p [L]
w_p^{IV}, w_p''', w_p'', w_p'	=	values of fourth, third, second, and first derivatives, respectively, of deflection w with respect to the depth x at depth x_p
w_s	=	lateral (uniform) soil movement, or shaft displacement [L]
w_t	=	pile-head settlement, or lateral deflection [L]
w_{t1}	=	settlement of a single pile under unit head load [L]
x, x_p, $\overline{x}$, $\overline{x}_p$	=	depth below ground level, slip depth of plastic zone, $\overline{x} = \lambda x$, $\overline{x}_p = \lambda x_p$
x_i	=	depth measured from point O on the sliding interface [L]
x_{max}	=	depth at which maximum bending moment occurs ($x_{max} = x_p + z_{max}$ when $\psi(\overline{x}_p) \geq 0$; $x_{max} = x_{max}$ when $\psi(\overline{x}_p) < 0$) [L]
x_{maxi}	=	depth of maximum bending moment measured from point O in plastic zone [L]

x_p, x_{pi}	=	slip depth from the elastic to the plastic state; or x_p from the elastic-plastic boundary to point O [L]
x_s	=	thickness of the zone, in which pile deflection exceeds soil movement [L]
$y(z)$	=	pile body displacement at depth z, or simply written as y [L]
Y_i	=	Bessel functions of the second kinds and of order i (i = 0, 1)
YRP	=	yield at rotation point
y_o	=	pile deflection at sand surface [L]
z	=	depth measured from the mudline [L]
$z, \bar{z}$	=	$x - x_p$, depth measured from the slip depth [L] and $\bar{z} = \lambda z$, respectively
z_c	=	critical depth [L]
z_i	=	$x_i - x_{pi}$, depth measured from the slip depth, x_{pi} [L]
z_m	=	depth of maximum bending moment [L]
$z_{\max2}$	=	depth of $M_{\max2}$ measured from the slip depth, x_{p2} [L]
$z_{\max i}$	=	depth of $M_{\max i}$ measured from the sliding interface [L]
$z_o(z_1)$	=	slip depth initiated from mudline (pile-base) [L]
z_r	=	depth of rotation point [L]
z_t	=	an infinite small depth [L]
z_*	=	slip depth z_o at the moment of the tip-yield [L]

Greek

α	=	average pile–soil adhesion factor in terms of total stress
α, β	=	stiffness factors for elastic solutions (lateral piles) [L^{-1}]
α_c	=	nondimensional creep parameter for standard linear model, or ratio of the shear modulus over the undrained strength, G/s_u
α_E	=	0.0231RQD – 1.32 ≥ 0.15, and RQD in percentage
α_g	=	shear modulus factor for ground surface
α_{ij}	=	pile-pile interaction factor between pile i and pile j (vertical loading)
α_m	=	shaft friction factor
α_n	=	consolidation factor
α_N, β_N	=	α/λ, and β/λ, normalized α and β by λ, respectively
$\alpha_o(\bar{\alpha}_o)$	=	an equivalent depth to account for ground-level limiting force with $\bar{\alpha}_o = \alpha_o\lambda$
α_r	=	a reduction factor (related to q_u) for shaft friction

α_{si}	=	a factor correlating maximum shear force to bending moment (i = 1, 2) of passive piles
α_{pP}, α_{pM}	=	the interaction factor between the *i*-th pile and *j*-th pile, reflecting increase in deflection due to lateral load and moment loading, respectively
$\alpha_{\theta P}$, $\alpha_{\theta M}$	=	the interaction factor between the *i*-th pile and *j*-th pile, reflecting increase in rotation due to lateral load and moment loading, respectively
β	=	average pile–soil adhesion factor in terms of effective stress
β_r	=	a factor correlated to the discontinuity spacing in the rock mass
γ ($\dot{\gamma}$)	=	shear strain (shear strain rate)
γ_b	=	load transfer factor
γ_j ($\dot{\gamma}_j$)	=	shear strain (shear strain rate) for elastic spring *j*
γ_m	=	effective unit weight of the rock mass [FL^{-3}]
$\gamma_{r\theta}$, $\gamma_{\theta z}$, γ_{rz}	=	shear strain within the r-θ plane, θ-z plane, and the r-z plane, respectively
$\gamma_s(\gamma'_s)$	=	unit weight of the overburdened soil (effective γ_s) [FL^{-3}]
γ_w	=	the unit weight of water [FL^{-3}]
δ	=	factor used for displacement prediction
δ	=	interface frictional angle, being consistent with that measured in simple shear tests
$\delta\theta$	=	mean total stress
$\delta\sigma'_r$ ($\delta\sigma'_\theta$)	=	increments of the effective stress during consolidation in radial and circumferential directions
$\delta\sigma'_z$	=	increments of the effective stress during consolidation in depth direction
ε_{ij}	=	strain components within the surrounding soil
ε_r, ε_θ, ε_z	=	strain in the radial, circumferential, and depth directions
ε_v	=	the volumetric strain
ζ, ζ_j	=	shaft load transfer factor, nonlinear measure of the influence of load transfer for spring *j* (*j* = 1, 2) within the creep models
ζ_c	=	a nondimensional creep function
ζ_g	=	shaft load transfer factor for a pile in a two-pile group

η	=	group efficiency correlated to the ratio ρ of the shaft (skin) load, P_{fs} over the total capacity, P_u (i.e., $\rho = P_{fs}/P_u$)
η_s, η_b	=	the efficiency factors of shaft and base
η_s'	=	geometric efficiency
$\eta_{\gamma i}(\eta)$	=	shear viscosity for the dash at strain γ_i ($i = 2, 3$)
θ	=	power of the shear stress distribution, nonhomogeneity factor (Chapter 4)
θ	=	an angle between the interesting point and loading direction (Chapter 7)
θ_B	=	pile rotation angle at base level
θ_g, θ_t	=	rotational angle at groundline (or point O), the angle at pile-head level
θ_o	=	pile-head rotation angle, or differential angle between upper and lower layer at point O (Chapter 12)
θ_w	=	rotation angle of pile at the point of e_w (Chapter 11)
λ	=	relative stiffness ratio between pile Young's modulus and the initial soil shear modulus at just above the base level, E_p/G_L (vertical loading)
λ, λ_i	=	reciprocal of characteristic length, $\lambda = \sqrt[4]{k/(4E_pI_p)}$ (lateral loading), λ for i-th layer [L^{-1}]
λ	=	factor to correlate shaft friction to mean effective overburden pressure, and undrained cohesion
λ_n	=	the n-th root for the Bessel functions
λ_s	=	Lame's parameter
μ	=	degree of pile–soil relative slip
$\nu_p(\nu_s)$	=	Poisson's ratio of a pile (soil)
ξ	=	shaft stress softening factor, when $w > w_e$ (Chapter 4)
ξ	=	outward radial displacement of the soil around a pile [L] (Chapter 5)
ξ	=	a factor to capture impact of pile-head constraints and soil resistance, etc., on the resistance zone of a passive pile (Chapter 12)
ξ_b	=	pile base shear modulus nonhomogeneous factor, G_L/G_b
ξ_c	=	a yield stress ratio for cyclic loading
ξ_{max}, ξ_{min}	=	the maximum and minimum values of the factor ξ (Chapter 12)
π_1, π_1^*	=	normalized pile displacement and local limiting of π_1

π_2	=	normalized depth with pile length
π_3	=	normalized pile–soil relative stiffness factor
π_4	=	normalized pile–soil relative stiffness for plastic case
π_{2p}	=	normalized depth with slip length
π_v	=	pile–soil relative stiffness
ρ	=	pile-head deformation [L]
ρ_g	=	ratio of the average soil shear modulus over the pile embedded depth to the modulus at depth *L*
σ_{ho}	=	horizontal stress [FL^{-2}]
σ_r, σ_θ, σ_z	=	radial, circumferential, and vertical stress in the surrounding soil, respectively [FL^{-2}]
σ'_v	=	effective overburden pressure [FL^{-2}]
σ'_{vb}	=	effective overburden pressure at the toe of the pile [FL^{-2}]
$\sigma'_{vs}(\bar{\sigma}'_{vs})$	=	effective overburden pressure over the pile shaft (average σ'_{vs}) [FL^{-2}]
$\tau(\tau_f)$	=	local shear stress (limiting τ) [FL^{-2}]
$\dot{\tau}$	=	shear stress rate for spring 1 in the creep model
τ_f^c	=	new limiting shaft stress under cyclic loading [FL^{-2}]
$\tau_j(\tau_{fj})$	=	local shear stress on elastic spring *j* with *j* = 1, 3 (maximum τ_j) [FL^{-2}]
$\tau_o, \tau_o(t^*)$	=	shear stress on pile–soil interface, and τ_o at the time of t*[FL^{-2}]
$(\tau_o)_{ave}$	=	average shear stress on a pile–soil interface over all the entire pile length [FL^{-2}]
τ_{oj}	=	shear stress on pile–soil interface at elastic spring *j* (*j* = 1, 2) [FL^{-2}]
τ_s	=	(average) shaft friction along a pile shaft [FL^{-2}]
τ_{ultj}	=	ultimate (soil) shear stress for spring *j* (*j* = 1, 3), respectively [FL^{-2}]
$\phi(\phi')$	=	angle of friction of soil (effective ϕ)
ϕ	=	ultimate moment reduction factor (Chapter 10)
$\phi(r)$	=	attenuation of soil displacement at *r* from the pile axis
ϕ'_1	=	undisturbed friction angle at the pile toe before pile installation
ϕ'_{cv}	=	critical frictional angle at constant volume
ϕ_p	=	frictional angle under plane strain conditions
ϕ_r	=	residual angle of friction of soil

ϕ_{tr}	=	frictional angle under axisymmetric conditions
χ_v	=	a ratio of shaft and base stiffness factors for vertical loading
ψ	=	factor to correlate adhesion factor α to unconfined compressive strength q_u
ψ_j	=	$(\tau_{oj}R_{fj}/\tau_{ultj})$, nonlinear stress level on pile–soil interface for spring j (j = 1, 2)
ω	=	water content (Chapter 1)
ω	=	a pile-base shape and depth factor (Chapters 4 and 5)
ω	=	a rotation angle (in radian) of a lateral pile
ω_g	=	base shape and depth factor for a pile in a two-pile group
ω_h, ω_{oh}	=	a coefficient for estimating "ω", accounting for the effect of H/L and the value of ω_h at a ratio of $H/L = 4$
ω_v, ω_{ov}	=	a coefficient for estimating "ω", accounting for the effect of ν_s and the value of ω_v at a ratio of $\nu_s = 0.4$

Principal Subscript (Piles Only)

A	=	upper plastic zone (lateral piles)
B	=	lower elastic zone (lateral piles)
b	=	pile-base
max, m	=	maximum
p	=	pile
t	=	pile-head
u	=	ultimate

Chapter 1

Strength and stiffness from in situ tests

In the design of pile foundations, two critical soil parameters of shear modulus and limiting friction along pile–soil interface are generally required. Because they vary from site to site, they must be determined for each location. This can be very costly if done through laboratory and/or in situ tests. In practice, these parameters may be gained through some correlations with in situ tests. Typical correlations are reviewed here concerning two common types of Standard Penetration Tests (SPT) and Cone Penetration Tests (CPT).

1.1 STANDARD PENETRATION TESTS (SPT)

The Standard Penetration Test was developed in the late 1920s, but the test procedure was not standardized until 1958 (see ASTM D1586), with periodic revisions to date. The test consists of the following procedures: (a) driving the standard split-barrel sampler a distance of 460 mm into the soil at the bottom of the boring, (b) counting the number of the blows to drive the sampler for the last two 150 mm distance (total = 305 mm) to obtain the blow counts number N, (c) using a 63.5-kg driving mass (or hammer) falling "free" from a height of 760 mm onto the top of the boring rods. The SPT test is normally halted if (a) 50 blows are required for any 150 mm increment, (b) 100 blows are obtained (to drive the required 300 mm), and (c) 10 successive blows produce no advance. The raw SPT data must be corrected to yield the real (corrected) value of SPT, as outlined next.

1.1.1 Modification of raw SPT values

Skempton (1986) conducted an extensive review about the impact of some key factors on efficiency of SPT tests. The review provides the

- Values of hammer efficiency for four typical methods of releasing the hammer

- Values of rod energy ratio, which is defined as the ratio of the actual hammer energy to sampler and the input (free-fall) energy
- Correction factor of less than unity for a rod length less than 10 m, as a result of loss in delivered energy
- Tentative correction factors of 1.05 and 1.15 for 150 mm and 200 mm boreholes, respectively, as lower *N* values are normally gained using the boreholes than those from 115 mm boreholes
- Impact of overburden stress associated with sand density

In particular, the energy ratio is equal to 0.3 to 1.0 (Kovacs and Salmone 1982; Riggs 1983). A standard rod energy ratio of 60% is recommended to normalize all measured blow account *N* values, by simple proportion of energy ratios, to this standard, and the normalized values are denoted as N_{60}. Note standard penetration resistance is sometimes normalized to a rod energy ratio of 70% (E_{r70} = 70). The associated standard below counts of N_{70} (the subscript of "70" refers to E_{r70} = 70) is estimated using $E_{r70}N_{70} = E_{r60}N_{60}$. The blow counts, N_{60} of SPT data using the energy E_{60} is equal to another blow counts, N_{70}, of the energy E_{70}. For instance, given N_{70} = 20, the value of N_{60} with E_{r60} = 60 is calculated as 23.3. The efficiency, the ratio, and correction factors may be multiplied directly with a measured blow count *N* to gain a corrected *N* value, as detailed next.

1.1.1.1 Method A

In light of these influence factors listed, the N_{60} may be revised as (Skempton 1986):

$$N_{60} = \frac{E_m C_B C_S C_R N}{0.6} \tag{1.1}$$

where N_{60} = SPT *N* value corrected for field procedures; E_m = hammer efficiency; C_B = borehole diameter correction; C_s = sample correction; C_R = rod length correction; and *N* = measured SPT *N* value. In addition, the SPT data should be adjusted to accommodate the effect of effective stress and field procedures. This leads to a corrected SPT N'_{60} value of

$$N'_{60} = C_N N_{60} \tag{1.2}$$

where C_N is a modification factor for overburden stress (see Figure 1.1) with $C_N = 2/(1+\sigma'_v/p_a)$ (normally consolidated fine sands), $C_N = 3/(2+\sigma'_v/p_a)$ (normally consolidated coarse sands), and $C_N = 1.7/(0.7+\sigma'_v/p_a)$ (overconsolidated fine sands) (Skempton 1986); p_a = 100 kPa, atmospheric pressure; and σ'_v = vertical effective stress at the test location (see Figure 1.1).

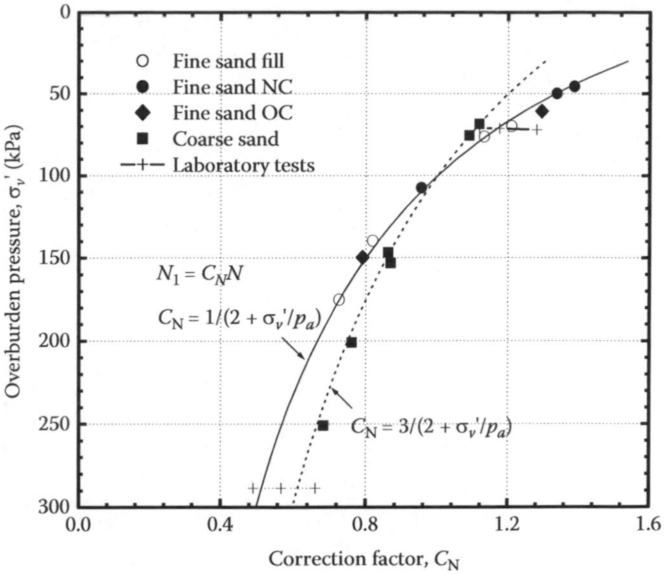

Figure 1.1 SPT N correction factor for overburden pressure. (After Skempton, A. W., *Geotechnique* 36, 3, 1986.)

1.1.1.2 Method B

When the SPT test is carried out in very fine sand or silty sand below the water table, the measured N value, if greater than 15, should be corrected for the increased resistance due to negative excess pore water pressure set up during driving. In this circumstance, the resulting SPT blow counts N' is given by (Terzaghi and Peck 1948)

$$N' = 15 + 0.5(N - 15) \tag{1.3}$$

Generally, the standard penetration resistance (N), relative density (D_r) may be related to effective overburden pressure (σ_v') by (Meyerhof 1957)

$$\frac{N}{D_r^2} = a + b\sigma_v' \tag{1.4}$$

Values of the parameters a and b for more than 50 field tests in sand have been determined (Skempton 1986), with typical examples being provided in Figure 1.2 (and σ'_v in units of kPa).

1.1.2 Relative density

The SPT value has been used to classify the consistency or relative density of a soil. Skempton (1986) added appropriate values of N'_{60} to Terzaghi and Peck's (1948) classification of relative density, as shown in Table 1.1.

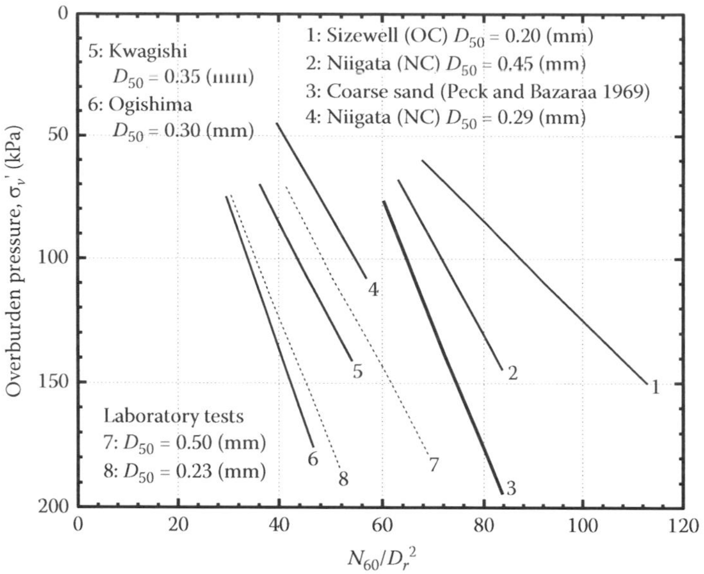

Figure 1.2 Effect of overburden pressure. (After Skempton, A. W., *Geotechnique* 36, 3, 1986.)

The value of D_r can be assessed in terms of SPT values (Kulhawy and Mayne 1990):

$$D_r = \sqrt{\frac{N'_{60}}{C_p C_A C_{OCR}}} \tag{1.5}$$

where $C_p = 60 + 25\log D_{50}$, $C_A = 1.2 + 0.05\log(0.01t)$ and $C_{OCR} = OCR^{0.18}$; D_r = relative density (in decimal form); C_p = grain-size correction factor; C_A = aging correction factor; C_{OCR} = overconsolidation correction factor; D_{50} = grain size at which 50% of the soil is finer (mm); t = age of soil (time since deposition)(years); and OCR = overconsolidation ratio.

Table 1.1 Soil relative density description

N'_{60} (Terzaghi et al. 1996)	*Relative density (%)*	*Descriptive term*	*Penetration resistance N (Blows/305 mm) (BSI 1981)*
0–3	0–15	Very loose	0–4
3–8	15–35	Loose	4–10
8–25	35–65	Medium	10–30
25–42	65–85	Dense	30–50
42–58	85–100	Very dense	>50

1.1.3 Undrained soil strength vs. SPT *N*

It is useful to have a correlation between SPT values and soil strength. Many empirical expressions were proposed and are dependent of plasticity index (I_p). The I_p is normally used to describe the state of clay and silt, both alone and in mixtures with coarser material. The soil plasticity is normally classified in terms of liquid limit (LL) (BSI 1981), as low (LL < 35%), intermediate (LL = 35~50%), high (LL = 50~70%), very high (LL = 70~90%), or extremely high (LL > 90%)

Stroud (1974) presented the variation of N value with undrained shear strength (s_u) for London clay. As replotted in Figure 1.3, the s_u is about (4–5) N kPa for clays of medium plasticity and (6–7)N kPa or more for soil with a plasticity index (I_P) less than 20. Tezaghi and Peck (1967) reported a high value of $s_u = 12.8N$. Sower (1979) attributed the variations to clay plasticity, and $s_u = 7.2N$ (low plasticity), 14.2N (medium), and 24N (high plasticity), respectively. The high ratio is also noted as $s_u = 29N^{0.27}$ (Hara et al. 1974). On the other hand, low strength of s_u = (1~4)N kPa is commonly adopted in Southeast Asia region: an average s_u = ~4N kPa for the weathered Kenny Hill Formation from eight test sites within Kuala Lumpur (Wong and Singh 1996); and a strength $s_u \approx N$ for clayey silt, and $s_u \approx$ (2~3)N for silty clay in Malaysia (Ting and Wong 1987). The Chinese hydraulic engineering code (SD 128-86) recommends a cohesion of (5.5~7)N (N = 3~13), and (3.5~5.2) N (N = 17~31) for alluvial and diluvial clay, as detailed in Table 1.2.

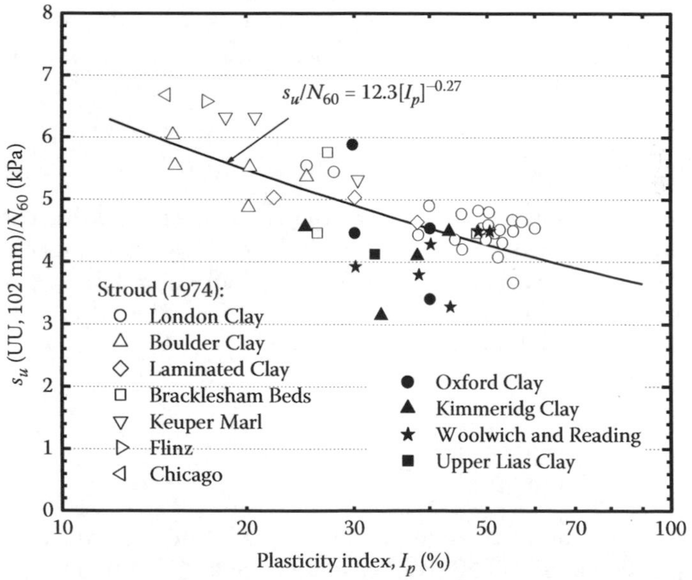

Figure 1.3 Plasticity versus ratio of undrained shear strength over N_{60}. (Data from Stroud, M. A., *Proc European Symp on Penetration Testing*, Stockholm, 1974.)

Table 1.2 Correlation between *N* value and friction angle or cohesion

$N_{63.5}$ (manual)	**3**	**5**	**7**	**9**	**11**	**13**	**15**
Friction angle, ϕ (°)	17.7	19.8	21.2	22.2	23.0	23.8	24.3
Cohesion, *c* (kPa)	17		49	59	66	72	78
$N_{63.5}$ (manual)	**17**	**19**	**21**	**25**	**29**	**31**	
Friction angle, ϕ (°)	24.8	25.3	25.7	26.4	27.0	27.3	
Cohesion, *c* (kPa)	83	87	91	98	103	107	

Source: Chinese Hydaulic Engineering Code (SD128-86), *Manual for Soil Testing*, 2nd ed., Chinese Hydraulic-Electric Press, Bejing, 1987.

Vane shear strength for normally consolidated clay was correlated to the overburden stress σ'_v and the plasticity index I_p by (Skempton 1957)

$$s_u / \sigma'_v = 0.11 + 0.0037 I_p \tag{1.6}$$

This expression agrees with that observed from Marine clays, but for a reverse trend of decrease in s_u/σ'_v with plasticity index I_p of 0~350 for soils with thixotropic behavior (dilate during shear) (Osterman 1959). Most of remolded clays have a s_u/σ'_v ratio of 0.3 ± 0.1.

It must be emphasized that the ratio of s_u/N depends on stress level to a large extent. In design of slope, dam, lateral piles, or predicting foundation response owing to liquefaction of sand, the clay or sand may "flow" around these structures or foundations. The associated undrained shear strength has been correlated with the N values, as illustrated in Figure 1.4. Indeed the observed ratio of s_u/N is much lower than those mentioned above in relation to foundation design at "pre-failure" stress level.

1.1.4 Friction angle vs. SPT *N*, D_r, and I_p

The angle of friction of soil has been correlated empirically with SPT blow counts by various investigators. Peck et al. (1953) suggested the curve in Figure 1.5, which is well represented with $\phi' = 0.3N_{60} + 26°$ to a *N* value of 40. The impact of effective overburden stress σ'_v on the angle has been incorporated into the estimation (e.g., the Japan railway). JSCE (1986) suggests

$$\phi' = 1.85^\circ \left[\frac{N}{0.7 + (\sigma'_v / p_a)} \right]^{0.6} + 26^\circ \tag{1.7}$$

This expression offers essentially similar values to those obtained from the Peck's expression to a *N* of 50 (Peck et al. 1953) at a $\sigma'_v < 100$ kPa, and otherwise results in an angle ~7 degrees smaller. The former prediction is close to the "extended" curve of the Chinese code (SD 128-86) of $0.3N + 19°$

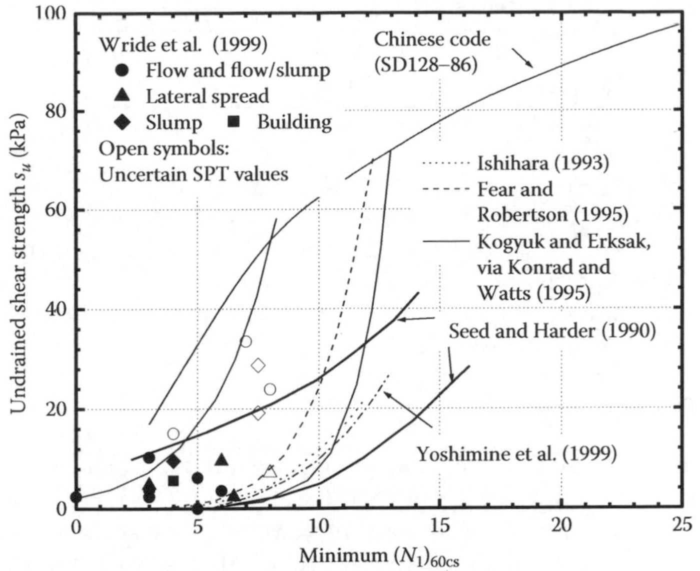

Figure 1.4 Undrained shear strength versus N_{60} (clean sand). (Data from Wride, C. E., E. C. McRoberts, and P. K. Robertson, Can. *Geotech. J.* 36, 1999.)

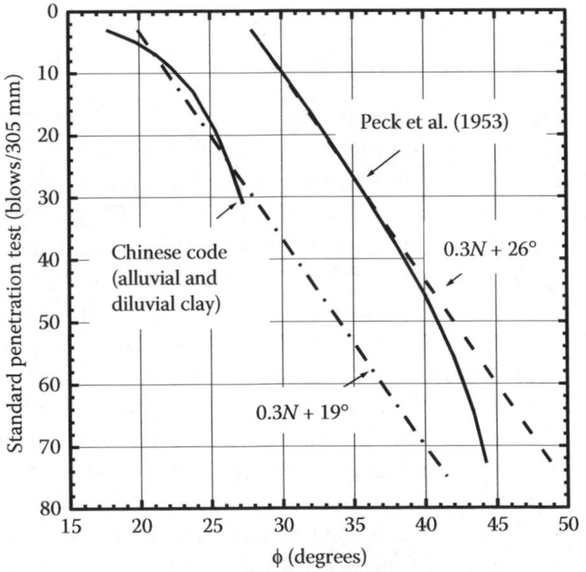

Figure 1.5 SPT blow counts versus angle of internal friction.

Table 1.3 Effective friction angle (ϕ'_{cv}) for clay soils

Plasticity Index (%)	15	30	50	80
ϕ'_{cv} (degrees)	30	25	20	15

Source: BSI, BS 8002, *BSI Standards*, 1994.

(see Figure 1.4), for example, at $N = 100$, Equation 1.7 offers $\phi' = 42.2°$ ($\sigma'_v = 200$ kPa) and 39.4° ($\sigma'_v = 300$ kPa). The low angle (for clay), in fact, characterizes the impact of high plasticity in comparison with the Peck's suggestion for low "plasticity" sand.

Using the relative density of Equation 1.5, the frictional angle of soil may also be obtained by (Meyerhof 1959)

$$\phi = 28 + 0.15D_r \ (D_r = \%) \tag{1.8}$$

In design of retaining structures in clay soils, in the absence of reliable laboratory test data, the conservative values (Table 1.3) of the constant volume (critical) angle of shearing resistance, ϕ_{cv} may be used with cohesion $c = 0$ (BSI 1994). The friction angle reduces by 10 degrees as the plasticity index I_p increases from 15% to 50%. The suggestion works well for laterally loaded piles (see Chapter 9, this book). On the other hand, some typical test data (Kenney 1959; Terzaghi et al. 1996) are plotted in Figure 1.6, which indicate the angle of friction (degrees) may be well correlated with plasticity index I_p by

$$\phi' = 56.2(I_p)^{-0.223} \tag{1.9}$$

1.1.5 Parameters affecting strength

Some correlations among overburden pressure, soil strength (density), and SPT blow counts are reported previously: (a) The $\sigma'_v \sim \phi \sim N_{60}$ plots (Schmertmann 1975), which provide a rough estimate of the angle ϕ (not suitable to a depth less than <2 m); (b) The $\sigma'_v \sim D_r \sim N_{60}$ plots (Holtz and Gibbs 1979), which offer the relative density D_r; (c) Gradation and maximum particle size on the angle ϕ (Lambe and Whitman 1979).

The impact of relative density D_r on friction angle ϕ of a soil is presented in Figure 1.7 (Bolton 1986), and that of plasticity index I_p against clay fraction was previously provided (Skempton 1964). The friction angle is also appreciably affected by initial void ratio and test (stress) conditions (e.g., intermediate principle stress). For instance, the angle of internal friction from a triaxial test (ϕ_{tr}) is 1~5 degrees smaller than the angle ϕ_p gained from a plane strain test. A higher difference is noted for dense specimens, which gradually diminishes for "loose" specimens, depending on the critical state friction angle (Bolton 1986). Figure 1.8 shows the measured friction angles of ϕ_{tr} and ϕ_p (Schanz and Vermeer 1996), which may be correlated by

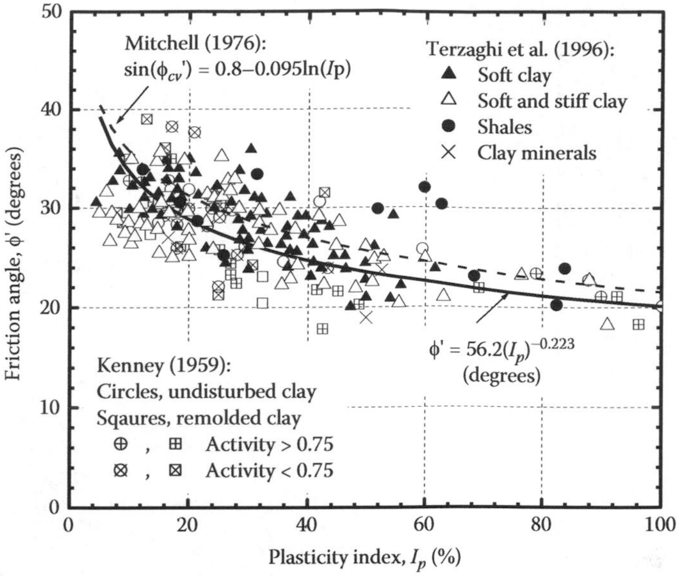

Figure 1.6 Variations of friction angle with plasticity index data. (Data from Terzaghi, K., B. P. Ralph and G. Mesri, *Soil mechanics in engineering practice*, John Wiley & Sons, New York, 1996; Kenney, T. C., *Proceedings*, ASCE 85, SM3, 1959.)

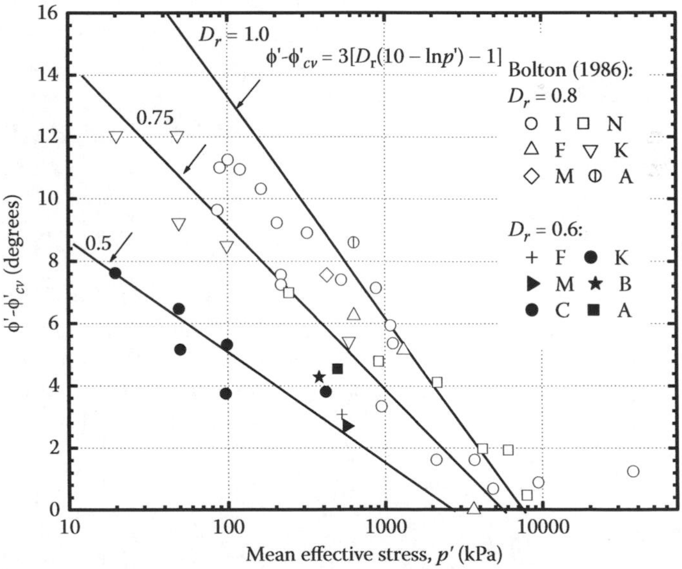

Figure 1.7 Shear resistance, ϕ'_{max}, ϕ'_{cv}, relative density D_r, and mean effective stress p'. (Data from Bolton, M. D., *Geotechnique* 36, 1, 1986.)

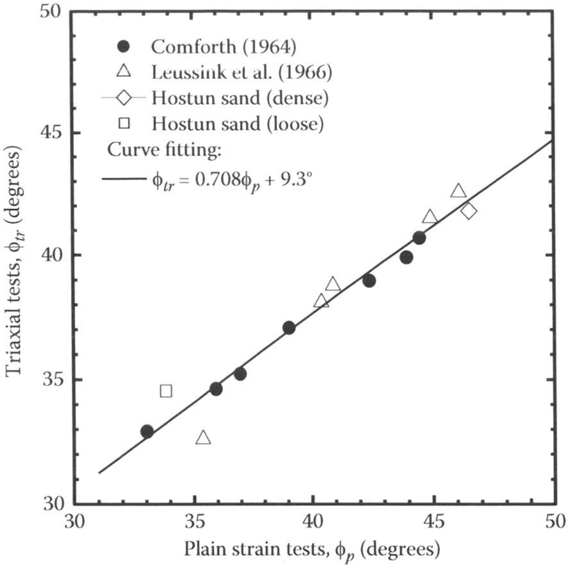

Figure 1.8 Results of regular and plane-strain triaxial tests. (Data from Schanz, T. and P. A. Vermeer, *Geotechnique* 46, 1, 1996.)

$$\phi_{tr} = 0.708\phi_p + 9.3° \text{ (for } \phi_{tr} > 30°) \quad \text{and} \quad \phi_{tr} = \phi_p \text{ (for } \phi_{tr} < 30°) \tag{1.10}$$

where ϕ_p = frictional angle under plane strain conditions, as would occur beneath a very long spread footing or a long wall leaning forward under later soil pressure; ϕ_{tr} = frictional angle under axisymmetric conditions (the intermediate principal stress = minor principal stress), as would occur at the tip of a pile or beneath a square footing. The adjustment of test conditions should be limited to no more than 5°.

1.2 CONE PENETRATION TESTS

A cone penetration test (CPT) is carried out by pushing the standard cone into ground at a rate of 10 to 20 mm/s, and recording the side friction resistance, q_s, and point resistance, q_c, at desired depths. CPT is suitable for fine to medium sand deposits, but not well adapted to gravel deposits or to stiff/hard cohesive deposits. Normally, friction ratio, f_r, is computed using the point and sleeve friction (side friction):

$$f_r = q_s / q_c (\%) \tag{1.11}$$

The soil sensitivity S_t may be estimated by (Robertson and Campanella 1983)

$$S_t = 10/f_r \tag{1.12}$$

where f_r is a percentage. The point resistance [also called cone tip (bearing) resistance] q_c is a very useful parameter. It has been adopted widely to indirectly determine soil strength, frictional angle, and SPT value.

1.2.1 Undrained shear strength

Undrained shear strength s_u may be correlated with the cone bearing resistance q_c by

$$s_u = \frac{q_c - \sigma_v}{N_k} \tag{1.13}$$

where $\sigma_v = \gamma_s z$ = overburden pressure at the depth z at which q_c is measured, and σ_v and q_c are of an identical unit. N_k = cone factor (a constant for each soil) ranging from 5 to 75, and largely between 15 and 20 (Mesri 2001) depending on values of plasticity index (Powell and Quarterman 1988).

Soil internal friction angle may be estimated using the cone bearing pressure q_c (MPa) by

$$\phi = 29° + \sqrt{q_c}\ (+5°\text{ for gravel; } -5°\text{ for silty sand}) \tag{1.14}$$

The frictional angle is affected by the effective overburden pressure (Robertson and Campanella 1983) and was presented in charts of q_c~σ'_v~ϕ.

1.2.2 SPT blow counts using q_c

The cone resistance q_c of a CPT test may be linearly correlated to a SPT N-blow count in both cohesive and cohesionless materials by

$$q_c = kN \tag{1.15}$$

where q_c is in units of MPa and the coefficient k varies from 0.1 to 1.0, as detailed in Table 1.4 (Sutherland 1974; Ramaswamy 1982), in which the N is N'_{60}.

The relationship between mean grain size (D_{50} or D_{10}) and q_c/N ratio has been widely reported. Typical results against D_{50} are as plotted in Figure 1.9. The q_c/N ratio relies significantly on soil particle size, compared to an early simple ratio q_c/N_{55} of 0.4 (MPa) (Meyerhof 1956).

1.3 SOIL STIFFNESS

Many empirical expressions are proposed to obtain soil shear modulus from in situ tests including SPT blow counts and cone resistance from CPT

Table 1.4 Relationship between CPT and SPT results

Soil type	q_c/N_{60}*(MPa)*
Silts, sandy silts, and slightly cohesive silt-sand mixtures	0.1~0.2 (Ramaswamy 1982)
Clean fine to medium sands and slightly silty sands	0.3~0.4
Coarse sands and sands with little gravel	0.5~0.7
Sandy gravel and gravel	0.8~1.0
Sandy silts	0.25 (Sutherland 1974)
Fine sand and silty fine sand	0.4
Fine to medium sand	0.48
Sand with some gravel	0.8

tests. To assess settlement of bored piles, Poulos (1993) summarized some empirical equations for estimating the modulus (low strain level). They are extended and presented in Table 1.5. Values of modulus were deduced using closed-form solutions against loading tests of laterally loaded piles in clay in Chapter 9, this book. They are plotted in Figure 1.10 against the uniaxial compressive strength.

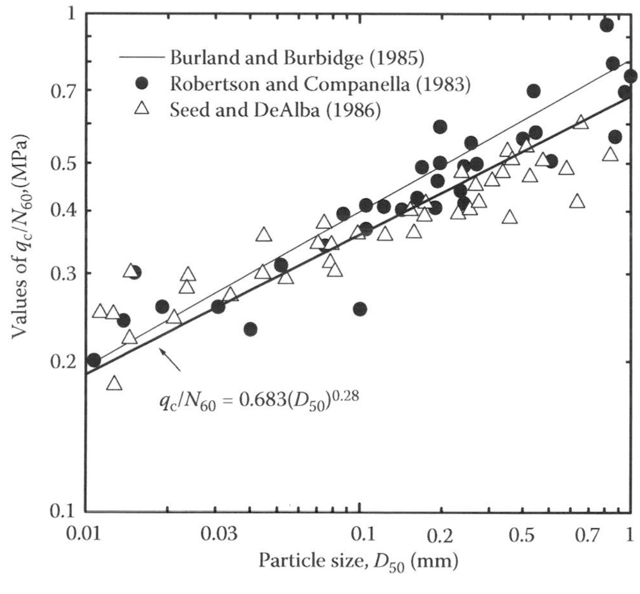

Figure 1.9 Cone penetration resistance q_c over standard penetration resistance N_{60} versus particle size D_{50} of sand. (Extended from Terzaghi, K., B. P. Ralph and G. Mesri, *Soil mechanics in engineering practice*, John Wiley & Sons, New York, 1995.)

Table 1.5 Near–pile drained soil modulus, *E*

Soil type	*Correlations*	*Remarks*
Clay	$E = 1.4N$ MPa	Hirayama (1991)
Clay	$E = (150–400)\ s_u$	Hirayama (1991), Poulos and Davis (1980)
Clay	$E = 7.3N^{0.72}$ MPa	Hara et al. (1974), using $E_s = 250s_u$
Clay	$E = 10\ q_c$	Christoulas and Frank (1991)
Silica sand	$E = (2.5–3.5)q_c$	Half the value for driven piles
Residual	$E = 3N$ MPa	Decourt et al. (1989)
Various	$E = 0.75N$ ($N < 14$) MPa	Christoulas and Frank (1991)
	$E = 7.5N–94.5$ MPa ($N \geq 14$)	
	$G/p_a = 12N^{0.8}$	USA Naval Facilities Engineering Command
	$G/p_a = 40N^{0.77}$	Fleming et al. (1992)
	$G/p_a = 10N$	Randolph (1981)
	$G/p_a = 8.25N$	Kenny Hill formation (Singapore)

Source: Revised from Poulos, H. G., *Proc., BAP II, Ghent*, Balkema, Rotterdam, 103–117, 1993.

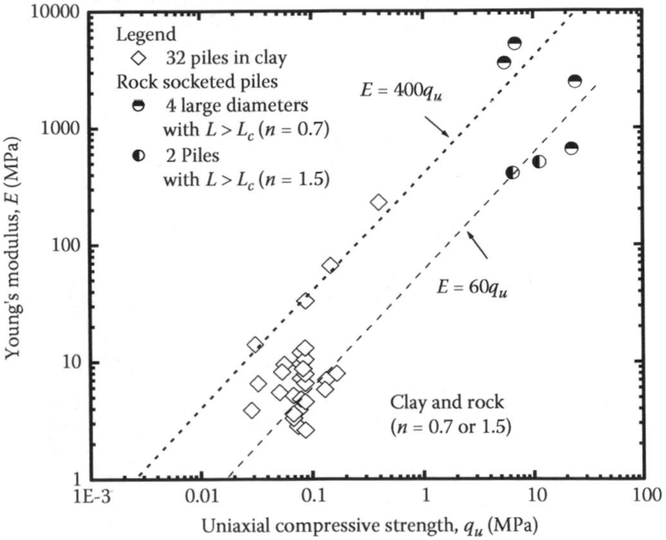

Figure 1.10 Moduli deduced from shafts in rock and piles in clay (q_u, E in MPa). (After Guo, W. D., *J Geotech Geoenviron Engrg*, ASCE, 2012a. With permission.)

1.4 STIFFNESS AND STRENGTH OF ROCK

1.4.1 Strength of rock

The engineering properties of rock pertinent to design are controlled by the extent and orientation of the bedding planes and joints within the rock mass together with the water pressures on the discontinuity planes. Rock mass strength (cohesion c and frictional angle ϕ) can be related to the unconfined compression strength q_u and the RQD (rock quality designation) value, as shown in Table 1.6.

The values of frictional angle for intact rock (Wyllie 1999) may be determined from Table 1.7, and higher values may be utilized for design of retaining structure as per Equation 1.10. The angles may be slightly conservative compared to those shown in Table 1.8, recommended by the British Standard. The rocks can be conservatively treated as composed of granular fragments. In other words, the rocks are closely and randomly jointed or otherwise fractured, having a RQD value close to zero. For instance, Guo and Wong (1987) reported clay-like shale materials with a residual shear strength of $c_r = 0$, $\phi_r = 21\sim24°$. It was classified as clayey silt (based on grain analysis). During slaking

Table 1.6 Strength of rock

RQD (%)	q_u	c	ϕ
0–70	0.33 q_{ui}[a]	0.1 q_{ui}	30°
70–100	0.33–0.8 q_{ui}	0.1 q_{ui}	30–60°

Source: Kulhawy, F. H., and R. E. Goodman, Design of foundations on discontinuous rock. *Proceedings of the International Conference on Structural Foundations on Rock*, Sydney, Australia, 1980.

[a] Unconfined compressive strength of intact rock

Table 1.7 Friction angle values for intact rock

Classification	*Type*	*Friction angle (degrees)*
Low friction	Schists (high mica content) Shale, Marl	20–27
Medium friction	Sandstone, Siltstone Chalk, Gneiss	27–34
High friction	Basalt, Granite, Limestone, Conglomerate	34–40

Source: Wyllie, D. C., *Foundation on Rock*. E & FN Spon, London and New York, 1999.

Table 1.8 Effective friction angle for rock

Stratum	*φ′ (degrees)*	*Notes*
Chalk	35	• The presence of a preferred orientation of joints, bedding or cleavage in a direction near that of a possible failure plane may require a reduction in all values, especially if the discontinuities are filled with weaker materials. • Chalk is defined here as weathered medium to hard, rubbly to blocky chalk, grade III
Clayey marl	28	
Sandy marl	33	
Weak sandstone	42	
Weak siltstone	35	
Weak mudstone	28	

Source: BSI, BS 8002, Clause 2.2.6, *BSI Standards*, 1994.

tests, it disintegrates quickly when contacted with water. The bond strength between grout and the fractured shale is ~340 kPa, and between grout and interbedded sandstone shale is 380 kPa. Both strength values are only fractions of the uniaxial compression strengths of sandstone (q_{ui} = 10 ~ 70 MPa) and shale (q_{ui} = 0.1 ~ 10 MPa). It should be stressed that the rock strength is generally dependent on moisture content, see Figures 1.11a and 1.12a (Clayton 1978; Dobereiner and Freitas 1986), and dry density (see Figure 1.11b).

1.4.2 Shear modulus of rock

The shear modulus G_m of an isotropic rock mass can be calculated from deformation modulus E_m and Poisson's ratio, v_m by $G_m = E_m/2(1 + v_m)$. For most rock masses, the Poisson's ratio v_m may be taken as 0.25 with v_m = 0.10 ~ 0.35 (U.S. Army Corps of Engineers 1994). The value of E_m is difficult, if not impossible, to determine in the laboratory as it is highly dependent on sample size.

Practically, the value of E_m is obtained by empirical correlations with rock mass classification or property indices including Rock Mass Rating (RMR), uniaxial compressive strength of intact rock (q_{ui}), Geology Strength Index (GSI), Rock Quality Designation (RQD), joint conditions, and modulus of intact rock, E_r. Five typical methods are listed in Table 1.9 as M1 ~ M4 and in Table 1.10 as M5. The RMR and GSI may be estimated as follows:

- The rock mass rating RMR (Bieniawski 1989), referred to as RMR_{89} hereafter, may be calculated by $RMR_{89} = R_A + R_B + R_C + R_D + R_E -$ Adj, where R_A, R_B, R_C, R_D, R_E = subratings for q_{ui}, RQD, spacing of discontinuities, conditions of discontinuities, and groundwater, respectively; and Adj = adjustment for orientation of discontinuities.
- The GSI may be roughly evaluated using GSI = RMR_{89} – 5, assuming completely dry conditions and a very favorable joint orientation (Hoek et al. 1995).

Figure 1.10 also provides the deduced Young's modulus versus undrained shear strength q_u from lateral pile tests (see Chapter 10, this book). They are higher than the deformation modulus for sandstone (Dobereiner and Freitas 1986), as shown in Figure 1.12. It should be stressed that the E_m may differ by 100 times using different methods of measurement. Zhang (2010) indicates the strength and modulus of various types of rocks may be empirically related by

$$\frac{q_u}{q_{ui}} = \left(\frac{E_m}{E_r}\right)^{0.7} \tag{1.16}$$

where q_{ui} = unconfined compression strength of an intact rock. The modulus of intact rock, E_r is (100~1000)q_{ui} (Reese 1997).

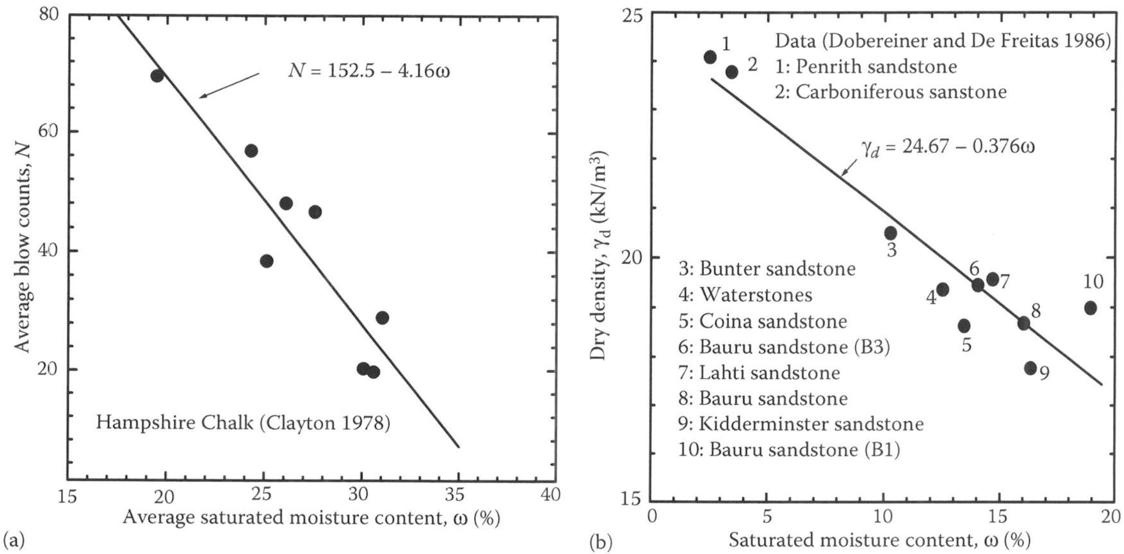

Figure 1.11 (a) SPT versus moisture content (density). (From Clayton, C. R. I., *Geotechnique* 28, 1, 1978); (b) Moisture content (density) versus strength. (From Dobereiner, L. and M. H. Freitas, *Geotechnique* 36, 1, 1986.)

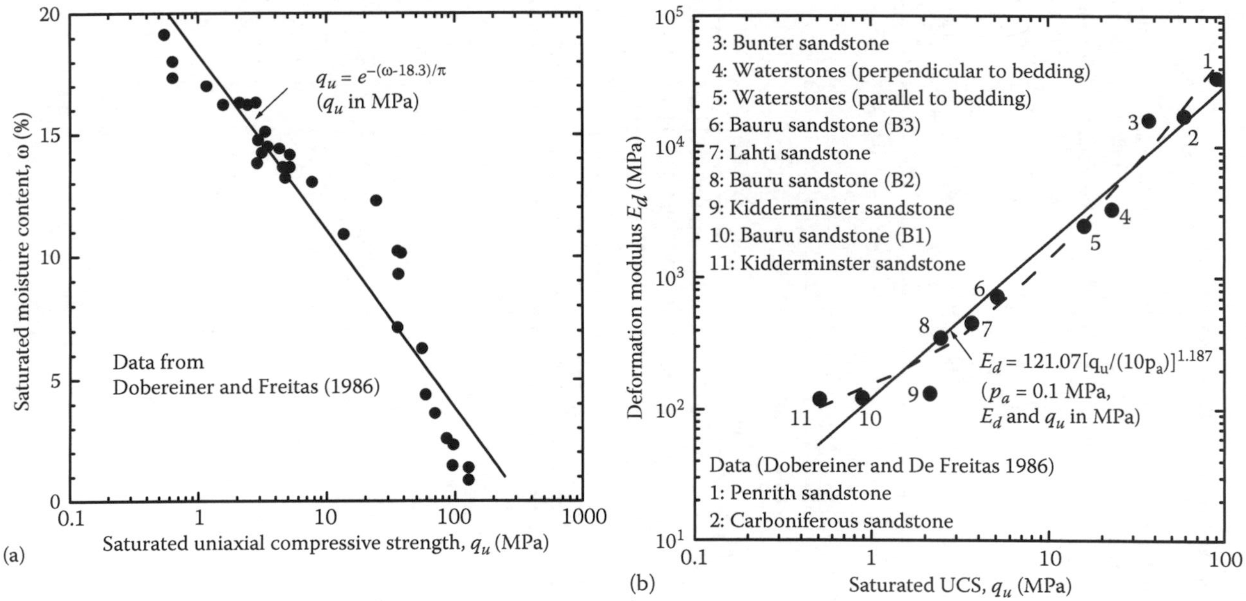

Figure 1.12 (a) Moisture content (density) versus strength. (b) Modulus versus uniaxial compressive strength. (From Dobereiner, L. and M. H. Freitas, *Geotechnique* 36, 1, 1986.)

Table 1.9 Empirical expressions for deformation modulus E_m

Method	*Reference*	*Empirical equation*	*Applicable conditions*	*Derived conditions*
M1	(Bieniawski 1989)	E_m (GPa) = 2 RMR − 100	RMR > 50	Mining support structures
M2	(Serafim and Pereira 1983)	E_m (GPa) = $10^{(RMR-10)/40}$	RMR = 20~85	Dam foundations
M3	(Rowe and Armitage 1984)	E_m (kPa) = $215\sqrt{q_{ui}}$	Any rock	Axially loaded drilled piles
M4	(Hoek 2000)	$E_m(GPa) = \sqrt{\frac{q_{ui}}{100}} 10^{(GSI-10)/40}$	$q_{ui} < 100$ MPa	Underground excavation
M5	(Sabatini et al. 2002)	Refer to Table 1.10	Any rock	Axially loaded drilled piles

Table 1.10 M5: Estimation of E_m based on RQD

	E_m/E_r		
RQD (%)	*Closed joints*	*Open joints*	*Remarks*
100	1.00	0.6	Values intermediate between tabulated entry values may be obtained by linear interpolation.
70	0.7	0.10	
50	0.15	0.10	
20	0.05	0.05	

Source: Sabatini, P. J., R. C. Bachus, P. W. Mayne, J. A. Schneider, and T. E. Zettler, Geotechnical Engineering Circular No. 5: Evaluation of Soil and Rock Properties, Rep. No. FHWA-IF-02-034, 2002.

Chapter 2

Capacity of vertically loaded piles

2.1 INTRODUCTION

Piles are commonly used to transfer superstructure load into subsoil and a stiff bearing layer. Piles are designed to ensure the structural safety of the pile body, an adequate geotechnical capacity of piles, and a tolerable settlement/displacement of piles. Under vertical loading, the design may be achieved by predicting nonlinear response of the pile using shaft friction and base resistance (see Chapter 4, this book). In practice, pile capacity is routinely obtained, assuming a rigid pile and a full mobilization of shaft friction. This is discussed in this chapter, along with pertinent issues and methods for estimating the negative skin friction and the capacity of pile groups.

2.2 CAPACITY OF SINGLE PILES

The total capacity of a single pile is customarily estimated using

$$P_u = q_b A_b + \tau_s A_s \tag{2.1}$$

where P_u = total ultimate capacity; q_b, τ_s = pressure on pile base and friction along pile shaft, respectively; and A_b and A_s = areas of the pile base and shaft, respectively. The capacity P_u consists of shaft component of $\tau_s A_s$ and base component of $q_b A_b$. It may be attained at a shaft displacement of 0.5%~2% and base displacement of ~20% pile diameter, depending on pile–soil relative stiffness.

2.2.1 Total stress approach: Piles in clay

2.2.1.1 α *Method* ($\tau_s = \alpha s_u$ *and* q_b)

With respect to piles in clay, the q_b and τ_s may be correlated to the undrained shear strength at pile tip level (c_{ub}) and the average shear strength along pile shaft (s_u) by $q_b = \omega c_{ub} N_c$, and $\tau_s = \alpha s_u$. The ω is a reduction factor that relates the ratio of the full-scale strength to the smaller sample strength, ω = 0.6~0.9 (Rowe 1972), or ω = 0.75 for bored piles in fissured clays (Whitaker and Cooke 1966).

Theoretically, the N_c only depends on the pile slenderness ratio L/d (L = pile length, d = diameter). However, higher values of N_c could be appropriate for displacement piles. On the basis of Skempton's work, $N_c = 9$ is generally used for all deep foundations in clay if $L/d > 4$, otherwise, it is reduced appropriately (e.g., $N_c = 5.14$ for $L/d = 0$; Craig 1997). In the case where piles penetrate into stiffer clays underlying soft clays, a low N_c may be adopted to reflect impact of base failure beyond the stiff clay layer. This may be assessed through a critical penetration ratio, $(L/d)_{crit}$ (Meyerhof 1976). With $L/d < (L/d)_{crit}$, any point failure is entirely contained within that layer and not influenced by the softer upper layer (Das 1990). The point bearing capacity in the lower layer could be assumed to increase linearly from the full bearing capacity of the softer overlying soil at the surface of the underlying stiffer soil to the full bearing capacity of the stiffer soil at the critical penetration ratio. On the other hand, higher values of N_c normally would be associated with driven piles. As an extreme case, the N_c reaches 15 to 21 for cone penetration tests (Mesri 2001). However, piles may be founded in stiff or very stiff clays, which tend to be less sensitive to disturbance effects, with negligible time-dependent variation of excess porewater pressures and strength.

The effects of pile installation, disturbance, porewater pressure development, and dissipation and consolidation are significant for soft clays, which will be discussed in Chapter 5, this book. An adhesion factor α of about unity or even higher may be selected to compensate for the immediate loss of strength (from undisturbed to remolded values) due to driving of the pile and the subsequent consolidation of the soil. However, an adhesion factor α of less than unity would be appropriate for sensitive soils, as full compensation for the immediate loss of strength may not occur even after the subsequent consolidation.

This adhesion factor α is determined empirically from the results of both full-scale and model pile load tests. Values of α are provided for driven piles in firm to stiff clay, sands underlined by stiff clay, and soft clay underlined by stiff clay, with respect to impact of pile length (Tomlinson 1970). The α is 0.3~0.6 for bored piles (Tomlinson 1970) or $\alpha = 0.45$ for piles in London clay (Skempton 1959) and 0.3 for short piles in heavily fissured clay. The shaft adhesion should not exceed about 100 kPa. The variation of α with s_u may be estimated using Equation 2.2, as synthesized from 106 load tests on drilled shafts in clay (Kulhawy and Jackson 1989)

$$\alpha = 0.21 + 0.26 p_a / s_u \quad (\alpha \le 1) \tag{2.2}$$

where s_u = undrained shear strength; p_a = the atmospheric pressure (≈ 100 kPa); and p_a and s_u are of same units. The values of α with undrained shear strength are recommended by API (1984), Peck et al. (1974), and Bowles (1997), to name a few. The variation is dependent of overburden pressure. Assumed α = 1.0 for normally consolidated clay, the α may be given for the general case by two expressions of (Randolph 1983) (see Figure 2.1):

$$\alpha = (s_u / \sigma'_v)^{0.5}_{nc} (s_u / \sigma'_v)^{-0.5} \quad (s_u / \sigma'_v \le 1) \tag{2.3a}$$

$$\alpha = (s_u / \sigma'_v)^{0.5}_{nc} (s_u / \sigma'_v)^{-0.25} \quad (s_u / \sigma'_v > 1) \tag{2.3b}$$

where σ'_v = effective overburden pressure. The "nc" subscript in these expressions refers to the normally consolidated state of soil. The shaft friction (thus α) is related to pile slenderness ratio and overburden pressure by (Kolk and Velde 1996)

$$\alpha = 0.55(s_u / \sigma'_v)^{-0.3}[40/(L/d)]^{0.2} \tag{2.4}$$

Given a slenderness ratio L/d of 10–250, Equation 2.4 covers the large range of measured data (see Figure 2.1).

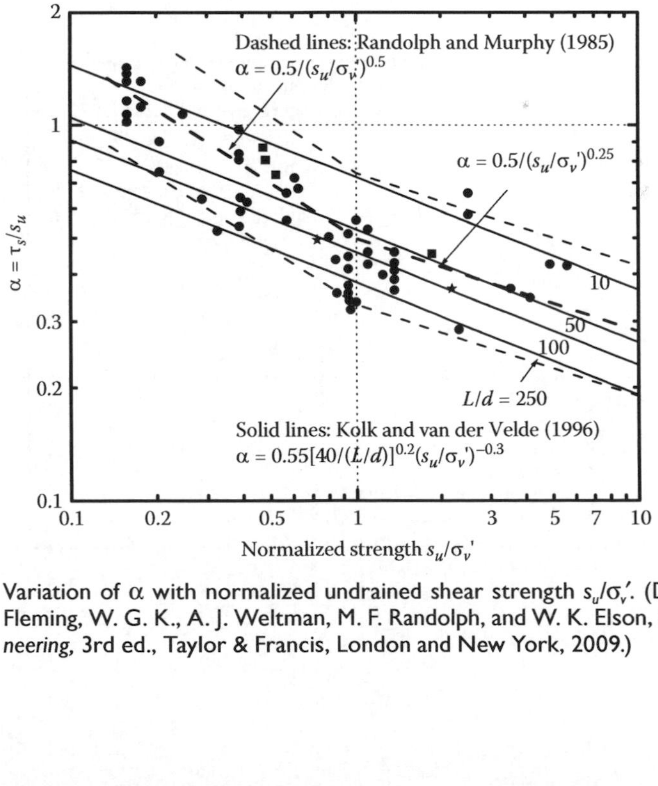

Figure 2.1 Variation of α with normalized undrained shear strength s_u/σ'_v. (Data from Fleming, W. G. K., A. J. Weltman, M. F. Randolph, and W. K. Elson, *Piling engineering*, 3rd ed., Taylor & Francis, London and New York, 2009.)

2.2.1.2 λ Method: Offshore piles

An alternative empirical (λ) method (Vijayvergiya and Focht 1972) has been proposed for estimating the shaft resistance of long steel pipe piles installed in clay. This method is generally used in the design of offshore piles (with an embedment L >15 m) that derive total capacity mainly from the ultimate shaft friction, τ_s (and negligible end-bearing component):

$$\tau_s = \lambda(\tilde{\sigma}'_{vs} + 2\tilde{s}_u) \tag{2.5}$$

where $\tilde{\sigma}'_{vs}$ and $\tilde{s}_u$ = the mean (denoted by "~") effective overburden pressure and undrained cohesion along the pile shaft, respectively. The λ factor is a function of pile penetration and decreases to a reasonably constant value for very large penetrations. Two typical expressions are shown in Figure 2.2. For instance, Kraft et al. (1981) suggested λ = 0.296 – 0.032ln(L/0.305) (normal consolidated soil) for $\tilde{s}_u/\sigma'_v < 0.4$; otherwise λ = 0.488 – 0.078ln(L/0.305) (overconsolidated soil), and L in m. Equations 2.2 and 2.5 show an inversely proportional reduction in shaft friction with the undrained shear strength s_u. It is interesting to note that the α deduced from Equation 2.5 compares well with the previously measured values of α, as shown in Figure 2.2. The predicted curves of $\alpha \sim s_u/\sigma'_v$ (σ'_{vs} is simply written as σ'_v) using a pile length of 10 and 50 m in Equation 2.5 bracket the measured data well using Vijayvergiye and Focht's λ factor.

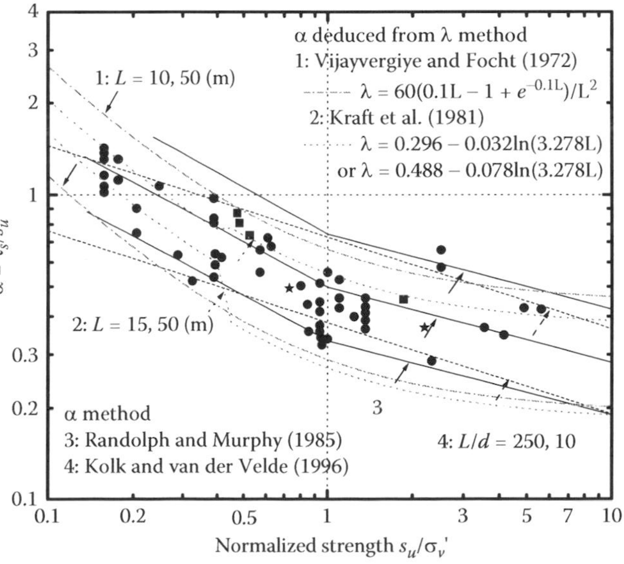

Figure 2.2 Variation of α as deduced from λ method.

2.2.2 Effective stress approach

2.2.2.1 β *Method for clay* ($\tau_s = \beta\sigma'_{vs}$)

The unit shaft resistance τ_s and end-bearing stress q_b are linearly related to effective overburden pressures σ'_{vs} and σ'_{vb}, respectively (Chandler 1968): $\tau_s = \beta\sigma'_{vs}$, and $q_b = \sigma'_{vb}N_q$, in which $\beta = K\tan\delta$, δ = interface frictional angle, being consistent with that measured in simple shear tests; N_q = bearing capacity factor; σ'_{vb} = effective overburden pressure at the toe of the pile; K = average coefficient of earth pressure on pile shaft, for driven pile, lying between $(1 - \sin\phi')$ and $\cos^2\phi'/(1 + \sin^2\phi')$, and close to $1.5K_o$ (K_o = static earth pressure coefficient); and ϕ' = effective angle of friction of soil. This effective strength approach is suitable for soft clays but not quite appropriate for stiff clays with cohesion.

For normally consolidated clay, assuming $K = K_o = 1 - \sin\phi'$ and $\delta = \phi'$, Burland (1973) obtained $\beta = (1 - \sin\phi')\tan\phi'$. The ϕ' may be taken as a residual friction angle ϕ'_{cv} and calculated using $\sin\phi'_{cv} \approx 0.8 - 0.094\ln I_p$, in which I_p = the plasticity index (Mitchell 1976; see Chapter 1, this book). The use of residue ϕ for pile design compares well with the values derived from a range of tests conducted on driven piles in soft, normally consolidated clays. However, slightly higher values of β (>0.3) were deduced from measured data, probably due to the actual K value being greater than K_o caused by pile installation. The β may be estimated using $\beta = \sin\phi'\cos\phi'/(1 + \sin^2\phi')$ (Parry and Swain 1977a). Given $21.5° \leq \phi' \leq 39°$, the equation yields $0.30 \leq \beta \leq 0.35$, which fit the test data well.

Assuming $K = K_o$, the effective stress β approach provides a lower and upper bound to the test results obtained for driven and bored piles in stiff London clay, respectively. Meyerhof (1976) demonstrated $K \approx 1.5K_o$ for driven piles in stiff clay and $K \approx 0.75K_o$ for bored piles, in which $K_o = (1 - \sin\phi')\sqrt{OCR}$, and OCR is the overconsolidation ratio of the clay. As deduced from 44 test pile data, the β may be correlated to pile embedment L (Flaate and Selnes 1977) by $\beta = 0.4\sqrt{OCR}\,(L + 20)/(2L + 20)$ (L in m).

2.2.2.2 β *Method for piles in sand* ($\tau_s = \beta\sigma'_{vs}$)

In calculating τ_s using $\beta = K\tan\delta$, the parameters K and δ may be estimated separately. The value of K must be related to the initial horizontal earth pressure coefficient, K_o ($= 1 - \sin\phi$). As with piles in clay, capacity may be estimated by assuming $K = 0.5$ for loose sands or $K = 1$ for dense sands. Kulhawy (1984) and Reese and O'Neill (1989) recommend the values of K/K_o in Table 2.1 to cater for impact of pile displacement and construction methods.

The angle of friction between the pile shaft and the soil δ is a function of pile surface roughness. It may attain the value of frictional angle of soil ϕ' on a rough-surface pile, and the pile–soil relative shearing occurs entirely within the soil. However, it is noted that the angle ϕ' ($\approx \phi'_{cv}$) is generally

Table 2.1 Values of average coefficient of earth pressure on pile shaft over that of static earth pressure K/K_o

Pile type	K/K_o	*Construction method (Bored piles)*	K/K_o
Jetted piles	1/2 ~ 2/3	Dry construction with minimal sidewall disturbance and prompt concreting	1.0
Drilled shaft, cast-in-place	2/3 ~ 1	Slurry construction—good workmanship	1.0
Driven pile, small displacement	3/4 ~ 5/4	Slurry construction—poor workmanship	2/3
Driven pile, large displacement	1 ~ 2	Casing under water	5/6
References	(Kulhawy 1984)	(Reese and O'Neill 1989)	

noted on a smooth-surface pile, and the shearing takes place on a pile surface. There have been many suggestions as to the values of δ and δ/ϕ'. Practically, for instance, Broms (1966) suggested $\delta = 20°$ for steel piles and a ratio of $\delta/\phi' = 0.75$ and 0.66 for concrete and timber piles respectively. Kulhawy (1984) recommends the ratios of δ/ϕ' in Table 2.2, again to capture the impact of pile displacement and construction methods.

Theoretically, Bolton (1986) demonstrates $\phi' = \phi'_{cv} + 0.8\psi$ for sand, in which ψ = dilatancy angle. It seems that volume changes cause the increase in shear strength, whether sand or clay. Chen et al. (private communication, 2010) conducted shear tests on clay and concrete interface. The roughness of the concrete was R_{max} = 1~10 mm, and the remolded Zhejiang clay has a plasticity index of 26.5 and liquid limit of 53.1. The tests indicate an angle of $\psi = 5°$ compared to $\phi'_{cv} = 6°\text{~}9°$. The shear strength indeed vary with the material on the pile surface, as shaft stress increases ~100% (from 30 kPa to 62 kPa) when the bentonite slurry was changed to a liquid polymer in constructing drilled shaft (Brown 2002).

Potyondi reported similar ratios of δ/ϕ' to what is mentioned above and indicated wet sand is associated with high ratios for steel and timber piles (Potyondy 1961). The soil-shaft interface friction angle for bored piles may drop by 30% depending on construction methods (see Table 2.2; Reese and O'Neill 1989).

During driving, the shaft resistance behind the pile tip is progressively fatigued. Randolph (2003) indicates the K varies with depth and may be captured by

$$K = K_{min} + (K_{max} - K_{min})e^{-0.05h/d} \tag{2.6}$$

where K_{min} = 0.2~0.4, K_{max} = (0.01~0.02)q_c/σ'_v, q_c = cone tip resistance from cone penetration test (CPT), and h = distance above pile tip level. The

Table 2.2 Values of angle between pile and soil over friction angle of soil δ/ϕ'

Pile material	δ/ϕ'	*Construction method (Bored piles)*	δ/ϕ'
Rough concrete (cast-in-place)	1.0	Open hole or temporary casing	1.0
Smooth concrete (precast)	0.8~1.0	Slurry method—minimal slurry cake	1.0
Rough steel (corrugated)	0.7~0.9	Slurry method—heavy slurry cake	0.8
Smooth steel (coated)	0.5~0.7	Permanent casing	0.7
Timber (pressure-treated)	0.8~0.9		
References	(Kulhawy 1984)	(Reese and O'Neill 1989)	

impact of fatigue coupled with dilation may be captured by (Lehane and Jardine 1994; Schneider et al. 2008):

$$K = 0.03A_r^{0.3}[\max(h/d,2)]^{-0.5}q_c/\sigma_v' \tag{2.7}$$

where $A_r = 1 - (d_i/d_o)^2$, d_i and d_o = inner diameter and outside diameter.

Recent study on model piles has revealed the impact of dilation on shear strength (shaft friction). Average shaft stress τ_s was measured under various overburden pressure σ'_{vo} (Lehane et al. 2005) on piles (with a diameter of 3~18 mm) in sand having a D_{50} of 0.2 mm and $\tan\phi'_{cv} = 0.7$ (ϕ'_{cv} = critical frictional angle at constant volume). The measured value of τ_s is normalized as $\tau_s/[\sigma_{ho}\tan\phi'_{cv}]$ (σ_{ho}= horizontal stress). The potential strength on pile shaft is normalized as $K_o\sigma'_{vo}\tan\phi'_{cv}/p_a$. The pair of normalized values for each test is plotted in Figure 2.3, which offers $\tau_s/(\sigma'_{ho}\tan\phi'_{cv}) = \psi(K_o\sigma'_{vo}\tan\phi'_{cv}/p_a)^{-0.5}$ with ψ = 1.0~2.2. This effect of dilatancy on the adhesion factor for shaft friction is compared late with that observed in piles in clay and rock sockets (Kulhawy and Phoon 1993).

Some measured variations of β with depth (Neely 1991; Gavin et al. 2009) are provided in Figure 2.4, which will be discussed later.

Values of $K\tan\delta$ for driven, jacked, and bored piles (Meyerhof 1976) are illustrated in Figure 2.5. The calculation uses an undisturbed value of friction angle of soil ϕ'_1 (see Figure 2.5a) but an interface friction angle δ (= $0.75\phi' + 10°$) for bored piles in Figure 2.5b. The angle ϕ' should be calculated using $\phi' = 0.75\,\phi'_1 + 10$ for driven piles and $\phi' = \phi'_1 - 3$ for bored piles (Kishida 1967). These values are combined in the Meyerhof method with the full calculated effective overburden pressure. A similar relationship was also deduced using model test results on steel piles (Vesić 1969; Poulos and Davis 1980). The values given in this figure may be conservative for other (rougher) surface finishes.

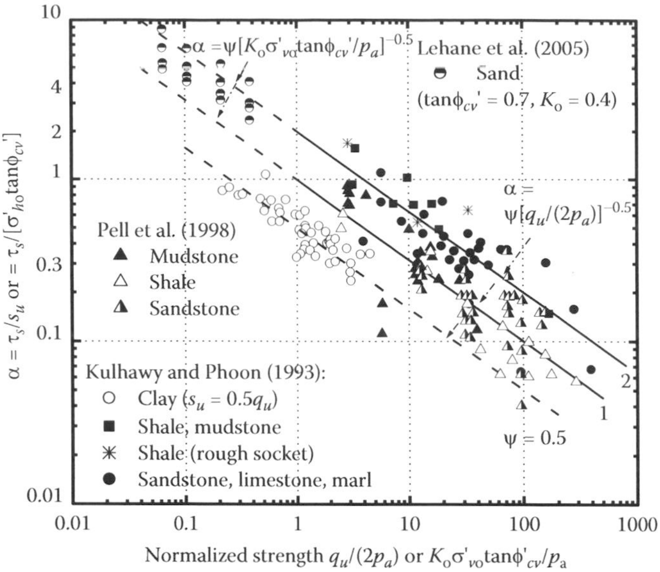

Figure 2.3 α versus the normalized uniaxial compressive strength q_u; p_a = 100 kPa. (Data from Kulhawy, F. H. and K. K. Phoon, *Proc Conf on Design and Perform Deep Foundations: Piles and Piers in Soil and Soft Rock*, ASCE, 1993; Lehane, B. M., C. Gaudin and J. A. Schneider, *Geotechnique* 55, 10, 2005; Pells, P. J. N., G. Mostyn and B. F. Walker, *Australian Geomechanics* 33, 4, 1998.)

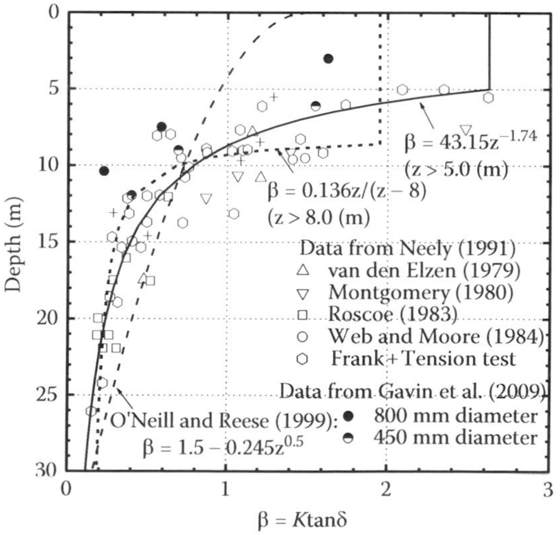

Figure 2.4 Variation of β with depth.

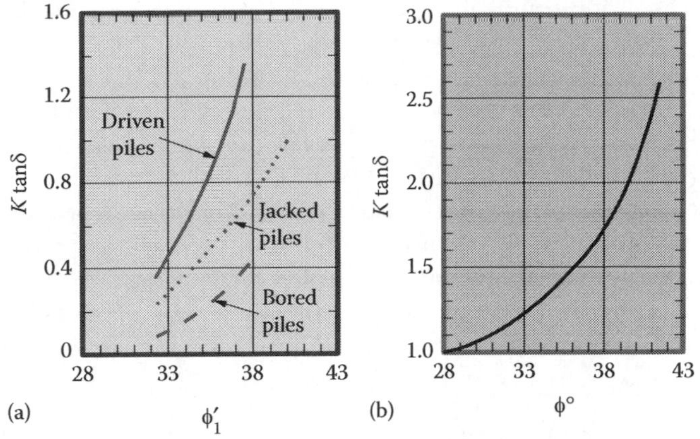

Figure 2.5 Ktan δ proposed by (a) Meyerhof (1976), (b) Vesić (1967). (After Poulos, H. G. and E. H. Davis, *Pile foundation analysis and design*, John Wiley & Sons, New York, 1980.)

2.2.2.3 *Base resistance* q_b (= $N_q\sigma'_{vb}$)

The bearing stress q_b for piles in sand is equal to $N_q\sigma'_{vb}$. Meyerhof (1951, 1976) stipulated a proportional increase in the effective overburden pressure at the pile base σ'_{vb} in a granular soil with depth. On the other hand, Vesić (1967), on the basis of model tests, concluded that the vertical effective pressure reaches a limiting value at a certain (critical or limiting) depth, z_c, beyond which it remains constant. The limiting vertical (and therefore horizontal) stress effect has been attributed to arching in the soil and particle crushing.

2.2.2.3.1 *Unlimited* σ'_{vb}

Meyerhof (1951, 1976) shows bearing capacity factor N_q depends on the penetration ratio L/d related to a critical value, $(L/d)_{crit}$. For instance, at $\phi = 40°$, it follows: $(L/d)_{crit} = 17$ and $N_q = 330$. The N_q is calculated using in situ undisturbed values of ϕ or $(0.5\sim0.7)\phi'$ if loosening of the soil is considered likely. If the critical embedment ratio is not attained, the N_q values must be reduced accordingly. For layered soils, where the pile penetrates from a loose soil into a dense soil, Meyerhof suggested use of the full bearing capacity of the denser soil if the penetration exceeds 10 pile widths ($L/d > 10$). Presuming the total pile embedment still satisfies the aforementioned critical ratios, the unit point bearing pressure q_b should be reduced by

$$q_b = q_{ll} + \frac{(q_{ld} - q_{ll})h_d}{10d} \le q_{ld} \tag{2.8}$$

where q_{ll}, q_{ld} = limiting unit base resistances of $\sigma'_{vb}N_q$ in the loose and dense sand, respectively; h_d = the pile penetration into the dense sand; and d = pile width or diameter. The σ'_{vb} is the (unlimited) effective overburden pressure, but the end bearing pressure $N_q\sigma'_{vb}$ is subject to the limiting value with $q_b \leq 50N_q \tan\phi$ (e.g., at $\phi = 45°$ [very dense] $q_b \leq 45$ MPa). In general practice, the q_b is less than 15 MPa.

Kulhawy (1984) suspects the existence of the limiting depth and attributed it to the impact of overconsolidation on the capacity of piles of typical (< 20 m) length and the model scale of pertinent tests. For instance, he proposed that the full overburden pressure should be used. For a square shape with width B (drained case), the unit tip resistance is computed as:

$$q_b = 0.3B\gamma'_s N_\gamma \zeta_{\gamma r} + qN_q \zeta_{qr}\zeta_{qs}\zeta_{qd} \tag{2.9}$$

where γ'_s= effective soil unit weight; $N_q = \tan^2(45 + 0.5\phi)e^{\pi\tan\phi}$; $N_\gamma = 2(N_q + 1)\tan\phi$; and q = vertical effective stress at embedment depth. The ζ modifiers reflect the impact of soil rigidity denoted by subscript "r", foundation shape by "s" and foundation depth by "d", respectively. For $L/d > 4$~5, the first term becomes less than 10% of the second term and can be ignored. The dimensionless portions of both terms are given by Kulhawy in graphical form against friction angle, ϕ. To use these graphs, it is necessary to estimate the so-called rigidity index, I_r. Examples of these calculations can be found in Fang's *Foundation Engineering Handbook* (1991).

2.2.2.3.2 Limited σ'_{vb}

Poulos and Davis (1980) have developed the normalized critical depth, z_c/d, for a given effective angle of friction ϕ' obtained using the undisturbed friction angle ϕ'_1 at the pile toe before pile installation. The value of σ'_{vb} is equal to the vertical effective overburden pressure if the toe of the pile is located within the critical depth z_c, otherwise $\sigma'_{vb} = z_c\gamma'_s$.

The mean of the curves between N_q and ϕ' is shown in Figure 2.6 (Berezantzev et al. 1961) for piles in granular soils, as they are nearly independent of the embedment ratio (L/d). The friction angle ϕ' is calculated from the undisturbed ϕ'_1 (before pile installation). For instance, at $\phi'_1 = 40°$, the post-installation friction angle, ϕ' is 40° for driven piles and 37° for bored piles. It thus follows $N_{q\,driven} \approx 200$ and $N_{q\,drilled} \approx 110$. Berezantev's method is used along with limiting depth theory and effective overburden pressures.

Finally, as is controversial for calculating the effective pressure σ'_{vb}, the value of σ'_{vs} may be calculated as effective overburden pressure (Meyerhof 1976; Kulhawy 1984), regardless of the critical depth z_c or limited effective overburden pressure below the critical depth.

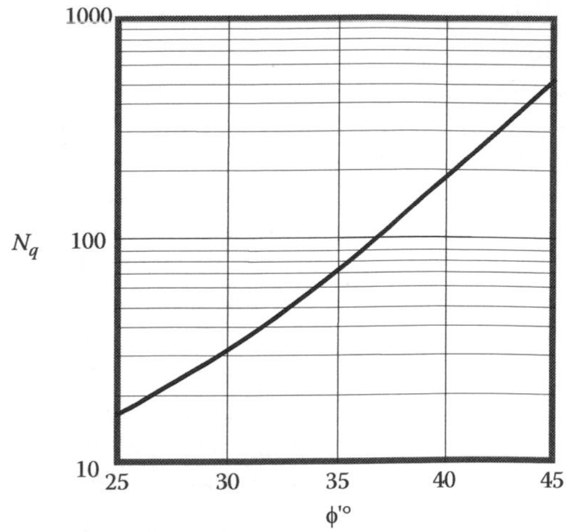

Figure 2.6 N_q. (After Berezantzev, V. G., V. S. Khristoforv and V. N. Golubkov, *Proc 5th Int Conf on Soil Mech and Found Engrg, speciality session 10*, Paris, 1961.)

2.2.3 Empirical methods

Meyerhof suggests $\tau_s = 2N$ (in kPa) and $\tau_s = N$ for high and low displacement piles based on SPT N values. Reese and O'Neill suggest the following variation of β with depth (O'Neill and Reese 1999):

$$\beta = 1.5 - 0.245\sqrt{z} \; (0.25 \le \beta \le 1.2) \tag{2.10}$$

where z = depth from the ground surface (m). This expression underestimates the values of β to a depth of 6–8 m (see Figure 2.4). The existing data may be represented by either $\beta = 43.15z^{-1.74}$ ($z > 5$ m) or $\beta = 0.136z/(z - 8)$ ($z > 8$ m).

The net unit end-bearing capacity q'_b for drilled shafts in cohesionless soils will be less than that for other piles (Reese and O'Neill 1989):

$$q'_b = 0.6 p_a N_{60} \le 4.5MPa \quad (d < 1200 \text{ mm}) \tag{2.11}$$

and

$$q'_b = 4.17\frac{d_r}{d} q'_b \quad (d \ge 1200 \text{ mm}) \tag{2.12}$$

where p_a = 100 kPa; d_r = reference width, 0.3 m; d = base diameter of drilled shaft; and N_{60} = mean SPT N value for the soil between the base of the shaft and a depth equal to $2d$ below the base, without an overburden correction. If the base of the shaft is more than 1200 mm in diameter, the value of q'_b from Equation 2.11 could result in settlements greater than

25 mm, which would be unacceptable for most buildings. Equation 2.12 should be used instead.

Some theoretical development has been made in correlating shaft friction and base resistance with cone resistance q_c (Randolph et al. 1994; Jardine and Chow 1996; Lee and Salgado 1999; Foray and Colliat 2005; Lehane et al. 2005), which is not discussed here. In contrast, estimating shaft friction using N (SPT blow counts) is still largely empirical (Fleming et al. 1996). The Meyerhof correlations are adopted in China and Southeast Asia (see Figure 2.7) as well (e.g., for bored piles in Malaysian soil, the shaft friction resistance τ_s is generally limited to [1~6]N kPa as per instrumented pile loading tests) (Buttling and Robinson 1987; Chang and Goh 1988). Typical correlations between base pressure and SPT blow counts are provided in Figure 2.8, for bored pile and driven piles in sand and clay. Low values of q_b/N than Meyerhof's suggestions are noted.

2.2.4 Comments

The β values given for Vesić's method are about twice those of Meyerhof's (1976). The former may be used for driven piles and the latter for bored piles (Poulos and Davis 1980).

Side support system is adopted when installing bored piles in granular soils. These design recommendations are valid when casing is utilized. A more common technique is to use bentonite slurry to support the drilled hole during excavation. The bentonite is displaced upward (from the base of the pile) by pouring the concrete through a tremie pipe. Any slurry caked on the sides should be scoured off by the action of the rising concrete; however, this may not always occur in practice. A 10~30% reduction in the ultimate shaft resistance is recommended (Sliwinski and Fleming 1974), as is reflected in the reduced ratio of δ/ϕ' in Table 2.2, although no reduction in the resistance is noted in some cases (Touma and Reese 1974). A special construction procedure means the need of modifying the empirical expression, such as use of a full length of permanent liner around large diameter piles (Lo and Li 2003).

The design methods outlined here are applicable to granular soils comprising silica sands. They are not suitable to calcareous sands predominated in the uncemented and weakly cemented states. An angle of friction of 35°~45° may be measured for undisturbed calcareous sand, but normal stresses on a driven pile shaft are noted as extremely low (as low as 5 kPa). The driving process tends to significantly crush the sand particles. Bored-and-grouted and driven-and-grouted piles should be used in such soil.

It is customary to deduce the correlation ratio of shaft friction over SPT (τ_s/N) or over the cone resistance (τ_s/q_c) without accounting for the impact of pile–soil relative stiffness. The ratios deduced may be sufficiently accurate in terms of gaining ultimate capacity. Nevertheless, consideration of

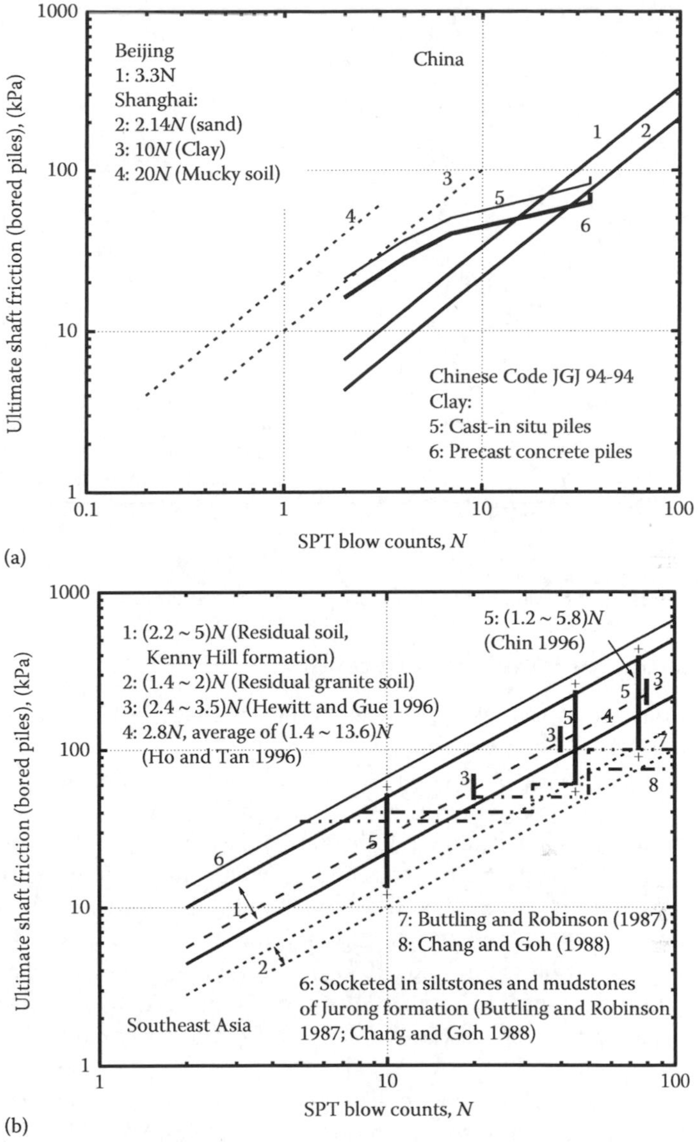

Figure 2.7 **Variation of ultimate shaft friction with SPT blow counts. (a) China. (b) Southeast Asia.**

friction fatigue on shaft friction renders the importance of considering displacement and pile stiffness. Ideally, the impact may be determined by matching theoretical solutions (see Chapter 4, this book) with measured pile-head load and displacement curve, assuming various form of shaft stress and shear modulus distribution profiles.

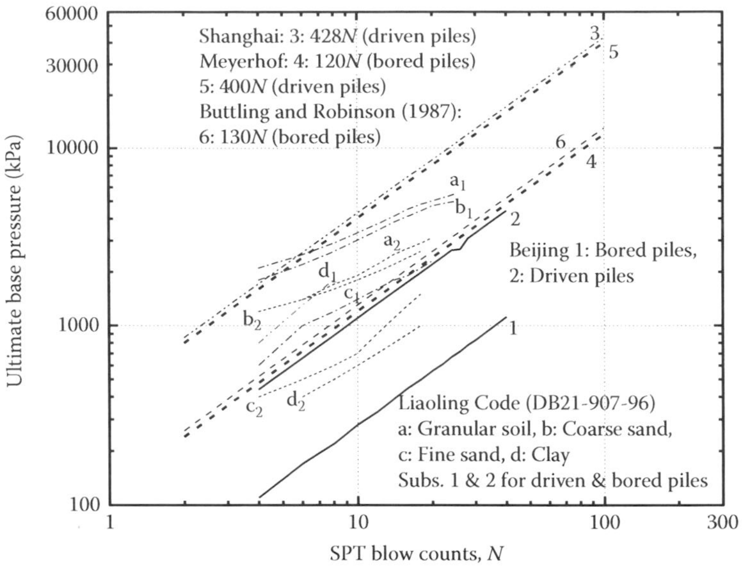

Figure 2.8 Variation of ultimate base pressure with SPT blow counts.

2.2.5 Capacity from loading tests

The most reliable capacity should be gained using loading test. The detailed load test procedure can be found in relevant code (e.g., ASTM D 1143, BS8004: clause 7.5.4). With a predominantly friction pile, the pile-head force may reach a maximum and decrease with larger penetrations, or the highest force may be maintained with substantially no change for penetrations between 1% and 5% of the shaft diameter, depending on the soil stiffness as shown in Figure 2.9 (Guo and Randolph 1997a). With an end-bearing pile, the ultimate bearing capacity may be taken as the force at which the penetration is equal to 10% of the diameter of the base of the pile. In other instances, a number of factors will need to be taken into account, as shown in the BS 8004: Clause 7.5.4 and explored in Chapters 4 and 5, this book.

Generally, pile capacity may be decomposed into shaft friction and end-bearing components using the empirical approach (Van Weele 1957), as the pile-head force is taken mostly by skin friction until the shaft slip is sufficient to mobilize the limiting value. When the limiting skin resistance is mobilized, the point load increases nearly linearly until the ultimate point capacity is reached. Chapter 4, this book, provides an analytical solution, which allows the components of base contribution and shaft friction to be readily resolved concerning a power law distribution of the shaft friction.

Among many other methods, the following methods are popularly adopted in practice for analyzing pile loading tests:

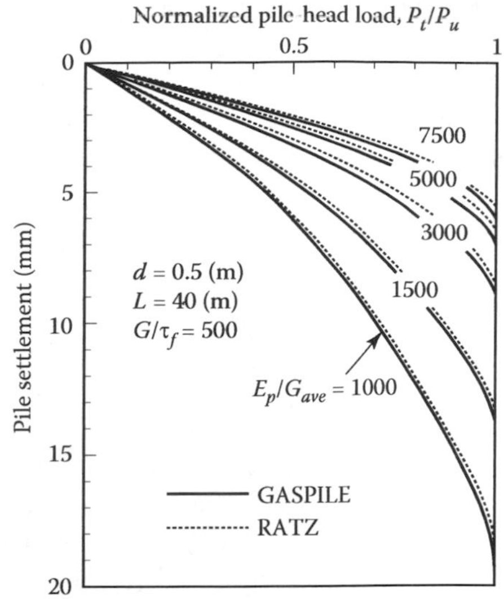

Figure 2.9 Effect of soil stiffness on pile-head load versus settlement relationship. (After Guo, W. D. and M. F. Randolph, *Int J Numer and Anal Meth in Geomech* 21, 8, 1997a.)

1. Chin's method (1970) based on hyperbolic curve fitting to the measured pile-head load-settlement curve to determine ultimate capacity.
2. Davisson's approach (1972), assuming ultimate capacity at a head settlement w_t of $0.012d_r + 0.1d/d_r + P_tL/A_pE_p$, where d_r = 0.3m, P_t = applied load; E_p = Young's modulus of an equivalent solid pile; A_p = cross-sectional area of an equivalent solid pile.
3. Van der Veen's approach (1953), based on a plot of settlement w_t versus ln $(1 - P_t/P_u)$ curve.
4. The logarithmic plot approach, based on plots of log P_t versus log w_t, and log T versus w_t (T = loading time). The log P_t versus log w_t plot is normally well fitted using two lines, and the load at the intersection of the lines is taken as the ultimate pile-head load, P_u. The log T~w_t curves are straight lines for load < ultimate P_u, but turn to curves after P_u. This load P_u is verified using log T versus w_t plot, as at and beyond the P_u, a curve plot of log T versus w_t is observed.

A comparison among nine popular methods (Fellenius 1980) shows values of capacity differening by ~38% for the very same load-displacement curve. This means a rigorous method is required to minimize uncertainty. Correlating load with displacement, the closed-form solutions discussed in Chapters 4 and 5, this book, are suitable for estimating the capacity.

2.3 CAPACITY OF SINGLE PILES IN ROCK

Equation 2.1 may be used to get a rough estimate of the capacity of a rock socket pile. The ultimate skin friction along rock socket shaft τ_s may be estimated using uniaxial compressive strength of rock q_u (Williams and Pells 1981)

$$\tau_s = \alpha_r \beta_r q_u \tag{2.13}$$

where q_u = uniaxial compressive strength of the weaker material (rock or concrete), τ_s and q_u in MPa, α_r = a reduction factor related to q_u, β_r = a factor correlated to the discontinuity spacing in the rock mass. Briefly, (a) τ_s = minimum value of (0.05~0.1)f_c and (0.05~0.1)q_u (f_c = 28-day compressive strength of concrete piles) for moderately fractured, hard to weak rock; (b) τ_s/q_u = 0.03~0.05 for highly fractured rocks (e.g., diabsic breccia); and (c) τ_s/q_u = 0.2~0.5 for soft rock (q_u < 1.0 MPa) (Carrubba 1997). Other suggestions are highlighted in Table 2.3. A united parameter ψ (see Figure 2.3) has been introduced to correlate the shaft friction with the undrained shear strength s_u (= 0.5q_u) of clay or rock (via q_u in MPa) by (Kulhawy and Phoon 1993; Pells et al. 1998)

$$\frac{\tau_s}{p_a} = \psi \left(\frac{0.5 q_u}{p_a} \right)^n \tag{2.14}$$

Table 2.3 Estimation of shaft friction of pile in rock using Equations 2.13 and 2.14

References	ψ *(n = 0.5)*	*References*	$\alpha_r \beta_r$ *for Equation 2.13*
Rosenberg and Journeaux (1976)	1.52	Thorne (1977)	0.1
Hobbs and Healy (1979)	1.8 (Chalk; q_u > 0.5 MPa)		0.05
Horvath et al. (1979, 1980)	1.1	Reynolds and Kaderbeck (1980)	0.3
Rowe and Armitage (1987)	2.0~2.7	Gupton and Logan (1984)	0.2
Carter and Kulhawy (1988)	0.9	Toh et al. (1989)	0.25
Reese and O'Neill (1989)	0.9	Osterberg (1992)	0.3–0.5 (q_u = 0.33~3.5 MPa)
Benmokrane et al. (1994)	0.9~1.3		0.1–0.3 (q_u = 3.5~14 MPa)
Carrubba (1997)	0.6~1.1		0.03–0.1 (q_u = 14~55 MPa)

Source: Revised from Seidel, J. P., and B. Collingwood, *Can Geotech J*, 38, 2001.

Some low adhesion factors from the Australian rock sockets (Pells et al. 1998) were also added to Figure 2.3. This offers a ψ for rock sockets from 0.9 to 2.9 with an average of ~1.7. The several-times disparity in shaft friction mirrors the coupled impact of socket roughness (via asperity heights and chord lengths), rock stiffness (via Young's modulus), shear dilation (varying with socket diameter, initial normal stress between concrete and rock), and construction practices. The critical factor among these may be the dilation, as revealed by the micromechanical approach (Seidel and Haberfield 1995; Seidel and Collingwood 2001), and also inferred from the shear tests on model piles in sand (Lehane et al. 2005). The impact of the listed factors on shaft friction needs to considered, and may be predicted using a program such as ROCKET, as is substantiated by large-scale shear tests on rock-concrete interface and in situ measured data (Haberfield and Collingwood 2006).

The maximum end-bearing capacity $q_{b\max}$ (in kPa) may be correlated to the unconfined compressive strength q_{ui} (kPa) of an intact rock by (after Zhang 2010)

$$\frac{q_{b\max}}{p_a} = 15.52\left(\frac{0.5q_{ui}}{p_a}\right)^{0.45}(\alpha_E)^{0.315} \tag{2.15}$$

where $\alpha_E = 0.0231\text{RQD} - 1.32 \geq 0.15$; and RQD in percentage.

The deformation modulus of rock mass E_m may be correlated to that of intact rock E_r by $E_m = \alpha_E E_r$ as well (Gardner 1987), as discussed in Chapter 1, this book. It is useful to refer to measured responses of typical rock socketed piles in Australia (Williams and Pells 1981; Johnston and Lam 1989) and in Hong Kong (Ng et al. 2001).

2.4 NEGATIVE SKIN FRICTION

When piles are driven through strata of soft clay into firmer materials, they will be subjected to loads caused by negative skin friction in addition to the structural loads, if the ground settles relative to the piles, as illustrated in Figure 2.10. Such settlement may be due to the weight of superimposed fill, to groundwater lowering, or as a result of disturbance of the clay caused by pile driving (particularly large displacement piles in sensitive clays leading to reconsolidation of the disturbed clay under its own weight). The depth of neutral plane L_n is (0.8~0.9)L for frictional piles and around 1.0L for end-bearing piles, which may vary with time. Above the plane, the soil subsides more than the pile compression.

Dragloads are dominated by vertical soil stress, as pile–soil relative slip generally occurs along the majority of the pile shaft. Therefore, the β method is preferred to calculate total downdrag load P_{ns}:

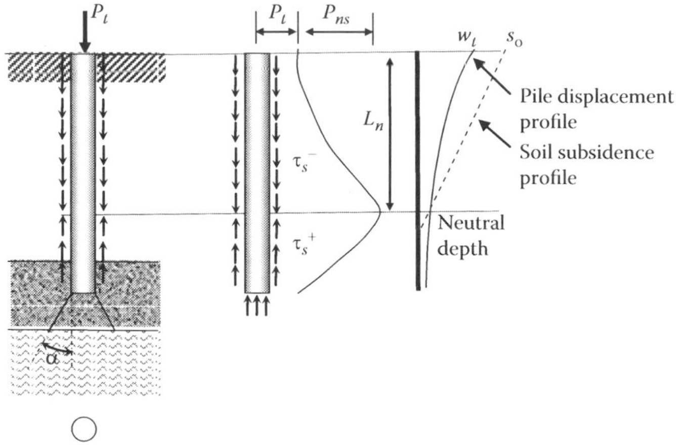

Figure 2.10 Schematic of a single pile with negative skin friction (NSF). (After Shen, R. F. Negative skin friction on single piles and pile groups. Department of Civil Engineering, Singapore, National University of Singapore. PhD thesis, 326.)

$$P_{ns} = A_s \beta \sigma'_v \text{ and } \beta = \eta K \tan\phi' \quad (2.16)$$

where

- $\eta = 0.7\sim0.9$; with tubular steel pipe piles, $\eta = 0.6$ (open-ended) $\eta = 1.0$ (closed-end).
- $\beta = 0.18\sim0.25$ with $\beta = 0.25$ for very silty clay, 0.20 for low plasticity clays, 0.15 for clays of medium plasticity, and 0.10 for highly plastic clays (Bjerrum 1973).

Results of measurements of negative skin friction on piles have been presented by numerous investigators (Zeevaert 1960; Brinch Hansen 1968; Fellenius and Broms 1969; Poulos and Mattes 1969; Sawaguchi 1971; Indraratna et al. 1992). Bjerrum (1973) drew attention to the fact that the negative adhesion depends on the pile material, the type of clay, the time elapsed between pile installation and test, the presence or lack of other material overlying the subsiding layer, and the rate of relative deformation between the pile and the soil. At a high rate of relative movement, Equation 2.16 may be used to estimate the magnitude of the negative skin friction forces. The β approach compares well with measured data and numerical analysis incorporating pile–soil relative slip. It appears that a small relative movement of ~10 mm is sufficient to fully mobilize negative skin friction.

This additional loading due to negative skin friction forces may be illustrated via an early case (Johannessen and Bjerrum 1965). A hollow steel pile was driven through ~40 m of soft blue clay to rock. After 10 m of fill was placed around the pile for 2.5 years, the ground surface settled nearly

2 m and the pile shortened by 15 mm. The compressive stress induced at the pile point reached 200 MPa, owing to the negative skin friction. The adhesion developed between the clay and the steel pile was about the value of the undrained shear strength of the clay, measured by the in site van test, before pile driving. It is equal to 0.2 times the effective vertical stress.

Increase in axial loads on pile-head reduces dragloads but increases pile settlement and negative skin friction (downdrag). The reduction is small for end-bearing piles compared to frictional piles (Jeong et al. 2004). Some researchers believe that transient live loads and dragload may never occur simultaneously, thus only dead load and dragload need to be considered in calculating pile axial capacity. However, in the case of short stubby piles found on rock, elastic compression may be insufficient to offset the negative skin friction. A conservative design should be based on the sum of dead load, full live load, and negative skin friction.

The group interaction reduces the downdrag forces by 20%, 40%, and 55% in two-pile, four-pile, and nine-pile groups, respectively (Chow et al. 1990). To incorporate the reduction, a number of expressions were developed. In particular, Zeevaert (1960) assumed the reduced effective overburden pressure among a pile group is equal to average unit negative skin friction mobilized on pertinent number of piles (n). The unit friction is presented in term of the reduced effective stress using β method. This results in a dimensional ordinary differential equation for resolving the reduced stress. Given boundary condition on ground level, the reduced stress is obtained (Zeevaert 1960). The stress in turn allows the unit shaft friction to be integrated along pile length to gain total shaft downdrag.

Shibata et al. (1982) refines the concept of "effective pile number n" in Zeevaert's method to when a pile under consideration is placed at the center of a square with a width of $2s$ (s = pile center to center spacing), and n is the number of piles enclosed by the square (see Figure 2.11). He obtained the total downdrag of a single pile in a group P_{ns}:

$$P_{ns} = \pi d \beta L_n \left[p\pi \left(\frac{1-e^{-\chi}}{\chi} \right) + L_n \gamma'_s \left(\frac{\chi + e^{-\chi} - 1}{\chi^2} \right) \right] \tag{2.17}$$

where p = surcharge loading on ground surface, L_n = depth of neutral plane, and

$$\chi = 4n\pi\beta \left(L_n / d \right) / \left[16 \left(\frac{L}{d} \right)^2 - n\pi \right].$$

Typically $n = 2.25$ for pile no. 1, 4, 13, and 16; $n = 3.0$ for pile no. 2, 3, 5 etc; and $n = 4.0$ for pile no. 6, 7, 10, and 11 (see Figure 2.11). Once the total number of piles in a group exceeds 9, the n value stays constant. This "constant" and its impact are consistent with recent centrifuge results (Shen 2008).

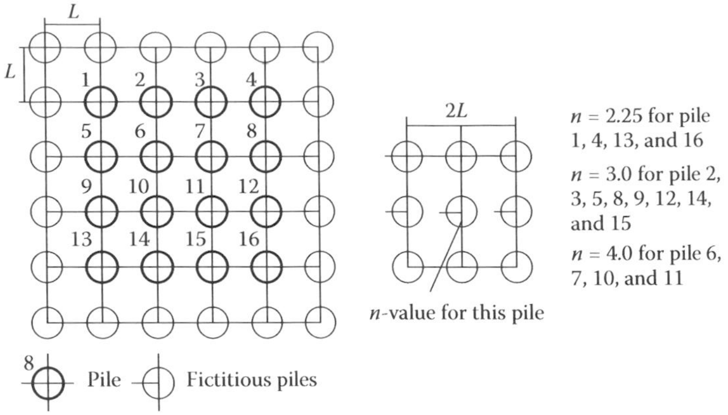

Figure 2.11 Schematic of a pile in a group with negative skin friction (NSF).

To reduce negative skin friction, a number of measures were attempted (Broms 1979), including driving piles inside a casing with the space between pile and casing filled with a viscous material and the casing withdrawn, and coating the piles with bitumen (Bjerrum et al. 1969).

2.5 CAPACITY OF PILE GROUPS

Single piles can be used to support isolated column loads. They are more commonly used as a group at high load levels, which are generally linked by a raft (or pile-cap) above ground level or embedded in subsoil.

A vertically loaded pile imposes mainly shear stress around the subsoil, which attenuates with distance way from the pile axis from maximum on the pile surface (see Chapter 6, this book). The stress in the soil may increase owing to installing piles nearby, which reduce the capacity of the soil by principally increasing the depth of stress influence (see Figure 2.12). On the other hand, installation of displacement piles may increase soil density and lateral stresses locally around a pile, although, for example, driving piles into sedimentary rocks can sometimes lead to significant loss of end-bearing (known as relaxation).

The pile group capacity is conventionally calculated as a proportion of the sum of the capacities of the individual piles in the group by

$$P_{ug} = \eta n_g P_u \tag{2.18}$$

where P_{ug} = ultimate capacity of the pile group; n_g = number of piles in the pile group; and η = a group efficiency. With frictional piles, increase in capacity due to densification effects is normally not considered, although the NAVFC DM-7.2 code refers to η >1.0 for cohesionless soils at the usual spacings of 2 to 3 pile diameter. As for end-bearing piles, the capacity of

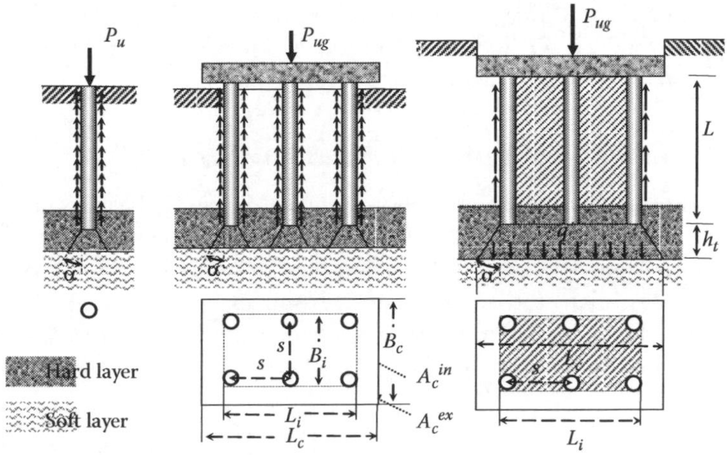

Figure 2.12 Schematic modes of pile failure.

the pile group is a simple sum of the capacity of each individual pile in the group.

2.5.1 Piles in clay

A group of piles may fail as a block under a loading less than the sum of the bearing capacity of the individual piles (Whitaker and Cooke 1966). A block failure occurred for pile spacings of the order of two diameters. For pile groups in clay, the capacity mainly derives from the shaft resistance component. The component can be significantly reduced by the proximity of other piles, depending on whether the piles are freestanding or capped at the ground surface. The end-bearing resistance is largely unaffected.

A single efficiency factor η is adopted in practice to calculate the total pile resistance including both shaft and base components. Two efficiency factors of η_s and η_b are also used to distinguish shaft and base resistance, respectively.

2.5.2 Spacing

Upheaval of the ground surface should be minimized by controlling minimum spacing during driving of piles into dense or incompressible material. A too-large spacing, on the other hand, may result in uneconomic pile caps. The driving of piles in sand or gravel should start from the center of a group and then progressively work outwards to avoid difficulty with tightening up of the ground. CP 2004 suggests the minimum pile spacing is the *perimeter* of the pile for frictional piles, *twice* the least width for end-bearing piles, and 1.5 times diameter of screw blades for screw piles. The Norwegian

Code of Practice on Piling recommends a minimum pile spacing of $3d$ (for $L < 12$ m in sand) or $4d$ (for $L < 12$ m in clay), increasing by one diameter for 12–24 m piles and by two diameters for $L > 24$ m.

2.5.3 Group interaction (free-standing groups)

Free-standing pile groups refer to those groups for which the cap capacity is negligible. This is noted when a pile cap is above soil surface or the ground resistance to a pile cap cannot be relied on. A few expressions have been proposed to estimate the capacity of a group using the group efficiency concept. For instance, the efficiency η is correlated to the ratio ρ of the shaft (skin) load P_{fs} over the total capacity P_u (i.e., $\rho = P_{fs}/P_u$) by (Sayed and Bakeer 1992):

$$\eta = 1 - (1 - \eta_s' K_g)\rho \tag{2.19}$$

where η_s' = geometric efficiency; K_g = 0.4~0.9, group interaction factor, with higher values for dense cohesionless or stiff cohesive soils, and 0.4~1.0 for loose or soft soils. The values of K_g may be determined according to relative density of the sand or consistency of the clay. It is noted that $\rho = 0$ for end-bearing piles and 1.0 for frictional piles. A value of $\rho > 0.6$ is usually experienced for a friction (floating) pile in clay. For a group of circular piles, the geometric efficiency η_s' may be estimated by

$$\eta_s' = 2 \times \left\{ \frac{\left[(n-1)\times s + d\right] + \left[(m-1)\times s + d\right]}{\pi \times m \times n \times d} \right\} \tag{2.20}$$

where m = number of rows of piles, n = number of piles in each row, and s = center to center spacing of the piles. In the case of square piles, π and d are replaced by 4 and b (b = width of the pile), respectively. The η_s' increases with an increase in the pile spacing-to-diameter ratio, s/d, and it normally lies between 0.6 and 2.5. For a configuration other than rectangular or square, it may be evaluated by

$$\eta_s' = Q_g / \sum Q_p \tag{2.21}$$

where Q_g = perimeter of the pile group (equivalent large pile) and ΣQ_p = summation of the perimeters of the individual piles in the group.

Equations 2.18 and 2.20 seem to be consistent with model test results (Whitaker 1957; O'Neill 1983). Tomlinson suggested an efficiency ratio of 0.7 (for $s = 2d$) to 1.0 (for $s = 8d$). The action of driving pile groups into granular soils will tend to compact the soil around the piles. The greater equivalent width of a pile group as compared to a single pile will render an increase in the ultimate failure load. For these reasons, the efficiency ratio of a group of piles in granular soils may reach 1.3 to 2.0 for spacing at 2~3 pile widths, as revealed in model tests (Vesić 1969), and about unity at a large pile spacing.

Driving action could loosen dense and very dense sand, but this situation is not a concern, as it is virtually impossible to drive piles through dense sand. As for bored piles in granular soils, the shaft contributes relatively small component to total resistance. The use of an efficiency factor of one should not cause major overestimation of capacity. This use is also common for end bearing (driven and bored) piles in granular soils. However, due allowance for any loosening during pile formation should be given in assessing individual capacity of bored piles.

2.5.4 Group capacity and block failure

2.5.4.1 Free-standing groups

A pile group may fail together as a block defined by the outer perimeter of the group (Terzaghi and Peck 1967). The block capacity consists of the shear around the perimeter of the group defined by the plan dimensions, and the bearing capacity of the block dimension at the pile points. (The only exception is point-bearing piles founded in rock where the group capacity would be the sum of the individual point capacities.) Terzaghi and Peck's block failure hypothesis offers a total failure load of the group, P_{BL} (see Figure 2.12):

$$P_{BL} = N_c A_b^{in} s_u + A_s \tilde{s}_u \tag{2.22}$$

where N_c = bearing capacity factor; $A_b^{in} = B_i \times L_i$, base area enclosed by the pile group; s_u = undrained shear strength at base of pile group; $A_s = 2(B_i + L_i)L$, perimeter area of pile group; $\tilde{s}_u$= average s_u around the perimeter of the piles; B_c, *and* L_c= the width and the length of block; and L = the depth of the piles.

The transition from individual pile failure (slip around individual piles) to block failure of a group (slip lines around the perimeter of the pile group) is confirmed by model pile tests (Whitaker 1957) as the pile spacing decreases. At a small spacing, "block" failure mode governs the capacity of the group. At the other extreme, the capacity of individual piles predominates group capacity at a very large spacing. As indicated in Figure 2.13, at a small spacing, free-standing and capped groups exhibit very similar responses. At a large spacing, the rate of increase in efficiency of the free-standing pile groups is smaller than that for the capped groups.

Model tests on lower cap-pile groups (Liu et al. 1985) indicate the limit spacing for shear block failure and individual failure is about (2~3)d, and may increase to (3~4)d for driven piles in sand. A smooth transition from

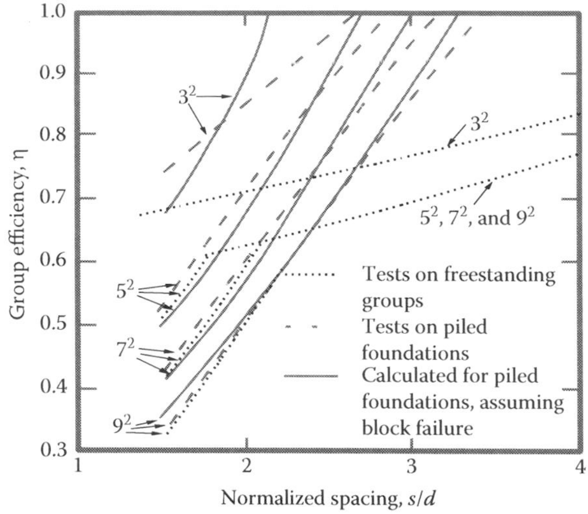

Figure 2.13 Comparison of η~s/d relationships for free-standing and piled groups. (After Poulos, H. G. and E. H. Davis, *Pile foundation analysis and design*, John Wiley & Sons, New York, 1980.)

individual to block failure modes is approximately captured using Equation 2.19. Increasing the number of piles in the group beyond a critical number would gain little in total capacity.

2.5.4.2 Capped pile groups versus free-standing pile groups

A rigid pile cap may bear directly on subsoil that provides reliable support (see Figure 2.12). This cap enables the piles to work as a group with rigid boundary condition at the pile heads, and the group is termed a capped group. The ultimate load capacity of the group P_{ug} is the lesser of mode (a) the sum of the capacity of the block containing the piles, P_{BL}, plus the capacity of that area A_c^{ex} of the cap lying outside the block, P_{cap}^{ex}; and mode (b) the sum of the individual pile capacities, $n_g P_u$, plus the bearing capacity of the pile cap on the bearing stratum, P_{cap}. The latter is a sum of P_{cap}^{ex} and P_{cap}^{in} deduced from both areas A_c^{ex} and A_c^{in}, with $A_c^{ex} = B_c L_c - B_i L_i$, and $A_c^{in} = B_i L_i - nA_b$.

The capped group capacity P_{ug} may be calculated using the shaft (P_{fs}) and base (P_{fb}) resistances of a single pile, the efficiency factors of shaft (η_s) and base (η_b), and the capacity of the cap. For instance, the Chinese design code JGJ94-94 recommends to calculate P_{ug} by (Liu et al. 1985)

$$P_{ug} = (\eta_s P_{fs} + \eta_b P_{fb}) n_g + P_{cap}^{ex} + P_{cap}^{in} \tag{2.23}$$

where P_{fs}, P_{fb} = ultimate shaft and base resistance, respecitvely, P_{cap}^{ex} and P_{cap}^{in} are given by (Liu 1986),

$$P_{cap}^{ex} = cN_c A_c^{ex} \eta_c^{ex} \tag{2.24}$$

$$P_{cap}^{in} = cN_c A_c^{in} \eta_c^{in} \tag{2.25}$$

and N_c = bearing capacity factor incorporating depth and shape effect, c = cohesion relevant to the bearing of the cap, $\eta_c^{ex} = 0.125[2 + (s/d)]$, and $\eta_c^{in} = 0.08(s/d)(B_c/L)^{0.5}$. It should be cautioned that in estimating η_c^{in}, the ratio of cap width over pile length B_c/L is taken as 0.2 for soft clay, regardless of the cap dimension and pile length; and taking $B_c/L = 1.0$ for other soil if $B_c/L > 1.0$. The product of cN_c is the ultimate capacity of the subsoil underneath the cap. The shaft component P_{fs} and base component P_{fb} of the capped pile capacity is calculated as they were for a single pile. The factors for shaft η_s and base η_b are different between sand and silt/clay.

Clay:

$$\eta_s = 1.2\left(1-\frac{d}{s}\right)\left\{1+0.12\sqrt{\frac{s}{d}}-0.5\left[\frac{B_c}{L}-\ln\left(0.3\frac{B_c}{L}+1\right)\right]\right\} \tag{2.26}$$

$$\eta_b = 6\left(\frac{d}{s}\right)\ln\left[1.718+\frac{1}{6}\left(\frac{s}{d}\right)\right]\left[1+0.1\ln\left(0.5\frac{B_c}{L}+1\right)\right] \tag{2.27}$$

Silt, sand:

$$\eta_s = \alpha_s\left\{1+0.1\left(\frac{s}{d}\right)-0.8\left[\frac{B_c}{L}-\ln\left(0.2\frac{B_c}{L}+1\right)\right]\right\} \tag{2.28}$$

$$\eta_b = \frac{1}{6+\dfrac{s}{d}}\left[10+\frac{s}{d}\sqrt{\frac{B_c}{L}}\right] \tag{2.29}$$

where $\alpha_s = 0.4 + 0.3(s/d)$ ($s/d = 2$~3), otherwise $\alpha_s = 1.6 - 0.1(s/d)$ ($s/d = 3$~6).

Equations 2.24 and 2.25 may overestimate the cap capacity without the reduction factors η_c^{ex}, and η_c^{in} for the enclosed areas. The η_c^{ex} increases from 0.63 to 1.0 as the s/d increases from 3 to 6. The η_c^{in} varies as follows: (1) 0.11~0.24 (s/d = 3), (2) 0.14~0.32 (s/d = 4), (3) 0.18~0.40 (s/d = 5) and (4) 0.21~0.48 (s/d = 6), respectively.

In the case of liquefaction, backfill, collapsible soil, sensitive clay, and/or under consolidated soil, the cap capacity may be omitted by taking $\eta_c^{ex} = \eta_c^{in} = 0$. In other words, Equations 2.23 through 2.25 are also directly used for free-standing pile groups. In this case, the impact of the parameter in the brackets with B_c/L is negligible, and the η_s and η_b reduce to the following forms:

$$\text{Clay: } \eta_s = 1.2\left(1 - \frac{d}{s}\right), \quad \text{and} \quad \eta_b = 6\left(\frac{d}{s}\right)\ln\left[1.718 + 6\left(\frac{s}{d}\right)\right] \tag{2.30}$$

$$\text{Silt, sand: } \eta_s = \alpha_s, \quad \text{and} \quad \eta_b = 10 / \left(6 + \frac{s}{d}\right) \tag{2.31}$$

Finally, Equation 2.23 may be rewritten as Equation 2.32 to facilitate comparison with Equation 2.19 (free-standing):

$$\eta = \eta_s \frac{P_{fs}}{P_u} + \eta_b \frac{P_{fb}}{P_u} \tag{2.32}$$

2.5.5 Comments on group capacity

Equation 2.23 is deduced empirically from 27 in situ model tests on 17 pile groups in silt and 10 groups in clay. It works well against measured data from piles in clay and silt for 15 high-rise buildings. The accuracy of Equation 2.19 for estimating capacity is largely dependent of the value of K_g. A comparison between Equations 2.19 and 2.32 indicates $K_g = \eta_s/\eta_s'$. In light of Equation 2.30 or 2.31 for η_s and Equation 2.20 for η_s', the values of K_g were obtained for typical model tests and are shown in Figure 2.14. The obtained curves of K_g agree well (but for the reverse trend with n_g) with those reported by Sayed and Bakeer (1992) for piles in clay, whereas the values of K_g (rewritten as K_{g1} for $s/d < 3$, otherwise as K_{g2}) only provide low bounds for the piles in sand. The values of K_{g1} and K_{g2} were increased by 2.5 times (= $2.5\eta_s/\eta_s'$) and are plotted in Figure 2.14 as $2.5K_{g1}$ and $2.5K_{g2}$. The latter agree well with higher value of measured data for all the piles in sand. This seems logical, as Equations 2.26 and 2.28 are deduced from model bored piles, which normally offer ~50% resistance mobilized along driven piles.

2.5.6 Weak underlying layer

Care must be taken, of course, to ensure that no weaker or more compressible soil layers occur within the zone of influence of the pile group, in particular for end-bearing piles (with tip founded in sand, gravel, rock, stiff clay, etc.). The underlying softer layer in Figure 2.12 may not affect the capacity of a single pile, but may enter the zone of influence of the pile group and punch through the bearing layer. The bearing capacity of the group may be estimated using footing analyses (see Figure 2.12, in which $\alpha = 30°$). The stress q′ on the top of the weaker layer is

$$q' = \frac{P_{ug}}{(B_c + h_t)(L_c + h_t)} \tag{2.33}$$

The stress of q′ should be less than $3s_u$ to avoid block failure. When driving piles into dense granular soils, care must be taken to ensure that

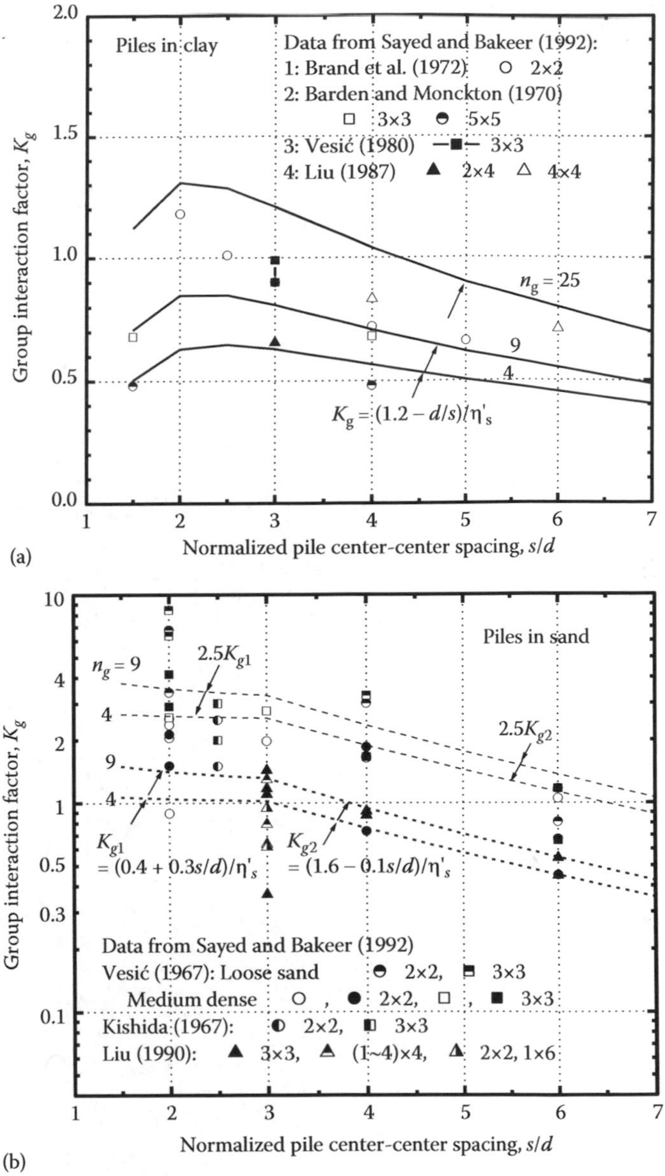

Figure 2.14 Group interaction factor K_g for piles in (a) clay and (b) sand.

previously driven piles are not lifted by the driving of other piles. For further discussion on group action, reference can be made to Tomlinson (1970) and Whitaker (1957).

Chapter 3

Mechanism and models for pile–soil interaction

Pile behavior may be predicted using various numerical and analytical methods, one of the most popular of which is the load transfer model. They are discussed in this chapter together with pertinent concepts and mechanisms, which will be used in subsequent chapters to develop solutions for piles.

3.1 CONCENTRIC CYLINDER MODEL (CCM)

Load transfer analysis is an uncoupled analysis, which treats the pile–soil interaction along the shaft and at the base as independent springs (Coyle and Reese 1966). The stiffness of the elastic springs, expressed as the gradient of the local load transfer curves, may be correlated to the soil shear modulus by load transfer factors (Randolph and Wroth 1978). The load transfer factors are significantly affected by (a) nonhomogeneous soil profile, (b) soil Poisson's ratio, (c) pile slenderness ratio, and (d) the relative ratio of the depth of any underlying rigid layer to the pile length. They are nearly constant for a given pile–soil system. Continuum-based FLAC (Itasca 1992) analysis has been used previously to calibrate the load transfer factors (Guo and Randolph 1998), which are recaptured next.

3.1.1 Shaft and base models

Load transfer approach is applied to a typical pile–soil system shown in Figure 3.1, which is characterized by shear modulus varying as a power of depth, z

$$G = A_g z^n \tag{3.1}$$

where n = power for the profile; A_g = a constant giving the magnitude of the shear modulus; $\xi_b = G_L/G_b$, base shear modulus jump (referred to as the end-bearing factor); and G_L, G_b = values of shear modulus of the soil

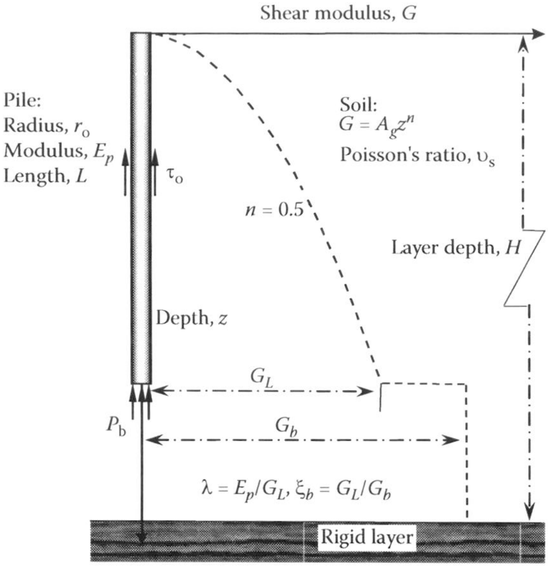

Figure 3.1 Schematic analysis of a vertically loaded pile.

just above the level of the pile tip and beneath the pile tip. The impact of ground-level modulus is incorporated in Chapter 4, this book.

The shaft displacement, w_s, is related to the local shaft stress, τ_o (on pile surface), and shear modulus, G, by the concentric cylinder approach (Randolph and Wroth 1978)

$$w_s = \frac{\tau_o r_o}{G}\zeta \tag{3.2}$$

where r_o = pile radius and ζ = the shaft load transfer factor.

The pile base settlement is estimated through the solution for a rigid punch acting on an elastic half-space (Randolph and Wroth 1978)

$$w_b = \frac{P_b(1-\nu_s)\omega}{4r_o G_b} \tag{3.3}$$

where w_b = base settlement; P_b = mobilized base load; ω= base load transfer factor; and ν_s = Poisson's ratio for the soil.

Assuming a constant with depth of the shaft load transfer factor ζ, closed-form solutions were established (see Chapter 4, this book) that encompass the displacement, w, and load, P, of a pile at any depth, z (Guo 1997); the pile-head stiffness of $P_t/(G_L w_t r_o)$ (P_t, w_t = pile-head load and settlement, respectively); and the ratio of pile base load (P_b) over the pile-head load (P_t). The solutions (see Chapter 4, this book) were used to examine the validity of the load transfer approach against FLAC analysis (Guo and Randolph 1998).

3.1.2 Calibration against numerical solutions

Numerical FLAC analysis (Guo and Randolph 1998) was conducted on piles in a soil with a shear modulus G following Equation 3.1 to entire depth H, and also with soil modulus given by $G_L/G_b = 1$ below pile base (see Figure 3.1), in which H = depth to underlying hard layer; L = pile embedment; $\lambda = E_p/G_L$; E_p = Young's modulus of an equivalent solid cylinder pile; and ν_p = Poisson's ratio of the pile. For piles with L/r_o = 10~80, λ = 300~10,000, H/L = 1.2~6, n = 0 ~1.0, and ν_p = 0.2, the head stiffness, ratios of w_t/w_b, and P_b/P_t are tabulated in Tables 3.1 through 3.3 for G strictly following Equation 3.1. The impact of G_L/G_b is discussed in Chapter 4, this book. The impact of ratio ν_p on this study is negligible (see Table 3.1). Table 3.2 shows comparisons among boundary element analysis (BEM; Randolph and Wroth 1978), variational method (VM; Rajapakse 1990), and the FLAC analysis for single piles in homogeneous soil (n = 0, H/L = 4). The FLAC and BEM analyses were based on a Poisson's ratio of soil ν_s of 0.4, while the VM analysis adopts a ν_s of 0.5. As a higher Poisson's ratio generally leads to a higher stiffness (see Chapter 4, this book), the stiffness from FLAC analysis is slightly higher than other predictions. Table 3.3 shows a further comparison with FEM analysis (Randolph and Wroth 1978), for both homogeneous (n = 0) and Gibson soil (n = 1). The reduction from n = 0 to n = 1 in the stiffness of 32~38% (FLAC) or 40~45% (FEM) is primarily attributed to the reduction in average value of modulus over the pile embedment, as elaborated in Chapter 4, this book.

Table 3.1 Effect of Poisson's ratio on the pile (ν_s = 0.49, L/r_o = 40, λ = 1,000)

n	*0*	*0.25*	*0.5*	*0.75*	*1.0*
$\frac{P_t}{G_L w_t r_o}$	59.08[a] / 59.04[b]	51.93 / 51.91	46.64 / 46.63	42.62 / 42.61	39.56 / 39.55
$\frac{w_t}{w_b}$	1.64 / 1.64	1.60 / 1.60	1.58 / 1.58	1.55 / 1.55	1.53 / 1.53
$\frac{P_b}{P_t}$	7.08 / 7.09	8.75 / 8.76	10.5 / 10.51	12.07 / 12.08	13.68 / 13.69

Source: Guo, W. D., and M. F. Randolph, *Computers and Geotechnics*, 23, 1–2, 1998.

[a] numerator for ν_p = 0.2
[b] denominator for ν_p = 0

Table 3.2 FLAC analysis versus other approaches ($n = 0$)

$\dfrac{P_t}{G_L w_t r_o}$	40	FLAC	69.70	64.38	53.60	36.51
		BEM	65.70	61.3	52.00	36.80
		VM	72.2[a]	65.1	54.9	38.7
	80	FLAC	109.0	85.0	61.6	36.2
		BEM	102.2	85.2	61.6	38.0
$\dfrac{W_t}{W_b}$	40	FLAC	1.05	1.18	1.55	2.92
		BEM	1.05	1.12	1.49	2.66
		VM	–	1.19	1.59	3.25
	80	FLAC	1.18		2.96	7.99
		BEM	1.16	1.54	2.68	6.75
L/r_o	$\lambda(=E_p/G_L)$		10,000	3,000	1,000	300

Source: Guo, W. D., and M. F. Randolph, *Computers and Geotechnics*, 23, 1–2, 1998.

[a] rigid pile; $v_s = 0.5$ for VM analysis, $v_s = 0.4$ for BEM and FLAC analyses, and H/L = 4 for FLAC analysis.

The FLAC analysis is utilized to deduce the base and shaft load transfer factors. The base factor ω was directly back-figured by Equation 3.3, in light of the base load P_b (estimated through the base stress) and the base displacement w_b (the base node displacement). The profile of shaft load transfer factor ζ and its average value were deduced:

1. With the profiles of local shaft shear stress τ_o and displacement w_s along a pile obtained by FLAC analysis, the profile of the factor ζ was back-figured using Equation 3.2, which is plotted as "FLAC" in Figure 3.2a.
2. Taking ζ as a constant with depth, the value of ζ was deduced by matching FLAC analysis and closed-form solutions for pile-head stiffness and the load ratio between pile base and head loads, P_b/P_t, respectively.

Figure 3.2a indicates the shaft factor ζ is approximately a constant with depth. With a constant ζ, the predicted profiles of load and displacement

Table 3.3 $P_t/(G_L w_t r_o)$ from FEM and FLAC analyses

FLAC	43.95	56.84	63.89	0	Note:
FEM	41.5	53.6	65.3		$v_s = 0.4$, $\lambda = 1,000$ for all analyses,
FLAC	29.89	37.21	39.53	1.0	but H/L = 2 for FEM and H/L = 2.5
FEM	25.0	34.8	35.8		for FLAC analyses.
L/r_o	20	40	80	n	

Source: Guo, W. D., and M. F. Randolph, *Computers and Geotechnics*, 23, 1–2, 1998.

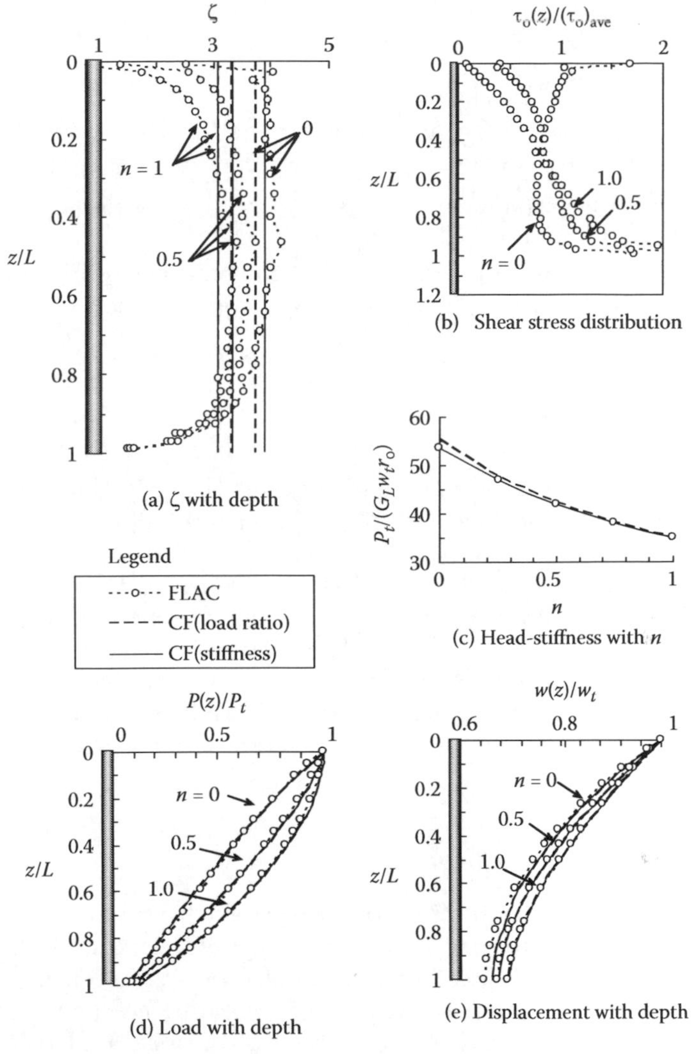

Figure 3.2 Effect of back-estimation procedures on pile response (L/r_o = 40, ν_s = 0.4, H/L = 4). (a) ζ with depth. (b) Shear stress distribution. (c) Head-stiffness ratio. (d) Load with depth. (e) Displacement with depth. (After Guo, W. D., and M. F. Randolph, *Computers and Geotechnics* 23, 1–2, 1998.)

using closed-form solutions (Chapter 4, this book) are very close to those from the FLAC analysis, as seen in Figure 3.2d and e, respectively.

3.1.2.1 Base load transfer factor

The base load transfer factor was synthesized as

$$\omega = \frac{\omega_h}{\omega_{oh}}\frac{\omega_v}{\omega_{ov}}\omega_o \tag{3.4}$$

where ω_h, ω_v = the parameters to capture the effect of H/L and soil Poisson's ratio; $\omega_{oh} = \omega_h$ at H/L = 4, and $\omega_{ov} = \omega_v$ at $\nu_s = 0.4$. They are given as:

- $\omega_o = 1/[0.67 - 0.0029(L/r_o) + 0.15n]$ ($L/r_o < 20$), otherwise $\omega_o = 1/[0.6 + 0.0006(L/r_o) + 0.15n]$ ($L/r_o \geq 20$)
- $\omega_v = 1/[1/\omega_o + 0.3(0.4 - \nu_s)]$ ($\nu_s \leq 0.4$, compressible), otherwise $\omega_v = 1/[1/\omega_o + 1.2(\nu_s - 0.4)]$ ($\nu_s > 0.4$, nearly incompressible)

$$\omega_h = \frac{1}{0.1483n + 0.6081}\left(1 - \exp(1 - \frac{H}{L})\right)^{-0.1008n+0.2406}$$

- The effect of pile–soil relative stiffness may be ignored over a practical range of the stiffness ratio, λ, between 300 and 3000.

The inverse of ω, $1/\omega$ reflects the base stiffness of $P_b(1 - \nu_s)/(4G_b r_o w_b)$. It is presented in Figure 3.3 to be consistent with the pile-head stiffness. It is equal to 0.6~0.95, with a base-load transfer parameter, ω, of 1.05~1.7. Practically, a unit value of ω may still be used to compensate the smaller load near pile base gained using a constant ζ (see Figure 3.2) under same amount of displacement. In addition, the effect of ω on the overall pile-head stiffness is extremely small.

3.1.2.2 Shaft load transfer factor

Back-figured shaft load transfer factors are presented in Figures 3.4 through 3.6 using pile-head stiffness or load ratio from FLAC analysis. The deduced values of ζ should be identical for a given pile–soil system if load transfer analysis is exact, otherwise the impact of neglecting layer interaction is detected. Figure 3.4 shows the variation of ζ with pile-slenderness ratio and soil nonhomogeneous profile described by Equation 3.1. Figures 3.5 and 3.6 show the impact of Poisson's ratio and the finite layer ratio on the factor ζ, respectively.

The impact on the shaft load transfer factor by a combination of pile slenderness ratio L/r_o, the soil nonhomogeneity factor n, and the soil Poisson's ratio ν_s can be approximated by (Cooke 1974; Frank 1974; Randolph and Wroth 1978)

$$\zeta = \ln\left[\frac{r_m}{r_o}\right] \tag{3.5}$$

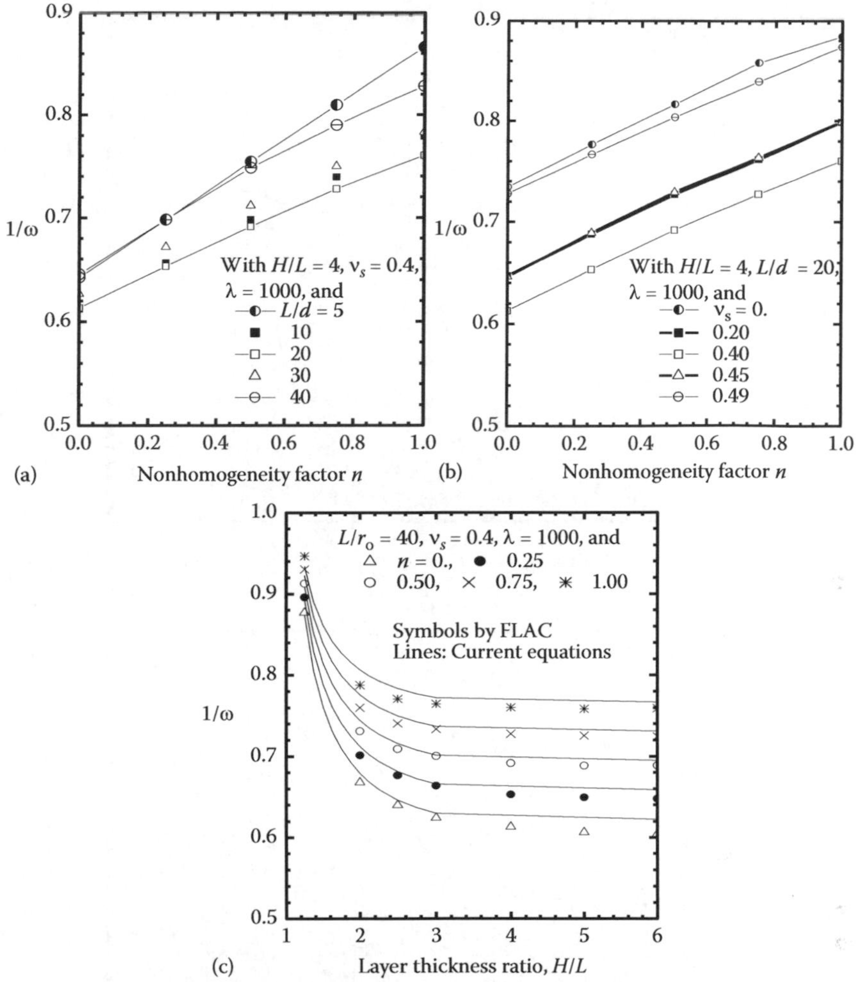

Figure 3.3 $1/\omega$ versus (a) slenderness ratio, L/d; (b) Poisson's ratio, ν_s; (c) layer thickness ratio, H/L. (After Guo, W. D., and M. F. Randolph, *Computers and Geotechnics* 23, 1–2, 1998.)

The maximum radius of influence of the pile r_m beyond which the shear stress becomes negligible was expressed in terms of the pile length, L (Guo and Randolph, 1998)

$$r_m = A(1-\nu_s)\rho_g L + Br_o \tag{3.6}$$

where $\rho_g = 1/(1+n)$. The parameters A and B were estimated through fitting Equation 3.5 to the values of ζ obtained by matching pile-head stiffness. This offers an A given by

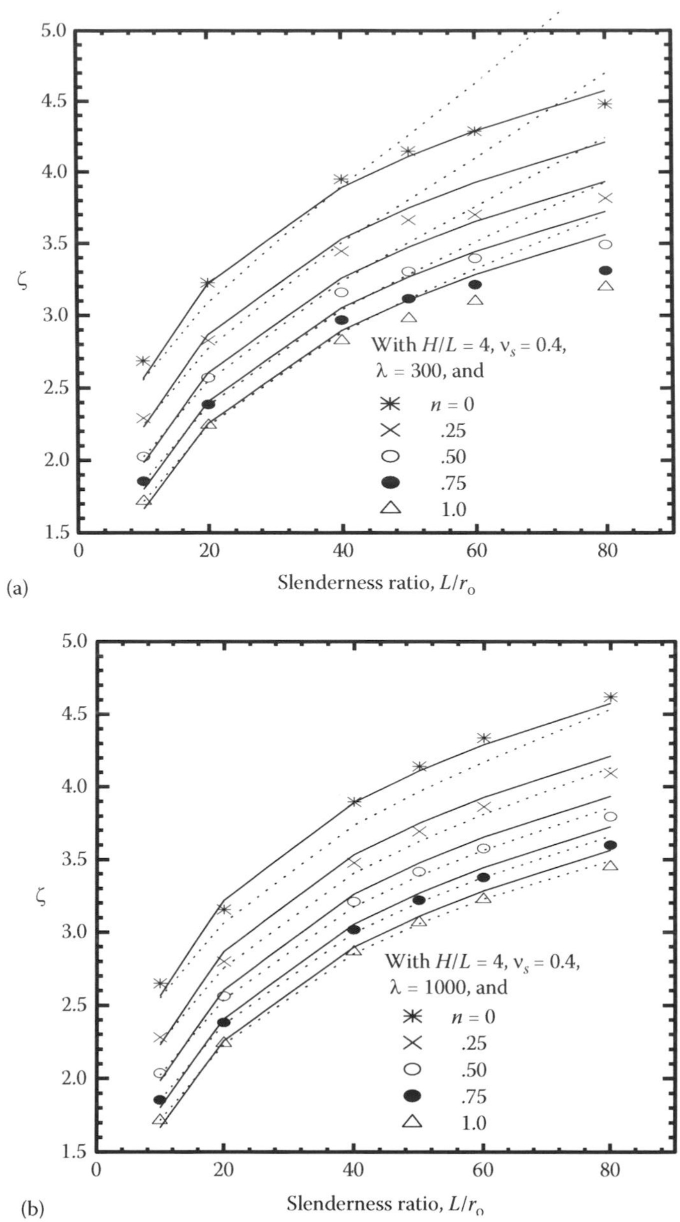

Figure 3.4 Load transfer factor versus slenderness ratio (H/L = 4, ν_s = 0.4). (a) λ = 300. (b) λ = 1,000. (c) λ = 10,000. (After Guo, W. D., and M. F. Randolph, *Computers and Geotechnics* 23, 1–2, 1998.)

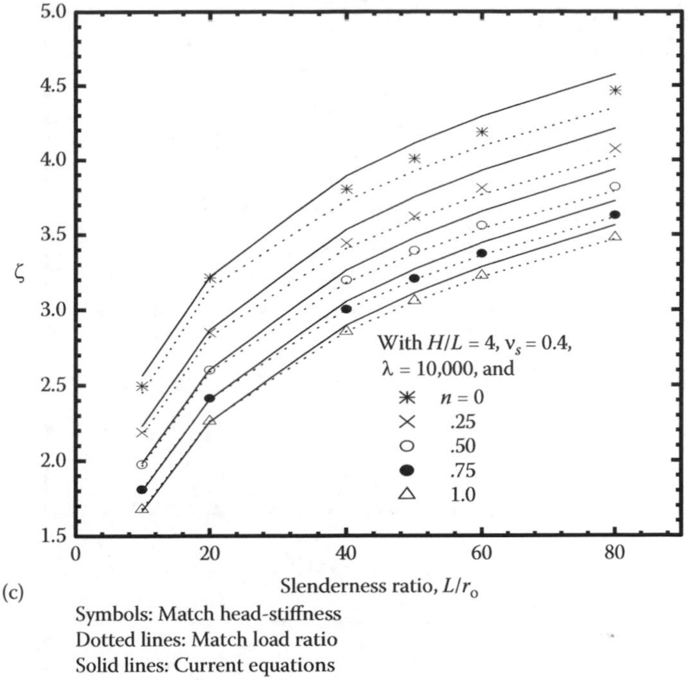

Figure 3.4 (Continued) Load transfer factor versus slenderness ratio (H/L = 4, ν_s = 0.4). (a) λ = 300. (b) λ = 1,000. (c) λ = 10,000. (After Guo, W. D., and M. F. Randolph, *Computers and Geotechnics* 23, 1–2, 1998.)

$$A = \frac{A_h}{A_{oh}}[\rho_g\left(\frac{0.4-\nu_s}{n+0.4}+\frac{2}{1-0.3n}\right)+C_\lambda(\nu_s-0.4)] \tag{3.7}$$

where $C_\lambda = 0$, 0.5, and 1.0 for $\lambda = 300$, 1,000, and 10,000, respectively, and with negligible impact but for short piles, as shown in Figure 3.7; $A_{oh} = A_h$ at a ratio of $H/L = 4$, and A_h is given by

$$A_h = 0.124e^{2.23\rho_g}\left(1-e^{1-\frac{H}{L}}\right)+1.01e^{0.11n} \tag{3.8}$$

Equation 3.6, without physical implication, compares well with that back-figured from FLAC analysis, as illustrated in Figures 3.4 through 3.6. It adopts a modifier of 1 – exp (1 – H/L) recommended by Lee (1991).

3.1.2.3 Accuracy of load transfer approach

The values of A deduced by matching either load ratio or head stiffness are different, in particular, for a homogeneous soil profile or a higher slenderness

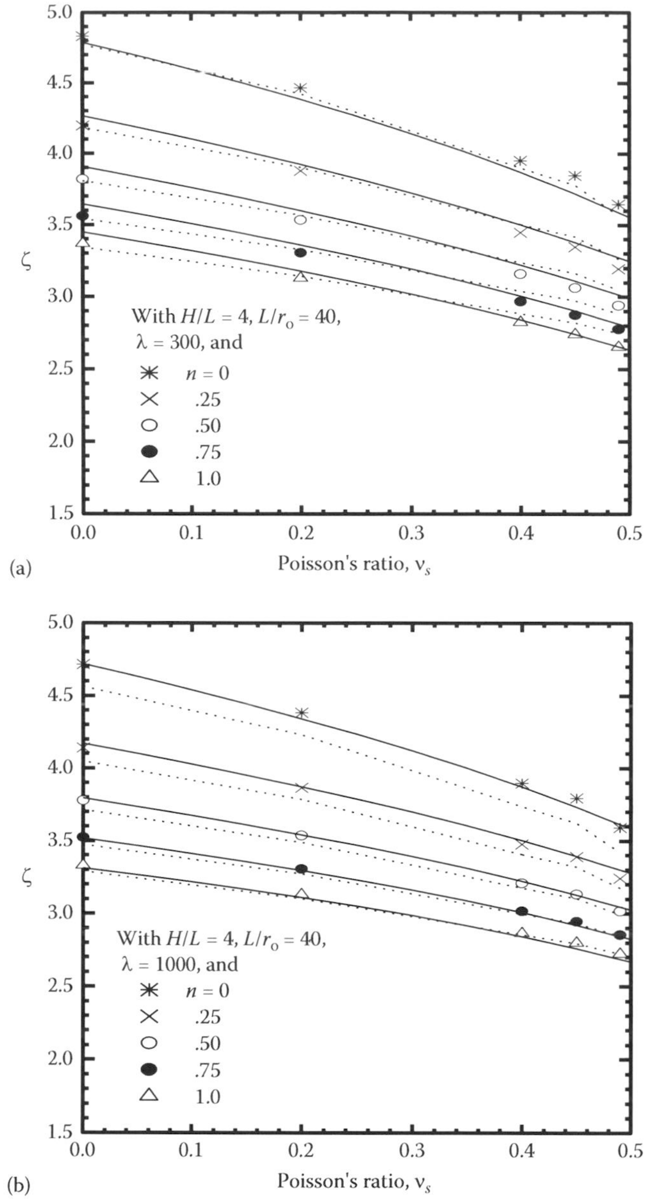

Figure 3.5 Load transfer factor versus Poisson's ratio relationship (H/L = 4, L/r_o = 40). (a) λ = 300. (b) λ = 1,000. (c) λ = 10,000. (After Guo, W. D., and M. F. Randolph, *Computers and Geotechnics* 23, 1–2, 1998.)

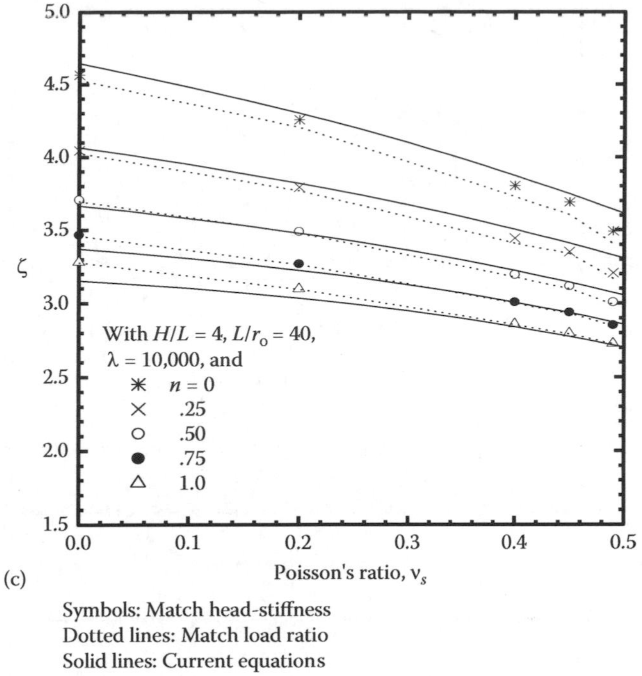

Figure 3.5 (Continued) Load transfer factor versus Poisson's ratio relationship (H/L = 4, L/r_o = 40). (a) λ = 300. (b) λ = 1,000. (c) λ = 10,000. (After Guo, W. D., and M. F. Randolph, *Computers and Geotechnics* 23, 1–2, 1998.)

ratio together with a lower stiffness (e.g., λ = 300) (Figures 3.4 and 3.7). This implies less accuracy of the load transfer approach for the cases.

The two parameters that most affect ζ are the soil layer depth ratio, H/L, and the nonhomogeneity factor, n. The shaft load transfer parameter, ζ, may be estimated from Equation 3.5, taking B = 1 and A given approximately by (Guo and Randolph 1998):

$$A \approx 1 + 1.1e^{-\sqrt{n}}\left(1 - e^{1-\frac{H}{L}}\right) \tag{3.9}$$

Figure 3.8 shows a comparison of Equation 3.9 with results from the FLAC analyses. For deep layers, a limiting value of A = 2.1 is noted along with B = 1 (for an infinite, homogeneous soil n = 0). This A is lower than A = 2.5 (B = 0) (Randolph and Wroth 1978; Guo and Randolph 1996). The discrepancy arises from the higher head-stiffness of FLAC analysis than the boundary element approach. A 30% difference in choice of A value would lead to less than 10% difference in the predicted pile-head stiffness. The accuracy of A, however, becomes important in modeling pile–pile interaction factors (Guo and Randolph 1996; Guo 1997).

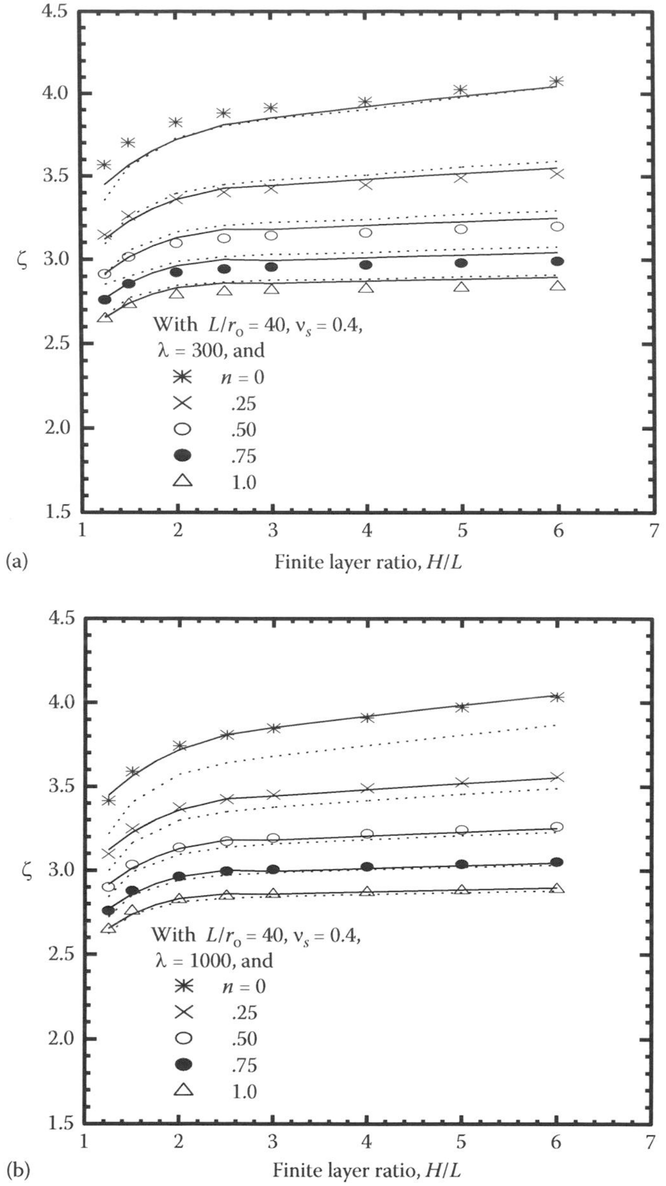

Figure 3.6 Load transfer factor versus H/L ratio (L/r_o = 40, ν_s = 0.4). (a) λ = 300. (b) λ = 1,000. (c) λ = 10,000. (After Guo, W. D., and M. F. Randolph, *Computers and Geotechnics* 23, 1–2, 1998.)

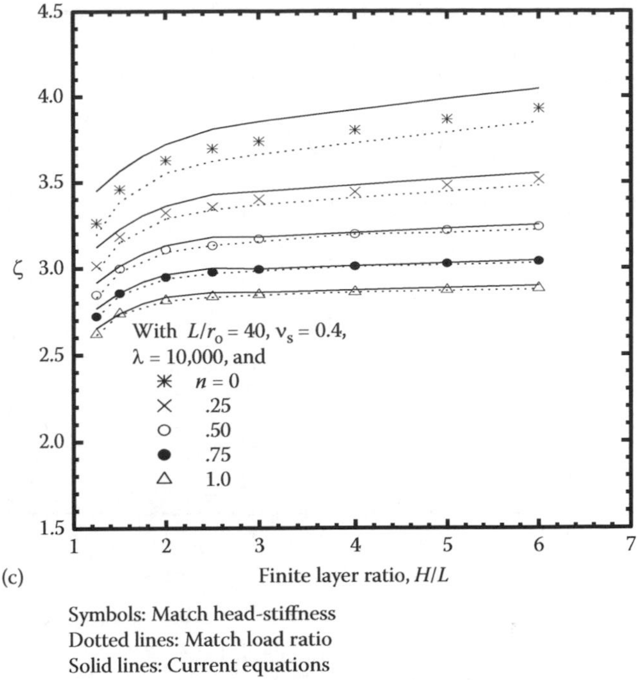

Figure 3.6 (Continued) Load transfer factor versus *H*/*L* ratio (L/r_o = 40, ν_s = 0.4). (a) λ = 300. (b) λ = 1,000. (c) λ = 10,000. (After Guo, W. D., and M. F. Randolph, *Computers and Geotechnics* 23, 1–2, 1998.)

The base contribution to the pile-head stiffness is generally less than 10%. Therefore, taking ω as unity will result in a less than 7% difference in the predicted pile-head stiffness. Figure 3.9 shows a comparison of the back-figured values of ζ using the simple ω = 1 and the more precise values of ω for two extreme cases of higher L/r_o of 80 and lower L/r_o of 10, together with the prediction by Equation 3.5.

3.2 NONLINEAR CONCENTRIC CYLINDER MODEL

The concentric cylinder load transfer model is next extended to incorporate nonlinear pile–soil interaction using a hyperbolic law to model the soil stress-strain relationship for pile shaft and base.

3.2.1 Nonlinear load transfer model

In comparison with Equation 3.5 for elastic case, a hyperbolic stress-strain curve would render the load transfer factor, ζ, to be recast as in Randolph (1977) (Kraft et al. 1981)

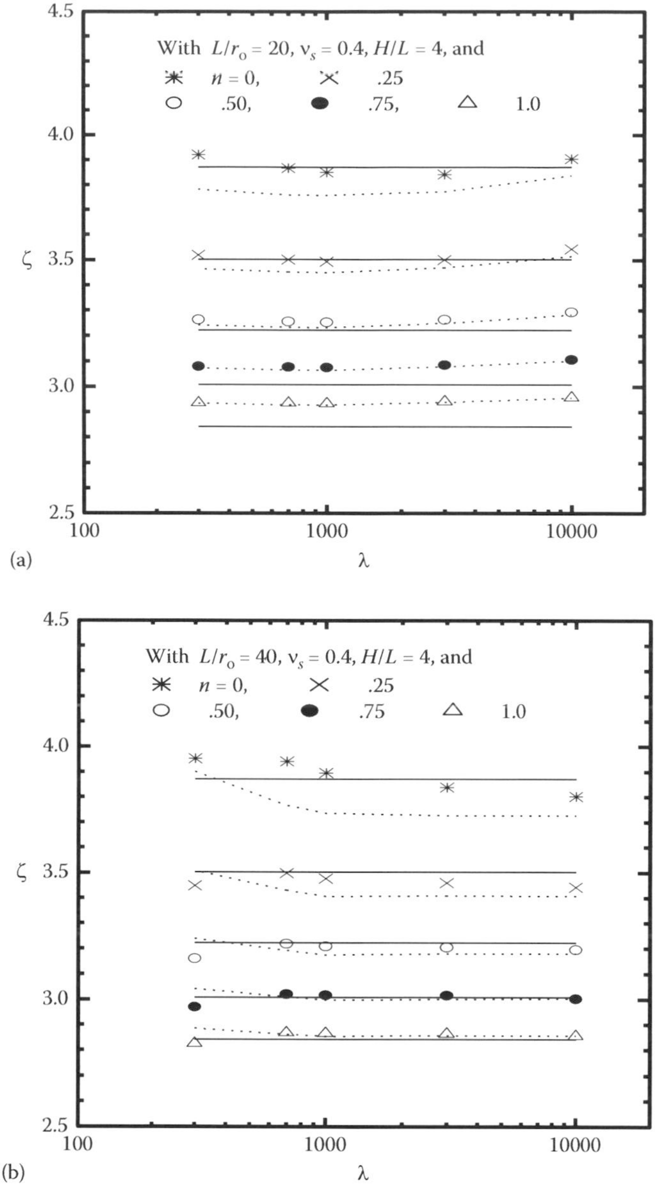

Figure 3.7 Load transfer factor versus relative stiffness (ν_s = 0.4, *H*/*L* = 4). (a) L/r_o = 20. (b) L/r_o = 40. (c) L/r_o = 60. (After Guo, W. D., and M. F. Randolph, *Computers and Geotechnics* 23, 1–2, 1998.)

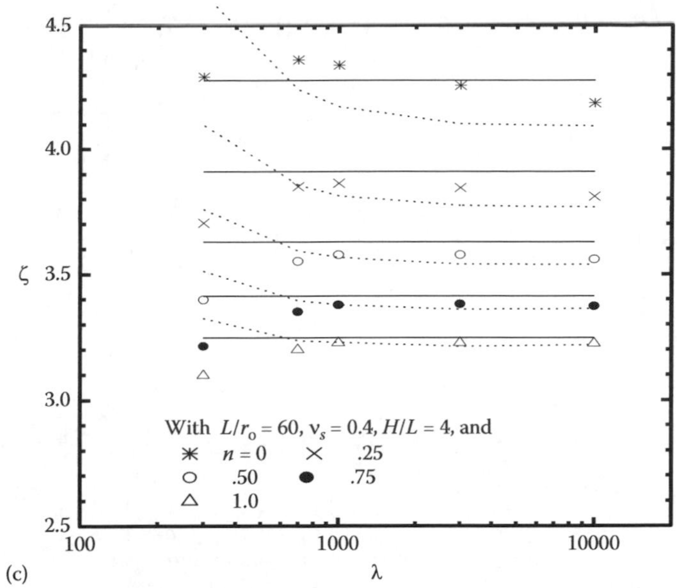

(c)

Symbols: Match head-stiffness
Dotted lines: Match load ratio
Solid lines: Current equations

Figure 3.7 (Continued) Load transfer factor versus relative stiffness ($\nu_s = 0.4$, $H/L = 4$). (a) $L/r_o = 20$. (b) $L/r_o = 40$. (c) $L/r_o = 60$. (After Guo, W. D., and M. F. Randolph, *Computers and Geotechnics* 23, 1–2, 1998.)

$$\zeta = \ln[(\frac{r_m}{r_o} - \psi)/(1 - \psi)] \tag{3.10}$$

where $\psi = R_f \tau_o / \tau_f$, the nonlinear stress level on the pile–soil interface, and the limiting stress is higher than but is taken as the failure shaft stress τ_f; $R_f = 0$~1.0, a flexible parameter controlling the degree of nonlinearity, $R_f = 0$ corresponds to a linear elastic case, and

$$\tau_f = A_v z^\theta \tag{3.11}$$

where A_v, θ = constants, which may be estimated using SPT blow counts or CPT profiles (see Chapter 1, this book). As the pile-head load increases, the mobilized shaft shear stress τ_o will reach the limiting value τ_f. This occurs at a local limiting displacement w_e, which in turn, in light of Equations 3.1 and 3.2 and $\theta = n$, can be expressed as

$$w_e = \zeta r_o \frac{A_v}{A_g} \tag{3.12}$$

Thereafter, as the pile–soil relative displacement exceeds the limiting value, the shear stress stays as τ_f (underpinned by an ideal elastic perfectly plastic

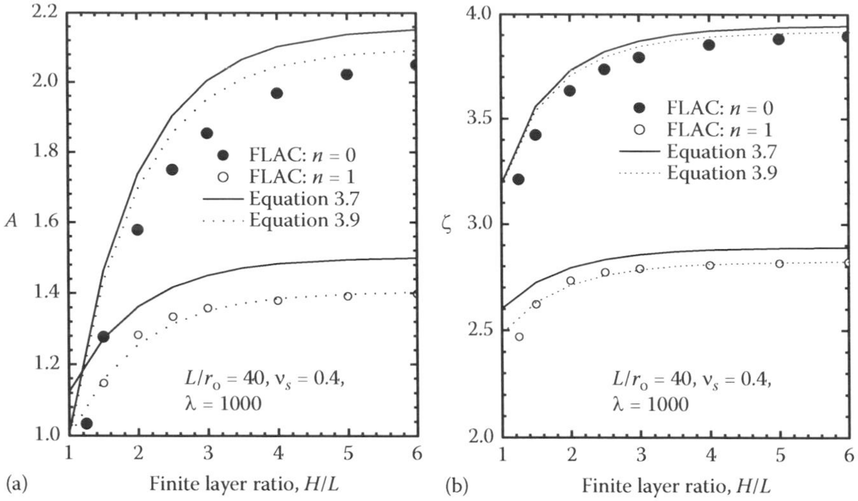

Figure 3.8 Effect of soil-layer thickness on load transfer parameters A and ζ (L/r_o = 40, ν_s = 0.4, λ = 1000). (a) Parameter A. (b) Parameter ζ. (After Guo, W. D., and M. F. Randolph, *Computers and Geotechnics* 23, 1–2, 1998.)

load transfer curve). The limiting shaft displacement w_e is a constant down the pile length.

Equation 3.10 is dependent on stress level via ψ and is referred to as nonlinear (NL) analysis. The load transfer curve may be simplified as simple linear analysis (SL) by assuming a constant value of ψ. As an example,

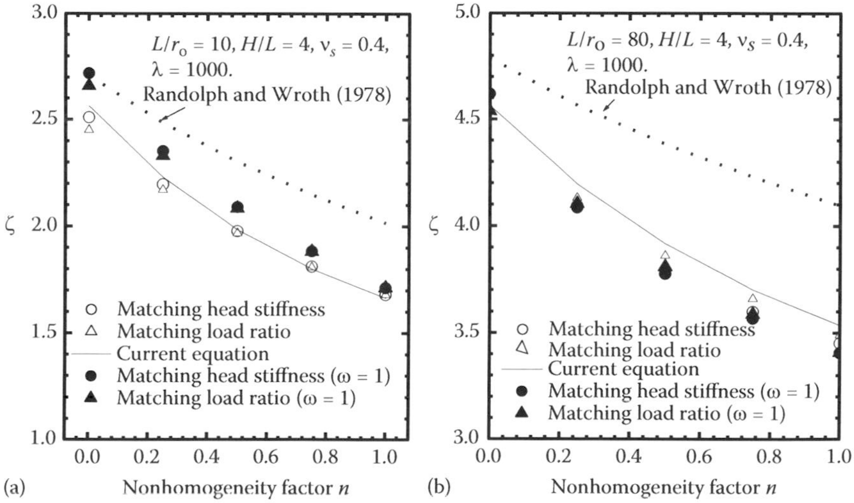

Figure 3.9 Variation of the load transfer factor due to using unit base factor ω and realistic value. (a) L/r_o = 10. (b) L/r_o = 80. (After Guo, W. D., and M. F. Randolph, *Computers and Geotechnics* 23, 1–2, 1998.)

Figure 3.10a shows the nondimensional shear stress versus displacement relationship obtained for τ_f/G_i = 350 (subscript i denotes initial subsequently), L/r_o = 100, and ν_s = 0.5 using NL (R_f = 0.9) and SL (ζ = constant by ψ = 0.5) analyses. It indicates that shaft friction is fully mobilized at a displacement w_e of 1%–2% of the pile radius, which accords well with model tests (Whitaker and Cooke 1966).

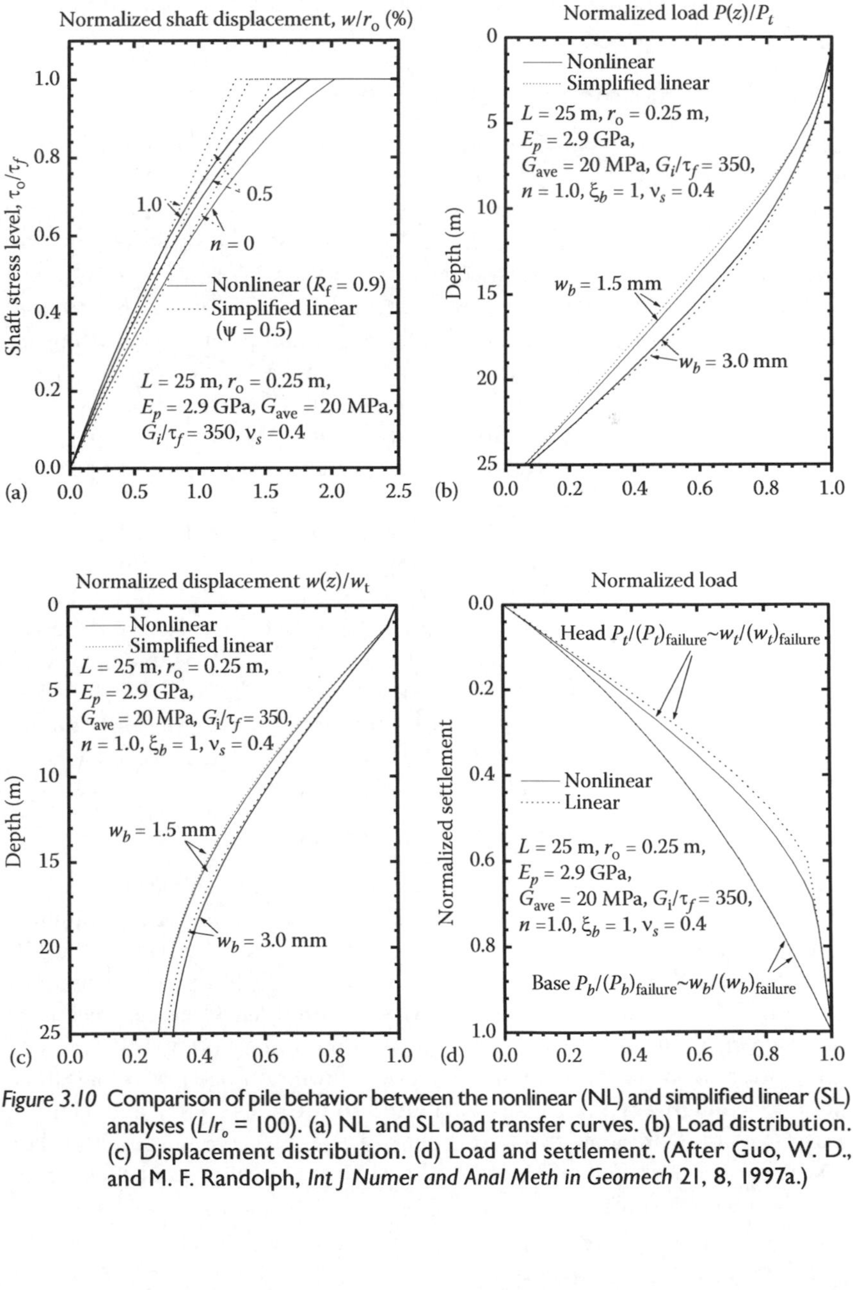

Figure 3.10 Comparison of pile behavior between the nonlinear (NL) and simplified linear (SL) analyses (L/r_o = 100). (a) NL and SL load transfer curves. (b) Load distribution. (c) Displacement distribution. (d) Load and settlement. (After Guo, W. D., and M. F. Randolph, *Int J Numer and Anal Meth in Geomech* 21, 8, 1997a.)

Compared to Equation 3.3 for elastic base, nonlinear base load displacement relationship was written as (Chow 1986b):

$$w_b = \frac{P_b(1-\nu_s)\omega}{4r_oG_{ib}}\frac{1}{(1-R_{fb}\,P_b/P_{fb})^2} \tag{3.13}$$

where P_{fb} = the limiting base load and R_{fb} = a parameter controlling the degree of nonlinearity.

3.2.2 Nonlinear load transfer analysis

A program operating in Microsoft Excel called GASPILE has been developed to allow analysis of pile response in nonlinear soil. The analytical procedure resembles that for computing load-settlement curves of a single pile under axial load (Coyle and Reese 1966). A pile is discretised into elements, and each element is connected to a soil load transfer spring. Load transfer is realized via the shaft model of Equation 3.2 along with 3.10, and base model of Equation 3.13, respectively. The input parameters include (a) limiting pile–soil friction distribution down the pile (e.g., Equation 3.11); (b) initial shear modulus distribution down the pile (e.g., Equation 3.1); (c) the end-bearing factor and soil Poisson's ratio; and (d) the dimensions and Young's modulus of the pile. Comparison shows the consistency between GASPILE and RATZ predictions (Randolph 2003a).

GASPILE was used to analyze a typical pile–soil system to examine the effect of the nonlinear soil model. The pile has a length L of 25 m, a radius r_o of 0.25 m and an equivalent Youngs' modulus E_p of 2.9 GPa (for a solid cylindric pile). The pile is discretized into 20 segments (with little different results using 10 segments). The soil has G_{ave} of 20 MPa (subscript "ave" denotes "average value over pile embedment"), Poisson's ratio, $\nu_s = 0.4$. The pile–soil system has a ratio of modulus and strength, G_i/τ_f of 350; an end-bearing factor ξ_b of 1; and the ultimate base load P_{fb} of 1.2 MN.

3.2.2.1 *Shaft stress-strain nonlinearity effect*

The nonlinear model (NL, R_f = constant) and the simple linear model (SL, ψ = constant), as shown in Figure 3.10a, were used, along with an identical base soil models featured by R_{fb} = 0.9. The difference between the NL (R_f = 0.9) and SL (ψ = 0.5) models is generally small and consistent, as shown in Figure 3.10, in terms of nondimensional load and displacement distributions down the pile at a base displacement, w_b of 1.5, and 3.0 mm; and the pile-head response owing to degree of nonhomogeneity n of 0, 0.5, and 1.0 (see Figure 3.11). This is attributed to the consistency between the shaft NL and SL models for a stress level to ~0.6 (see Figure 3.10a). For most realistic cases, the effect of nonlinearity is expected to be significant

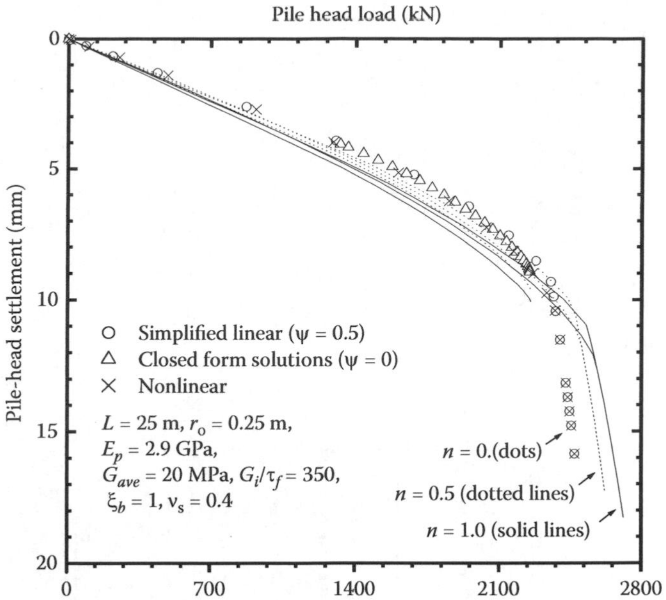

Figure 3.11 Comparison of pile-head load settlement relationship among the nonlinear and simple linear (ψ = 0.5) GASPILE analyses and the CF solution (L/r_o = 100). (After Guo, W. D., and M. F. Randolph, *Int J Numer and Anal Meth in Geomech* 21, 8, 1997a.)

only at load levels close to failure (e.g., Figure 3.10d with $R_f = 0.9$). The SL from a constant value of ζ is generally sufficiently accurate.

3.2.2.2 *Base stress-strain nonlinearity effect*

The influence of base stress level is obvious only when a significant settlement occurs (Poulos 1989). If the base settlement, w_b, is less than the local limiting displacement, w_e, the base soil is generally expected to behave elastically. The exception is when the underlying soil is less stiff than the soil above the pile base level (ξ_b > 1). This case is fortunately associated with a relatively small base contribution. As a result, an elastic consideration of the base interaction before full shaft slip is generally adequate.

3.3 TIME-DEPENDENT CCM

The accuracy of the load transfer model prompts the extension into time-dependent case. Load transfer functions for the shaft may be derived from the stress-strain response of the soil using the concentric cylinder approach. As depicted in Figure 3.12b, the approach is based on a simple $1/r$ variation of shear stress, τ_{rz}, around the pile (where r is the distance from pile

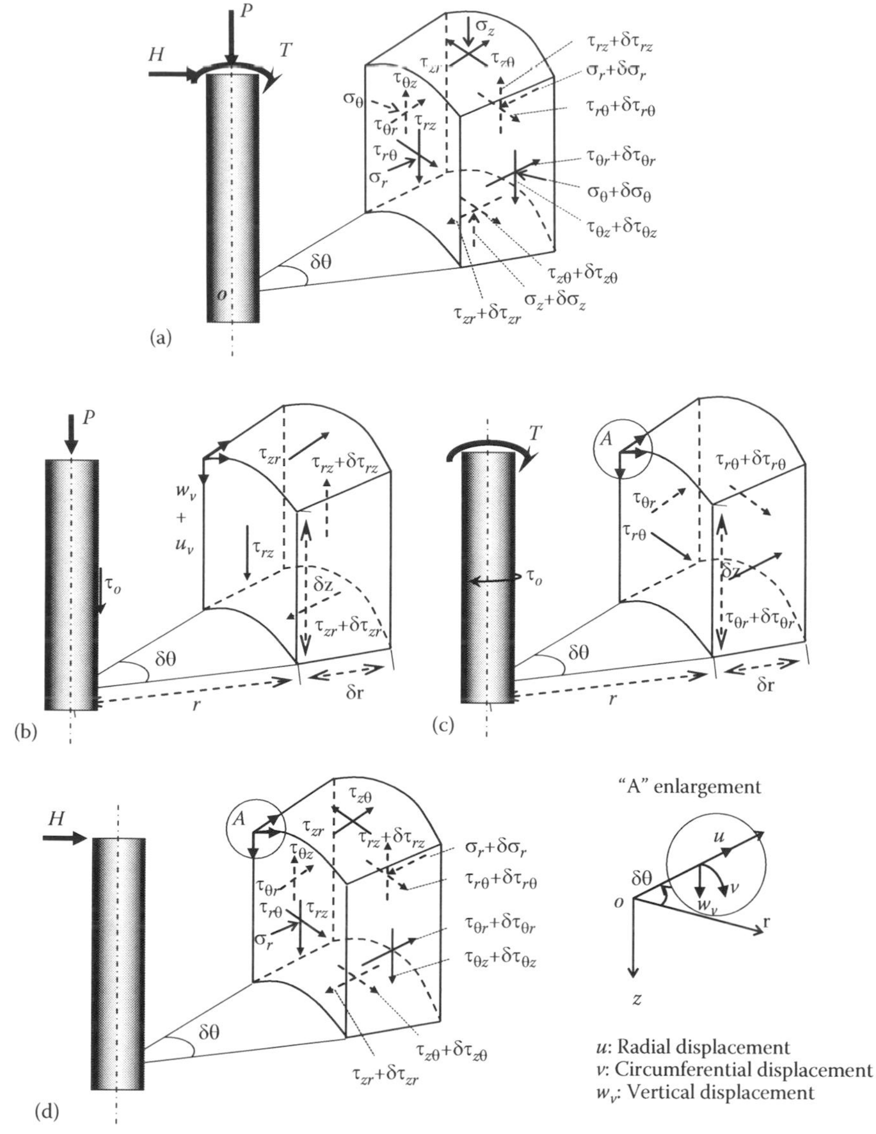

Figure 3.12 Stress and displacement fields underpinned load transfer analysis. (a) Cylindrical coordinate system with stresses. (b) Vertical (P) loading. (c) Torsional (T) loading. (d) Lateral (H) loading. (After Guo, W. D., and F. H. Lee, *Int J Numer and Anal Meth in Geomech* 25, 11, 2001.)

axis) (Cooke 1974; Frank 1974; Randolph and Wroth 1978). The treatment below extends those functions to allow for visco-elastic response of the soil.

3.3.1 Nonlinear visco-elastic stress-strain model

A pile in clay under a sustained load usually undergoes additional (creep) settlement, owing to time-dependent stress-strain behavior (Mitchell and Solymar 1984). The creep settlement occurs in the soil surrounding the pile as well as on the pile–soil interface itself (Edil and Mochtar 1988). A model consisting of Voigt and Bingham elements in series can account well for the creep behavior of several soils (Komamura and Huang 1974), but for the difficulty in determining the slider threshold value for the Bingham element. An alternative is to adopt a hyperbolic stress-strain curve as shown by experiment (Feda 1992). This alternative use leads to a modified intrinsic time dependent nonlinear creep model (Figure 3.13a) (Guo 2000b)

$$\gamma = \gamma_1 + \gamma_2 \tag{3.14}$$

$$\tau_j = \gamma_j G_j k_j \tag{3.15}$$

$$\tau_3 = \eta_{\gamma 3} \dot{\gamma}_3 \tag{3.16}$$

$$\tau_1 = \tau_2 + \tau_3 \tag{3.17}$$

where γ_j = shear strain for the elastic spring 1 and 2 and dashpot 3 (j = 1, 2, and 3), respectively; γ = total shear strain; G_j = instantaneous and delayed initial elastic shear modulus (j = 1, 2), respectively; $\dot{\gamma}_3$ = shear strain rate for the dashpot ($\gamma_3 = \gamma_2$); $\eta_{\gamma 3}$ = shear viscosity at a strain rate of $\dot{\gamma}_3$; τ_j = shear stress acted on spring 1 and 2 and dashpot 3 (j = 1, 2, and 3), respectively;

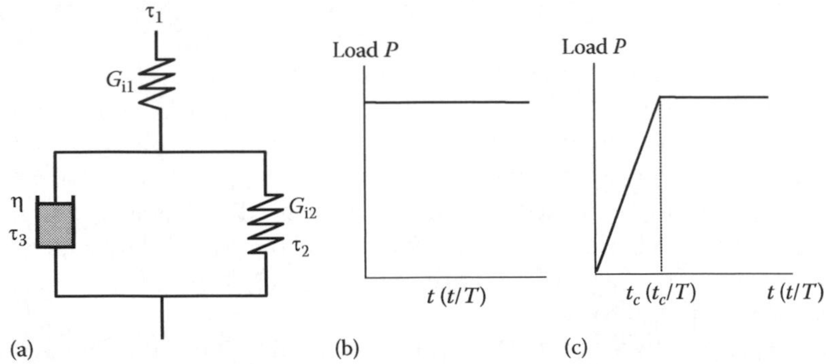

Figure 3.13 Creep model and two kinds of loading adopted in this analysis. (a) Visco-elastic model. (b) One-step loading. (c) Ramp loading. (After Guo, W. D., *Int J Numer and Anal Meth in Geomech* 24, 2, 2000b.)

and k_j = coefficient for considering nonlinearity of elastic springs 1 and 2 (j = 1, 2), respectively.

The shear strain rate, $\dot{\gamma}$, is developed and related to absolute temperature and/or deviatoric shear stress using rate process theory (Murayama and Shibata 1961; Christensen and Wu 1964; Mitchell 1964; Mitchell et al. 1968). However, they did not account for the nonlinearity of the soil creep. A nonlinear hyperbolic model can fit well with measured stress-strain relationship at different times (Feda 1992). This may be realized via Equations 3.14 to 3.17 and the coefficient k_j given by

$$k_j = 1 - \psi_j = G_j / G_{ij} \tag{3.18}$$

where $\psi_j = R_{fj}\tau_j / \tau_{fj}$ (j = 1, 2); R_{fj} is originally defined as τ_{fj}/τ_{ultj} (τ_{ultj}, τ_{fj} = ultimate and failure local shear stress for spring j, respectively) (Duncan and Chang 1970); and i denotes "initial."

During a creep process, the coefficient k_j is a constant. Equations 3.14 to 3.17 may be converted to

$$\tau_1 J + \frac{\eta_{\gamma 3}}{G_{\gamma 2}} \frac{1}{G_{\gamma 1}} \dot{\tau}_1 = \gamma + \frac{\eta_{\gamma 3}}{G_{\gamma 2}} \dot{\gamma} \tag{3.19}$$

where $J = 1/G_{\gamma 1} + 1/G_{\gamma 2}$; $G_{\gamma j} = G_j k_j$, instantaneous and delayed elastic shear modulus at a strain of γ_j (j = 1, 2) respectively; $\dot{\tau}_1$, $\dot{\gamma}$= shear stress rate and shear strain rate respectively. Equation 3.19 is of an identical form to that for linear visco-elastic material. It captures the impact of nonlinearity of the modulus, G_j, through the simple reduction factor, k_j. Equation 3.19 is integrated with respect to time, and with the initial condition of $\gamma = 0$ at $t = 0$, the total shear strain is deduced as

$$\gamma = \frac{\tau_1}{G_{\gamma 1}} \left(1 + \frac{G_{\gamma 1}}{G_{\gamma 2}} \frac{G_{\gamma 2}}{\eta_{\gamma 3}} \int_0^t \frac{\tau_1(t^*)}{\tau_1} \exp\left(-\frac{G_{\gamma 2}}{\eta_{\gamma 3}} (t - t^*) \right) dt^* \right) \tag{3.20}$$

where τ_1, $\tau_1(t^*)$ = soil shear stress at times t and t^*, respectively; t^* = a variable for the integration. The shear strain of Equation 3.20, underpinned by the nonlinear soil model of Equations 3.14 to 3.17, is characterized by instantaneous elasticity (G_1) and delayed elasticity (G_2). At the onset of loading, only elastic shear strain is initiated. Creep displacement gradually appears (via delayed elasticity) on and/or around the pile–soil interface and is dominated by the ratio of strength over modulus, $\tau_1/G_{\gamma 2}$; the relaxation time, $\eta_{\gamma 3}/G_{\gamma 2}$; and the loading path via $\tau_1(t^*)/\tau_1$.

The stress initially taken by the dashpot redistributes to the elastic spring 2 (Figure 3.13a) until finally all the stress is transferred. During the transferring process, spring 2 will yield upon attaining the failure stress τ_{f2}. A larger fraction of the stress subsequently has to be endured by the dashpot, which could lead to a nonterminating creep or eventually trigger a failure. The stress τ_{f2} (= τ_{ult2}) is a long-term value and is lower than τ_{f1} (τ_{ult1}) (Geuze and Tan 1953; Murayama and Shibata 1961; Leonardo 1973), (e.g., τ_{f2}/τ_{f1} = 0.71) (Murayama and Shibata 1961). Soil strength [thus $\tau_{f2}(\tau_{ult2})$, $\tau_{f1}(\tau_{ult1})$] reduces linearly with the logarithmic time elapsed (Casagrande and Wison 1951), as also formulated by Leonardo (1973), and reduces logarithmically with increase in water content.

The magnitude of the creep parameters, for either disturbed or undisturbed clays, are as follows (Guo 1997):

1. The relaxation time, $\eta_{\gamma 3}/G_{\gamma 2}$, a constant for a given clay is 0.3~5 ($\times 10^5$ second) varying from site to site (Lo 1961; Qian et al. 1992). The compressibility index ratio, $G_{\gamma 1}/G_{\gamma 2}$, depending on water content, varies from 0.05 to 1.5 (Lo 1961).
2. The individual values of $G_{\gamma 1}$, $G_{\gamma 2}$, and $\eta_{\gamma 2}$, however, vary with load (stress) level.

The two factors of $\tau_1/G_{\gamma 2}$ and $\eta_{\gamma 3}/G_{\gamma 2}$ may be estimated through interface (pile–soil) shear test or backfigured through field or laboratory pile tests (see Example 3.1). They may be similar to the above-mentioned values.

Generally speaking, secondary compression of all remolded and undisturbed clays obtained by odometer tests can be sufficiently accurately predicted by the model of Equation 3.20 for the elastic case ($\psi_1 = \psi_2 = 0$). The model is adequate to piles, as remolding of the soil is generally inevitable during pile installation.

3.3.2 Shaft displacement estimation

3.3.2.1 *Visco-elastic shaft model*

Local shaft displacement can be predicted through the concentric cylinder approach, which itself is based on elastic theory (Randolph and Wroth 1978; Kraft et al. 1981; Guo and Randolph 1997a; Guo and Randolph 1998). The correspondence principle (Lee 1955; Lee 1956; Lee et al. 1959) states that the analysis of stress and displacement field in a linear visco-elastic medium can be treated in terms of the analogous linear elastic problem having the same geometry and boundary conditions. A shaft model reflecting nonlinear visco-elastic response thus has to be deduced directly from the generalized visco-elastic stress strain relationship of Equation 3.20, with suitable shear modulus (Guo 2000b).

Model pile tests show that load transfer along a model pile shaft leads to a nearly negligible volume change (or consolidation) in the surrounding soil (Eide et al. 1961). Approximately, the vertical displacement u_v (subscript v for vertical loading; see Figure 3.12b) along depth z ordinate may be ignored (Randolph and Wroth 1978). Therefore, it follows that

$$\gamma = \frac{\partial u_v}{\partial z} + \frac{\partial w_v}{\partial r} \approx \frac{\partial w_v}{\partial r} \tag{3.21}$$

where w_v = local displacement of shaft element at time t. Based on the concentric cylinder approach, the shaft displacement w_s is obtained by integration from the pile radius, r_o, to the maximum radius of influence, r_m

$$w_s = \int_{r_o}^{r_m} \frac{\partial w_v}{\partial r} dr = \int_{r_o}^{r_m} \frac{\tau}{G} dr \tag{3.22}$$

The shear stress, τ_1 (= τ_{rz}) at a distance of r away from the pile axis may be deduced using force equilibrium on the stress element in vertical direction (see Figure 3.12b), which gives $\tau_1 = \tau_o r_o / r$. The shear stress and shear strain γ of Equation 3.20 were substituted into Equation 3.22, which gives us

$$w_s = \int_{r_o}^{r_m} \frac{\tau_o r_o}{r} \frac{1}{G_{\gamma 1}} dr + \frac{G_{\gamma 2}}{\eta_{\gamma 3}} \int_{r_o}^{r_m} \int_0^t \frac{\tau_o(t^*) r_o}{r} \frac{1}{G_{\gamma 2}} \exp(-\frac{G_{\gamma 2}}{\eta_{\gamma 3}}(t - t^*)) dt^* \, dr \tag{3.23}$$

where τ_o, $\tau_o(t^*)$ = shear stress on the pile soil interface at time t and t^* respectively. $G_{\gamma j}$ becomes the shear modulus at distance r away from the pile axis for elastic spring j (j = 1, 2). Although the shear modulus and the viscosity parameter are functions of the stress level, the relaxation time $\eta_{\gamma 3}/G_{\gamma 2}$ may be taken as a constant (Guo 1997). Hence, it is replaced with T ($T = \eta/G_2$, η = the value of $\eta_{\gamma 3}$ at strain $\gamma_3 = 0\%$).

The inverse linear reduction of shear stress away from a pile (i.e., $\tau_1 = \tau_o r_o / r$) along with Equation 3.18 allow a variation of shear modulus with distance r to be determined as

$$G_{\gamma j} = G_{ij}(1 - \frac{r_o}{r}\psi_j) \tag{3.24}$$

where $\psi_j = R_{fj}\tau_{oj}/\tau_{fj}$; nonlinear stress level on the pile–soil interface for elastic spring j (j = 1, 2); and τ_{oj} = shear stress on pile–soil interface (j = 1, 2). The shear modulus variation of Equation 3.24 allows Equation 3.23 to be simplified as

$$w_s = \frac{\tau_o r_o}{G_1} \zeta_1 \zeta_c \tag{3.25}$$

where

$$\zeta_c = 1 + \frac{\zeta_2}{\zeta_1} \frac{G_1}{G_2} A(t) \tag{3.26}$$

Equation 3.25 is a nonlinear visco-elastic load transfer (t-z) model. The shear measure of influence ζ is equal to a product of two entities ζ_1 and ζ_c.

The displacement calculation embracing nonlinear visco-elastic behavior still retains the simplicity and pragmatism of Equation 3.5 (Randolph and Wroth 1978; Kraft et al. 1981). The radial shear influence is the same as Equation 3.10, and for spring j,

$$\zeta_j = \ln\left(\frac{r_m/r_o - \psi_j}{1-\psi_j}\right) \tag{3.27}$$

Normally, ζ_1 gradually becomes higher than ζ_2, as stress level ψ_j increases, because ψ_2 is gradually higher than ψ_1 (with the failure stress, $\tau_{ult2} < \tau_{ult1}$).

The time dependent part $A(t)$ is related to stress level–time relationship by

$$A(t) = \frac{1}{T}\int_o^t \frac{\tau_o(t^*)}{\tau_o}\exp(-\frac{(t-t^*)}{T})\,dt^* \tag{3.28}$$

Most practical loading tests follow "ramp type" loading, which is a combination of constant rate of loading during addition of load (i.e., $t < t_c$ in Figure 3.13c; t_c = the time at which a constant load commences) and sustained loading ($t > t_c$, a creep process). Within the elastic stage, the shear stress at any time, t^* (in between 0 and t_c) should follow a similar pattern of time dependency to the loading, thereby

$$\tau_o(t^*)/\tau_o(t) = t^*/t \tag{3.29}$$

Afterwards, when $t^* > t_c$, the stress ratio stays at unity. Therefore, if the total loading time t exceeds t_c, Equation 3.28 may be integrated, allowing $A(t)$ to be written as

$$A(t) = \frac{t_c}{t}\exp\left(-\frac{t-t_c}{T}\right) - \frac{T}{t}\left(\exp\left(-\frac{t-t_c}{T}\right) - \exp\left(-\frac{t}{T}\right)\right) + 1 - \exp\left(-\frac{t-t_c}{T}\right) \tag{3.30}$$

Otherwise, if $t \le t_c$, $A(t)$ is given

$$A(t) = 1 - \frac{T}{t}\left(1-\exp\left(-\frac{t}{T}\right)\right) \tag{3.31}$$

At the extreme case of $t_c = 0$ (i.e., one-step loading, Figure 3.13b), $A(t)$ is given by

$$A(t) = 1 - exp(-t/T) \tag{3.32}$$

The non-dimensional local displacement and stress level for nonlinear visco-elastic case (NLVE) is illustrated in Figure 3.14a for a typical pile

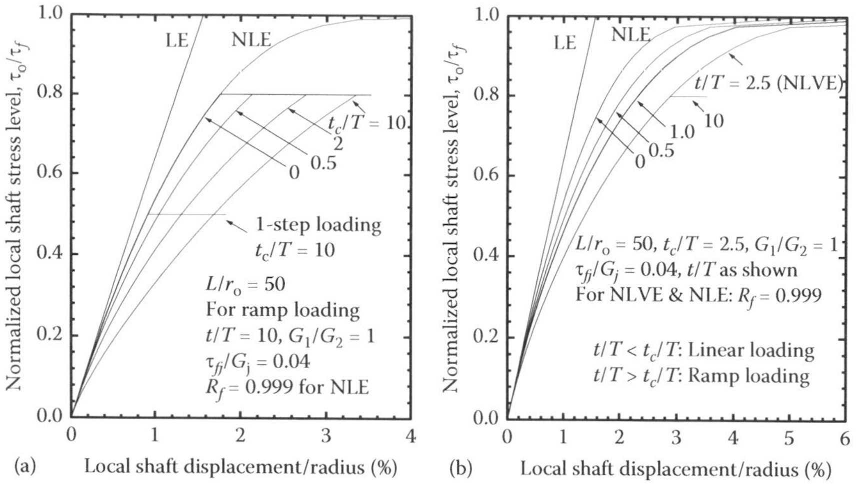

Figure 3.14 Normalized local stress displacement relationships for (a) 1-step and ramp loading, and (b) ramp loading. (After Guo, W. D, *Int J Numer and Anal Meth in Geomech* 24, 2, 2000b.)

of $L/r_o = 50$ in a clay with $\tau_{fj}/G_j = 0.04$ ($j = 1, 2$), $\nu_s = 0.5$, $n = 1$, $G_1/G_2 = 1$, and pile–soil interaction factors of $\zeta_2/\zeta_1 = 1$ and $\xi_b = 1$. First, the linear elastic (LE) and nonlinear elastic (NLE) load transfer curves were gained and are illustrated in Figure 3.14a as reference. The creep is then illustrated in Figure 3.14a for one-step loading initiated at stress level τ_o/τ_{f1} of 0.5 and ramp loading initiated at the beginning ($\tau_o/\tau_{f1} = 0$) and held at a prolonged load level of 0.8 from time t_c. Figure 3.14a indicates a significant effect of the relative ratio of the duration of constant rate of loading, t_c, over total loading time, t, on the stress-displacement response, as is evident in Figure 3.14b concerning a constant of t_c/T but varying t/T.

3.3.2.2 Nonlinear creep displacement

Figure 3.15a shows ~200% increase of the parameter ζ_1 with the shear stress level, ψ, approaching 0.9. At failure, the secant stiffness of the load transfer curve is approximately half the initial tangent value for values of R_f in the region of 0.9. The whole shape of the curve may be approximated closely by a parabola (Randolph 1994). The elastic behavior of pile (corresponding to $\zeta_c = 1$) under working load may be predicted by the solutions (Guo and Randolph 1997a).

To predict secondary deformation of clay, the ratio of ζ_2/ζ_1 may be taken as unity (identical shaft failure stress for both springs 1 and 2), in light of the correspondence principle for linear visco-elastic media. Generally, at a large load level, the stress level on spring 2 may exceed that on spring 1 at the same degree of shaft displacement mobilization, owing to the limiting shear stress $\tau_{ult2} < \tau_{ult1}$. The parameter ζ_2 estimated by Equation 3.27 is larger than ζ_1. A higher shaft displacement (via high ζ_c) is expected as per Equation 3.25. At this stage, the pile would not yield, but has significant creep displacement, particularly for long piles.

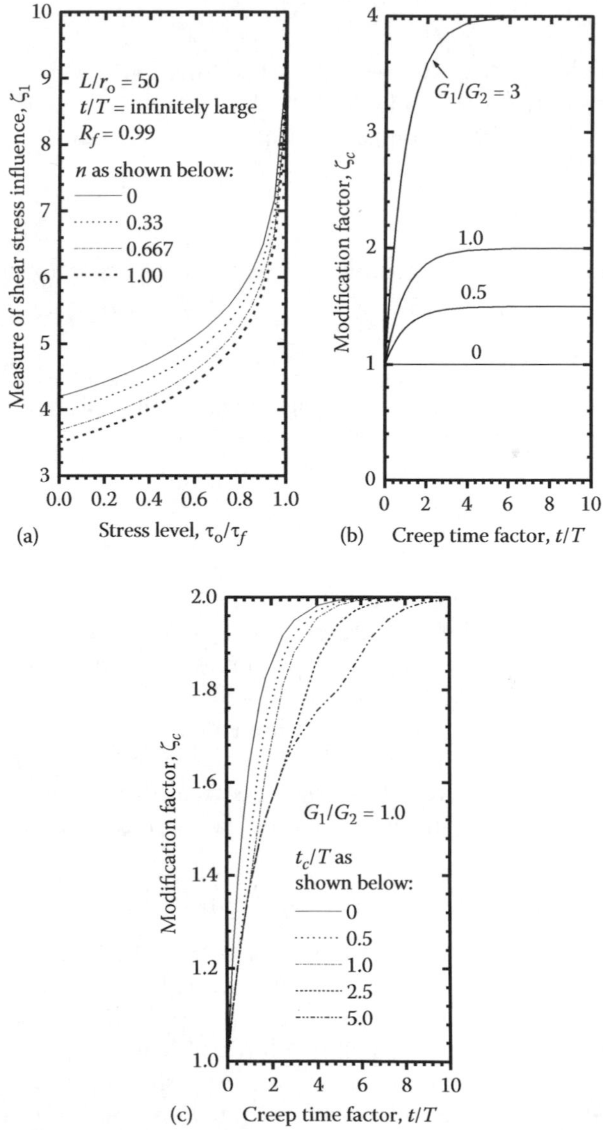

Figure 3.15 (a) Variation of nonlinear measure with stress level. (b) Modification factor for step loading. (c) Modification factor for ramp loading. (After Guo, W. D., *Int J Numer and Anal Meth in Geomech* 24, 2, 2000b.)

The creep modification factor, ζ_c, varies with nondimensional time t/T and depends on modulus ratios, G_1/G_2, as shown in Figure 3.15b for step loading. It depends on the t_c/t of the ramp loading, as illustrated in Figure 3.15c for $G_1/G_2 = 1$. Given a sufficient time, it leads to $\zeta_c = (G_1/G_2) + 1$. Equation 3.32 may be converted to a creep function $J(t)$ with

$$J(t) = A_c + B_c e^{-t/T} \tag{3.33}$$

where $A_c = 1/G_1 + \zeta_2/G_2\zeta_1$; $B_c = -\zeta_2/G_2\zeta_1$. The function is equivalent to that adopted previously (Booker and Poulos 1976) and will be used in Chapter 5, this book. In addition, comparing Equation 3.26 with Equation 3.33 indicates

$$J(t) = \zeta_c/G_1 \tag{3.34}$$

This relationship enables Equation 3.25 to be written as a function of $J(t)$ as well:

$$w_s = \tau_o r_o \zeta_1 J(t) \tag{3.35}$$

Equations 3.25 and 3.26 allow the creep component of displacement (step loading) to be expressed as

$$w_c = \frac{\tau_o r_o}{G_2}\zeta_2 A(t) = \frac{\psi_2 \tau_{f2} r_o}{R_{f2} G_2}\zeta_2\left(1 - \exp(-\frac{t}{T})\right) \tag{3.36}$$

where w_c = local creep displacement at time t. Equation 3.36 implies that the rate of creep displacement of a frictional pile is proportional to the diameter of the pile and the stress ratio. This agrees well with theoretical and/or empirical findings (Mitchell 1964; Edil and Mochtar 1988). The impact of pile slenderness ratio, the shaft nonhomogeneity factor, and Poisson's ratio on the w_e is captured through ζ_2. The time-displacement relationship of Equation 3.36 is different from the empirical expression by Edil and Mochtar (1988), but it matches well with experimental data shown next.

3.3.2.3 *Shaft model versus model loading tests*

The shaft displacement can be easily gained from Equation 3.25, which includes the nonlinear elastic component obtained by using $\zeta_c = 1$ and the creep component (e.g. by Equation 3.36 for step loading). The nonlinear model was theoretically verified (Randolph and Wroth 1978; Kraft et al. 1981; Guo and Randolph 1997a). Next, only the creep component of Equation 3.36 is checked against experimental data, concerning the impact of the initial elastic and delayed shear moduli, the ultimate (≈ failure) shaft shear stress for the springs 1 and 2, the relaxation time, and the geometry and elastic property of the pile.

As discussed in Chapter 2, this book, appropriate values of τ_{f1} (τ_{ult1}) may be estimated using the shear strength of the soil, the effective overburden stress, the CPT, or the SPT tests. The variation of the shear stress, τ_{fj}, due to reconsolidation may be estimated by the relevant elastic or visco-elastic consolidation theory to be discussed in Chapter 5, this book.

The modulus G_1 may be deduced by fitting Equation 3.25 with the measured local shear stress-displacement relationship. The equivalent modulus to evaluate a pile settlement of 1% of the pile radius may be initially taken as $3G_{1\%}$ ($G_{1\%}$ = shear modulus at a shear strain of 1%) (Kuwabara 1991), otherwise a smaller value should be taken for a large settlement. With normally consolidated clays, the shear modulus at a shear strain of 1%, $G_{1\%}$ and the initial modulus G_1 may be taken as (Kuwabara 1991) $G_{1\%} = (80\sim90)s_u$ and $G_1 = (400\sim900)s_u$, respectively. Use of nonlinear elastic or elastic form ($\psi = 0$) of Equation 3.25 generally results in little discrepancy of the overall pile response over a loading level between 0 and 0.75 (Guo and Randolph 1997a). The initial shear modulus, G_1, can generally be chosen as 1~3 times the corresponding shear modulus gained from field measurement or empirical formulas (Fujita 1976).

The rate factor, $1/T$, should be ascertained for a range of relevant loading levels. Three laboratory tests (Edil and Mochtar 1988) provide time-dependent settlement relationships for short, "rigid" model piles. They were well fitted using Equation 3.36 and the parameters in Table 3.4, as shown in Figure 3.16a. The poor fits to measured data during initial stage are noted irrespective of using Equation 3.36 or Edils and Mochtar's statistical formula. This indicates nonlinear elastic displacement in the creep tests and reflects the hydrodynamic period of consolidation process (Lo 1961).

The creep parameters for a given load may also be deduced from measured settlement versus time relationship of a loading test. Equations 3.25 and 3.36 show the creep settlement rate may be expressed as

$$\frac{dw_c}{dt} = \left[\frac{1}{T}\exp\left(-\frac{t}{T}\right)\right]\left[\frac{\tau_o r_o}{G_2}\zeta_2\right] \tag{3.37}$$

Table 3.4 Curve fitting parameters for creep tests in Figure 3.16a

Test No.	G_2/τ_{f2}	G_2/η (10^{-5}/sec)	Length L (mm)	Diameter d (mm)	Stress Level ψ ($R_f\tau_o/\tau_f$)
38	175.	0.5	115.6	10.1	0.91
32	175.	0.55	90.4	17.0	0.69
12	500.	2.67	77.5	26.7	0.68

Source: Guo, W. D., *Int J Numer and Anal Meth in Geomech* 24, 2, 2000b.

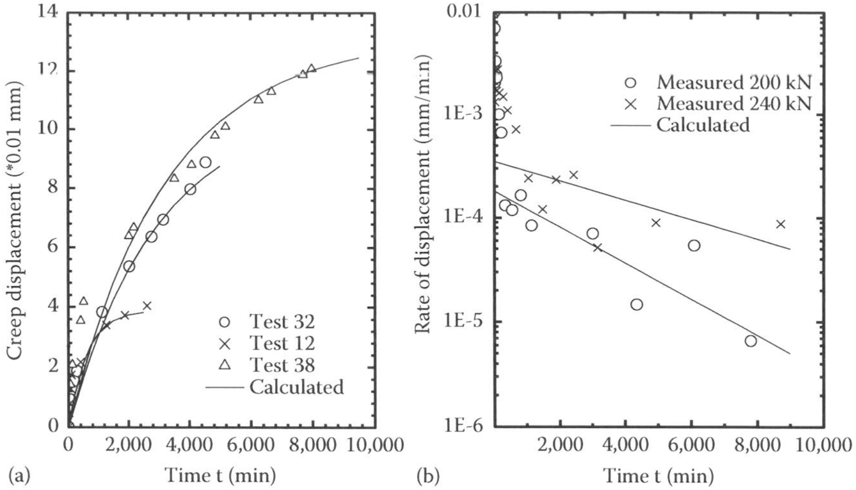

Figure 3.16 (a) Predicted shaft creep displacement versus test results (Edil and Mochtar 1988). (b) Evaluation of creep parameters from measured time settlement relationship (Ramalho Ortigão and Randolph 1983). (After Guo, W. D., *Int J Numer and Anal Meth in Geomech* 24, 2, 2000b.)

Under a sustained load at the pile top, the logarithmic creep settlement rate of $\log(dw_c/dt)$ may be obtained using loading test. The plot against time (normally a straight line) can be fitted by Equation 3.37, which allows the parameters to be readily determined. The use of this method is elaborated in Example 3.1.

Example 3.1 Determining creep parameters from loading tests

Ramalho Ortigão and Randolph (1983) reported a pile tested in clay until failure in an increment sustained tensile loading pattern. The closed-ended steel pipe pile was 203 mm in diameter and 6.4 mm in wall thickness and had an equivalent Young's modulus of 2.1×10^5 MPa. The pile was driven 9.5 m into stiff, overconsolidated clay. The shear modulus of the clay was 12 MPa, the failure shaft friction was 41.5 kPa, as deduced from the load settlement curve.

To estimate creep parameters, an average pile stress level is used to gain the nonlinear elastic load transfer measure, ζ. The loading tests offer an ultimate load of 280 kN. The stress (load) levels were 0.714 and 0.857, respectively, for the pile-head loads of 200 and 240 kN. With the pile geometry, a soil Poisson's ratio of 0.3, the nonlinear elastic measure, ζ, as per Equation 3.27, is calculated as 6.35 [i.e., $(\xi_2)_1$] at 200 kN and as 7.04 [i.e., $(\xi_2)_2$] at 240 kN, respectively. The tests offer the plots of the log creep settlement rate and time, which are shown in Figure 3.16b. The two lines for loading of 200 and 240 kN offer relaxation times, $1/T_1$ of 6.64×10^{-6}/s and $1/T_2$ of 3.6×10^{-6}/s, respectively.

The intersections in the creep settlement rate ordinate for the two loading levels are 0.00018 and 0.00035 mm/min. At t = 0, Equation 3.37 for 200 kN and 240 kN is written as

$$(\xi_2)_1\left(\frac{\tau_o r_o}{G_1}\right)_1\left(\frac{G_1}{G_2}\right)_1\frac{1}{T_1}=0.00018\ (\text{mm/min}) \tag{3.38a}$$

$$(\xi_2)_2\left(\frac{\tau_o r_o}{G_1}\right)_2\left(\frac{G_1}{G_2}\right)_2\frac{1}{T_2}=0.00035\ (\text{mm/min}) \tag{3.38b}$$

Equation 3.38a, along with $\psi_1 = 0.714$, $(\xi_2)_1 = 6.35$, $r_o = 101.5$ mm, $1/T_1 = 6.64 \times 10^{-6}$/s, and $\tau_f = 41.5$ kPa offer $G_1/G_2 = 0.2839$, $(G_2)_1 = 42.27$ MPa. Equation 3.38b together with $\psi_2 = 0.857$, $(\xi_2)_2 = 7.04$, $r_o = 101.5$ mm, $1/T_2 = 3.6 \times 10^{-6}$/s, and $\tau_f = 41.5$ kPa lead to $G_1/G_2 = 0.7653$, $(G_2)_2 = 15.68$ MPa.

The initial shear modulus, G_1, generally increases with soil consolidation, but it can be regarded as a constant upon completion of the primary consolidation. As mentioned before, the creep parameters, G_2 and η, normally vary with the loading (stress) level (Figure 3.16b). The ratio of delayed shear moduli $(G_2)_1/(G_2)_2$ is 2.69, indicating the reduction in G_2 with the increase in load level. However, the ratios of G_1/G_2 and G_2/η are nearly constants to a working load level, say 70% (= τ_{f2}/τ_{f1}) of failure load level. Afterwards, the G_1/G_2 may become higher. The ratio G_2/η influences the duration of creep time rather than the final pile-head response and may roughly be taken as a constant over general working load.

Use of Equation 3.37, in essence, is identical to that proposed by Lee (1956). The $1/T$ of $(0.36{\sim}0.664) \times 10^{-5}$/s [or $T = (1.5{\sim}2.78) \times 10^5$ second] deduced for the loading tests is consistent with laboratory tests mentioned earlier. Thus, lab tests on soil may be used for crude estimation of the time process, though interface tests on pile and soil interface are recommended. An average value of $1/T$ over a range of working load may be employed in Equation 3.25, as it only slightly affects the time-dependent process but not the final pile response.

3.3.3 Base pile–soil interaction model

The nonlinear visco-elastic (time-dependent) response of the pile-base load and displacement is still estimated using Equation 3.13, in which the base shear modulus of G_{ib} is replaced with the following time-dependent shear modulus $G_b(t)$:

$$G_b(t)=\frac{G_{b1}}{1+A(t)G_{b1}/G_{b2}} \tag{3.39}$$

where G_{b1}, G_{b2} = shear modulus just beneath the pile tip level for springs 1 and 2, respectively, with $G_{b1}/G_{b2} \approx G_1/G_2$.

3.3.4 GASPILE for vertically loaded piles

The GASPILE program was extended to incorporate the visco-elastic response using Equations 3.25 and 3.39. Using the Mechant's model, the conclusions described below have been directly adopted in this program (Guo 1997; Guo and Randolph 1997b; Guo 2000b): (a) τ_{ultj} (limiting shaft stress) = τ_{fj} (failure stress on the pile–soil interface); (b) $\tau_{ult1} = \alpha s_u$ or $\tau_{ult1} = \beta\sigma_v'$; (c) $\tau_{f2} = 0.71\tau_{f1}$ if not available; (d) Nonlinear and creep responses to stress are captured via instantaneous elasticity ($G_{\gamma 1}$) and delayed elasticity ($G_{\gamma 2}$). At the onset of loading, the stress-strain response is a nonlinear hyperbolic curve. Under a specified stress, a creep process is displayed. The base settlement is based on Equation 3.13 and time-dependent modulus of Equation 3.39. Overall, secondary deformation is generally sufficiently accurately modeled for piles in "remolded" clays (see Chapter 5, this book).

3.3.5 Visco-elastic model for reconsolidation

Installation of piles in clay will induce pore water pressure. The maximum pore pressure $u_o(r)$ immediately following driving may approximately equal or even exceed the total overburden pressure in overconsolidated soil (Koizumi and Ito 1967; Flaate 1972). The pore pressure decreases rapidly with distance r from the pile wall and becomes negligible at a distance of 10~20r_o. This dissipation is well simulated using the one-dimensional, cylindrical cavity expansion analogy (Randolph and Wroth 1979) or the strain path method (Baligh 1985) (see Figure 3.17a). The former theory has been extended for visco-elastic soil through a model depicted in Figure 3.17b, which is elaborated on Chapter 5, this book.

3.4 TORQUE-ROTATION TRANSFER MODEL

Torsional loading on piles may occur due to an eccentric lateral loading. It is important for some lateral piles (Guo and Randolph 1996). Numerical and analytical solutions have been published for piles subjected to torsion, for which the soil is elastic with either homogeneous modulus, or modulus proportional to depth (Poulos 1975; Randolph 1981a), or a modulus varying with a simple power law of depth (see Equation 3.1). The power law works for homogeneous (n = 0) and proportionally varying cases as well and is more accurate to capture the impact of modulus profiles on the pile-head stiffness. This is important, as load transfer is generally concentrated in the upper portion of a torsional pile (Poulos 1975). In the context of load transfer model, the torsional behavior may be modeled by a series of "torsional" springs distributed along the pile shaft. The model allows elastic

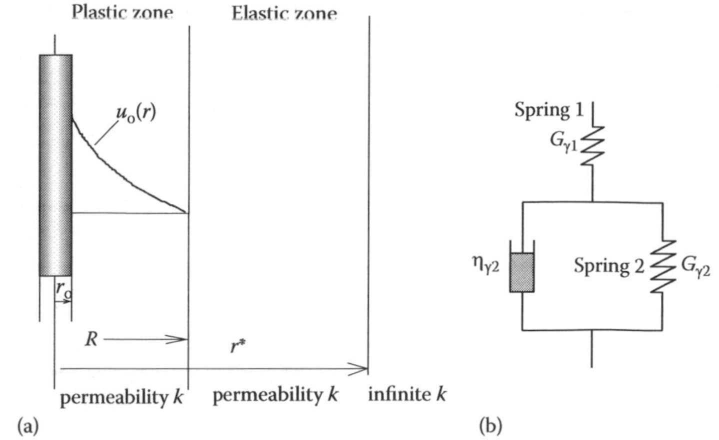

Figure 3.17 Modeling radial consolidation around a driven pile. (a) Model domain. (b) Visco-elastic model. (After Guo, W. D., *Computers and Geotechnics* 26, 2, 2000c.)

and elastic-plastic solutions to be developed (Guo and Randolph 1996, Guo et al. 2007). The solutions are not presented in this book, but it is necessary to review the torsional model (Guo and Randolph 1996) to facilitate understanding the load-transfer model.

3.4.1 Nonhomogeneous soil profile

In Equation 3.1, the G is replaced with the initial (tangent) shear modulus G_i at depth z. The limiting shaft friction τ_f follows Equation 3.11, and the parameters A_v and θ are replaced with a new gradient of A_t and power t ("t" for torsion). In addition, our model is again restricted to a similar profile between the shear modulus and shaft friction by taking $n = t$ and a constant A_g/A_t over pile embedment.

3.4.2 Nonlinear stress-strain response

Equations 3.15 and 3.18 indicate the secant shear modulus, G_r, for a hyperbolic stress-strain law, which is given by

$$G_r = G_i\left(1 - \frac{\tau}{\tau_{ult}}\right) \tag{3.40}$$

Note again that the limiting shear stress in the soil is identical to the failure pile–soil shaft friction. The concentric cylinder approach may be used to estimate the radial variation of shear stress around a pile subjected to torsion (Frank 1974; Randolph 1981a). The longitudinal stress gradients (parallel to the pile, Figure 3.12c) are small compared to radial stress gradients

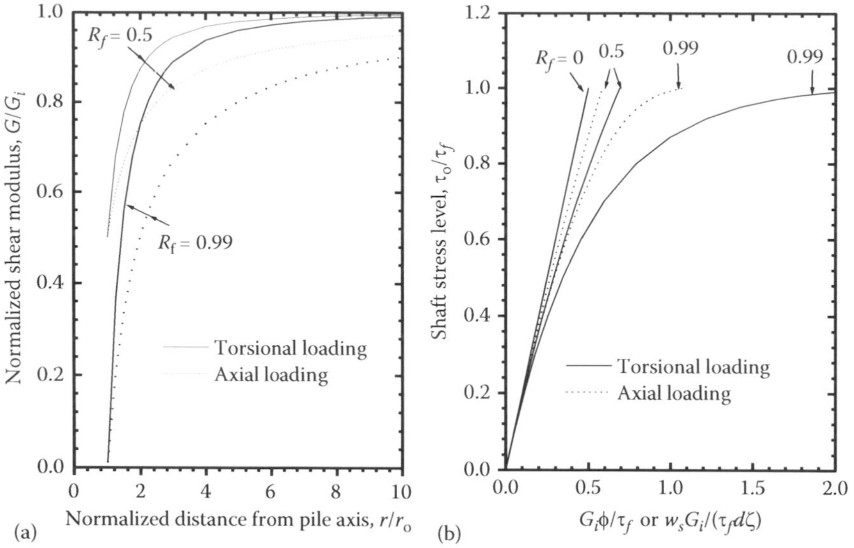

Figure 3.18 Torsional versus axial loading: (a) Variation of shear modulus away from pile axis. (b) Local load transfer behavior. (After Guo, W. D., and M. F. Randolph, *Computers and Geotechnics* 19, 4, 1996.)

and the shear stress (formally $\tau_{r\theta}$, but the double subscript is omitted here) at any radius, r, is given by

$$\tau = \tau_o \frac{r_o^2}{r^2} \tag{3.41}$$

Substituting this into Equation 3.40 gives the radial variation of secant shear modulus as

$$G_r = G_i\left(1 - \frac{r_o^2}{r^2}\psi\right) \tag{3.42}$$

These relationships are similar to those derived for axial loading of a pile of Equation 3.24 (Kraft et al. 1981). However, the effect of nonlinearity is much more localized around a torsional pile, as shown by Figure 3.18a where the normalized shear modulus, G_r/G_i, is plotted as a function of radius, r/r_o, for the two types of loading.

3.4.3 Shaft torque-rotation response

The shear strain, $\gamma_{r\theta}$, around a pile subjected to torsion may be written as (Randolph 1981a) (see Figure 3.12c)

$$\gamma_{r\theta} = \frac{\tau_{r\theta}}{G_r} = \frac{1}{r}\frac{\partial u}{\partial \theta} + r\frac{\partial}{\partial r}\left(\frac{v}{r}\right) \tag{3.43}$$

where u = radial soil deformation, v = circumferential deformation, and θ = angular polar coordinate. From symmetry, $\partial u/\partial\theta$ is zero and Equation 3.43 is combined with Equation 3.41 to give

$$\frac{\partial}{\partial r}\left(\frac{v}{r}\right) = \frac{\tau_o r_o^2}{G_r r^3} \tag{3.44}$$

Substituting Equation 3.42 and integrating this with respect to r from r_o to yields the angle of twist ϕ at the pile as

$$\phi = \left(\frac{v}{r}\right)_O = \frac{\tau_o}{2G_i}\left(\frac{-\ln(1-\psi)}{\psi}\right) \tag{3.45}$$

This equation may be rewritten as

$$\phi = \frac{\tau_f}{G_i}\frac{1}{2R_f}[-\ln(1-\psi)] \tag{3.46}$$

which shows that the angle of twist depends logarithmically on the relative shear stress level. Again, the form of the torque-twist relationship is similar to that for axial loading, as seen from Equation 3.10. However, as shown in Figure 3.18b, the degree of nonlinearity (for a given R_f value) is somewhat more evident for the torsional case (w_s and ζ in the figure for vertical loading).

3.5 COUPLED ELASTIC MODEL FOR LATERAL PILES

A number of simple solutions were developed for laterally loaded piles (see Figure 3.19) using an empirical load transfer [p-y (w) curve] model (Matlock 1970) or a two-parameter model (Sun 1994) in which p = force per unit length and y (w) = displacement. As with vertical loading, pile–soil interaction is modeled using independent elastic springs along the shaft and may be referred to as an uncoupled model. The two-parameter model caters to the coupled impact among the springs through a single factor and may be referred to as a coupled model (Guo and Lee 2001). The accuracy of these solutions essentially relies on the estimation of the (1-D) properties of the elastic springs that represent the 3-D response of the surrounding soil.

The coupled two-parameter (or Vlasov's foundation) model (Jones and Xenophontos 1977; Nogami and O'Neill 1985; Vallabhan and Das 1988) mimics well the effect of the surrounding soil displacement (at Poisson's ratio ν_s < 0.3), through modulus of subgrade reaction k (= p/w) for the

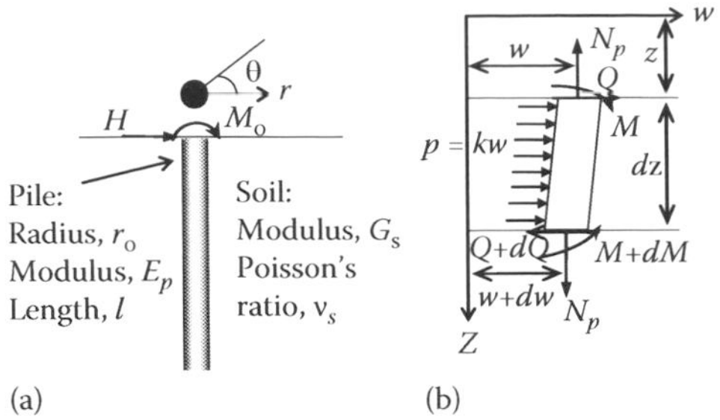

Figure 3.19 Schematic of a lateral pile–soil system. (a) Single pile. (b) Pile element. (After Guo, W. D., and F. H. Lee, *Int J Numer and Anal Meth in Geomech* 25, 11, 2001.)

independent springs, and a fictitious tension N_p of a stretched membrane used to tie together the springs. The model was developed using an assumed displacement field and variational approach. Unfortunately, the ratio ν_s generally exceeds 0.3. The modelling for $\nu_s > 0.3$ was circumvented by incorporating the effect of ν_s into a new shear modulus (Guo and Lee 2001) and using a rational stress field (see Figure 3.12d). The new model is termed a "theoretical" load transfer model, and the parameters k and N_p were obtained for some typical head and base conditions by Guo and Lee (2001), as discussed next.

3.5.1 Nonaxisymmetric displacement and stress field

As depicted in Figure 3.19a, a circular pile is subjected to horizontal loading (H, M_o) at the pile-head level. The pile is of length, L, and radius, r_o, and is embedded in an elastic, homogeneous, and isotropic medium. The pile response is characterized by the displacement, w; the bending moment, M; and the shear force, Q. The displacements and stresses in the soil element around the pile are described by a cylindrical coordinate system r, θ, and z as depicted in Figure 3.12d.

The displacement field around the laterally loading pile is nonaxisymmetric and is normally dominated by radial u and circumferential displacement v, whereas the vertical displacement w_v is negligible. It is expressed as a Fourier series (as shown in Chapter 7, this book). The effect of Poisson's ratio may be captured using the modulus G^* (Randolph 1981b) with $G^* = (1 + 3\nu_s/4)G$, G = an average shear modulus of the soil over the effective length L_c beyond which pile response is negligible (see next section). Taking $\nu_s = 0$ (thus Lame's constant = 0), the displacement and stress fields, as represented by one component of Fourier series (Chapter 7, this book) (see Figure 3.12d), are as follows:

$$u = w(z)\phi(r)\cos\theta \quad v = -w(z)\phi(r)\sin\theta \quad w_v = 0 \tag{3.47}$$

$$\sigma_r = 2Gw(z)\frac{d\phi}{dr}\cos\theta \quad \sigma_\theta = \sigma_z = 0$$
$$\tau_{r\theta} = -Gw(z)\frac{d\phi}{dr}\sin\theta \quad \tau_{\theta z} = -G\frac{dw}{dz}\phi\sin\theta \quad \tau_{zr} = G\frac{dw}{dz}\phi\cos\theta \tag{3.48}$$

where σ_r = radial stress; σ_θ, σ_z = circumferential stress and vertical stress, which are negligible; w and dw/dz = local lateral displacement, and rotational angle of the pile body at depth z; θ = an angle between the loading direction and the line joining the center of the pile cross-section to the point of interest; r = a radial distance away from the pile axis and $w(z)$ = the pile displacement at depth z. The radial attenuation function $\phi(r)$ was resolved as modified Bessel functions of the second kind of order zero, $K_o(\gamma_b)$ (Sun 1994; Guo and Lee 2001):

$$\phi(r) = K_o(\gamma_b r/r_o)/K_o(\gamma_b) \tag{3.49}$$

where r_o = radius of an equivalent solid cylinder pile; γ_b = load transfer factor. The coupled interaction between pile displacement, $w(z)$ and the displacements u, v of the soil around is achieved through the load transfer factor, γ_b [or $\phi(r)$]. The calculation of the factor γ_b is discussed in the next section.

The modulus of subgrade reaction k [FL^{-2}] of p/w is given by

$$k = 1.5\pi G\{2\gamma_b K_1(\gamma_b)/K_o(\gamma_b) - \gamma_b^2[(K_1(\gamma_b)/K_o(\gamma_b))^2 - 1]\} \tag{3.50}$$

The fictitious tension, N_p [F] of the membrane linking the springs is determined by

$$N_p = \pi r_o^2 G[(K_1(\gamma_b)/K_o(\gamma_b))^2 - 1] \tag{3.51}$$

3.5.2 Short and long piles and load transfer factor

Response of a lateral pile becomes negligible as the pile–soil stiffness, (E_p/G), exceeds a critical value $(E_p/G)_c$. It depends on pile-base and head conditions, as illustrated in Figure 3.20, such as free-head (*FreH*), clamped base pile (*CP*) under a later load H and moment loading (M_o), and fixed-head (*FixH*), floating pile (*FP*). The $(E_p/G)_c$ may be estimated by

$$\frac{E_p}{G} \approx 0.05(L/r_o)^4 \tag{3.52}$$

Conversely, a critical length, L_c, for a given pile–soil relative stiffness, E_p/G, may be approximated by

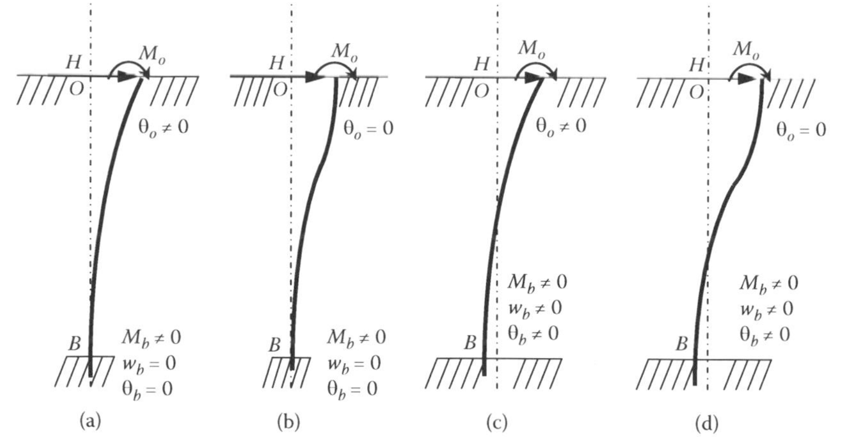

Figure 3.20 Schematic of pile-head and pile-base conditions (elastic analysis). (a) FreHCP (free-head, clamped pile). (b) FixHCP (fixed-head, clamped pile). (c) FreHFP (free-head, floating pile). (d) FixHFP (fixed-head, floating pile). (After Guo, W. D., and F. H. Lee, *Int J Numer and Anal Meth in Geomech* 25, 11, 2001.)

$$L_c \approx 1.05dE_p/G)^{0.25} \tag{3.53}$$

Note in all equations G is used. With $L < L_c$ or $(E_p/G) > (E_p/G)_c$, or $(E_p/G^*) > (E_p/G^*)_c$, the piles are referred to as short piles, otherwise referred to as long piles. In the cases of $FreHCP(M_o)$ and $FixHFP(H)$, the critical stiffness, $(E_p/G)_c$, should be increased to $4(E_p/G)_c$. As shown in Chapter 7, this book, "short" piles are defined herein, are not necessarily "rigid" piles. Most lateral piles used in practice behave as if infinitely long.

The load transfer factor, γ_b, is illustrated in Figure 3.21 for the typical slenderness ratios against pile–soil relative stiffness. In general, it may be estimated by

$$\gamma_b = k_1 \left(\frac{E_p}{G^*}\right)^{k_2} \left(\frac{L}{r_o}\right)^{k_3} \tag{3.54}$$

where k_1, k_2, and k_3 = coefficients given in Table 3.5 for short piles. For long piles regardless of base constraints, (a) free-head requires $k_2 = -0.25$, $k_3 = 0$, and $k_1 = 1.0$, or 2.0 for the lateral load, H, or the moment, M_o, respectively; (b) fixed-head (translation only) needs $k_1 = 0.65$, $k_2 = -0.25$, and $k_3 = -0.04$ due to load; whereas $k_1 = 2.0$, $k_2 = -0.25$, and $k_3 = 0.0$ due to moment. These values of γ_b were estimated for either the load H or the moment M_o. Two different γ_b will be adopted for the pile analysis, subjected to both the load and the moment simultaneously.

The γ_b for a combined loading should lie between the extreme values of γ_b for the load H and the moment M_o, from which the maximum difference

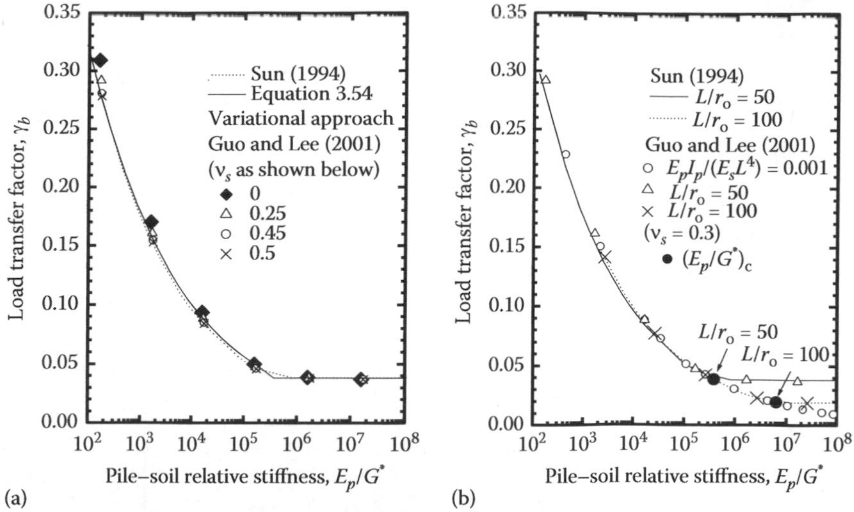

Figure 3.21 Load transfer factor γ_b versus E_p/G^* (clamped pile, free-head). (After Guo, W. D., and F. H. Lee, *Int J Numer and Anal Meth in Geomech* 25, 11, 2001.)

in the response of the pile and the soil is readily assessed. As the maximum difference in the moduli of subgrade reaction, k for the H and the M_o is generally less than 40% (particularly for rigid piles), an average k is likely to be within 20% of a real k. A 20% difference in the k will, in turn, give rise to a much smaller difference in the predicted pile response. Therefore, the superposition using the two different γ_b (thus k) may be roughly adopted for designing piles under the combined loading, although a single k for the combined loading may still be used (shown next).

Using Equations 3.50 and 3.51, the parameters k and N_p were estimated due to either the moment (M_o) or the lateral load (H) and are plotted in Figure 3.22. For the typical slenderness ratio of $L/r_o = 50$, the figure shows that the increase in the pile–soil relative stiffness renders an increase in the fictitious tension (Figure 3.22b), but decrease in the "modulus of subgrade reaction" (Figure 3.22a). Also, the critical stiffness for the moment loading is higher than that for the other cases.

Table 3.5 Parameters for estimating the factor γ_b for short piles

	Free-head due to H		*Free-head due to M_o*		*Fixed-head due to H*	
Item	*Clamped*	*Floating*	*Clamped*	*Floating*	*Clamped*	*Floating*
k_1	1.9	2.14	2.38	3.8	1.5	0.76
k_2	0	0	−0.04	0	−0.01	0.06
k_3	−1.0	−1.0	−0.84	−1.0	−0.96	−1.24

Source: Guo, W. D., and F. H. Lee, *Int J Numer and Anal Meth in Geomech,* 25, 11, 2001.

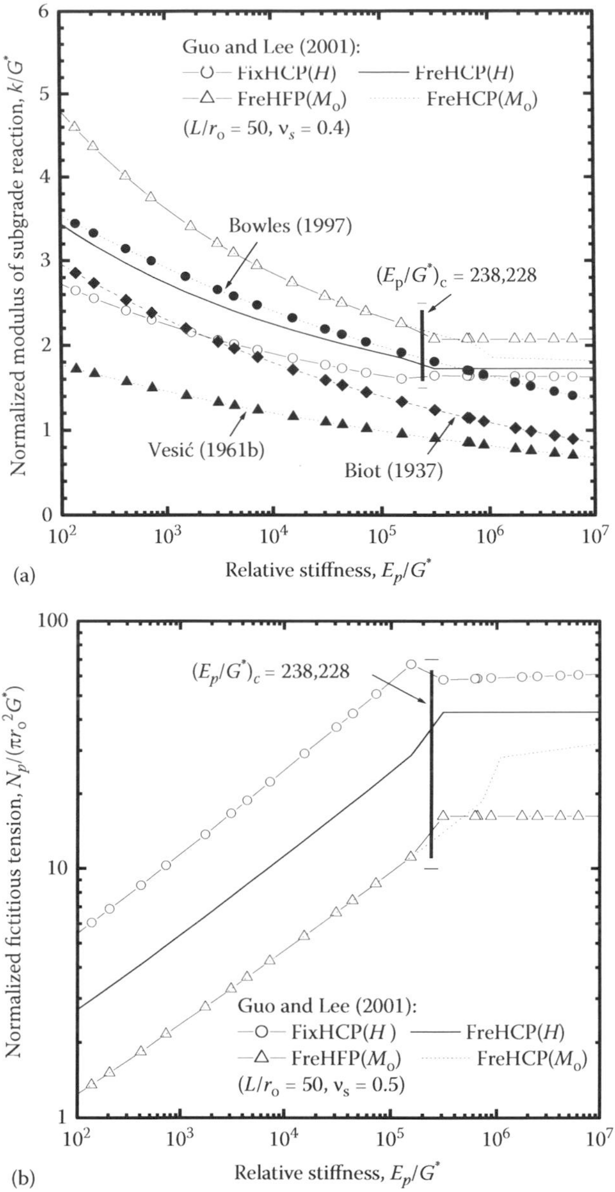

Figure 3.22 (a) Normalized modulus of subgrade reaction. (b) Normalized fictitious tension. (After Guo, W. D., *Proc 8th Int Conf on Civil and Structural Engrg Computing*, paper 112, Eisenstadt, Vienna, 2001a.)

Example 3.2 γ_b for Rigid and Short Piles

The critical relative stiffness, and pile length are approximated by Equations 3.52 and 3.53. Given free-head (*FreH*), long piles with clamped base (*CP*), or floating base (*FP*) [denoted as *FreHCP*(*H*), *FreHFP*(*H*), and *FreHFP*(M_o)], the load transfer factor at critical length should offer $(E_p/G^*)_c^{0.25} = L_c/(2.27r_o)$ with $\nu_s = 0.49$, in light of Equation 3.53. This allows Equation 3.54 to be simplified as

$$\gamma_b \approx 2.27 r_o / L_c \tag{3.55}$$

With Equation 3.54 and Table 3.5 for short piles [of the same base and head conditions, e.g., *FreHCP*(*H*), and *FreHFP*(*H*)], the factor is approximated by

$$\gamma_b \approx (1.9 \sim 3.8) r_o / L \tag{3.56}$$

The difference of the "γ_b" gained between Equations 3.55 and 3.56 is expected. As in between long flexible and short rigid piles, there are "transitional" nonrigid short piles for which the "γ_b"varies between those given by Equations 3.55 and 3.56.

3.5.3 Subgrade modulus

In the conventional, uncoupled (Winkler) model, the modulus of subgrade reaction was deduced through fitting with relevant rigorous numerical solutions (Biot 1937; Vesić 1961a; Baguelin et al. 1977; Scott 1981), as summarized below:

- Biot (1937) compared maximum moments between continuum elastic analysis and the Winkler model for an infinite beam (with concentrated load) resting on an elastic medium, and suggested the following "*k*" (termed as Biot's *k*):

$$\frac{k}{G} \approx \frac{1.9}{(1-\nu_s)}\left(\frac{128}{\pi(1-\nu_s)}\frac{G}{E_p}\right)^{0.108} \tag{3.57}$$

- Vesić (1961a), in a similar idea to Biot's approach, matched the maximum displacement of the beam and proposed the following expression (Vesić's *k*):

$$\frac{k}{G} \approx \frac{1.3}{(1-\nu_s)}\left(\frac{128}{\pi}\frac{G(1+\nu_s)}{E_p}\right)^{1/12} \tag{3.58}$$

The difference between Equations 3.57 and 3.58 implies that the uncoupled model is not sufficiently accurate for simulating beam-soil interaction, although it is used for lateral loaded piles, pipeline, and even slab (Daloglu and Vallabhan 2000). The "k" of the empirical Equation3.58) was doubled and used for piles (Kishida and Nakai 1977; Bowles 1997) (termed as Bowles' k). These suggestions are indeed empirical, owing to significant difference in soil deformation pattern between a laterally loaded pile and a beam as demonstrated experimentally (Smith 1987; Prasad and Chari 1999). The "k" is dependent of loading properties, pile-head and base conditions, pile–soil relative stiffness, and so on (Guo and Lee 2001).

To estimate the modulus for lateral piles, a comparison of different derived moduli was made and is shown in Figure 3.22a. It indicates that for long flexible piles, the values of Bowles' k and Biot's k are close to the current "k" for free-head clamped piles, *FreHCP* (H), and fixed-head clamped piles, *FixHCP* (H), respectively. The Vesić's k is lower than all other suggestions. For short piles, all the available "k" are significantly lower than the current k given by Equation 3.50. The current solution indicates significant variations of the k with the load properties and pile slenderness ratio (Figure 3.23a), apart from pile–soil relative stiffness considered in previous expressions. More comparisons among different k are presented in Chapter 7, this book.

The impact of different values of k was examined using the closed-form solutions for lateral piles (Guo and Lee 2001). The adequacy of using the available expressions (Biot 1937; Vesić 1961a; Kishida and Nakai 1977; Bowles 1997; Guo and Lee 2001) for modulus of subgrade reaction has been examined against finite element solutions (Guo 2001a). As shown in Chapter 7, this book, the comparison demonstrates that

- The Biot's k and Vesić's k for beam are not suitable for pile analysis, while the Bowles' k is conditional valid to flexible piles. The second parameter N_p is generally required to gain sufficiently accurate analysis of lateral piles. The conventional "Winkler" model ($N_p = 0$) is sufficiently accurate only for free-head pile due to moment loading [i.e., *FreH* (M_o)].
- Only by using both k and N_p would a consistent pile response be predicted against relevant numerical simulations at any pile–soil relative stiffness.

The impact of loading eccentricity (e) on k/G may be seen from Figure 3.24a and b for rigid and flexible piles at $e = 0$ and infinitely large. The k/G at any e may be given by

$$k/G = \left[8.4 + \frac{\bar{e}}{0.119 + 0.2427\bar{e}}\right]\left(\frac{L}{d}\right)^{-0.45[1+\frac{\bar{e}}{1+5\bar{e}}]} \quad \text{(Rigid piles)} \tag{3.59}$$

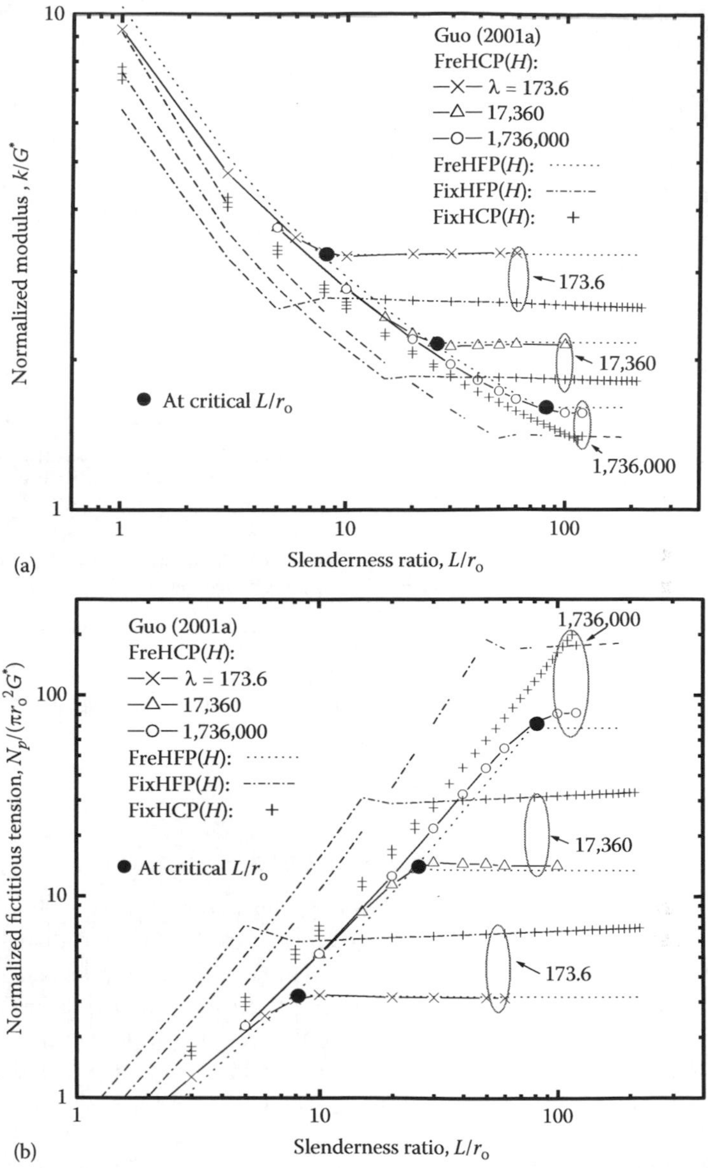

Figure 3.23 (a) Modulus of subgrade reaction. (b) Fictitious tension for various slenderness ratios. (After Guo, W. D., *Proc 8th Int Conf on Civil and Structural Engrg Computing*, paper 112, Eisenstadt, Vienna, 2001a.)

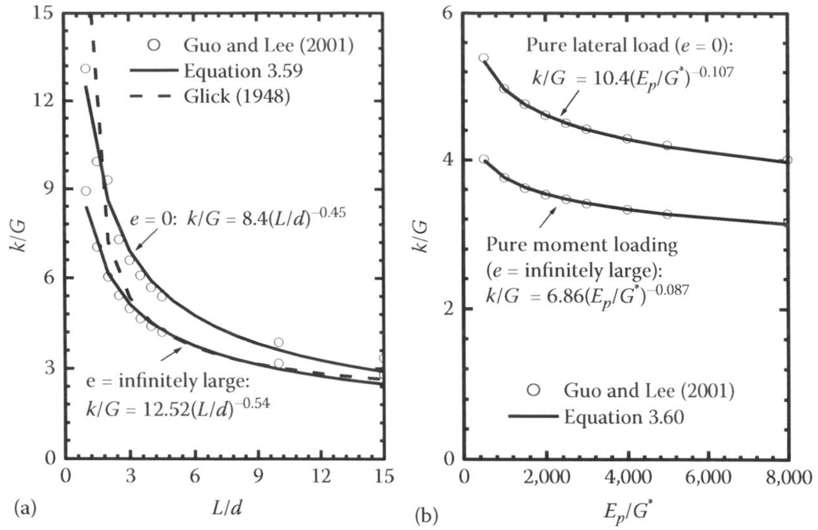

Figure 3.24 Subgrade modulus for laterally loaded free-head (a) rigid and (b) flexible piles. (Revised from Guo, W. D., *J Geotech Geoenviron Engrg*, ASCE, 2012a. With permission from ASCE.)

$$k/G = [6.86 + \frac{\bar{e}}{0.1458 + 0.2834\bar{e}}](\frac{E_p}{G^*})^{-(0.087+\frac{\bar{e}}{11.49+50\bar{e}})} \quad \text{(Flexible piles) (3.60)}$$

where $\bar{e} = e/L$.

3.5.4 Modulus *k* for rigid piles in sand

The pile–soil interaction may be modeled by a series of springs distributed along the pile shaft, as the model captures well the stress distribution around a rigid pile (see Figure 3.25). In reality, each spring has a limiting force per unit length p_u at a depth z $[FL^{-1}]$ with $p_u = A_r zd$ (Figure 3.26b). If less than the limiting value p_u, the on-pile force (per unit length), p $[FL^{-1}]$ at any depth is proportional to the local displacement, u $[L]$, and to the modulus of subgrade reaction, kd $[FL^{-2}]$ (see Figure 3.26b), which offers:

$$p = kdu \text{ (Elastic state)} \quad (3.61)$$

The gradient k $[FL^{-3}]$ may be written as $k_o z^m$ $[k_o, FL^{-m-3}]$, with $m = 0$ and 1 being referred to as constant k and Gibson k hereafter.

The magnitude of a constant k may be related to the shear modulus G by Equation 3.50, in which the k (= p/w) should be replaced with kd (= p/u). The average modulus of subgrade reaction concerning a Gibson k is $k_o(l/2)d$, for which the modulus G in Equation 3.50 is replaced with an average $\tilde{G}$

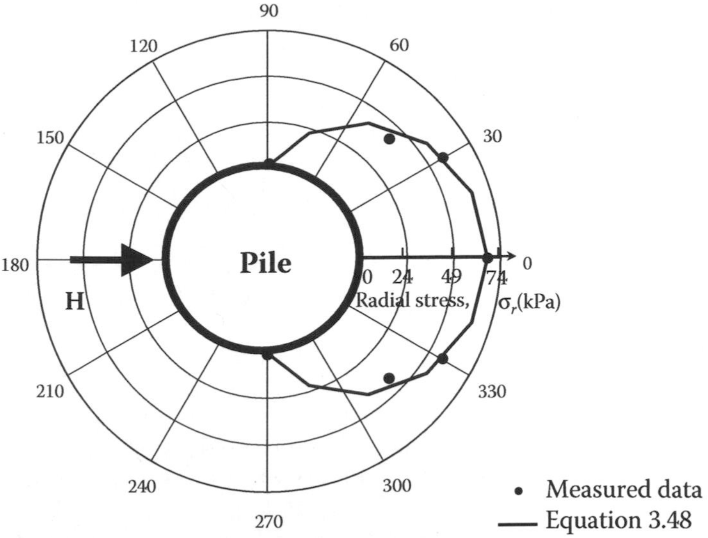

Figure 3.25 Comparison between the predicted and the measured (Prasad and Chari 1999) radial pressure, σ_r, on a rigid pile surface. (After Guo, W. D., *Can Geotech J* 45, 5, 2008.)

(bar "~" denotes average) over the pile embedment. This allows Equation 3.50 to be rewritten more generally as

$$k_o\left(\frac{l}{2}\right)^m d = \frac{3\pi\tilde{G}}{2}\left\{2\gamma_b\frac{K_1(\gamma_b)}{K_o(\gamma_b)} - \gamma_b^2\left[\left(\frac{K_1(\gamma_b)}{K_o(\gamma_b)}\right)^2 - 1\right]\right\} \quad (m = 0 \text{ or } 1) \qquad (3.62)$$

Use of Equations 3.50 and 3.62 will be addressed in later chapters. For rigid piles (see Figure 3.26), the following points are worthy to be mentioned.

1. The diameter d is incorporated into Equation 3.61 and thus appears on the left-hand side of Equation 3.62. This is not seen for flexible piles (Guo and Lee 2001), but to facilitate the establishment of new solutions in Chapter 7, this book.
2. The G and $\tilde{G}$ are not exactly "proportional" to the pile diameter (width), due to the dependence of the right-hand side on L/r_o (see Equation 3.62, via γ_b). For instance, given $l = 0.621$ m, $r_o = 0.0501$ m, the factor γ_b was estimated as 0.173~0.307 using Equation 3.54 and $k_1 = 2.14$~3.8 and then revised as 0.178 given $e = 150$ mm. $K_1(\gamma_b)/K_0(\gamma_b)$ was computed to be 2.898. As a result, the kd (constant k) is evaluated as 3.757G, and $0.5k_0dl$ (Gibson k) = 3.757$\tilde{G}$. Conversely, shear modulus G may be deduced from $G = kd/3.757$, or $\tilde{G} = 0.5k_odl/3.757$, as discussed in Chapter 7, this book.

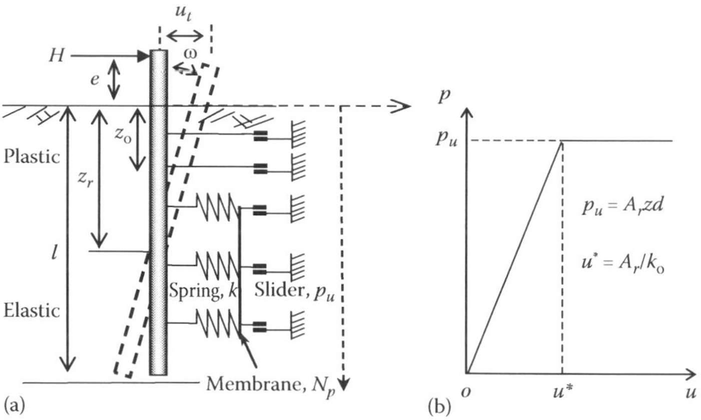

Figure 3.26 Schematic analysis for a rigid pile (a) pile–soil system (b) load transfer model ($p_u = A_r dz$ and $p = k_o dzu$). (Revised from Guo, W. D., *Can Geotech J* 45, 5, 2008.)

3. Equation 3.62 is approximate. The average k (= $0.5k_o l$) and $\tilde{G}$ refer to those at the middle embedment of a pile, with linear increase from zero to k (= $k_o l$) and from zero to G_L (pile-tip shear modulus), respectively.
4. The values of k_m (due to pure moment loading) and k_T (pure lateral loading) were calculated using Equations 3.50 and 3.54 and are plotted in Figure 3.24a. The modulus ratio k_m/k_T is provided in Table 3.6. The calculation shows the ratio k_m/k_T reduces from 1.56 to 1.27 with increase in the slenderness ratio l/r_o from 1 to 10. The fictitious upper limit is 3.153 at $l/r_o = 0$, as is seen from Figure 3.24a. The ratio k/k_T reduces from k_m/k_T ($e = \infty$) to 1 ($e = 0$) as the free length e decreases (e.g., from 1.35 to 1.0 for $l/r_o = 5$). The ratio k/G may be underestimated by 30~40% (for l/r_o = 3~8) if the impact of high eccentricity is neglected.

Given a pile-head load exerted at $e > 0$, the displacement is conservatively overestimated using $k = k_T$ compared to that obtained using a real k ($k_T<k<k_m$). Influence of the e on the k is catered for by selecting the k_1 in determining γ_b via Equation 3.54, as is seen in Equation 3.59. The outmost difference can be obtained by comparing with the upper bound solutions capitalized on $k = k_m$. Equation 3.59 may be used to estimate the k.

Table 3.6 Reduction of k_m/k_T with Increase in l/r_o of rigid piles

l/r_o	1	2	3	4	5	6	7	8	9	10
k_m/k_T[a]	1.56	1.47	1.41	1.37	1.35	1.32	1.31	1.29	1.28	1.27

Source: Guo, W. D., *Can Geotech J*, 45, 5, 2008.

[a] k_m/k_T = 3.153 (l/r_o = 0); 1.27~1.22 (10~20); 1.22~1.19(20~30); 1.14~1.12 (100~200).

In the context of the load transfer approach, it is critical to capture stress development along the pile and in the surrounding soil. As noted earlier, the stress distribution is unsymmetrical, and in three-dimensions (3D), which results in the non-uniform distribution of force (pressure) on the pile surface in radial (Prasad and Chari 1999) and longitudinal dimensions; and the alteration of modulus of subgrade reaction due to non-uniform soil displacement field around the pile. Load transfer approach (Guo and Lee 2001) well captures these two features, (e.g., the stress distribution; see Figure 3.25). The approach requires much less computing effort than numerical modeling and is thus utilized to model elastic-plastic response.

3.6 ELASTIC-PLASTIC MODEL FOR LATERAL PILES

As schematically depicted in Figure 3.26 (rigid piles), a laterally loaded pile is embedded in a nonhomogenous elastoplastic medium. The free length (eccentricity) measured from the point of applied load, H, to the ground surface is written as e. An uncoupled model indicated by the p_u is utilized to represent the plastic interaction, and the coupled load transfer model indicated by the k, and N_p ($N_p = 0$ for rigid piles) to portray the elastic pile–soil interaction, respectively. The two interactions occur respectively in regions above and below the "slip depth" z_o. In other words, the following hypotheses are adopted:

1. Each spring has an idealized elastic-plastic p-y curve (y being written as u, Figure 3.26b).
2. Equivalent, homogenous, and isotropic elastic properties (modulus and Poisson's ratio) are used to estimate the elastic k and the N_p.
3. In plastic state, the interaction among the springs is negligible by taking $N_p = 0$.
4. Pile–soil relative slip develops down to the slip depth z_o where the displacement, u^*, is equal to p_u/k and net resistance per unit length p_u is fully mobilized. (The p_u and k may be constant or increase with depth z [e.g., $p_u = A_r zd$ for sand]; and A_r is given in Table 3.7).
5. The slip (or yield) z_o occurs from ground level and progress downwards, and another slip (z_1) may be developed from the rigid pile tip.

The five assumptions (see Table 3.8) are adopted for model flexible piles as well, for which no slip occurs at pile base (owing to infinite length). As shown in Figure 3.27, the z_o is replaced with x_p, u with w, u^* with p_u/k, etc. These conditions allow the nonlinear closed-form solutions examined in Chapters 8 through 13, this book, to be established. Concepts of mobilizing shaft friction and capacity are discussed next, with respect to rigid and fixed-head piles.

Table 3.7 Capacity of lateral piles based on limit states

Mtds	$H_o/(A_r dl^2)$	A_r (kN/m³)	*References*
A	Equilibrium about rotating point	$K_{qz}\gamma_s'^{a}$	Brinch Hansen (1961)
B	$l/[6(1+e/l)]$	$3K_p\gamma_s'$	Broms (1964a)
C	$l/[2.13(1+1.5e/l)/l]^{b}$	28 ~ 228 kPa [b]	McCorkle (1969)
D	$0.5[2(z_r/l)^2-1]$	$(3.7K_p-K_a)\gamma_s'$	Petrasovits and Award (1972)
E	$l/(1+1.4e/l)$	$F_b S_{bu}(K_p-K_a)\gamma_s'{}^{c}$	Meyerhof et al. (1981)
F	$l/[8(1+1.5e/l)/l]^{b}$	80 ~ 160 kPa [b,d]	Balfour Beatty Construction (1986)
G	$\dfrac{1-d/l}{[1-0.333\ln(l/d)]e/d}$	$4.167\gamma_s'$	Dickin and Wei (1991)
H	$0.51\dfrac{z_r}{l}\left(1.59\dfrac{z_r}{l}-1\right)$	$0.8\gamma_s' 10^{1.3\tan\phi'+0.3}$	Prasad and Chari (1999)
I	$0.1181/(1+1.146e/l)$	$K_p^2\gamma_s'$	Guo (2008)

Source: Guo, W. D., *Can Geotech J*, 45, 5, 2008.

[a] K_{qz} = passive pressure coefficient as well, but relying on l/d ratio, etc.
[b] Dimensional expressions with a uniform p_u, and A_r in kPa.
[c] F_b = lateral resistance factor, 0.12 for uniform soil; S_{bu} = a shape factor varying with the length l and the angle ϕ'.
[d] Smaller values in presence of water.

3.6.1 Features of laterally loaded rigid piles

3.6.1.1 Critical states

A rigid pile is subject to a lateral load, H, at an eccentricity, e, above mudline. The net limiting force per unit length or p_u (LFP) on the pile may be stipulated as linear with depth z (i.e., $p_u = A_r dz$), upon which the *on-pile force per unit length profile* (p profile) alters (see the solid lines in Figure 3.28a$_1$ to c$_1$). The linear LFP (Broms 1964a) is indicated by the two dashed lines (independent of pile displacement), while the p profile characterizes the mobilization of the resistance along the unique p_u profile. The displacement u of the pile varies linearly with depth z (see Figure 3.28a$_2$ to c$_2$), and is described by $u = \omega z + u_g$, in which ω and u_g are rotation and mudline displacement of the pile, respectively. Typical stress states between pile and soil are noted (Guo 2008):

1. *Pre-tip yield and tip-yield states*: The force per unit length at pile tip just attains the limiting p_u (Figure 3.28a$_1$). This is known as tip-yield state. Up to the tip-yield state, the on-pile force profile follows the positive p_u down to a depth of z_o (the slip depth), below which the p is governed by

Table 3.8 Features of laterally loaded piles

Item	*Assumptions*	*Features*
p-y(w) curve and x_p	• The springs are characterized by an idealized elastic-plastic *p-y* curve (*y* rewritten as *w* in Figure 3.29f). The uncoupled and coupled load transfer models are utilized to portray the plastic and elastic zones.	• Plastic and elastic zones encountered at slip depth x_p
G for k and N_p	• Average *G* over a depth of L_c is used to estimate the *k* and N_p for elastic state	• *k* and N_p deduced from a soil deflection mode.
LFP	• Average soil properties over a maximum slip depth are employed to estimate the LFP for plastic state.	• LFP given by N_g, α_o, and *n*
Plastic zone	• Interaction among the springs is negligible (i.e., $N_p = 0$, Figure 3.29b). Pile deflection *w*(*x*) exceeds w_p (see Figure 3.29f), for which resistance per unit length, *p*, is fully mobilized.	$N_p = 0$ $w(x) > w_p$ $p = p_u$.
w_p	• At the slip depth, x_p, the pile deflection *w*(*x*) equals w_p (see Figure 3.29f). Below the x_p, the deflection w(z) < w_p; and the resistance *p* (< p_u) is linearly proportional to the *k*.	• $w_p = p_u/k$, shown in Figure 3.29f • $p = k\,w(z)$
Slip	• Pile–soil relative slip [e.g., value of $w(x) - w_p$] can only be initiated from mudline and can only move downwards.	• Allowing closed-form solutions to be generated

Notes: (1) $G = (25\sim340)s_u$ with an average of $92.3s_u$. $G = (0.25\sim0.62)N$ (MPa) with an average of $0.5N$ (MPa). The parameters s_u and *N* are calculated as the average values over the critical pile length L_c

(2) p_u is effective to maximum slip depth x_p (initially taken as 8*d* for prototype piles, and 20*d* for model piles), and s_u (undrained shear strength), *N* (blow count of SPT), and ϕ' (effective angle of friction of soil) are average values over the x_p.

(3) Free-head piles in cohesive soil: α_o = 0.05~0.2 m, N_g = 0.6~4.8 (average 1.6), and *n* = 0.7~1.7. n = 0.7 for a uniform strength profile; whereas n = 1.5~1.7 for a sharp increase strength profile (e.g., in stiff clay), otherwise, *n* is in the middle range (e.g., for multi-layered soil). Given $n = \alpha_o = 0$, the p_u reduces to s_uN_gd, with N_g = 2~4 (Viggiani 1981).

(4) Free-head piles in sand: $N_g = s_gK_p^2$, $\alpha_o = 0$, and *n* = 1.7, where $K_p = \tan^2(45° + \phi'/2)$, and s_g = 0.3~2.0 (average 1.12). High s_g is used to cater for dilatancy, which is less possible for FixH piles.

elastic interaction. At the slip depth z_o, the pile displacement *u* attains a threshold u^* $(= \omega z_o + u_g)$. With the p_u, the threshold u^* is also given by $u^* = A_r/k_o$, and $u^* = A_r z/k$, respectively, in light of a force per unit length *p* of $k_o z du$ (Gibson *k*, see Equation 3.62) or *kdu* (constant *k*).

2. *Post-tip yield state*: A higher load beyond the tip yield state enables the limiting force to be fully mobilized from the tip (depth *l*) as well and to a depth of z_1 ($z_1 = l$ for tip-yield state), at which the local displacement *u* just touches $-u^*$ with $-u^* = \omega z_1 + u_g$, or $u = -u^* z_1/z_o$ (constant *k*). A new portion of the on-pile force profile appears as the negative LFP over the depth *l* to z_1. The overall profile resembles that

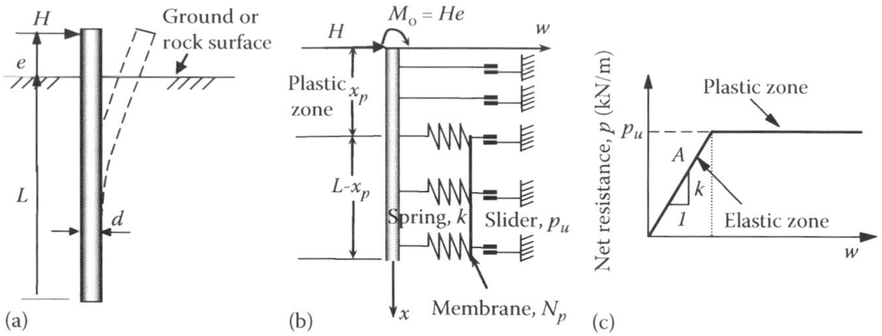

Figure 3.27 Coupled load transfer analysis for a laterally loaded free-head pile. (a) The problem addressed. (b) Coupled load transfer model. (c) Load transfer (*p-y(w)*) curve. (From Guo, W. D. *Proc 8th Int Conf on Civil and Structural EngrgComputing*, paper 108: Civil-Comp Press, Stirling, UK, 2001b; Guo, W. D., *Computers and Geotechnics* 33, 11, 2006.)

illustrated in Figure 3.28b$_2$. Upon the pile-tip yield (i.e., $z_1 = l$), the z_o is later rewritten as z_*. The pile rotates about a depth z_r (= $-u_g/\omega$), at which no displacement u (= $\omega z_r + u_g = 0$) occurs.

3. Further increase in the load brings the depths z_o and z_1 closer to each other and to the limiting depth of rotation z_r (with the impossible state of $z_o = z_r = z_1$, termed as yield at rotation point, YRP). To the depth z_r, the on-pile force profile now follows the positive LFP from the head; and the negative LFP from the tip. This impossible ultimate on-pile force profile, fully plastic (ultimate) state, is depicted in Figure 3.28c$_1$ and c$_2$. It had been adopted by some investigators (Brinch Hansen 1961; Petrasovits and Award 1972).

Similar features for a constant p_u to those mentioned above for a Gibson p_u are noted, as detailed in Chapter 8, this book.

3.6.1.2 Loading capacity

Available expressions for estimating capacity H_o (i.e., H at a defined state) of a lateral pile are generally deduced in light of force equilibrium on the pile and bending moment equilibrium about the pile-head or tip. They were deduced using a postulated on-pile force profile that unfortunately varies with load levels (Brinch Hansen 1961; Broms 1964a; Petrasovits and Award 1972; Meyerhof et al. 1981; Prasad and Chari 1999). Typical expressions are outlined in Table 3.7, which are broadly characterized by either the normalized rotational depth z_r/l (Methods A, D and H), or the normalized eccentricity e/l (Methods B, C, E, F, and G). Nevertheless, the capacity (alike to H_o) relies on both facets of e/l and z_r/l, in addition to the critical value A_r and pile dimensions.

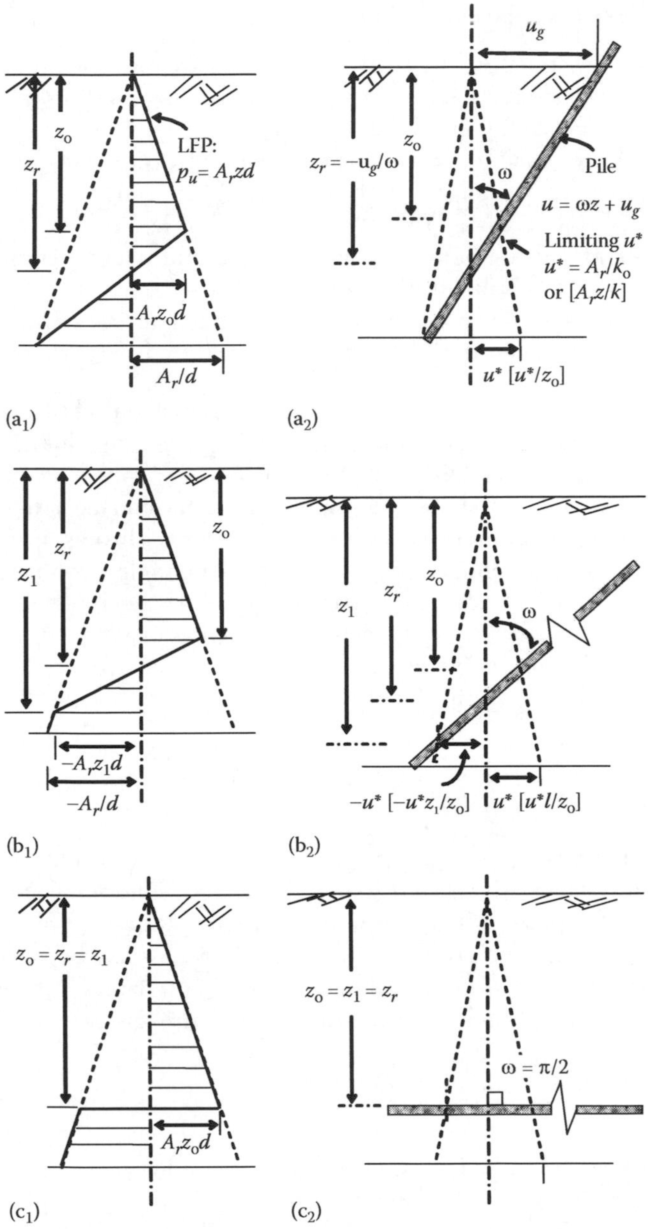

Figure 3.28 Schematic profiles of limiting force, on-pile force, and pile deformation. (a$_i$) Tip-yield state. (b$_i$) Post-tip yield state. (c$_i$) Impossible yield at rotation point (YRP) (i = 1, 2 for p, p_u profiles and displacement profiles, respectively). (Revised from Guo, W. D., *Can Geotech J* 45, 5, 2008.)

Salient features of the expressions are as follows:

1. In the denominator, the relative weight of free length over the constant component in general varies from e/l (Broms 1964a) to $1.5e/l$ (McCorkle 1969; Balfour-Beatty-Construction 1986).
2. Calculated capacities of H_o using each expression may agree with measured data via adjusting the gradient A_r, as seen from the diversity of the A_r. For example, the low A_r for McCorkle's method is 20~30% that suggested for Balfour-Beatty method and may be compensated by the 3.76 times higher ratio of $H_o/(A_r dl^2)$.
3. The expressions are not explicitly related to magnitude of the displacement or rotation angle and may correspond to different stress states.

Any solutions underpinned by a single stipulated on-pile force profile unfortunately would not guarantee compatibility with the altering profiles (see Figure 3.28a_1 to c_1) recorded at different loading levels for a single pile. Capacity is defined as the load at a specified displacement, say, 20% pile diameter upon a measured load-displacement curve (Broms 1964a). It is also taken as the load inducing a certain rotation angle upon a measured load-rotation curve (Dickin and Nazir 1999). These two independent definitions would not normally yield identical capacities even for the same test. This artificial impingement has been resolved using displacement-based solutions (see Chapter 8, this book).

3.6.2 Generic net limiting force profiles (LFP) (plastic state)

Response of single or pile groups largely depends on profile of limiting force per unit length (LFP) and its depth of mobilization. In contrast to the linear p_u for piles in sand, a number of expressions are available for constructing the LFP (p_u profile) (e.g., Figure 3.29). Typical expressions are tabulated in Table 3.9. In particular, a generic expression for p_u has been proposed (Guo 2006) and is written in the following forms

$$p_u = \tilde{s}_u N_g d[(x+\alpha_o)/d]^n \text{ (Cohesive soil),} \tag{3.63}$$

$$p_u = \gamma_s' N_g d^2[(x+\alpha_o)/d]^n \text{ (Cohesionless soil), and} \tag{3.64}$$

$$p_u = N_g q_u^{1/n} d^{1+n}[(x+\alpha_0)/d]^n \text{ (Rock)} \tag{3.65}$$

where x = depth below ground level (note, z is reserved for elastic zone only, see Chapter 9, this book); α_o = an equivalent depth to include the p_u at mudline level; n = a power showing the increase in the p_u with depth; $\tilde{s}_u$ or

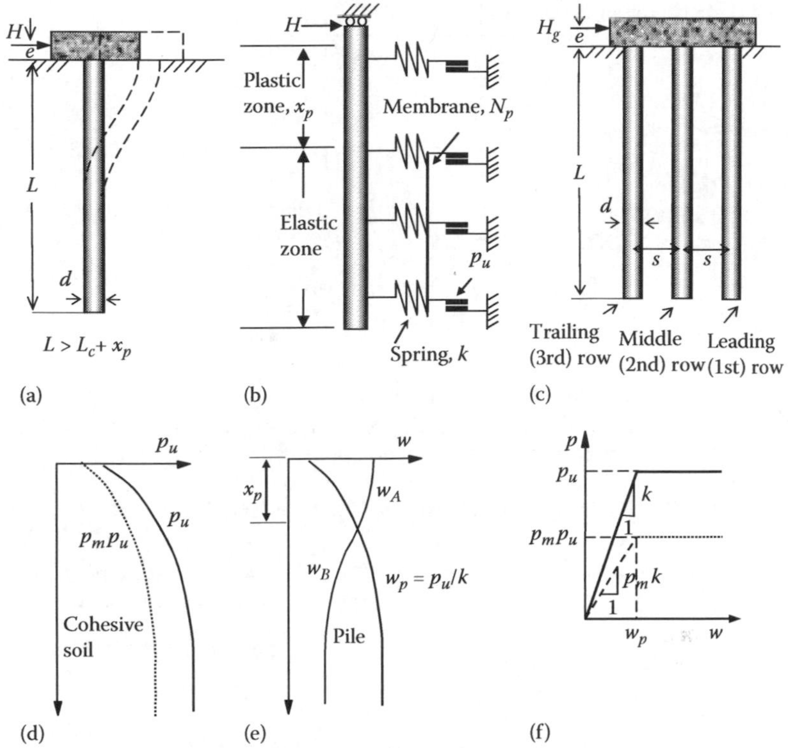

Figure 3.29 Schematic limiting force and deflection profiles. (a) A fixed-head pile. (b) Model of single pile. (c) Piles in a group. (d) LFPs. (e) Pile deflection and w_p profiles. (f) *p-y* curve for a single pile and piles in a group. (After Guo, W. D., *Int J Numer and Anal Meth in Geomech* 33, 7, 2009.)

q_u = average undrained shear strength s_u of the soil, or uniaxial compressive strength, [FL^{-2}] over the max x_p (i.e., maximum depth of mobilization of the p_u); N_g = a gradient correlated clay strength or sand density to the limiting p_u, which depends on pile-head constraints; γ_s' = effective unit weight of the overburden soil [FL^{-3}] (i.e., dry weight above water table, and buoyant weight below). Equations 3.63 through 3.65 are only effective to the max x_p. Magnitudes of the parameters α_o, N_g, and n are independent of load levels over the entire loading regime.

Equation 3.63 indicates at $n = 0$, p_u for cohesive soil is reduced to $\tilde{s}_u N_g d$; and at mudline, p_u becomes $\tilde{s}_u N_{co} d$, or $\alpha_o = (N_{co}/N_g)^{1/n} d$ (n ≠ 0). An equivalent $\tilde{N}_c$ [$= \tilde{p}_u/(\tilde{s}_u d)$] is defined using an average limiting force per unit length $\tilde{p}_u$ over the maximum slip depth. A maximum $\tilde{N}_c$ of 9.14~11.94 for a uniform s_u with depth is expected (Randolph and Houlsby 1984), although a superficially high value of $\tilde{N}_c$ may result by using an average $\tilde{s}_u$ for an increasing strength profile with depth (Murff and Hamilton 1993).

Table 3.9 Typical limiting force profiles

Name	*Expressions*
Broms LFP (Broms 1964a)	$N_g = 3K_p,\ \alpha_o = 0,\ n = 1$
Matlock LFP (Matlock 1970)	$N_g = \gamma_s' d/\tilde{s}_u + 0.5,\ \alpha_o = 2d/N_g,\ n = 1$
Reese LFP(S) for sand (Reese et al. 1974)	$p_u = \gamma_s' x\{f_{r1} x + [\frac{\tan\beta}{\tan(\beta-\phi)} - K_a]d\}$ $(x < x_t)$ and $p_u = \gamma_s' x[K_a d(\tan^8\beta - 1) + K_o \tan\phi \tan^4\beta]$ $(x \ge x_t)$ $f_{r1} = \frac{\tan\beta}{\tan(\beta-\phi)}(\frac{K_o\tan\phi\cos\beta}{\cos\alpha} + \tan\beta\tan\alpha) + K_o\tan\beta(\tan\phi\sin\beta - \tan\alpha)$ $K_o = 1 - \sin\phi'$, $\alpha = 0.5\phi$, $\beta = 45 + 0.5\phi$, $K_a = \tan^2(45 - 0.5\phi)$, and x_t = critical depth. A modification of the calculated p_u is used.
Reese LFP(C) (Reese et al. 1975)	$p_u = (2 + x\frac{\gamma_s'}{s_u} + 2.83\frac{x}{d})s_u d$ *and* ($p_u \le 11 s_u d$) (API for clay)
API (1993) sand	$p_u = \min\begin{bmatrix}(C_1x + C_2d)\gamma_s' x\\ C_3 d\gamma_s' x\end{bmatrix}$ An approximate version of Reese's LFP(S), with ϕ = 25 to 40°, C_1 = 1.2~5, C_2 = 2.0~4.3, and C_3 = 0.8~5.3.
Hansen LFP (Brinch Hansen 1961)	$p_u = (K_q^D\gamma_s' x + K_c^D c)d$ where $K_q^D = (K_q^o + \alpha_q K_q^\infty \frac{x}{d})/(1+\alpha_q\frac{x}{d})$, $K_c^D = (K_c^o + \alpha_c K_c^\infty\frac{x}{d})/(1+\alpha_c\frac{x}{d})$, $\alpha_q = K_q^o K_o \sin(\phi)/[(K_q^\infty - K_q^o)\sin(45^\circ+0.5\phi)]$, $\alpha_c = 2\sin(45^\circ+0.5\phi)K_c^o/[(K_c^\infty - K_c^o)]$ $K_q^o = e^{(0.5\pi+\phi)\tan\phi}\cos\phi\tan(45^\circ+0.5\phi) - e^{-(0.5\pi-\phi)\tan\phi}\cos\phi\tan(45^\circ-0.5\phi)$ $K_c^o = [e^{(0.5\pi+\phi)\tan\phi}\cos\phi\tan(45^\circ+0.5\phi) - 1]\mathrm{ctan}\phi$ $K_c^\infty = (1.58 + 4.09\tan^4\phi)[e^{\pi\tan\phi}\tan(45^\circ+0.5\phi) - 1]\mathrm{ctan}\phi$, $K_q^\infty = K_c^\infty(1-\sin\phi)\tan\phi$
Guo LFP (Guo 2006)	Clay: $A_L = \tilde{s}_u N_g d^{1-n}$, with α_o = 0.05~0.2 (m), n = 0.7 (but for n = 1.5 for stiff clay), and N_g = 2~4 Sand: $A_L = \gamma_s' N_g d^{2-n}$, with α_o = 0, n = 1.7, and $N_g = s_g K_p^2$ Rock: $A_L = q_u^{1/n} N_g d^{1-n}$, with α_o = 0, n = 0.7~2.5, and N_g (see Chapter 10, this book) *Note* s_g = an integral factor to cater for all sorts of influence; and $K_p = \tan^2(45^\circ+\phi/2)$.
Reese LFP(R) (Reese 1997)	$p_u = \alpha_r q_u d(1 + 1.4x/d)$ $(0 \le x \le 3d)$ Rock, α_r = 0~1.0
Zhang LFP	$p_u = [\gamma_m x + q_u(\frac{m_b\gamma_m x}{q_u} + s)^m + g_s\sqrt{q_u}]d$ (p_u in MPa, g_s = 0.2 and 0.8 for smooth and rough shaft sockets, respectively) $m_b = m_i e^{(GSI-100)/28}$, and $s = e^{(GSI-100)/9}$, m = 0.5 (GSI > 25); or s = 0, m = 0.65–GSI/200 (GSI<25) where m_i = material constant for the intact rock, dependent on rock type.

Equations 3.63 through 3.65 along with $L > L_c$ + max x_p (infinitely long piles) require full mobilization of the p_u from mudline (or rock surface) to the slip depth, x_p (see Figure 3.27), and the x_p increases with lateral loads.

The parameters N_g, N_{co}, and n may be determined using the following options deduced from Table 3.9.

- *Option 1*. The LFP for a cohesionless soil (rigid piles), as suggested by Broms (1964a) and termed as *Broms'* LFP, is given by

$$N_g = 3K_p, \quad N_{co} = 0, \quad n = 1 \tag{3.66}$$

where $K_p = \tan^2 (45° + \phi'/2)$, the passive earth pressure coefficient; ϕ' = effective friction angle of the soil.

- *Option 2*. The LFPs for a cohesive soil (Matlock 1970; Reese et al. 1975) may be represented by

$$N_g = \gamma_s' d/\tilde{s}_u + J, \quad N_{co} = 2, \quad n = 1 \tag{3.67}$$

where $J = 0.5$~3 (Matlock 1970). The LFP with $J = 0.5$ and $J = 2.8$ are referred to as *Matlock LFP* and *Reese LFP (C)*, respectively.

- *Option 3*. The LFP for flexible piles in sand employed in COM624P [i.e., *Reese LFP(S)*], was underpinned by $N_g = K_p^2$, $N_{co} = 0$, and n = 1.7 (Guo and Zhu 2004).
- *Option 4*. Values of p_u may be acquired from measured *p-y* curves for each depth to generate the LFP.
- *Option 5*. The LFP for a layered soil profile may be constructed by the following steps. (1) The entire soil is assumed as the clay or the sand, the *Reese LFP(C)* or *LFP(S)* is obtained. (2) The obtained p_u within a zone of 2*d* above or below an sand-clay interface should be increased in average by ~40% for a weak (clay) layer adjoining a stiff (sand) layer; and decreased by ~30% for a stiff layer adjoining a weaker layer (Yang and Jeremic 2005). (3) The increased and decreased p_u of the two adjacent layers is averaged, using a visually gauged n, as an exact shape of the LFP (thus n) and makes little difference to the final predictions (see Chapter 9, this book). Finally, an LFP is created for a two-layered soil. This procedure is applicable to a multiple layered soil by choosing n (thus the LFP) to fit the overall limiting force profile. Any layer located more than 2*d* below the maximum slip depth is excluded in this process.
- *Option 6*. The parameters N_g, N_{co}, and n may be backfigured through matching predicted with measured responses of a pile, as elaborated in Chapter 9, this book, for piles in soil and in Chapter 10, this book, for rock-socket piles, respectively.

Use of options 1~5 and 6 by Guo (2006) is amply demonstrated in Chapter 9, this book, for piles in soil and in Chapter 10, this book, for rock socket piles (e.g., parameters n, N_g). For rigid piles in sand, $n = 1$, Equation 3.64 is consistent with the experimental results (Prasad and Chari 1999), and the pertinent recommendation (Zhang et al. 2002), in contrast to other expressions for the A_r provided in Table 3.7.

The p_u profiles vary with pile-head constraints (Guo 2009). The determination of the parameters may refer to Table 3.8 for free-head and ad hoc guidelines G1–G5 in Table 3.10 for capped piles. They are synthesized from study on 70 free-head piles (32 piles in clay, 20 piles in sand, and other piles in layered soil), and capped piles (7 single piles and 27 pile groups). Capitalized on average soil parameters, the impact of soil nonhomogeneity and layered properties on nonlinear pile response is well simulated by adjusting the factor n, owing to the dominance of plastic interaction, and negligible effect of using an average k. Typical cases are elaborated in Chapter 9, this book, for single piles and in Chapter 11, this book, for pile groups, respectively. Using Equation 3.63 or 3.64 to construct the LFP, a high n (= 1.3~2.0) is noted for a sharply varying strength with depth, whereas a moderate n (= 0.5~0.7) is seen for a uniform strength profile. The N_g, n, and α_o together (see Table 3.8) are flexible to replicate various p_u profiles. The total soil resistance on the pile in plastic zone T_R is given by

$$T_R = \int_0^{\max x_p} p_u\, dx = \frac{A_L}{1+n}[(\max x_p + \alpha_o)^{n+1} - \alpha_o^{n+1}] \tag{3.68}$$

where $A_L = \tilde{s}_u N_g d^{1-n}$ (cohesive soil), $A_L = \gamma_s' N_g d^{2-n}$ (cohesionless soil); max x_p is calculated per G4. The difference between the resistance T_R and the load H_{max} renders head restraints to be detected (G3). The input parameters where possible should be deduced from measured data of a similar project (Guo 2006). The net limiting force per unit length along a pile in groups may be taken as $p_u p_m$ (p_u for free-head, single piles). Group interaction can be catered for by p-multiplier p_m (see Figure 3.29). The p_m may be estimated by

$$p_m = 1 - a(12 - s/d)^b \tag{3.69}$$

where s = pile center-to-center spacing; $a = 0.02 + 0.25\,Ln(m)$; $b = 0.97(m)^{-0.82}$; and m = row number. For instance, the second row (with $m = 2$) has a = 0.193 and b = 0.55. The row number m should be taken as 3 for the third and subsequent rows. The p_m varies with row position, spacing, configuration of piles in a group, and pile installation method. The values of p_m for 24 pile groups were deduced by gaining most favorable predictions against measured data (Guo 2009) and are plotted in Figure 3.30a and b. Insights about the p_m as gleaned herein are provided in Table 3.10. The p-multipliers

Table 3.10 Guidelines G–G5 for selecting input parameters

G1 for G_s	• A value of (0.5~2.0)G may be used for FixH piles with G being determined from Table 3.8 using $\tilde{s}_u$ or SPT. Given identical G_s, the k is higher for FreH piles than FixH ones (Gazetas and Dobry 1984; Syngros 2004), • Diameter effect is well considered by the expression correlating G to k (Table 3.8), otherwise, 5~10 times the real G (Guo 2008) may be deduced from a given k. For large diameter piles, a small maximum x_p is seen in gaining the $\tilde{s}_u$ (Guo and Ghee 2004; Guo and Zhu 2004) • G for piles in a group may increase by 2~10 times in sand or may reduce by 50% in clay, compared to single piles.		
G2	FixH single piles	Group piles	Semi-fixed head group
Sand	n_s^{FixH} = 1.1~1.3, $N_g^{FixH} = K_p^2$		n = 0.5~0.65 (= $0.5n_s^{FixH}$), $N_g = K_p^2$ using FixH solutions
Silt or clay	$n = n_c^{FreH}$ = 0.5~0.7, independent of head restraints	$n = 1.7n_c^{FreH}$ if soil strength reduces; Otherwise $n = n_c^{FreH}$	n = 0.85 regardless of FreH or FixH.
	N_g = 0.6~4.8, α_o = 0.05~0.2 m and considering reduction as described in the note.		$N_g = 0.25N_g^{FreH}$ or $N_g = N_g^{FreH}$ using FixH or FreH solutions, respectively
Note for n and N_g	• n is independent of head restraints for piles in silt/clay, but reduced for FixH piles in sand. A high n is used for disturbed clay within a pile group; whereas a low n is seen for homogenizing sand among a pile group, owing to constraints imposed by nearby piles and sand dilatancy. • N_g, n, and α_o are selected by referring to Table 3.8 to cater for non-uniform resistance with depth and justified using G5, especially for piles in layered soil. Over a maximum depth x_p, a low value N_g of $0.25N_g^{FreH}$ as mobilized along a FixH pile gradually increases (e.g. to $N_g = 0.7N_g^{FixH}$) (Yan and Byrne 1992) towards that for a FreH pile, as the fixed-head restraint degrades.		
G3 for T_R	• A close T_R leads to similar predictions, insensitive to the values of α_o, n, and N_g. • The result of $T_R > P_{max}$ indicates a FixH condition, whereas $T_R < P_{max}$ implies a FreH condition. An additional resistance of $T_R - P_{max}$ is induced for FixH pile owing to restraining rotation.		
G4 for x_p	The maximum x_p for FixH piles are initially taken as 16d for prototype piles and 20d for model piles.		
G5 for p_m	The p_m reduces as number of piles in a group increases, such as by 50% with 9~16 piles in a group (Ilyas et al. 2004). The p_m is barely to slightly affected by the pile-head restraints.		

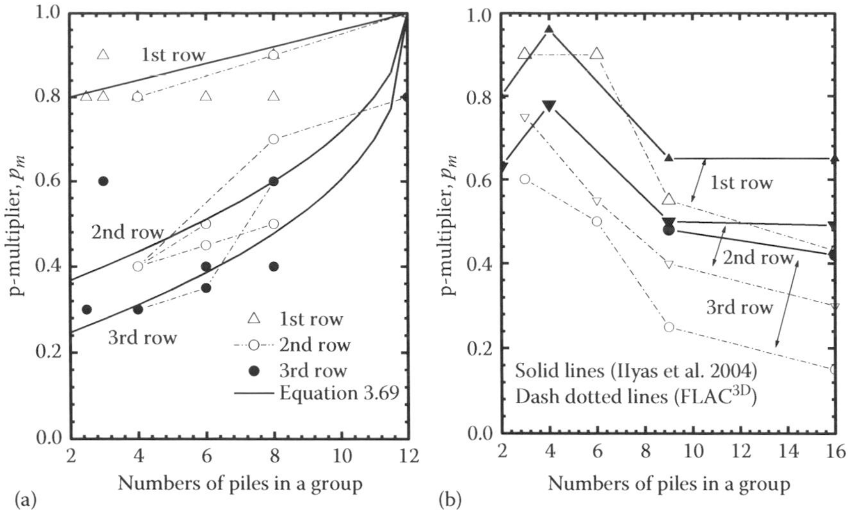

Figure 3.30 Determination of p-multiplier p_m. (a) Equation 3.69 versus p_m derived from current study. (b) Reduction in p_m with number of piles in a group. (After Guo, W. D., *Int J Numer and Anal Meth in Geomech* 33, 7, 2009.)

are higher than other suggestions at large spacing (Mokwa and Duncan 2005; Rollins et al. 2006), despite of good agreement with recent suggestions (Dodds 2005). This also implies that the calculated p_m using Equation 3.69 should be reduced with more number of piles in a group (see Figure 3.30b). Furthermore, the p_m varies from one pile to another in a row to characterize response of individual piles.

Chapter 4

Vertically loaded single piles

4.1 INTRODUCTION

"Hybrid analysis" has been developed to analyze large group of piles. It combines numerical and analytical solutions to perform a complete analysis (Chow 1987; Guo 1997) and is the most efficient approach to date. To enhance the approach, the impact of soil nonhomogeneity on the pile–soil–pile interaction and group pile behavior needs to be incorporated into the analysis in an efficient manner (Poulos 1989; Guo and Randolph 1999). This prompts the development of closed-form solutions for piles in nonhomogeneous soil (Rajapakse 1990).

The available closed-form solutions for vertically loaded piles are, strictly speaking, largely limited to homogenous soil (Murff 1975; Motta 1994). Guo and Randolph (1997a) developed new solutions for nonhomogeneous soil characterized by shear modulus as a power of depth. Guo (2000a) extended the solutions to accommodate the impact of nonzero modulus at ground surface and further to cater for impact of strain softening (Guo 2001c). The closed-form solutions for a vertically loaded pile in a nonhomogeneous, elastic-plastic soil (Guo and Randolph 1997) are expressed in modified Bessel functions of non-integer order. Numerical estimates of the solutions are performed by either Mathcad™ and/or a purposely designed spreadsheet program. These solutions are generally sufficiently accurate for modeling normally consolidated and overconsolidated soil.

The closed-form solutions are generally based on the load transfer approach (Coyle and Reese 1966; Kraft et al. 1981; Guo and Randolph 1997a; Guo and Randolph 1998), treating the soil as independent springs. The approach is underpinned by load transfer factors recaptured in Chapter 3, this book, and is able to cater for impact of nonzero input of shear modulus (Guo and Randolph 1998). The solutions are sufficiently accurate against various rigorous numerical solutions. This chapter presents the solutions and their application in predicting load displacement response, loading capacity of a pile in strain-softening soil, and safe cyclic load amplitude for a vertical pile.

4.2 LOAD TRANSFER MODELS

4.2.1 Expressions of nonhomogeneity

As an extension to that presented in Chapter 3, this book, for zero modulus at ground level (i.e., $\alpha_g = 0$), a more generalized soil profile is addressed herein. Pertinent profiles and nondimensional parameters are briefly described below.

The soil shear modulus, G, along a pile is stipulated as a power function of depth

$$G = A_g(\alpha_g + z)^n \tag{4.1}$$

where n, α_g, A_g = constants; z = depth below the ground surface. Below the pile-base level, shear modulus is taken as a constant, G_b, which may differ from the shear modulus at just above the pile-base level, G_L, as reflected by the ratio, ξ_b (= G_L/G_b Figure 3.1, Chapter 3, this book). The variation of limiting shear stress, τ_f, associated with Equation 4.1 is assumed to be (note τ_f is different from the average τ_s in Chapter 2, this book)

$$\tau_f = A_v(\alpha_v + z)^\theta \tag{4.2}$$

where θ, α_v, A_v = constants. To date, it is assumed that $\alpha_v = \alpha_g$ and $\theta = n$. The ratio of modulus to shaft-limiting stress is thus equal to A_g/A_v, independent of the depth.

The nonhomogeneity factor ρ_g is defined as the ratio of the average soil shear modulus over the pile length and the modulus at the pile-base level, G_L (see Equation 3.6 in Chapter 3, this book)

$$\rho_g = \frac{1}{1+n}\left(1 + \frac{\alpha_g}{L} - \frac{\alpha_g}{L}\left(\frac{\alpha_g/L}{1+\alpha_g/L}\right)^n\right) \tag{4.3}$$

The pile–soil relative stiffness factor, λ, is defined as the ratio of pile Young's modulus, E_p, to the shear modulus at pile-base level, G_L,

$$\lambda = E_p/G_L \tag{4.4}$$

4.2.2 Load transfer models

Closed-form solutions are developed within the framework of load transfer models for nonhomogeneous soil (see Chapter 3, this book). In the shaft model, the shaft displacement, w, is correlated to the local shaft stress and shear modulus by (Randolph and Wroth 1978)

$$w = \frac{\tau_o r_o}{G_i}\zeta \tag{4.5}$$

where

$$\zeta = \ln\left(\frac{r_m}{r_o}\right) \tag{4.6}$$

where τ_o = local shaft shear stress; r_o = pile radius; r_m = maximum radius of influence of the pile beyond which the shear stress becomes negligible, and may be expressed in terms of the pile length, L, and "n" as (Guo 1997; Guo and Randolph 1997a; Guo and Randolph 1998)

$$r_m = A\frac{1-\nu_s}{1+n}L + Br_o \tag{4.7}$$

where ν_s = Poisson's ratio of the soil and B may generally be taken as 5. As discussed in Chapter 3, this book, A is dependent on the ratio of the embedded depth of underlying rigid layer, H, to the pile length, L (Figure 4.1a), Poisson's ratio, ν_s, and nonhomogeneity factor, n. With the modulus distribution of Equation 4.1, the A may still be estimated using Equation 3.7 (Chapter 3, this book) by replacing the factor "n" with an equivalent nonhomogenous factor "n_e" (Figure 4.1) (Guo 2000a, such that

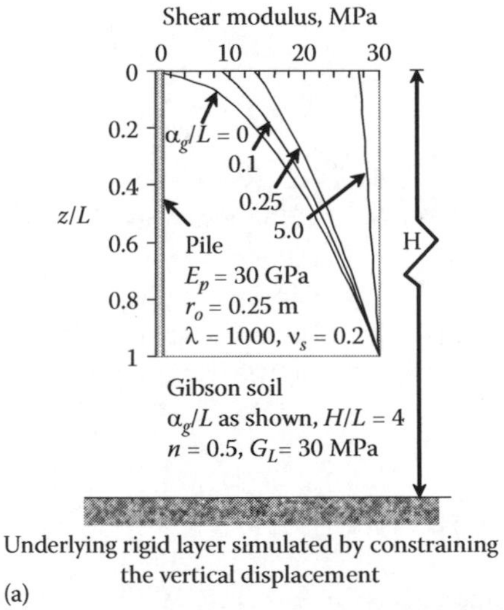

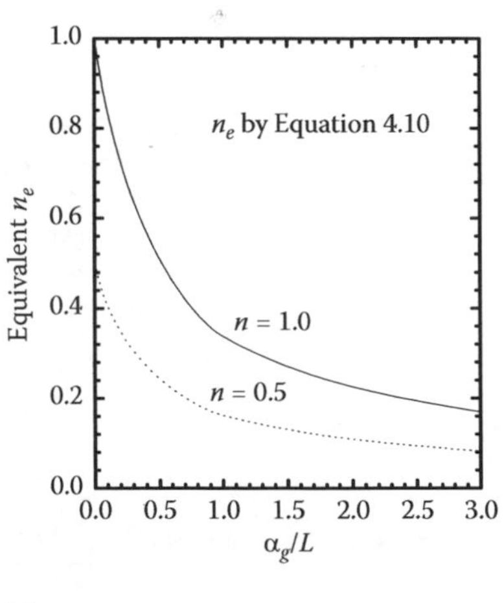

Figure 4.1 (a) Typical pile–soil system addressed. (b) α_g and equivalent n_e. (After Guo, W. D., *Int J Numer and Anal Meth in Geomech* 24, 2, 2000b.)

$$A = \frac{A_b}{A_{ob}}\left(\frac{1}{1+n_e}\left(\frac{0.4-\nu_s}{n_e+0.4}+\frac{2}{1-0.3n_e}\right)+C_\lambda(\nu_s-0.4)\right) \tag{4.8}$$

where $C_\lambda = 0$, 0.5, and 1.0 for $\lambda = 300$, 1,000, and 10,000. A_{ob} is A_b at a ratio of $H/L = 4$, A_b is given by

$$A_b = 0.124e^{2.23\rho_g}\left(1-e^{1-\frac{H}{L}}\right)+1.01e^{0.11n_e} \tag{4.9}$$

where

$$n_e = 1/\rho_g - 1 \tag{4.10}$$

The mobilized shaft shear stress on the pile surface will reach the limiting value, τ_f, once the local pile–soil relative displacement, w, attains the local limiting displacement, w_e, which with $n = \theta$ and $\alpha_g = \alpha_v$ is obtained as

$$w_e = \zeta r_o \frac{A_v}{A_g} \tag{4.11}$$

Thereafter, the shear stress stays as τ_f with $w > w_e$ (i.e., an ideal elastic-plastic load transfer model is adopted).

The base settlement can be estimated through the solution for a rigid punch acting on a half-space:

$$w_b = \frac{P_b(1-\nu_s)\omega}{4r_oG_b} \tag{4.12}$$

where P_b = mobilized base load and ω = pile-base shape and depth factor. The ω is taken as unity (Randolph and Wroth 1978). This only incurs a < 7% (Guo and Randolph 1998) difference in predicted pile-head stiffness but may have a higher impact on predicting load and/or displacement distribution profiles down a pile. Generally, the more accurate expressions presented in Chapter 3, this book (Guo 1997; Guo and Randolph 1998), are adopted herein except where specified.

4.3 OVERALL PILE–SOIL INTERACTION

Generally, a pile behaves elastically under vertical loading. With constant diameter and Young's modulus, the governing equation for pile–soil interaction is as follows (Randolph and Wroth 1978)

$$\frac{d^2u(z)}{d^2z} = \frac{2\pi r_o \tau_o}{E_p A_p} \tag{4.13}$$

where A_p = cross-sectional area of an equivalent solid cylinder pile, $u(z)$ = axial pile deformation.

4.3.1 Elastic solution

Within the elastic range, shaft stress, τ_o in Equation 4.13 is related to the local displacement, w as prescribed by Equation 4.5. The basic differential equation governing the axial deformation of a pile fully embedded in the nonhomogeneous soil, described by Equation 4.1, is thus deduced as

$$\frac{d^2u(z)}{d^2z} = \frac{A_g}{E_p A_p}\frac{2\pi}{\zeta}(\alpha_g + z)^n w(z) \tag{4.14}$$

The axial pile displacement, $u(z)$, is equal to the pile–soil relative displacement, $w(z)$, neglecting any external soil subsidence. The load transfer factor ζ is a constant with depth (see Chapter 3, this book). Equation 4.14 is thus resolved using Bessel functions of non-integer order as

$$w(z) = w_b\left(\frac{z+\alpha_g}{L+\alpha_g}\right)^{1/2}\left(\frac{C_3(z)+\chi_v C_4(z)}{C_3(L)}\right) \tag{4.15}$$

$$P(z) = k_s E_p A_p w_b (z+\alpha_g)^{n/2}\left(\frac{z+\alpha_g}{L+\alpha_g}\right)^{1/2}\left(\frac{C_1(z)+\chi_v C_2(z)}{C_3(L)}\right) \tag{4.16}$$

where $w(z)$, $P(z)$ = displacement and load at a depth of z ($0 < z \le L$), respectively; $C_i(z)$ (i = 1 to 4) is expressed as the modified Bessel functions, K, and I of non-integer order, m and m–1,

$$\begin{aligned} C_1(z) &= -K_{m-1}I_{m-1}(y) + K_{m-1}(y)I_{m-1} \\ C_2(z) &= K_m I_{m-1}(y) + K_{m-1}(y)I_m \\ C_3(z) &= K_{m-1}I_m(y) + K_m(y)I_{m-1} \\ C_4(z) &= -K_m I_m(y) + K_m(y)I_m \end{aligned} \tag{4.17}$$

where $m = 1/(n+2)$; I_m, I_{m-1}, K_{m-1}, and K_m are the values of the Bessel functions at $z = L$; and the variable y is given by

$$y = 2mk_s(\alpha_g + z)^{1/(2m)} \tag{4.18}$$

and

$$k_s = \frac{\alpha_g + L}{r_o}\sqrt{\frac{2}{\lambda\zeta}}\left(\frac{1}{\alpha_g + L}\right)^{1/(2m)} \tag{4.19}$$

The ratio χ_v is given by

$$\chi_v = \frac{2\sqrt{2}}{\pi(1-\nu_s)\omega\xi_b}\sqrt{\frac{\zeta}{\lambda}} \tag{4.20}$$

At any depth, z, the stiffness can be calculated using

$$\frac{P(z)}{G_L w(z) r_o} = \sqrt{2\pi}\sqrt{\frac{\lambda}{\zeta}}C_v(z) \tag{4.21}$$

where

$$C_v(z) = \frac{C_1(z) + \chi_v C_2(z)}{C_3(z) + \chi_v C_4(z)}\left(\frac{\alpha_g + z}{\alpha_g + L}\right)^{n/2} \tag{4.22}$$

At the ground surface ($z = 0$), it is necessary to take the limiting value of $C_v(z)$ as z approaches zero. Equations 4.15 and 4.16 allow base settlement to be written as a function of pile-head load and displacement. The accuracy of these closed-form solutions (hereafter denoted by *CF*) has been checked by Mathcad™ and corroborated by continuum-based finite difference analysis, as shown later.

4.3.2 Verification of the elastic theory

The closed-form solutions outlined here are underpinned by the uncoupled load transfer analysis. As shown in Chapter 3, this book, the solutions compare well with more rigorous numerical approaches for $\alpha_g = 0$ (Guo 1997; Guo and Randolph 1998) and $\alpha_g \neq 0$ (Guo 2000a). Typically, a continuum-based numerical analysis is conducted using the finite-difference program FLAC (Itasca 1992; Guo and Randolph 1997a) on the pile shown in Figure 4.1a, in which $H/L = 4$; Young's modulus of the pile, $E_p = 30$ GPa; and Poisson's ratio, $\nu_p = 0.2$. Given constant values, say, of $\lambda = 1{,}000$, $G_L = 30$ MPa, regardless of n ($n > 0$), the increase in α_g renders reduction in value of A_g in Equation 4.1. At a sufficiently high α_g, the soil approaches a homogenous medium (see Figure 4.1a and b). The associated pile-head stiffness (see Figure 4.2) then approaches the upper limit for homogeneous case.

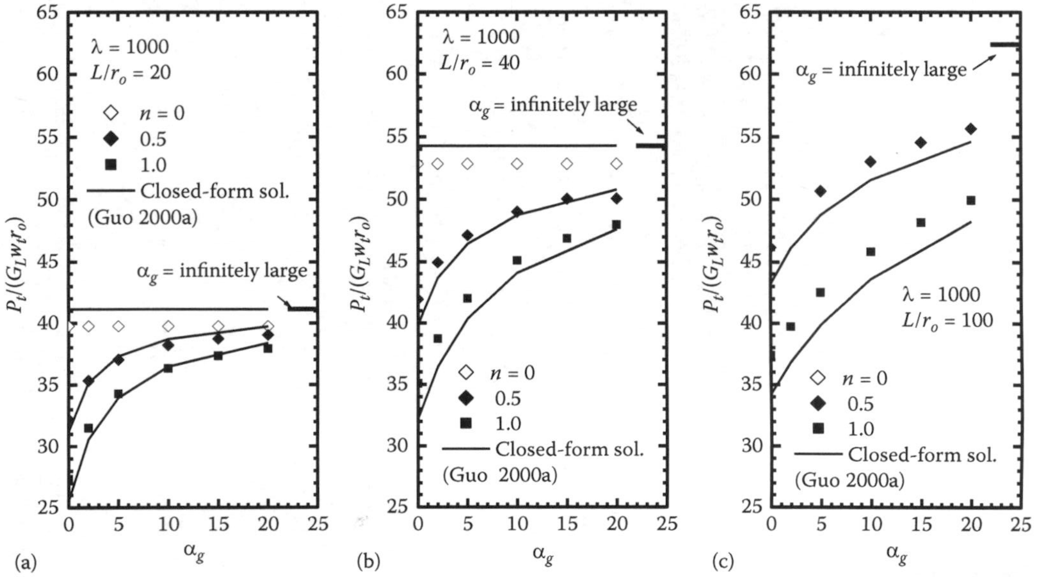

Figure 4.2 Effect of the α_g on pile-head stiffness for a L/r_o of (a) 20, (b) 40, and (c) 100. (After Guo, W. D., *J Geotech and Geoenvir Engrg*, ASCE 126, 2, 2000a. With permission from ASCE.)

Table 4.1 Estimation of n_e and parameter A for the CF analysis

L/r_o	20		40		100	
α_g	n = 0.5	n = 1.0	n = 0.5	n = 1.0	n = 0.5	n = 1.0
0	0.50[a]	1.00	.50	1.00	.50	1.00
	1.57[b]	1.43	1.57	1.43	1.57	1.43
2	.265	.556	.341	.714	.417	.862
	1.72	1.54	1.66	1.49	1.61	1.45
5	.160	.333	.238	.50	.341	.714
	1.81	1.67	1.74	1.57	1.66	1.49
10	0.97	.20	.160	.333	.265	.556
	1.88	1.77	1.81	1.67	1.72	1.54
20	.055	.111	.097	.200	.184	.385
	1.93	1.86	1.88	1.77	1.79	1.63

Source: Guo, W. D., *Proceedings of the Eighth International Conference on Civil and Structural Engineering Computing*, Paper 112, Civil-Comp Press, Stirling, UK, 2001a.

[a] numerator: value of n_e estimated by Equation 4.10
[b] denominator: value of A by Equation 4.8. In the estimation, $H = 4L$, $\lambda = 1{,}000$, and $\nu_s = 0.4$.

The pile-head stiffness ($z = 0$) was estimated using Equation 4.21, in which A was estimated from Equation 4.8 using $H/L = 4$, $\nu_s = 0.4$, $\xi_b = 1$, $\omega = 1$, and $B = 1$. It was obtained for a few typical values of α_g and L/r_o. The corresponding values of n_e and A are tabulated in Table 4.1 for each case. A low value of $B = 1$ (Guo and Randolph 1998) was adopted in view of the slight

Table 4.2 $P_t/(G_L w_t r_o)$ from FLAC and CF analyses

L/r_o	20		40		100	
α_g	n = 0.5	n = 1.0	n = 0.5	n = 1.0	n = 0.5	n = 1.0
0	32.2[a]	27.4	41.9	35.1	462	37.4
	31.2[b]	25.6	39.9	32.2	432	34.2
2	35.3	31.5	44.9	38.7	--	39.7
	35.1	30.6	43.7	36.4	46.1	36.8
5	37.1	34.3	47.1	42.0	50.7	42.5
	37.3	33.9	46.4	40.4	48.8	39.9
10	38.2	36.3	49.0	45.1	53.0	45.8
	38.7	36.4	48.8	44.1	51.6	43.6
20	39.0	37.9	50.0	48.0	55.7	50.0
	39.7	38.4	50.8	47.6	54.6	48.3

Source: Guo, W. D., *Proceedings of the Eighth International Conference on Civil and Structural Engineering Computing*, Paper 112, Civil-Comp Press, Stirling, UK, 2001a.

[a] numerator from FLAC analysis ($H = 4L$, $\lambda = 1{,}000$, and $\nu_s = 0.4$);
[b] denominator from closed-form solutions (CF).

overestimation of pile-head stiffness from the FLAC analysis (compared to other approaches). The values of the pile-head stiffness obtained are shown in Table 4.2, which are slightly higher than the FLAC predictions for short piles and slightly lower for long piles (particularly, at $n = 1$), with a difference of ~5%.

4.3.3 Elastic-plastic solution

As the vertical pile-head load increases, pile–soil relative slip is stipulated to commence from the ground surface and progressively develop to a depth called transition depth (L_1), at which the shaft displacement, w, is equal to the local limiting displacement, w_e. As shown in Figure 4.3a, the upper portion of the pile above the transition depth is in plastic state, while the lower portion below the depth is in an elastic state. Within the plastic state, the shaft shear stress in Equation 4.13 should be replaced by the limiting shaft stress from Equation 4.2 (i.e., an increasing τ_f with depth, see Figure 4.3b). Pile-head load is thus a sum of the elastic component represented by letters with subscript of "e," and the plastic one:

$$P_t = w_e k_s E_p A_p (\alpha_g + L)^{n/2} C_v(\mu L) + \frac{2\pi r_o A_v[(\alpha_v + \mu L)^{1+\theta} - \alpha_v^{1+\theta}]}{1+\theta} \qquad (4.23)$$

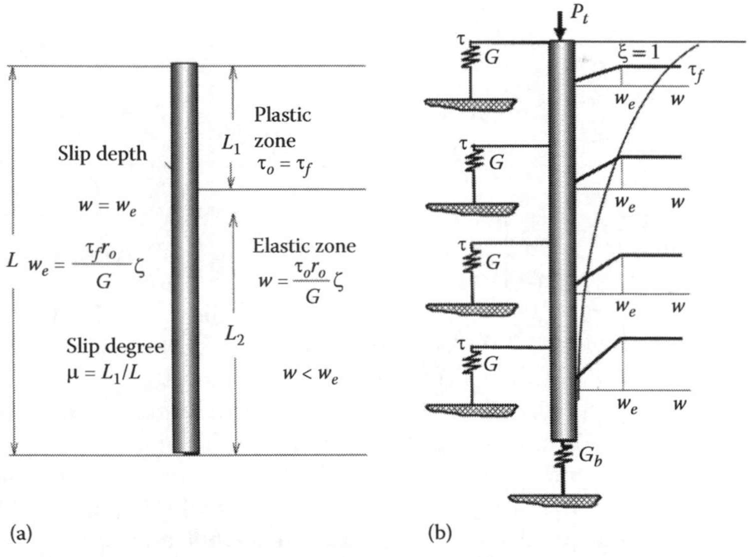

Figure 4.3 Features of elastic-plastic solutions for a vertically loaded single pile. (a) Slip depth. (b) τ_o~w curves along the pile.

where $\mu = L_1/L$ is defined as degree of slip ($0 < \mu \le 1$), and L_1 = the length of the upper plastic part; P_e, w_e = the pile load $P(L_1)$, and displacement $w(L_1)$, at the transition depth L_1, estimated from Equation 4.21. Likewise, the pile-head settlement is expressed as

$$w_t = w_e[1 + \mu k_s L(\alpha_g + L)^{n/2} C_v(\mu L)] + \frac{2\pi r_o A_v}{E_p A_p} \frac{\alpha_v^{2+\theta} + (\alpha_v + \mu L)^{1+\theta}[(1+\theta)\mu L - \alpha_v]}{(1+\theta)(2+\theta)} \tag{4.24}$$

These solutions provide three important results: (a) By specifying a slip degree, μ, the pile-head load and settlement are estimated by Equations 4.23 and 4.24, respectively; repeating the calculation for a series of slip degrees, a full pile-head load-settlement relationship is obtained. (b) For a specific pile-head load, the corresponding degree of slip of the pile can be deduced from Equation 4.23. (c) The distribution profiles of load and displacement can be readily obtained at any stage of the elastic-plastic development. Within the upper plastic portion, at any depth of z, the load, $P(z)$, can be predicted by

$$P(z) = P_e + \frac{2\pi r_o A_v[(\alpha_v + \mu L)^{1+\theta} - (\alpha_v + z)^{1+\theta}]}{1+\theta} \tag{4.25}$$

and the displacement, $w(z)$, can be obtained by

$$w(z) = w_e + \frac{P_e(\mu L - z)}{E_p A_p} + \frac{2\pi r_o A_v}{E_p A_p} \frac{(\alpha_v + z)^{2+\theta} + (2+\theta)(\mu L - z)(\alpha_v + \mu L)^{1+\theta} - (\alpha_v + \mu L)^{2+\theta}}{(1+\theta)(2+\theta)} \tag{4.26}$$

The current analysis is limited to $\alpha_v = \alpha_g$ and $n = \theta$. They are all preserved in the equations to indicate the physical implications of n and α_g to elastic state and α_v and θ to plastic state.

Equations 4.23 to 4.26 are as accurate against the GASPILE program (Guo and Randolph 1997b). An example is provided here for a pile of L/r_o = 100, embedded in a soil with $n = \theta = 0.5$, $\lambda = 1000$, $A_g/A_v = 350$, $\xi_b = 1.0$, $\nu_s = 0.4$, and $H/L = 4.0$. Using Equations 4.23 and 4.24, the prediction adopts ω = 1 and values of A estimated using n_e of 0.5 ($\alpha_g = 0$), 0.238 ($\alpha_g = 12.5$), and 0 ($\alpha_g = \infty$), respectively. The predicted pile-head load displacement relationships are plotted in Figure 4.4a. The GASPILE analyses were conducted using 20 segments for the pile and the aforementioned parameters, with nonlinear ζ (see Equation 3.10, Chapter 3, this book), an advanced version of Equation 4.5. The predicted results are also

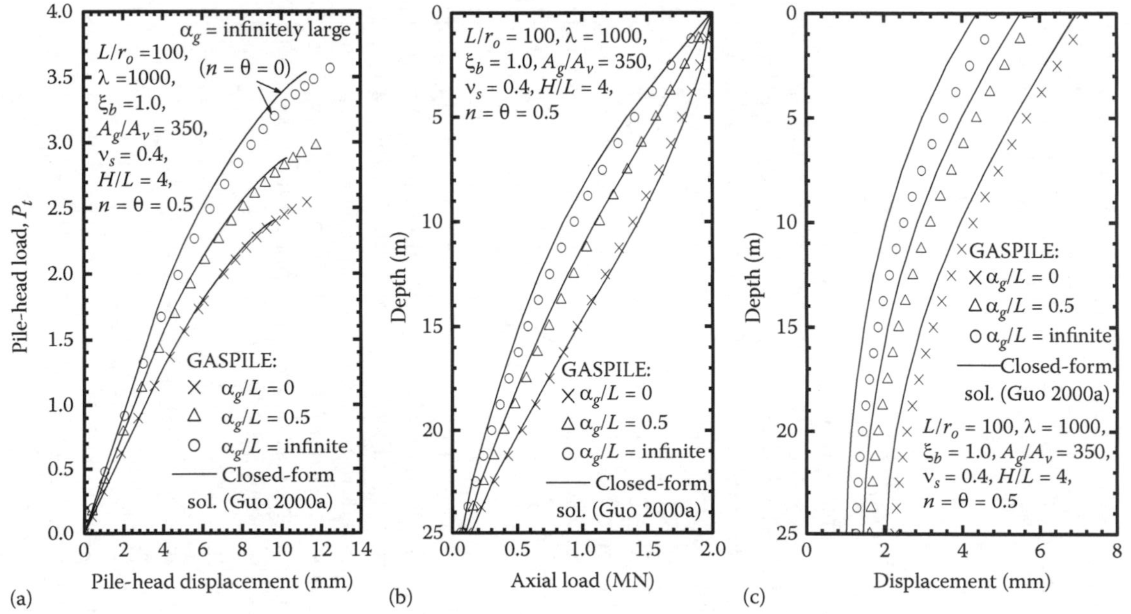

Figure 4.4 Effect of the α_g on pile response (L/r_o = 100, λ = 1000, $n = \theta = 0.5$, A_g/A_v = 350, α_g = as shown). (a) P_t–w_t, (b) Load profiles. (c) Displacement profiles. (After Guo, W. D., *J Geotech and Geoenvir Engrg*, ASCE 126, 2, 2000a. With permission from ASCE.)

Table 4.3 Critical values at P_t = 2 MN from CF analysis

α_g	0	2.5	12.5	∞
$\mu \mid P_e$ (kN)	0.527∣1141	0.443∣1211	0.306∣1323	0.126∣1575
w_e (mm)∣w_b (mm)	3.25∣2.04	3.28∣1.80	3.33∣1.45	3.43∣1.02

Source: Guo, W. D., *Proceedings of the Eighth International Conference on Civil and Structural Engineering Computing*, Paper 112, Civil-Comp Press, Stirling, UK, 2001a.

shown in Figure 4.4a. The comparison demonstrates the limited effect of the nonlinear stress on the pile-head load and displacement, as is noted in load-deformation profiles (Guo and Randolph 1997a).

Load and displacement distributions below and above the transition depth are estimated respectively using the elastic and elastic-plastic solutions. The depth of the transition (= μL) or the degree of slip μ is first estimated using Equation 4.23 (i.e., by using Mathcad) for a given pile-head load. Below the transition point, the distributions are estimated by Equations 4.15 and 4.16, respectively. Otherwise, they are evaluated by Equations 4.25 and 4.26, respectively. Any load beyond full shaft resistance (full slip) should be taken as the base load (see later Example 4.2). For P_t = 2 MN, the load and displacement profiles were predicted using Equations 4.16 and 4.25 and Equations 4.15 and 4.26, respectively. These profiles are presented in Figure 4.4b and c, together with the GASPILE analyses. They may be slightly different from a continuum-based numerical approach, as the ω is equal to 1.36, 1.437, and 1.515 for the α_g of 0, 12.5, and ∞ [from Guo and Randolph's (1998) equation in Chapter 3, this book, in which the "n" is replaced with n_e]. The critical values of μ, w_e, and w_b for the P_t were obtained from the CF solutions and are shown in Table 4.3. The results demonstrate that an increase in the α_g renders a decrease in the degree of slip, μ, and base settlement, w_b, but an increase in the load, P_e, and limiting displacement, w_e.

Example 4.1 Solutions for homogeneous soil

4.1.1 Elastic solution

In an ideal nonlinear homogeneous soil (n = 0), the coefficients in Equation 4.17 can be simplified as

$$\begin{aligned} C_1(z) = C_4(z) &= \frac{1}{k_s L}\sqrt{\frac{L}{z}}\sinh k_s(L-z) \\ C_2(z) = C_3(z) &= \frac{1}{k_s L}\sqrt{\frac{L}{z}}\cosh k_s(L-z) \end{aligned} \tag{4.27}$$

The C_i (i = 1~4) allows Equation 4.15 for shaft displacement, $w(z)$, and Equation 4.16 for axial load, $P(z)$, of the pile body at depth of z to be simplified as

$$w(x) = w_b[\cosh k_s(L-x) + \chi_v \sinh k_s(L-x)], \tag{4.28}$$

$$P(x) = k_s E_p A_p w_b[\chi_v \cosh k_s(L-x) + \sinh k_s(L-x)] \tag{4.29}$$

With Equation 4.29, the load acting on pile-head P_t is related to the force on the pile base P_b, the base settlement, w_b, and the shaft (base) settlement ratio by

$$P_b \cosh\beta + k_s E_p A_p w_b \sinh\beta = P_t \tag{4.30}$$

where $\beta = k_s L$. From Equation 4.28, the head settlement w_t (at z = 0) can be expressed as

$$w_t = w_b(\cosh\beta + \chi_v \sinh\beta) \tag{4.31}$$

With Equation 4.21, the nondimensional relationship between the head load P_t [hence deformation $w_p = P_t L/(E_p A_p)$] and settlement w_t is deduced as

$$\frac{w_p}{w_t} = \beta(\tanh\beta + \chi_v)/(\chi_v \tanh\beta + 1) \tag{4.32}$$

Equation 4.32 can be expanded to the previous expression (Randolph and Wroth 1978), for which "μL" (their symbol) = β and χ_v replaced with Equation 4.20.

4.1.2 Elastic-plastic solution

Within the elastic-plastic stage, Equations 4.11 and 4.32 allow the pile load at the transition depth to be derived as

$$P_e = \frac{\pi d \tau_f L}{\beta}\left(\frac{\tanh\bar{\beta} + \chi_v}{\chi_v \tanh\bar{\beta} + 1}\right) \tag{4.33}$$

where $\bar{\beta} = \alpha L_2 = \beta(1-\mu)$ for plastic zone ($0 \le z \le L_1$). Equations 4.23 and 4.33 permit the head load to be related to slip degree by

$$P_t = \pi d \tau_f L\left(\mu + \frac{1}{\beta}\frac{\tanh\bar{\beta} + \chi_v}{\chi_v \tanh\bar{\beta} + 1}\right) \tag{4.34}$$

In terms of Equations 4.11 and 4.24, the pile-head settlement can also be rewritten as

$$w_t = w_e\left(1 + \frac{\beta^2\mu^2}{2} + \beta\mu\frac{\tanh\bar{\beta} + \chi_v}{\chi_v \tanh\bar{\beta} + 1}\right) \tag{4.35}$$

Example 4.2 Shaft friction and base resistance at full slip state

A pile head-load P_t may be resolved into the components of total shaft friction and base resistance at the full slip state. In comparison with Equations 4.25 and 4.26, the axial load $P(z)$ at depth z, the pile-head load P_t and displacement w_t can alternatively be expressed as

$$P(z) = P_t - 2\pi r_o \frac{(\alpha_v + z)^{1+\theta} - \alpha_v^{1+\theta}}{1+\theta} \tag{4.36}$$

$$w_t = w_b + \frac{1}{A_p E_p} \int_0^L P(z)\,dz \tag{4.37}$$

The base displacement w_b of Equation 4.12 may be correlated to the pile cross-sectional properties by $w_b = P_b L/(R_b E_p A_p)$ and a newly defined base deformation ratio R_b of

$$R_b = \frac{L}{A_p E_p} \frac{4 r_o G_b}{(1-\nu_s)\omega} \tag{4.38}$$

The base resistance P_b is the $P(z)$ at $z = L$ using Equation 4.36. With the w_b obtained and Equation 4.36, the w_t of Equation 4.37 is simplified, which offers the following linear load and displacement relationship at and beyond full shaft slip state

$$P_t = \frac{R_b}{1+R_b} \frac{A_p E_p}{L} w_t + \frac{P_s}{1+R_b} \{1 + R_b \alpha_m\} \tag{4.39}$$

where P_s = total shaft friction along entire pile length and α_m = dimensionless shaft friction factor, which for a τ_f distribution of Equation 4.2 is given by

$$\alpha_m = \frac{1}{1 - \left(\dfrac{\alpha_v}{\alpha_v + L}\right)^{1+\theta}} \left\{ \frac{1}{2+\theta} \left[1 + \frac{\alpha_v}{L} - \left(\frac{\alpha_v}{\alpha_v + L}\right)^{1+\theta} \frac{\alpha_v}{L} \right] - \left(\frac{\alpha_v}{\alpha_v + L}\right)^{1+\theta} \right\} \tag{4.40}$$

where $\alpha_m = 0.5$ for a uniform τ_f ($\theta = 0$), $\alpha_m = 0.4$ for a distribution by $\theta = 0.5$ and $\alpha_v = 0$, and $\alpha_m = 1/3$ for a Gibson profile ($\theta = 1$) at $\alpha_v = 0$

$$P_t = \frac{1}{1-\alpha_m} \frac{A_p E_p}{L} w_t \tag{4.41}$$

It is more straightforward to draw the line P_t–w_t of Equation 4.41 together with the line P_t–w_t of Equation 4.39 and a measured P_t~w_t curve. The value of "P_t" at the intersection of the two lines is shaft friction P_s, as is illustrated in Example 4.3. Equation 4.39 provides the theoretical base for the earlier empirical approach (Van Weele 1957).

4.4 PARAMATRIC STUDY

In the form of modified Bessel functions, numerical evaluations of the current solutions (Guo 1997; Guo and Randolph 1997a) have been performed through a spreadsheet program (using a macro sheet in Microsoft Excel), with the shaft load transfer factor of Equation 4.7 and the base load transfer factor of Equation 3.4 (see Chapter 3, this book), except for using $\omega = 1$ to compare with the FLAC analyses. All the following CF solutions result from this program.

4.4.1 Pile-head stiffness and settlement ratio (Guo and Randolph 1997a)

The closed-form solution for the pile, which is later referred to as "CF," is underpinned by the load transfer parameter, ζ (thus, the ψ_o and r_m/r_o). Assuming $A = 2$, $B = 0$ (see Chapter 3, this book), $\nu_s = 0.4$, $\xi_b = 1$, $\psi_o = 0$, and $\omega = 1$ (to compare with FLAC analysis), the solutions were obtained and are presented here. The pile-head stiffness predicted by Equation 4.21 is plotted against FLAC analyses in Figure 4.5, along with the simple analysis (SA; Randolph and Wroth 1978). Note the latter uses $A = 2.5$ (see Chapter 3, this book). The results show that:

1. The CF approach is reasonably accurate, with a underestimation of the stiffness by ~10% in comparison with the FLAC results.
2. The SA analysis progressively overestimates the stiffness by ~20% with either increase in nonhomogeneity factor n (particularly, $n = 1$), or decrease in pile–soil relative stiffness factor.
3. For a pile in homogenous soil ($n = 0$), the CF and SA approaches are exactly the same (see Example 4.1). The discrepancy in the head stiffness between the two approaches is because of different values of A.

The ratios of pile-head and base settlement estimated by Equation 4.15 are compared with those from the FLAC analyses in Figure 4.6. With extremely compressible piles, the CF solutions diverge from the FLAC results as the displacement prediction becomes progressively more sensitive to the neglecting of the interactions between each horizontal layer of soil, which is also observed in the deduced values of ζ in Figures 3.4a and 3.7c (Chapter 3, this book).

4.4.2 Comparison with existing solutions (Guo and Randolph 1998)

Table 4.4 shows that the pile-head stiffness predicted by Equation 4.21 and the ratio of pile-base and head load by Equation 4.16 generally lie between

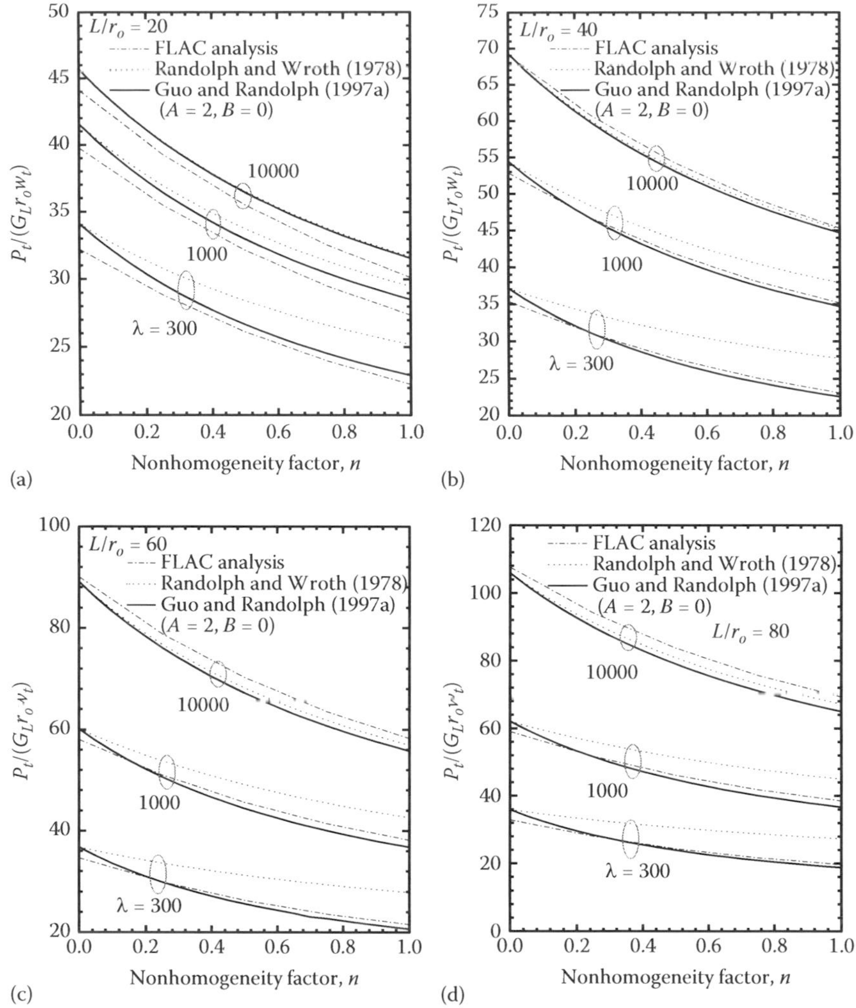

Figure 4.5 Comparison of pile-head stiffness among FLAC, SA (A = 2.5), and CF (A = 2) analyses for a L/r_0 of (a) 20, (b) 40, (c) 60, and (d) 80. (After Guo, W. D. and M. F. Randolph, *Int J Numer and Anal Meth in Geomech* 21, 8, 1997a.)

those obtained from the VM analysis (Rajapakse 1990) and the current FLAC analyses.

Figure 4.7a shows a nonlinear increase in pile-head stiffness with slenderness ratio for a pile in a homogeneous, infinite half space (Banerjee and Davies 1977; Chin et al. 1990). As expected, the present CF solution yields slightly higher head stiffness than those by other approaches. The solution using a value of $A = 2.5$ [i.e., CF ($A = 2.5$)] was also conducted and is shown in Figure 4.7. It agrees well with those from the more rigorous numerical

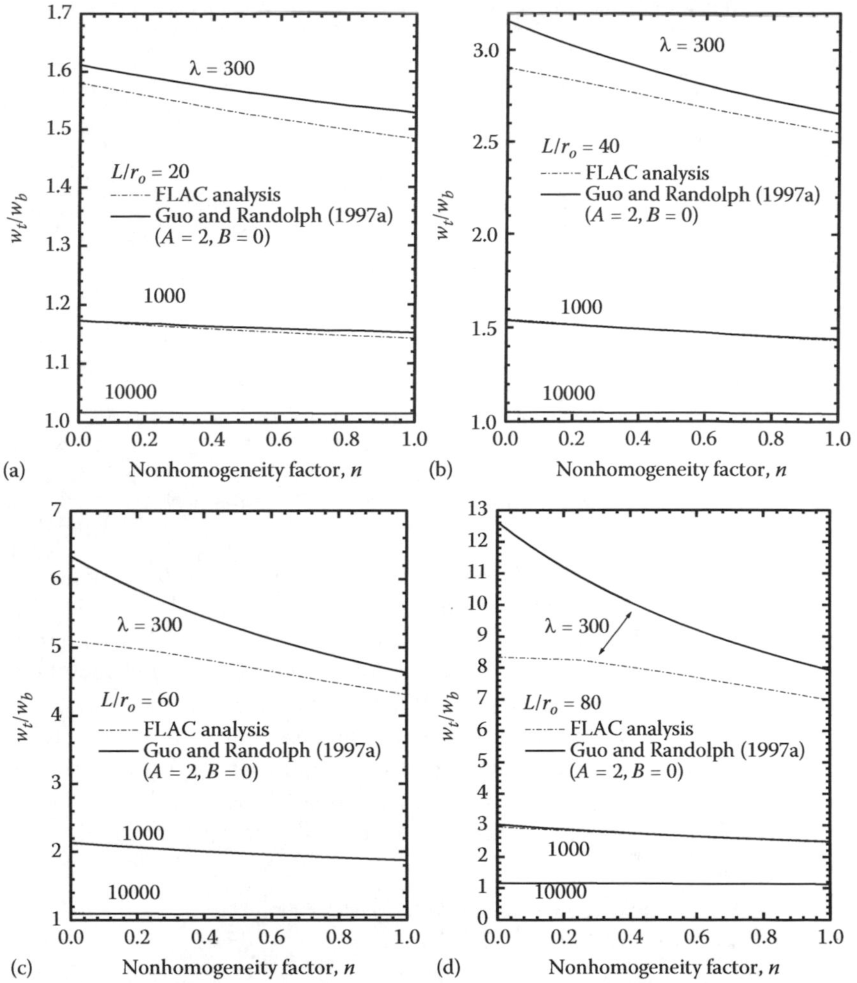

Figure 4.6 Comparison between the ratios of head settlement over base settlement by FLAC analysis and the CF solution for a L/r_o of (a) 20, (b) 40, (c) 60, and (d) 80. (After Guo, W. D., and M. F. Randolph, *Int J Numer and Anal Meth in Geomech* 21, 8, 1997a.)

approaches shown there. As for a pile in a Gibson soil ($n = 1$), Figure 4.7b also indicates a good comparison of the head stiffness between the CF solution and the numerical results (Banerjee and Davies 1977; Chow 1989). Again, an increase in slenderness ratio causes an increase in pile-head stiffness until a critical ratio is reached.

Figure 4.8a_1 through 4.8c_1 indicates the significant impact of soil nonhomogeneity and finite layer ratio, H/L, on the head stiffness. Poisson's ratio (ν_s) reflects the compressibility of a soil. Figure 4.8 shows that more incompressible

Table 4.4 FLAC analysis versus the VM approach

$\frac{P_t}{G_L w_t r_o}$ $(n = 0)$	FLAC	74.82	68.71	56.73	38.09
	VM	72.2[a]	65.1	54.9	38.7
	CF	71.68	66.75	56.59	38.68
$\frac{P_t}{G_L w_t r_o}$ $(n = 1.0)$	FLAC	52.44	48.08	39.13	24.7
	VM	44.46	40.38	–	22.2
	CF	49.79	45.93	38.13	24.3
$\frac{P_b}{P_t}(\%)$ $(n = 0)$	FLAC	8.47	8.01	6.9	4.47
	VM	5.4	5.2	4.6	3.1
	CF	6.63	6.38	5.69	4.0
	$\lambda(=E_p/G_L)$	10,000	3,000	1,000	300

Source: Guo, W. D., and M. F. Randolph, *Computers and Geotechnics* 23, 1–2, 1998.

[a] rigid pile; VM and CF analyses (ν_s = 0.5) and FLAC analysis (ν_s = 0.495 and H/L = 4)

(higher ν_s) soil may have ~25% higher pile-head stiffness (at $\nu_s = 0.5$) compared to that at $\nu_s = 0$. With $L/r_0 = 40$, the increase in finite-layer ratio H/L (within effective pile slenderness ratio) from 1.25 to 4 incurs approximately a 15% reduction in the head stiffness but only a slight decrease in base load (not shown), as reported previously (Valliappan et al. 1974; Poulos and Davis 1980). The effect of the ratio of H/L can be well represented for other slenderness ratios by the current load transfer factors, as against the numerical results (Butterfield and Douglas 1981). Equation 3.4 (see Chapter 3, this book) for ω and Equation 4.7 for ζ are sufficiently accurate for load transfer analysis.

4.4.3 Effect of soil profile below pile base (Guo and Randolph 1998)

The analysis in the last section is generally capitalized on the shear modulus of Equation 4.1 through the entire soil layer of depth, H. A constant value of shear modulus below the pile-tip level may be encountered (see Figure 3.1, Chapter 3, this book), for which modified expressions for the parameter A in Equation 4.7 (Guo 1997) were presented previously. This situation induces a slightly softer pile response compared with Equation 4.1, as is pronounced for shorter piles ($L/r_0 < 30$). It may result in ~10% difference in pile-head stiffness, particularly in soil with a significant strength increase with depth (high n values).

4.5 LOAD SETTLEMENT

Equations 4.23 and 4.24 offer good prediction of a load-settlement relationship against continuum-based analyses. This is illustrated next for a

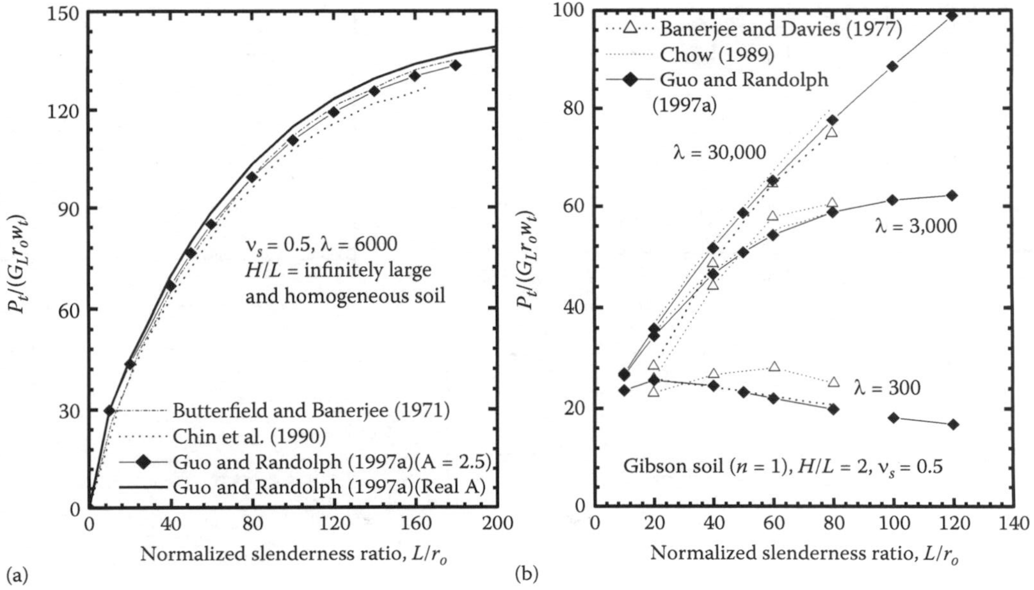

Figure 4.7 Pile-head stiffness versus slenderness ratio relationship. (a) Homogeneous soil (n = 0). (b) Gibson soil (n = 1). (Guo, W. D., and M. F. Randolph, *Computers and Geotechnics* 23, 1–2, 1998.)

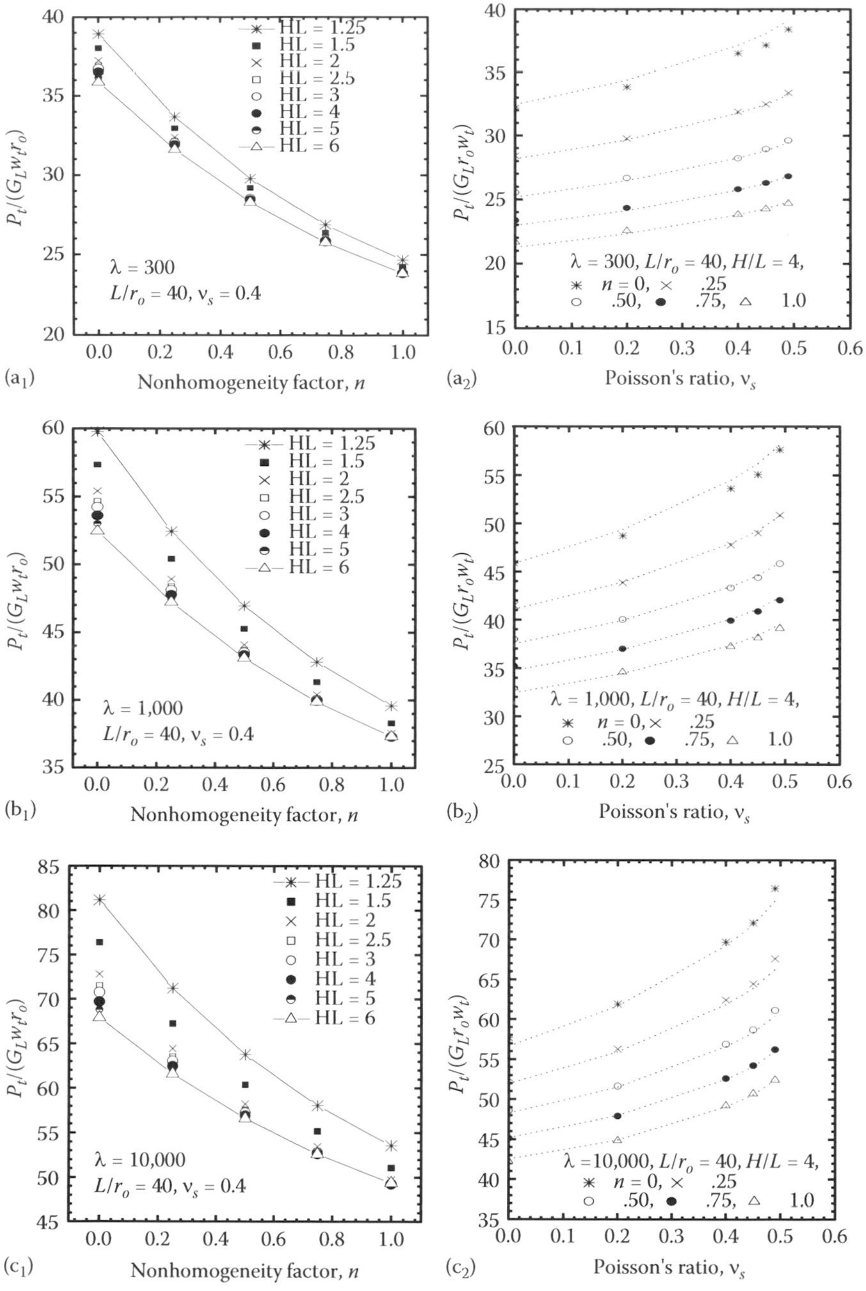

Figure 4.8 Pile-head stiffness versus (a_1–c_1) the ratio of *H*/*L* relationship (ν_s = 0.4), (b_2–c_2) Poisson's ratio relationship (*H*/*L* = 4). (Guo, W. D., and M. F. Randolph, *Computers and Geotechnics* 23, 1–2, 1998.)

particular set of pile and soil parameters, concentrating on the elastic-plastic response.

4.5.1 Homogeneous case (Guo and Randolph 1997a)

A pile of 30 m in length, 0.75 m in diameter, and 30 GPa in Young's modulus was installed in a homogeneous soil layer 50 m deep. The soil has an initial tangent modulus (for very low strains) of 1056 MPa and a Poisson's ratio of 0.49. The constant limiting shaft resistance was 0.22 MPa over the pile embedded depth. The numerical analyses by GASPILE and the closed-form solutions offer the load settlement curves depicted in Figure 4.9. They are compared with a finite element analysis involving a nonlinear soil model (Jardine et al. 1986), boundary element analysis (BEA utilizing an elastic-plastic continuum-based interface model), and BEA with a hyperbolic continuum-based interface model, respectively (Poulos 1989). A good agreement of the load transfer analysis with other approaches is evident. Nevertheless, as noted previously (Poulos 1989), the response of very stiff piles (e.g., $E_p = 30{,}000$ GPa), obtained using an elastic, perfectly plastic soil response, can differ significantly from that obtained using a more gradual nonlinear soil model.

Base contribution is generally limited to the pile-head response except for short piles. An evidently nonlinear base behavior will be observed only when local shaft displacement at the base level exceeds the limiting displacement, w_e, as determined from the difference between the nonlinear GASPILE and linear (closed form) analyses.

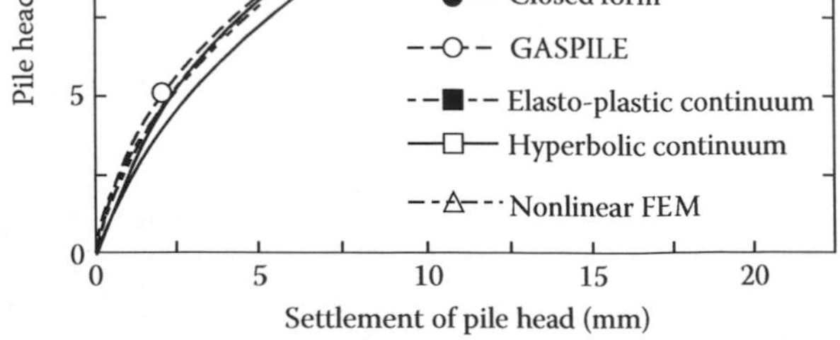

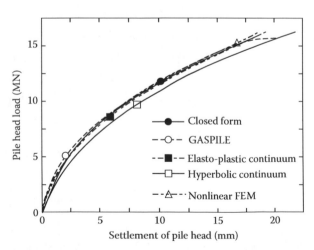

Figure 4.9 Comparison between various analyses of single pile load-settlement behavior. (After Guo, W. D., and M. F. Randolph, *Int J Numer and Anal Meth in Geomech* 21, 8, 1997a.)

4.5.2 Nonhomogeneous case (Guo and Randolph 1997a)

Previous analyses (Banerjee and Davies 1977; Poulos 1979; Rajapakse 1990) indicate a substantial decrease in pile-head stiffness as the soil shear modulus non-homogeneity factor (n) increases. This is partly owing to use of a constant modulus at the pile-tip level, together with reduction in the average shear modulus over the pile length as n increases (see Figure 4.1). By changing the nonhomogeneity factor ($\theta = n$) and maintaining the average shaft shear modulus, the closed-form prediction by $\psi_o = 0$ (linear elastic-plastic case) was obtained and is shown in Figure 4.10a. Only ~30% difference (this case) is observed due to variation in the n (within the elastic stage). This mainly comes from the load variations at low load levels and displacement variations at higher load levels, as demonstrated in Figure 4.10b and c. At the same slip degrees, the w_t is only slightly affected until

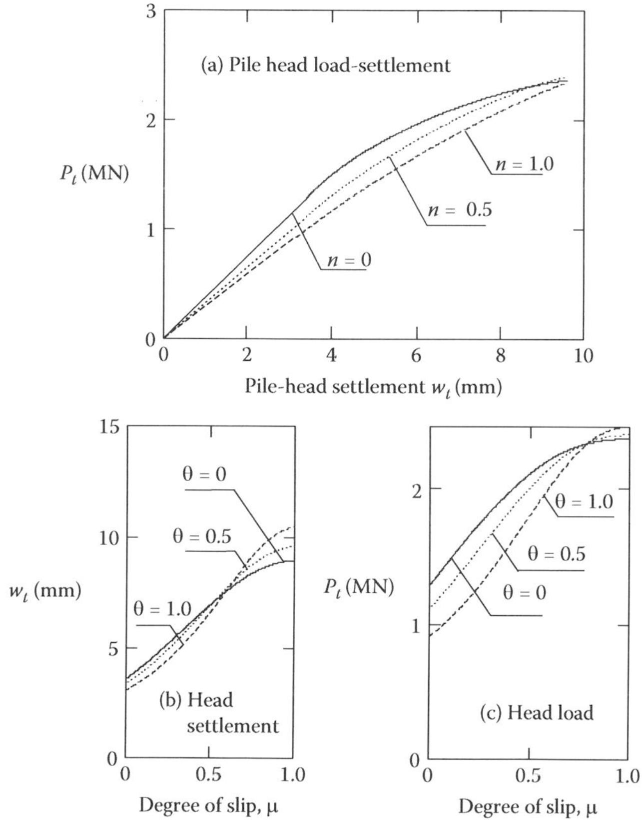

Figure 4.10 Effect of slip development on pile-head response (n = θ). (After Guo, W. D., and M. F. Randolph, *Int J Numer and Anal Meth in Geomech* 21, 8, 1997a.)

$\mu > 0.6$. These features are important to assessing cyclic capacity. With an identical average shaft friction for all three cases, the P_t~μ curves should converge towards an identical pile head load at full slip ($\mu = 1$), but for the difference in the base.

Example 4.3 Analysis of a loading test on an instrumented pile

An analysis is presented here to show the impact of nonhomogeneous soil property and the pile–soil relative slip on pile response. Gurtowski and Wu (1984) detailed the measured response of an instrumented pile. The pile was a 0.61-m-wide octagonal prestressed concrete hollow pile with a plug at the base and was driven to a depth of about 30 m. The input parameters (Poulos 1989) include E_p = 35 GPa, E (soil; Young's modulus) = 4*N* MPa (*N* = SPT value, with *N* = 2.33 *z*/m to a depth of 30 m, *z* in meters), τ_f (shaft) = 2*N* kPa, τ_{fb} (base) = 0.4 MPa, and ν_s = 0.3. The pile-head and base load settlement curves were predicted by GASPILE (with R_{fs} = 0.9, G_i/τ_f = 769.2 R_{fb} = 0.9) and by the closed form solutions (with $\psi_o = 0.5$). They are plotted in Figure 4.11a, which compare well with boundary element analysis (Poulos 1989), in view of the difference at failure load levels caused by the assumed τ_{fb} (base).

The shaft friction and base resistance were obtained using Equations 4.39 and 4.41 and are plotted in Figure 4.11b. The α_m was 1/3 using $\theta = 1$, $\alpha_v = 0$ in Equation 4.40. The R_b was calculated as 0.55, with $\omega = 1.0$, $\nu_s = 0.3$, $G_b/G_L = 1$, and $\zeta = 4.455$. The shaft resistance P_s was determined as 4.019 MN.

The load and displacement distributions are predicted using the closed-form solutions and the nonlinear GASPILE analysis. The degrees of slip at P_t = 1.8 MN are 0.058, 0.136, 0.202, 0.258, and 0.305 for *n* = 0, 0.25, 0.5, 0.75, and 1.0 respectively, and at P_t = 3.45 MN, μ = 0.698, 0.723, 0.743, 0.758, and 0.771 accordingly. At P_t = 4.52 MN, full slip occurs and the base takes 1.07 MN. For the three soil profiles of *n* = 0, 0.5, and 1.0, Figure 4.12 shows the predicted settlement (only at two load levels) and load distribution profiles. Typical predictions of the closed form solutions (*n* = 0, 0.5, and 1) and the GASPILE (*n* = 1 only) are shown in Figure 4.13, along with BEM results (Poulos 1989) and the measured data. The figures show the linear soil strength and shear modulus (*n* = 1) yield reasonable predictions of the axial force against the measured data. The accuracy of the CF solutions and GASPILE is evident.

4.6 SETTLEMENT INFLUENCE FACTOR

The settlement influence factor, *I*, is defined as the inverse of a pile-head stiffness

$$I = \frac{G_{iL} w_t r_o}{P_t} \tag{4.42}$$

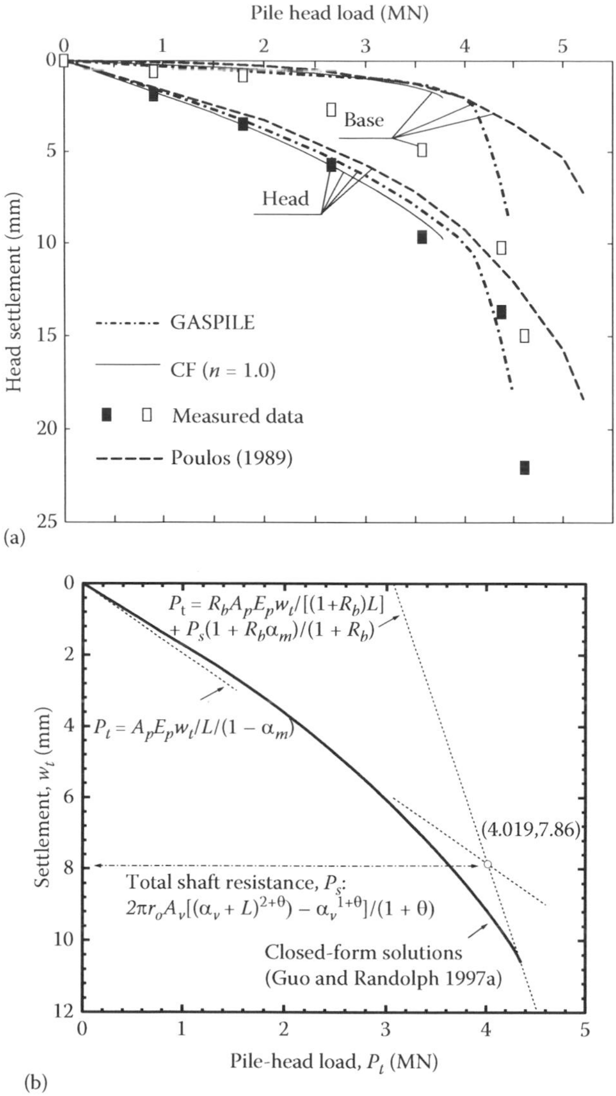

Figure 4.11 (a) Comparison among different predictions of load settlement curves (measured data from Gurtowski and Wu 1984) (Guo and Randolph 1997a); (b) Determining shaft friction and base resistance.

where G_{iL} = initial shear modulus of soil at pile-base level. The factor can be derived directly from Equation 4.21 for the elastic stage as

$$I = \frac{1}{\pi\sqrt{2}C_{vo}}\sqrt{\frac{\zeta}{\lambda}} \tag{4.43}$$

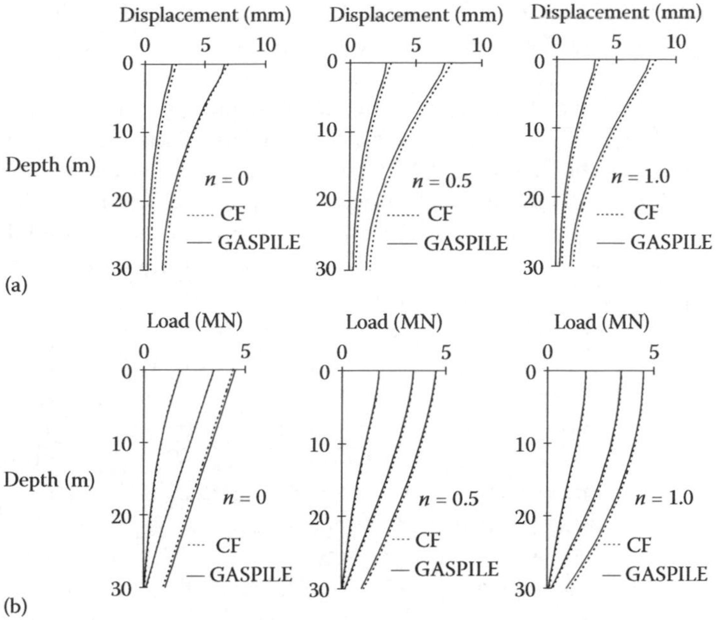

Figure 4.12 Comparison between the CF and the nonlinear GASPILE analyses of pile (a) displacement distribution and (b) load distribution. (After Guo, W. D., and M. F. Randolph, *Int J Numer and Anal Meth in Geomech* 21, 8, 1997a.)

It is straightforward to obtain the factor for an elastic-plastic medium using Equations 4.23 and 4.24. The settlement influence factor is primarily affected by pile slenderness ratio, pile–soil relative stiffness factor, the degree of the nonhomogeneity of the soil profile, and the degree of pile–soil relative slip. Our attention is subsequently confined to elastic medium, to compare the factor with those of pile groups in Chapter 6, this book,

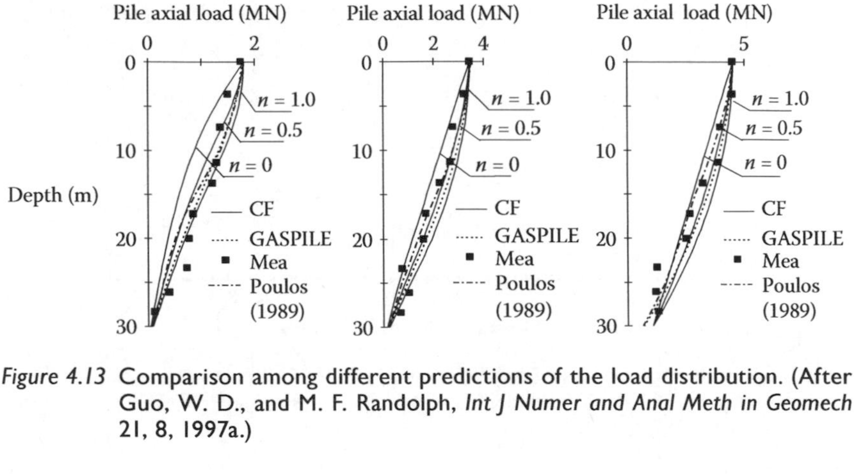

Figure 4.13 Comparison among different predictions of the load distribution. (After Guo, W. D., and M. F. Randolph, *Int J Numer and Anal Meth in Geomech* 21, 8, 1997a.)

to estimate settlement of pile groups. The elastic factor estimated using Equation 4.43 (termed as Guo and Randolph 1997a) is presented in Figures 4.14 and 4.15.

Figure 4.14 shows the settlement influence factor of piles in a Gibson soil with different slenderness ratios at a constant relative stiffness factor (λ = 3,000), together with the BEM analysis based on Mindlin's solution (Poulos 1989), BEM analysis of three dimensional solids (Banerjee and Davies 1977), and the approximate closed-form solution (Randolph and Wroth 1978). The results of Equation 4.43 are generally consistent with those provided by the other approaches.

Figure 4.15 shows the impact of relative stiffness λ on the settlement influence factor for four different slenderness ratios at n = 0 and 1, in comparison with the boundary element (BEM) analysis (Poulos 1979) and the FLAC analysis. The BEM analysis is for the case of H/L = 2, while this CF solution corresponds to the case of H/L = 4. As presented in Figure 4.8, an increase in the value of H/L reduces the pile-head stiffness and increases the settlement influence factor (e.g., with L/r_o = 50, λ = 26,000; E_p/G_L = 10,000), n = 0, an increase in H/L from 2 to infinity raises the settlement factor by ~21% (Poulos 1979). In view of the H/L effect, the closed-form solutions are generally quite consistent with the numerical analysis.

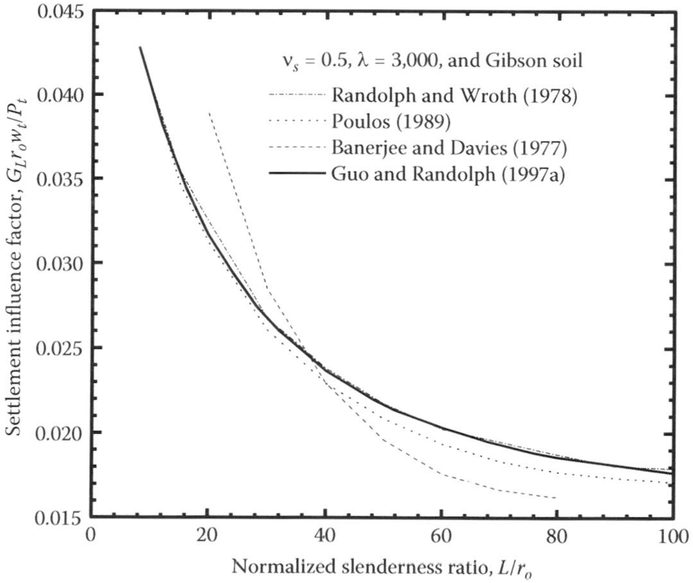

Figure 4.14 Comparison of the settlement influence factor (n = 1.0) by various approaches. (After Guo, W. D., and M. F. Randolph, *Int J Numer and Anal Meth in Geomech* 21, 8, 1997a.)

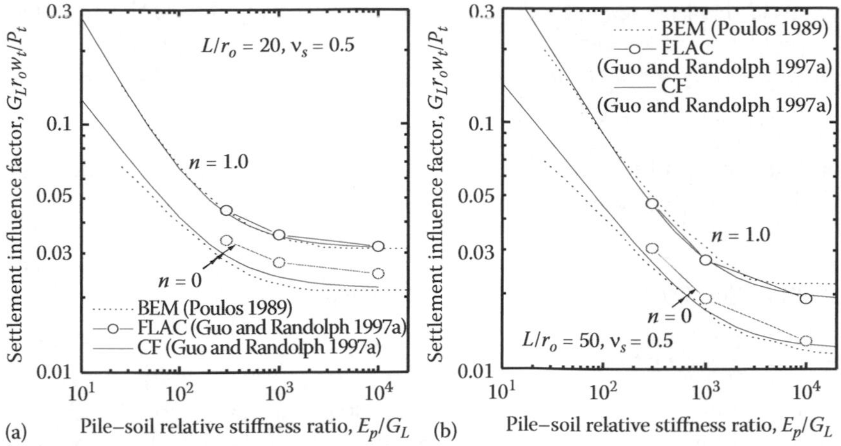

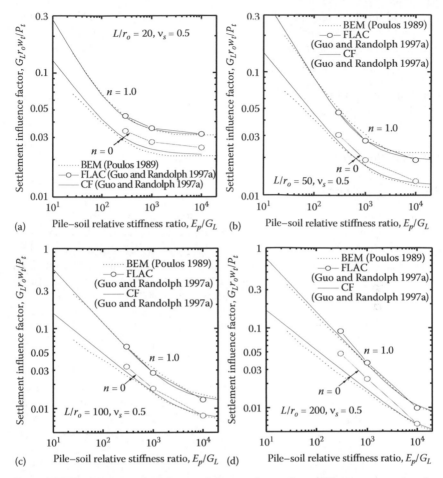

Figure 4.15 Comparison of settlement influence factor from different approaches for a L/r_o of: (a) 20, (b) 50, (c) 100, and (d) 200. (After Guo, W. D., and M. F. Randolph, *Int J Numer and Anal Meth in Geomech* 21, 8, 1997a.)

4.7 SUMMARY

The following conclusions can be drawn:

1. Nonlinear and simplified elastic-plastic analyses offer slightly different results. The closed-form solutions underpinned by the simplified elastic-plastic model are sufficiently accurate.
2. The influence of n on pile-head stiffness or settlement influence factor is largely attributed to the alteration of average shear modulus over the pile length. The nonhomogeneity factor, n, may be adjusted to fit the general trend of the modulus with depth for a complicated shear

modulus profile. The solution presented here may thus still be applied with reasonable accuracy

3. The evolution of the pile–soil relative slip on load-settlement behavior is readily simulated by the closed-form solutions.

The last conclusion is useful to gain capacity of piles in strain-softening soil, as highlighted subsequently.

4.8 CAPACITY FOR STRAIN-SOFTENING SOIL

4.8.1 Elastic solution

To facilitate determining capacity of piles in strain-softening soil, the elastic-plastic solutions are recast in nondimensional form (Guo 2001c), with the same definitions of parameters (except where specified). The axial pile displacement, u, is equal to the pile–soil relative displacement, w. Within elastic state, the governing equation for the axial deformation of a pile fully embedded in the soil addressed is as follows (Murff 1975)

$$\frac{d^2w}{dz^2} = \frac{\pi d}{E_p A_p} \frac{w}{w_e} \tau_f \tag{4.44}$$

where $d = 2r_o$, diameter of the pile. Under cyclic loading, α_v and α_g may be taken as zero. Introducing nondimensional parameters, Equation 4.44 is transformed into (see Figure 4.16a), with $\alpha_g = \alpha_v = 0$.

$$\frac{d^2\pi_1}{d\pi_2^2} = \pi_v^{1/m}\pi_2^{\theta}\pi_1 \tag{4.45}$$

where $\pi_1 = w/d$, $\pi_2 = z/L$ $(0 < z \leq L)$. The pile–soil relative stiffness, π_v, is a constant along the pile. With $w_e = A_v r_o \zeta / A_g$ (see Equation 4.11), it is given by

$$\pi_v = \left(\frac{2}{\lambda\zeta} \left(\frac{L}{r_o} \right)^2 \right)^m \tag{4.46}$$

Therefore, Equation 4.45 may be solved in terms of modified Bessel functions, I and K, of the second kind of non-integer order m and $m - 1$, using the pile-head load, P_t, and the base load, P_b. The P_b is correlated to the base displacement, w_b, by the base settlement ratio R_b. The R_b of Equation 4.38, with Equation 4.12, may be rewritten as

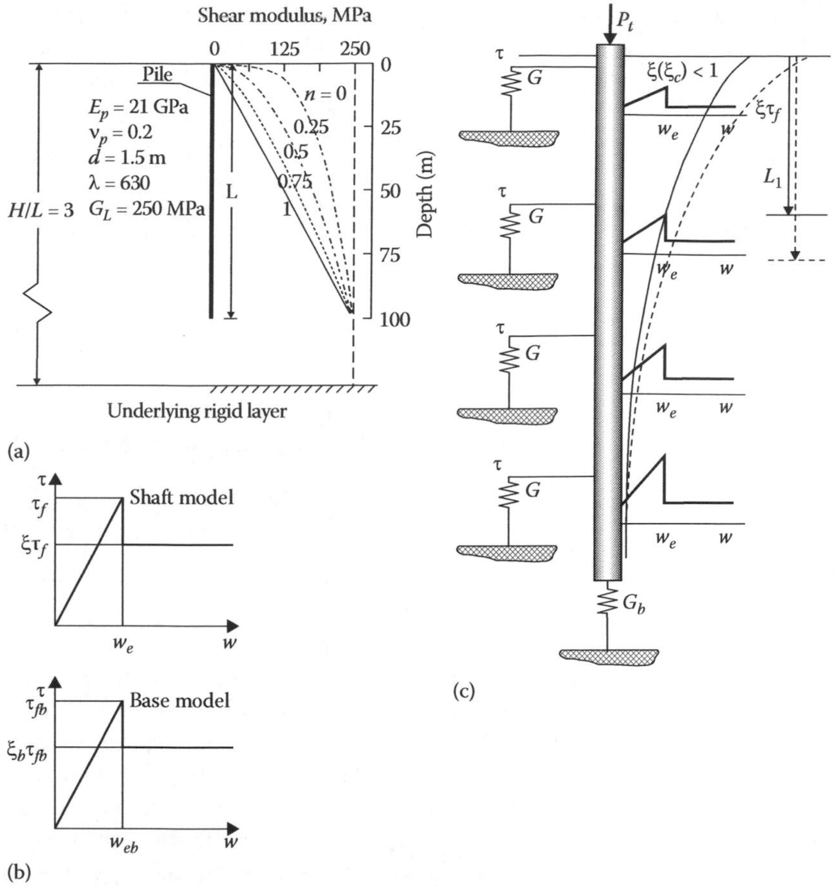

Figure 4.16 Schematic pile–soil system. (a) Typical pile and soil properties. (b) Strain-softening load transfer models. (c) Softening $\xi\tau_f$ ($\xi_c\tau_f$) with depth.

$$R_b = \frac{4}{(1-\nu_s)\pi\omega}\frac{1}{\lambda\xi_b}\frac{L}{r_o} \tag{4.47}$$

At depth z, the pile stiffness, compared to Equation 4.21, may be expressed as

$$\frac{P(z)}{G_L w(z) r_o} = \frac{2\pi}{\zeta}\left(\frac{L}{r_o}\right)\frac{1}{\pi_v^{1/2m}}C_v(z) \tag{4.48}$$

A new form of $C_v(z)$ is as follows:

$$C_v(z) = \frac{C_1(z) + C_2(z)\,R_b/\pi_v^{1/2m}}{C_3(z) + C_4(z)\,R_b/\pi_v^{1/2m}}(\pi_2)^{n/2} \tag{4.49}$$

Within the elastic state, the shaft displacement at the ground level ($z = z_t$), $w(z_t)$ equals the head settlement, w_t. For a rigid pile (i.e., with a displacement of w_t at any depth), the total shaft load, P_{fs}, is obtained by integration of Equation 4.2 ($\alpha_v = 0$) over the pile length

$$P_{fs} = \pi d A_v\ L^{1+\theta}/(1+\theta) \tag{4.50}$$

As $A_v = w_t G/(\zeta r_o L_\theta)$, in light of Equations 4.2 and 4.5, Equation 4.50 can be rewritten as

$$\frac{P_{fs}}{G_L w_t r_o} = \frac{2\pi}{\zeta}\left(\frac{L}{r_o}\right)\frac{1}{1+\theta} \tag{4.51}$$

Mobilization of the shaft capacity (for a pile of any rigidity) may be quantified by a capacity ratio n_p ($= P_t/P_{fs}$) deduced from Equations 4.48 and 4.51:

$$n_p = \frac{P_t}{P_{fs}} = \frac{(1+\theta)C_v(z_t)}{\pi_v^{1/2m}} \tag{4.52}$$

4.8.2 Plastic solution

A rigid pile sitting on a soft layer may yield initially at the pile base before it does from the pile top. In the majority of cases, as load on the pile top increases, plastic yield may be assumed to initiate at the ground surface and propagate down the pile. As shown in Figure 4.16c, the yield transfers to a transitional depth, L_1, at which the soil displacement, w, equals w_e, above which the soil resistance is in a plastic state, below which it is in an elastic state. Strain softening renders the limiting stress reduce to $\xi\tau_f$ from τ_f (Figure 4.16b, subscript "b" for pile base), and the w_e reduces to $\xi\ w_e$, and the depth L_1 increases to the "dash line" position (Figure 4.16c). The pile–soil interaction is governed by the following differential equation

$$\frac{d^2\pi_1}{d\pi_{2p}^2} = \pi_4 \pi_{2p}^{\theta} \tag{4.53}$$

where $\pi_{2p} = z/L_1$, L_1 = length of the upper plastic zone; and $\pi_4 = \xi\pi A_v L_1^{2+\theta}/(E_p A_p)$. The π_4 is positive where the pile is in compression and negative where it is in tension. Integration of Equation 4.53 offers two constants, which are determined using the boundary conditions of load P_t at the top of plastic zone ($\pi_{2p} = 0$) and the displacement π_1^* ($= w_e/d$) at the transition depth ($\pi_{2p} = 1.0$). This offers the following

$$\pi_1 = \frac{\pi_4}{(1+\theta)(2+\theta)}(\pi_{2p}^{2+\theta} - 1) + \frac{P_t L_p}{E_p A_p d}(\pi_{2p} - 1) + \pi_1^* \tag{4.54}$$

4.8.3 Load and settlement response

To assess pile capacity, the plastic solutions are now transformed into functions of the capacity ratio n_p defined by Equation 4.52. The pile-head deformation from Equation 4.54 may be rewritten as

$$\frac{w_t}{w_e} = 1 - \frac{\xi \pi_v^{1/m}}{(1+\theta)(2+\theta)} \left(\frac{L_1}{L}\right)^{2+\theta} + \frac{n_p \pi_v^{1/m}}{1+\theta} \left(\frac{L_1}{L}\right) \tag{4.55}$$

At the transition depth of the elastic-plastic interface ($z = L_1$), firstly, as with Equation 4.50, the pile load, P_e, is deduced as

$$P_e = P_t - \frac{\pi d A_v \xi}{1+\theta} L_1^{1+\theta} = \frac{\pi d A_v L^{1+\theta}}{1+\theta} \left(n_p - \left(\frac{L_1}{L}\right)^{1+\theta} \xi \right) \tag{4.56}$$

and, secondly, Equation 4.53 allows the displacement, w_e, to be related to P_e by

$$\frac{w_e}{d} = \frac{P_e L}{E_p A_p d \pi_v^{1/2m}} \frac{1}{C_v(L_1)} \tag{4.57}$$

Therefore, the capacity ratio n_p for an elastic-plastic case is ($n = \theta$)

$$n_p = \xi \left(\frac{L_1}{L}\right)^{1+\theta} + \frac{1+\theta}{\pi_v^{1/2m}} C_v(L_1) \tag{4.58}$$

Equation 4.58 is used to examine the effect of the slip development on the pile capacity. As slip develops ($L_1 > 0$), pile capacity, n_p may increase due to further mobilization of the shaft stress in the lower elastic portion of the pile or decrease because of the strain softening ($\xi < 1$) in the upper plastic portion (see Figure 4.16c). The incremental friction is the difference between the reduced shaft friction of $(1 - \xi)\tau_f \pi d L_1$ and possible shaft friction increase in elastic zone. This is evident in Figure 4.17a for a pile in a soil with stiffness factors of $\lambda = 1{,}000$ ($\pi_v = 1.1$ given $n = 0$ or 1) and $R_b = 0$. The figure shows that as long as $\xi > 0.75$, the overall increase for the pile is greater than the decrease. Thus, the capacity ratio, n_p, at any degree of slip is higher than that at the incipient of slip (P_e/P_{fs}). Particularly for the case of $n = 1$, irrespective of the softening factor, ξ, the upper portion (e.g., ~20% pile length) of the pile may be allowed to slip to increase the capacity ratio. The portion extends to a critical degree of slip, μ_{max}, at which the capacity ratio n_p reaches maximum n_{max}. This maximum can be viewed in another angle from Figure 4.17c, showing the capacity ratio, n_p (given by Equation 4.58), and the normalized pile-head displacement w_t/w_e (by Equation 4.55).

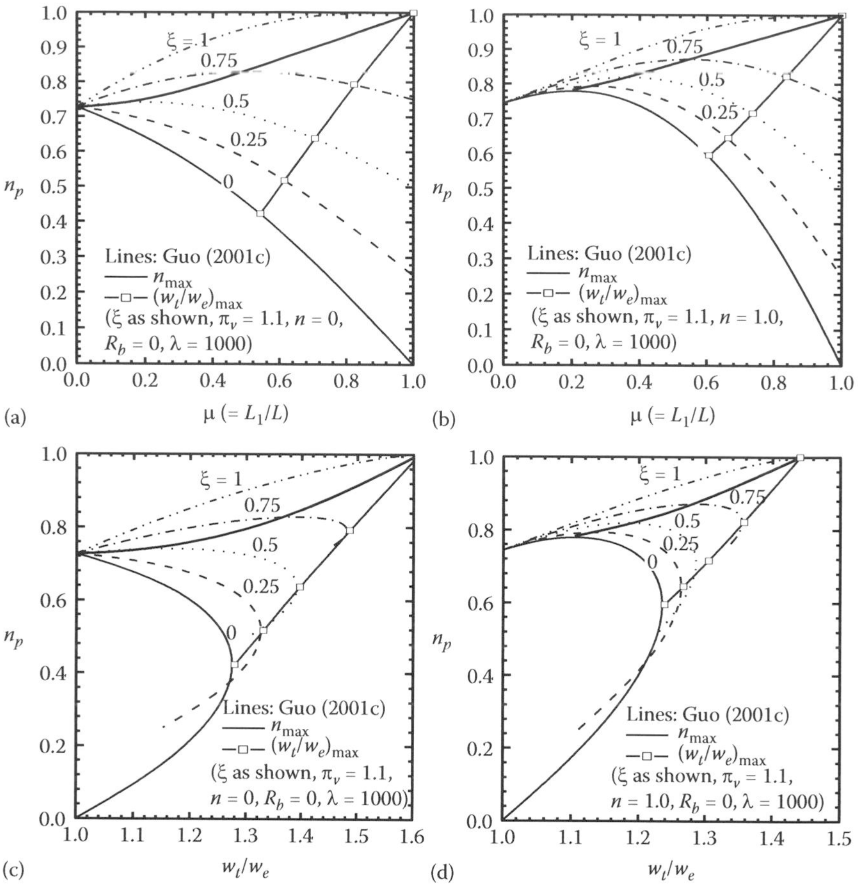

Figure 4.17 Capacity ratio versus slip length. (a) n = 0. (b) n = 1.0; or normalized displacement. (c) n = 0. (d) n = 1.0. (After Guo, W. D., *Soils and Foundations* 41, 2, 2001c.)

The figure indicates the existence of a maximum value of the displacement, w_t/w_e [written as $(w_t/w_e)_{max}$]. As the displacement, w_t/w_e, reaches $(w_t/w_e)_{max}$, it will increase indefinitely, although the capacity ratio, n_p, will stay at a constant (written as n_w). This is not shown in the figure, but instead the unloading curve is given, which indicates the w_t/w_e returns to unity upon a complete unloading ($P_t = 0$).

With Equation 4.58, the maximum, n_{max} (see Table 4.5), may be mathematically determined through setting the first derivative of n_p (with respect to μ) as zero:

$$\frac{dn_p}{d\mu} = (1+\theta)\left(\mu^{\theta}\xi + L\frac{d}{dL_1}\left(\frac{C_v(L_1)}{\pi_v^{1/2m}}\right)\right) \tag{4.59}$$

Table 4.5 Maximum capacity ratio (n_{max}) and degree of slip (μ_{max})

ξ	n	*0.1*[a]	*0.25*	*0.5*	*0.75*	*1.0*	*1.5*	*2.0*
0	0	< 0[b]	< 0	< 0	< 0	< 0	< 0	< 0
		.1000[c]	.2499	.4820	.6526	.7616	.8742	.9242
	0.5	.752	.514	.271	.141	.075	.024	.009
		.1397	.3176	.5417	.6898	.7851	.8857	.9312
	1.0	.781	.578	.367	.243	.168	.091	.055
		.1768	.3802	.6053	.7389	.8196	.9030	.9408
0.25	0	.868	.671	.342	.012	< 0	< 0	< 0
		.3037	.3842	.5184	.6526	.7616	.8742	.9242
	0.5	.870	.686	.422	.238	.131	.042	.017
		.3284	.4373	.5921	.7099	.7930	.8872	.9316
	1.0	.873	.698	.47	.318	.222	.121	.074
		.3518	.4834	.6488	.7583	.8290	.9057	.9419
0.5	0	.912	.780	.559	.339	.119	< 0	< 0
		.5266	.5666	.6332	.6998	.7664	.8742	.9242
	0.5	.913	.782	.576	.394	.250	.092	.037
		.5393	.5954	.6807	.7537	.8128	.8912	.9326
	1.0	.913	.785	.592	.434	.317	.178	.110
		.5519	.6217	.7193	.7933	.8467	.9111	.9439
0.75	0	.945	.863	.725	.588	.451	.176	< 0
		.7588	.7720	.7940	.8160	.8380	.8820	.9242
	0.5	.945	.863	.728	.597	.473	.264	.131
		.7631	.7821	.8123	.8402	.8657	.9078	.9348
	1.0	.945	.864	.731	.607	.495	.320	.210
		.7672	.7917	.8285	.8601	.8867	.9255	.9497

Source: Guo, W. D., *Soils and Foundations*, 41, 2, 2001c.

[a] $1/\pi_v^{1/2m}$
[b] Numerator: μ_{max}
[c] denominator: n_{max}; and μ_{max} taken as if < 0

With n_p of Equation 4.58, the pile-head displacement of Equation 4.55 may be rewritten as

$$\frac{w_t}{w_e} = 1 + \frac{\xi \pi_v^{1/m}}{(2+\theta)} \left(\frac{L_1}{L} \right)^{2+\theta} + \pi_v^{1/2m} \left(\frac{L_1}{L} \right) C_v(L_1) \tag{4.60}$$

As with that for obtaining n_{max}, Equation 4.60 allows maximum values of the displacement ratio, $(w_t/w_e)_{max}$, to be gained through setting the first derivative of w_t/w_e (with respect to μ) as zero:

$$\pi_v^{1/2m}\left(\frac{L_1}{L}\right)^{1+\theta}\xi+C_v(L_1)+L_1\frac{dC_v(L_1)}{dL_1}=0 \tag{4.61}$$

Two possible values of the capacity ratio are seen in Figure 4.17c and d for a given pile-head displacement (w_t/w_e). The displacement ratio, $(w_t/w_e)_{max}$, was thus obtained for various values of pile–soil stiffness and softening factor and is detailed in Table 4.6 together with the degree of slip, μ_w, and the capacity ratio, n_w, at which the displacement ratio occurs. Generally, the critical displacement reduces as the softening becomes more severe (ξ gets smaller).

Example 4.4 Use of equation 4.58 to calculate n_p

Equation 4.58 states that as L_1 ($R_b = 0$) approaches the pile length, n_p reduces to ξ. As L_1 ($R_b = 0$) approaches zero, Equation 4.58 reduces to Equation 4.52, n_p becomes the ratio of the limiting load (P_e) divided by the ultimate load (P_{ult}), P_e/P_{ult}. Given $n = \theta = 0$, Equation 4.58 reduces to Equation 4.62 given previously by Murff (1980),

$$n_p=\xi\left(\frac{L_1}{L}\right)+\frac{1}{\pi_v}\frac{R_b/\pi_v+\tanh[(1-L_1/L)\pi_v]}{1+(R_b/\pi_v)\tanh[(1-L_1/L)\pi_v]} \tag{4.62}$$

Given $L_1 = 0$, and replacing the stiffness, π_v, with an equivalent stiffness, $\pi_v^{1+n/2}$, in Equation 4.62, the n_p for the nonhomogeneous case may be approximated by

$$n_p=\frac{1}{\pi_v^{1/2m}}\frac{R_b/\pi_v^{1/2m}+\tanh(\pi_v^{1/2m})}{1+(R_b/\pi_v^{1/2m})\tanh(\pi_v^{1/2m})} \tag{4.63}$$

The validity of Equation 4.63 is discussed in Example 4.6.

Example 4.5 Use of equation 4.59 to estimate n_{max}

The n_{max} for various values of $1/\pi_v$, at $n = 0$ and 1 ($R_b = 0$, $\xi = 0.75$), was estimated using Equation 4.59 and is illustrated in Figure 4.18. It forms the envelope lines of the capacity ratio gained using Equation 4.58 for various degrees of slip μ. The figure indicates that the capacity ratio (n_p) generally increases at lower values of $1/\pi_v$ (< 1) for "flexible" piles, with evolution of the slip until the n_p attains the n_{max}. Subsequently, the n_p generally decreases, especially for higher values of $1/\pi_v$ (i.e., "rigid" piles), until $n_p = \xi$ at $\mu = 1$. The n_{max} attains at higher degrees of slip for lower values of $1/\pi_v$ ("flexible" piles), but at approximately zero degrees of slip for high values of $1/\pi_v$ ("rigid" piles). The n_{max} is generally lower for flexible piles.

Capacity of offshore (normally "flexible") piles may reduce due to strain softening and degradation of soil strength and stiffness upon cyclic loading.

Table 4.6 Capacity ratio (n_w), degree of slip (μ_w) at maximum $(w_t/w_e)_{max}$

ξ	n	*0.1*[a]	*0.25*	*0.5*	*0.75*	*1.0*	*1.5*	*2.0*
0	0	8.766[b]	3.334	1.797	1.394	1.232	1.107	1.061
		.825[c]	.698	.598	.556	.535	.517	.510
		.094[d]	.209	.333	.399	.434	.467	.481
	0.5	8.760	3.330	1.719	1.350	1.205	1.094	1.053
		.833	.714	.625	.589	.571	.557	.551
		.093	.204	.431	.502	.538	.569	.582
	1.0	8.211	3.034	1.655	1.315	1.183	1.084	1.048
		.840	.728	.647	.615	.601	.588	.584
		.172	.349	.506	.576	.610	.639	.651
0.25	0	18.028	4.422	2.000	1.472	1.273	1.124	1.071
		.887	.773	.675	.631	.609	.590	.582
		.303	.373	.455	.500	.525	.548	.557
	0.5	15.421	3.948	1.859	1.404	1.233	1.106	1.06
		.888	.780	.691	.653	.635	.620	.613
		.327	.423	.524	.574	.601	.624	.634
	1.0	13.650	3.614	1.758	1.355	1.204	1.092	1.052
		.890	.786	.705	.672	.657	.644	.639
		.350	.467	.579	.631	.657	.679	.688
0.5	0	28.270	5.721	2.256	1.573	1.326	1.147	1.083
		.922	.837	.758	.720	.700	.683	.676
		.526	.562	.604	.628	.641	.654	.659
	0.5	23.233	4.894	2.040	1.474	1.272	1.121	1.069
		.825	.840	.766	.732	.716	.703	.698
		.094	.589	.643	.672	.687	.700	.706
	1.0	19.863	4.335	1.893	1.407	1.232	1.104	1.059
		.922	.842	.774	.744	.730	.718	.713
		.551	.613	.677	.708	.723	.737	.743
0.75	0	39.235	7.228	2.579	1.706	1.398	1.177	1.10
		.952	.900	.853	.831	.820	.810	.806
		.759	.770	.783	.790	.793	.796	.798
	0.5	31.727	6.022	2.276	1.571	1.322	1.143	1.081
		.952	.901	.856	.853	.826	.818	.814
		.763	.779	.797	.805	.810	.814	.816
	1.0	26.72	5.217	2.073	1.480	1.271	1.121	1.068
		.952	.902	.859	.840	.831	.824	.821
		.767	.788	.809	.819	.825	.829	.831

continued

Table 4.6 (Continued) Capacity ratio (n_w), degree of slip (μ_w) at maximum $(w_t/w_e)_{max}$

Source: Guo, W. D., *Soils and Foundations*, 41, 2, 2001c.

a $1/\pi_v^{1/2m}$

b $(w_t/w_e)_{max}$

c μ_w,

d n_w at $(w_t/w_e)_{max}$

To avoid the reduction, the degree of slip may be controlled (Randolph 1983). By specifying a degree of slip (μ) (via depth of softening layer), the design value of $1/\pi_v$ (thus, pile dimensions) may be taken as that ensuing n_{max}, since further increases in rigidity via $1/\pi_v$ (μ = constant) render little increase in the pile capacity. Under a given n_p, a lower degree of slip may be induced for a flexible pile (low $1/\pi_v$). For instance, as is evident in Figure 4.18, given $\mu = 0.75$ ($\xi = 0.75$, $R_b = 0$), n_p reaches the n_{max} of 0.794, and 0.828 at values of $1/\pi_v$ of ~0.5, 0.63 for $n = 0$, and 1, respectively. With $1/\pi_v$ increased further to 2.0, the n_{max} exhibits 2.2% and 3.5% increases for $n = 0$ and 1, respectively. With a lower degree of slip of $\mu = 0.5$, the n_{max} then occurs at higher values of $1/\pi_v \approx 0.9$ ($n = 0$) and 1.1 (1.0), respectively.

With Equation 4.59, the influence of the softening factor, ξ, on the maximum ratio of P_t/P_{fs} (n_{max}) was obtained for various values of relative stiffness π_v and is shown in Figure 4.19. The figure illustrates that the capacity ratio, n_{max}, is the ratio of P_e/P_{ult} for $\xi = 0$ and is less affected by the strain-softening factor, at high values of $1/\pi_v$ ("rigid" piles). These capacity ratios of n_{max} occur at the degrees of slip, μ_{max} shown in Figure 4.20 as also tabulated in Table 4.5.

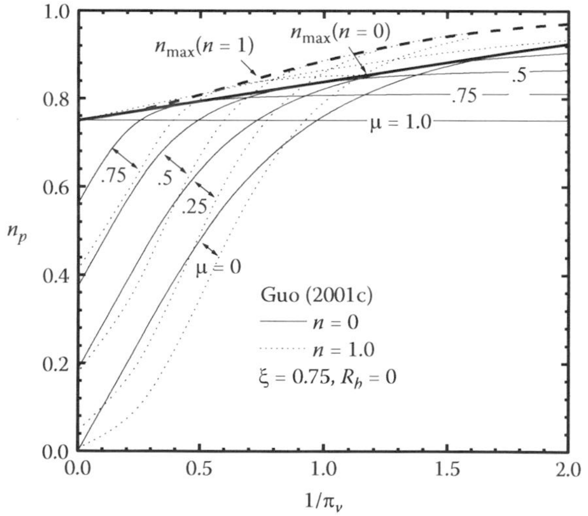

Figure 4.18 Development of load ratio as slip develops. (After Guo, W. D., *Soils and Foundations* 41, 2, 2001c.)

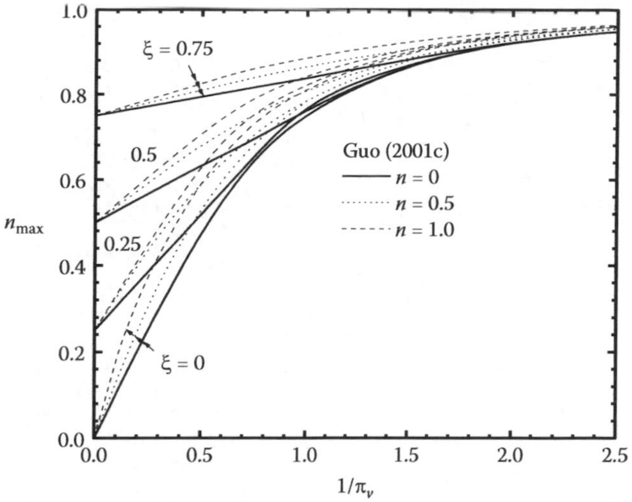

Figure 4.19 n_{max} versus π_v relationship (ξ as shown and R_b = 0). (After Guo, W. D., *Soils and Foundations* 41, 2, 2001c.)

Example 4.6 μ_{max} for $n = 0$

At $n = 0$, the μ_{max} may be simply given by (Murff 1980)

$$\mu_{max} = 1 + \frac{1}{2\pi_v}\ln\left(\frac{R_b + \pi_v}{R_b + \pi_v}\left(1 - \frac{2}{\xi} + \frac{2}{\xi}\sqrt{1-\xi}\right)\right) \quad (4.64)$$

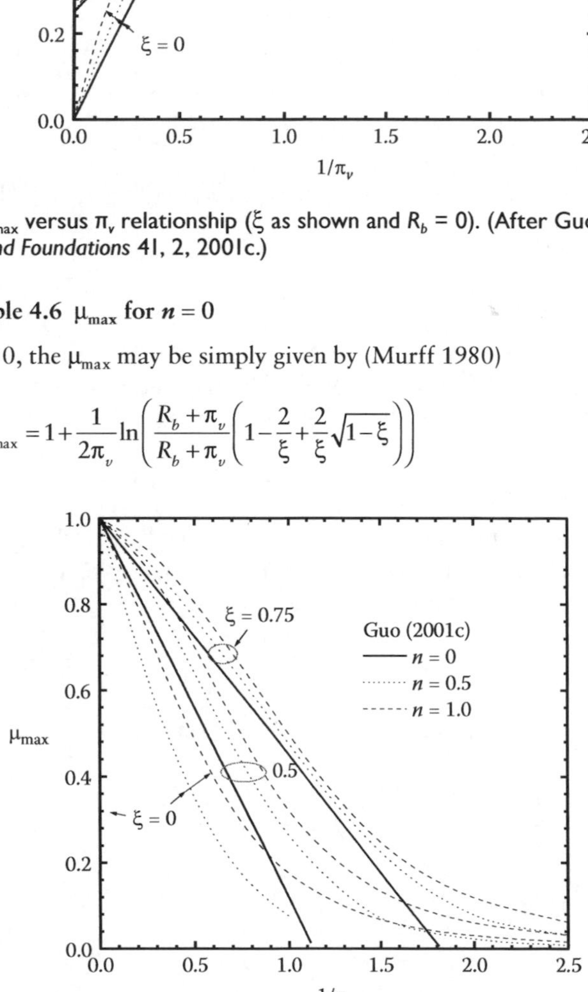

Figure 4.20 μ_{max} versus π_v relationship (ξ as shown and R_b = 0). (After Guo, W. D., *Soils and Foundations* 41, 2, 2001c.)

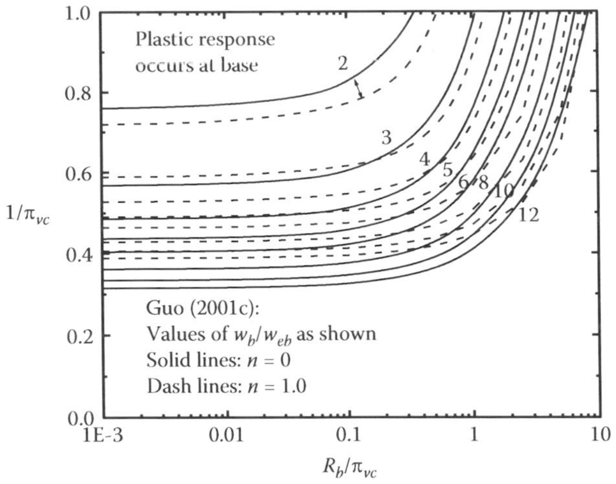

Figure 4.21 Critical stiffness for identifying initiation of plastic response at the base prior to that occurring at ground level. (After Guo, W. D., *Soils and Foundations* 41, 2, 2001c.)

As the value of $1/\pi_v$ increases (toward rigid piles), the degree of slip, μ_{max} gradually reduces to zero (when $\mu_{max} < 0$, μ_{max} is taken as zero). Thus, Equation 4.63 compares well with Equation 4.58 for $n = 0$ and $n = 1$ (not shown herein, Guo 2001c).

Finally, it should be stressed that all the solutions are intended for the initiating of slip from ground level. In rare case, the base slip occurs prior to the slip from ground level, which occurs when $\pi_v < \pi_{vc}$ (π_{vc} = a critical value of π_v), for which the current solutions are not applicable. As shown in Figure 4.21, the critical ratio π_{vc} depends on the base limiting displacement w_{eb} ($= \tau_{fb} r_o \zeta / G_b$).

4.9 CAPACITY AND CYCLIC AMPLITUDE

The normalized limiting load, P_e, by the P_{fs} at the elastic-plastic transition depth, L_1, is determined from Equation 4.52 as

$$\frac{P_e}{P_{fs}} = \frac{1+\theta}{\pi_v^{1/2m}} C_v(L_1) \tag{4.65}$$

As mentioned previously, mobilization of pile capacity is dominated by pile–soil relative slip and strain-softening properties of the soil. In the circumstance where piles are in a homogenous soil with a lower softening factor (e.g., $\xi < 0.5$; Figure 4.17a), no slip is allowed even at the ground surface (Randolph 1983). In contrast, with a nonhomogeneous soil profile, a small, upper portion of slip (see Figure 4.17b) is beneficial to increase pile capacity.

Under a cyclic loading, the limiting shaft stress τ_f on the pile reduces to the residue skin friction, $\xi\tau_f$, at a lower limiting local shaft stress, τ_f^c (Randolph 1988a; Randolph 1988b)

$$\tau_f^c = \alpha_c \tau_f \tag{4.66}$$

where $\alpha_c = 0.5(1 + \xi_c)$ under one-way cyclic loading between zero and τ_f^c; $\alpha_c = (1 + \xi_c)/(3 - \xi_c)$ under two-way (symmetric) cyclic loading between $-\tau_f^c$ and τ_f^c. ξ_c = a yield stress ratio for cyclic loading, which may be approximately taken as ξ for the monotonic loading described before (Randolph 1988b). This limiting shaft stress, τ_f^c, renders a reduced limiting shaft displacement w_e^c

$$w_e^c = \alpha_c \frac{\tau_f r_o}{G} \zeta = \alpha_c w_e \tag{4.67}$$

Equation 4.58 is modified to gain cyclic load amplitude n_c.

$$n_c = \xi \left(\frac{L_1}{L} \right)^{1+\theta} + \frac{\alpha_c (1+\theta)}{\pi_v^{1/2m}} C_v(L_1) \tag{4.68}$$

where n_c = safe cyclic load amplitude. Figure 4.22 shows the effect of the yield stress ratio, ξ_c, on the amplitude of n_c. The impact of the discrepancy between the yield stress ratio, ξ_c, and the residue stress ratio, ξ (= 0.5), is determined as: (1) n_c increases with the stress ratio, ξ_c, at high values of $1/\pi_v$, as the slip never occurs for rigid piles, with $L_1 \approx 0$ in Equation 4.68; (2) given $\xi_c = \xi$, the difference in the values of n_c obtained from $\mu = 0$ and $\mu > 0$ (directly from Equation 4.68) is significant at low values of $1/\pi_v$, indicating shaft resistance from residue strength in the slip zone. This resistance, nevertheless, vanishes for piles in calcareous sediments, with a rather low ξ (Randolph 1988b). The effect of the residue value of ξ on the capacity ratio, n_{max}, may also be determined from Figure 4.23, apart from Figure 4.19.

Example 4.7 Capacity and cyclic amplitude

A typical offshore pile is studied here to gain capacity and cyclic amplitude. The pile had $L = 100$ m, $d = 1.5$ m, wall thickness = 50 mm, and Young's modulus $E_p = 7.037 \times 10^8$ kPa. It was installed in a soil with $A_v = 150$ kPa/m, $\xi = 0.5$, and $w_e = 0.01d$. The predicted capacity of the

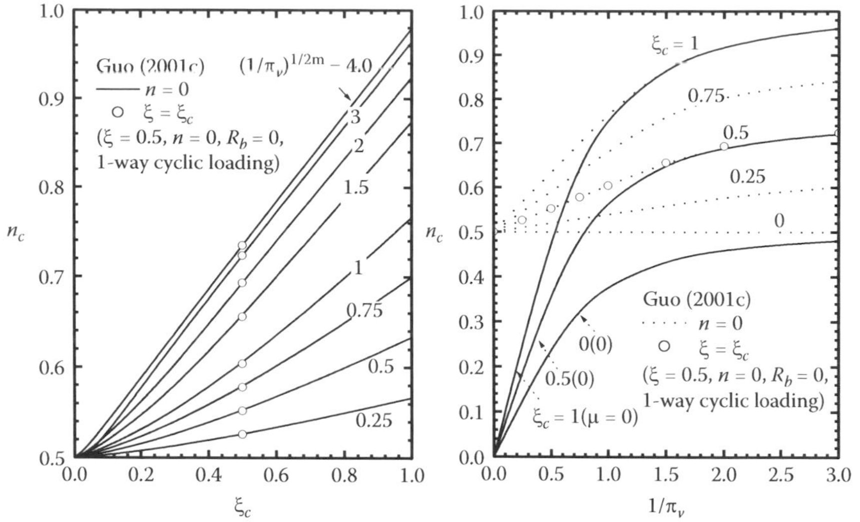

Figure 4.22 Effect of yield stress level (ξ_c) on the safe cyclic amplitude (n_c) ($\xi = 0.5$, $n = 0$, $R_b = 0$, one-way cyclic loading). (After Guo, W. D., *Soils and Foundations* 41, 2, 2001c.)

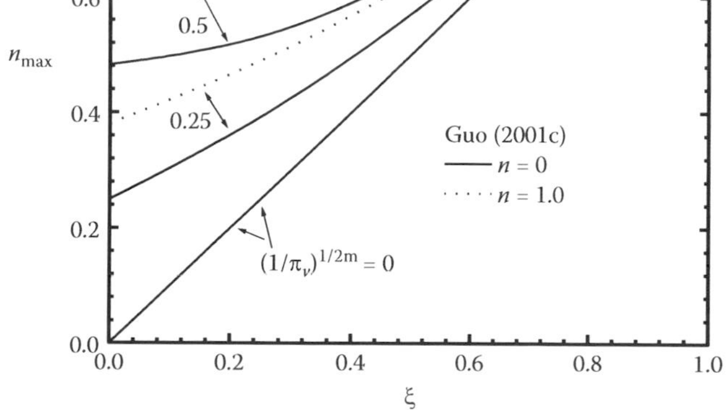

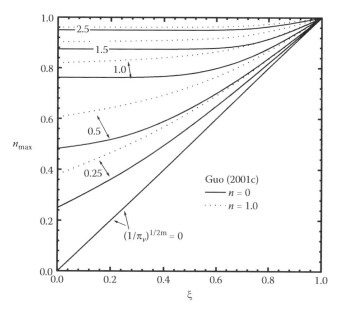

Figure 4.23 Effect of strain-softening factor on ultimate capacity ratio (n_{max}) ($R_b = 0$). (After Guo, W. D., *Soils and Foundations* 41, 2, 2001c.)

Table 4.7 Analysis of an offshore pile in strain-softening soil

n	$1/\pi_v^{1/2m}$	P_e/P_{fs}* (MN) (P_{fs} = constant)	At n_{max} State $\frac{n_{max}}{\mu_{max}}$	At n_{max} State $\frac{(w_t/w_e)}{P_{max}}$	At $(w_t/w_e)_{max}$ State n_w/μ_w	At $(w_t/w_e)_{max}$ State $(w_t/w_e)_{max}$
0	0.3194	$\frac{22.48}{70.69}$	$\frac{.5851}{.719}$	$\frac{3.856}{41.35}$	$\frac{.5757}{.809}$	3.961
0.5	0.2608	$\frac{16.97}{70.69}$	$\frac{.5993}{.773}$	$\frac{4.512}{42.17}$	$\frac{.5917}{.835}$	4.594
1.0	0.2259	$\frac{14.14}{70.69}$	$\frac{.6111}{.805}$	$\frac{4.968}{42.35}$	$\frac{.6045}{.853}$	5.037

Source: Guo, W. D., *Soils and Foundations*, 41, 2, 2001c.

pile is provided in Tables 4.7 and 4.8 for three typical soil profiles. As the factor n increases, maintaining ultimate pile capacity, P_{fs}, as 70.69 MN would reduce the ratio n_{max} by only ~4.5% (see Table 4.7), but reduce the limiting load, P_e at ground level by ~37%. Maintaining a constant stiffness of 0.3194 would increase the ratio n_{max} by ~11.3% (see Table 4.8), reduce the limiting load P_e by ~51%, and alter slightly the ratio P_e/P_{fs}.

The capacity of 41.35 MN (at $n = 0$) obtained here is lower than 52 MN based on a gradually softening soil model (Randolph 1983). The limiting load, P_e (for initiating slip at ground level, $L_1 = 0$), at $n = 1$ is about half of that at $n = 0$. As degradation of pile capacity under cyclic loading is less severe at $n = 1$ than at $n = 0$ (Poulos 1981). It appears appropriate to allow a limiting load for the case of $n > 0$ not less than that for $n = 0$. A slip degree μ of 0.1556 and 0.255 for $n = 0.5$ and 1, respectively, may be allowed to reach a value of $P_e/P_{fs} = 0.3183$ for $n = 0$. In the case of two-way cyclic loading ($\alpha_c = 1/3$) and $\xi_c = 0$, the safe cyclic load amplitude n_c is estimated as 4.13 MN.

Table 4.8 Analysis of an offshore pile in strain-softening soil

n	$1/\pi_v^{1/2m}$ (= constant)	P_e/P_{fs} (MN)	At n_{max} state $\frac{n_{max}}{\mu_{max}}$	At n_{max} state $\frac{(w_t/w_e)}{P_{max}}$	At $(w_t/w_e)_{max}$ state n_w/μ_w	At $(w_t/w_e)_{max}$ state $(w_t/w_e)_{max}$
0	0.3194	$\frac{22.48}{70.69}$	$\frac{.5851}{.719}$	$\frac{3.856}{41.35}$	$\frac{.5757}{.809}$	3.961
0.5		$\frac{14.30}{47.12}$	$\frac{.6202}{.724}$	$\frac{3.351}{29.22}$	$\frac{.6073}{.813}$	3.447
1.0		$\frac{10.92}{35.34}$	$\frac{.6512}{.729}$	$\frac{3.01}{23.02}$	$\frac{.6355}{.817}$	3.099

Source: Guo, W. D., *Soils and Foundations*, 41, 2, 2001c.

Chapter 5

Time-dependent response of vertically loaded piles

Two types of time-dependent response of vertically loaded piles are presented in this chapter, owing to (a) visco-elastic load transfer behavior, and (b) the variations of limiting shaft friction and soil stiffness with reconsolidation.

5.1 VISCO-ELASTIC LOAD TRANSFER BEHAVIOR

Piles are installed largely to reduce settlement of a building or a structure. Two kinds of settlement-based design methodologies (Randolph 1994) are *creep piling* and *optimizing pile-raft analysis*. The creep piling approach (Hanbo and Källström 1983) allows each pile to operate at a working load of about 70–80% of its ultimate bearing capacity. The optimising pile-raft analysis (Clancy and Randolph 1993) is proposed to control the differential settlement of a foundation by reducing the number of piles and concentrating piles in the middle of a foundation. Either methodology will generally lead to piles operating at a high load level at which creep can induce significant pile-head movement at constant load, and even a gradual reduction in shaft capacity.

Creep or viscoelastic response of the soil leads to variations in stiffness and capacity depending on the time-scale of loading. The loading history may be a major concern for predicting settlement of a pile foundation when using different time-scale loading tests, including maintained loading tests and constant rate loading tests. The tests may be simulated sufficiently accurately by one-step loading and ramp (linear increase followed by sustained) loading, respectively.

Numerical solutions for axial pile response, based on elasticity, have incorporated the impact of nonhomogeneity of the soil (Banerjee and Davies 1977; Poulos 1979), relative slip between pile and soil (Poulos and Davis 1980), and visco-elastic response of soil (Booker and Poulos 1976). However, the analysis can be more efficiently conducted using the unified compact closed-form solutions in the context of the load transfer approach (see Chapter 4, this book) with the visco-elastic solutions to be presented in this chapter.

The hyperbolic approach of pile analysis (Fleming 1992) was extended (England 1992) to cater to time-dependent pile response, with separate hyperbolic laws being used to describe the time dependency of the (average) shaft and base response. This phenomenological approach is limited by the difficulty in linking the parameters required for the model to fundamental and measurable properties of the soil.

Visco-elastic shaft and base load transfer models have been presented in Chapter 3, this book, for the step and ramp type loading respectively. With the models, the closed-form solutions for a pile in an elastic-plastic nonhomogeneous media (Guo and Randolph 1997a) are extended herein to account for visco-elastic response. The visco-elastic solutions are subsequently compared with the numerical analysis (Booker and Poulos 1976) for the case of one-step loading. They are used to examine the overall pile response for the two commonly encountered loading types. Finally, two example analyses are compared with measured pile responses to illustrate the validity of the proposed theory to practical applications.

5.1.1 Model and solutions

5.1.1.1 Time-dependent load transfer model

Local shaft displacement for visco-elastic soil is described in Chapter 3, this book, where it shows

$$w = \frac{\tau_o r_o}{G_1} \zeta_1 \zeta_c \tag{5.1}$$

where

$$\zeta_c = 1 + \frac{\zeta_2}{\zeta_1} \frac{G_1}{G_2} A(t) \tag{5.2}$$

If the total loading time t exceeds t_c, $A(t)$ is given by

$$A(t) = \frac{t_c}{t} \exp\left(-\frac{t-t_c}{T}\right) - \frac{T}{t}\left(\exp\left(-\frac{t-t_c}{T}\right) - \exp\left(-\frac{t}{T}\right)\right) + 1 - \exp\left(-\frac{t-t_c}{T}\right) \tag{5.3}$$

Otherwise, if $t \le t_c$, it follows that

$$A(t) = 1 - \frac{T}{t}\left(1 - \exp\left(-\frac{t}{T}\right)\right) \tag{5.4}$$

where t_c = the time at which a constant load commences. If $t_c = 0$, Equation 5.3 is reduced to one-step loading (Figure 3.13b, Chapter 3, this book), which can be simply described by

$$A(t) = 1 - exp(-t/T) \tag{5.5}$$

For this case, the creep function $J(t)$ (Booker and Poulos 1976) is inferred as

$$J(t) = A_c + B_c e^{-t/T} \tag{5.6}$$

where $A_c = 1/G_1 + \zeta_2/G_2\zeta_1$, $B_c = -\zeta_2/G_2\zeta_1$, and

$$w = \tau_o r_o \zeta_1 J(t) \tag{5.7}$$

The base load-settlement relationship is given by

$$w_b = \frac{P_b(1-\nu_s)\omega}{4r_o G_{b1}} \frac{1 + A(t)G_{b1}/G_{b2}}{(1 - R_{fb}\, P_b/P_{fb})^2} \tag{5.8}$$

5.1.1.2 Closed-form solutions

Closed-form solutions are presented in Chapter 4, this book, for a pile in an elastic-plastic nonhomogeneous soil. Under the circumstance of a constant of ζ_c with depth, these solutions can be readily extended to account for visco-elastic response of soil by simply replacing the non-linear elastic load transfer, ζ_1, with the new load transfer factor, $\zeta_c\zeta_1$, and replacing the base shear modulus, G_b, with the time-dependent $G_b(t)$ (Equation 3.39, Chapter 3, this book). This is highlighted next for load ratio, settlement influence factor, and pile-head load displacement. The load ratio of pile base and head (see Equation 4.16, Chapter 4, this book) can be predicted by

$$\frac{P_b}{P_t} = \frac{4r_o G_b(t)}{(1-\nu_s)\omega} \frac{1}{k_s E_p A_p z_t^{n/2} \left(\dfrac{z_t}{L}\right)^{1/2} \left(\dfrac{C_1(z_t) + \chi_v C_2(z_t)}{C_3(L)}\right)} \tag{5.9}$$

where z_t = depth for pile-head, taken as an infinitesimal value; and P_t = pile-head load.

$$\begin{aligned} C_1(z) &= -K_{m-1} I_{m-1}(y) + K_{m-1}(y) I_{m-1} \\ C_2(z) &= K_m I_{m-1}(y) + K_{m-1}(y) I_m \\ C_3(z) &= K_{m-1} I_m(y) + K_m(y) I_{m-1} \\ C_4(z) &= -K_m I_m(y) + K_m(y) I_m \end{aligned} \tag{5.10}$$

The modified Bessel functions $I_m(y)$, $I_{m-1}(y)$, $K_{m-1}(y)$, and $K_m(y)$ are written as I_m, I_{m-1}, K_{m-1}, and K_m at $z = L$; $m = 1/(2 + n)$. The ratio χ_v becomes

$$\chi_v = \frac{2\sqrt{2}}{\pi(1-\nu_s)\omega\xi_b}\sqrt{\frac{\zeta_c\zeta_1}{\lambda}} \tag{5.11}$$

where $\lambda = E_p/G_L$. The variable y and the stiffness factor, k_s, are given, respectively, by

$$y = 2mk_s z^{1/2m} \tag{5.12}$$

and

$$k_s = \frac{L}{r_o}\sqrt{\frac{2}{\lambda\zeta_c\zeta_1}}\left(\frac{1}{L}\right)^{1/2m} \tag{5.13}$$

The settlement influence factor, I, is recast as

$$I = \frac{G_L w_t r_o}{P_t} = \frac{1}{\pi C_v(z_t)}\sqrt{\frac{\zeta_c\zeta_1}{2\lambda}} \tag{5.14}$$

where

$$C_v(z_t) = \frac{C_1(z_t)+\chi_v C_2(z_t)}{C_3(z_t)+\chi_v C_4(z_t)}\left(\frac{z_t}{L}\right)^{n/2} \tag{5.15}$$

As the pile-head load increases, the shaft shear stress will reach the limiting shaft stress, τ_f

$$\tau_f = A_v z^{\theta} \tag{5.16}$$

where A_v, θ = a gradient, and a constant for limit shear stress distribution ($\theta = n$), respectively. This τ_f is attained at a local limiting displacement, w_e, of

$$w_e = r_o\zeta_c\zeta_1 A_v/A_g \tag{5.17}$$

In light of Equations 4.23 and 4.24 for $\alpha_g = \alpha_v = 0$ (see Chapter 4, this book), pile-head load, P_t, and settlement, w_t, can be expressed as the slip degree, $\mu = L_1/L$ (L_1 = slip length) by

$$P_t = w_e k_s E_p A_p L^{n/2} C_v(\mu L) + \pi d A_v \frac{(\mu L)^{1+\theta}}{1+\theta} \tag{5.18}$$

$$w_t = w_e[1+\mu k_s L^{n/2+1} C_v(\mu L)] + \frac{\pi d A_v}{E_p A_p}\frac{(\mu L)^{2+\theta}}{2+\theta} \tag{5.19}$$

These solutions are referred to as "Guo (2000b)" or "closed-form solutions" in the subsequent figures. Under high stress levels and/or at a higher slenderness ratio, ζ_c is no longer a constant, which invalidates Equations 5.18 and 5.19. This case may be simulated using numerical analysis (e.g., the GASPILE program) (Guo and Randolph 1997b).

5.1.1.3 Validation

Booker and Poulos (1976), through incorporating a linear visco-elastic model into Mindlin's solution, showed the impact of the following three variables on the settlement influence factor and the ratio of base and head loads: the pile–soil relative stiffness, λ; the ratio of long-term and short-term soil response, $J(\infty)/J(o)$ (Chapter 3, this book); and the nondimensional time, t/T. The numerical analysis is sufficiently rigorous, despite neglecting the effect of viscosity on the Poisson's ratio. It agrees with the closed-form prediction (Guo 2000) concerning the settlement influence factor for two different relative stiffnesses at a ratio of $J(\infty)/J(o) = 2$, see Figure 5.1a; and regarding the ratio of base and head load predicted by Equation 5.9, both with and without the base creep (see Figure 5.1b). The base creep significantly affects the load ratio, although with limited impact on the settlement influence factor as shown in Figure 5.2. At a higher ratio of $J(\infty)/J(o)$, such as 10 (i.e., $G_1/G_2 = 9$), the difference is evident between the predicted I by Equation 5.14 and the numerical solution (Booker and Poulos 1976) (Figure 5.2). This is fortunately not a concern, as the ratio of G_1/G_2 is normally lower (Lo 1961) and generally less than 5 as backfigured from a few different field tests.

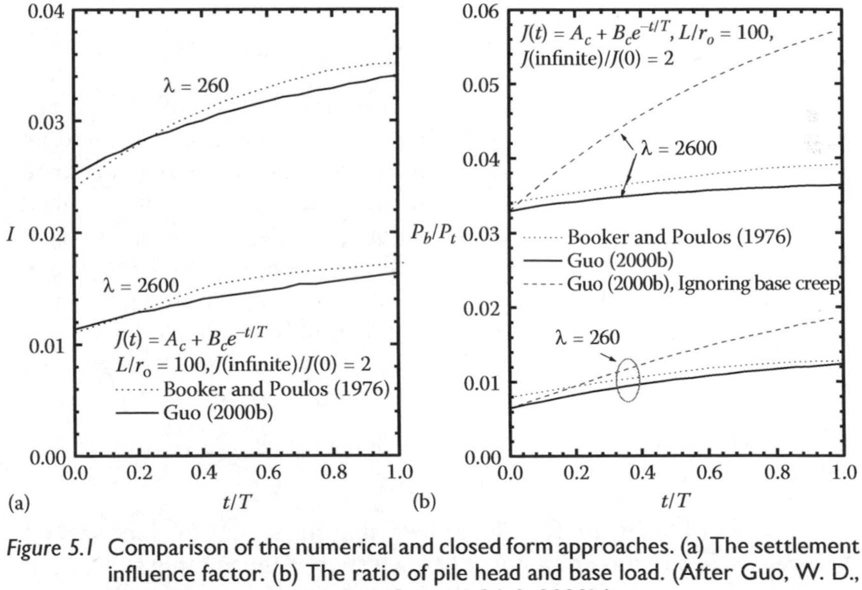

Figure 5.1 Comparison of the numerical and closed form approaches. (a) The settlement influence factor. (b) The ratio of pile head and base load. (After Guo, W. D., *Int J Numer and Anal Meth in Geomech* 24, 2, 2000b.)

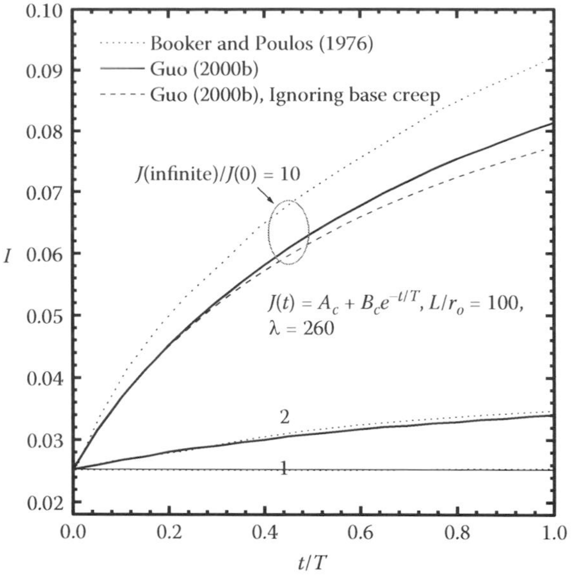

Figure 5.2 Comparison of the settlement influence factor. (After Guo, W. D., *Int J Numer and Anal Meth in Geomech* 24, 2, 2000b.)

Figure 5.3 shows the predicted pile-head load and settlement curves using Equations 5.18 and 5.19, respectively. They are consistent with the numerical prediction by GASPILE (Guo and Randolph 1997b) (see Chapter 3, this book).

5.1.2 Effect of loading rate on pile response

The impacts of one-step and ramp loading on the settlement influence factor was obtained using the closed-form solution, Equation 5.14, and are shown in Figure 5.4a and b, respectively. They demonstrate that step loading is associated with a larger settlement. Increasing the time t_c (hence reducing the loading rate) can reduce secondary pile settlement to a large extent. Likewise, step loading induces a slightly higher proportion of base load over the head load (as per Equation 5.9), in comparison with that for the ramp loading, as is evident in Figure 5.5a and b.

5.1.3 Applications

Two kinds of time-dependent loading tests on piles are frequently conducted in practice:

1. A series of loading tests are performed at different time intervals following installation of a pile. For each step of the loading tests, a sufficient time is given.

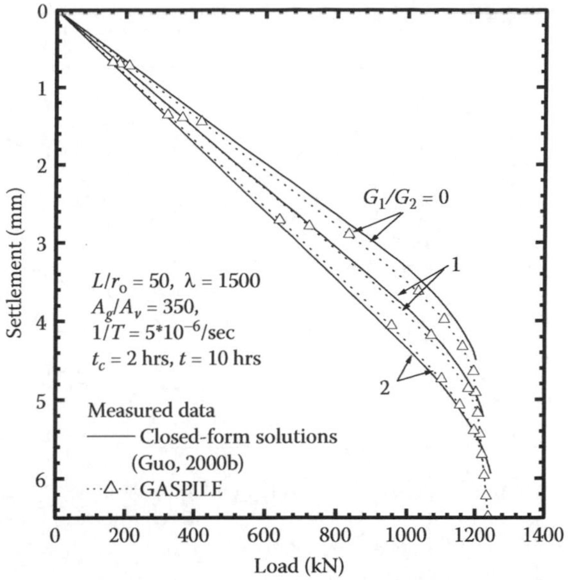

Figure 5.3 Comparison between closed-form solutions (Guo 2000b) and GASPILE analyses for different values of creep parameter G_1/G_2. (After Guo, W. D., *Int J Numer and Anal Meth in Geomech* 24, 2, 2000b.)

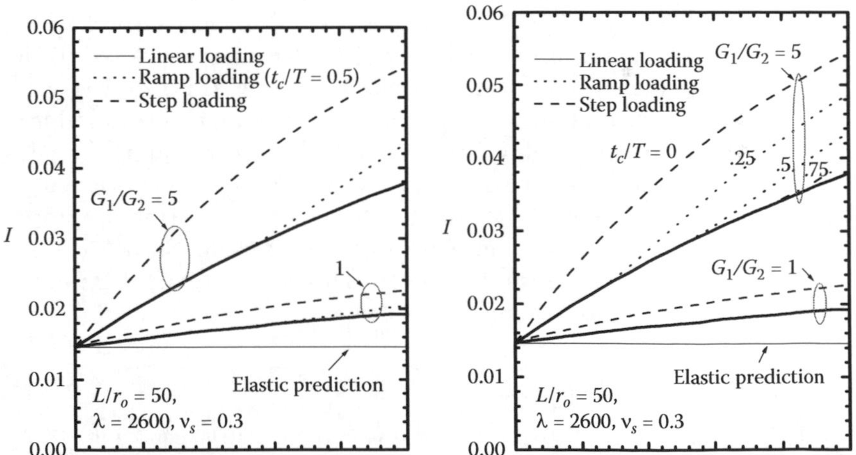

Figure 5.4 Loading time t_c/T versus settlement influence factor. (a) Different loading. (b) Influence of relative ratio of t_c/T. (After Guo, W. D., *Int J Numer and Anal Meth in Geomech* 24, 2, 2000b.)

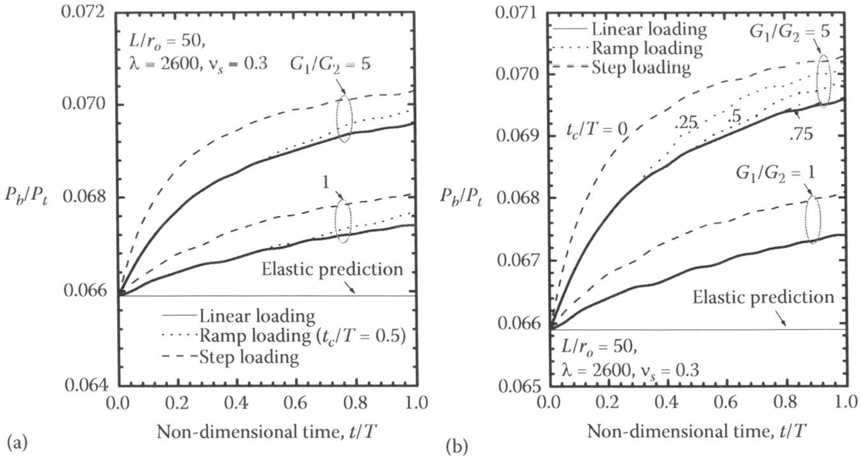

Figure 5.5 Loading time t_c/T versus ratio of P_b/P_t. (a) Comparison among three different loading cases. (b) Influence of relative t_c/T. (After Guo, W. D., *Int J Numer and Anal Meth in Geomech* 24, 2, 2000b.)

2. Only one loading test is performed after full reconsolidation of the destructed soil around the pile. Each step of loading is maintained at a specified interval of time.

The first kind of test may capture the recovery of the soil strength (modulus) with reconsolidation (see examples in section 5.2, Visco-Elastic Consolidation). The second kind of test reflects purely the pile response due to time-dependent loading and may be simulated by either the closed-form solutions of Equations 5.18 or 5.19 or the numeric GASPILE program. Normally, if the test time for each step loading is less than that required for a 90% degree of consolidation t_{90} of the soil, the soil surrounding the pile may behave elastically, otherwise the extra deformation mainly originates from visco-elastic response. Unfortunately, the current criterion for stopping each step of loading uses settlement rate (e.g., Maintained Loading Test) rather than degree of consolidation t_{90}. This criterion is not helpful to classify the consolidation and creep settlements.

Example 5.1 Tests reported by Konrad and Roy

Konrad and Roy (1987) reported the results of an instrumented pile loaded to failure at intervals after driving. The closed-ended steel pipe pile was 0.219 m in outside radius, 8.0 mm in wall thickness, with a Young's modulus of 2.07×10^5 MPa and a cross-sectional area of 53.03 cm^2 (an equivalent pile modulus of 29,663 MPa). The pile was jacked vertically to a depth of 7.6 m. The test site consists of topsoil (0.4 m in thickness); weathered clay crust (1.2 m); soft silty, marine clay (8.2 m); very soft clayey silt (4.0 m); and a deep layer of dense sand extending from a depth z of 13.7 m to more than 25 m. The undrained shear strength, s_u, increased approximately nearly linearly from 18 kPa (at z = 1.8 m) to 28 kPa (at z = 9 m). The pile was loaded to failure in 10 to 15 increments of 6.67 kN. Each increment was maintained for a period of 15 min. The immediate elastic response (t = 0) measured at different days is plotted in Figure 5.6a, and the total settlement in Figure 5.6b. Assuming a soil shear modulus of 260 s_u, and $G/\tau_f = 270$ (see Table 5.1), the load-settlement relationship was predicted using GASPILE and the closed-form solutions as shown in Figure 5.6a and b. Figure 5.6a indicates, at a load level higher than about 70%, a nonlinear relationship between the initial load and settlement prevails with increasing curvature as failure approaches. This nonlinearity principally reflects the effect of the base nonlinearity, since by simply using a nonlinear base model ($R_{fb} = 0.95$ in Equation 5.8), an excellent prediction using GASPILE is achieved. Time-dependent creep settlements for the test at 33 days after completion of the driving were predicted by the visco-elastic analysis, with $G_1/G_2 = 2$. As shown in Figure 5.6b, the current closed-form solutions agree with those measured at a number of time intervals, 0, 15, and 90 minutes to a load of ~40 kN. Nevertheless, at higher load levels, the factor ζ_2 is no longer a constant as adopted in the prediction, or else the effect of the base nonlinearity become evident. Both factors render the divergence between the closed form solution and measured data.

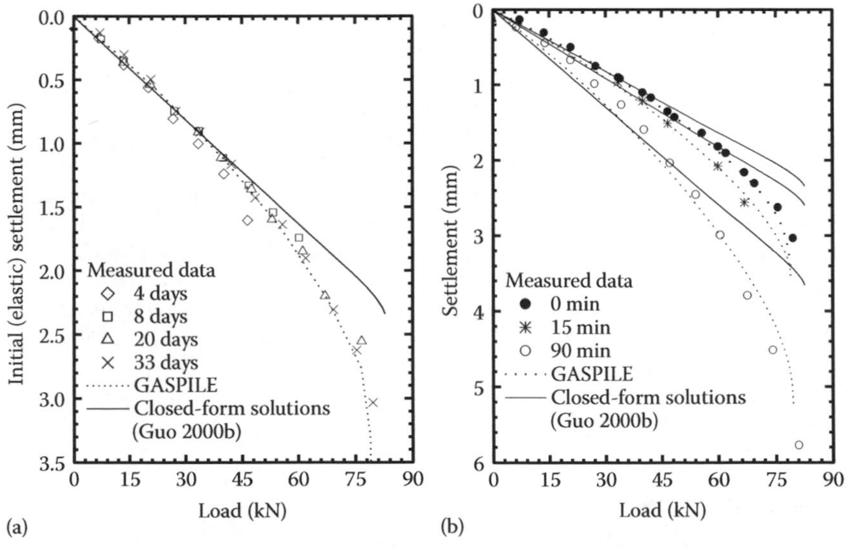

Figure 5.6 Comparison between the measured and predicted load and settlement relationship. (a) Initial settlement. (b) Total settlement for the tests (33 days) (Konrad and Roy 1987). (After Guo, W. D., *Int J Numer and Anal Meth in Geomech* 24, 2, 2000b.)

Table 5.1 Parameters for creep analysis

Examples	G_1/s_u	G_1/τ_{f1}	ζ_1	ζ_c	G_1/G_2	*Note*
5.1	260	270	4.60	1./1.13/1.666 for 0/15/90 minutes	___	ω = 1.0, ξ_b = 1.0,
5.2	47.5	80	6.27	1.025	0.025	ν_s = 0.4

Source: Guo, W. D., *Int J Numer and Anal Meth in Geomech*, 24, 2, 2000b.

Note: τ_{f2}/τ_{f1} was taken as unity; ψ_{oj} was taken as 0.5.

Example 5.2: Visco-elastic behavior under compressive loading

Bergdahl and Hult (1981) conducted tests on two driven wooden piles in a site (mainly of postglacial organic clay) about 20 km west of Stockholm. The undrained shear strength increased almost linearly from 9 kPa at a depth z of 4~5 m to 25 kPa at z = 14 m. The square piles B_1 and B_2 were 100 mm on each side and 15 m in length. The creep behavior was monitored by maintained load tests, with the load increased every 15 minutes in steps of 1/16 of the estimated bearing capacity of the pile. The measured values of load-settlement and load-creep settlement are plotted in Figure 5.7a and c, respectively. The two piles gave consistent results, therefore only pile B_1 is analysed herein. The analysis adopts a Young's modulus of the piles of 10^4 MPa, and these parameters tabulated in Table 5.1. In particular, an equivalent shear modulus distribution of $\tilde{G}$ = 755.6 kPa, n = 0.75 is employed in the closed-form predictions. The nonlinear elastic-plastic predictions (using ψ_{oj} = 0.5 in estimating ζ_i) of load-settlements by numeric GASPILE program and the closed-form solutions (Chapter 4, this book) are compared with the measured data in Figure 5.7a. Nonlinear visco-elastic (NLVE) and nonlinear elastic (NLE) predictions were made, respectively, using ψ_{oj} = 0.5 in estimating the factor ζ_j, and ψ_{oj} = 0 in estimating ζ_1 (see Equation 3.10, Chapter 3, this book). Their difference is the creep displacement, which is shown in Figure 5.7c, together with the measured creep displacement. The corresponding load distribution down the pile is illustrated in Figure 5.7b. In this instance, the secondary deformation due to the viscosity of the soil is sufficiently accurately predicted by a visco-elastic analysis over a loading level of 75~85% of the ultimate bearing capacity (determined by constant rate of penetration [CRP]test). Afterwards, considerable creep occurs as shown in the tests, which implies failure of spring 2 (Figure 3.13a, Chapter 3, this book), and may be accounted for by taking τ_{f2}/τ_{f1} = 0.75~0.85.

5.1.4 Summary

The shaft and base pile–soil interaction models capture well the nonlinear visco-elastic soil behavior at any stress levels. Capitalized on the models, the closed-form solutions or the GASPILE program well model the overall pile response under one-step and ramp-type loading. Numerical analysis (e.g.,

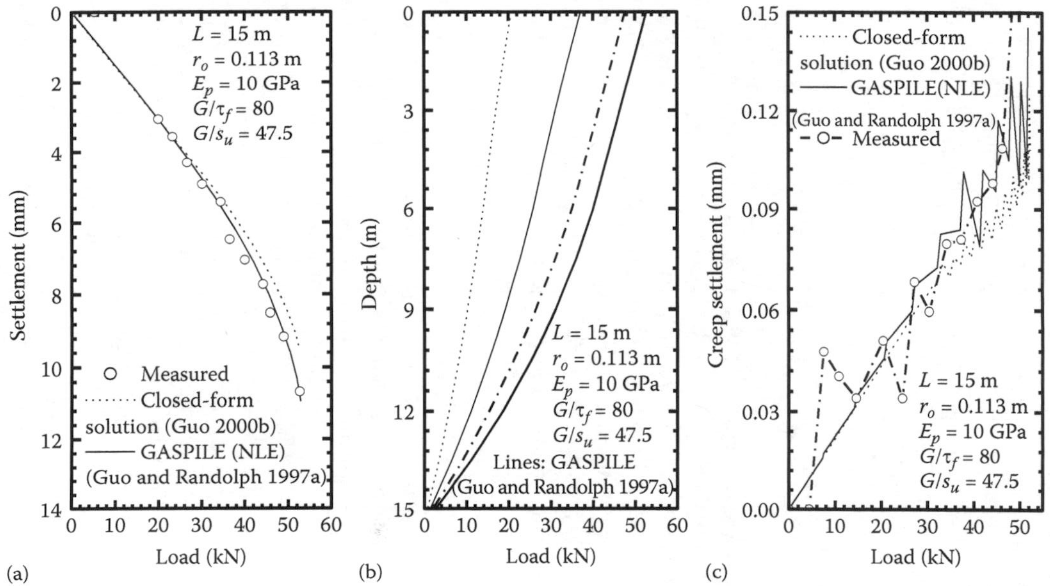

Figure 5.7 Creep response of pile B1. (a) Load settlement. (b) Load distribution. (c) Creep settlement (measured from Bergdahl and Hult 1981). (After Guo, W. D., *Int J Numer and Anal Meth in Geomech* 24, 2, 2000b.)

GASPILE analysis) is sufficiently accurate to predict creep behavior, while the closed-form solutions are generally valid to normal working loads of ~70% of ultimate load level by adopting a constant load transfer factor, ζ_c. Input values of the shear modulus, and failure shear stress with depth, might be simply obtained from empirical formulas or, more accurately, by field tests. A suitable control of the ramp type loading can avoid excessive secondary settlement.

The solutions hitherto are limited to invariable values of limiting stress (τ_f) and shear modulus G. To allow for the variation of load-displacement response owing to reconsolidation, the variations of τ_f and G are modeled next.

5.2 VISCO-ELASTIC CONSOLIDATION

Driving a pile into ground generally will remold the surrounding soil. This will alter the pile capacity (see Chapter 2, this book). In particular, some loss in the strength and an increase in pore water pressure are observed in the adjacent clay around the pile. Subsequent to driving, a gradual increase in the bearing capacity of the pile in clay is seen (Seed and Reese 1955), with increase in strength (e.g., s_u) with time, until a final soil strength equal to or greater than the initial value (Orrje and Broms 1967; Flaate 1972; Fellenius and Samson 1976; Bozozuk et al. 1978).

The maximum pore pressure occurs immediately following driving. Koizumi and Ito (1967) and Flaate (1972) demonstrate that the pressure may approximately equal or exceed the total overburden pressure in over-consolidated soil. It decreases rapidly with distance from the pile wall and becomes negligible at a distance of 5 to 10 pile diameters. This distribution is well simulated using the (one-dimensional) cylindrical cavity expansion analogy (Randolph and Wroth 1979a) and the strain path method (Baligh 1985). The former analogy with sufficient accuracy against the latter has been extended to visco-elastic soil (Guo 2000c).

Soderberg (1962) and Randolph and Wroth (1979a), in light of a radial consolidation theory, show the consistency between the rates of development of pile capacity in soft clay and pore-pressure dissipation. The prediction of time-dependent capacity for an impervious pile may thus be scaled from dissipation of the excess pore pressures. The dissipation also causes a gradual raise in the stiffness of the surrounding soil (Eide et al. 1961; Flaate 1972; Flaate and Selnes 1977; Bergdahl and Hult 1981; Trenter and Burt 1981). This raise needs to be incorporated to design settlement reduction piles (Olson 1990; Fleming 1992; Randolph 1994), apart from the time-dependent bearing capacity following pile installation.

Consolidation problems have generally been modeled using diffusion theory (Terzaghi 1943; Rendulic 1936) and elastic theory (Biot 1941). The diffusion theory is generally less rigorous than the elastic theory, as it stipulates a constant of mean total stress and may result in a different value of coefficient of consolidation from rigorous elastic theory (Murray 1978). However, the diffusion theory is mathematically much simpler to apply and is readily extended to cater to the impact of soil visco-elasticity and soil shear modulus nonhomogeneity. A coefficient from elastic theory may be used to replace that in the diffusion theory, allowing the diffusion solution to be converted into a rigorous solution.

The behavior of a driven pile is characterized by generation of the pore water pressure during installation of a pile, resembling the cylindrical cavity expansion and the radial reconsolidation subsequent to pile installation (Randolph and Wroth 1979a). This behavior has been well modeled by elastic theory (Randolph and Wroth 1979a). However, as noted earlier, the viscosity effect may become appreciable on load-settlement response. This is incorporated into elastic solutions via establishing (Guo 2000b): (1) a volumetric strain and excess pore water pressure relationship; (2) a governing equation for radial reconsolidation of a visco-elastic medium, for which general solution is established; and (3) solutions for radial consolidation for a given logarithmic variation of initial pore pressure. The visco-elastic solutions are elaborated next, together with three examples.

5.2.1 Governing equation for reconsolidations

The model on effect of driving a pile into clay using expansion theory is underpinned by plain strain expansion of a long cylindrical cavity under undrained conditions in an ideal elastic, perfectly plastic material, characterized by the shear moduli and the undrained shear strength. The plane strain expansion is verified experimentally for the middle portion of the pile (Clark and Meyerhof 1972). The soil properties and the stress state immediately following pile driving have been simplified and illustrated in Figure 3.17 (Chapter 3, this book), which offer the initial stress and boundary conditions for the reconsolidation investigated herein.

5.2.1.1 Visco-elastic stress-strain model

Reconsolidation of a visco-elastic soil is simulated using the model illustrated in Figure 3.17b (also called Mechant's model). Under a prolonged constant loading, the model offers a creep compliance, $F(t)$, of (Booker and Small 1977)

$$F(t) = \frac{1}{G_{\gamma 1}}\{1 + m_2[1 - \exp(-t / T_2)]\} \tag{5.20}$$

Table 5.2 Summary of relaxation factors for creep analysis

Authors	*Lo (1961)*	*Lo (1961)*	*Qian et al. (1992)*	*Ramalho Ortigão and Randolph (1983)*
$G_{\gamma 2}/\eta_{\gamma 2}$ ($\times$ $10^{-5}s^{-1}$)	0.2~0.4	0.5~2.67	1.71~3.29	0.36~0.664
Description	Odometer test	Creep test on model piles	Vacuum preloading	Field pile test

Source: Guo, W. D., *Computers and Geotechnics*, 26, 2, 2000c.

where $m_2 = G_{\gamma 1}/G_{\gamma 2}$; $1/T_2 = G_{\gamma 2}/\eta_{\gamma 2}$; $\eta_{\gamma 2}$ = the shear viscosity at visco element 2; $G_{\gamma j}$ = an average shear modulus, at the outset of the reconsolidation, for each of the elastic spring, j for the concerned domain. More generally, in the model shown in Figure 3.17, a Voigt element may be added to simulate some special soil behavior (Lo 1961; Guo 1997). It must stress that prior to the reconsolidation (i.e., just after installation of a pile), the soil parameters should correspond to large deformation (strain γ_j) level, as indicated by the subscript "γ_j" for spring j; during the process of reconsolidation, the modulus, $G_{\gamma j}$, and shear viscosity, $\eta_{\gamma 2}$, remain constants; and subsequent to completion of the reconsolidation, the soil parameters for modeling pile loading test should account for the effect of the time-dependent stress-strain relationship as discussed previously.

The magnitude of the relaxation time, T_2, is provided in Table 5.2, as collected from relevant publications (Ramalho Ortigão and Randolph 1983; Edil and Mochtar 1988; Qian et al. 1992). Lo (1961), through model tests, showed that the rate factor, $G_{\gamma 2}/\eta_{\gamma 2}$, is generally a constant for a given type of clay; the compressibility index ratio, $G_{\gamma 1}/G_{\gamma 2}$, generally lies between 0.05 and 0.2 (depending on water content), except for a soil of loose structure; and the individual values of $G_{\gamma 1}$, $G_{\gamma 2}$, and $\eta_{\gamma 2}$ vary with load (stress) level. Clark and Meyerhof (1972) conducted field tests and showed that during a loading test, the change in pore water pressure along the shaft of a pile is insignificant and the magnitudes of the total and effective radial stress surrounding a pile are primarily related to the stress changes brought about during the pile driving and subsequent consolidation. Their variations during loading tests are insignificant relative to the initial values. The ratio of $G_{\gamma 1}/\tau_{f1}$ may be stipulated as a constant during a loading test and be deduced by matching the measured load-settlement curve with the theoretical solution (see Chapter 4, this book). Soil stress-strain nonlinearity has only a limited effect on such a back-analysis (Guo 1997), as further demonstrated in later examples.

5.2.1.2 Volumetric stress-strain relation of soil skeleton

The volumetric effective stress-strain relationship for an elastic medium is obtained using the plane strain version of Hooke's law as

$$\varepsilon_r = \frac{1}{2G}[(1-\nu_s)\delta\sigma_r' - \nu_s\delta\sigma_\theta'] = -\frac{\partial\xi}{\partial r}$$
$$\varepsilon_\theta = \frac{1}{2G}[-\nu_s\delta\sigma_r' + (1-\nu_s)\delta\sigma_\theta'] = -\frac{\xi}{r}$$
$$\varepsilon_z = 0$$
$$\text{Also} \quad \varepsilon_r - \varepsilon_\theta = r\frac{\partial\varepsilon_\theta}{\partial r} \tag{5.21}$$

where ξ = outward radial displacement of the soil around a pile; r = distance away from the pile axis; ν_s = Poisson's ratio of the soil; ε_r, ε_θ, ε_z = strains in the radial, circumferential, and depth directions, respectively; and G = elastic soil shear modulus; $\delta\sigma_r'$, $\delta\sigma_\theta'$, $\delta\sigma_z'$ = increments of the effective stresses during consolidation in the radial, circumferential, and depth directions, with $\delta\sigma_z' = \nu_s(\delta\sigma_r' + \delta\sigma_\theta')$. Combining Equation 5.21 and the effective stress principle, the volumetric effective stress-strain relationship for plain strain cases may be written as

$$\varepsilon_v = \frac{1-2\nu_s}{G}[\delta\theta - (u - u_o)] \tag{5.22}$$

where ε_v = volumetric strain; $\delta\theta = 0.5(\delta\sigma_r + \delta\sigma_\theta)$, total mean stress change; u = excess pore pressure; and u_o = initial value following driving (Randolph and Wroth 1979a). Radial equilibrium leads to

$$\frac{\partial\delta\sigma_r'}{\partial r} + \frac{\partial\delta\sigma_r' - \partial\delta\sigma_\theta'}{r} + \frac{\partial(u-u_o)}{\partial r} = 0 \tag{5.23}$$

As with that by McKinley (1998), Equations 5.21 and 5.23 offer

$$\delta\sigma_r' + \delta\sigma_\theta' = -\frac{1}{1+\nu_s}(u - u_o) \tag{5.24}$$

Equation 5.24 is valid to a domain with a sufficiently faraway outer boundary (i.e., $r > r^*$). Beyond the critical radius r^*, the pore pressure, u, and the stresses, $\delta\sigma_r'$, $\delta\sigma_\theta'$, $\delta\sigma_z'$, are all zero at any time. Thus, the total mean stress change may be expressed as

$$\delta\theta = \frac{1-2\nu_s}{2(1-\nu_s)}(u - u_o) \tag{5.25}$$

and Equation 5.22 may be rewritten as

$$\varepsilon_v = -\frac{1-2\nu_s}{2(1-\nu_s)G}(u - u_o) \tag{5.26}$$

Equation 5.26 is derived for an elastic medium. It is transformed into Equation 5.27 for visco-elastic media by using the correspondence principle (Mase 1970) and applying the inverse Laplace transform (Guo 2000c):

$$\varepsilon_v = -\frac{1-2\nu_s}{2(1-\nu_s)G_{\gamma 1}}\left((u-u_o)+G_{\gamma 1}\int_0^t (u-u_o)\Big|_\tau \frac{dF(t-\tau)}{d(t-\tau)}d\tau\right) \tag{5.27}$$

At $t = 0$, ε_v is equal to 0. The Poisson's ratio was regarded as a constant in this derivation, as is adopted in numerical analysis (Booker and Poulos 1976). Equation 5.27 renders the changing rate of volumetric strain as

$$\frac{\partial \varepsilon_v}{\partial t} = -\frac{(1-2\nu_s)}{2(1-\nu_s)}\frac{1}{G_{\gamma 1}}\left(\frac{\partial u}{\partial t}+G_{\gamma 1}\int_0^t \frac{\partial u}{\partial \tau}\frac{dF(t-\tau)}{d(t-\tau)}d\tau\right) \tag{5.28}$$

Equation 5.28 satisfies both radial equilibrium and Hooke's law.

5.2.1.3 Flow of pore water and continuity of volume strain rate

The volumetric strain rate is deduced next by considering the flow of pore water and continuity of volume. The pore water velocity is related to the pressure gradient by Darcy's law. The rate of volumetric strain for continuity is related to the flow of pore water into and out of any region by (Randolph and Wroth 1979a)

$$\frac{\partial \varepsilon_v}{\partial t} = \frac{k}{r_w}\frac{1}{r}\frac{\partial}{\partial r}\left(r\frac{\partial u}{\partial r}\right) \tag{5.29}$$

where k = permeability of the soil and γ_w = unit weight of water. Equations 5.28 and 5.29 are combined to yield

$$c_v \frac{1}{r}\frac{\partial}{\partial r}\left(r\frac{\partial u}{\partial r}\right) = \frac{\partial u}{\partial t}+G_{\gamma 1}\int_0^t \frac{\partial u}{\partial \tau}\frac{dF(t-\tau)}{d(t-\tau)}d\tau \tag{5.30}$$

where c_v = coefficient of soil consolidation under plain strain condition (Randolph and Wroth 1979a), given by

$$c_v = \frac{kG_{\gamma 1}}{\gamma_w}\frac{2(1-\nu_s)}{1-2\nu_s} \tag{5.31}$$

The subscript "γ_1" in $G_{\gamma 1}$ will be dropped subsequently, for convenience. Equation 5.30 is the governing equation for radial consolidation. For an elastic soil, Equation 5.20 offers $dF(t-\tau)/d(t-\tau) = 0$, which in turn allows Equation 5.30 to reduce to elastic case as

$$c_v \frac{1}{r}\frac{\partial}{\partial r}\left(r\frac{\partial u}{\partial r}\right)=\frac{\partial u}{\partial t} \tag{5.32}$$

5.2.1.4 Comments and diffusion equation

Equation 5.25 indicates the dependenence of total mean stress, $\delta\theta$, on the pore pressure, u, during the process of consolidation, as is noted in other kinds of consolidation (Mandel 1957; Cryer 1963; Murray 1978). Generally speaking, it is difficult to gain a similar expression to Equation 5.25 to correlate $\delta\theta$ with u. Instead, the $\delta\theta$ is taken as a constant or zero (Rendulic 1936; Terzaghi 1943; Murray 1978). With $\delta\theta = 0$, a new similar equation to Equation 5.30 was derived that yields a different coefficient, c_v (Guo 1997). It naturally does not warrant radial equilibrium and is a diffusion equation (Murray 1978). The diffusion equation is readily resolved and compares well with a corresponding rigorous solution with a proper c_v (Christian and Boehmer 1972; Davis and Poulos 1972; Murray 1978). In fact, a few popular theories are underpinned by this assumption (Murray 1978), including the sand drain problem (Barron 1948).

5.2.1.5 Boundary conditions

Randolph and Wroth (1979a) elaborated the boundary conditions for radial consolidation of an elastic medium around a rigid, impermeable pile. These conditions as described below are generally valid for the visco-elastic medium as well:

$$u\big|_{t=0} = u_o(r)(t=0,\ r\geq r_o) \tag{5.33a}$$

$$\left.\frac{\partial u}{\partial r}\right|_{r=r_o} = 0\ (t\geq 0) \tag{5.33b}$$

$$u\big|_{r\geq r^*} = 0\ (t\geq 0) \tag{5.33c}$$

$$\mathrm{u} = 0\ as\ t \to \infty\ (r \geq r_o) \tag{5.33d}$$

where r_o = the pile radius (r^* is defined in the note for Equation 5.24). Initially, $u = 0$ for $r \geq R$ (R is the width of the plastic zone). However, during consolidation, outward flow of pore water will give rise to excess pore pressures in the region $r > R$, and generally it is necessary to take r^* as 5 to 10 times R.

5.2.2 General solution to the governing equation

Equation 5.30 is resolved by separating the variables for time-dependent and independent parts through

$$u = wT(t) \tag{5.34}$$

With a separation constant of λ_n^2, Equation 5.30 is transformed into

$$\frac{\partial^2 w}{\partial r^2}+\frac{1}{r}\frac{\partial w}{\partial r}+\lambda^2 w=0 \tag{5.35}$$

$$\frac{dT(t)}{dt}+G_{\gamma 1}\int_o^t \frac{dT(t)}{d\tau}\frac{dF(t-\tau)}{d(t-\tau)}d\tau+\alpha_n^2 T(t)=0 \tag{5.36}$$

where

$$\alpha_n^2=c_v\lambda_n^2 \tag{5.37}$$

The parameter, λ_n, is one of the infinite roots satisfying Equation 5.35, which is governed by Bessel functions through

$$w_n(r)=A_nJ_o(\lambda_n r)+B_nY_o(\lambda_n r) \tag{5.38}$$

where A_n = constants determined using boundary conditions; J_o, Y_o, J_1, and Y_1 = Bessel functions of zero order and first order; and J_i, Y_i = Bessel functions of the first and second kinds, respectively.

Cylinder functions, $V_i(\lambda_n r_o)$, of i-th order are expressed as (McLachlan 1955)

$$V_i(\lambda_n r)=J_i(\lambda_n r)-\frac{J_1(\lambda_n r_o)}{Y_1(\lambda_n r_o)}Y_i(\lambda_n r) \tag{5.39}$$

The boundary condition of Equation 5.33b renders $B_n=-A_n\,J_1(\lambda_n r_o)/Y_1(\lambda_n r_o)$. Equations 5.38 and 5.33b are simplified, respectively, as

$$w_n(r)=A_nV_o(\lambda_n r) \tag{5.40}$$

$$\left.\frac{dw_n(r)}{dr}\right|_{r-r_o}=A_nV_1(\lambda_n r)\big|_{r=r_o}=0 \tag{5.41}$$

Also, Equation 5.33c of $u = 0$ for $r \geq r^*$ provides

$$V_o(\lambda_n r^*)=J_o(\lambda_n r^*)-\frac{J_1(\lambda_n r_o)}{Y_1(\lambda_n r_o)}Y_o(\lambda_n r^*)=0 \tag{5.42}$$

Equations 5.41 and 5.42 define the cylinder functions. There are an infinite number of roots of λ_n satisfying these equations.

The time-dependent solution for the standard linear visco-elastic model (Figure 3.17, Chapter 3, this book) is

$$T_n(t)=\frac{(\omega_1(n)-\alpha_c)\exp(-\omega_1(n)t)-(\omega_2(n)-\alpha_c)\exp(-\omega_2(n)t)}{\omega_1(n)-\omega_2(n)} \tag{5.43}$$

where

$$\omega_1(n) = \frac{\alpha_c + \alpha_n^2}{2} + \frac{1}{2}\sqrt{(\alpha_c + \alpha_n^2)^2 - 4\alpha_n^2 / T_2} \tag{5.44}$$

$$\omega_2(n) = \frac{\alpha_c + \alpha_n^2}{2} - \frac{1}{2}\sqrt{(\alpha_c + \alpha_n^2)^2 - 4\alpha_n^2 / T_2} \tag{5.45}$$

and

$$\alpha_c = (1 + m_2) / T_2 \tag{5.46}$$

This time factor, $T_n(t)$, is essentially identical to the previous one (Christie 1964) using a similar model for one-dimensional consolidation. For the elastic case of $\alpha_c = 0$, and $m_2 = 0$, Equations 5.44 and 5.45 reduce to $\omega_1(n) = \alpha_n^2$ and $\omega_2(n) = 0$, respectively; and Equation 5.43 reduces to

$$T_n(t) = e^{-\alpha_n^2 t} \tag{5.47}$$

The full expression for pore pressure, u, will be a summation of all the possible solutions

$$u = \sum_{n=1}^{\infty} A_n V_o(\lambda_n r) T_n(t) \tag{5.48}$$

Normally use of the first 50 roots of the Bessel functions offers a sufficiently accurate value of u. With Equations 5.33a and 5.48, it follows

$$A_n = \int_{r_o}^{r^*} u_o(r) V_o(r\lambda_n) r\, dr \Bigg/ \int_{r_o}^{r^*} V_o^2(r\lambda_n) r\, dr \tag{5.49}$$

These visco-elastic solutions may be readily obtained, in terms of the elastic solutions, by the correspondence principle (Guo 1997).

5.2.3 Consolidation for logarithmic variation of u_o

For a cavity expansion from zero radius to a radius of r_o (pile radius), the initial stress state for radial consolidation of a visco-elastic medium around a rigid, impermeable pile is taken as that of an elastic medium (Guo 2000c), which, as reported previously (Randolph and Wroth 1979a), is adopted herein along with $G_{\gamma 1}$ at a suitable strain level.

(1) During the cavity expansion from zero radius to a radius of r_o (pile radius), the stress change, $\delta\theta$, within the plastic zone ($r_o \le r \le R$)

follows Equation 5.50. Note R = the radius, beyond which the excess pore pressure is initially zero (see Figure 3.17, Chapter 3, this book)

$$\delta\theta = s_u[\ln(G_{\gamma 1} / s_u) - 2\ln(r / r_o)] \tag{5.50}$$

where s_u = the undrained shear strength of the soil. The width of the plastic zone is given by

$$R = r_o(G_{\gamma 1} / s_u)^{1/2} \tag{5.51}$$

(2) Under undrained conditions and an invariable mean effective stress, the initial excess pore pressure distribution away from pile wall varies according to

$$\begin{aligned} u_o(r) &= 2s_u \ln(R / r) \quad & r_o \le r \le R \\ u_o &= 0 & R < r < r^* \end{aligned} \tag{5.52}$$

The initial pore pressure distribution of Equation 5.52 allows the coefficients, A_n, to be simplified as

$$A_n = \frac{4s_u}{\lambda_n^2} \frac{V_o(\lambda_n r_o) - V_o(\lambda_n R)}{r^{*2} V_1^2(\lambda_n r^*) - r_o^2 V_o^2(\lambda_n r_o)} \tag{5.53}$$

These values of A_n permits the pore pressure to be estimated using Equation 5.48. The estimations were readily conducted using a spreadsheet program operating in Microsoft Excel™.

The rate of consolidation may be quantified by the nondimensional time factor, T (Soderberg 1962),

$$T = c_v t / r_o^2 \tag{5.54}$$

The visco-elastic effect may be represented by the factor, N (Christie 1964),

$$N = \frac{r_0^2}{T_2} \frac{\gamma_w}{G_{\gamma 1} k} \tag{5.55}$$

A parametric study was undertaken using Equation 5.48. In Figure 5.8, the predicted dissipation of pore water pressure is presented as $(u_o - u)/u_o$ (u_o, u is the pore pressure on the pile–soil interface). Figure 5.8a illustrates the effect of viscosity, η, by the factor N on the dissipation process. The "extreme" curves for $N = 0$ and ∞ all behave as Terzaghi's consolidation, except for the time factor being as a multiplier of the factor, T (Christie 1964). As the factor N increases, the effect on the dissipation process will start earlier. Figure 5.8b shows the slowdown in dissipation of pore pressure at a later stage of the process, owing to $G_{\gamma 2} < G_{\gamma 1}$ (i.e., secondary shear modulus, $G_{\gamma 2}$, is lower than the primary modulus).

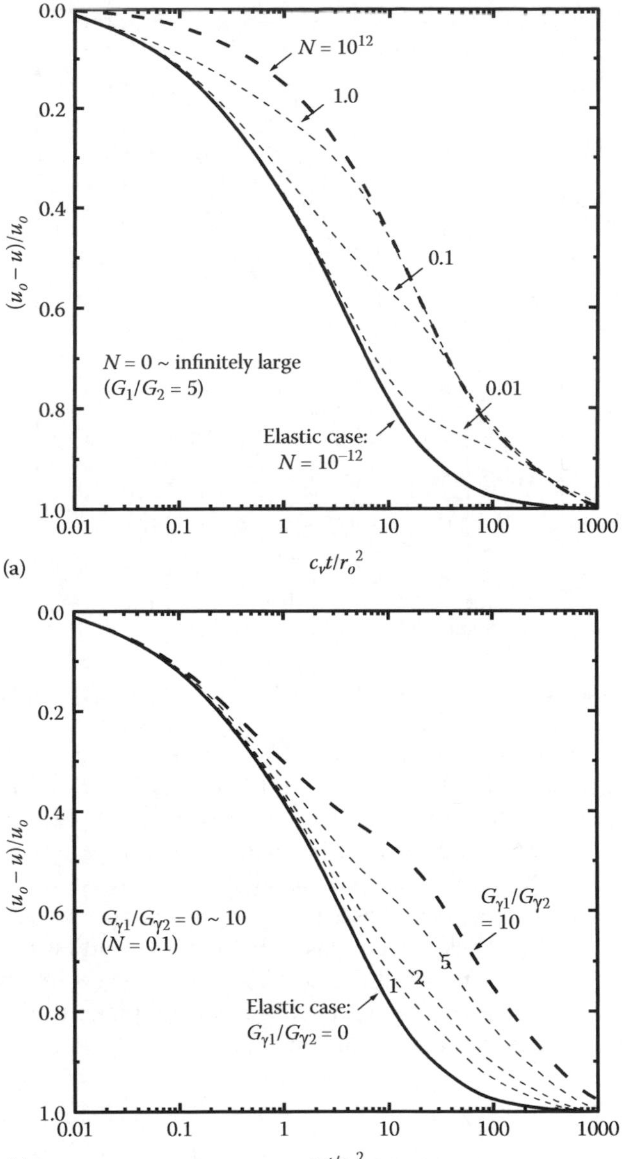

Figure 5.8 Influence of creep parameters on the excess pore pressure. (a) Various values of N. (b) Typical ratios of $G_{\gamma 1}/G_{\gamma 2}$. (After Guo, W. D., *Computers and Geotechnics* 26, 2, 2000c.)

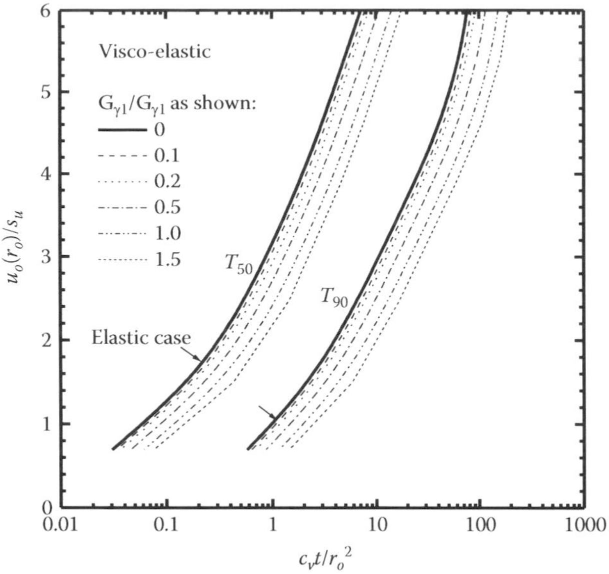

Figure 5.9 Variations of times T_{50} and T_{90} with the ratio $u_o(r_o)/s_u$. (After Guo, W. D., *Computers and Geotechnics* 26, 2, 2000c.)

The time factor T at 50% and 90% degree (of consolidation) for the dissipation of pore pressure, $(u_o - u)/u_o$, is denoted as T_{50} and T_{90}. This degree of consolidation using pore pressure dissipation is not equal to that based on settlement (Christie 1964; Booker and Small 1977) for a visco-elastic medium. Figure 5.9 shows a set of plots of the factor, T_{50} and T_{90}, at different values of $u_o(r_o)/s_u$ and $G_{\gamma1}/G_{\gamma2}$ ($u_o[r_o]$ = initial pore pressure on pile–soil interface immediately following pile installation). Increase in the ratio of $G_{\gamma1}/G_{\gamma2}$, (i.e., including the secondary consolidation) would render higher values of T_{50} and T_{90}, thus longer consolidation times and higher displacements compared to elastic analysis ($G_{\gamma1}/G_{\gamma2} = 0$). Figures 5.8 and 5.9 are applicable for both cases of constant total stress and plane strain deformation (but for different coefficients of consolidation, c_v).

5.2.4 Shaft capacity

The shaft resistance was estimated in terms of total and effective stress using the following expressions (see Chapter 2, this book):

$$\tau_{f1} = \alpha s_u \tag{5.56}$$

and

$$\tau_{f1} = \beta\sigma'_v \tag{5.57}$$

where σ'_v = effective overburden pressure; τ_{f1} = limiting shaft stress, α = average pile–soil adhesion factor in terms of total stress; and β = average pile–soil adhesion factor in terms of effective stress.

5.2.5 Visco-elastic behavior

Recovery of the soil strength and modulus with reconsolidation was investigated by a series of loading tests conducted at different time intervals following installation of the piles (Trenter and Burt 1981). Each measured load-settlement curve allows a time-dependent soil strength and modulus to be back-figured by matching with the theoretical solution (via the GASPILE program; see Chapters 3 and 4, this book) (Guo 2000a; Guo 2000b; Guo 2000c). As illustrated next, only the initial shear modulus $G_{\gamma 1}$ needs to be changed to match a measured load-settlement curve with GASPILE analysis for a set of ratio of creep moduli, $G_{\gamma 1}/G_{\gamma 2}$, and load transfer factor, ζ_j. Imposing identical values of $G_{\gamma 1}/G_{\gamma 2}$ and ζ_j for each pile analysis at different times eliminates their impact on the time-dependent normalized values of limiting shaft stress τ_{f1}, shear modulus, α, and β obtained.

Example 5.3 Test reported by Trenter and Burt (1981)

Trenter and Burt (1981) conducted loading tests on three driven open-ended piles in Indonesia, mainly by maintained load procedure. The basic pile properties are shown in Table 5.3, along with a Young's modulus of 29,430 MPa. The undrained shear strength of the subsoil at the site varies basically according to $s_u = 1.5z$ (s_u, kPa; z, depth, m). The ratio, $G_{\gamma 1}/s_u$ is back-analyzed from the test data, with $G_{\gamma 1}/G_{\gamma 2}$ = 0.15 and $G_{\gamma 2}/\eta_{\gamma 2} = 0.5 \times 10^{-5}$ (s^{-1}).

The load transfer factor, ζ_1, should be limited to $L/r_o = 180$, against numerical solutions (Guo 1997; Guo and Randolph 1998), as the real slenderness ratio exceeds the critical ratio of $3\sqrt{\lambda}$ [λ = ratio of pile Young's modulus to the soil shear modulus at pile tip level (Fleming et al. 1992)]. To avoid the recursive calculation of the critical pile slenderness ratio, the load transfer factor is simply estimated using the real pile slenderness ratios. With the input parameters tabulated in Table 5.3 using GASPILE analysis, the relevant average values are back-figured and shown in Table 5.4 for piles 4, 3, and 2. At ultimate capacities, the measured pile-head displacements were 4% and 6.3% of the diameter for piles 4 and 3, respectively, which agree with the lower value of $G_{\gamma 1}/\tau_{f1}$ for pile 3 against pile 4 (Table 5.4) (Kuwabara 1991).

The theoretical solutions using GASPILE were adopted to match the measured data of pile 2, pile 4, and pile 3, individually. For pile 4 at 1.7 and 10.5 days, the following analyses were undertaken (Guo 2000c): nonlinear elastic (NLE, with $G_{\gamma 1}/G_{\gamma 2} = 0$ and $\psi_{oj} = 0.5$), nonlinear visco-elastic (NLVE, using $\psi_{oj} = 0.5$ in estimating the ζ_j and $G_{\gamma 1}/G_{\gamma 2} = 0.15$), linear elastic (LE, with $G_{\gamma 1}/G_{\gamma 2} = 0$ and $\psi_{oj} = 0$), and linear

Table 5.3 Parameters for the analysis of the tests

Pile No.	*Penetration (m)*	ζ_1 *(ψ_{o1} = 0)*	ζ_1 *(ψ_{o1} = 0.5)*	ω/ξ_b	*Note*
2a & b	24/30.3	4.5/4.73	5.2/5.4	1.0/2	Diameter: 400 mm
3	53.5/54.5	5.31	6.0	1.0/2	Wall thickness: 12 mm
4	43.3	4.4	5.08	1.0/2	

Source: Guo, W. D., *Computers and Geotechnics*, 26, 2, 2000c.

Table 5.4 Parameters for analysis of test piles 4, 3, and 2

Time (days)	$G_{\gamma 1}$ *(MPa)*	τ_{f1} *(kPa)*	$G_{\gamma 1}/\tau_{f1}$	*Example 5.3*
1.7	6.11	20.18	303	Pile 4:
10.5	7.64	25.28	302	$G_{\gamma 2}/\eta_{\gamma 2}t$ = 12.96
20.5	8.43	27.12	311	$G_{\gamma 1}/G_{\gamma 2}$ = 0.15
32.5	8.43	27.5	306	
2.3	3.62	20.717	175	Pile 3:
3.0	3.69	21.13	175	$G_{\gamma 2}/\eta_{\gamma 2}t$ = 12.96
4.2	3.9	22.308	175	$G_{\gamma 1}/G_{\gamma 2}$ = 0.15
Infinite (L = 24 m)	8.46	20.225	418	Pile 2a & b: $G_{\gamma 2}/\eta_{\gamma 2}$ = 0.5 × $10^{-5}(s^{-1})$
Infinite (L = 30.3 m)	7.75	23.865	308	$G_{\gamma 1}/G_{\gamma 2}$ = 0.15

Source: Guo, W. D., *Computers and Geotechnics*, 26, 2, 2000c.

visco-elastic (LVE, with $G_{\gamma 1}/G_{\gamma 2}$ and $\psi_{oj} = 0$). It was found that with a low value of $G_{\gamma 1}/G_{\gamma 2} = 0.15$, the difference amongst these analyses (NLE, NLVE, LVE, and LE) is negligibly small and the final settlement and the settlement rate under a given load depend on $G_{\gamma 1}/G_{\gamma 2}$ and $G_{\gamma 2}/\eta_{\gamma 2}$, respectively.

The reported parameters α and β (Trenter and Burt 1981) for piles 3 and 4 are tabulated in Table 5.5. Their normalized values (by those at 1.7 days) turn out quite consistent with those of the shear modulus (Table 5.5) against time. More generally, strength increases

Table 5.5 Parameters for empirical formulas

Pile No.	4				3		
Time (days)	*1.7*	*10.5*	*20.5*	*32.5*	*2.3*	*3.0*	*4.2*
α	0.63	0.81	0.87	0.87	0.51	0.53	0.55
β	0.16	0.20	0.22	0.22	0.13	0.13	0.14
α/α_o [a]	1.0	1.286	1.381	1.381	1.0	1.039	1.0784
β/β_o [a]	1.0	1.25	1.375	1.375	1.0	1.0	1.077
G_i/G_{io} [a]	1.0	1.25	1.38	1.38	1.0	1.02	1.077

Source: Guo, W. D., *Computers and Geotechnics*, 26, 2, 2000c.

[a] α_o, β_o, G_{io} = the values of α, β, G_i at 1.7 days

logarithmically with time (Bergdahl and Hult 1981; Sen and Zhen 1984), but is obviously capped by the "original" undisturbed strength. This example demonstrates that the pile–soil interaction stiffness increases simultaneously as soil strength regains, and secondary compression of clay only accounts for a small fraction of the settlement of the pile, with a low $G_{\gamma 1}/G_{\gamma 2}$ and a short period of consolidation.

5.2.6 Case study

Pore pressure, u, was estimated using Equation 5.48 for plane strain deformation for each case study discussed next. The dissipated pressure, $u_o - u$, (u_o initial value given by Equation 5.52 at $r = r_o$) is then normalized by the initial value u_o and compared with (a) the normalized difference of the measured (if available) pore pressure, $u_o - u$, by the initial pressure, u_o; (b) the normalized, back-figured shear modulus by the modulus at t_{90}; (c) the normalized, back-figured limiting shear strength by the strength at t_{90}; and (d) the normalized, measured time-dependent pile capacity by the capacity at t_{90}. The shear modulus and limiting strength with time were deduced in a similar manner described in Example 5.3, using the measured load-settlement curves at different times following pile driving. Also, the values at t_{90} were obtained through linear interpolation.

Example 5.4 Tests by Seed and Reese

Seed and Reese (1955) assessed the change in pile-bearing capacity with reconsolidation of soil following pile installation by conducting loading tests on an instrumented pile at intervals after driving. The pile, of radius 0.0762 m, was installed through a sleeve, penetrating the silty clay from a depth of 2.75 to 7 m. The Young's modulus is 2.07×10^5 MPa and the cross-sectional area is 9.032 cm^2 (i.e., the equivalent pile modulus = 10,250 MPa). Through fitting the measured load-settlement response by elastic analysis using the GASPILE program (Figure 5.10), values of $G_{\gamma 1}$, τ_{f1} were back-figured from each load-settlement curve. These values are tabulated in Table 5.6.

With Poisson's ratio, $\nu_s = 0.49$, permeability $k = 2 \times 10^{-6}$ m/s and $c_v = 0.0529$ m^2/day, the 90% degree of reconsolidation (elastic case) is estimated to occur at $t_{90} = 8.76$ days ($T_{90} = 74.43$, $G_{\gamma 1}/\tau_{f1} = 350$). At the time of t_{90}, Table 5.6 shows a shear modulus of ~3.55 MPa, which is less than 90% of 4.5 MPa (i.e., the maximum value at 33 days), and a shaft friction of ~11.6 kPa (the final pile–soil friction = 12.7 kPa), which is a fraction of the initial soil strength of 18 kPa (due to soil sensitivity). The back-figured shear modulus and the limiting strength were normalized by the values at t_{90} and are plotted in Figure 5.11a together with the normalized pile capacity and elastic prediction of dissipation of pore water pressure by Equation 5.48.

Taking $G_{\gamma 1}/G_{\gamma 2} = 1$, and the parameters for elastic analysis, the visco-elastic analysis offers a new $t_{90} = 16.35$ days ($T_{90} = 148.85$,

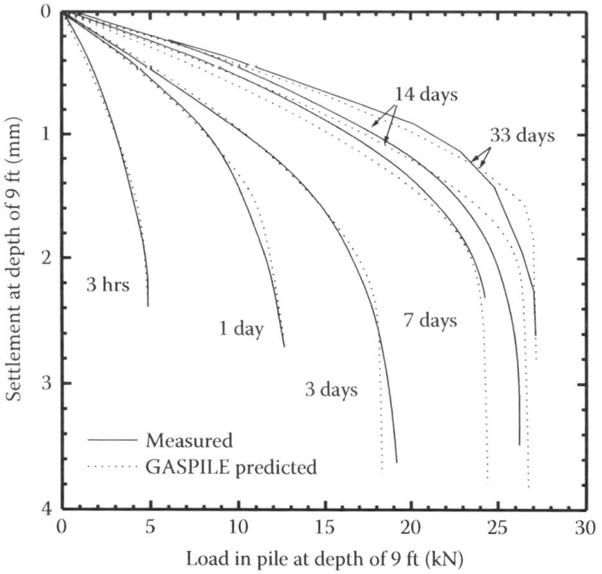

Figure 5.10 Comparison between the calculated and measured (Seed and Reese 1955) load-settlement curves at different time intervals after driving. (After Guo, W. D., *Computers and Geotechnics* 26, 2, 2000c.)

$G_{\gamma 1}/\tau_{f1} = 350$). At this new t_{90}, Table 5.6 indicates a shear modulus of 4.06 MPa (≈90% of the final value, 4.5 MPa) and a shaft friction of 12.54 kPa. The normalized back-figured shear modulus and the limiting strength by the values at this new t_{90} are shown in Figure 5.11b, together with the normalized pile-capacity (by the value at this new t_{90}) and the visco-elastic prediction of the dissipation of pore pressure.

This example shows that both elastic and visco-elastic analysis can well simulate variation of shear strength or pile capacity but for the difference in consolidation time.

Example 5.5 Tests reported by Konrad and Roy

Konrad and Roy (1987) reported an instrumented pile loaded to failure at intervals after driving. The closed-ended pile, of outside radius 0.219 m and wall thickness 8.0 mm, was jacked to a depth of 7.6 m. It has a Young's modulus of 2.07×10^5 MPa and a cross-sectional area of 53.03 cm^2 (i.e., equivalent pile modulus = 29,663 MPa). The normalized increase in pile (shaft) capacity with consolidation (Konrad and Roy 1987) by which at 2 years after installation seems to agree with the normalized dissipation of pore pressure measured at three depths of 3.05, 4.6, and 6.1 m (see Figure 5.12a and b).

The initial load-settlement response measured at various time intervals after pile installation allows $G_{\gamma 1}/s_u = 270$ to be deduced using elastic analysis (Guo 1997). The associated final load-settlement

Table 5.6 Back-figured parameters in Examples 5.4 and 5.5

Example 5.4	*From measured data*					
Time (days)	.125	1	3	7	14	33
τ_{fl} (kPa)	2.26	5.71	8.4	11.3	12.52	12.68
$G_{\gamma l}$(MPa)	.6	1.6	2.1	3.4	4	4.5

Example 5.5	*From measured data*				
Time (days)	4	8	20	33	730
τ_{fl} (kPa)	5.58[a]/12.93[b]	6.56/19.49	7.75/23.0	8.06/23.93	8.63/25.61
$G_{\gamma l}$ (MPa)	1.07/3.19	1.44/4.29	1.65/4.9	1.73/5.15	1.9/5.64

Sources: Guo, W. D., *Computers and Geotechnics*, 26, 2, 2000c; Seed, H. B., and L. C. Reese, *Transactions, ASCE*, 122, 1955; Konrad, J.-M., and M. Roy, *Geotechnique*, 37, 2, 1987.

[a] numerators for ground level
[b] denominators for the pile base level

relationships offer $G_{\gamma l}/s_u$ = 210–230, deduced using $G_{\gamma l}/G_{\gamma 2}$ = 2 and visco-elastic GASPILE analysis. Using linear distributions, the values of shear modulus and strength back-figured are tabulated in Table 5.6.

The visco-elastic analysis offers a slightly better agreement with the measured response at low load levels of ~70% ultimate load (Figure 5.13a) than elastic analysis (Figure 5.13b). At high load levels, the prediction may be further improved by incorporating nonlinear base response and variation of ζ_1 with load level. Nevertheless, the current analysis is deemed sufficiently accurate for gaining the variations of shear strength and modulus with reconsolidation.

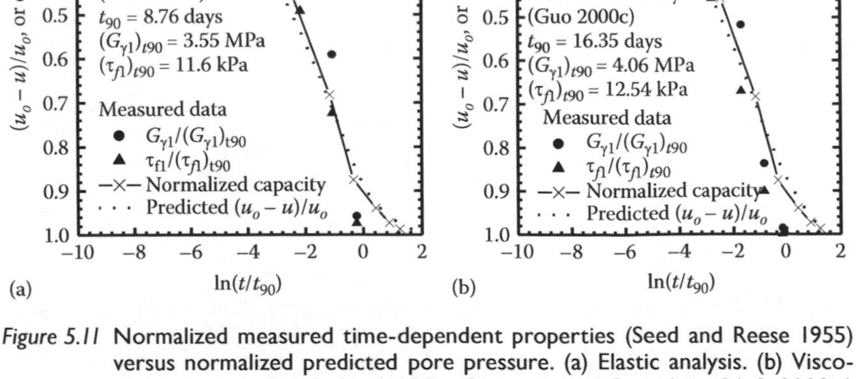

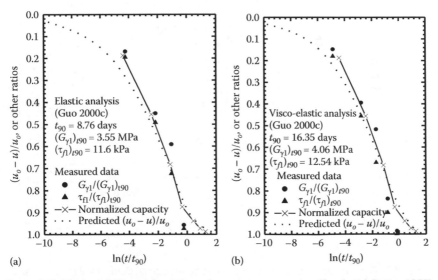

Figure 5.11 Normalized measured time-dependent properties (Seed and Reese 1955) versus normalized predicted pore pressure. (a) Elastic analysis. (b) Visco-elastic analysis. (After Guo, W. D., *Computers and Geotechnics* 26, 2, 2000c.)

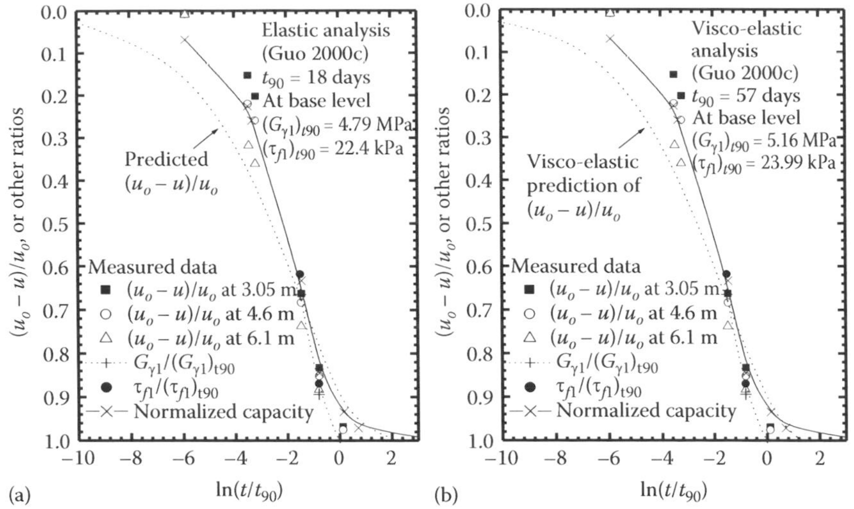

Figure 5.12 Normalized measured time-dependent properties (Konrad and Roy 1987) versus normalized predicted pore pressure. (a) Elastic analysis. (b) Visco-elastic analysis. (After Guo, W. D., *Computers and Geotechnics* 26, 2, 2000c.)

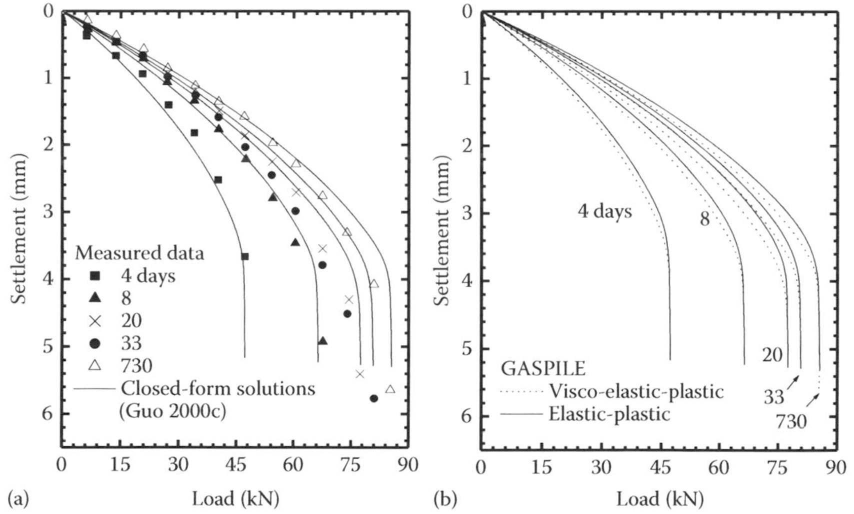

Figure 5.13 Comparison of load-settlement relationships predicted by GASPILE/closed-form solution with the measured (Konrad and Roy 1987): (a) Visco-elastic-plastic. (b) Elastic-plastic and visco-elastic-plastic solutions. (After Guo, W. D., *Computers and Geotechnics* 26, 2, 2000c.)

The coefficient of consolidation c_v was 0.0423 m²/day (ν_s = 0.45) (Konrad and Roy 1987). Elastic analysis offers the time factor T_{90} of 65, with $G_{\gamma 1}/s_u$ = 230 from Figure 5.9; hence, t_{90} ≈18 days. At the time of t_{90}, Table 5.6 shows a shear strength at the pile base level of 22.4 kPa, which agrees with 23.0 kPa (= 90% of limiting stress, τ_{f1} of 25.61 kPa); and a shear modulus of 4.79 MPa, which is slightly lower than 5.08 MPa (= 90% of the maximum modulus 5.64 MPa). Using visco-elastic analysis, and $G_{\gamma 1}/G_{\gamma 2} = 2$, $T_{90} = 205.1$, the t_{90} is estimated as 57 days, at which $\tau_{f1} = 23.99$ kPa and $G_{\gamma 1} = 5.16$ MPa (by interpolation from Table 5.6). With these estimated values at t_{90}, the normalized variations are plotted in Figure 5.12a and b, respectively, for elastic and visco-elastic analyses, together with the theoretical curves of dissipation of pore water pressure.

The predicted pore pressure by Equation 5.48 at the initial stage is lower than the measured data, indicating the impact of the radial nonhomogeneity of shear modulus (Guo 1997) or soil stress-strain nonlinearity (Davis and Raymond 1965). Radial nonhomogeneity of modulus can reduce the ratio of $(u - u_o)/u_o$ and retard the regain in the average shear modulus (at some distance away from the pile axis). Thus, a comparatively higher new value of t_{90} for the nonhomogenous modulus regain is anticipated than the currently predicted t_{90}. Using the values at this new t_{90} to normalize the rest of the measured data or back-estimated values, a lower trend of the curves is obtained than the current presentation in the figure, which will be more close to the prediction by Equation 5.48.

Soil strength may increase due to reconsolidation or decrease due to creep. The latter may continue until it attains a long-term strength (e.g., ~70% of the soil strength) (Geuze and Tan 1953; API 1993; Guo 1997). The effect of reconsolidation and creep on the soil strength may offset in this particular case.

5.2.6.1 Comments on the current predictions

The back-figured values of $G_{\gamma 1}/G_{\gamma 2}$ for the two field studies are higher than those based on confined compression (oedometer) tests (Lo 1961). The current radial consolidation theory is based on a homogeneous medium. Radial nonhomogeneity can alter the shape of the time-dependent curve at the initial stage and increase the time for regain of shear modulus.

5.2.7 Summary

A method is provided to predict overall response of a pile following driving rather than just the pile capacity. A number of important conclusions can be drawn:

1. Visco-elastic solutions can be obtained by the available elastic solutions using the correspondence principle.

2. The viscosity of a soil can significantly increase the consolidation time and the pile-head settlement, despite its negligible effect on soil strength or pile capacity.
3. During reconsolidation, the normalized pile–soil interaction stiffness (or soil shear modulus) increase, pore pressure dissipation on the pile–soil interface, and soil strength regain observe similar time-dependent path. These time-dependent properties can be sufficiently accurately predicted by the radial consolidation theory. They in turn allow load-settlement response to be predicted at any times following driving by either GASPILE analysis or the previous closed-form solutions (Guo 1997; Guo 2000a; Guo 2000b).

Chapter 6

Settlement of pile groups

6.1 INTRODUCTION

When piles are in a group, the settlement will increase. This increase, as revealed previously using various numerical analyses, is dependent on at least five factors: pile center-center spacing, the number of piles in a group, pile–soil relative stiffness, the depth of the underlying rigid layer, and the profile of shear modulus both vertically and horizontally.

A large body of information is available for analyzing pile groups. Generally, the settlement of pile groups can be predicted by the following procedures:

1. Empirical methods (Terzaghi 1943; Skempton 1953; Meyerhof 1959; Vesić 1967);
2. Shallow footing analogy or imaginary footing method;
3. Load transfer approaches, based on either simple closed-form solutions (Randolph and Wroth 1978; Randolph and Wroth 1979b; Lee 1993a) or discrete layer approach (Chow 1986b; Lee 1991);
4. Elastic continuum-based methods such as boundary element analysis (Poulos 1968; Butterfield and Banerjee 1971; Chin et al. 1990), infinite layer approach (Guo et al. 1987; Cheung et al. 1988), and FEM analysis (Valliappan et al. 1974; Ottaviani 1975; Pressley and Poulos 1986);
5. Hybrid load transfer approach (O'Neill et al. 1977; Chow 1986a; Clancy and Randolph 1993; Lee 1993b), which takes advantage of both numerical and closed-form approaches and provides the possibility of analyzing large group piles.

In view of the five influencing factors, empirical methods are perhaps the least reliable, and the shallow footing analogy is overly simplistic. Elastic methods of analysis are perhaps the most suitable method for the majority of engineering problems, as the pile–soil interaction may generally be in elastic state. For more critical structures, finite element or boundary element analyses may be more appropriate.

It is also important to note that the zone of influence of a pile group extends considerably further below the pile bases than that of the individual piles. If a soft layer lies within the extended zone, the compression of the layer must be considered (see Chapter 2, this book).

6.2 EMPIRICAL METHODS

A few nondimensional parameters have been introduced to describe pile group behavior. A commonly used settlement ratio, R_s, is defined as the ratio of the average group settlement, w_g, to the settlement of a single pile, w_t, carrying the same average load (Poulos 1968). Skempton correlated the group settlements, w_g, with the single pile settlement, w_t, by

$$\frac{w_g}{w_t} = \left(\frac{4B_c + 3}{4 + B_c}\right)^2 \tag{6.1}$$

in which w_g = settlement of a pile group of width B_c in meter, and w_t = observed settlement of a single pile at the same load intensity. Later, Meyerhof (1959) revised Equation 6.1 to Equation 6.2 to account for the geometry of the pile group (Meyerhof 1959):

$$\frac{w_g}{w_t} = \frac{s(5 - s/3)}{(1 + 1/m)^2} \tag{6.2}$$

where s = ratio of spacing to diameter; and m = number of rows for a square group.

The empirical formulas were generally established by comparing full-scale or model test results between the settlements of pile groups and those of single piles in sands (Skempton 1953; Meyerhof 1959). Kaniraj (1993) introduced a new settlement ratio as a ratio of the settlement of a pile group over that of a single pile *under an identical average stress on their respective load transmitting area*. The load transmitting area is, at the pile-base level, estimated through the dispersion angle (≈7°, Berezantzev et al. 1961) as illustrated in Figure 6.1. A semi-empirical equation was proposed for the ratio and compared with the measured values, which show slight improvement over Equations 6.1 and 6.2. These empirical formulas do not account for all five of the factors mentioned in introduction.

6.3 SHALLOW FOOTING ANALOGY

Settlement of a pile group may be estimated using shallow footing analogy or "imaginary footing method." The pile group presumably acts as a block,

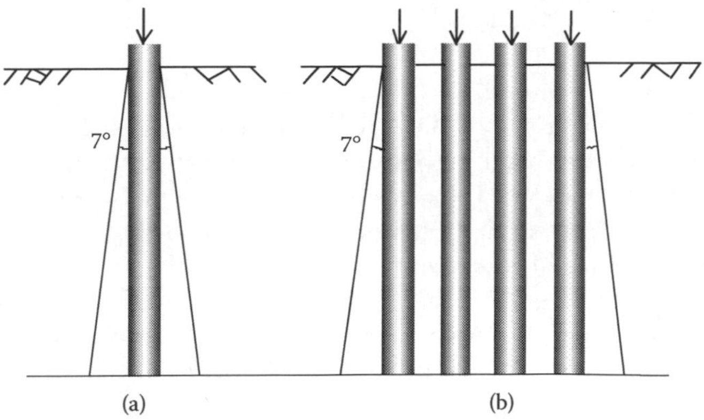

Figure 6.1 Load transmitting area for (a) single pile, (b) pile group. (After Kaniraj, S. R., *Soils and Foundations*, 33, 2, 1993.)

the dimensions of which are defined by the perimeter of the group and a depth z_f equal to either the depth of the group for end-bearing piles or to two thirds the depth of the group for friction piles (see Figure 6.2). Assume the load beneath the imaginary footing spreads out over a 30° frustum. The increase in the vertical stress in the subsoil at the depth z_f is:

$$\Delta\sigma'_v = \frac{P_g}{(B_c + 2z_f \tan 30^\circ)(L_c + 2z_f \tan 30^\circ)} \tag{6.3}$$

where $\Delta\sigma'_v$ = increase in the vertical effective stress; P_g = downward load acting on a pile group; B_c, L_c = width and length of pile group (cap), which equal to width and length of the imaginary footing, respectively (see Figure 2.12, Chapter 2, this book); and z_f = depth below bottom of imaginary footing. The settlement of the imaginary footing may be readily estimated by the methods for shallow foundations. Thereafter, the elastic compression w_e of the piles should be added to the settlement of the footing to gain the total settlement of the pile groups:

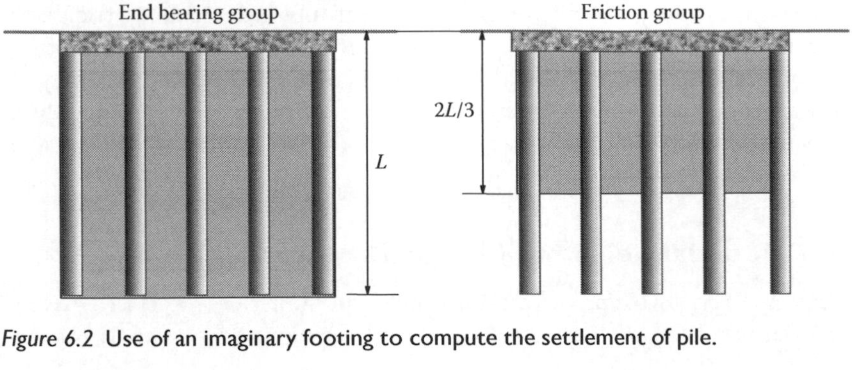

Figure 6.2 Use of an imaginary footing to compute the settlement of pile.

$$w_e = \frac{P_t z_i}{A_p E_p} \tag{6.4}$$

where w_e = settlement due to elastic compression of piles; P_t = downward load on each pile; z_i = depth to bottom of imaginary footing; A_p = equivalent cross-sectional area of a single solid pile; E_p = modulus of elasticity of a pile, = 2×10^5 MPa for steel; = $4806.6\sqrt{f_c'}$ MPa for concrete, and f_c' = 28-day compressive strength of concrete.

Assuming that the compressibility of the soil beneath the imaginary footing is constant or increases with depth, Meyerhof (1976) developed the following expressions for calculating settlement of piles in cohesionless soils:

$$w_g = \frac{q_e'}{1.2N_{60}'}\sqrt{B_c} \tag{6.5}$$

$$w_g = \frac{B_c q_e' I}{2q_c} \tag{6.6}$$

$$I = 1 - \frac{z_i}{8B_c} \geq 0.5 \tag{6.7}$$

where w_g = settlement of pile group (mm); $q_e' = P_g/(B_c L_c)$, equivalent net bearing pressure (kPa); B_c = width of pile group (m); L_c = length of pile group block (m); N_{60}' = SPT N-value within a depth of z_i to $z_i + B_c$ with overburden correction; q_c = CPT cone bearing within a depth of z_i to $z_i + B_c$. The "shallow footing analogy" method ignores the load-transfer interaction between piles and soil, which depends on the five factors mentioned earlier. The block behavior may never exhibit before the piles fail individually, which renders the block analogy inappropriate.

6.4 NUMERICAL METHODS

Finite element and boundary element analyses, especially for a 3-dimensional analysis, are sophisticated to predicting pile group settlements, particularly for complex boundary conditions. These analyses are warranted for settlement sensitive structures, which are justified economically. The accuracy of these methods is critically dependent on the failure criterion adopted, and particularly the constitutive laws that govern the pile–soil interface. Typical numerical methods are briefed next.

6.4.1 Boundary element (integral) approach

Using the BEM (BI) approaches, Butterfield and Banerjee (1971) extensively explored pile-head stiffness for different pile groups of rigid cap at various pile slenderness ratios and pile–soil relative stiffnesses. Poulos (1968) introduced the pile–soil–pile interaction factor, as mentioned earlier, and generated values of the settlement ratio, R_s, and load distribution within a group considering the influence of pile spacing, pile length, type of group, depth of a rigid layer, and Poisson's ratio of the soil. Chin et al. (1990) reported pile–soil–pile interaction factors in terms of Chan's solution (Chan et al. 1974) for various pile spacing, relative stiffness, and slenderness ratios.

6.4.2 Infinite layer approach

Guo et al. (1987) and Cheung et al. (1988) proposed an infinite layer approach. The stress analysis for a single pile embedded in layered soil was performed through a cylindrical coordinate system. Each soil layer was represented by an infinite layer element and the pile by a solid bar. The displacements of the soil layer were given as a product of a polynomial and a double series. The strain-displacement and stress-strain relations were established from the displacement fields; therefore, the total stiffness matrix could be readily formed.

The interaction between piles 1 and 2 (see Figure 6.3) is simulated through the following procedure:

1. Replacing pile 2 with a soil column of the surrounding soil properties. The settlement of pile 1 as well as the soil due to the action of unit load on the pile is then computed by the single pile model. Ignoring the change in the displacement field due to the existence of pile 2, the force acting along pile 2 can be readily calculated by multiplying the displacement vector and the stiffness matrix of pile 2. The differences

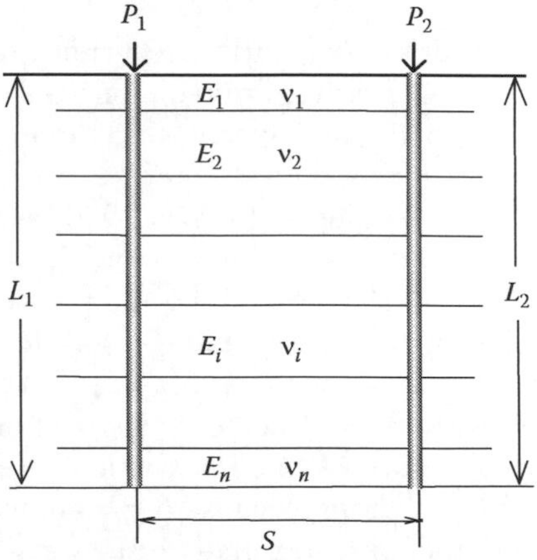

Figure 6.3 Model for two piles in layered soil. (From Guo, D. J., L. G. Tham, and Y. K. Cheung, *Computers and Geotechnics* 3, 4, 1987.)

between the forces on the pile 2 and those computed from the infinite layer model are regarded as residual forces, which are applied in the opposite direction along pile 2 to maintain the equilibrium of the whole system.

2. Likewise, if the forces are applied to pile 2, pile 1 is replaced by a soil column. The soil movement and residual forces induced in pile 1 are computed.
3. The whole procedure (1) to (2) is repeated by applying the residual forces of each step on pile 1 and pile 2 accordingly until the changes in the displacement of both piles due to the loading are negligible. By this analysis, the resulting interaction factors for two identical piles embedded in homogenous soil generally agree with those by Poulos (1968).

6.4.3 Nonlinear elastic analysis

Trochanis et al. (1991) studied the response of a single pile and pairs of piles by undertaking a 3-dimensional FE analysis using an elastoplastic model. The results demonstrated that as a result of the nonlinear behavior of the soil, the pile–soil interface interaction, especially under axial loading, is reduced markedly compared to that for an elastic soil bonded to piles. The commonly used elastic methods for evaluating pile–soil–pile interaction can substantially overestimate the degree of interaction in realistic situations. In load transfer analysis, this nonlinear effect may be modeled using elastic interaction factors and adding nonlinear components afterwards (Randolph 1994; Mandolini and Viggiani 1997).

6.4.4 Influence of nonhomogeneity

Vertical soil nonhomogeneity significantly affects pile group behavior (Guo and Randolph 1999), although it has limited effect on single pile response (Chapter 4, this book) under same average shear modulus along the pile depth. Figure 6.4a and b shows the difference in the interaction factors between homogeneous soil (Poulos and Davis 1980) and Gibson soil (Lee 1993b). The differences in the pile-pile interaction factor may be attributed to the variation of the average shear modulus, and the twice as large a value for a homogeneous soil compared with a Gibson soil (Randolph and Wroth 1978).

Horizontal nonhomogeneity considered so far has been limited to the shear modulus alteration caused by pile installation (Randolph and Wroth 1978; Poulos 1988). This alteration leads to a significant change in the load transfer factor and, therefore, normally results in a lower value of the pile-pile interaction factor as noted experimentally (O'Neill et al. 1977) and obtained numerically (Poulos 1988). Horizontal nonhomogeneity may

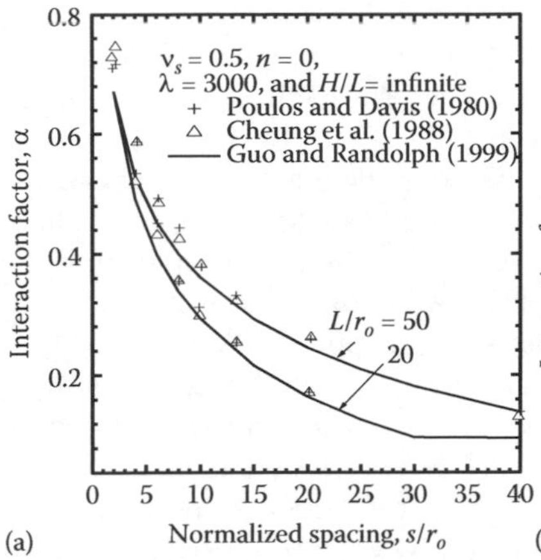

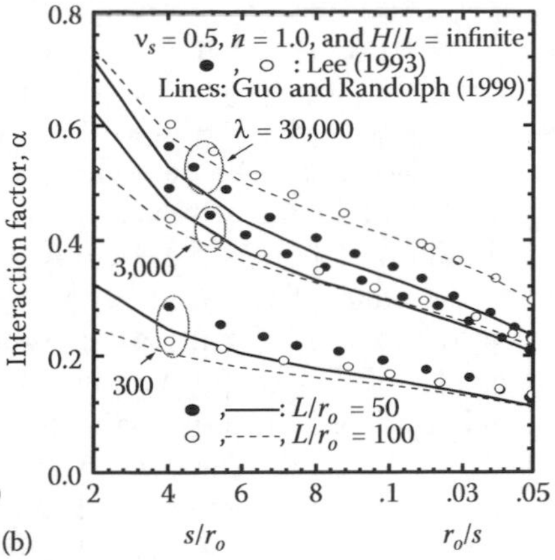

Figure 6.4 α~*s*/*d* relationships against (a) slenderness ratio, (b) pile–soil relative stiffness.

be readily incorporated into the "ζ" (refer to Equation 4.6, Chapter 4, this book) (Randolph and Wroth 1978). Therefore, closed-form solutions as shown next may be directly used to account for the nonhomogeneity.

6.4.5 Analysis based on interaction factors and superposition principle

Pile group settlement may be estimated using design charts developed through various numerical approaches and for a vast range of boundary conditions (Poulos and Davis 1980).

The influence of the displacement field of a neighboring identical pile was represented by an interaction factor between pairs of incompressible piles (Poulos 1968). The interaction factor α is defined as the ratio of the additional settlement of a pile due to the displacement field of a similarly loaded neighboring pile. The factor may be expressed using pile-head stiffness as

$$\alpha_{ij} = \frac{\text{Pile-head stiffness of a single pile}}{\text{Pile-head stiffness of a pile in a group of two}} - 1 \qquad (6.8)$$

where α_{ij} = interaction factor between pile i and pile j. The interaction factor originally defined for two identical piles is then extended to unequally loaded piles. Instead of pile-head stiffness, pile shaft displacement increase due to a displacement field of a similarly loaded neighboring pile may be represented by a shaft interaction factor between pile i and pile j, and base displacement by a base interaction factor (Lee 1993a). In fact, the interaction factor can be defined in other forms depending on the manner of estimating displacements. For instance, the settlement of a single pile δ_s is a sum of elastic shortening of a pile (w_s) and the base settlement owing to shaft (w_{pp}) and base (w_{ps}), respectively. The components w_{pp} and w_{ps} may be estimated using the pile-pile interaction factors α^{ij}_{ps} and α^{ij}_{pp}, which capture an increase in the settlement at the base of pile j due to the load transmitted along the shaft and at the base of pile i, respectively.

With a known displacement field or pile–soil–pile interaction factors, the behavior of a pile in a group can be readily evaluated using the principle of superposition. The results generally agree with those when analyzing the entire pile group (Clancy and Randolph 1993). This is illustrated later in various figures. The validity of the superposition approach both to the estimation of the pile settlement and to the determination of the load carried by each pile was confirmed through a number of field tests (Cooke et al. 1980).

It should be noted that all the solutions herein are intended for a free-standing group with the pile cap located above the principal founding layer and not contributing to the overall group performance. Otherwise, the cap-pile interaction should be considered (Randolph 1994).

6.5 BOUNDARY ELEMENT APPROACH: GASGROUP

GASGROUP (Generalized **A**nalytical **S**olutions for Vertically Loaded Pile **Group**s) is a numerical program using closed-form solutions (Guo and Randolph 1997a; Guo and Randolph 1999) designed specially for predicting behavior of vertically loaded single piles and pile groups. The solutions were developed based on the load transfer approach (see Chapter 4, this book), and the program compares well with other more rigorous numerical approaches. The distinct advantage of this program is that it is very efficient, particularly in analyzing large pile groups. As a matter of fact, for large pile groups, a rigorous numerical approach may become impractical or impossible in many cases, while GASGROUP offers an efficient and sufficiently accurate approach.

6.5.1 Response of a pile in a group

6.5.1.1 Load transfer for a pile

Closed-form solutions for a pile in a nonhomogeneous soil have been developed previously (Guo and Randolph 1997a) for the case where the elastic shear modulus of the soil varies with power law of depth (see Equation 4.1, Chapter 4, this book, and Figure 6.5).

In order to allow for the presence of neighboring piles, load transfer factors for a pair of piles is revised as (Randolph and Wroth 1979b)

$$\zeta_g = \ln(r_{mg}/r_o) + \ln(r_{mg}/s) \tag{6.9}$$

where s = pile spacing; r_{mg} = a radius of the "group," which is given by

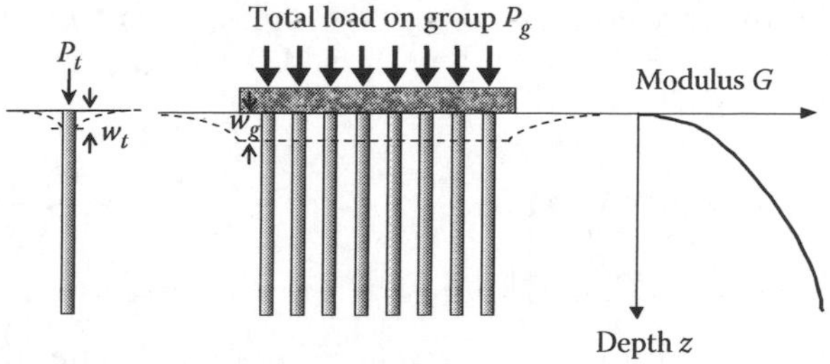

Figure 6.5 Pile group in nonhomogeneous soil. n_g = total number of piles in a group; R_s = settlement ratio; $w_g = R_s w_t$ (rigid cap); $P_t = P_g/n_g$ (flexible cap); $G = A_g(\alpha_g + z)^n$ (see Chapter 4, this book).

$$r_{mg} = r_m + \left[1 - \exp\left(1 - \frac{H}{L}\right)\right] r_g \tag{6.10}$$

where H = depth to the underlying rigid layer; r_g = about one-third to one-half of the pile spacing (Randolph and Wroth 1979b; Lee 1991). More generally, r_g may be estimated by $r_g = (0.3 + 0.2n)s$, with the s being taken as the lesser of s and r_m. The multiplier of r_g is adopted to account for the effect of a finite layer on reduction in the value of the group radius, which resembles that for the parameter A (see Equation 3.8, Chapter 3, this book).

The base-load transfer factor may be expressed as

$$\omega_g = \omega(1 + 2r_o/s\pi) \tag{6.11}$$

Generally, in the present analysis, the factors r_m and ω have been estimated using Equations 3.6 and 3.4, respectively. The analysis based on these values is termed later as "A by Guo and Randolph's equation"' or "Real A." Previously, for a pile in an infinite layer, the values of A and ω were reported as 2.5 and 1, respectively (Randolph and Wroth 1978). Analyses using these simple values for A and ω are given later as well for $H/L = \infty$, and the results are referred to as "$A = 2.5$."

6.5.1.2 Pile-head stiffness

The solutions for a single pile can be readily extended to a pile in a group, through replacement of the load transfer factors, ζ, ω for a single pile in Equations 4.5 and 4.12 (see Chapter 4, this book) with the factors, ζ_g, ω_g for a pile in a group. Therefore, the ratio of load, P, and settlement, w, at any depth, z, may be expressed as (Guo and Randolph 1997a)

$$\left[\frac{P(z)}{G_L w(z) r_o}\right]_g = \sqrt{2}\pi\sqrt{\frac{\lambda}{\zeta_g}} C_{vg}(z) \tag{6.12}$$

where the subscript g refers to a pile in a group; λ = relative stiffness ratio between pile Young's modulus, E_p, and the soil shear modulus at just above the base level, G_L; the function, $C_{vg}(z)$, is given by

$$C_{vg}(z) = \frac{C_1(z) + \chi_{vg} C_2(z)}{C_3(z) + \chi_{vg} C_4(z)} \left(\frac{\alpha_g + z}{\alpha_g + L} \right)^{n/2} \tag{6.13}$$

The individual functions $C_j(z)$ are given by modified Bessel functions of fractional order of Equation 4.17 (see Chapter 4, this book) for which the argument y is replaced with y_g

$$y_g = 2mk_{sg}(\alpha_g + z)^{1/2m} \tag{6.14}$$

$$k_{sg} = \frac{\alpha_g + L}{r_o} \sqrt{\frac{2}{\lambda \zeta_g}} \left(\frac{1}{L + \alpha_g} \right)^{1/2m} \tag{6.15}$$

where $m = 1/(2 + n)$. The ratio χ_{vg} is given by

$$\chi_{vg} = \frac{2\sqrt{2}}{\pi(1 - \nu_s)\omega_g \xi_b} \sqrt{\frac{\zeta_g}{\lambda}} \tag{6.16}$$

Note that the surface value of C_{vg} must be taken as a limit, as z approaches zero. The ζ_g and ω_g may be replaced with the $\zeta_g\zeta_c$ and $\omega_g[1 + A(t)G_{b1}/G_{b2}]$ to incorporate creep effect (see Chapter 5, this book).

6.5.1.3 Interaction factor

Influence of the displacement field of a neighboring identical pile may be captured by the interaction factor of Equation 6.8, which is rewritten as

$$\alpha_{ij} = \frac{P_t/(G_L w_t r_o)}{[P_t/(G_L w_t r_o)]_g} - 1 \tag{6.17}$$

where P_t, w_t = pile-head load and settlement, respectively; $P_t/(G_L r_o w_t)$ = pile-head stiffness of a single pile and α_{ij} = the conventional interaction factor, which can be expressed explicitly from Equations 6.12 and 6.13:

$$\alpha_{ij} = \frac{C_v}{C_{vg}} \sqrt{\frac{\zeta_g}{\zeta}} - 1 \tag{6.18}$$

where C_{vg} and C_v = limiting values of the function, $C_{vg}(z)$ in Equation 6.13 as z approaches zero, with values of ζ_g, ω_g and ζ, ω respectively.

6.5.1.4 Pile group analysis

The settlement of any pile in a group can be predicted using the superposition principle together with appropriate interaction factors. For a symmetrical group, the settlement w_i of any pile i in the group can be written as

$$w_i = w_{t1} \sum_{j=1}^{n_g} P_j \alpha_{ij} \tag{6.19}$$

where w_{t1} = settlement of a single pile under unit head load; α_{ij} = interaction factor between pile i and pile j (for $i = j$, $\alpha_{ij} = 1$) estimated by Equation 6.18; and n_g = total number of piles in the group. The total load applied to the pile group is the sum of the individual pile loads, P_j.

For a perfectly flexible pile cap, each pile load will be identical and so the settlement can be readily predicted with Equation 6.19. For a rigid pile cap with a prescribed uniform settlement of all the piles in a group, the loads may be deduced by inverting Equation 6.19. This procedure for solving Equation 6.19 has been designed in the GASGROUP program. By assuming pile cap as rigid, the GASGROUP program gives a very good prediction of settlement in comparison with other numerical approaches and field measured data. In the present analysis, estimation of settlement of a single pile under unit head load and the interaction factors are based on closed-form solutions. Therefore, the calculation is relatively quick and straightforward (e.g., for a 700-pile group, the calculation only takes about five minutes on an IBM 486 personal computer).

6.5.2 Methods of analysis

To facilitate calculation of group settlement, the settlement ratio is rewritten as

$$R_s = \frac{\text{average group settlement}}{\text{settlement of a single pile in the group with the same average load}} \tag{6.20}$$

For groups with less than 16 piles, the R_s can be calculated using the pile-head stiffness $P_g/(sG_L w_g n_g^{.5})$ in Figures 6.6 through 6.8. It has negligible variation with the shape of the group. The figure may thus be used for other similar shapes of groups with the same number of piles. As the bases of the piles get closer to the rigid layer, the pile-head stiffness increases and interaction factor decreases. For groups containing more than 16 piles, the group stiffness or settlement influence factors I_g (= $w_g d E_L / P_g$) and I (= $w_t r_o G_L / P_t$, Chapter 4, this book) may be gained using Figure 6.9, which allow the R_s to be estimated using

$$R_s = \frac{I_g n_g}{I \text{ (Influence factor for a single pile)}} \frac{1}{2(1+\nu_s)} \tag{6.21}$$

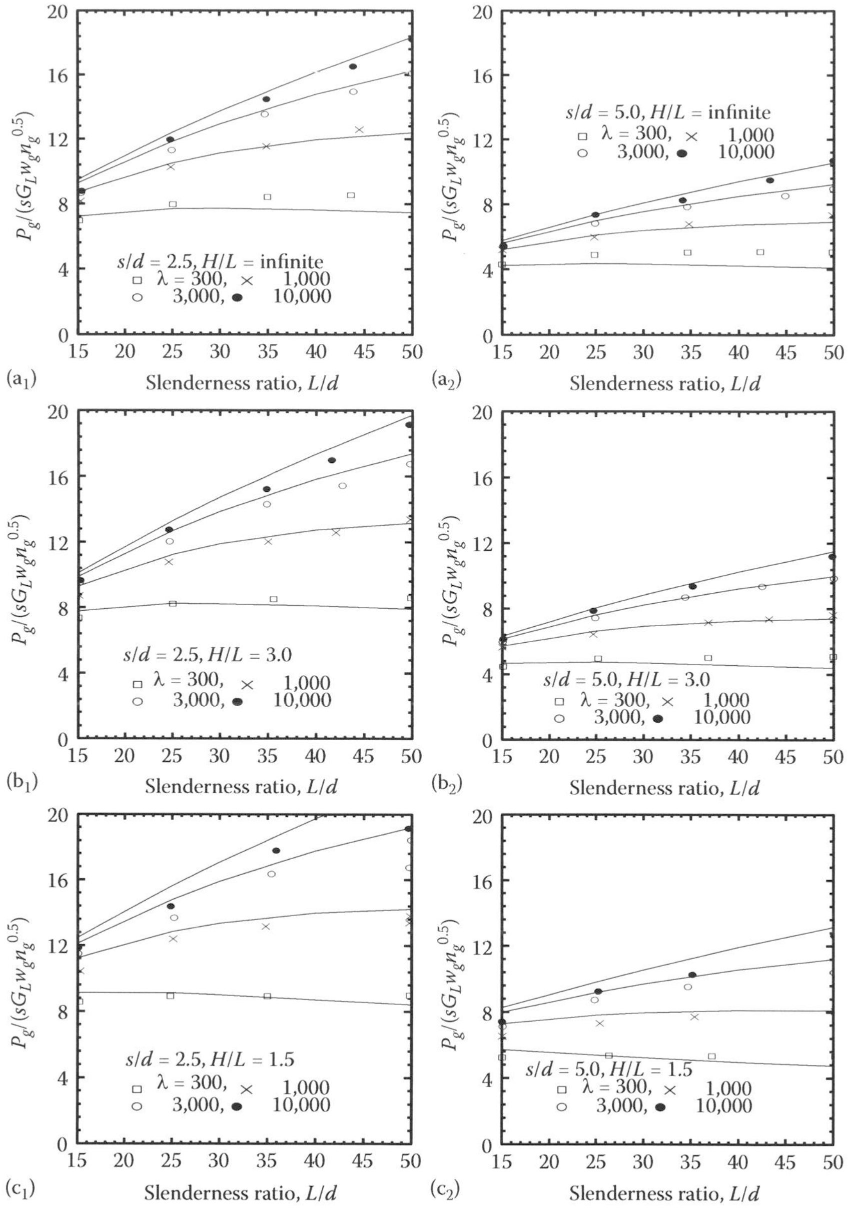

Figure 6.6 Group stiffness for 2×2 pile groups. (a_i) $H/L = \infty$. (b_i) $H/L = 3$. (c_i) $H/L = 1.5$ ($i = 1$, $s/d = 2.5$ and $i = 2$, $s/d = 5$) (lines: Guo and Randolph 1999; dots: Butterfield and Douglas 1981).

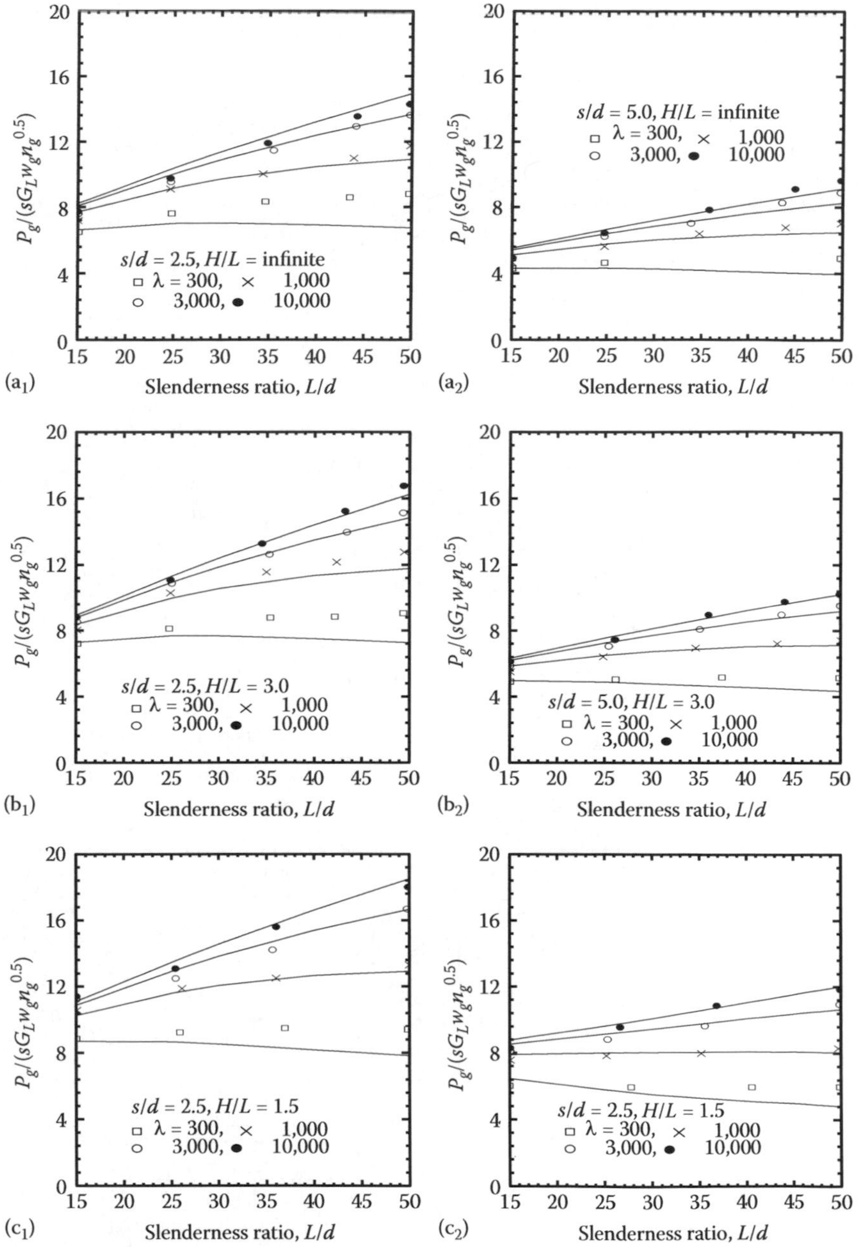

Figure 6.7 Group stiffness for 3×3 pile groups (a_i) $H/L = \infty$. (b_i) H/L = 3. (c_i) H/L = 1.5 (i = 1, s/d = 2.5 and i = 2, s/d = 5) (lines: Guo and Randolph 1999; dots: Butterfield and Douglas 1981).

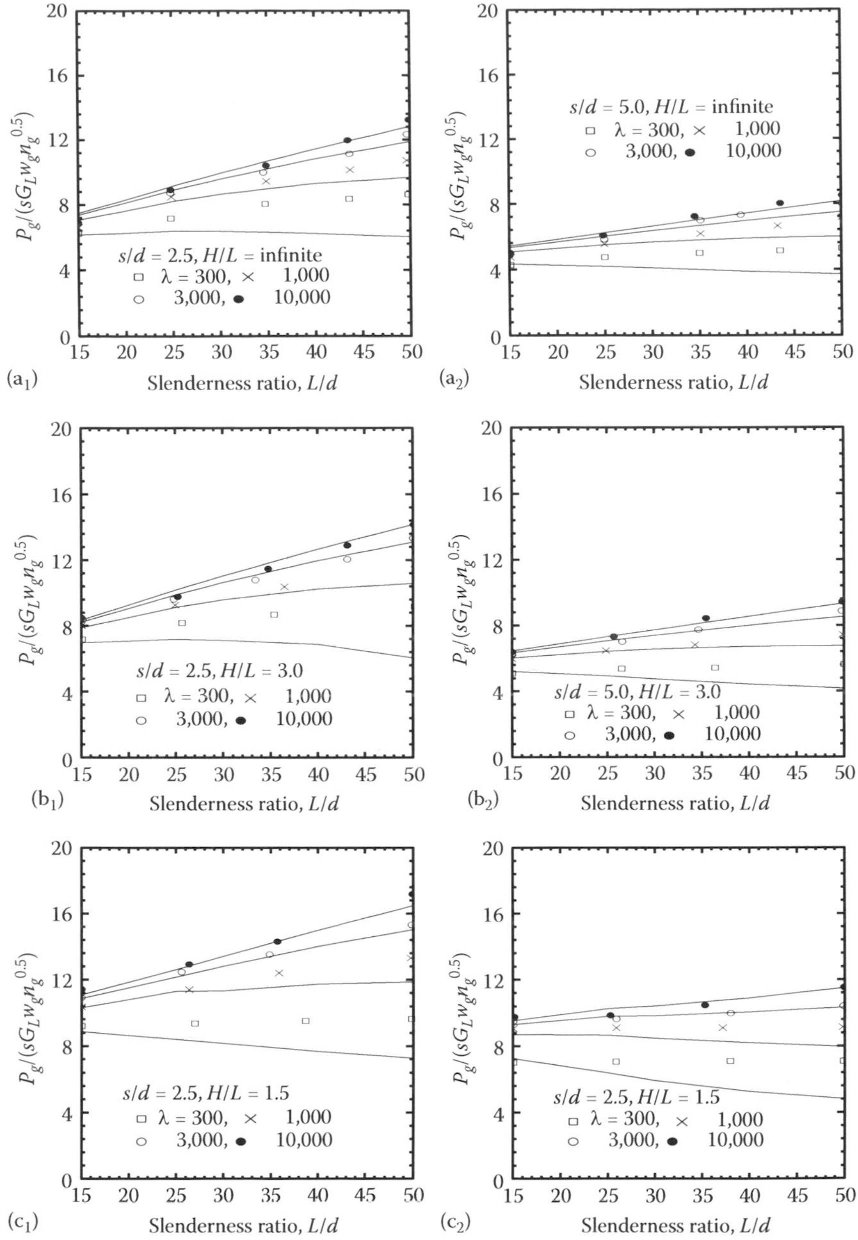

Figure 6.8 Group stiffness for 4×4 pile groups (a_i) $H/L = \infty$. (b_i) $H/L = 3$. (c_i) $H/L = 1.5$ ($i = 1$, $s/d = 2.5$ and $i = 2$, $s/d = 5$). (lines: Guo and Randolph 1999; dots: Butterfield and Douglas 1981).

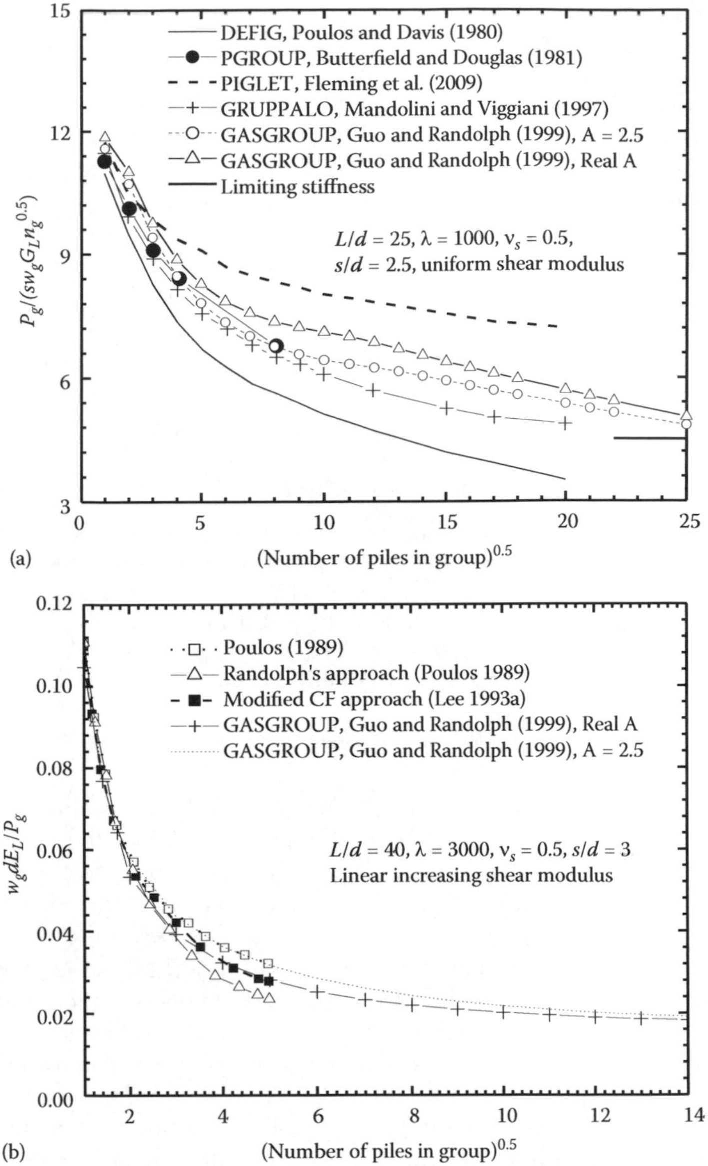

Figure 6.9 (a) Group stiffness factors. (b) Group settlement factors.

The R_s presented herein captures the impact of the five factors (mentioned earlier) on the group settlement w_g, and allows the settlement w_g (= $R_s w_t$) to be calculated for a rigid pile-cap (see Figure 6.5). Figure 6.9a (for L/d = 40) indicates that the increase in the number of piles in a group can reduce the group settlement. However, this becomes increasingly less effective as the number exceeds, say, 81, as is evident in Figure 6.9a for more closely spaced piles. For example, with $s/d = 3$, the value of $w_g dE_L/P_g$ for a 2×2 group is 0.0582, and 0.0393 for a 3×3 group. The reduction in settlement is 67.5% by more than doubling the pile numbers. If the group is further increased to 5×5 piles, the value of $w_g dE_L/P_g$ would decrease to 0.0317, giving a total settlement 54.5% that of 4 piles, but the pile cap size (width and breadth) will increase dramatically as well.

The group settlement may be reduced, as the depth to the rigid layer H decreases or the size of the pile group increases (Poulos 1968). An increase in stiffness of frictional piles leads to an increase in R_s, whereas a reverse trend is noted for end bearing piles. For instance, with $L/d = 50$ and an infinitely large H ($H/L > 4$), the stiffness $P_g/(sG_L w_g n_g^{.5})$ is about 19.0, 15.0, and 12.8 for 4, 9, and 16 piles in a group, respectively, whereas the stiffness becomes 20.0, 18.2, and 16.5 for the groups in soil with $H/L = 2$.

Example 6.1 A comparison of head stiffness

Given a pile slenderness ratio L/d of 40, and a normalized pile center-to-center spacing ratio s/d of 3, a 4×4 group will have a breadth B_i of 10d. The ratio B_i/L (see Figure 2.12, Chapter 2, this book) of 0.25 leads to $P_g/(G_L w_g B_i) = 9.0$ (Randolph 2003b). On the other hand, Figure 6.9 indicates $w_g dE_L/P_g = 0.033$ for the group, which can be converted to $P_g/(G_L w_g B_i) = 9.0$ (with $\nu_s = 0.5$) as well.

6.5.3 Case studies

Input parameters for each analysis include (a) soil shear modulus distribution down the pile, Poisson's ratio, and the ratio of H/L; (b) the dimensions and Young's modulus of the pile; (c) the number of piles in the foundation and pile centre-centre space. In the prediction of R_s, there is no practical difficulty in using the exact center-center spacing for each pair of piles. However, for convenience, an equivalent average pile spacing may be assessed for practical prediction. First, irregular plans of large groups are converted to equivalent rectangular plans. Second, a mean area per pile is obtained, with the total plan area of a pile group being divided by the number of piles in the group. Finally, taking the mean area as a square, the average pile "spacing" is thus the length of the side of the square. Ten published cases were studied using the current approach. The input parameters are summarized in Tables 6.1 and 6.2 (assuming $\alpha_g = 0$ for convenience); typical cases are examined in the examples that follow.

Example 6.2 19-story R. C. building

The 19-story building (Koerner and Partos 1974) was founded on 132 permanently cased driven piles, covering an approximately rectangular area, about 24 m by 34 m. The piles were cased 0.41 m in diameter and 7.6 m in

Table 6.1 Summary of parameters for group pile analysis

Example	6.2	6.3	6.4	6.5	6.6	6.7	6.8a and b	6.9	6.10[a]	6.11[a]
L (m)	7.6	16.0	15.0	4.5	13.0	13.1	20.0	9.15	27.0	13.4
r_o (m)	0.205	0.4	0.2425	0.084	0.225	0.137	0.30	0.1365	0.25[b]	0.26
E_p(GPa)	30.0	9.8	30.0	210.0	30.0	13.44	23.0	13.4	26.0	30.0
n	0.5	0.8	0.4	0.5	0.4	0.65	1.0	0.0	1.0	0.0
ν_s	–	0.3	0.5	0.5	0.5	0.5	0.3	0.3	0.45	0.4
G_L (MPa)		92.2	50.0	38.0	38.72	151.0	145~147	43.43	13.6	28.6
G_L/G_b		1.0	1–3	1–3	1–3	1–3	1–3	3.0	1.0	1.0
Spacing (m)	2.48	4.41	1.5	0.505	1.6	0.82	2.14	0.82	2.0	2.02
H/L	2.2	2.70	1.9	4.0	1–4	4.0	1.75~2.7	1.5	4.0	2.5
w_t (mm)	3.3	5.0	1.17	0.23	0.85	2.85	1.3	4.56	5.0	3.15

a Guo and Randolph (1999).

b width of a square cross-section (m), converted to r_o = 0.141 m.

Table 6.2 Comparison between predicted and calculated settlement of group piles

Example	*Description*	*Modulus (MPa)*[a]	*Number of piles*	*Observed/ predicted* w_g *(mm)*
6.2	19-story concrete building	$G = 16.34z^{0.5}$	132	64/65.5
6.3	5-story building	$G = 10.03z^{0.8}$	20	10~20(14)/12.6
6.4	Dashwood house (OCR > 1)	$G = 18.8z^{0.4}$	462	33/37.98
6.5	Jacked piles in London Clay	$G = 20.53z^{0.5}$	3	0.38/0.41
6.6	Stonebridge Park	$G = 13.8z^{0.4}$	351	18/17.7
6.7	Driven piles in Houston clay (OCR > 1)	$G = 49.5z^{0.65}$	9	4.55/6.68
6.8	Napoli: (a) Office Tower	$G = 7.195z$	314	20.8/22.0
	(b) Holiday Inn	$G = 7.367z$	323	30.9/30.15
6.9	San Francisco 5-pile	$G = 43.43$	5	5.75/5.56
6.10[b]	Molasses tank	$G = 0.504z$	55	29~30/28.7~29.2
6.11[b]	Ghent terminal silos	$G = 28.6$	697	180/186.3

[a] Depth z in meters
[b] Guo and Randolph (1999)

length, with an expanded base of 0.76 m. Each of the total 132 piles shares a mean area of 6.18 m^2 and has a pile "spacing" of 2.48 m. The SPT variation indicates $n = 0.5$. The shear modulus follows $G(\text{MPa}) = 16.43z^{0.5}$, with the ratio of H/L = 2.2. The single pile-loading test shows a secant stiffness P_t/w_t of 350 kN/mm. Young's modulus of the pile was measured as 30 GPa. With these parameters, the GASGROUP predicts a settlement ratio (R_s) of 19.85, a single pile settlement of 3.3 mm under the average load of 1.05 MN, and an average group settlement of 65.5 mm. This prediction agrees with the measured average of ~64 mm, ranging from a maximum of 80 mm near the center to 37 mm near the corners of the building.

Example 6.3 5-story building

A piled raft foundation has been constructed in Japan for a five-story building. The plan area was measured 24 m by 23 m. The total working load of 47.5 MN was supported by a total of 20 piles to reduce the settlement (Yamashita et al. 1993). The piles were 16 m in length, 0.7 and 0.8 m in diameter, and were at a pile center-to-center spacing of 6.3 to 8.6 times the pile diameter.

The shear modulus profile (Yamashita et al. 1993) may reasonably be approximated by $G\ (MPa) = 10.03z^{0.8}$. Assuming E_p = 9.8 GPa, the GASGROUP analysis was conducted as shown in Table 6.3. The settlement ratio, R_s, was estimated to be 2.516. Under the average working load of 2.4 MN, the single pile settlement was ~5.0 mm. Because of this, the settlement of the pile group was predicted as 12.6 mm. This agrees well with the average settlement of ~14 mm (ranging between 10 and 20 mm) measured at completion of the building.

Table 6.3 A samplc calculation by GASGROUP

Method I (A by Guo and Randolph's equation):
INPUT DATA FILE ofYamashit.dat

Five-story building (Yamashita et al., 1993) Non-rectangular
16.,0.4,9800000.,20,0.8,0.3,92210.,1.0 20,0.,6.,12.,18.,24.,0.,6.,12.,18.,24.,0.,6.,12.,18., 24.,0.,6.,12.,18.,24.
20,0.,0.,0.,0.,0.,6.,6.,6.,6.,6., 11.,11.,11.,11.,11.,23.,23.,23.,23.,23. 2.7 Rigid 1.
A by Guo and Randolph's equation 5.0

OUTPUT FROM PILE GROUP ANALYSIS PROGRAM - GASGROUP
(Version dated October 1999)
Project name: Five-story building (Yamashita et al., 1993)
Input file name = yamashit.DAT

SOIL AND PILE PROPERTIES

Pile length, L (m) = 16.00
Pile Young's modulus, E_p (kPa) = 0.98E+07
Nonhomogeneity factor, n = 0.80
Poisson ratio of the soil = 0.30
Pile radius, r_o (m) = 0.40
Number of piles in the group, n_g = 20
Modulus distribution factor, A_g (kPa/m^n) = 10034.18
Modulus at just above the base level, G_L (kPa) = 92210.00
End-bearing non-homogeneity factor, G_L/G_b = 1.00
Coordinates for selected piles in the group (m)

Pile No.	X	Y	Pile No.	X	Y	Pile No.	X	Y	Pile No.	X	Y
1	0.000—	0.000	2	6.000—	0.000	3	12.000—	0.000	4	18.000—	0.000
5	24.000—	0.000	6	0.000—	6.000	7	6.000—	6.000	8	12.000—	6.000
9	18.000—	6.000	10	24.000—	6.000	11	0.000—	11.000	12	6.000—	11.000
13	12.000—	11.000	14	18.000—	11.000	15	24.000—	11.000	16	0.000—	23.000
17	6.000—	23.000	18	12.000—	23.000	19	18.000—	23.000	20	24.000—	23.000

Ratio of the depth to the underlying rigid layer and L, H/L : 2.70
CONDITIONS FOR THE ANALYSIS
Pile cap is assumed as rigid
Uzing Guo and Randolph's (1998) expression of "A" to estimate the load transfer factor
Run commenced on 10/27/99 At 23:12:33.75

OUTPUT OF THE ANALYSIS

Pile-pile interaction factor for selected piles, (I,J)

Pile I—J,	Alpha(I,J)	Pile I—J,	Alpha(I,J)	Pile I—J,	Alpha(I,J)	Pile I—J,	Alpha(I,J)
1—1	1.000000	1—2	0.108655	1—3	0.073560	1—4	0.073484
1—5	0.073446	1—6	0.108655	1—7	0.086154	1—8	0.073536
1—9	0.073477	1—10	0.073443				

Load transfer parameter, A = 1.45
Base load transfer factor = 1.26
Settlement ratio, R_s = 2.516
Settlement factor, I_g = 0.048
Normalized pile-pile spacing (s) by pile diameter (d), s/d = 7.500
Normalized group stiffness, $P_g/(w_g s G_L n_g^{0.5})$ = 1.614
Group stiffness, $P_g/(w_g r_o G_L)$ = 108.257
Input settlement of a zingle pile (mm) = 5.00
Settlement of the pile group, w_g (mm) = 12.579
Run completed on 10/27/99 At 23:12:36.66

Example 6.4 Dashwood house

Dashwood House (Green and Hight 1976) is a 15-story, 61-m-high building located close to Liverpool Street Station in London. The building has a single-story basement resting on a rectangular piled raft foundation of 33.8 × 32.6 m. The overall load on the foundation was 279 MN, and the measured settlement was 33 mm. The pile group consisted of 462 (21 × 22) bored piles in a grid of 1.5 m square spacing. Each had a diameter of 0.485 m, a length of 15 m, and a Young's modulus, E_p, of 30 GPa. The subsoil profile beneath the building was 1 m placed compacted gravel, followed by approximately 29 m of London clay, and 10 m of Woolwich and Reading beds. The shear modulus G (London clay) was estimated as: $G = 30 + 1.33z$ (MPa). Poisson's ratio, ν_s, was taken as 0.5.

The equivalent n was deduced as 0.2~0.4 in terms of the constant of P_t/w_t. Under an average working load of 604 kN per pile, settlement of a single pile was estimated as 1.17 mm. The results from the two trial analyses are as follows: (1) $n = 0.2$, $A_g = 26.1$ MPa/m$^{0.2}$, $G_L = 44.92$ MPa, $P_g/(G_L w_g r_o) = 668.8$, $R_s = 32.68$, and w_g is evaluated accordingly as 38.2 mm; (2) $n = 0.4$, $A_g = 18.8$ MPa/m$^{0.4}$, $G_L = 55.51$ MPa, $P_g/(G_L w_g r_o) = 543.6$, $R_s = 32.47$, and $w_g = 38$ mm. The estimated settlement of 38~38.2 mm is within 15% of the measured value. This is quite satisfactory in view of 462 piles in the group, and the sensitivity to any discrepancy between the calculated single pile settlement of 1.17 mm and the actual one.

Example 6.5 Jacked piles in London clay

A series of field tests was conducted on pile groups embedded in London clay (Cooke et al. 1980). Each tubular steel pile has an external radius of 84 mm and a wall thickness of 6.4 mm. The piles were embedded to a depth of 4.5 m and at a spacing of three pile diameters. The clay extends from the mudline to a depth of about 30 m. Shear modulus increases linearly from 15.6 MPa (at the surface) to 38.0 MPa (at the pile base, G_L), taking Poisson's ratio as 0.5. The measured load-settlement curve for pile A (the middle pile) shows a settlement of 0.23 mm at an average working load of 33.3 kN. Settlement of the three-pile group loaded with a rigid pile-cap was recorded as 0.38 mm under a total load P_g of 100 kN.

The measured P_t/w_t of 145 kN/mm allows $G_L = 43.6$ MPa and $n = 0.5$ ($A_g = 20.53$ MPa/m$^{0.5}$) to be back-calculated. The group results are $P_g/(G_L w_g r_o) = 100.1$ and $R_s = 1.79$. The predicted group settlement is 0.41 mm, which is close to the measured value of 0.38 mm.

Example 6.6 Stonebridge Park apartment building

The 16-story Stonebridge Park apartment building (Cooke et al. 1981) located in the London borough of Brent has a building foundation of 43.3 m long by 19.2 m wide that consists of a heavily reinforced concrete raft 0.9 m thick resting on 351 cast in situ bored concrete piles. The piles are 0.45 m in diameter and 13 m in length, formed at a square spacing of 1.6 m. The shear modulus distribution of the subsoil is described by $G = 20 + 1.44z$ (MPa).

The average working load per pile was 565 kN as any load carried by the raft was ignored. The settlement of building was measured ~10.5 mm at the end of construction and increased to 18 mm after 4 years. Incremental quick loading tests were carried out on the site prior to construction. They provide higher bounds of P_t/w_t and shear modulus G_L of 38.72 MPa (assuming $n = 0.4$). With $A_g = 13.8$ MPa/m$^{0.4}$, GASGROUP offers $P_g/(G_L w_g\ r_o) = 399.3$, $R_s = 36.7$, and $w_g = 31.9$ mm. This estimated settlement of the pile group is larger than the measured value of 19 mm. If the underlying rigid layer exist at a depth of, say, 20~26 m (thus $H/L = 1.5$~2.0), the group settlement will be well predicted as 17.7~25.2 mm.

Example 6.7 Driven piles in overconsolidated clay

A series of tests were conducted on single piles and a pile group at the University of Houston (O'Neill et al. 1982). Each steel pipe pile was 273 mm diameter and 9.3 mm in wall thickness. The piles were driven (closed ended) 13.1 m into stiff overconsolidated clay. A group was installed in a 3×3 configuration (nine-pile group) at a center-to-center spacing of $6r_o$ and connected to a rigid reinforced concrete block. Two separate piles were installed 3.7 m from the center of the group and on opposite sides.

The two single piles and the nine-pile group were loaded to failure. At the average load of 550 kN per pile, the settlement was recorded as 2.85 mm (or $P_t/w_t = 260$ kN/mm) from single pile tests. The settlement of the 3×3 pile group was measured at 4.55 mm. The shear modulus was deduced as increase from 47.9 MPa (at surface) to 151 MPa (at the pile base) ($\nu_s = 0.5$).

Young modulus of an equivalent solid pile was estimated using $E_p = E_{steel}\ [r_o^2 - (r_o - t)^2]/r_o^2$, where E of steel = 200 GPa, r_o and t = the outside radius and wall thickness of the pile. G_L was deduced as 263.57 MPa using $P_t/w_t = 269$ kN/mm and $n = 0.65$ (from reported distribution). The group was featured by $n = 0.65$, $A_g = 49.5$ MPa/m$^{0.65}$, $G_L = 263.57$ MPa, $P_g/(G_L w_g\ r_o) = 26.28$, and $R_s = 2.34$. The group settlement was estimated as 6.68 mm and ~2 mm higher than the measured value. This may be attributed to neglecting the ground-level modulus of the highly overconsolidated clay; to the gradual underestimation of the modulus at greater depths by using $n = 0.65$; and the existence of any underlying rigid layer. All these factors could reduce the calculated settlement, which is not fully considered herein.

Example 6.8 Napoli Holiday Inn and office tower

Parts of the new Directional Center of Napoli were built on rigid foundation slabs with a thickness of 1.2~3.5 m (Mandolini and Viggiani

1997). They in turn rest on 314 (Office Tower) and 323 (Holiday Inn) auger piles of the "PressoDrill" type to support ~200 MN building load. Each pile had a diameter of 0.6 m, a length of 20 m, and a Young's modulus of 23 GPa (back-calculated), which caters to an average permanent load of 670 kN.

The site subsoil from the surface downwards is man-made ground, volcanic ashes/organic soils; stratified sands, cohesionless pozzolana, and volcanic tuff. The upper soils extend about 16 m and encounter a peat layer (with very small values of q_c). The thickness of stratified sands increases from 5 to 20 m from the Office Tower towards the Inn, being "parallel" to the deeper pyroclastic formation from 22 to 37 m below the surface.

The piles are arranged in rectangular patterns of 15×20 (Office Tower) and 15×21 (Holiday Inn). The Office Tower site is 1,308 m^2, thus the average single pile area was ~4.36 m^2 with a pile spacing of ~2.18 m for both sites.

6.8.1 Office tower

The site subsoil is underlaid by a volcanic tuff layer ~35 m below the surface (i.e., H/L = 1.75). The G_L was deduced as 145.2 MPa using a P_t/w_t = 600 kN/mm (measured from single pile tests). The calculated results are: G_L = 147.9 MPa (slightly different from single pile using G_L/G_b = 3.0), $P_g/(G_L w_g r_o)$ = 253.4, and R_s = 16.96. The settlement of the group is thereby evaluated as 22.0 mm.

6.8.2 Holiday Inn

The rigid layer of volcanic tuff dramatically increases with depth across the site with H > 54 m. Taking H/L = 2.7 and G_L/G_b = 2.0, a G_L of 143.9 MPa was deduced from the measured P_t/w_t. These allow $P_g/(G_L w_g r_o)$ = 185.8, R_s = 23.19, and a group settlement of 30.15 mm.

Calculations using a ratio G_L/G_b between 1.0 and 3.0 induce a less than 1 mm difference in the computed settlement. The predicted 22.0 mm agrees well with the average of 20.8 mm at the Office Tower, with 9.2 mm (at the edge) and 29.1 mm (at the center of the foundation). The prediction of 31.15 mm is also confirmed by the measured settlements of 26.9~32.7 mm (an average of 30.9 mm) at the Holiday Inn.

Example 6.9 San Francisco five-pile group

Load tests to failure were performed on a single pile and a five-pile group installed in hydraulic-fill sand at a site in San Francisco (Briaud et al. 1989). The closed-end steel pipe piles were 273 mm in diameter and 9.3 mm in wall thickness. All piles were driven to a depth of 9.15 m below the ground surface (through a 300-mm predrilled hole to a depth of 1.37 m) at spacing of $6r_o$. The piles were connected by 1.8-m, rigid, thick reinforced concrete cap. The subsoil at the site was sandy gravel fill (to a depth 1.37 m below the surface), followed by a hydraulic-fill layer of clean sand (down to 12.2 m), and further down bedrock (at a depth 14.3 m). Results of CPT indicate a uniform distribution of the modulus ($n = 0$).

At an average load of 311.5 kN per pile, the single pile settlement was measured as 4.56 mm, and the group settlement was 5.75 mm. This offers P_t/w_t of 222 kN/mm. With $E_p = 13.4$ GPa, it follows $n = 0$, and $G_L = A_g = 43.43$ MPa; and furthermore, $P_g/(G_L w_g r_o) = 152.8$, $R_s = 1.22$, and $w_g = 5.56$ (mm). This estimated 5.56 mm well matches the measured value of 5.75 mm.

Finally, the impact of the shear modulus profiles of the subsoil on the settlement predictions were examined previously through the "Molasses tank" and "Ghent terminal silos" building (Guo and Randolph 1999), which are provided here in Tables 6.1 and 6.2 as Examples 6.10 and 6.11, respectively.

6.6 COMMENTS AND CONCLUSIONS

A wide variety of geological conditions were encountered for the ten cases investigated so far. A comparison between the observed and calculated settlements for all cases is provided in the Table 6.2. For normal clay, the predictions are within 7% to the measured data. Lack of information may be complemented through parametric analysis, such as the affect of the H/L in Example 6.6. Below are some key findings in this study.

- Using average pile center-to-center spacing has negligible effect on the predicted settlement.
- Variation of G_L/G_B from 1 to 3 has limited effect on the predicted settlement.
- A shallow rigid layer of $H/L < 2$ will have obvious effect on the prediction.
- Existing distribution of shear modulus, and also back-estimated one (from single pile tests) should be used in calculating settlement of pile groups.
- Inaccuracies can creep into the back-estimation method due to the selection of P_t/w_t from a plot, as misinterpretation of the initial gradient may occur. Prediction of settlement of a single pile is necessary. A nonzero ground-level modulus should be adopted for overconsolidated clays to improve the accuracy of the predicted settlement, as observed in Examples 6.4 and 6.7.
- The program is efficient with all calculations taking under two min, even for a 697-pile group that caters to all important factors such as expanded pile bases, varying subsoil profiles, and depths to rigid layers.

Chapter 7

Elastic solutions for laterally loaded piles

7.1 INTRODUCTION

Analyzing response of laterally loaded piles and pile groups so far has generally been recourse to numerical approaches (Poulos 1971; Banerjee and Davies 1978; Randolph 1981b; Chow 1987). These approaches are more rigorous, but their practical applications are often limited to small pile groups (Guo 1997), and their accuracy is difficult to justify without analytical solutions (Guo 2010). Approaches using an empirical load transfer model (Hetenyi 1946; Matlock and Reese 1960) and a two-parameter model (Sun 1994) have been proposed by modeling pile–soil interaction using a series of independent elastic springs along the shaft.

The elastic springs are generally characterized by the modulus of subgrade reaction k (Hetenyi 1946). The modulus was gained through fitting with relevant rigorous solutions (Vesić 1961a; Baguelin et al. 1977; Scott 1981) but is different between matching deflection and moment (Scott 1981). The modulus for p-$y(w)$ curve is generally based on an empirical fitting to relevant measurements (Matlock and Reese 1960). None of them seems to incorporate properly the impact of the soil response in radial direction on the modulus, which is apparent for vertically loaded piles (Randolph and Wroth 1978; Scott 1981; Guo 1997; Guo and Randolph 1997a; Guo and Randolph 1998; Guo and Randolph 1999; Guo 2000b). To resolve the problem, the soil displacement of $w(z)\phi(r)$, at depth z and a distance r from pile axis is correlated to the pile displacement, $w(z)$, by displacement reduction factor $\phi(r)$ (see Equation 7.1). The functions $w(z)$ and $\phi(r)$ are resolved via total energy of a pile–soil system and linked together via a load transfer factor, γ_b (Scott 1981; Sun 1994; Guo and Lee 2001; Basu and Salgado 2008). The solutions provide unique values of modulus of subgrade reaction k and fictitious tension N_p (which ties together the independent springs) using the properly determined factor γ_b (Jones and Xenophontos 1977; Nogami and O'Neill 1985; Vallabhan and Das 1988; Vallabhan and Das 1991; Sun 1994).

The potential energy is gained using the radial displacement *u*, circumferential displacement *v*, and their compatible stress components in the pile and the soil (Vallabhan and Mustafa 1996). The energy, in light of variational approach, allows governing equations and boundary conditions to be established. This generally results in semi-analytical solutions (Vallabhan and Das 1991; Vallabhan and Mustafa 1996). Moreover, by ignoring higher order components, the stress field may be simplified (termed as a "load transfer model"), allowing compact and sufficiently accurate closed-form solutions to be obtained (Randolph and Wroth 1978; Randolph 1981a) against more rigorous numerical approaches (Guo 1997; Guo and Randolph 1998; Guo and Randolph 1999). The use of the load transfer model is preferred for a laterally loaded pile, as the two-parameter (or Vlasov's foundation) model (Nogami and O'Neill 1985; Sun 1994) is inherently unreasonable at a high Poisson's ratio of $\nu_s \geq 0.3$.

Hetenyi (1946) first developed compact-form solutions for a free-end beam of infinite and finite length (underpinned by the subgrade modulus k) that is subjected to transverse loading and external axial force N (= N_p). Sun (1994) provided solutions for fixed- and free-head piles for various base conditions using the two-parameter model. Apart from the difference in boundary conditions, the latter is more rigorous in determining the two parameters, k and N_p, using total energy, while the early solutions require advanced mathematical skills to obtain compact expressions.

In this chapter, the inherent disadvantage of the two-parameter model is circumvented using a simplified stress field. Applying the variational approach on total energy of the lateral pile–soil system, a theoretical load transfer model is developed. The model is underpinned by the modulus of subgrade reaction k and the fictitious tension N_p, which in turn are based on Bessel functions of the load transfer factor γ_b. The pile response and the factor are presented in compact expressions, respectively, along with those for estimating critical pile length, maximum bending moment, and the depth at which the moment occurs. The stress field and the closed-form solutions are verified by the available results for a rigid disc (Baguelin et al. 1977) and various numerical approaches (Banerjee and Davies 1978; Randolph 1981b; Chow 1987), respectively.

7.2 OVERALL PILE RESPONSE

7.2.1 Nonaxisymmetric displacement and stress field

The response of a circular pile subjected to horizontal load H and moment M_o at the pile-head level (see Figure 7.1a) includes the displacement, w, the bending moment, M, and the shear force, Q (see Figure 7.1b). The pile with

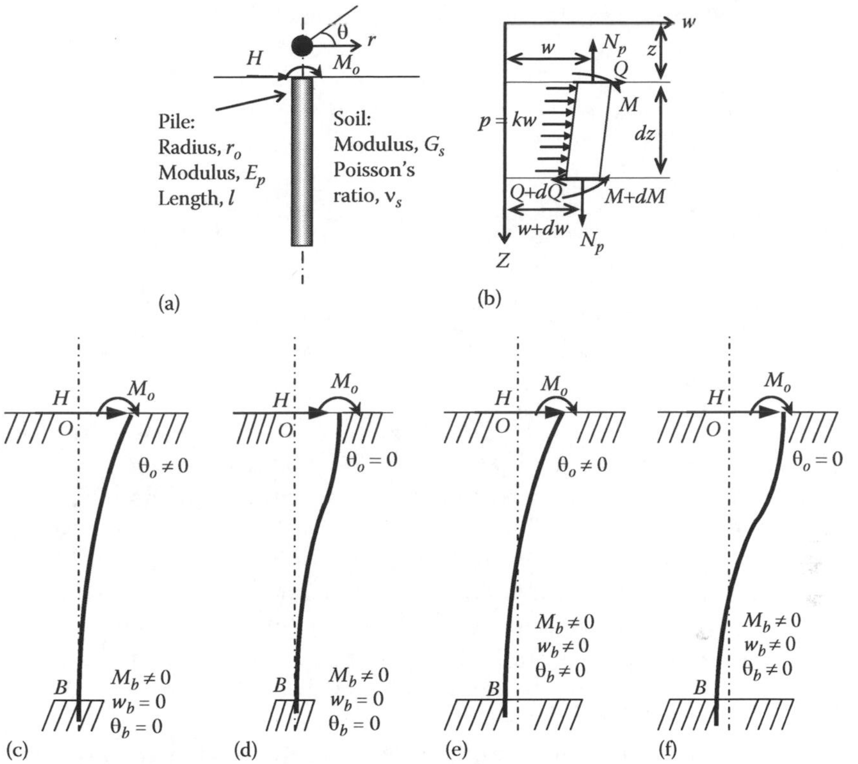

Figure 7.1 Schematic of (a) single pile, (b) a pile element, (c) free-head, clamped pile, (d) fixed-head, clamped pile, (e) free-head, floating pile, and (f) fixed-head, floating pile. (After Guo, W. D., and F. H. Lee, *Int J Numer and Anal Meth in Geomech* 25, 11, 2001.)

a length l and radius r_o is embedded in a linear elastic, homogeneous, and isotropic medium. The displacement and stress fields in the soil around the pile are described by a cylindrical coordinate system r, θ, and z as depicted in Figure 3.12a in Chapter 3, this book.

A nonaxisymmetric displacement field around the pile is observed with dominant radial u and circumferential v displacements and negligible vertical (w_s) displacement. The field may be expressed in Fourier series (Sun 1994; Cook 1995)

$$u = \sum_{n=0}^{\infty} w_n(z)\phi_n(r)\cos n\theta \quad v = -\sum_{n=0}^{\infty} w_n(z)\phi_n(r)\sin n\theta \quad w_s = 0 \tag{7.1}$$

where $w_n(z)$ = the n-th component of the pile body displacement at depth, z, and in the direction of the n-th loading component; $\phi_n(r)$ = the n-th component of the attenuation function of soil displacement at a radial distance,

r, from the pile axis; and θ = angle between the line joining the center of the pile cross-section to the point of interest and the direction of the n-th loading component. Applying elastic theory (Timoshenko and Goodier 1970) on Equation 7.1, the stresses in the soil surrounding the pile are deduced as (Guo and Lee 2001)

$$\sigma_r = (\lambda_s + 2G)\sum_{n=0}^{\infty} w_n \frac{d\phi_n}{dr}\cos n\theta \quad \sigma_\theta = \sigma_z = \lambda_s \sum_{n=0}^{\infty} w_n \frac{d\phi_n}{dr}\cos n\theta$$
$$\tau_{r\theta} = -G\sum_{n=0}^{\infty} w_n \frac{d\phi_n}{dr}\sin n\theta \quad \tau_{\theta z} = -G\sum_{n=0}^{\infty}\frac{dw_n}{dz}\phi_n \sin n\theta \quad \tau_{zr} = G\sum_{n=0}^{\infty}\frac{dw_n}{dz}\phi_n \cos n\theta \tag{7.2}$$

where λ_s = Lame's constant; G = shear modulus of the soil. With an equivalent concentrated load, H, and moment, M_o, at the pile-head (see Figure 7.1), only the $n = 1$ term of the series in Equations 7.1 and 7.2 exist (Sun 1994; Cook 1995), which offer identical displacement and stress fields to those adopted previously (Sun 1994). Other terms (e.g., $n = 2, 3$) are needed to cater for load and/or moment components in multi-directions "θ." A relevant solution for each term (n) may be obtained by the procedure detailed in this chapter for $n = 1$ and then superimposed together to yield the final results, seen in Equation 7.1, for the displacements (Cook 1995).

The displacements along with the derived stresses are directly used to establish solutions for laterally loaded piles (Sun 1994) and beams (Vallabhan and Das 1988; Vallabhan and Das 1991) or along with simplified stresses by neglecting less important components (Nogami and O'Neill 1985). The two-parameter model indicates a remarkable impact of a higher Poisson's ratio ν_s (> 0.3) on pile response, which is, however, not seen in numerical analysis (Poulos 1971; Sun 1994). The impact can be readily incorporated by using a modulus, G^* [= $(1 + 3\nu_s/4)G$)] (Randolph 1981b) and by a simplified stress field deduced using $\nu_s = 0$ (i.e., Lame's constant, $\lambda_s = 0$) in Equation 7.2 (Guo and Lee 2001). This offers Equation 3.48 shown in Chapter 3, this book.

The assumed stresses of Equation 7.2 are compared with finite element analysis and a simplified (intact model) solution (at the directions of $\theta = 0$ and $\pi/2$) for a rigid disc (Baguelin et al. 1977). The stresses σ_r and $\tau_{r\theta}$ exhibit similar trends among the predictions (see Figure 7.2 for $\nu_s = 0$). The constantly low σ_r compared to the rigid disc reflects the alleviating (coupled) interaction from neighboring springs along the pile shaft. The zero circumferential stress σ_θ also matches well with the average stress ($\nu_s = 0.33$) over radial direction of the rigid disc solution. This stress is, however, significantly overestimated by the two-parameter model using the actual λ_s.

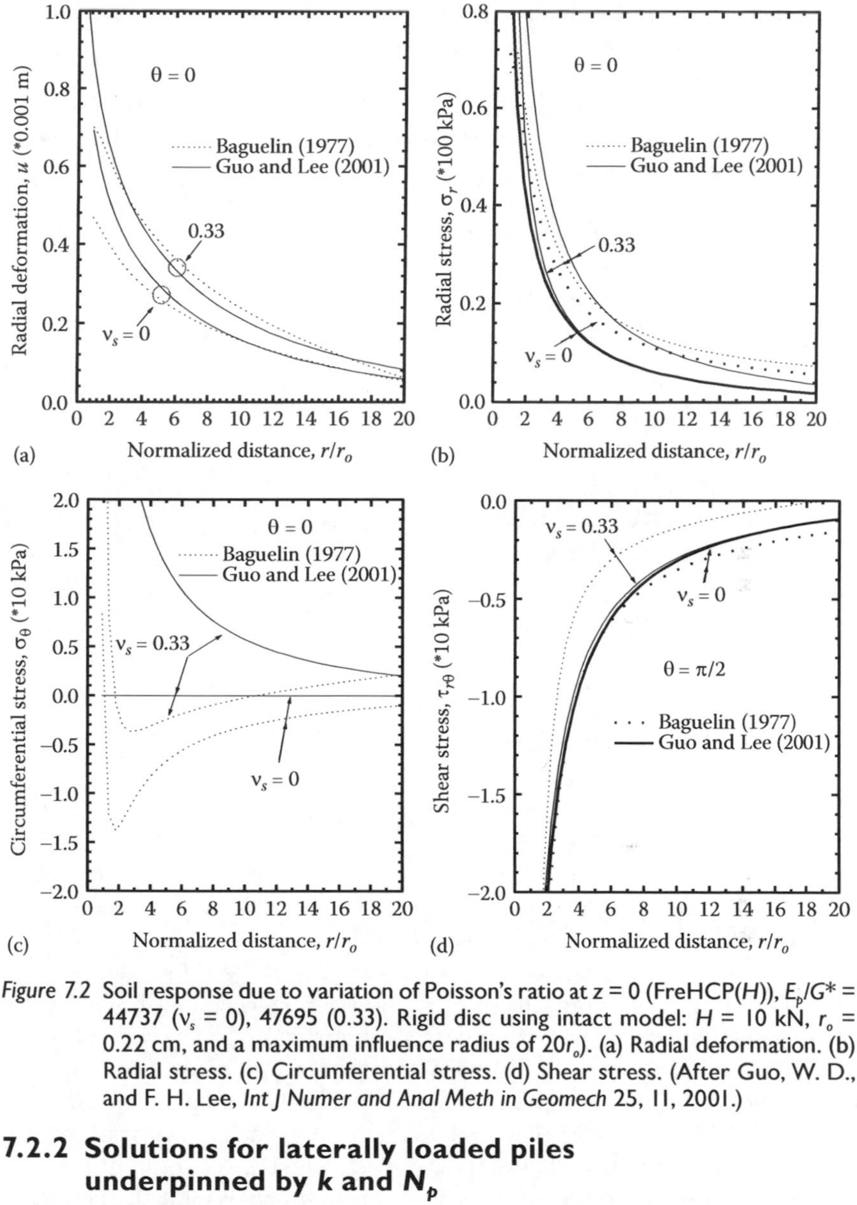

Figure 7.2 Soil response due to variation of Poisson's ratio at z = 0 (FreHCP(*H*)), E_p/G^* = 44737 (ν_s = 0), 47695 (0.33). Rigid disc using intact model: *H* = 10 kN, r_o = 0.22 cm, and a maximum influence radius of $20r_o$). (a) Radial deformation. (b) Radial stress. (c) Circumferential stress. (d) Shear stress. (After Guo, W. D., and F. H. Lee, *Int J Numer and Anal Meth in Geomech* 25, 11, 2001.)

7.2.2 Solutions for laterally loaded piles underpinned by *k* and N_p

Potential energy of the lateral pile–soil system *U* is obtained using the displacement field of Equation 7.1 and the stress field of Equation 7.2. The variation of the energy, δU, may be expressed as

$$\delta U = E_p I_p \int_0^l \frac{d^2w}{dz^2}\delta\left(\frac{d^2w}{dz^2}\right)dz + \pi r_o^2 \int_L^\infty G\frac{dw}{dz}\delta\left(\frac{dw}{dz}\right)dz + \iiint \sigma_{ij}\,\delta\varepsilon_{ij}\, r\,dr\,d\theta\,dz \tag{7.3}$$

where E_p, I_p = Young's modulus and moment of inertia of an equivalent solid cylinder pile, respectively; r_o = radius of an equivalent solid cylinder pile; σ_{ij} and ε_{ij} = stress and strain components in the surrounding soil of the pile, respectively (see Guo and Lee 2001). The virtual work, δW, done by the load H and the moment M_o due to a small displacement δy, and rotation $\delta(dw/dz)$, may be expressed as

$$\delta W = H\,\delta w\big|_{z=0} + M_o\,\delta(dw/dz)\big|_{z=0} \tag{7.4}$$

where H and M_o = the concentrated load and moment, respectively, exerted at the pile-head level. Equilibrium of the pile–soil system leads to

$$\begin{aligned}\delta U = (EI)_p &\left\{\left[\frac{d^2w}{dz^2}\delta\left(\frac{dw}{dz}\right) - \frac{d^3w}{dz^3}\delta w\right]_o^l + \int_o^l \frac{d^4w}{dz^4}\delta w dz\right\}\\ &+ \pi r_o^2 G\left\{\left[\frac{dw}{dz}\delta w\right]_l^\infty - \int_l^\infty \frac{d^2w}{dz^2}\delta w\,dz\right\} + k\int_o^\infty w\,\delta w\,dz + \left[2U_m\frac{d\phi}{dr}\delta\phi\right]_{r_o}^\infty\\ &- \int_{r_o}^\infty\left[2U_m\left(\frac{d\phi}{dr} + r\frac{d^2\phi}{dr^2}\right) - 2\phi U_n\right]\delta\phi dr + N_p\left\{\left[\frac{dw}{dz}\delta w\right]_o^\infty - \int_o^\infty \frac{d^2w}{dz^2}\delta w\,dz\right\}\end{aligned} \tag{7.5}$$

The last four components are energy in the surrounding soil, and

$$k = 2U_k = 3\pi G\int_{r_o}^\infty r\left(\frac{d\phi}{dr}\right)^2 dr \quad N_p = 2U_N = 2\pi G\int_{r_o}^\infty r\phi^2(r)\,dr \tag{7.6}$$

$$U_m = \frac{3}{2}\pi Gr\int_o^\infty w^2\,dz \quad U_n = \pi Gr\int_o^\infty\left(\frac{dw}{dz}\right)^2 dz \tag{7.7}$$

where G = shear modulus of the soil; U_k and U_N = energy of unit pile length for unit pile movement (w = 1) and unit pile rotation (dw/dz = 1), respectively; U_m and U_n = energy of unit radial length for unit radial rotation ($d\phi/dr$ = 1) and unit radial variation (ϕ = 1), respectively. The pairs U_k and U_m and U_N and U_n denote the potential energy due to the stress variations in radial direction and vertical direction, respectively. U_k, U_N could have been defined as k and N_p, respectively, but for inconsistency with the classical elastic theory (Timoshenko and Goodier 1970). Using Equations 7.4 and 7.5, the condition of $\delta W + \delta U = 0$ allows pertinent conditions and governing equations to be deduced (Guo and Lee 2001), which are recaptured in Table 7.1.

For instance, the radial attenuation function, $\phi(r)$, is governed by the first equation in Table 7.1. As $r \to \infty$, $\phi(\infty) \to 0$, and at $r = r_o$, $\phi(r_o) = 1$, the $\phi(r)$

Table 7.1 Expressions for governing equations and various conditions

Name	*Expressions*	*In Equation 7.5, collecting*
Radial function, $\phi(r)$	$r^2\frac{d^2\phi}{dr^2}+r\frac{d\phi}{dr}-\left(\frac{\gamma_b}{r_o}\right)^2 r^2\phi=0$	Coefficients of $\delta\phi$ for $r_o \le r < \infty$
Factor γ_b	$\gamma_b = r_o\sqrt{U_n/U_m}$	
Pile $w(z)$	$(EI)_p\frac{d^4w}{dz^4}-N_p\frac{d^2w}{dz^2}+kw=0$	Coefficients of δw for $0 \le z < l$
Free-head (at z = 0)	$(EI)_p\frac{d^3w}{dz^3}-N_p\frac{dw}{dz}-H=0$	Coefficients for δw
	$(EI)_p\frac{d^2w}{dz^2}+M_o=0$	Coefficients for $\delta(dw/dz)$
Floating-base (at z = *l*)	$(EI)_p\frac{d^2w}{dz^2}-N_p\frac{dw}{dz}-\sqrt{kN_p}w=0$	Coefficients for δw
	$\frac{d^2w}{dz^2}=0$	Coefficients for $\delta(dw/dz)$
Fixed-head and clamped base	$\frac{dw}{dz}=0$ (at z = 0, fixed head)	
	$w=0$ *and* $\frac{dw}{dz}=0$ (z = *l*, clamped base)	
Base shear force, Q_B	$-N_p\frac{d^2w}{dz^2}+kw=0 \quad Q_B=-\sqrt{kN_p}w(l)$	Coefficients of δw for $l \le z < \infty$, at $r = 0$
Pile $w(z)$, $w'(z)$, $w''(z)$, $w'''(z)$, $w^{IV}(z)$	$w(z)=(C_1\cos(\beta z)+C_2\sin(\beta z))e^{\alpha z}+(C_5\cos(\beta z)+C_6\sin(\beta z))e^{-\alpha z}$ $w(z)=e^{-\alpha z}(C_5\cos\beta z+C_6\sin\beta z)$ $w'(z)=e^{-\alpha z}[(-\alpha C_5+\beta C_6)\cos\beta z+(-\beta C_5-\alpha C_6)\sin\beta z]$ $w''(z)=e^{-\alpha z}\{[(\alpha^2-\beta^2)C_5-2\alpha\beta C_6]\cos\beta z+[2\alpha\beta C_5+(\alpha^2-\beta^2)C_6]\sin\beta z\}$ $w'''(z)=e^{-\alpha z}[-\alpha(\alpha^2-3\beta^2)C_5+\beta(3\alpha^2-\beta^2)C_6]\cos\beta z$ $+e^{-\alpha z}[-(3\alpha^2-\beta^2)\beta C_5-\alpha(\alpha^2-3\beta^2)C_6]\sin\beta z$ $w^{IV}(z)=e^{-\alpha z}[(\alpha^4-6\alpha^2\beta^2+\beta^4)C_5+4\alpha\beta(-\alpha^2+\beta^2)C_6]\cos\beta z$ $+e^{-\alpha z}[4\alpha\beta(\alpha^2-\beta^2)C_5+(\alpha^4-6\alpha^2\beta^2+\beta^4)C_6]\sin\beta z$	

is resolved as modified Bessel functions of the second kind of order zero, $K_o(\gamma_b)$ (McLachlan 1955) underpinned by the load transfer factor γ_b

$$\gamma_b = r_o\sqrt{U_n/U_m} \tag{7.8}$$

$$\phi(r) = K_o(\gamma_b r/r_o)/K_o(\gamma_b) \tag{7.9}$$

The $U_n = 0$ and $\gamma_b = 0$ are noted for a two-dimensional rigid disc. The associated $\phi(r)$ reduces to a logarithmic function that agrees with previous findings (Baguelin et al. 1977). Using Equation 7.9, Equation 7.6 is simplified as

$$k=\frac{3\pi G}{2}\left(2\gamma_b\frac{K_1(\gamma_b)}{K_o(\gamma_b)}-\gamma_b^2\left(\left(\frac{K_1(\gamma_b)}{K_o(\gamma_b)}\right)^2-1\right)\right)\quad N_p=\pi r_o^2 G\left(\left(\frac{K_1(\gamma_b)}{K_o(\gamma_b)}\right)^2-1\right) \tag{7.10}$$

These expressions were also provided in Chapter 3, this book.

7.2.3 Pile response under various boundary conditions

The pile displacement, $w(z)$, is governed by (see Table 7.1),

$$(EI)_p\frac{d^4w}{dz^4}-N_p\frac{d^2w}{dz^2}+kw=0 \tag{7.11}$$

where $w(z)$ is measured in the direction of the applied loading H, and/or M_o. Equation 7.11 is identical to that for a straight bar under simultaneous axial and transverse loading (Hetenyi 1946; Scott 1981), but for the concepts of the modulus of subgrade reaction k and the fictitious tension N_p. Only the case of $N_p < 2\sqrt{kE_pI_p}$ is of practical interest (Scott 1981; Yin 2000), for which $w(z)$ is solved from Equation 7.1 in the form of four unknown coefficients C_i ($i = 1\sim4$) as shown in Table 7.1. For long piles, the coefficients C_1 and C_2 are zero, whereas the C_3 and C_4 are determined using the boundary conditions in Table 7.1. The pile body displacement, $w(z)$, at depth z may be written as

$$w(z)=\frac{H}{E_pI_p\delta}\left(H(z)+\frac{\sqrt{kN_p}}{E_pI_p}B(z)\right)+\frac{M_o}{E_pI_p\delta}\left(I(z)+\frac{\sqrt{kN_p}}{E_pI_p}C(z)\right) \tag{7.12}$$

where $H(z)$ and $I(z)$ are functions used to capture pile-head boundary conditions (with subscript "o"), see Figure 7.1, while $B(z)$ and $C(z)$ reflect pile-base conditions (with "b"). They, along with the factor δ, are provided in Tables 7.2, 7.3, 7.4, and 7.5 for free-head, clamped-base piles (FreHCP), fixed-head, clamped-base piles (FixHCP), free-head, floating-base piles (FreHFP), and fixed-head, floating-base piles (FixHFP), respectively. These conditions are illustrated in Figure 7.1, in which M = bending moment, and θ = rotation angle. The solutions are featured by the parameters α and β:

$$\alpha=\sqrt{\sqrt{\frac{k}{4E_pI_p}}+\frac{N_p}{4E_pI_p}}\ \text{and}\ \beta=\sqrt{\sqrt{\frac{k}{4E_pI_p}}-\frac{N_p}{4E_pI_p}} \tag{7.13}$$

The derivatives of $w(z)$, as provided in Table 7.1, offer the profiles of pile rotation, bending moment, and shear force.

Table 7.2 Expressions for free-head, clamped-base piles (FreHCP)

$B(z) = C(z) = 0, z' = l - z$, and

$$H(z) = \beta sh(\alpha z')[2\alpha\beta ch(\alpha l)\cos(\beta z) + (\alpha^2 - \beta^2)sh(\alpha l)\sin(\beta z)] - \alpha\sin(\beta z')[2\alpha\beta ch(\alpha z)\cos(\beta l) + (\alpha^2 - \beta^2)sh(\alpha z)\sin(\beta l)]$$

$$I(z) = \beta(\alpha^2 + \beta^2)sh(\alpha z')[-\beta sh(\alpha l)\cos(\beta z) + \alpha ch(\alpha l)\sin(\beta z)] + \alpha(\alpha^2 + \beta^2)\sin(\beta z')[\beta sh(\alpha z)\cos(\beta l) - \alpha ch(\alpha z)\sin(\beta l)]$$

$$\delta = (\alpha^2 + \beta^2)[(\alpha^2 - \beta^2)(\alpha^2\sin^2(\beta l) + \beta^2 sh^2(\alpha l)) + 2\alpha^2\beta^2(ch^2(\alpha l) + \cos^2(\beta l))]$$

At the pile-head level ($z = 0$) and base level ($z = l$), it follows

$H(l) = I(l) = 0$, and $H'(l) = I'(l) = 0$

$H''(0) = 0 \quad H(0) = \alpha\beta[\beta sh(2\alpha l) - \alpha\sin(2\beta l)]$,

$H'(0) = I(0) = -(\alpha^2 + \beta^2)[\beta^2 sh^2(\alpha l) + \alpha^2\sin^2(\beta l)]$

$I'(0) = \alpha\beta(\alpha^2 + \beta^2)[\alpha\sin(2\beta l) + \beta sh(2\alpha l)]$

Table 7.3 Expressions for fixed-head, clamped-base piles (FixHCP)

$I(z) = B(z) = C(z) = 0, z' = l - z$, and

$$H(z) = sh(\alpha z')[\alpha\beta ch(\alpha l)\sin(\beta z) + \beta^2 sh(\alpha l)\cos(\beta z)] - \sin(\beta z')[\alpha\beta sh(\alpha z)\cos(\beta l) + \alpha^2 ch(\alpha z)\sin(\beta l)]$$

$$\delta = \alpha\beta(\alpha^2 + \beta^2)[\alpha\sin(2\beta l) + \beta sh(2\alpha l)]$$

At the pile-head level ($z = 0$) and base level ($z = l$), it follows $H(l) = H'(l) = 0$

$H'(0) = 0 \quad H(0) = \beta^2 sh^2(\alpha l) - \alpha^2\sin^2(\beta l) \quad H''(0) = -(\alpha^2 + \beta^2)(\beta^2 sh^2(\alpha l) + \alpha^2\sin^2(\beta l))$

The shear force Q_B at the base of a floating pile was gained as $-\sqrt{kN_p}w(l)$ (see Table 7.1, $z \to \infty$, $w \to 0$). It is generally rather small and may be ignored in estimating $H(z)$, $I(z)$, $B(z)$, $C(z)$, and δ, (by taking $\sqrt{kN_p} = 0$, referred to as "FP excluding base," although the real values of the k and N_p are still used in calculating parameters α and β). By setting $B(z) = 0$ and $M_o = 0$, Equation 7.12 does reduce to the solution for a free-end beam of finite length with a concentrated load at one end (Hetenyi 1946).

All the expressions presented have been verified using Maple™ and Mathcad™. For floating-base piles, some errors were found in the previous solutions (Sun 1994). The solution for the case of $N_p \geq 2\sqrt{kE_pI_p}$ may be obtained in a similar way. However, it is not presented herein because of its minor relevance to practical design.

7.2.4 Load transfer factor γ_b

The coupled interaction between the pile and soil [e.g., the displacement, $w(z)$, and the radial attenuation function, $\phi(r)$] is captured through the load transfer factor, γ_b. Equation 7.8 is rewritten as

Table 7.4 Expressions for free-head, floating-base piles (FreHFP)

$$H(z) = (\alpha^2+\beta^2)[\alpha(\alpha^2-3\beta^2)\sin(\beta l)C_h(z)+\beta(3\alpha^2-\beta^2)sh(\alpha l)C_h(z')]$$

where $C_h(z) = (\alpha^2-\beta^2)sh(\alpha z)\sin(\beta(l-z))+2\alpha\beta ch(\alpha z)\cos(\beta(l-z))$, $z' = l - z$

$$B(z) = 2\alpha\beta sh(\alpha l)[(\alpha^2-\beta^2)ch(\alpha z')\sin(\beta z)+2\alpha\beta sh(\alpha z')\cos(\beta z)] - 2\alpha\beta\sin(\beta l)[(\alpha^2-\beta^2)sh(\alpha z)\cos(\beta z')-2\alpha\beta ch(\alpha z)\sin(\beta z')]$$

$$I(z) = \alpha(\alpha^2+\beta^2)^2(\alpha^2-3\beta^2)\sin(\beta l)[-\beta sh(\alpha z)\cos(\beta z')+\alpha ch(\alpha z)\sin(\beta z')] + \beta(\alpha^2+\beta^2)^2(3\alpha^2-\beta^2)sh(\alpha l)[-\beta sh(\alpha z')\cos(\beta z)+\alpha ch(\alpha z')\sin(\beta z)]$$

$$C(z) = 2\alpha\beta(\alpha^2+\beta^2)sh(\alpha l)[-\beta ch(\alpha z')\cos(\beta z)+\alpha sh(\alpha z')\sin(\beta z)] + 2\alpha\beta(\alpha^2+\beta^2)\cos(\beta l)[\beta sh(\alpha z)\cos(\beta z')-\alpha ch(\alpha z)\sin(\beta z')]$$

$$\delta = (\alpha^2+\beta^2)^2[\beta^2(3\alpha^2-\beta^2)^2 sh^2(\alpha l)-\alpha^2(\alpha^2-3\beta^2)^2\sin^2(\beta l)] + \frac{\sqrt{kN_p}}{(EI)_p}\alpha\beta(\alpha^2+\beta^2)[\alpha(\alpha^2-3\beta^2)\sin(2\beta l)+\beta(3\alpha^2-\beta^2)sh(2\alpha l)]$$

At the pile-head level ($z = 0$), it follows $H''(0) = B''(0) = 0$, and

$$H(0) = \alpha\beta(\alpha^2+\beta^2)[\beta(3\alpha^2-\beta^2)sh(2\alpha l)+\alpha(\alpha^2-3\beta^2)\sin(2\beta l)]$$

$$H'(0) = I(0) = -(\alpha^2+\beta^2)^2[\beta^2(3\alpha^2-\beta^2)sh^2(\alpha l)-\alpha^2(\alpha^2-3\beta^2)\sin^2(\beta l)]$$

$$B'(0) = C(0) = -\alpha\beta(\alpha^2+\beta^2)[\beta sh(2\alpha l)+\alpha\sin(2\beta l)]$$

$$I'(0) = \alpha\beta(\alpha^2+\beta^2)^2[\beta(3\alpha^2-\beta^2)sh(2\alpha l)-\alpha(\alpha^2-3\beta^2)\sin(2\beta l)]$$

$$C'(0) = 4\alpha^2\beta^2(\alpha^2+\beta^2)[sh^2(\alpha l)-\cos^2(\beta l)]$$

Table 7.5 Expressions for fixed-head, floating-base piles (FixHFP)

$I(z) = C(z) = 0$, $z' = l - z$, and

$$H(z) = \alpha(\alpha^2+\beta^2)(3\beta^2-\alpha^2)\cos(\beta l)[\alpha ch(\alpha z)\cos(\beta z')-\beta sh(\alpha z)\sin(\beta z')] + \beta(\alpha^2+\beta^2)(3\alpha^2-\beta^2)sh(\alpha l)[\alpha ch(\alpha z')\sin(\beta z)+\beta sh(\alpha z')\cos(\beta z)] + \alpha(\alpha^2+\beta^2)^2[\alpha ch(\alpha z)\cos(\beta z)-\beta sh(\alpha z)\sin(\beta z)]$$

$$B(z) = 2\alpha\beta sh(\alpha l)[\beta ch(\alpha z')\cos(\beta z)+\alpha sh(\alpha z')\sin(\beta z)] - 2\alpha\beta\cos(\beta l)[\beta sh(\alpha z)\cos(\beta z')+\alpha ch(\alpha z)\sin(\beta z')]$$

$$\delta = \alpha\beta(\alpha^2+\beta^2)^2[\beta(3\alpha^2-\beta^2)sh(2\alpha l)+\alpha(3\beta^2-\alpha^2)\sin(2\beta l)] + 4\alpha^2\beta^2(\alpha^2+\beta^2)\frac{\sqrt{kN_p}}{(EI)_p}[sh^2(\alpha l)+\cos^2(\beta l)]$$

At the pile-head level ($z = 0$), it follows $H'(0) = B'(0) = 0$, and

$$H(0) = (\alpha^2+\beta^2)[4\alpha^2\beta^2-\alpha^2(3\beta^2-\alpha^2)\sin^2(\beta l)+\beta^2(3\alpha^2-\beta^2)sh^2(\alpha l)]$$

$$H''(0) = -(\alpha^2+\beta^2)^2[\alpha^2(3\beta^2-\alpha^2)\sin^2(\beta l)+\beta^2(3\alpha^2-\beta^2)sh^2(\alpha l)]$$

$$B(0) = \alpha\beta[\beta sh(2\alpha l)-\alpha\sin(2\beta l)] \quad B''(0) = -\alpha\beta(\alpha^2+\beta^2)[\beta sh(2\alpha l)+\alpha\sin(2\beta l)]$$

$$\frac{r_o^2}{3}\frac{2\int_o^l (dw/dz)^2\,dz+\sqrt{k/N_p}\,w^2(l)}{\int_o^l w^2\,dz+\sqrt{N_p/4k}\,w^2(l)}-\gamma_b^2=0 \tag{7.14}$$

The $w(z)$, k, and N_p are interrelated through γ_b. Nonlinear Equation 7.14 has been resolved numerically for these boundary conditions. The γ_b obtained is synthesised as a simple power form of pile–soil relative stiffness E_p/G^* and pile slenderness ratio, l/r_o (see Equation 3.54, Chapter 3, this book). In particular, with $w(l) \approx 0$ for long piles, the γ_b is dominated by the rotation, dw/dz and the lateral displacement, $w(z)$, over the whole length; with a combined load and moment simultaneously, the γ_b may be directly estimated by using Equation 7.14 with $w(z)$ by Equation 7.12. It should lie in between these for the load H and the moment M_o, and offers a single k in light of Equation 7.10. As mentioned in Chapter 3, this book, the superposition using the two different values of γ_b (thus k) may be roughly adopted for designing piles under the combined loading.

7.2.5 Modulus of subgrade reaction and fictitious tension

The current solutions (e.g., Equation 7.12) are underpinned by the two energy parameters k and N_p. Dependent of loading properties, pile slenderness ratio, and pile–soil relative stiffness, they were estimated using Equation 7.10, and are illustrated previously in Figures 3.22 and 3.23 (Chapter 3, this book), due to either the moment (M_o) or the lateral load (H). In particular, the k and N_p for short piles rely primarily on the slenderness ratio, as the total energy U, the displacement w, and the rotation dw/dz do.

The k originates from the stress variations in radial direction and reflects the coupling effect among the independent springs via the parameter γ_b. A constant w in Equation 7.2 for a two-dimensional disc implies that τ_{zr}, and $\tau_{\theta z}$ and γ_b reduce to zero. The ϕ (hence k, see Table 7.1) becomes independent of variations in vertical directions. Only then does the k become a ratio of the local, uncoupled reaction per unit area on the pile, p, over the pile deflection, w (Figure 7.1b) (i.e., the conventional modulus of subgrade reaction).

The fictitious tension, N_p, is due to the stress variations in the vertical direction (Equation 7.6) and causes the shear force, Q_B, at a pile base. It may be a pair of equilibrating external forces acting in the center of gravity of the end cross-sections of the pile (Hetenyi 1946); or a ratio of the modulus of subgrade reaction, k, over a "shear stiffness" of the pile (Cowper 1966; Yin 2000). Under lateral loading, real tensile force (e.g., in a steel bar of soil nails) (Pedley et al. 1990) is a multiplier of the pile rotation and the distance between the bar and the neutral axis (Yin 2000).

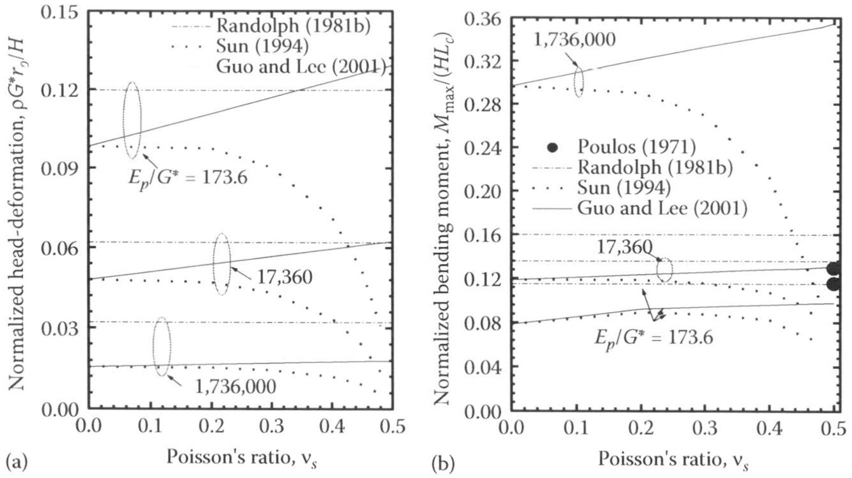

Figure 7.3 Comparison of pile response due to Poisson's ratio (L/r_o = 50). (a) Deflection. (b) Maximum bending moment. (After Guo, W. D., and F. H. Lee, *Int J Numer and Anal Meth in Geomech* 25, 11, 2001.)

7.3 VALIDATION

The current approach differs essentially from the two-parameter model (Sun 1994) in using the new stress field and the shear modulus, G^*. This is further validated in terms of normalized deformation $\rho G^* r_o/H$ (ρ = pile-head deformation) and normalized moment M_{max}/HL_c (M_{max} = the maximum bending moment; and L_c = a critical pile length defined next) in the loading direction ($\theta = 0$). The impact of Poisson's ratio, ν_s, on the pile response is investigated using Equation 7.12 and shown in Figure 7.3. The figure indicate: (1) an excellent agreement in head-displacement and moment at $\nu_s = 0$ between current predictions and the two-parameter model (Sun 1994), although both are slightly lower than the finite element predictions (Randolph 1981b); (2) similar bending moment (at $\nu_s = 0.5$) to the boundary element approach (Poulos 1971); and (3) significant underestimation of the displacement and moment using the two-parameter model (Sun 1994) for $\nu_s > 0.3$. The current solutions offer a better simulation of the response of a single pile–soil system than the two-parameter model at a Poisson's ratio higher than 0.3, as further highlighted next.

7.4 PARAMETRIC STUDY

7.4.1 Critical pile length

There exists a critical length, L_c, beyond which pile-head response and maximum bending moment stay as constants (Hetenyi 1946; Randolph 1981b). The modulus, k, may be as high as 10G, thus critical length L_c may be as low as

$$L_c \approx 2.1 r_o [(1+0.75\nu_s) E_p / G^*]^{0.25} \tag{7.15}$$

On the other hand, in Equation 7.15 the slenderness ratio of L_c/r_o allows a critical pile–soil stiffness, $(E_p/G^*)_c$, to be gained as

$$\left(\frac{E_p}{G^*}\right)_c \approx \frac{0.05}{(1+0.75\nu_s)}\left(\frac{L_c}{r_o}\right)^4 \tag{7.16}$$

Given $l < L_c$, or $(E_p/G^*) > (E_p/G^*)_c$, the piles are referred to as "short piles," otherwise as "long piles." It may be shown that the critical pile length for lateral loading is generally the shortest, in comparison with vertical or torsional loading (Randolph 1981a; Fleming et al. 1992; Guo and Randolph 1996); and "Short piles," as defined herein, are not necessarily equivalent to rigid piles, as stubby, rock-socketed piles are (Carter and Kulhawy 1992).

7.4.2 Short and long piles

The pile-head deformation and the maximum bending moment were estimated for various pile slenderness ratios, l/r_o, using Equation 7.12 and its second derivative. This allows a critical length L_c (i.e, Guo and Lee 2001) to be directly, albeit approximately, determined for each pile–soil relative stiffness ratio. The results compare well with those estimated by Equation 7.15 (see Figure 7.4), albeit generally slightly lower than those obtained from the finite element analysis (Randolph 1981b). The critical stiffness, $(E_p/G^*)_c$, increases to four times that of Equation 7.16 in the case of free-head clamped piles due to the moment loading, FreHCP(M_o), or fixed-head floating piles due to the lateral loading, FixHFP(H), as is seen in the load transfer factors or in later figures for the moment or rotation response. Thereby, with Equation 7.15, the critical pile length should be 41% longer, which, represented by 1.41*Equation 7.15, is illustrated in Figure 7.4 as the upper limit of the effective pile length. Overall, the critical length may well fall in those estimated by Equation 7.15 and 1.4*Equation 7.15.

7.4.3 Maximum bending moment and the critical depth

Simple expressions for the maximum bending moment, M_{max}, and the depth, z_m, at which the M_{max} occurs are developed using Equation 7.12 and relevant available solutions.

7.4.3.1 Free-head piles

For free-head piles (no restraints but soil resistance), the depth z_m and moment M_{max} were obtained directly using Equation 7.12. The

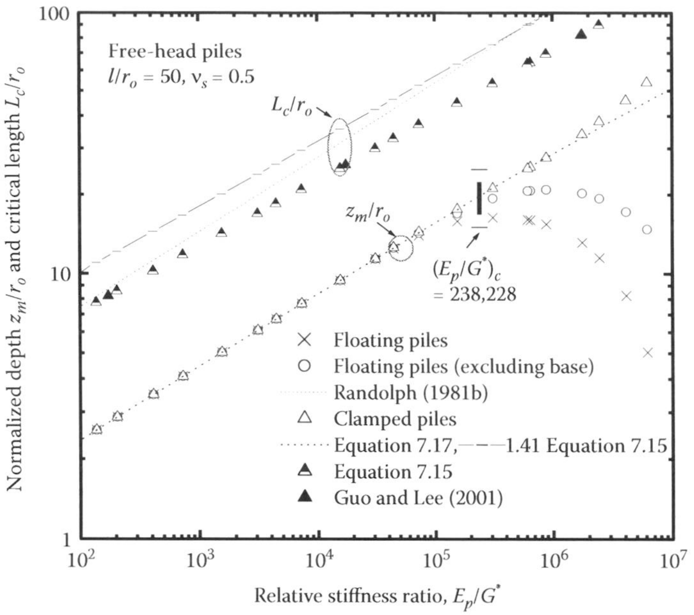

Figure 7.4 Depth of maximum bending moment and critical pile length. (After Guo, W. D. and F. H. Lee, *Int J Numer and Anal Meth in Geomech* 25, 11, 2001.)

corresponding normalized values are presented in Figures 7.4 and 7.5b as Guo and Lee, which show the following:

1. The depth z_m is 25~30% the critical length L_c (Randolph 1981b)
2. With $E_p/G^* < (E_p/G^*)_c$, the z_m and the M_{max} for long piles may be estimated using Equations 7.17 and 7.18a developed for a semi-infinitely long beam (Hetenyi 1946):

$$z_m = \frac{1}{\beta}\arctan\left(\frac{\beta}{\alpha}\right) \tag{7.17}$$

$$M_{max} = \frac{-H\sqrt{\alpha^2+\beta^2}}{(3\alpha^2-\beta^2)}e^{-\alpha z_m} \tag{7.18a}$$

The M_{max} may be estimated by Equation 7.18b proposed for an infinitely long beam (Scott 1981):

$$M_{max} = -H/(4\alpha) \tag{7.18b}$$

With α given by Equation 7.13, Equation 7.18b compares well with the FEM results (Randolph 1981b), although both are slightly higher than the current predictions.

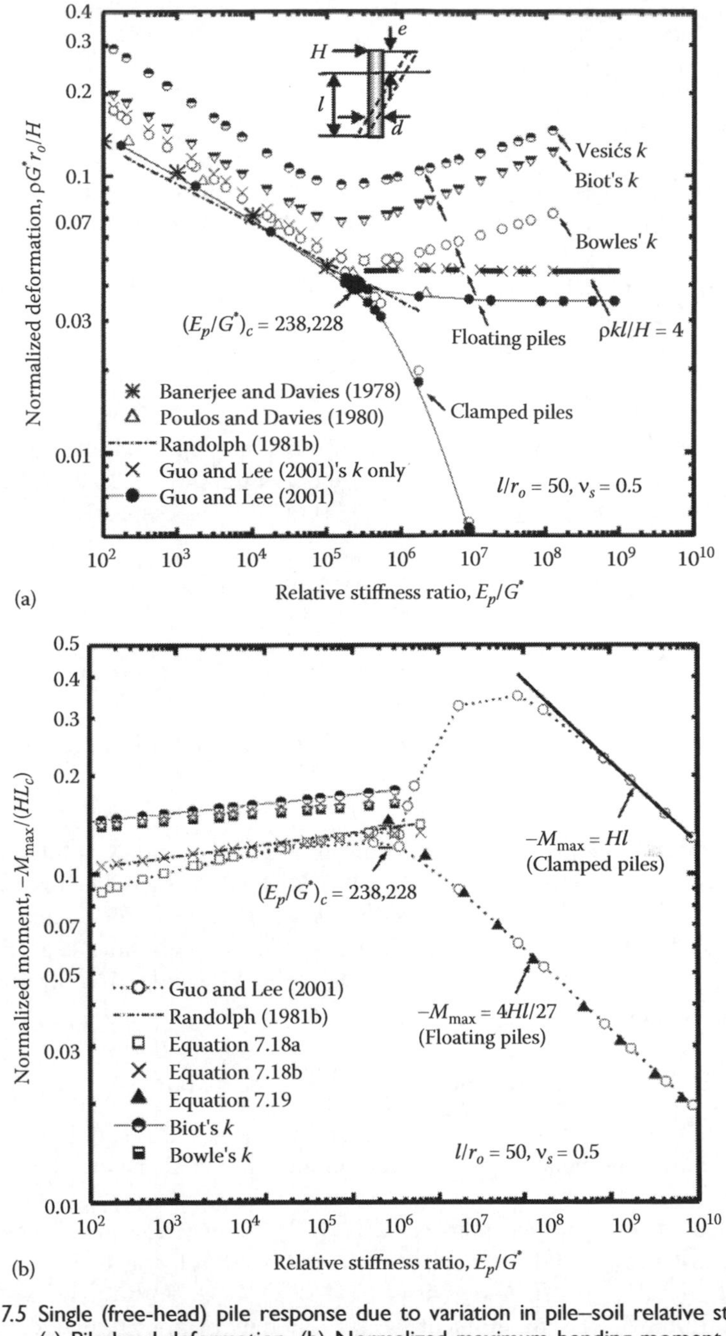

Figure 7.5 Single (free-head) pile response due to variation in pile–soil relative stiffness. (a) Pile-head deformation. (b) Normalized maximum bending moment. (After Guo, W. D., *Proc 8th Int Conf on Civil and Structural Engrg Computing*, paper 112, Eisenstadt, Vienna, 2001a.)

3. Otherwise, if $E_p/G^* > (E_p/G^*)_c$, the M_{max} for floating base, rigid piles may be estimated by Equation 7.19 (Scott 1981), whereas $M_{max} \approx -Hl$ for clamped base, rigid piles:

$$M_{max} = -4Hl/27 \tag{7.19}$$

   Free-head, short, floating piles [FreHFP(H)] may be taken as rigid, whereas short, clamped piles [FreHCP(H)] should not, only if, say, $E_p/G^* > 10^8$ at $l/r_o = 50$.
4. $-M_{max}/HL_c$ is 0.13 for flexible piles (Randolph 1981b) and 0.148 l/L_c for rigid piles.

7.4.3.2 *Fixed-head piles*

For fixed-head piles, the moment, M_{max}, occurs at ground level ($z_m = 0$). Equation 7.12 allows M_{max} to be simplified as Equations 7.20 and 7.21 for clamped [FixHCP(H)] and floating [FixHFP(P)] piles, respectively:

$$M_{max} = \frac{-H}{\alpha\beta}\left(\frac{\alpha^2 \sin^2(\beta l) + \beta^2 sh^2(\alpha l)}{\alpha \sin(2\beta l) + \beta sh(2\alpha l)}\right) \text{ (Clamped base)} \tag{7.20}$$

$$M_{max} = \frac{-H}{\delta}(\alpha^2 + \beta^2)^2[\alpha^2(3\beta^2 - \alpha^2)\sin^2(\beta l) + \beta^2(3\alpha^2 - \beta^2)sh^2(\alpha l)]$$
$$-\frac{H}{\delta}\frac{\sqrt{kN_p}}{E_p I_p}\alpha\beta(\alpha^2 + \beta^2)\left[\beta sh(2\alpha l) + \alpha \sin(2\beta l)\right] \text{ (Floating base)} \tag{7.21}$$

As shown in Figure 7.6b, using the parameters k and N_p of Equation 7.10, the M_{max} by either Equation 7.20 or 7.21 is ~7% smaller than finite element results for long piles (Randolph 1981b); and the difference in the M_{max} between Equations 7.20 and 7.21 is ~20% for rigid piles (e.g., $E_p/G^* \geq 10^7$, given $l/r_o = 50$), with an average M_{max} of 0.5Hl and 0.6Hl for floating piles and clamped piles, respectively.

7.4.4 Effect of various head and base conditions

The impact of the pile-head and base conditions on the response of maximum bending moment, M_{max}, and the pile-head displacement, ρ, is examined here.

For free-head, clamped piles [FreHCP(H)], Figure 7.5a and b show at a high stiffness E_p/G^*, a negligible normalized pile head-deformation $\rho G r_o/H$, and $M_{max} \approx Hl$. Also, with $E_p/G^* < (E_p/G^*)_c$, both the normalized deformation $\rho G r_o/H$ and moment $-M_{max}/HL_c$ may be approximated by relevant simple expressions Equations 7.18a and b for the moment M_{max}. Free-head, floating short piles [FreHFP(H)] observe $\rho kl/H = 4$ (Scott 1981). The

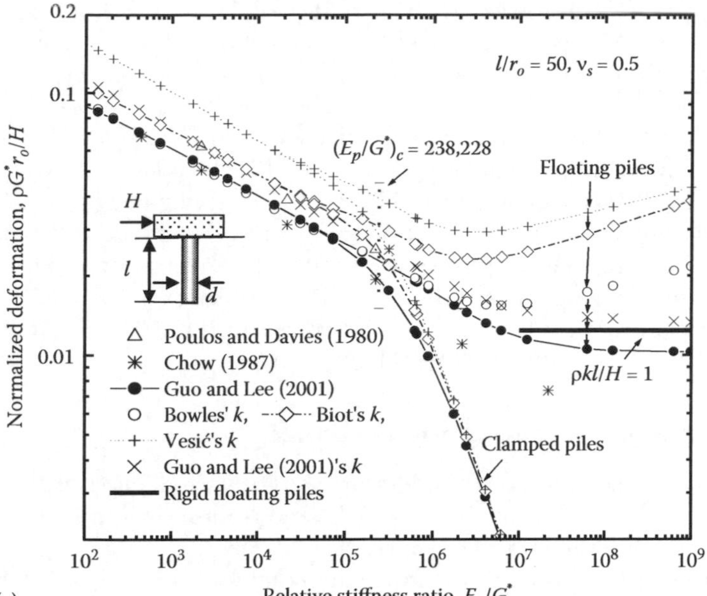

(a)

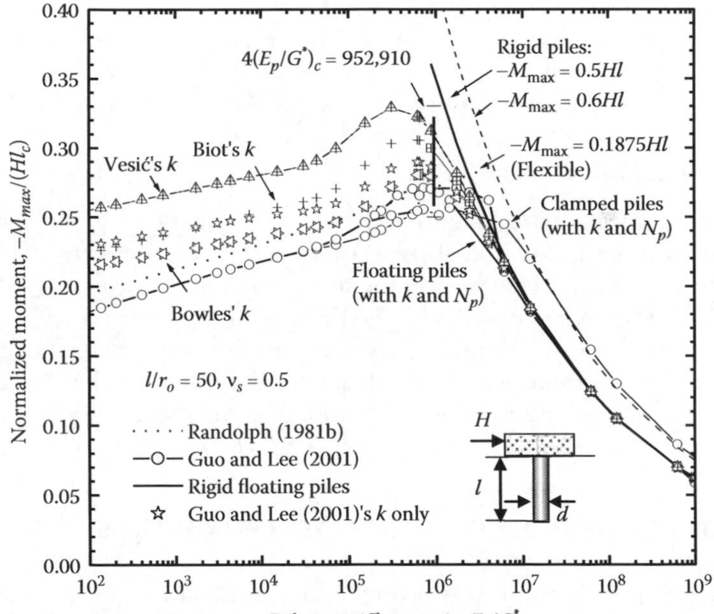

(b)

Figure 7.6 Single (fixed-head) pile response due to variation in pile–soil relative stiffness. (a) Deformation. (b) Maximum bending moment. (After Guo, W. D., *Proc 8th Int Conf on Civil and Structural Engrg Computing*, paper 112, Eisenstadt, Vienna, 2001a.)

ratio (see Figure 7.5a) agrees with current solution using k only and shows the impact of N_p against Guo and Lee's predictions (2001). The impact is not seen in the maximum moment (independent of the k) and well captured by Equation 7.19.

For fixed-head piles, comparison between Figures 7.5a and 7.6a shows the normalized pile-head deformation of FixHCP(H) piles over that of free-head FreHCP(H) pile ranges from 1/2~2/3 for long, clamped piles to 1/4 for short, floating piles. The deformation for the fixed-head [FixHFP(H)], rigid piles (see Figure 7.6a), compares well with other approaches. The maximum moment for rigid piles differs by 2~7 times due to head and base conditions, as mentioned before from ~0.6Hl (FixHCP) to Hl (FreHCP), and from 0.148Hl (FreHFP) to Hl (FreHCP).

7.4.5 Moment-induced pile response

With pure moment, M_o, on the pile-head, the normalized pile-head displacement was calculated using Equation 7.12 and is illustrated in Figure 7.7. It compares well with numerical approaches (Poulos 1971; Randolph 1981b) for long piles, but it is markedly lower than that for short rigid piles characterized by a limiting displacement, ρ of $-6M_o/(kl^2)$ (Scott 1981). The latter difference implies the accuracy of the simplified stress field, Equation 7.2, and the displacements fields, Equation 7.1. The maximum bending moment M_{max} (= M_o) occurs at $z_m = 0$, which is not the case for a combined load, H, and the moment, M_o. The latter has to be obtained by Equation 7.12.

7.4.6 Rotation of pile-head

The pile-head rotation θ_o due to either the lateral load H, or the moment M_o (at the ground level), has been estimated individually using Equation 7.12. As presented in Figure 7.7b, the normalized pile rotations $\theta_o G^* r_o^2/H$ and $\theta_o G^* r_o^3/M_o$ compare well with the FEM results (Randolph 1981b) with $E_p/G^* < 4(E_p/G^*)_c$. Otherwise, they are slightly lower than that gained from the limiting rotations, θ_o of $-6H/(kl^2)$, or higher than that calculated from $\theta_o = 12M_o/(kl^3)$, respectively, for rigid, floating piles due to lateral load H or moment M_o (Scott 1981).

7.5 SUBGRADE MODULUS AND DESIGN CHARTS

Normalized pile response was estimated using Equation 7.12, in light of existing formulae for subgrade modulus including Biot's k (1937), Vesić's k (1961b), and Bolwes' k (1997) (see Chapter 3, this book). The results are presented in form of $\rho G^* r_o/H$, $M_{max}/(HL_c)$, $\theta_o G^* r_o^2/H$ and $\theta_o G^* r_o^3/M_o$ in Figures 7.5 through 7.7 together with relevant numerical results. These figures indicate that

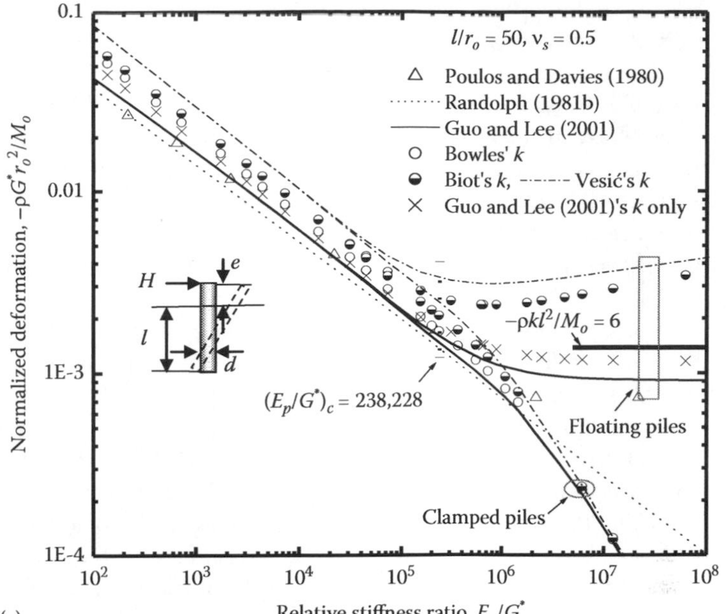

(a)

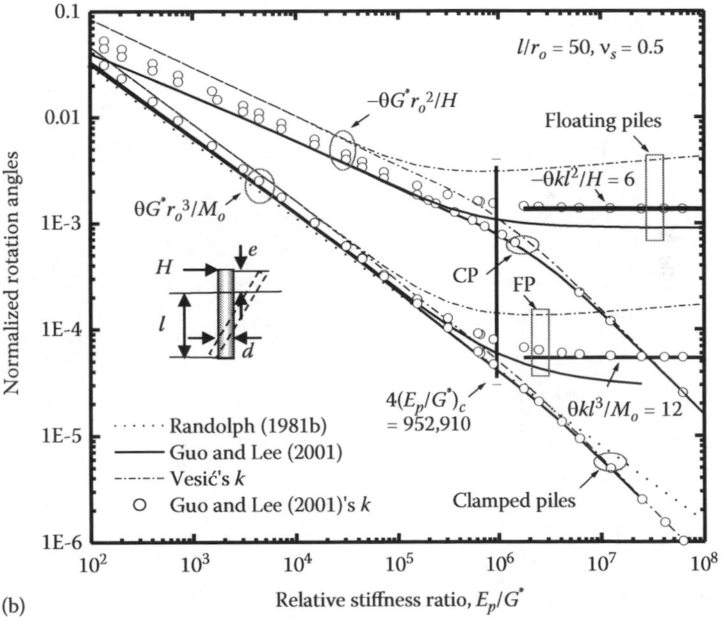

(b)

Figure 7.7 Pile-head (free-head) (a) displacement due to moment loading, (b) rotation due to moment or lateral loading. (After Guo, W. D., *Proc 8th Int Conf on Civil and Structural Engrg Computing*, paper 112, Eisenstadt, Vienna, 2001a.)

- For free- or fixed-head piles due to lateral load, the solutions using Vesić's k for beams offer invariably highest values than other results. The k is conservative for lateral pile analysis.
- The Bowles' k is close to the current k (Guo and Lee 2001), for fixed- and free-head flexible piles. Nevertheless, without the fictitious tension (N_p), both would generally overestimate the displacement and M_{max} by ~30% compared to relevant numerical results. The Biot's k is in between the Bowles' k and the current k.
- None of the available suggestions are suitable for rigid pile analysis.

Measured distributions of the mobilized front pressure around the circumference of a pile approximately follows the theoretical prediction (Baguelin et al. 1977; Guo 2008). Side shear contributes 88% of soil reaction from horizontal equilibrium (Smith 1987), rather than 100% as implicitly assumed previously (Kishida and Nakai 1977; Bowles 1997). Therefore, the current proposal is more rational than the available suggestions.

Design charts for typical pile slenderness ratios are provided for the normalized deflections $\rho G^*r_o/H$ and $-\rho G^*r_o^2/M_o$ (Figure 7.8), normalized bending moment, $M_{max}/(Hl)$ (Figure 7.9), and rotations $\theta_o G^*r_o^2/H$ and $\theta_o G^*r_o^3/M_o$ (Figure 7.10). The deflections for a combined lateral load and moment can be added together, such as Figure 7.11a and b for some typical loading eccentricities ($e = M_o/H$), and with clamped and floating base, respectively.

7.6 PILE GROUP RESPONSE

Pile-pile interaction was well predicted using expressions derived from Equation 7.9 against the FEM results (at various directions of θ) (Randolph 1981b), such as in Figure 7.12, and is affected by the pile-head constraints (see Figure 7.13).

7.6.1 Interaction factor

The increase in displacement of a pile due to a neighboring pile is normally estimated by an interaction factor, α (Poulos 1971; Randolph 1981b). Generally, the deformation of the i-th pile in a group of n_g piles may be written as (Poulos 1971)

$$w_i = \frac{\rho}{H}\sum_{j=1}^{n_g}\alpha_{ij}H_j \tag{7.22}$$

where α_{ij} = the interaction factor between the i-th pile and j-th pile; later it may be simply written as α ($\alpha_{ij} = 1$, $i = j$). Generally, four interaction factors

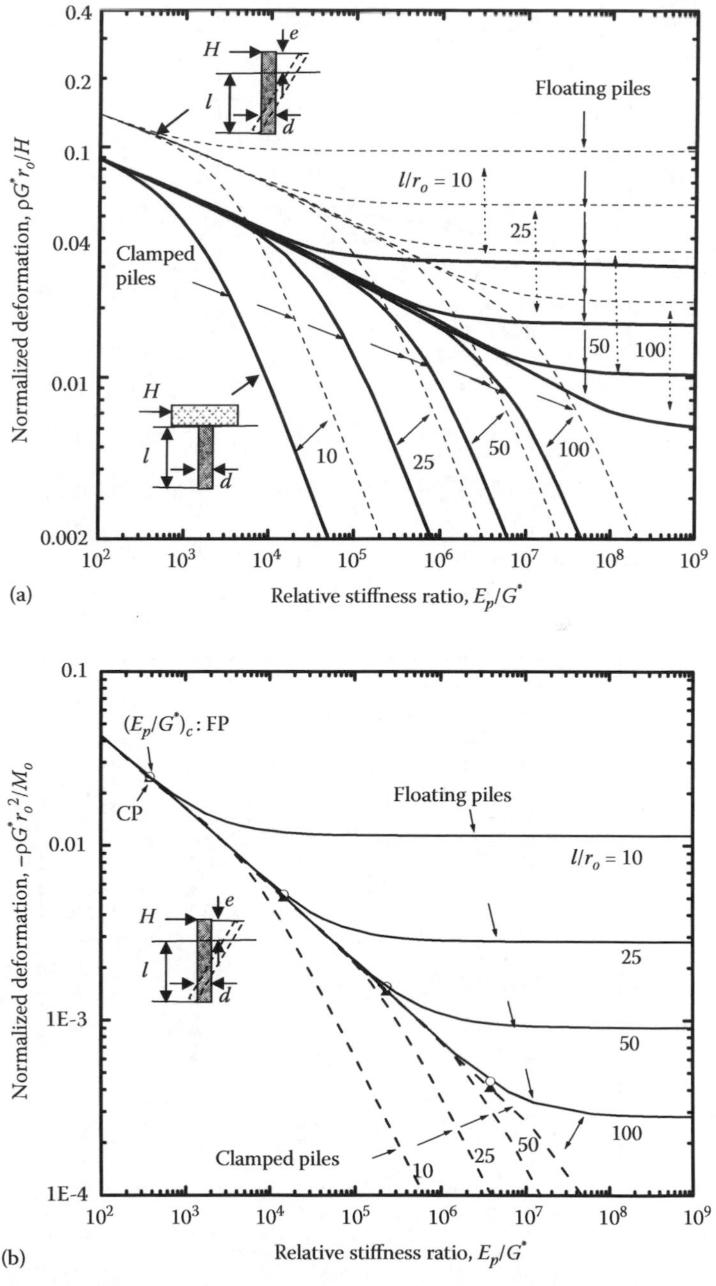

Figure 7.8 Impact of pile-head constraints on the single pile deflection due to (a) lateral load *H*, and (b) moment M_o.

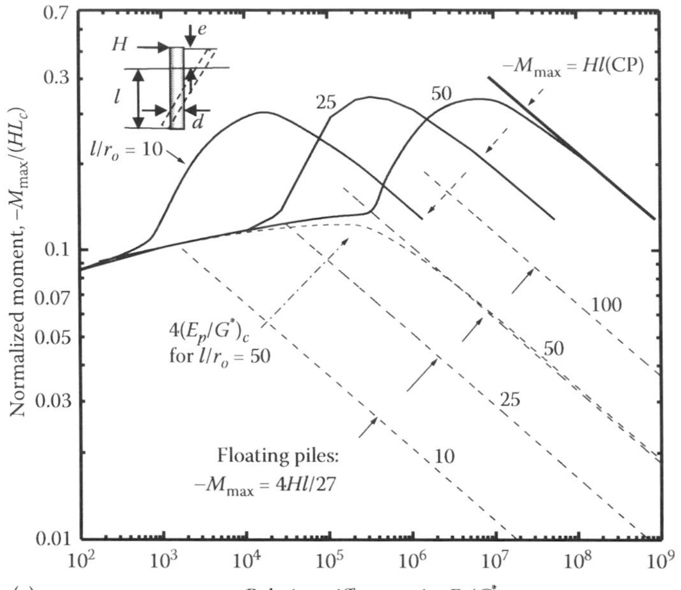

(a)

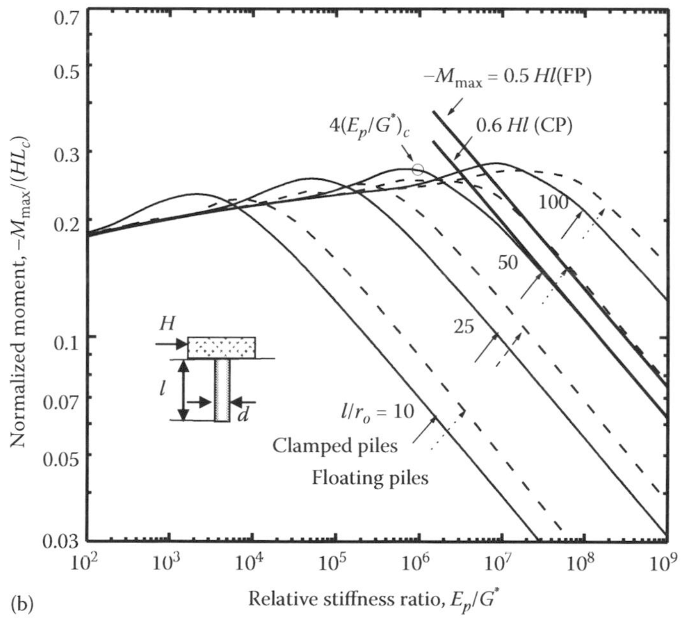

(b)

Figure 7.9 Impact of pile-head constraints on the single pile bending moment. (a) Free-head. (b) Fixed-head.

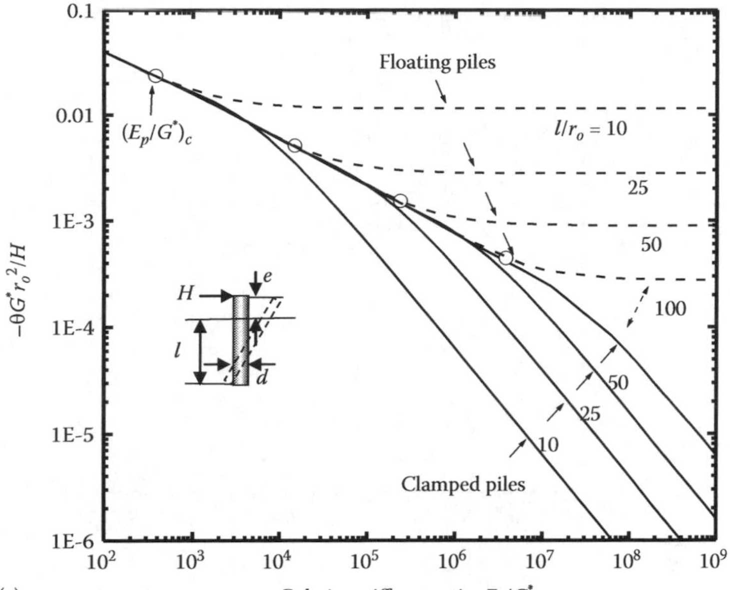

(a)

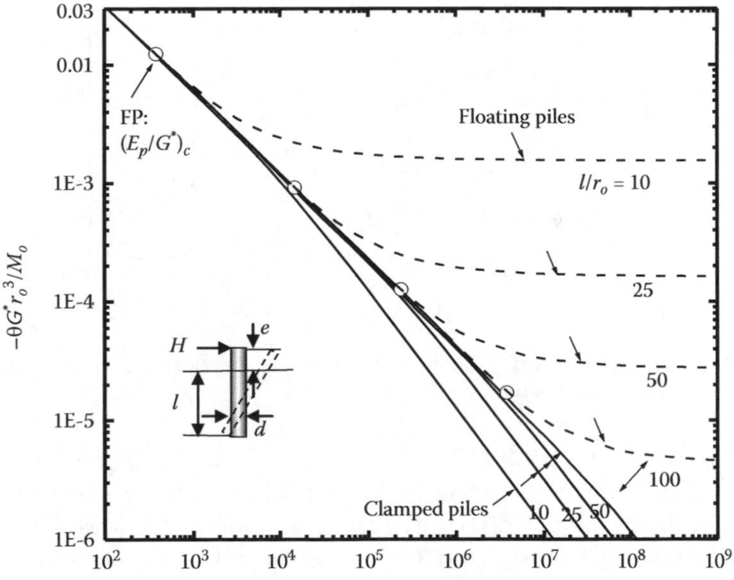

(b)

Figure 7.10 Normalized pile-head rotation angle owing to (a) lateral load, H, and (b) moment M_o.

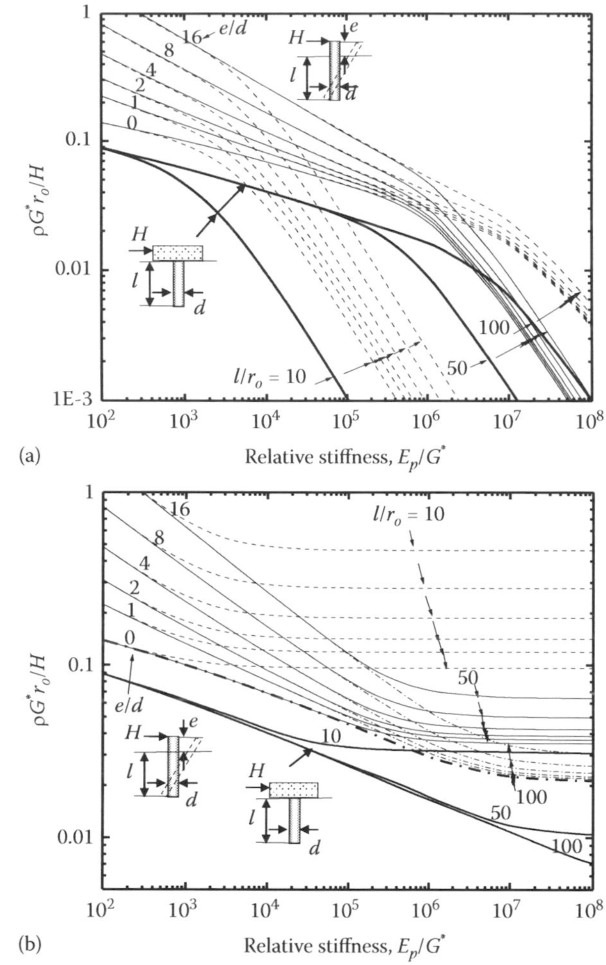

Figure 7.11 Pile-head constraints on normalized pile-head deflections: (a) clamped piles (CP) and (b) floating piles (FP).

are required (Poulos 1971): $\alpha_{\rho P}$ and $\alpha_{\rho M}$ reflect increase in deflection due to lateral load (for both free-head and fixed-head) and moment loading, respectively; $\alpha_{\theta P}$ and $\alpha_{\theta M}$ reflect increase in rotation due to lateral load and moment loading ($\alpha_{\rho M} = \alpha_{\theta P}$ from the reciprocal theorem), respectively. The factors for two identical piles at a pile center-to-center spacing, *s*, may be obtained using Equation 7.2. For instance, $\alpha_{\rho P}$ may be expressed as

$$\alpha_{\rho P} = w(z)\phi(s) \Bigg/ \left(w(z)\phi(s) \frac{Gw(z)\phi'}{\sqrt{\sigma_r^2 + \tau_{r\theta}^2}} \right) \tag{7.23}$$

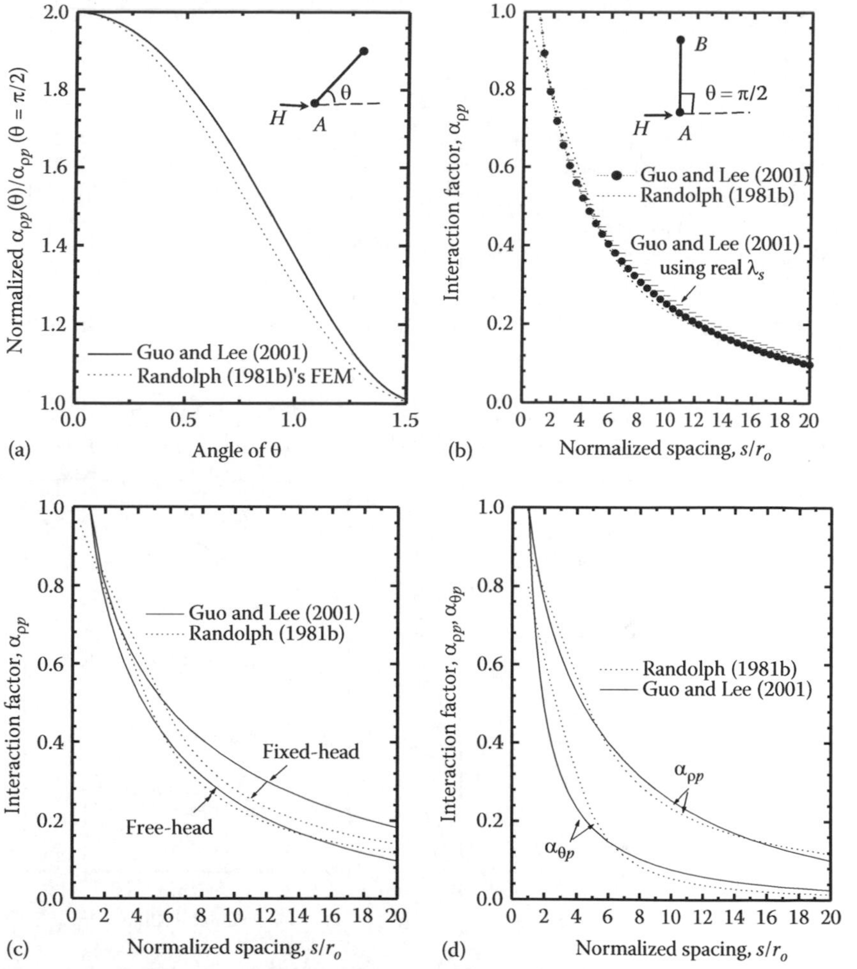

Figure 7.12 **Variations of the interaction factors (l/r_o = 50) (except where specified: ν_s = 0.33, free-head, E_p/G^* = 47695): (a) effect of θ, (b) effect of spacing, (c) head conditions, and (d) other factors.**

The expression may be further simplified as

$$\alpha_{\rho P} = \sqrt{\sin^2\theta + 4\cos^2\theta}\,\frac{K_o(\gamma_b r_o/s)}{K_o(\gamma_b)} \tag{7.24}$$

Figure 7.12a shows that Equation 7.24 well captures the effect of loading direction on the displacement factor, $\alpha_{\rho P}$, compared to the FEM (Randolph 1981b), which is also evident for the particular loading direction $\theta = \pi/2$ at various spacing (Figure 7.12b). At $\theta = 0$, the interaction for the overall pile response seems to be influenced mainly by radial stress σ_r. The normalized

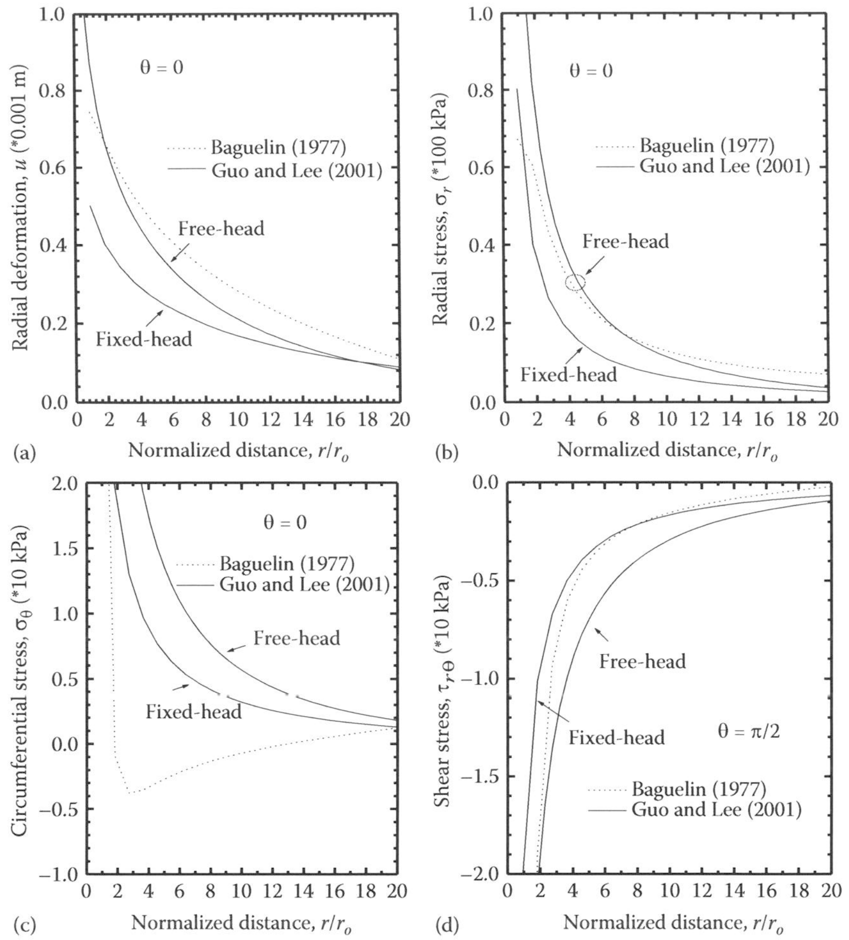

Figure 7.13 Radial stress distributions for free- and fixed-head (clamped) pile. For intact model: ν_s = 0.33, H = 10 kN, r_o = 0.22 cm, E_p/G^* = 47695, R = $20r_o$ (free-head), $30r_o$ (fixed-head). (a) Radial deformation. (b) Radial stress. (c) Circumferential stress. (d) Shear stress.

stresses (by the values at the pile–soil interface) such as Equation 7.24 predominately affect the neighboring pile. They are well predicted by the current analysis using $\lambda_s = 0$, rather than the two-parameter model at a Poisson's ratio $\nu_s > 0.3$, despite of underestimating the stress, σ_r, and overestimating the $\tau_{r\theta}$ (e.g., Figure 7.2). A larger pile-pile interaction (see Figure 7.12c) is associated with the fixed-head case than the free-head case, although the former involves lower stresses than the latter does (Figure 7.13). The negligible impact of Poisson's ratio on the interaction factor (not shown) is not well modeled using the two-parameter model (see Figure 7.2).

In a similar manner, thc factors $\alpha_{\theta P}$ and $\alpha_{\theta M}$ are deduced, respectively, as

$$\alpha_{\theta P} = \sqrt{\sin^2\theta + 4\cos^2\theta}\,\frac{K_1(\gamma_b r_o/s)}{K_1(\gamma_b)} \tag{7.25}$$

$$\alpha_{\theta M} = \sqrt{\sin^2\theta + 4\cos^2\theta}\,\frac{K_1(\gamma_b r_o/s)r_o/s - \gamma_b K_o(\gamma_b r_o/s)}{K_1(\gamma_b) - \gamma_b K_o(\gamma_b)} \tag{7.26}$$

Equation 7.25 compares well with FEM study (Randolph 1981b), as indicated in Figure 7.12d, but Equation 7.26 shows slight overestimation of the $\alpha_{\theta M}$ (not shown here).

7.7 CONCLUSION

This chapter provides elastic solutions for lateral piles, which encompass a new load transfer approach for piles in a homogenous, elastic medium; and compact closed-form expressions for the piles and the surrounding soil that are underpinned by modulus of subgrade reaction, k, and the fictitious tension, N_p, and the load transfer factor γ_b linking the response of the pile and the surrounding soil. The expressions compare well with more rigorous numerical approaches and well capture the effect of the pile-head and base conditions to any pile–soil relative stiffness but avoid the unreasonable predictions at high Poisson's ratio using the conventional two-parameter model.

The approach can be reduced physically to the available uncoupled approach for a beam using the Winkler model, and/or a two-dimensional rigid disc. The current parameters k and N_p may be used in the available Hetenyi's solutions to gain predictions for long flexible piles. Short piles of sufficiently high stiffness, or of free head, floating base, may be treated as "rigid piles." The maximum bending moment for the free-head rigid piles is ~7 times that for the fixed-head case.

Complicated loading may be decomposed into a number of components, which can be modeled using the current approach for a concentrated load and moment.

Chapter 8

Laterally loaded rigid piles

8.1 INTRODUCTION

Extensive in situ full-scale and laboratory model tests have been conducted on laterally loaded rigid piles (including piers and drilled shafts, see Figure 8.1). The results have been synthesized into various simple expressions (see Table 3.7, Chapter 3, this book) to estimate lateral bearing capacity. To assess nonlinear pile–soil interaction, centrifuge and numerical finite element (FE) modeling (Laman et al. 1999) have been conducted. These tests and modeling demonstrate a diverse range of values for the key parameter of limiting force per unit length along the pile (p_u profile, also termed as LFP). The divergence has been attributed to different stress levels and is theoretically incomparable (Guo 2008). With the development of mono-pile (rigid) foundation for wind turbine, it is imperative to have a stringent nonlinear model for modeling the overall nonlinear response at any stage. The model should also allow unique back-estimation of the parameters using measured nonlinear response.

In this chapter, elastic-plastic solutions are developed for laterally loaded rigid piles. They are presented in explicit expressions, which allow (1) nonlinear responses of the piles to be predicted; (2) the *on-pile force profiles* at any loading levels to be constructed; (3) the new yield (critical) states to be determined; and (4) the displacement-based pile capacity to be estimated. By stipulating a linear LFP or a uniform LFP, the study employs a uniform modulus with depth or a linearly increasing modulus.

A spreadsheet program was developed to facilitate numeric calculation of the solutions. Comparison with measured data and FE analysis will be presented to illustrate the accuracy and highlight the characteristics of the new solutions. A sample study will be elaborated to show all the aforementioned facets, apart from the impact of modulus profile, the back-estimation of soil modulus, and the calculation of stress distribution around a pile surface.

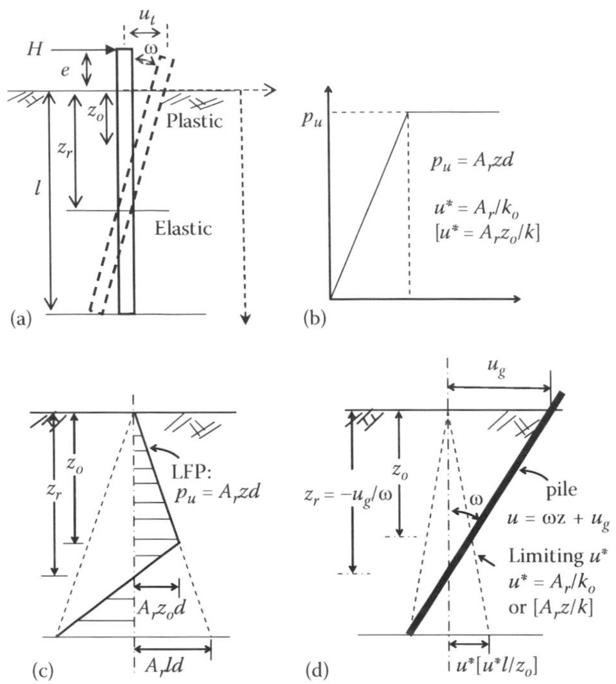

Figure 8.1 Schematic analysis for a rigid pile. (a) Pile–soil system. (b) Load transfer model. (c) Gibson p_u (LFP) profile. (d) Pile displacement features.

8.2 ELASTIC PLASTIC SOLUTIONS

8.2.1 Features of laterally loaded rigid piles

Under a lateral load, H, applied at an eccentricity, e, above groundline, elastic-plastic solutions for infinitely long, flexible piles have been developed previously (Guo 2006). Capitalized on a generic LFP, these solutions, however, do not allow pile–soil relative slip to be initiated from the pile tip. Thus, theoretically speaking, they are not applicable to a rigid pile discussed herein.

With a Gibson p_u (a linear LFP), the on-pile force (pressure) profile alters as presented previously in Chapter 3, this book, for constant or Gibson k. It actually characterizes the mobilization of the resistance along the unique p_u profile (independent of pile displacement). Given a uniform p_u (= $N_g s_u d$), and constant k [= $p/(du)$] the profile is shown in Figure 8.2a$_1$–c$_1$ as solid lines, which is mobilized along the stipulated constant LFP (Broms 1964) indicated by the two dashed lines. As with a Gibson p_u, three typical states of yield between pile and soil are noted. At the tip-yield state, the on-pile force profile follows the positive LFP down to a depth of z_o, below which it

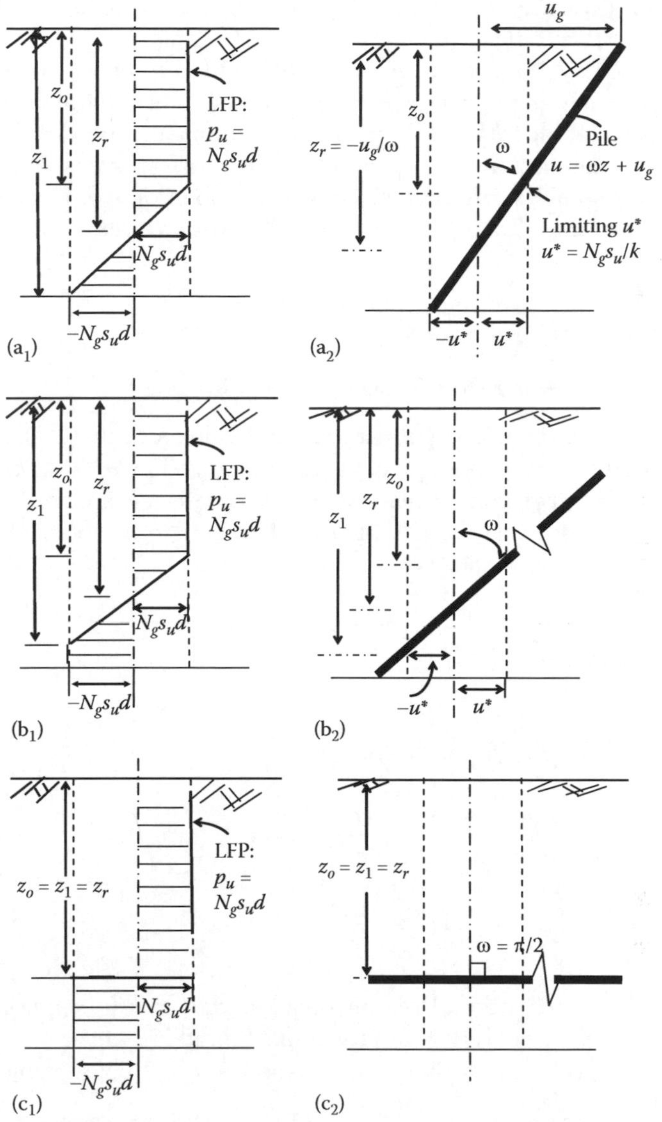

Figure 8.2 Schematic limiting force profile, on-pile force profile, and pile deformation. (a$_i$) Tip-yield state, (b$_i$) Post-tip yield state, (c$_i$) Impossible yield at rotation point (YRP) (piles in clay) (i = 1 p profiles, 2 deflection profiles).

is governed by elastic interaction ($u^* = N_g s_u/k$). In addition, a load beyond the tip-yield state enables the limiting force to be fully mobilized from the tip as well and to a depth of z_1. A maximum load may render the depths z_o, z_1 to merge with the depth of rotation z_r (i.e., $z_o = z_r = z_1$), which is practically unachievable, but for a fully plastic (ultimate) state (see Figure 8.2c$_1$).

Assuming a constant k, solutions for a rigid pile were deduced previously for a uniform p_u profile with depth (Scott 1981), and for a linear (Gibson) p_u profile (Motta 1997), respectively. With a Gibson p_u profile, closed-form solutions were developed and presented in compact form (Guo 2003) of the slip depths (see Table 8.1). The latter allows nonlinear responses to be readily estimated, along with responses at defined critical states (Guo 2008), such as the on-pile p_u profiles in Figure 8.3. The solutions for Gibson p_u and new solutions for constant p_u are presented in this chapter.

8.2.2 Solutions for pre-tip yield state (Gibson p_u, either k)

8.2.2.1 $\bar{H}$, $\bar{u}_g$, $\bar{\omega}$, and $\bar{z}_r$ for Gibson p_u and Gibson k

Given a Gibson k featured by $p = k_o dzu$, and Gibson p_u (= $A_r dz$), the solutions (Guo 2008) are elaborated next. Characterized by $u^* = A_r/k_o$ and $u = \omega z + u_g$, the unknown rotation ω and groundline displacement u_g of a rigid pile are expressed primarily as functions of the pile-head load, H, and the slip depth, z_o (see Figure 8.1).

$$\bar{H} = \frac{1}{6}\frac{1+2\bar{z}_o+3\bar{z}_o^2}{(2+\bar{z}_o)(2\bar{e}+\bar{z}_o)+3}, \tag{8.1}$$

$$\bar{u}_g = \frac{3+2[2+(\bar{z}_o)^3]\bar{e}+(\bar{z}_o)^4}{[(2+\bar{z}_o)(2\bar{e}+\bar{z}_o)+3](1-\bar{z}_o)^2}, \tag{8.2}$$

$$\omega = \frac{A_r}{k_o l}\frac{-2(2+3\bar{e})}{[(2+\bar{z}_o)(2\bar{e}+\bar{z}_o)+3](1-\bar{z}_o)^2}, \quad \text{and} \tag{8.3}$$

$$\bar{z}_r = \frac{-u_g}{\omega l} \tag{8.4}$$

where $\bar{H} = H/(A_r dl^2)$, normalized pile-head load; $\bar{u}_g = u_g k_o/A_r)$, normalized groundline displacement; ω = rotation angle (in radian) of the pile; $\bar{z}_r = z_r/l$, normalized depth of rotation, $\bar{z}_o = z_o/l$, and $\bar{e} = e/l$. These solutions are characterised by:

1. Two soil-related parameters, k_o and A_r, or up to three measurable soil parameters, as the k_o is related to G (see Equation 3.62, Chapter 3, this book); while A_r is calculated using the unit weight γ_s', and angle of soil friction, ϕ' with $A_r = \gamma_s' N_g$ (see Equation 3.68 and Table 3.7, Chapter 3, this book, with $A_r = A_L/d$, n =1).
2. The sole variable z_o/l to capture nonlinear response. Assigning a value to z_o/l, for instance, a pair of pile-head load H and groundline displacement u_g are calculated using Equation 8.1 and Equation 8.2, respectively.

Table 8.1 Solutions for a rigid pile at pre-tip and tip-yield states (constant k)

Constant p_u *(n = 0)*	*Gibson* p_u *(n = 1, Guo 2008)*	*Equation*
$u = \omega z + u_g$, $z_r/l = -u_g/(\omega l)$, $kd = \frac{3\pi\tilde{G}}{2}\{2\gamma_b \frac{K_1(\gamma_b)}{K_o(\gamma_b)} - \gamma_b^2[\left(\frac{K_1(\gamma_b)}{K_o(\gamma_b)}\right)^2 - 1]\}$ where $K_i(\gamma_b)$ (i = 0, 1) is modified Bessel functions of the second kind, and of order i. $\gamma_b = k_1(r_o/l)$; $k_1 = 2.14 + \bar{e}/(0.2+0.6\bar{e})$, increases from 2.14 to 3.8 as e increases from 0 to ∞.		
$p = kdu, p_u = N_g s_u d, u^* = N_g s_u / k$	$p = kdu, p_u = A_r dz, u^* = A_r z_o / k$	
$\bar{H} = 0.5(1+2\bar{z}_o)/(2+\bar{z}_o+3\bar{e})$	$\bar{H} = 0.5\bar{z}_o/(2+\bar{z}_o+3\bar{e})$	(8.1g)
$\bar{u}_g = \frac{2+3(1+\bar{z}_o^2)\bar{e}+\bar{z}_o^3}{(2+\bar{z}_o+3\bar{e})(1-\bar{z}_o)^2}$	$\bar{u}_g = \frac{(3\bar{e}+2)\bar{z}_o}{(2+\bar{z}_o+3\bar{e})(1-\bar{z}_o)^2}$	(8.2g)
$\bar{\omega} = \frac{-3(1+2\bar{e})}{(2+\bar{z}_o+3\bar{e})(1-\bar{z}_o)^2}$	$\bar{\omega} = \bar{z}_o \frac{\bar{z}_o^2-3+3\bar{e}(\bar{z}_o-2)}{(2+\bar{z}_o+3\bar{e})(1-\bar{z}_o)^2}$	(8.3g)
$\bar{z}_m = \bar{H}$ $\bar{z}_m = \frac{1+2\bar{z}_o^3+6\bar{e}\bar{z}_o^2}{3(1+2\bar{e})}$	$\bar{z}_m = \sqrt{2\bar{H}}$ $\bar{z}_m = \frac{1+\bar{z}_o(\bar{z}_o+3\bar{e})}{3(1+2\bar{e})-\bar{z}_o(\bar{z}_o+3\bar{e})}$	($z_m \le z_o$) (8.20g) ($z_m > z_o$) (8.21g)
$\bar{M}_m = \bar{H}(\bar{e}+0.5\bar{z}_m)$ ($z_m \le z_o$) $\bar{M}_m = \frac{[(1+2\bar{z}_m-3\bar{z}_o^2)\bar{e}+(\bar{z}_m-\bar{z}_o^3)](1-\bar{z}_m)^2}{2(2+\bar{z}_o+3\bar{e})(1-\bar{z}_o)^2}$ ($z_m > z_o$) and $\bar{M}_o\|_{e=\infty} = (1+2\bar{z}_o)/6$	$\bar{M}_m = \bar{H}(\bar{e}+2\bar{z}_m/3)$ $\bar{M}_m = \frac{(\bar{z}_m-\bar{z}_o)^3}{(1-\bar{z}_o)^2(2+\bar{z}_o)}\left(\bar{H}\bar{e}+\frac{\bar{z}_o^3}{3}\right) + \bar{H}(\bar{e}+\bar{z}_m)+\frac{1}{6}\bar{z}_o^2(2\bar{z}_o-3\bar{z}_m)$	($z_m \le z_o$) (8.22g) ($z_m > z_o$)

continued

Table 8.1 (Continued) Solutions for a rigid pile at pre-tip and tip-yield states (constant *k*)

Constant p_u *(n = 0)*	*Gibson* p_u *(n = 1, Guo 2008)*	*Equation*
	$+\frac{\bar{z}_o}{2}\left[\frac{(\bar{z}_m\bar{z}_o-2+\bar{z}_m)(\bar{z}_o-\bar{z}_m)^2}{(1-\bar{z}_o)(2+\bar{z}_o)}\right]$	
At tip-yield state		
$\bar{z}_* = -(1.5\bar{e}+0.5)+0.5\sqrt{3+6\bar{e}+9\bar{e}^2}$	$\bar{z}_* = -(1.5\bar{e}+0.5)+0.5\sqrt{5+12\bar{e}+9\bar{e}^2}$	(8.19g)
$\bar{M}_o = (1-2\bar{H}_o-2\bar{H}_o^2)/6$	$\bar{M}_o = (1-8\bar{H}_o-4\bar{H}_o^2)/12$	(8.29g)
$\bar{M}_m = (1-\bar{H}_o)^2/6$ or $\bar{M}_m = 0.5\bar{H}_o^2+\bar{M}_o$	$\bar{M}_m = \frac{1}{12}(1-8\bar{H}_o-4\bar{H}_o^2)+\frac{(2\bar{H}_o)^{1.5}}{3}$	(8.30g)
$\bar{u}_g = u_g k/N_g s_u$, $\bar{\omega} = \omega kl/N_g s_u$, $\bar{H} = H/(N_g s_u dl)$, $\bar{M}_m = M_m/(N_g s_u dl^2)$, and $\bar{M}_o = M_o/(N_g s_u dl^2)$	$\bar{u}_g = u_g k/A_r l, \bar{\omega} = \omega k/A_r, \bar{H} = H/(A_r dl^2)$, $\bar{M}_m = M_m/(A_r dl^3)$, and $\bar{M}_o = M_o/(A_r dl^3)$	

Note: $\bar{z}_o = z_o/l, \bar{z}_r = z_r/l = -\bar{u}_g/\bar{\omega}, \bar{z}_m = z_m/l, \bar{e} = e/l$

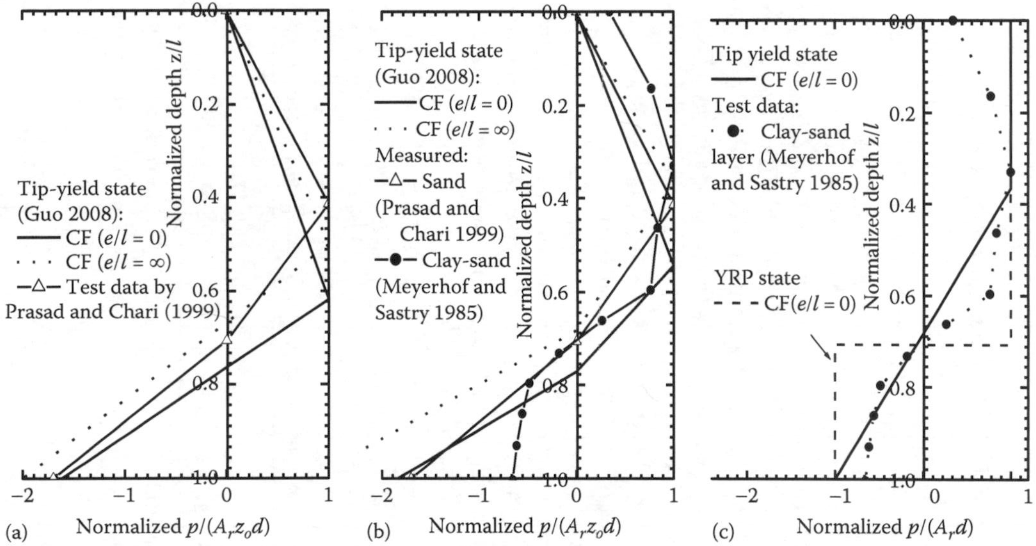

Figure 8.3 Predicted versus measured (see Meyerhof and Sastray 1985; Prasad and Chari 1999) normalized on-pile force profiles upon the tip-yield state and YRP: (a) Gibson p_u and constant k. (b) Gibson p_u and Gibson k. (c) Constant p_u and constant k.

3. A proportional increase of the H to the pile diameter (width) as per Equation 8.1. The u_g implicitly involves with the pile dimensions via the k_o that in turn is related to pile slenderness ratio (l/r_o) via the γ_b (Chapter 3, this book).
4. A free length e of the loading point (H) above ground level, or an $e = M_o/H$ to accommodate moment loading M_o at groundline level.

8.2.2.2 $\bar{H}$, $\bar{u}_g$, $\bar{\omega}$, and $\bar{z}_r$ for Gibson p_u and constant k

As for Gibson p_u and constant k (with $p = kdu$), the solutions are provided in Table 8.1 (right column). The equations or results are generally highlighted using numbers postfixed with "g" and/or are set in brackets. They have similar features to Gibson p_u and Gibson k. The difference is that a plastic (slip) zone for Gibson k is not initiated (i.e., $z_o > 0$) from groundline until the H exceeds $A_r dl^2/(24e/l + 18)$; whereas the slip for constant k is developed upon a tiny loading. The latter implies the need of elastic-plastic solutions rather than elastic solutions (Scott 1981; Sogge 1981) in practice.

Salient features of the current solutions are illustrated for two extreme cases of $e = 0$ and ∞. Assuming a Gibson p_u and Gibson k, the usage of relevant expressions for $z_o \le z_*$ ($z_* = z_o$ at tip-yield state) (for $e < 3.0$ practically) may refer to Table 8.2. Given $e = \infty$ (or practically $e \ge 3$), Equations 8.1 through 8.4 do reduce to those expressions in Table 8.3 gained for pure moment loading M_o (with $H = 0$), including the normalized ratios for the u_g, ω, and M_o. For instance, the normalized moment per Equation 8.1 degenerates to $\bar{M}_o[= M_o/(A_r dl^3)]$ and $M_o = He$ of:

$$\bar{M}_o = \frac{1 + 2\bar{z}_o + 3\bar{z}_o^2}{12(2 + \bar{z}_o)} \tag{8.5}$$

Table 8.2 Response at various states [e = 0, Gibson p_u and Gibson k/(constant k)]

Items	$H/(A_r dl^2)$	$u_g k_o/A_r$ [$u_g k/A_r l$]	$\omega k_o l/A_r$ [$\omega k/A_r$]	$M_m/(A_r dl^3)$
$\bar{z}_o \le \bar{z}_*$	*Equation* 8.1 [*Equation* 8.1g]	*Equation* 8.2 [*Equation* 8.2g]	*Equation* 8.3 [*Equation* 8.3g]	*Equation* 8.22 [*Equation* 8.22g or M_m *for* $z_m > z_o$]
Tip-yield[a]	0.113 [0.118]	3.383 [3.236]	−4.3831 [−4.2352]	0.036 [0.038]
YRP[b]	0.130 [0.130]	∞	$0.5\pi k_o l/A_r$ [$0.5\pi k/A_r$]	0.0442 [0.0442]

[a] $\bar{z}_* = 0.5437/[0.618]$, $\bar{z}_r = 0.772/[0.764]$, $\bar{z}_m = 0.4756/[0.4859]$, and $C_y = 0.296/[0.236]$
[b] At the YRP state, all critical values are independent of k distribution. Thus, $\bar{z}_o = \bar{z}_r = 0.7937/[0.7937]$, $\bar{z}_m = 0.5098/[0.5098]$. Also, $M_o = 0$.

Table 8.3 Response at various states ($e = \infty$, Gibson p_u, and Gibson k [constant k])

Items	$u_g k_o/A_r/[u_g k/A_r l]$	$\omega k_o l/A_r/[\omega k/A_r]$	$M_o/(A_r d l^3)$
$\bar{z}_o \le \bar{z}_*$	$\frac{2+\bar{z}_o^3}{(1-\bar{z}_o)^2(2+\bar{z}_o)}$ $[\frac{\bar{z}_o}{(1-\bar{z}_o)^2}]$	$\frac{-3}{(1-\bar{z}_o)^2(2+\bar{z}_o)}$ $[\frac{\bar{z}_o(\bar{z}_o-2)}{(1-\bar{z}_o)^2}]$	$\frac{1+2\bar{z}_o+3\bar{z}_o^2}{12(2+\bar{z}_o)}$ $[\bar{z}_o/6]$
Tip yield [a]	2.155/[2.0]	−3.1545/[−3.0]	0.0752/[0.0833]
YRP [b]	∞	$0.5\pi k_o l/A_r/[0.5\pi k/A_r]$	0.0976/[0.0976]

[a] $\bar{z}_* = 0.366/[0.50]$, $\bar{z}_r = 0.683/[0.667]$, $\bar{z}_m = 0/[0]$, and $C_y = 0.464/[0.333]$
[b] $\bar{z}_o = \bar{z}_r = 0.7071/[0.7071]$, and $\bar{z}_m = 0$ [0].
Also, $H = 0$, and $M_o = M_m$

Assuming a Gibson p_u and constant k, the corresponding features are provided in the Tables 8.2 and 8.3 as well.

8.2.3 Solutions for pre-tip yield state (constant p_u and constant k)

Given constant p_u (= $N_g s_u d$) and constant k (= p/du), typical normalized expressions were deduced and are shown in Table 8.1 (left column). For instance, $\bar{H}$, $\bar{u}_g$ are given by

$$\frac{H}{N_g s_u dl} = \frac{0.5(1+2\bar{z}_o)}{2+\bar{z}_o+3\bar{e}} \quad \text{and} \tag{8.6}$$

$$\frac{u_g k}{N_g s_u} = \frac{2+3(1+\bar{z}_o^2)\bar{e}+\bar{z}_o^3}{(2+\bar{z}_o+3\bar{e})(1-\bar{z}_o)^2} \tag{8.7}$$

The normalized ratios for the $\bar{u}_g$, $\bar{\omega}$, and $\bar{M}_o$ are provided in Table 8.1. Given $\bar{e} = 0$, the solutions reduce to those obtained previouly (Scott 1981). Similar features to those for Gibson p_u (see Table 8.1) are noted, such as

- A plastic (slip) zone is not initiated (i.e., $z_o > 0$) from groundline until the H exceeds $0.5A_r dl/(3e/l + 2)$.
- Given $e = \infty$, the solutions do reduce to those obtained for pure moment loading M_o (with $H = 0$). For instance, the moment M_m degenerates to M_o (= He). Its nondimensional form of $\bar{M}_o$ [= $M_o/(N_g s_u dl^2)$] is given by:

$$\bar{M}_o = (1+2\bar{z}_o)/6 \tag{8.8}$$

- Finally, the calculation of M_m depends on the relative value between z_m and z_o, although the simple expression M_m for $z_m \le z_o$ is generally sufficiently accurate.

8.2.4 Solutions for post-tip yield state (Gibson p_u, either k)

8.2.4.1 $\bar{H}$, $\bar{u}_g$, and $\bar{z}_r$ for Gibson p_u and Gibson k

Equations 8.1 through 8.4 for *pre-tip* and *tip-yield* states are featured by the yield to the depth z_o (being initiated from groundline only). As illustrated in Chapter 3, this book, at a sufficiently high load level, another yield zone to depth z_1 may be developed from the pile tip as well. Increase load will cause the two yield zones to move towards each other and to approach the practically impossible ultimate state of equal depths of $z_o = z_1 = z_r$ (see Figure 8.2c$_2$). This is termed the *post-tip yield state*, for which horizontal force equilibrium of the entire pile, and bending moment equilibrium about the pile-head (rather than the tip), were used to deduce the solutions. A variable C [$= A_r/(u_g k_o)$] is introduced as the reciprocal of the normalized displacement (see Equation 8.2). It must not exceed its value C_y at the tip-yield state (i.e., $C < C_y$, and u_g being estimated using Equation 8.2 and $z_o = z_*$) to induce the post-tip yield state. With this C, the equations or expressions for estimating z_r, H, and u_g (Gibson p_u and Gibson k) are expressed as follows:

1. The rotation depth z_r is governed by the C and the e/l

$$\bar{z}_r^3 + \frac{3+C^2}{2(1+C^2)}\bar{e}\bar{z}_r^2 - \frac{1}{4(1+C^2)}\left(2+3\bar{e}\right) = 0. \quad (8.9)$$

The solution of Equation 8.9 may be approximated by

$$\bar{z}_r = \sqrt[3]{A_0} + \sqrt[3]{A_1} - D_1/6 \quad (8.10)$$

$$A_j = (D_o/8 - D_1^3/216) + (-1)^j[(27D_o^2 - 2D_oD_1^3)/1728]^{1/2} \; (j = 0, 1) \quad (8.11)$$

$$D_1 = \frac{3+C^2}{1+C^2}\bar{e} \quad D_o = \frac{2+3\bar{e}}{1+C^2} \quad (8.12)$$

2. The normalized head-load, $H/(A_r dl^2)$ and the groundline displacement, u_g, are deduced as

$$\bar{H} = \left(1+\frac{C^2}{3}\right)\bar{z}_r^2 - 0.5 \quad (8.13)$$

$$u_g = A_r/(k_o C) \quad (8.14)$$

3. The slip depths from the pile head, z_o, and the tip, z_1, are computed using $z_o = z_r(1 - C)$ and $z_1 = z_r(1 + C)$, respectively.

8.2.4.2 $\bar{H}$, $\bar{u}_g$, and $\bar{z}_r$ for Gibson p_u and constant k

As for the constant k (Gibson p_u), a new variable C [$= A_r z_r/(u_g k)$] was defined using Equation 8.14g in Table 8.4 as the product of the reciprocal of the normalized displacement $u_g k/A_r l$ and the normalized rotation depth z_r/l. With this C, the counterparts for Equations 8.9 through 8.14 are provided in Table 8.4, and z_o [$= z_r/(1 - C)$] and z_1 [$= z_r/(1 + C)$] (see Table 8.5).

Table 8.4 Responses of piles with Gibson p_u and constant k (post-tip yield state)

Expressions	*Equations*
$\frac{H}{A_r dl^2} = 0.5\left[\frac{2}{1-C^2}\bar{z}_r^2 - 1\right]$ where $C = A_r z_r/(k\,u_g)$	[8.13g]
$u_g = A_r z_r/(kC)$	[8.14g]
The ratio z_r/l is governed by the following expression:	
$\bar{z}_r^3 + 1.5(1-C^2)\bar{e}\bar{z}_r^2 - (0.5+0.75\bar{e})(1-C^2)^2 = 0$	[8.9g]
Thus, the z_r/l should be obtained using	
$[-1.5\bar{e}C_g^2 - (0.5+0.75\bar{e})C_g^4]\bar{z}_r^4 + \bar{z}_r^3$	
$+[1.5\bar{e} + (1+1.5\bar{e})C_g^2]\bar{z}_r^2 - (0.5+0.75\bar{e}) = 0$ where $C_g = Cl/(kz_r)$.	
$\bar{z}_r$ may be approximated by the following solution (Guo 2003):	
$\bar{z}_r = 0.5(1-C^2)(\sqrt[3]{A_0} + \sqrt[3]{A_1} - \bar{e})$ (Iteration required)	[8.10g]
$A_j = \left(-\bar{e}_o^3 + \frac{2+3\bar{e}}{1-C^2}\right) + (-1)^j\left[\frac{2+3\bar{e}}{1-C^2}\left(-2\bar{e}_o^3 + \frac{2+3\bar{e}}{1-C^2}\right)\right]^{1/2}$ (j = 0, 1).	[8.11g]
It is generally ~5% less than the exact value of $\bar{z}_r$. Note that Equations 8.20, 8.23, and 8.24 are valid for this "Constant k" case.	

Source: Guo, W. D., *Can Geotech J*, 45, 5, 2008.

Table 8.5 Expressions for depth of rotation (Gibson p_u, either k)

u	*Depth of rotation*	*Slip depths*	*Figure*
$u = \omega z + u_g$	$z_r = -\frac{u_g}{\omega}$	$\bar{z}_o$ deduced using Equations 8.1~8.4	3.28a
		$\bar{z}_1 = (1+C)\bar{z}_r$ $[\bar{z}_1 = \bar{z}_r/(1-C)]$ $\bar{z}_o = \bar{z}_r(1-C)$ $[\bar{z}_o = \bar{z}_r/(1+C)]$	3.28b
	$\bar{z}_1 = \bar{z}_o = \bar{z}_r$		3.28c

Source: Guo, W. D., *Can Geotech J*, 45, 5, 2008.

Note: u_g, ω, z_r, z_o, z_1 refer to list of symbols. $C = A_r/(u_g k_o)$ (Gibson k), $C = A_r z_r/(u_g k)$ (Constant k).

Table 8.6 Solutions for $H(z)$ and $M(z)$ for piles (constant k)

Constant p_u	*Gibson* p_u *(Guo 2008)*
(1) $\bar{z}_* = -(1.5\bar{e}+0.5)+0.5\sqrt{3+6\bar{e}+9\bar{e}^2}$	$\bar{z}_* = -(1.5\bar{e}+0.5)+0.5\sqrt{5+12\bar{e}+9\bar{e}^2}$
(2) $\bar{z}_r = -\bar{e}+[\bar{e}^2+\bar{e}+0.5-\frac{1}{3}(\frac{C}{l})^2]^{0.5}$	$\bar{z}_r = 0.5(1-C^2)(\sqrt[3]{A_0}+\sqrt[3]{A_1}-\bar{e})$
where $A_j = (-\bar{e}^3+D_1)+(-1)^j[D_1(-2\bar{e}^3+D_1)]^{0.5}$ (j = 0, 1); $D_1 = (2+3\bar{e})/(1-C^2)$. Iteration required for $\bar{z}_r$, which is ~5% less than the exact value of $\bar{z}_r$.	
(3) $z_1 = z_r + C \quad z_o = z_r - C$	$z_1 = z_r/(1-C) \quad z_o = z_r/(1+C)$
(4) $\bar{H} = 2\bar{z}_r - 1$	$\bar{H} = 0.5\left[\frac{2}{1-C^2}\bar{z}_r^2 - 1\right]$
(5) $u_g = N_g s_u z_r/(kC)$, or $C = N_g s_u z_r/(k u_g)$	(6) $\omega = -u_g/z_r$
Equations 1–6 are used only for $C < C_y$.	
$0 < z \le z_o$: $\bar{M}_1(\bar{z}) = \bar{H}(\bar{e}+\bar{z})-0.5\bar{z}^2$	$0 < z \le z_o$: $\bar{M}_1(\bar{z}) = \bar{H}(\bar{e}+\bar{z})-\frac{1}{6}\bar{z}^3$
$\bar{H}_1(\bar{z}) = \bar{H}-\bar{z}$	$\bar{H}_1(\bar{z}) = \bar{H}-0.5\bar{z}^2$
$z_o < z \le z_1$:	$z_o < z \le z_1$:
$\bar{M}_2(\bar{z}) = \bar{H}(\bar{e}+\bar{z})-(\bar{z}_o\bar{z}-0.5\bar{z}_o^2)$ $-\frac{\omega l}{6u^*}(\bar{z}^3-3\bar{z}\bar{z}_o^2+2\bar{z}_o^3)-\frac{u_g}{2u^*}(\bar{z}-\bar{z}_o)^2$	$\bar{M}_2(\bar{z}) = \bar{H}(\bar{e}+\bar{z})-(0.5\bar{z}\bar{z}_o^2-\frac{\bar{z}_o^3}{3})$ $-\frac{\omega l}{6u^*}(\bar{z}^3-3\bar{z}\bar{z}_o^2+2\bar{z}_o^3)-\frac{u_g}{2u^*}(\bar{z}-\bar{z}_o)^2$
$\bar{H}_2(\bar{z}) = \bar{H}-\bar{z}_o-\frac{\omega l}{2u^*}(\bar{z}^2-\bar{z}_o^2)$ $-\frac{u_g}{u^*}(\bar{z}-\bar{z}_o)$	$\bar{H}_2(\bar{z}) = \bar{H}-0.5\bar{z}_o^2-\frac{\omega l\bar{z}_o}{2u^*}(\bar{z}^2-\bar{z}_o^2)$ $-\frac{u_g}{u^*}\bar{z}_o(\bar{z}-\bar{z}_o)$
$z_1 < z \le l$:	$z_1 < z \le l$:
$\bar{M}_3(\bar{z}) = \bar{H}(\bar{e}+\bar{z})-(\bar{z}_o\bar{z}-0.5\bar{z}_o^2)$ $-\frac{\omega l}{6u^*}(\bar{z}_1^3-3\bar{z}_1\bar{z}_o^2+2\bar{z}_o^3)$ $-\frac{u_g}{2u^*}(\bar{z}_1-\bar{z}_o)^2+0.5(\bar{z}-\bar{z}_1)^2$	$\bar{M}_3(\bar{z}) = \bar{H}(\bar{e}+\bar{z})-(0.5\bar{z}\bar{z}_o^2-\frac{1}{3}\bar{z}_o^3)$ $-\frac{\omega l}{6u^*}(\bar{z}_1^3-3\bar{z}_1\bar{z}_o^2+2\bar{z}_o^3)$ $-\frac{u_g\bar{z}_o}{2u^*}(\bar{z}_1-\bar{z}_o)^2+\frac{1}{6}(\bar{z}+2\bar{z}_1)(\bar{z}-\bar{z}_1)^2$
$\bar{H}_2(\bar{z}) = \bar{H}-\bar{z}_o-\frac{\omega l}{2u^*}(\bar{z}_1^2-\bar{z}_o^2)$ $-\frac{u_g}{u^*}(\bar{z}_1-\bar{z}_o)+(\bar{z}-\bar{z}_1)$	$\bar{H}_3(\bar{z}) = \bar{H}-0.5\bar{z}_o^2-\frac{\omega l\bar{z}_o}{2u^*}(\bar{z}_1^2-\bar{z}_o^2)$ $-\frac{u_g}{u^*}\bar{z}_o(\bar{z}_1-\bar{z}_o)+0.5(\bar{z}^2-\bar{z}_1^2)$

8.2.5 Solutions for post-tip yield state (constant p_u and constant k)

Given constant p_u and constant k, the expressions for estimating z_r, H, and u_g at this state are provided in Table 8.6 and highlighted in the following:

1. The normalized rotation depth $\bar{z}_r$ is governed by the C/l and the e/l

$$\bar{z}_r = -\bar{e} + \left[\bar{e}^2 + \bar{e} + 0.5 - \frac{1}{3}\left(\frac{C}{l}\right)^2\right]^{0.5} \tag{8.15}$$

where $C = z_r u^*/u_g$, C/l = ratio of the normalized rotation depth z_r/l over the normalized displacement u_g/u^*. The variable C must not exceed C_y (C at for the tip-yield state) to ensure the post-tip yield state. The C_y is obtained using u_g, which is calculated by substituting $z_o = z_*$ into Equation 8.7.

2. The normalized load, $H/(N_g s_u dl)$, and the groundine displacement, u_g, are given by

$$\bar{H} = 2\bar{z}_r - 1 \tag{8.16}$$

$$u_g = \frac{N_g s_u}{k}\frac{z_r}{C} \tag{8.17}$$

3. The slip depths from the pile head, z_o, and the tip, z_1, are computed using $z_o = z_r - C$ and $z_1 = z_r + C$, respectively.

The pile-head displaces infinitely as the C approaches zero, see Equation 8.14 or Equation 8.17, offering an upper bound featured by $z_r = z_o$. Equation 8.10 or Equation 8.15 provides the rotation depth, z_r (thus z_o and z_1), for each u_g. The z_r in turn allows the rotation angle ω ($= -u_g/z_r$) to be obtained. Other normalized expressions for the post-tip yield state are provided in Table 8.7, together with those for Gibson p_u.

8.2.6 $\bar{u}_g$, ω, and p profiles (Gibson p_u, tip-yield state)

Referring to Figure 3.28 (Chapter 3, this book), the expression of $-u^* = -A_r/k_o = \omega z_1 + u_g$ for tip-yield state ($z_o = z_*$) is transformed into the following form to resolve the z_*, by replacing u_g with that given by Equation 8.2, ω, with that by Equation 8.3, and z_1 with l:

$$\bar{z}_*^3 + (2\bar{e}+1)\bar{z}_*^2 + (2\bar{e}+1)\bar{z}_* - (\bar{e}+1) = 0 \tag{8.18}$$

The z_*/l was estimated for e/l = 0~100 using Equation 8.18, and it is illustrated in Figure 8.4a. With the values of z_*/l (the tip-yield state), the responses of $u_g k_o l^m/(A_r l)$ and $\omega k_o l^m/A_r$ (note $k = k_o z^m$, m = 0, and 1 for constant k and Gibson k) were calculated and are presented in Figure 8.4b and c, and in

Table 8.7 Response of piles at post-tip yield and YRP states (constant *k*)

Constant p_u	*Gibson* p_u	*Equation*
Post-tip yield state		
$\bar{z}_r = 0.5(1+\bar{H}_o)$	$\bar{z}_r = [(0.5+\bar{H}_o)(1-C^2)]^{0.5}$	
$\bar{M}_o = \frac{1}{4}(\bar{H}_o+1)^2 - \frac{1}{2} + \frac{1}{3}\left(\frac{C}{l}\right)^2$	$\bar{M}_o = \frac{2}{3\sqrt{1-C^2}}(0.5\bar{H}_o+1)^{1.5} - \frac{1}{3}$	
$\bar{z}_m = \bar{H}_o$	$\bar{z}_m = \sqrt{2\bar{H}_o}$	
$\bar{M}_m = 0.5\bar{H}_o^2 + \bar{M}_o$	$\bar{M}_m = \bar{M}_o + \frac{1}{3}(2\bar{H}_o)^{1.5}$	
Yield at rotation point (YRP) ($\bar{z}_1 = \bar{z}_o = \bar{z}_r, C = 0$)		
$\bar{z}_m = \bar{H}_o$	$\bar{z}_m = \sqrt{2\bar{H}_o}$	(8.20)
$\bar{z}_m = 2\bar{z}_r - 1$	$\bar{z}_m = [2\bar{z}_r^2 - 1]^{0.5}$	(8.23)
$\bar{z}_r = 0.5(1+\bar{H}_o)$	$\bar{z}_r = [\bar{H}_o + 0.5]^{0.5}$	
$\bar{M}_o = \frac{1}{4}(\bar{H}_o+1)^2 - \frac{1}{2}$	$\bar{M}_o = \frac{2}{3}(\bar{H}_o+0.5)^{1.5} - \frac{1}{3}$	(8.31g)
$\bar{M}_m = 0.5\bar{H}_o^2 + \bar{M}_o$	$\bar{M}_m = \bar{M}_o + \frac{1}{3}(2\bar{H}_o)^{1.5}$	(8.32g)
$\bar{M}_o$ and $\bar{M}_m$ at YRP for various loading directions		
$\bar{M}_{oi} = (-1)^{i+1}\left\{\frac{1}{4}\left(\bar{H}_{oi}(-1)^{j+1}+1\right)^2 - \frac{1}{2}\right\}$	$\bar{M}_{oi} = (-1)^{i+1}\frac{2}{3}[0.5\bar{H}_{oi}(-1)^{j+1}+1]^{1.5} - \frac{1}{3}$	
$\bar{M}_{mi} = (-1)^{i+1}0.5\bar{H}_{oi}^2 + \bar{M}_{oi}$	$\bar{M}_{mi} = (-1)^{i+1}\frac{1}{3}[2\bar{H}_{oi}(-1)^{j+1}]^{1.5} + \bar{M}_{oi}$	

Note: $\bar{z}_l = z_l/l; i = 1, 2, j = 1$; and $i = 3, 4, j = 2$. (1) $i = 1, j = 1$ $[\bar{H}_o^+, \bar{M}_o^+]$; (2) $i = 2, j = 1$ $[\bar{H}_o^+, \bar{M}_o^-]$; (3) $i = 3, j = 2$ $[\bar{H}_o^-, \bar{M}_o^+]$; and (4) $i = 4, j = 2$ $[\bar{H}_o^-, \bar{M}_o^-]$

Tables 8.2 and 8.3. The counterparts for constant k were obtained using z_*/l expression (right column in Table 8.6) and are provided in the square [] brackets for $e = 0$ and ∞ in Tables 8.2 and 8.3 as well. For instance, a Gibson p_u and constant k at $e = 0$ would have $z_* = 0.618l$ and $z_r = 0.764l$ (see Table 8.2); and at $e = \infty$, $z_* = 0.5l$ and $z_r = 0.667l$ (see Table 8.3). As the e increases from groundline (pure lateral loading) to infinitely large (pure moment loading), the u_g reduces by 36% [38%] and the ω reduces by 28% [29.2%]. These magnitudes of reduction of ~38% is associated with ~30% increase in the maximum bending moment for Gibson p_u and Gibson k (see Example 8.7)

Example 8.1 p **profiles at tip-yield state (Gibson** p_u**)**

The normalized on-pile force profiles of $p/(A_r z_o d)$ at the tip-yield state are constructed using estimated depths z_r and z_*. Given p_u and k profiles, a line is drawn from point (0, 0), to $(A_r d z_*, z_*)$, and then to $(0, z_r)$,

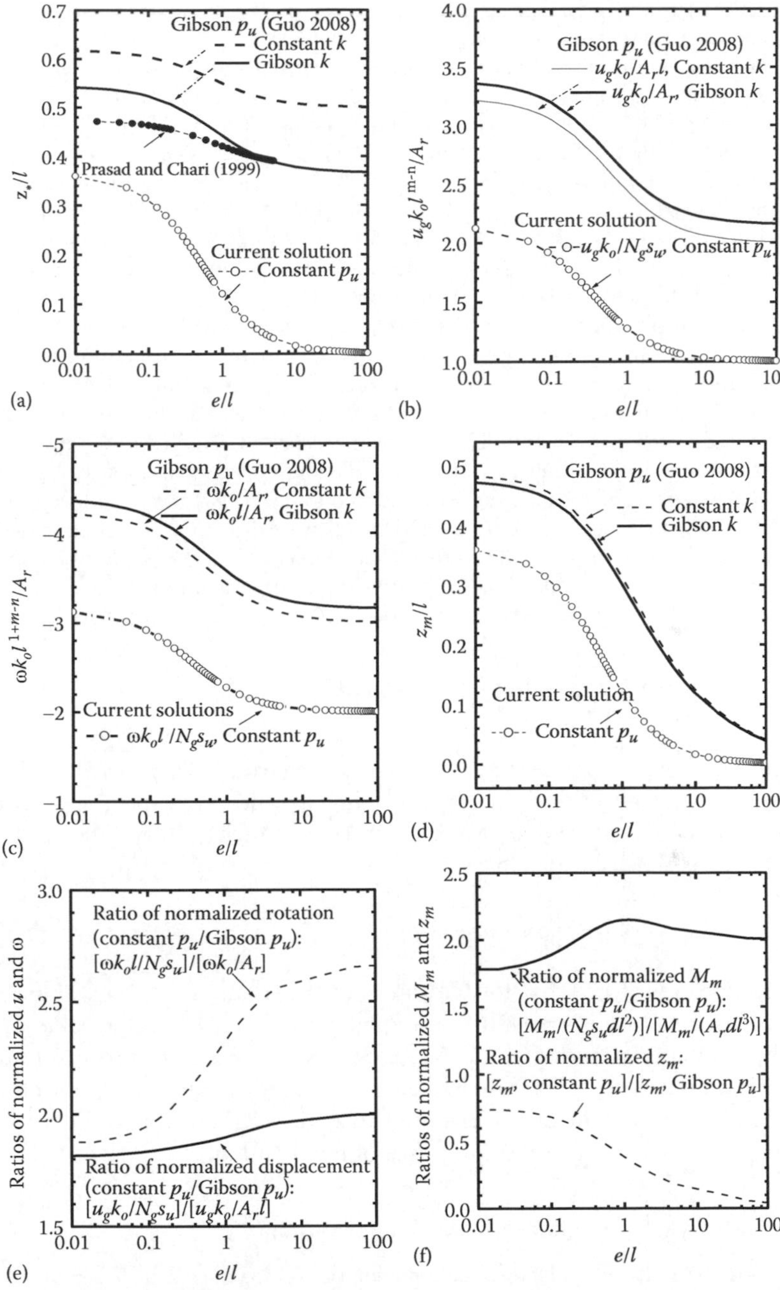

Figure 8.4 Response of piles at tip-yield state. (a) z_o/l. (b) $u_g k_o l^{m-n}/A_r$. (c) $\omega k_o l^{1+m-n}/A_r$. (d) z_m/l. (e) Normalized ratio of u, ω. (f) Ratios of normalized moment and its depth. $n = 0$, 1 for constant p_u and Gibson p_u; $k = k_o z^m$, $m = 0$ and 1 for constant and Gibson k, respectively.

and finally to $(-A_r dl, l)$ to produce the p profile. The first and the second coordinates may be normalized by $A_r d\, z_*$ and l, respectively. For instance, with Gibson p_u and constant k, the normalized profiles are constructed using $z_*/l = 0.618$ and $z_r/l = 0.764$ ($e = 0$), and $z_*/l = 0.5$ and $z_r/l = 0.667$ ($e = \infty$), respectively, as shown in Figure 8.3a. The pressure at the pile-tip level increases from $1.6A_r z_*$ to $2A_r z_*$, as the e increases from 0 to ∞. Likewise, with a Gibson p_u and Gibson k, the normalized profiles are obtained using either $z_* = 0.544l$ and $z_r = 0.772l$ ($e = 0$, Table 8.2) or $z_* = 0.366l$ and $z_r = 0.683l$ ($e = \infty$, Table 8.3), as shown in Figure 8.3b. The pile-tip pressure increases from $1.84A_r z_*$ to $2.73A_r z_*$, as the e shifts towards infinitely large from 0.

The on-pile force per unit length (p) profile is governed by the gradient A_r, the slip depths z_o (from the head) and z_1 (from the tip), and the rotation depth z_r. The constructed profiles at tip-yield state for $e = 0$ and ∞ (see Figure 8.3a and b) well-bracket the test data provided by Prasad and Chari (1999). The tri-linear feature of the "Test" profile is also captured using the Gibson k. The slight overestimation of z_o compared to the measured z_o (Prasad and Chari 1999) implies a pre-tip yield state (see Figure 8.4a), as is evident in the reported capacity shown later in Figure 8.8a. Other factors may affect the predictions such as k profile and its variation with loading eccentricity, lack of measured points about the z_o, and nonlinear elastic-plastic p-y curve.

8.2.7 $\bar{u}_g$, ω, and p profiles (constant p_u, tip-yield state)

Pile-tip yields upon satisfying $-u^* = \omega l + u_g$, regardless of p_u profile. This expression may be expanded, by replacing u_g and ω, respectively, with those gained from $\bar{u}_g$ and $\bar{\omega}$ (see Table 8.1), in which $z_o = z_*$. The $\bar{z}_*$ for a constant p_u is thus obtained as

$$\bar{z}_* = -\left(1.5\bar{e} + 0.5\right) + 0.5\sqrt{3 + 6\bar{e} + 9\bar{e}^2} \tag{8.19}$$

The $\bar{z}_*$ is different from that for Gibson p_u in the square root (see Table 8.1) and was estimated for $\bar{e} = 0$~100. The $\bar{z}_*$ allows normalized values of $\bar{u}_g$ and $\bar{\omega}$ to be calculated in terms of the expressions in Table 8.1. The results (indicated by constant p_u) are presented in Figure 8.4a through c. The impact of p_u profile on u_g and ω is illustrated in Figure 8.4e. Typical values are provided in Table 8.8 for extreme cases of $e = 0$ and ∞.

Example 8.2 p profiles at tip-yield state (constant p_u)

As with a Gibson p_u, the force per unit length p (along normalized depth) at the tip-yield state for constant p_u and constant k is constructed by drawing lines in sequence between adjacent points ($N_g s_u d$, 0), ($N_g s_u d$, $\bar{z}_*$), (0, $\bar{z}_r$), ($-N_g s_u d$, $\bar{z}_1$), and ($-N_g s_u d$, 1), resembling Figure 8.2b$_1$. A constant p_u is associated with $\bar{z}_* = 0.366$ and $\bar{z}_r = 0.683$ for $e = 0$. The

Table 8.8 Critical response at tip-yield state (constant *k*)

	For Constant p_u					
	z_*/l	$\frac{H}{N_g s_u dl}$	$\frac{u_g k}{N_g s_u}$	$\frac{\omega k l}{N_g s_u}$	z_r/l	C_y
e = 0	0.366	0.366	2.155	−3.155	0.683	0.194*l*
e = ∞	0	0	1.0	−2.0	0.5	0.5*l*
	For Gibson p_u					
	z_*/l	$\frac{H}{A_r dl^2}$	$\frac{u_g k}{A_r l}$	$\frac{\omega k}{A_r}$	z_r/l	C_y
e = 0	0.618	0.118	3.236	−4.235	0.764	0.236
e = ∞	0.5	0	2.0	−3.0	0.667	0.333

normalized on-pile $\bar{p}$ [$= p/(N_g s_u d)$] profile is plotted against the normalized depth in Figure 8.3c as the current "CF" (closed-form) solutions, in which $A_r = N_g s_u$ and normalized measured p on model piles in clay-sand layers (Meyerhof and Sastray 1985) is also provided.

8.2.8 Yield at rotation point (YRP, either p_u)

The pile-head displaces infinitely as the C approaches zero (see Equation 8.14), or z_o approaches z_r, the yield at rotation point (YRP). While being practically unachievable, the state provides a useful upper bound. At YRP state, with Gibson p_u, Equation 8.13 reduces to that proposed by Petrasovits and Award (1972), as the on-pile force profile does (see Figure 3.28c, Chapter 3, this book). The LFP is constructed by linking adjacent points (0, 0), ($A_r dz_{**}$, z_*), ($-A_r dz_{**}$, z_r), and ($-A_r dl$, l). Regardless of the modulus k profile, the z_r/l is estimated using Equation 8.10, $C = 0$, $D_o = 2 + 3e/l$, and $D_1 = 3e/l$ (from Equations 8.11 and 8.12) regarding $e \neq \infty$; otherwise, it is gained directly using Equation 8.9. For instance, at $e = 0$, $D_o = 2$, and $D_1 = 0$ (from Equation 8.12), $A_o = 0.5$, and $A_1 = 0$ (from Equation 8.11), and the z_r/l is evaluated as 0.7937; with $e = \infty$ in Equation 8.9, the z_r/l is directly computed as 0.7071.

As for a constant p_u, response of rigid piles at the YRP state (see Figure 8.2) is calculated using Equations 8.15 through 8.17 and $C = 0$. In the case of $e = 0$, we have $\bar{z}_o$ ($= \bar{z}_r$) = 0.707 and the on-pile force profile shown in Figure 8.3c.

8.2.9 Maximum bending moment and its depth (Gibson p_u)

8.2.9.1 Pre-tip yield ($z_o < \bar{z}_*$) and tip-yield ($z_o = \bar{z}_*$) states

The depth z_m at which the maximum bending moment occurs is given by Equation 8.20 if $z_m \leq z_o$, and otherwise by Equation 8.21.

$$\bar{z}_m = \sqrt{2\bar{H}} \ (z_m \le z_o) \tag{8.20}$$

$$\frac{z_m}{l} = \frac{3\bar{z}_o^3(\bar{z}_o + 2\bar{e}) + 1 + \sqrt{3}\sqrt{[\bar{z}_o^3(\bar{z}_o + 2\bar{e}) + 16\bar{e} + 11][3\bar{z}_o^3(\bar{z}_o + 2\bar{e}) + 1]}}{8(2 + 3\bar{e})} \ (z_m > z_o) \tag{8.21}$$

Equation 8.20 offers $H/(A_r dl^2) = 0.5(z_m/l)^2$. With $\bar{z}_*$ gained from Equation 8.18, the z_m/l at tip-yield state was computed for a series of e/l ratios and is plotted in Figure 8.4d against e/l. The calculation shows that the z_m generally follows Equation 8.20 as $z_m < z_*$ (regardless of e/l) is generally noted, but for a low load level (pre-tip yield state) and a small eccentricity e (which is generally not a concern). The z_m converges to zero, as the e/l approaches infinitely large (i.e., pure moment loading) (see Table 8.3).

The z_m/l from Equation 8.20 allows the maximum bending moment (M_m) (for $z_m \le z_o$) to be gained as

$$\bar{M}_m = (2\bar{z}_m/3 + \bar{e})\bar{H} \ (z_m \le z_o) \tag{8.22}$$

Otherwise, with z_m/l from Equation 8.21, the normalized M_m should be calculated using another expression (not shown herein) (Guo 2008). Equation 8.22 is generally valid, independent of pile–soil relative stiffness, as it is essentially identical to that for a flexible pile (Guo 2006). Likewise, the expressions for the z_m and M_m for Gibson p_u and constant k were derived as provided in Table 8.1. Note that Equations 8.20 and 8.22 are indepedent of k profile and are actaully valid from the tip-yield state through to the ultimate YRP state (with $z_m < z_o$). This will be shown in Section 8.5.

Example 8.3 Features of M_{max} and M_o at tip-yield state (Gibson p_u)

At the tip-yield state, the M_m was calculated using Equation 8.22 for Gibson p_u and Gibson k, and Gibson p_u and constant k, respectively, as $z_m < z_o$ (Figure 8.4a and d). It is presented in Figure 8.5a through c in form of $M_m/(A_r dl^3)$. The main features are as follows:

1. The normalized maximum moment M_m obtained is higher using constant k than Gibson k (see Figure 8.5b, Gibson p_u), which indicates the impact of a higher value of normalized H for constant k than Gibson k (shown later). At $e = 0$ (see Table 8.2), $M_m/(A_r dl^3) = 0.036$ [0.038], with $\bar{z}_* = 0.5473$ [0.618], and $H/(A_r dl^2) = 0.113$ [0.118]. At $e = \infty$ (see Table 8.3), $z_m = 0$ [0], $M_m = He$ (from Equation 8.22, either k), $M_m/(A_r dl^3) = 0.0752$ [0.0833] with $\bar{z}_* = 0.366$ [0.50], and $H = 0$ [0].
2. M_m gradually approaches M_o with increase in e, until $M_m \approx M_o$ at a ratio e/l >3. Typically, $M_m = 1.08M_o$ given $e/l = 2$ and Gibson k (see Figure 8.5a and b). The M_m and M_o have an identical upper limit (dotted line) of $e = \propto$ (for the corresponding k) obtained using Equation 8.5 (Gibson k) or $\bar{z}_*/6$ (constant k, Table 8.3).

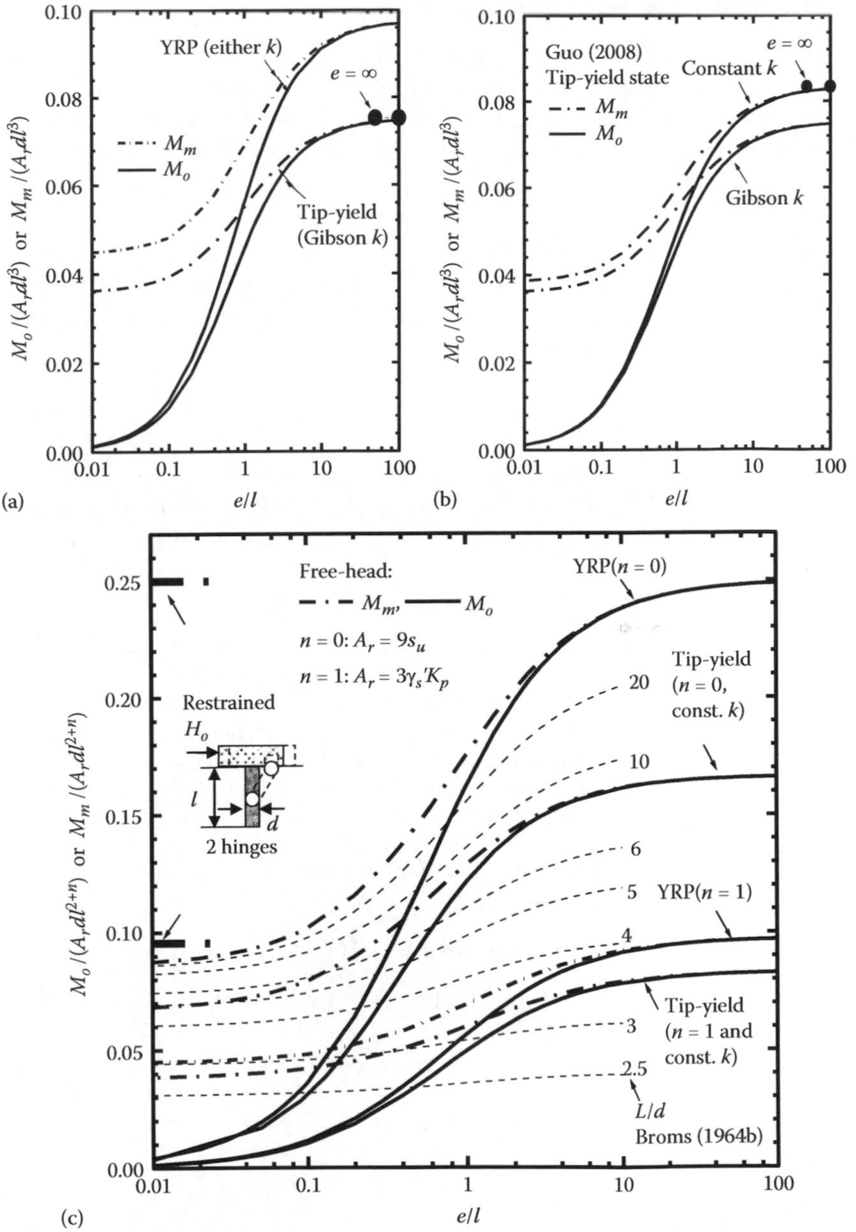

Figure 8.5 Normalized applied $\bar{M}_o$ (= $\bar{H}\,\bar{e}$) and maximum $\bar{M}_m$ (a), (b) Gibson p_u (= $A_r dz^n$, n = 1), (c) Constant p_u (= $N_g s_u dz^n$, n = 0). (Revised from Guo, W. D., *Can Geotech J* 45, 5, 2008.)

8.2.9.2 *Yield at Rotation Point (Gibson* p_u*)*

At the YRP state, the relationships of H versus z_m and H versus M_{max} are still govened by Equations 8.20 and 8.22 ($z_m \le z_o$, pre-tip yield), respectively, as mentioned earlier. For instance, $\bar{z}_m = (2\bar{H})^{0.5}$ is obtained by substituting $C = 0$ into Equation 8.13, as the depth z_m and the z_r relationship is correlated by

$$\bar{z}_m = [2\bar{z}_r^2 - 1]^{0.5} \tag{8.23}$$

where z_r/l is still calculated from Equation 8.9 or 8.10. Equation 8.23 was derived utilizing the on-pile force profile depicted in Chapter 3, this book, and shear force $H(z_m) = 0$ at depth z_m. With the $\bar{H}$, the normalized maximum bending moment is deduced as

$$\bar{M}_m = \bar{H}\bar{e} + \frac{1}{3}\bar{z}_m^3 \tag{8.24}$$

The normalized M_m was computed using Equation 8.24. It is plotted in Figure 8.5a as "*YRP* (either *k*)" as equations at YRP state are independent of the *k* profiles. The moment of M_o (= He) is also plotted in the figure.

Example 8.4 M_m response at YRP state (Gibson p_u)

At the YRP state, the z_o/l (= z_r/l) of 0.7937 ($e = 0$) and 0.707 ($e = \infty$) offer $z_m/l = 0.5098$ and 0; and accordingly $H/(A_r dl^2) = 0.130$ ($e = 0$) and 0 ($e = \infty$); also $M_m/(A_r dl^3) = 0.0442$ ($e = 0$) and 0.0976 ($e = \infty$) from Equation 8.24; and $M_m/(A_r dl^3) = [1 - 2(z_r/l)^3]/3$ in light of moment equilibrium about ground line and the on-pile force profile for $e = \infty$ (see Figure 3.28c$_1$, Chapter 3, this book).

8.2.10 Maximum bending moment and its depth (constant p_u)

8.2.10.1 *Pre-tip yield* ($z_o < \bar{z}_*$) *and tip-yield* ($z_o = \bar{z}_*$) *states*

In contrast to those for Gibson p_u, the depth z_m and maximum bending moment M_m are given by ($z_m \le z_o$):

$$\bar{z}_m = \bar{H} \ (z_m \le z_o) \tag{8.25}$$

$$\bar{M}_m = \bar{H}_t(\bar{e} + 0.5\bar{H}) \ (z_m \le z_o) \tag{8.26}$$

Example 8.5 $\bar{M}_m$ at tip-yield state (constant p_u)

Nonlinear response of the M_m and its depth z_m for constant p_u is captured by the slip depth z_o, as with Gibson p_u (see Table 8.1), and with a power-law increase p_u for flexible piles (Guo 2006). At tip-yield state, the normalized $\bar{M}_m$ and $\bar{z}_m$ were computed for a series of $\bar{e}$. Typical

Table 8.9 Critical response for tip-yield and YRP states

	Gibson p_u *and Gibson* k *(Constant* k*) (Guo 2008)*				
	z_o/l	z_m/l	$H_o/(A_r dl^2)$	$M_m/(A_r dl^3)$	M_m *Difference (%)*
$e = 0$	0.544[a]	0.475	0.113	0.0362	From tip-yield to YRP state, at $e = 0$ $M_m/(A_r dl^3)$ increases by 22.8%, or at $e = \infty$ by 24.9%
	[0.618][b]	[0.486]	[0.118]	[0.038]	
	0.794[c]	0.510	0.130	0.0442	
$e = \infty$	0.366[a]	0.	0.	0.0752	
	[0.500][b]	0.	0.	[0.0833]	
	0.707[c]			0.0976	
	From $e = 0$ to ∞, $M_m/(A_r dl^3)$ increases by 109% (tip-yield) and 120% (YRP)				
	Constant p_u *and Constant* k				
	z_o/l	z_m/l	$H_o/(N_g s_u dl)$	$M_m/(N_g s_u dl^2)$	M_m *Difference (%)*
$e = 0$	[0.366][b]	[0.366]	[0.366]	[0.067]	From tip-yield to YRP state, at $e = 0$ $M_m/(N_g s_u dl^2)$ increases by 28.4%, or at $e = \infty$ by 50%
	0.707[c]	0.414	0.414	0.086	
$e = \infty$	[0]	[0.]	[0.]	[0.1667]	
	0.5	0.	0.	0.25	
	From $e = 0$ to ∞, $M_m/(N_g s_u dl^2)$ increases by 149% (tip-yield) and 191% (YRP)				

a Gibson *k* (tip-yield state).
b Constant *k* (tip-yield state).
c either *k* for YRP state

values concerning $e = 0$ and ∞ are tabulated in Table 8.9, which shows (1) At $e = 0$, $M_m/(N_g s_u dl^2) = 0.067$, $\bar{z}_m = \bar{z}_* = 0.366$, and $H_o/N_g s_u dl = 0.366$ (H_o denotes H at tip-yield state, and also at the YRP state, or at onset of plastic state); and (2) At $e = \infty$, $\bar{M}_m = 0.1667$, with $z_m = 0$, $\bar{z}_* = 0$, and $H = 0$, or directly from Equation 8.8.

The $\bar{z}_m$ is plotted in Figure 8.4d against the $\bar{e}$. As with the remarks about Gibson p_u, Figure 8.4 together with Table 8.9 indicate that $\bar{z}_m$ is generally less than $\bar{z}_*$ (see Figure 8.4a and d), and may be computed using $\bar{z}_m = \bar{H}$ (see Table 8.1); and it does reduce to groundline, as the $\bar{e}$ approaches infinitely large. The normalized maximum moment $\bar{M}_m$ is plotted in Figure 8.5c, which shows the $\bar{M}_m$ of 0.067 ($e = 0$) ~0.1667 ($e = \infty$) for constant p_u exceeds 0.038~0.0833 for Gibson p_u, as the average p_u at the YRP state for the constant p_u may be twice that for the latter. The M_m is larger than M_o until $M_m \approx M_o$ at a higher $\bar{e}$ (> 3) (note that $M_m = 1.02 M_o$ at $\bar{e} = 2$).

8.2.10.2 Yield at rotation point (constant p_u)

The depth z_m (= $2z_r - l$) was deduced utilizing the on-pile force profile (see Figure 8.2c$_1$) along with $H_i(z_m) = 0$ (see Table 8.6). The depth z_m and maximum bending moment M_m for constant p_u are given by

$$\bar{z}_m = 2\bar{z}_r - 1 \tag{8.27}$$

$$\bar{M}_m = \bar{M}_o + 0.5\bar{H}_o^2 \tag{8.28}$$

Example 8.6 $\bar{M}_m$ at YRP state (constant p_u)

Equations 8.25 and 8.26 for calculating z_m and M_m are supposedly valid from the pre-tip yield state through to YRP state with $z_m \le z_o$ (Guo 2008), with which the normalized $\bar{M}_m$ at YRP state was estimated for a spectrum of $\bar{e}$. For instance, with $\bar{z}_o$ (= $\bar{z}_r$) = 0.707 and 0.5, $H/(N_g s_u dl)$ is estimated as 0.414 and 0 for e = 0 and ∞; and $M_m/(N_g s_u dl^2)$ as 0.086 ($\bar{e}$ = 0) and 0.25 (e = ∞), respectively. The normalized $\bar{M}_o$ (= $\bar{H}\,\bar{e}$) is then calculated. The $\bar{M}_m$ and $\bar{M}_o$ are plotted in Figure 8.5c together with those for tip-yield state.

Figure 8.5c demonstrates that the $\bar{M}_m$ increases by 28.4~50% (e = 0~100), as tip-yield state (see Figure 8.2a$_1$ and a$_2$) moves to YRP state (see Figure 8.2c$_1$ and c$_2$); and by 149~191% (at tip-yield to YRP state), as the free-length e ascends from 0 to $3l$. The eccentricity has 4~5 times higher impact on the values of $\bar{M}_m$ than yield states. These percentages of increase are 1.5~2 times more evident than those gained for Gibson p_u (see Table 8.9 and Example 8.7). The YRP state featured by $z_o = z_1 = z_r$ would never be attained. A pile may rotate about a depth z_r outside the depth (0.5~0.794)l and around a stiff layer or pile-cap. The depth z_r may be used as the optimum load attachment point for suction caisson. This does alter the M_m, M_o relationship, as illustrated later using lateral-moment loading locus.

Example 8.7 Impact of p_u, eccentricity, and yield states on $\bar{M}_m$

Figure 8.5c demonstrates that first, from the initiation of slip at pile base (tip-yield) to full plastic state (YRP), $M_m/(A_r dl^3)$ (Gibson p_u and Gibson k) increases by 22.8% (from 0.036 to 0.0442) at e = 0 (see Table 8.2) or by 29.9% (from 0.0752 to 0.0976) at e = ∞ (see Table 8.3). (A slightly smaller increase in percentage is observed using constant k.) The M_{max} increase is ~30% in general. Second, as pure lateral load (e = 0) shifts to pure moment loading (e = ∞), the M_m increases by ~120%, with 109% at tip-yield state (from 0.036 to 0.0752), and 120% at rotation-point yield state (from 0.0442 to 0.0976). The eccentricity yields ~4 times higher values of M_{max} than the states of yield. The impact of p_u on z_m and M_m is shown in Figure 8.4d and f, respectively.

8.2.11 Calculation of nonlinear response

Regardless of the p_u or k profiles, the response of rigid piles is characterized by two sets of expressions concerning pre- ($z_o < z_*$) and post- ($z_o > z_*$) tip yield states. For instance, the response for Gibson p_u and Gibson k may be calculated by two steps: (1) A slip depth z_o (< z_*, pre-tip yield state) is specified to calculate load (H), displacement (u_g), and rotation (ω) using Equations 8.1, 8.2, and 8.3, respectively; and furthermore, the moment (M_m) using Equation 8.22 or 8.24. (2) The calculation step 1 is repeated for a series of z_o (< z_*) to gain overall response prior to tip yield state. (3) A value of C ($0 \le C \le C_y$, post-tip yield state) is assigned to calculate a rotation depth z_r using Equation 8.10 ($e \neq \infty$) otherwise Equation 8.9 ($e = \infty$); a load and a displacement using Equations 8.13 and 8.14, respectively; and a rotation angle ω ($= -u_g/z_r$), and the M_m using Equation 8.22 or 8.24. (4) The calculation step 3 is repeated for a series of C (< C_y) to gain overall response after the tip-yield state. The calculation steps 1–4 allow entire responses of the pile-head load, displacement, rotation, and maximum bending moment to be ascertained. Likewise the calculation for other p_u and k profiles may be conducted.

Example 8.8 Nonlinear pile response (Gibson p_u either k)

Nondimensional responses were predicted for a pile having $l/r_o = 12$, and at six typical ratios of e/l, including $u_g k_o/A_r$ [$u_g k/A_r l$], $\omega k_o l/A_r$ [$\omega k/A_r$], $M_m/(A_r d l^3)$ and $H/(A_r d l^2)$, and those at tip-yield state ($z_o = z_*$), assuming a Gibson p_u ($= A_r d z^n$, $n = 1$) and Gibson k (also constant k). The ultimate moment at YRP state and $e/l = 0$ was also predicted using Equation 8.24. The response and moment are shown in Figure 8.6a$_1$ through c$_1$, which demonstrate a higher impact of k profile on the normalized u_g and ω than on the normalized M_{max} (prior to tip yield). This implies a good match between two measured responses [e.g., H~u_g and H~M_{max} (or ω) curves] and the current solutions will warrant unique values of A_r and k to be back-figured in a principle discussed for a flexible pile (Guo 2006) (see Chapter 9, this book), as is illustrated later in Section 8.5.

Example 8.9 Nonlinear pile response (impact of p_u profiles, constant k)

The rigid pile of Example 8.8 was predicted again using a constant p_u ($= N_g s_u d z^n$, $n = 0$) and is depicted in Figure 8.6a$_2$ through c$_2$. The average p_u over pile embedment was identical to that for the Gibson p_u [i.e., $A_r = 2p_u/(ld)$]. In particular, the resulted $\bar{H}$ and $\bar{M}_m$ (Gibson p_u and constant k) are twice those presented previously for $A_r = p_u/d$ (Guo 2008) with $\bar{H} = H/(p_u l)$ (constant p_u) and $\bar{H} = H/(2p_u l)$ (Gibson p_u). The effect of p_u profile on the response is evident in the normalized curves of H~u_g (or ω) and H~M_m. Consequently, it is legitimate to deduce N_g (or A_r thus p_u) and k by matching two measured responses with the current solutions underpinned by either constant p_u or Gibson p_u (Guo 2006; Guo 2008).

To gain profiles of shear force $H(z)$ and bending moment $M(z)$, the normalized expressions in Table 8.6 for either p_u profile may be used. Under the

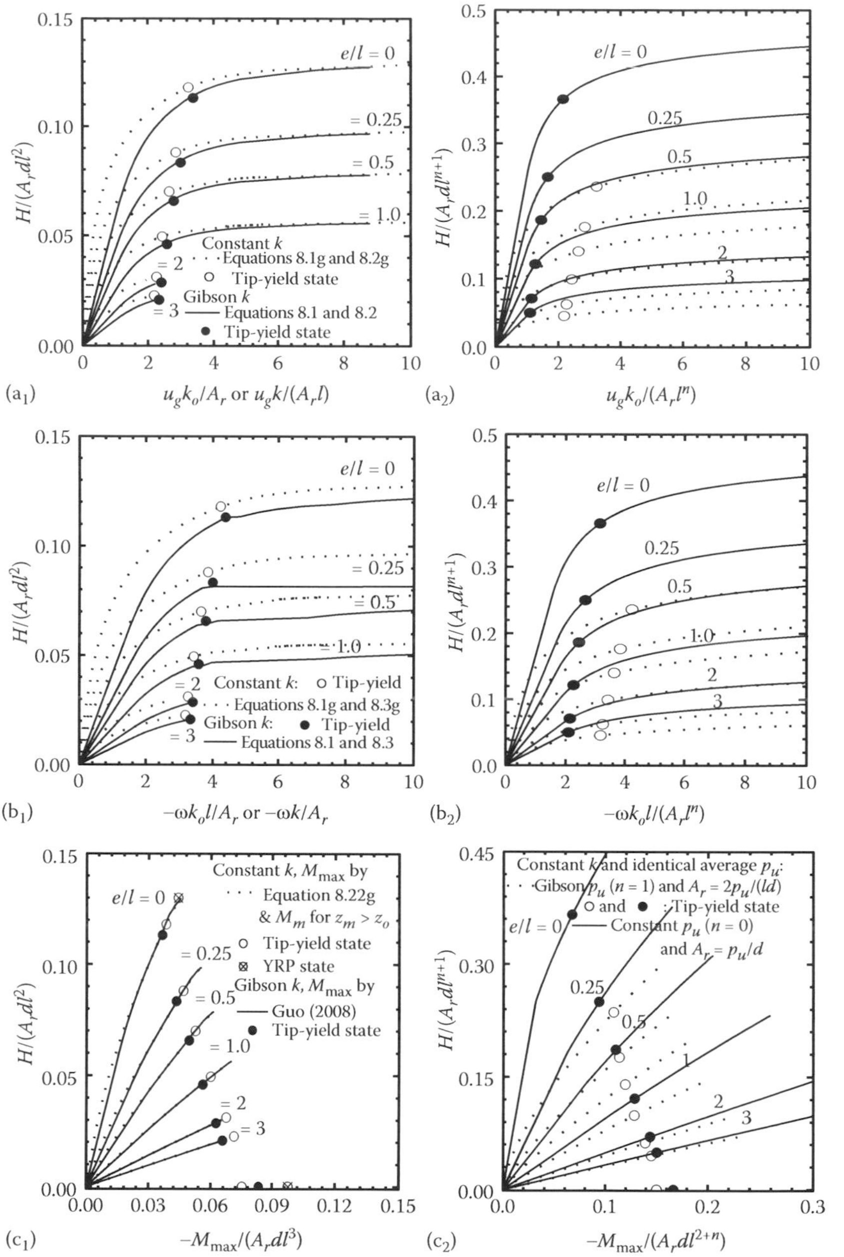

Figure 8.6 Normalized response of (a_i) pile-head load $\bar{H}$ and groundline displacement $\bar{u}_g$. (b_i) $\bar{H}$ and rotation $\bar{\omega}$. (c_i) $\bar{H}$ and maximum bending moment $\bar{M}_{max}$ (subscript i = 1 for Gibson p_u and constant or Gibson k, and i = 2 for constant k and Gibson or constant p_u). Gibson p_u (= $A_r dz^n$, n = 1), and constant p_u (= $N_g s_u dz^n$, n = 0). (Revised from Guo, W. D., *Can Geotech J* 45, 5, 2008.)

combined loading, and with the determined groundline displacement u_g, limiting u^*, and rotation ω, the distribution profiles of $H(z)$ and $M(z)$ can be ascertained as to zones covering depths $0\sim z_o$, $z_o\sim z_1$, and $z_1\sim l$, respectively. For instance, after tip-yield (with $z_o > z_*$) occurs, the response consists of three components of $H_1(z)$, $M_1(z)$ ($z \le z_o$), $H_2(z)$, $M_2(z)$ ($z_o < z \le z_l$), and $H_3(z)$, $M_3(z)$ ($z_1 < z \le l$), respectively.

Example 8.10 Pile response profiles

As an illustration, with respect to the tip-yield and the YRP states, the normalized distribution profiles along the pile of Figure 8.6a$_2$ through c$_2$ for five typical ratios of *e*/*l* were predicted. They are plotted in Figure 8.7a and b and Figure 8.7c and d, concerning constant p_u and Gibson p_u, respectively. The figures demonstrate that constant p_u (Figure 8.7a) at YRP state renders bilinear variation of shear force with depth, which is not observed for Gibson p_u (Figure 8.7c); and the normalized $\bar{H}(\bar{z})$ shifts more evidently from the tip-yield to the YRP states for constant p_u (Figure 8.7a) than it does for Gibson p_u (Figure 8.7c), so does the $\bar{M}(\bar{z})$ (see Figure 8.7b and d).

8.3 CAPACITY AND LATERAL-MOMENT LOADING LOCI

8.3.1 Lateral load-moment loci at tip-yield and YRP state

Given Gibson p_u (constant k), the $\bar{M}_o$ for the tip-yield state is obtained by replacing $\bar{z}_o$ in Equation 8.1 with $\bar{z}_*$ of Table 8.1:

$$\bar{M}_o = (1 - 8\bar{H}_o - 4\bar{H}_o^2)/12 \tag{8.29g}$$

where $\bar{H}_o = \bar{H}$ at tip-yield state. This $\bar{M}_o$ allows the $\bar{M}_m$ and $\bar{z}_m$ from Table 8.1 (right column) to be recast into

$$\bar{M}_m = \frac{1}{12}(1 - 8\bar{H}_o - 4\bar{H}_o^2) + \frac{(2\bar{H}_o)^{1.5}}{3} \quad \bar{z}_m = \sqrt{2\bar{H}_o} \tag{8.30g}$$

These expressions are provided in Table 8.1. At the YRP state (with C = 0 in the post-tip yield state), again for Gibson p_u, it is not difficult to gain the following:

$$\bar{M}_o = \frac{2}{3}(\bar{H}_o + 0.5)^{1.5} - \frac{1}{3} \tag{8.31}$$

$$\bar{M}_m = \bar{M}_o + \frac{1}{3}(2\bar{H}_o)^{1.5} \tag{8.32}$$

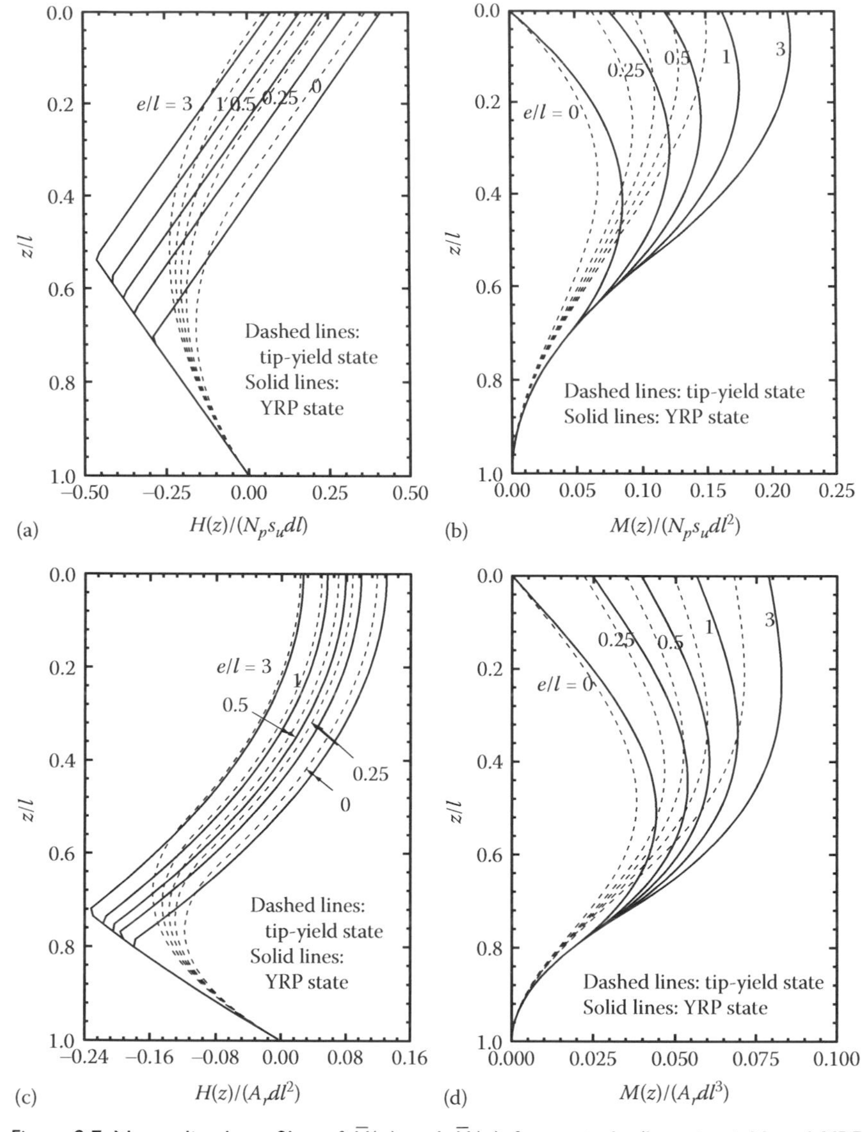

Figure 8.7 Normalized profiles of $\bar{H}(z)$ and $\bar{M}(z)$ for typical e/l at tip-yield and YRP states (constant k): (a)–(b) constant p_u; (c)–(d) Gibson p_u.

in view of $\bar{z}_m = [2\bar{z}_r^2 - 1]^{0.5}$, and $\bar{z}_m = \sqrt{2\bar{H}_o}$. These expressions are furnished in Table 8.7. Likewise, the solutions for $\bar{M}_o$ and $\bar{M}_m$ were deduced concerning constant p_u. They are presented in Tables 8.1 and 8.7 as well for either yield state. These solutions are subsequently compared respectively with existing solutions, experimental data, and numerical solutions.

8.3.2 Ultimate lateral load H_o against existing solutions

The ambiguity regarding H_o and A_r was highlighted previously in Chapter 3, this book, which may be removed by redefining the capacity H_o as the H at (a) tip-yield state or (b) yield at rotation point state. For instance, with a Gibson p_u and Gibson k, the H_o for tip-yield state is obtained by substituting z_* from Equation 8.18 for the z_o in Equation 8.1; whereas the H_o for the YRP state is evaluated using Equation 8.13, in which the z_r is calculated from Equation 8.10 and $C = 0$.

The normalized values of $\bar{H}_o$ for Gibson p_u at the two states (Guo 2008) are plotted in Figure 8.8a and b across the whole $\bar{e}$ spectrum. The sets of $\bar{H}_o$ for constant p_u are obtained using Equation 8.6 with z_* from Equation 8.19 (tip-yield state) and Equations 8.15 through 8.17 for YRP state. They are presented in the figures as well. Figure 8.8a also provides the normalized capacities of $\bar{H}_o$ gained from 1 g (g = gravity) model pile tests in clay (Meyerhof and Sastray 1985) (see Table 8.10) and in sand (Prasad and Chari 1999), and from centrifuge tests in sand (Dickin and Nazir 1999).

The figure demonstrates that (1) the highest measured values of $\bar{H}_o$ for rigid piles in sand tested in centrifuge ($\bar{e}$ = ~6) are just below the curve of tip-yield state (constant p_u and constant k) and may exceed those for Gibson p_u and k; (2) the highest measured values of $\bar{H}_o$ from the tests in clay ($\bar{e}$ = ~0.06) match up well with the YRP state; (3) the lowest measured values of $\bar{H}_o$ for piles in sand, perhaps obtained at pre-tip yield state, are slightly under the solutions based on Gibson p_u. These conclusions are also observed in our recent back-estimation against measured response of ~50 piles. Overall, constant p_u is seen on some rigid piles. For example, z_*/l at the tip-yield state (Gibson p_u, constant k) is obtained as 0.618 with $e = 0$. Regardless of the e, substituting 0.618 for the z_o/l in right Equation 8.1g, a new expression of $H_o/(A_r dl^2) = 0.1181/(1 + 1.146\,\bar{e})$ is developed (see Table 3.7, Chapter 3, this book), which is also close to the YRP state (not shown herein).

The groundline displacement may be accordingly estimated using Equation 8.2 (or Equation 8.7) concerning the tip-yield state, given k_o (or k); whereas it is infinitely large upon the YRP state.

Example 8.11 Comments on Broms' method

Broms (1964a) used simple expressions (see Table 8.10) to gain the ultimate M_m, H_o, and u_g concerning a set of e/d = 0, 1, 2, 4, 8, and 16. The normalized M_m, H_o are compared with current solutions for tip-yield state in Figures 8.9 and 8.10 for piles in clay and sand, respectively, and the normalized u_o (= u_g at tip-yield state) in Figure 8.11. The simple

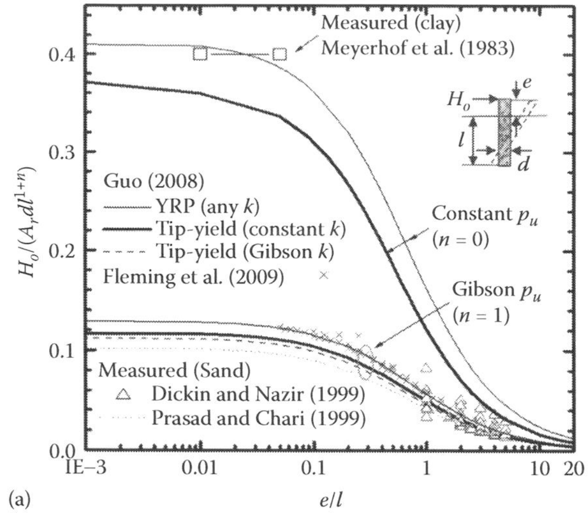

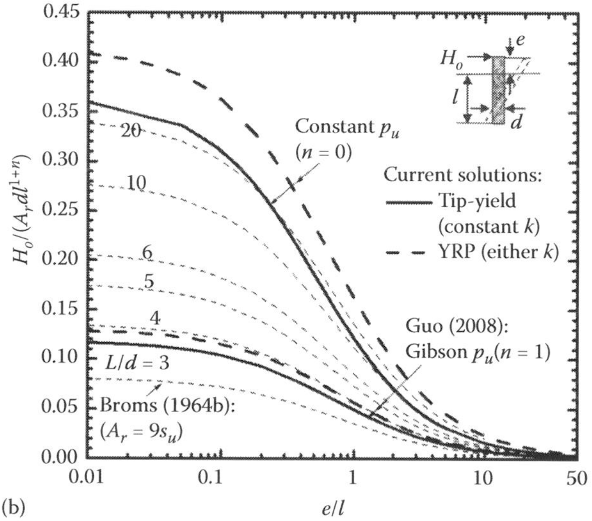

Figure 8.8 Predicted versus measured normalized pile capacity at critical states. (a) Comparison with measured data. (b) Comparison with Broms (1964b).

expressions (Broms 1964a) (see Table 8.10) were used to gain the ultimate M_m and H_o concerning a set of l/d = 3, 4, 5, 6, 10, and 20. The normalized ultimate $\bar{M}_m$ and $\bar{H}_o$ are plotted as dashed lines in Figures 8.5c and 8.8b, respectively. Comparing with current solutions using a constant p_u to a Gibson p_u, the figures indicate that Broms' solutions underpredict the $\bar{M}_m$ and $\bar{H}_o$ given $l/d < 3$, but offer similar $\bar{H}_o$ and $\bar{M}_m$ at l/d = 3~20.

Table 8.10 Typical Relationships for H_o, M_o and M_m (Independent of k Profiles)

Solutions	e = 0	e > 0
Current solutions Equations 8.6 and 8.26	Tip-yield state: $H_o = 0.427N_g s_u dl$ $M_m = (0.105 \sim 0.167)N_g s_u dl^2$	Pre-tip yield state: $\frac{H_o}{N_g s_u dl} = \frac{0.5(1+2\bar{z}_o)}{2+\bar{z}_o+3\bar{e}}$
Experiment results (Meyerhof and Sastray 1985)	$H_o = 0.4N_g s_u dl$ $M_o = 0.2N_g s_u dl^2$	$\frac{H_o}{N_g s_u dl} = \frac{0.4}{1+(1.4 \sim 1.9)\frac{e}{d}}$
Ultimate state solution (Broms 1964b)	$\frac{H_o}{N_g s_u dl} = \frac{9}{N_g}\frac{d}{l}\left\{\left[\left(\frac{l}{d}+2\frac{e}{d}+1.5\right)^2+\left(\frac{l}{d}-1.5\right)^2\right]^{0.5}-\left(\frac{l}{d}+2\frac{e}{d}+1.5\right)\right\}$ $\frac{z_m}{l} = 1.5\frac{d}{l}+\left(\frac{H_o}{N_g s_u dl}\right)\frac{N_g}{9}\ \ \frac{M_m}{N_g s_u dl^2} = \frac{2.25}{N_g}\left(1-\frac{z_m}{l}\right)^2$	
Ultimate state solution (Broms 1964a)	$\frac{H_o}{\gamma_s' K_p dl^2} = \left[2\left(1+\frac{d}{l}\frac{e}{d}\right)\right]^{-1}$ $\frac{M_{max}}{\gamma_s' K_p d^4} = \frac{H_o}{\gamma_s K_p d^3}\left\{\frac{e}{d}+\frac{2}{3}\frac{l}{d}\left[3\left(1+\frac{d}{l}\frac{e}{d}\right)\right]^{-0.5}\right\}$	

8.3.3 Lateral–moment loading locus

Yield loci are obtained for lateral (H_o) and moment (M_o) loading, and the induced maximum bending moment (M_m) to assess safety of piles subjected to cyclic loading.

8.3.3.1 *Impact of p_u profile (YRP state) on $\bar{M}_o$ and $\bar{M}_m$*

Yield loci $\bar{M}_o$ and $\bar{M}_m$ at the YRP state were obtained in light of the expressions of $\bar{M}_{oi}$ and $\bar{M}_{mi}$ in Table 8.7 (the subscript i = 1~4 indicates moment directions). With constant p_u, they are plotted in Figure 8.12, such as $\bar{M}_{o1}$, $\bar{M}_{o2}$, $\bar{M}_{o3}$, and $\bar{M}_{o4}$ for $f_c b'_c g_c$, $a_c b_c e_c$, $a_c b'_c e_c$; and $f_c b_c g_c$.; and $\bar{M}_{m1}$, $\bar{M}_{m2}$, $\bar{M}_{m3}$, and $\bar{M}_{m4}$ for $b'_c d_c$, $b_c d_c$, $b'_c d'_c$, and $b_c d'_c$., respectively. The figure shows the following features:

- Two values of $\bar{M}_m$ exisit for $\bar{H}_o > 0$ and $\bar{z}_r > 0.5$, (e.g., points e_c and d_c in Figure 8.12d and e. It is noted $\bar{M}_m = \bar{M}_o$ over the track $a_c e_c$ but for $\bar{M}_m$ (= 0.0625~0.25) > $\bar{M}_o$ on $b_c c_c$ ($\bar{H}_o \leq 0.414$). The latter indicates a possible bending failure owing to $\bar{M}_m$ rather than $\bar{M}_o$.
- Reversing H_o direction only would render the locus $a_c b_c e_c b'_c e'_c$ in the second and the fourth quadrants to relocate on $f_c b_c g_c b'_c f_c$ in the first

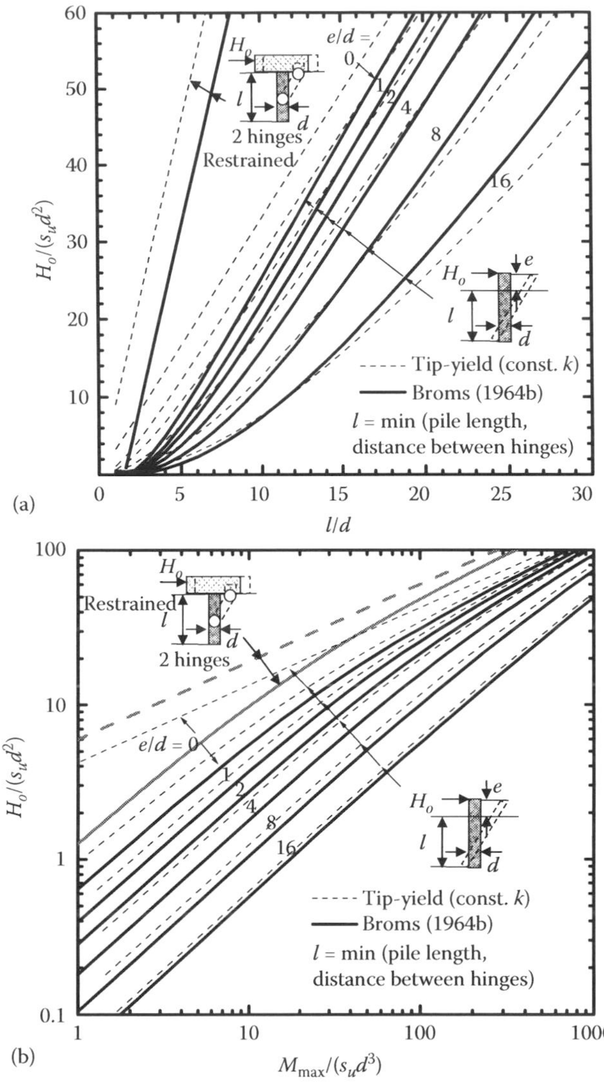

Figure 8.9 Tip yield states (with uniform p_u to l) and Broms' solutions (with uniform p_u between depths 1.5d and l). (a) Normalized $\overline{H}_o$. (b) $\overline{H}_o$ versus normalized moment ($\overline{M}_{max}$).

and the third quadrants, with typical points $a_c(-1.0, 0.5)$, $b_c(0, 0.5)$, $c_c(0.414, 0.086)$, $d_c(1, 0)$, and $e_c(1.0, -0.5)$.

- On-pile force profiles induced by $\overline{H}_o$ and $\overline{M}_o$ vary with rotation depth $\overline{z}_r$ that may be at a stiff layer or pile cap. Impossible to attain the YRP state of $z_o = z_1 = z_r$, a pile is likely to rotate about a depth outside (0.5~0.707)l for which $\overline{M}_o = \overline{M}_m$, otherwise $\overline{M}_m > \overline{M}_o$ with z_r = (0.5~0.707)l.

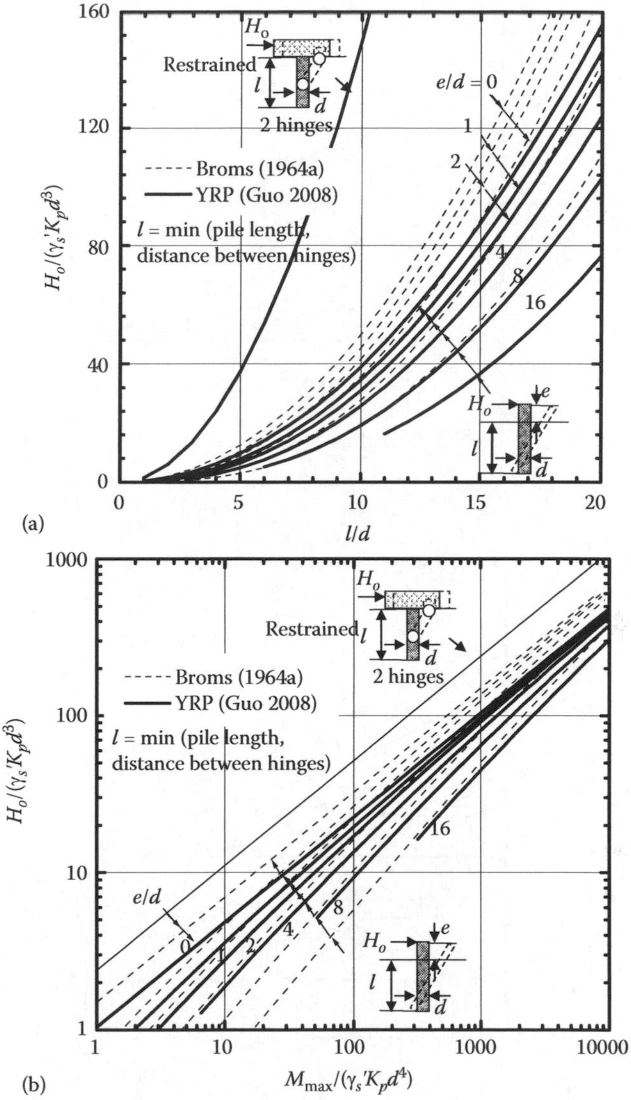

Figure 8.10 Yield at rotation point and Broms' solutions with $p_u = 3K_p\gamma_s'dz$. (a) Normalized $\bar{H}_o$. (b) $\bar{H}_o$ versus normalized moment ($\bar{M}_{max}$).

Likewise, the loci $\bar{M}_o$ and $\bar{M}_m$ at YRP state for Gibson p_u were constructed under the same average pressure over pile embedment as the constant p_u. They are plotted in Figure 8.13 along with those for constant p_u, in form of $\bar{M}_o = M_o/(p_u l^2)$ and $\bar{H}_o = H_o/(p_u l)$ (with the A_r and p_u relationsip in Example 8.9). The figure indicates similar features between either p_u profile but for the following:

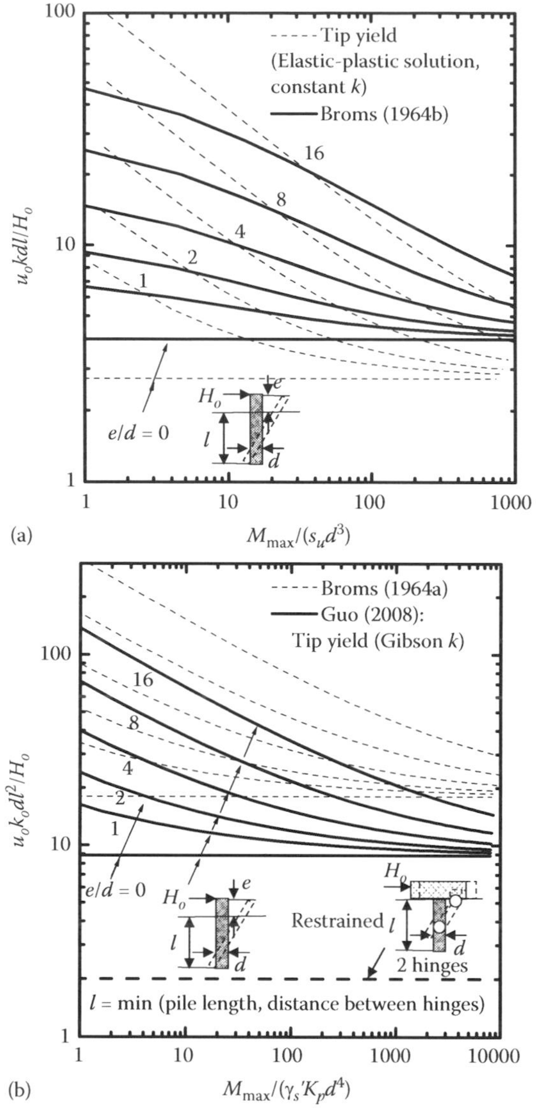

Figure 8.11 Tip-yield states and Broms' solutions. (a) Normalized stiffness versus normalized moment ($\bar{M}_{max}$). (b) $\bar{H}_o$ versus normalized moment ($\bar{M}_{max}$).

- $\bar{M}_m = \bar{M}_o$ along the track $a_g e_g$ but for $\bar{M}_m$ (= 0.0343~0.098) > $\bar{M}_o$ on track $b_g c_g$ at $\bar{H}_o < 0.171$, with $a_g(-1.0, 0.667)$, $b_g(0, 0.1952)$, $c_g(0.260, 0.0884)$, $d_g(1.0, 0)$, and $e_g(1.0, -0.667)$, twice higher values gained from $A_r = 2p_u/ld$.
- $\bar{M}_m > \bar{M}_o$ would not occur if piles rotate about a depth outside (0.707~0.794)l.

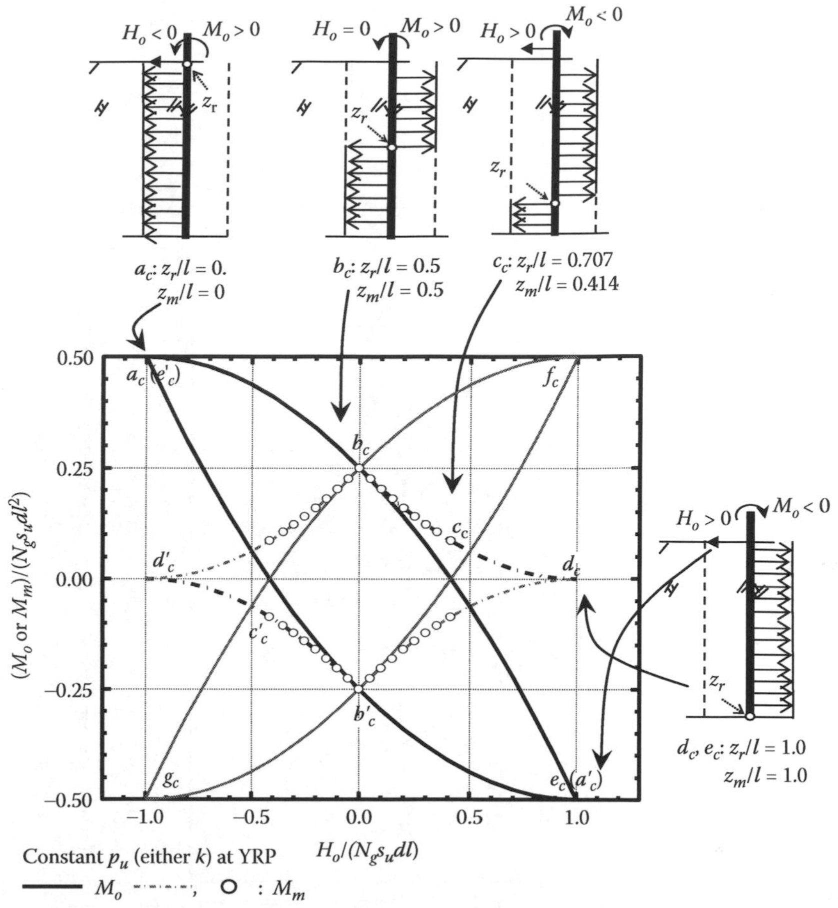

Figure 8.12 Normalized load $\bar{H}_o$-moment $\bar{M}_o$ or $\bar{M}_m$ at YRP state (constant p_u).

Point a_c (a_g) or point e_c (e_g) is unlikely to be attained as the resistance near the rotation point at ground level cannot be fully mobilized, as is observed in numerical results (Yun and Bransby 2007). The new $\bar{H}_o - \bar{M}_o$ loci reveal the insufficiency of $\bar{H}_o - \bar{M}_o$ loci (Poulos and Davis 1980). The loci for YRP state are independent of the k profile. Note the $\bar{H}_o - \bar{M}_o$ locus matches well with the quasi-linear relationship deduced from laboratory tests for piles in clay (Poulos and Davis 1980; Meyerhof and Yalcin 1984) and for piles in sand (Meyerhof et al. 1983).

8.3.3.2 Elastic, tip-yield, and YRP loci for constant p_u

Given constant k and p_u, new loci were generated concerning onset of plastic state and tip-yield states, respectively, and are plotted Figure 8.14, in which:

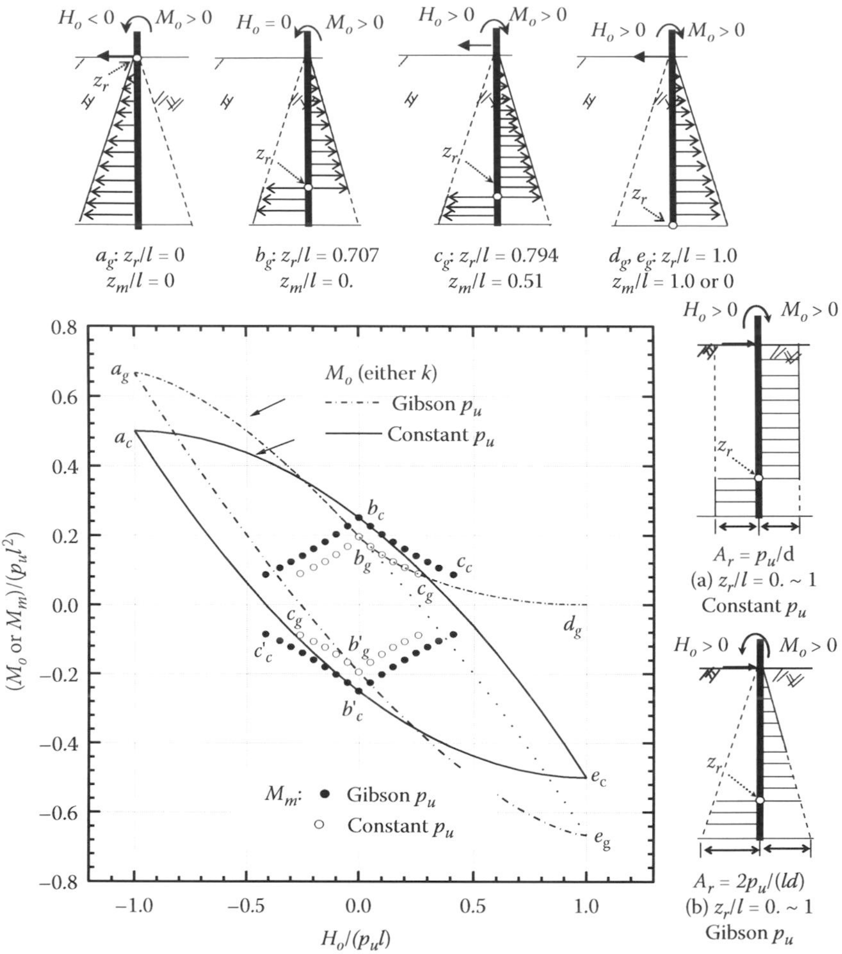

Figure 8.13 Impact of p_u profile on normalized load $\bar{H}_o$-moment $\bar{M}_o$ or $\bar{M}_m$ at YRP state.

- Onset of plastic state (see Figure 8.14a) is indicated by the diamond $e_1e_2e_3e_4$ (maximum elastic core). It was obtained using $\pm 2\bar{H}_o \pm 3\bar{M}_o = 0.5$ (to reflect four combinations of directions of loading $\bar{H}_o$ and $\bar{M}_o$), or Equation 8.6 with $\bar{z}_o = 0$.
- At tip-yield state, as shown in Figure 8.14a, (1) the $\bar{M}_m$ locus is the "hexagon" $t_1t_2t_3t_4t_5t_6$ gained by using the expression shown in Table 8.7; and (2) the $\bar{M}_o$ locus is a "baby leaf" consisting of the paths $h_ce_1i_c$ and $i_ce_3h_c$.
- Line Of_c (with an arrow) indicates a typical loading path of $M_o/(Hl) = 1$ and $\bar{e} = 1$, which would not cause plastic response until outside the diamond $e_1e_2e_3e_4$.

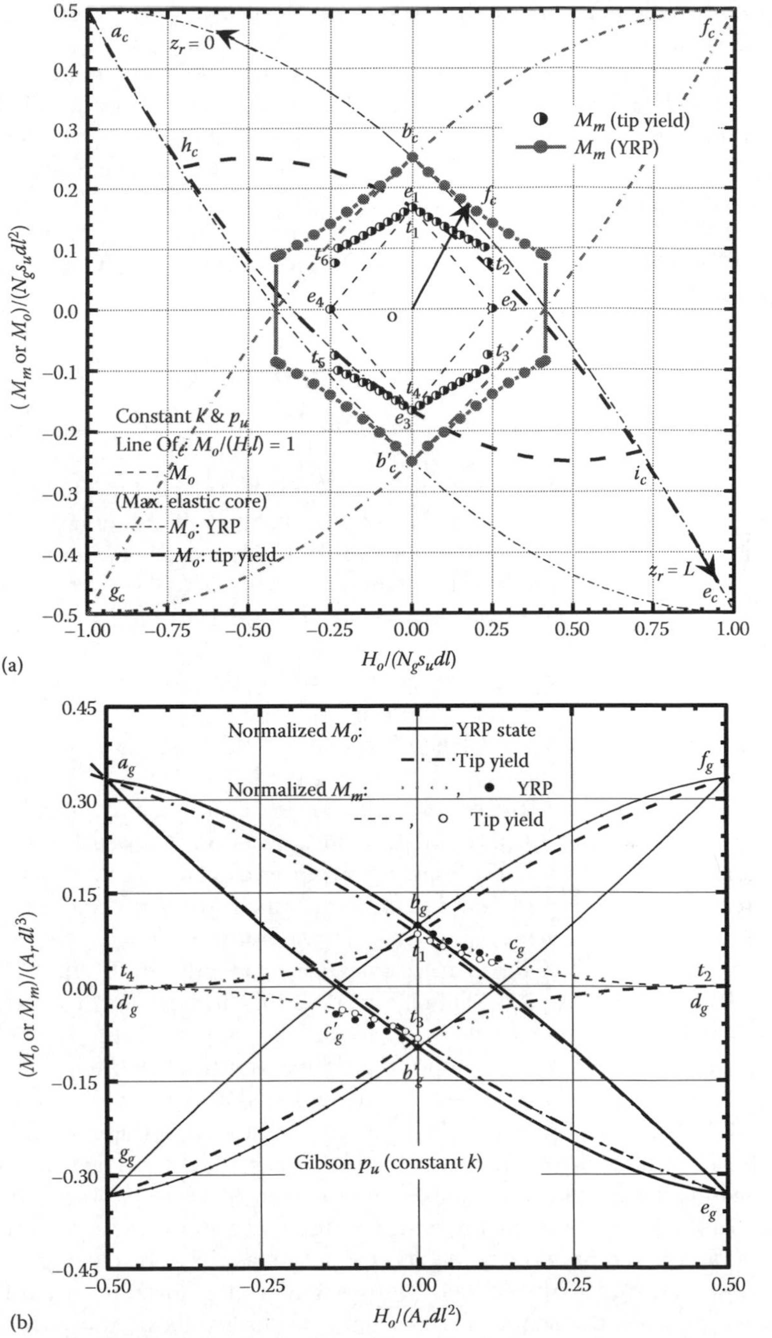

Figure 8.14 Normalized $\bar{H}_o$, $\bar{M}_o$, and $\bar{M}_m$ at states of onset of plasticity, tip-yield, and YRP. (a) Constant p_u and k. (b) Gibson p_u.

Likewise, with Gibson p_u, the yield loci $\bar{M}_o$ and $\bar{M}_m$ were obtained at the tip-yield state (note elastic core does not exist for either k). They are depicted in Figure 8.14b, together with those for YRP. The figure shows that the normalized $\bar{M}_o$ at the tip-yield state is equal to or slightly less than that at the YRP state, and the "open dots" on the sides t_1t_2 and t_3t_4 of the diamond indicate $\bar{M}_m$ of 0.038~0.0833 at the tip-yield state (see also Figure 8.5b).

In brief, $\bar{M}_m$ exceeds $\bar{M}_o$ once the piles rotate about the depth z_r of (0.5~0.707)l (constant p_u) or (0.707~0.794)l (Gibson p_u), which needs to be considered in pertinent design.

8.3.3.3 Impact of size and base (pile-tip) resistance

The shear resistance on the pile-tip is ignored in gaining the current expressions, but it is approximately compensated by using the factor N_g or A_r gained from entire pile–soil system. A unique yield locus for YRP state is obtained for a given p_u profile. Yun and Bransby (2007) conducted finite element (FE) analysis on footings with various dimensions. They show that yield loci may rotate slightly with changing ratios of l/d. The currently proposed single locus is, however, sufficiently accurate for practical design, as explained below.

The impossible YRP state requires the rigid pile lie horizontally. Furthermore, if a gap forms between the pile and soil on one side, the pile would actually resemble a "slender" footing. With respect to scoop and scoop-slide failure modes (see inserts of Figure 8.15) of the footing, charts of $\bar{H}_o$–$\bar{M}_o$ loci (using different normalizers) were drawn for typical sizes of l/d = 0.2, 0.5, and 1.0 (l = embedment depth and d = width for footing) (Yun and Bransby 2007). The locus rotates clockwise as l/d increases towards unity. The $\bar{H}_o$–$\bar{M}_o$ loci for l/d = 0.2 and 1 are replotted in Figure 8.15a and b, respectively, using the current normalizer $N_g s_u$ [N_g = 2.6 (constant p_u), or 2.7 (Gibson p_u), and $s_{uL} = s_u$ at pile-tip level]. The N_g was selected from 2.5~3.5 (Aubeny et al. 2003) for gap formed behind piles. The loci at YRP state were subsequently obtained using Equations 8.30g and 8.31g and are plotted in the figure as well. As anticipated, the current loci without base resistance are much skinnier than the footing ones. If 33% moment (due to base resistance) is added to the current solution for the point of pure moment loading (H_o = 0), the current $\bar{M}_o$ locus may match the footing one. This large portion of base resistance drops remarkably for piles with L/d > 2.5, which can be well catered to by N_g = 4.5 as noted previously. A larger yield locus than the current one is also observed under an inclined (with vertical and horizontal) loading, which is beyond the scope of this chapter. Nevertheless, the parabolic relationship between $\bar{M}_o$ and $\bar{H}_o$ (e.g., Equation 8.29g, Table 8.1) agrees with laboratory test results (Meyerhof et al. 1983).

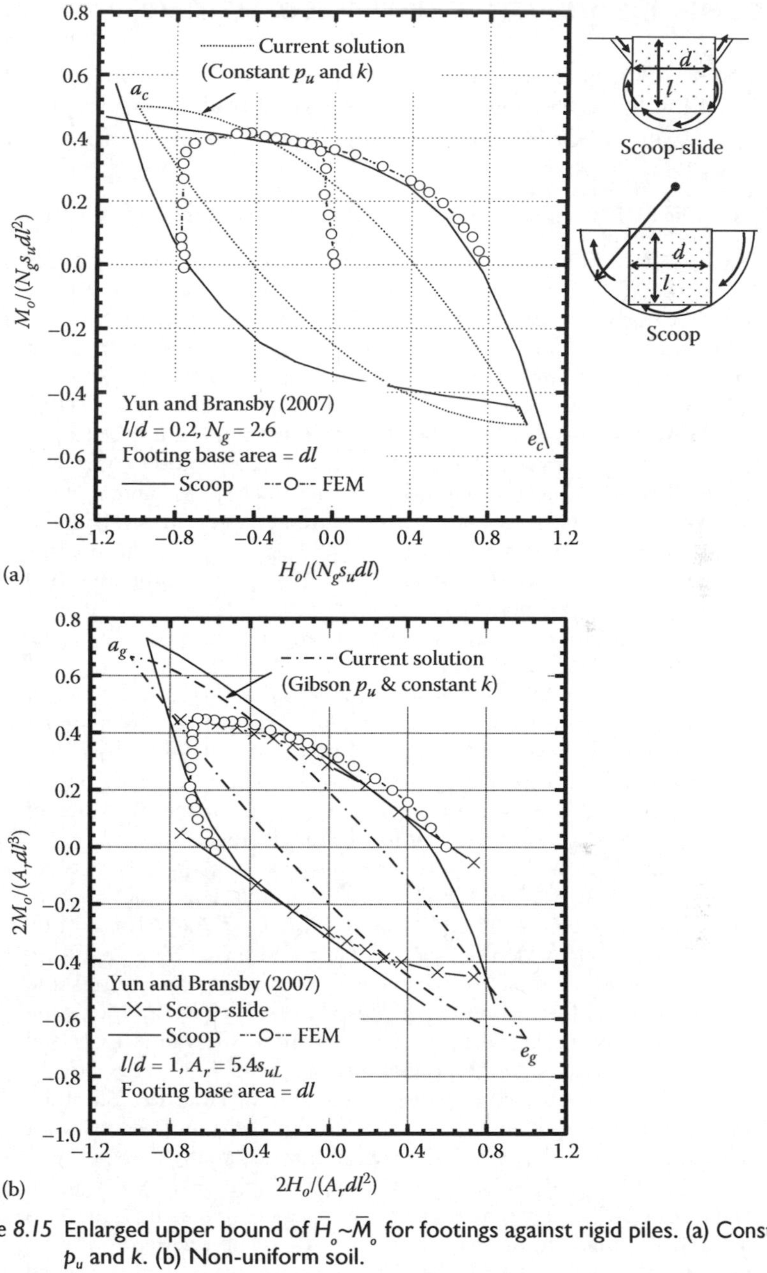

Figure 8.15 Enlarged upper bound of $\bar{H}_o \sim \bar{M}_o$ for footings against rigid piles. (a) Constant p_u and k. (b) Non-uniform soil.

8.4 COMPARISON WITH EXISTING SOLUTIONS

The current solutions have been implemented into a spreadsheet program called GASLSPICS operating in Windows EXCEL™. The results presented thus far and subsequently were all obtained using this program.

Example 8.12 Comparison with model tests and numerical solutions

Nazir (Laman et al. 1999) conducted three centrifugal tests: Test 1 at a centrifugal acceleration of 33 g on a pier with a diameter $d = 30$ mm and an embedment length $l = 60$ mm; Test 2 at 50 g on a pier with $d = 20$ mm and $l = 40$ mm; and Test 3 at 40 g on a pier with $d = 25$ mm and $l = 50$ mm, respectively. They are designed to mimic the behavior of a prototype pier ($d = 1$ m, $l = 2$ m, Young's modulus = 207 GPa, and Poisson's ratio = 0.25) embedded in dense sand (bulk density $\gamma_s' = 16.4$ kN/m^3 and frictional angle $\phi' = 46.1°$). The prototype pier was gradually loaded 6 m (= e) about groundline and to a maximum lateral load of 66.7 kN (i.e., $M_0 = 400$ kNm). In the 40 g test (Test 3), lateral loads were applied at a free-length (e) of 120 mm above groundline (Laman et al. 1999) on the model pier. It offers pier rotation angle (ω) under various moments (M_0) during the test, as is plotted in Figure 8.16a in prototype scale. Tests 1 and 2 show the modeling scale effect on the test results as plotted in Figure 8.16c.

Laman et al. (1999) conducted three-dimensional finite element analysis (FEA^{3D}) to simulate the tests, adopting a hyperbolic stress-strain model with stress-dependent initial and unloading-reloading Young's moduli. The predicted moment (M) is plotted against rotation (ω) in Figure 8.16a and c. It compares well with the median value of the three centrifugal tests, except for the initial stage.

The current predictions was made using an A_r of 621.7 kN/m^3 (= $\gamma_s' K_p^2$, $\gamma_s' = 16.4$ kN/m^3, and $\phi' = 46.1°$), and a modulus of subgrade reaction kd of 34.42 MPa ($d = 1$ m, Test 3), or kd of 51.63 MPa ($d = 1$ m, Test 2) (see Table 8.11). The kd was estimated using $kd = 3.02G$ (or $1.2E$) using Equation 3.62 (Chapter 3, this book), and initial and unloading-reloading Young's modulus, E, of 25.96 MPa, and 58.63 MPa, respectively. First assuming a constant k, the values of H and ω were estimated via right Equations 8.1g and 8.3g, respectively. The M_o (= He) obtained is plotted against ω in Figure 8.16a and c as "Current CF," which agrees well with the measured responses for Tests 2 and 3. The *tip-yield* occurred at a rotation angle of 3.8° (see Table 8.11). Second, assuming a Gibson k with $k_o = 17.5$ kN/m^3/m, and A_r of 621.7 kN/m^3, the M_0 and ω were calculated using Equations 8.1 and 8.3 and are plotted in Figure 8.16a. The prediction is also reasonably accurate. Third, the displacement u_g was calculated using Equation 8.2 and the right Equation 8.2g, respectively, and is plotted in Figure 8.16b against the respective H. The Gibson k solution is much softer than the uniform k, indicating the u_g is more sensitive to the k profile than the M_o is, as is noted for flexible piles (Chapter 9,

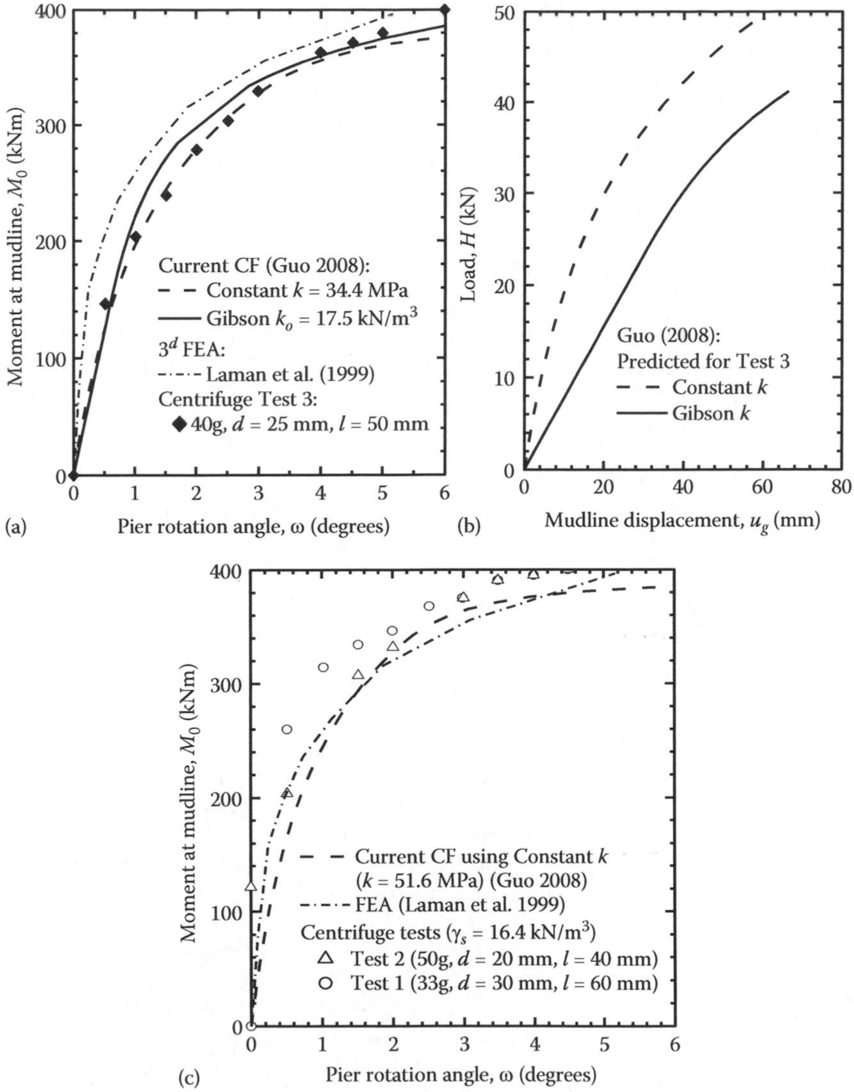

Figure 8.16 Comparison among the current predictions, the measured data, and FEA results (Laman et al. 1999). (a) M_o versus rotation angle ω (Test 3). (b) H versus mudline displacement u_g. (c) M_o versus rotation angle ω (effect of k profiles, tests 1 and 2). (Revised from Guo, W. D., *Can Geotech J* 45, 5, 2008.)

this book). A slightly higher A_r than 621.7 kN/m³ for Test 2 would achieve a better prediction against the measured (M_o~ω) curve (see Figure 8.16c). The current solutions are sufficiently accurate, in terms of capturing nonlinear response manifested in the tests and the 3-D FE analysis (FEA^{3D}).

Table 8.11 Pile in dense sand

Input parameters (l = 2 m, d = 1 m)			*Output for tip-yield state (Test 3)*		
A_r *(kN/m³)*	k *(MN/m³)*	γ_s' *(kN/m³)*	*Angle (deg.)*	z_r/l	M *(kNm)*
621.7	34.42[a]/51.63[b]	16.4	3.83	0.523	338.4

Source: Laman, M., G. J.W. King, and E.A. Dickin, *Computers and Geotechnics*, 25, 1999.

[a] Test 3
[b] Test 2

8.5 ILLUSTRATIVE EXAMPLES

Nonlinear response of rigid piles can be readily captured using the current solutions. This is illustrated next, following the procedures elaborated previously (Guo 2008).

Example 8.13 Model tests

Prasad and Chari (1999) conducted 15 tests on steel pipe piles in dry sand. Each model pile was 1,135 mm in length, 102 mm in outside diameter (*d*), and 5.6 mm in wall thickness. Each was installed to a depth (*l*) of 612 mm. The sand was made to three relative densities (D_r): a loose sand with $D_r = 0.25$, γ_s' (bulk densities) = 16.5 kN/m³, and ϕ (frictional angle) = 35°, respectively; a medium sand with $D_r = 0.5$, γ_s' = 17.3 kN/m³, and ϕ = 41°, respectively; and a dense sand with $D_r = 0.75$, $\gamma_s' = 18.3$ kN/m³, and $\phi = 45.5°$, respectively.

Lateral loads were imposed at an eccentricity of 150 mm on the piles until failure, which offers (1) distribution of σ_r across the pile diameter at a depth of ~0.276 m (shown in Figure 3.25, Chapter 3, this book), with a maximum σ_r = 66.85 kPa; (2) pressure profile (*p*) along the pile plotted in Figure 8.3 as "Test Data"; (3) normalized capacity $H_o/(A_r dl^2)$ versus normalized eccentricity (*e*/*l*) relationship, denoted as "Prasad and Chari (1999)" in Figure 8.8a; (4) lateral pile-head load (*H*) ~ groundline displacement (u_g) curves in Figure 8.17a; and (5) shear moduli at the pile-tip level (G_L for Gibson *k*) of 3.78, 6.19, and 9.22 MPa (taking ν_s as 0.3) for $D_r = 0.25$, 0.5, and 0.75, respectively, which, after incorporating the effect of diameter, were revised as 0.385 MPa (= 3.78*d*), 0.630 MPa (= 6.19*d*), and 0.94 MPa (= 9.22*d*). The revision is necessary, compared to 0.22~0.3 MPa deduced from a model pile of similar size (having *l* = 700 mm, *d* = 32 mm or 50 mm) embedded in a dense sand (Guo and Ghee 2005).

Responses 2 and 3 were addressed previously. Only the H~w_g, M_m, and z_m are studied in the subsequent Examples 8.14 through 8.16 to illustrate the use of the current solutions and the impact of the *k* profiles.

Example 8.14 Analysis using Gibson *k*

The measured H~u_g relationships (see Figure 8.14a) were fitted using the current solutions (Gibson *k*), following the procedure in "Calculation

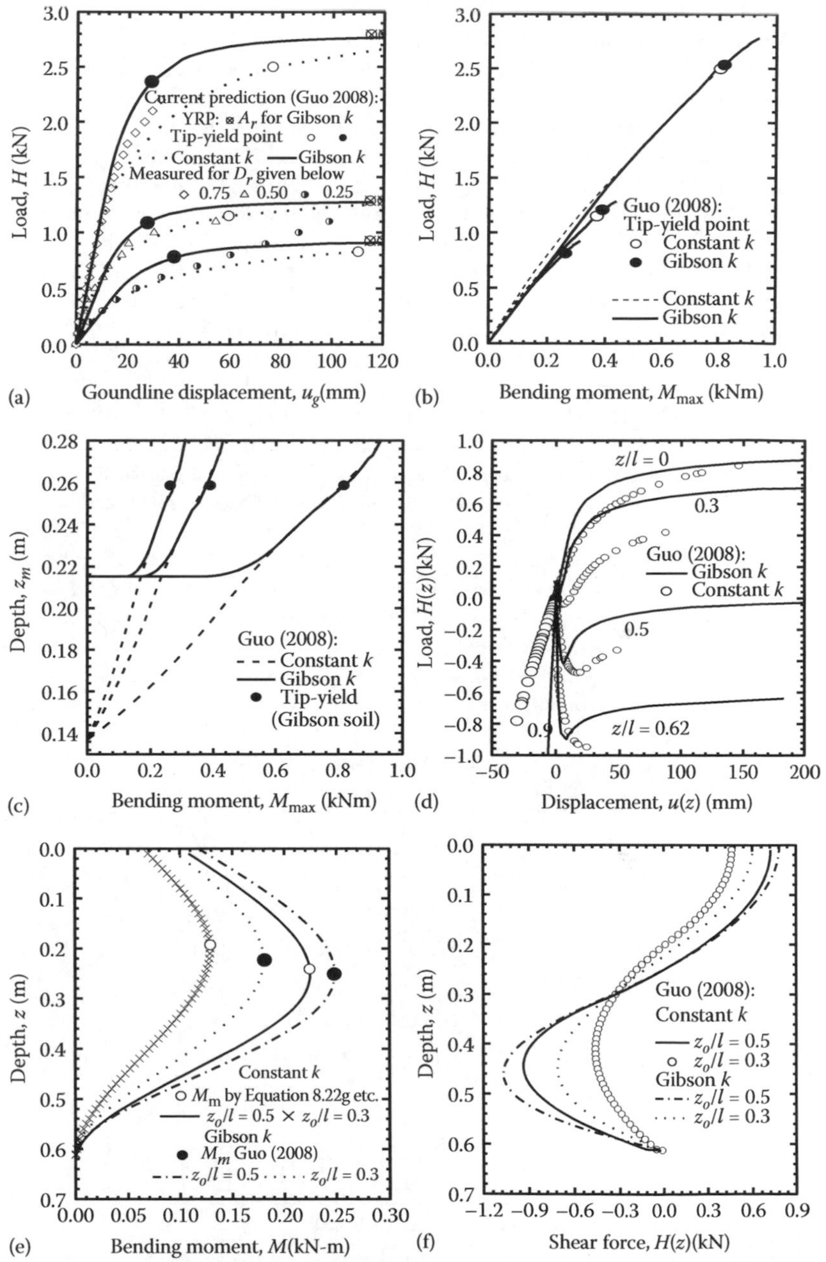

Figure 8.17 Current predictions of model pile (Prasad and Chari 1999) data: (a) Pile-head load H and mudline displacement u_g. (b) H and maximum bending moment M_{max}. (c) M_{max} and its depth z_m. (d) Local shear force~displacement relationships at five typical depths. (e) Bending moment profiles. (f) Shear force profiles. (After Guo, W. D., *Can Geotech J* 45, 5, 2008.)

of Nonlinear Response." This offers the parameters k_o and A_r (see Table 8.12) for each test in the specified D_r as deduced from matching the initial elastic gradient and subsequent nonlinear portion of the measured H~u_g curve. (Two measured curves would be ideal to deduce the two parameters.) The $k_o d$ and A_r permit shear modulus G_L, the maximum bending moment M_m and the depth z_m to be evaluated. The M_m is plotted in Figure 8.17b against H, and Figure 8.17c against z_m.

The study indicates (1) an A_r of 244.9 kN/m³, 340.0 kN/m³, and 739. kN/m³ for $D_r = 0.25$, 0.5, and 0.75 respectively, which is within ± ~15% of $\gamma_s' K_p^2$; (2) a G_L of 0.31 MPa, 0.801 MPa, and 1.353 MPa, which differ by –19.48%, 27.7%, and 43.9% from the revised measured moduli for $D_r = 0.25$, 0.5, and 0.75, respectively; (3) the tip-yield for $D_r = 0.5$ and 0.75 occurred around a displacement u_g of 0.2d (see Figure 8.17a), but for $D_r = 0.25$ occurred at a much higher displacement u_g of 39.5 mm than 0.2d, and a higher load H_o of 0.784 kN than the measured 0.529 kN observed at 0.2d; (4) the *tip-yield* associated with a rotation angle of 2.1~4.0 degrees (see Table 8.12), conforming to 2~6° (see Dickin and Nazir 1999) deduced from model piles tested in centrifuge.

The calculation for the test with $D_r = 0.25$ ($A_r = 244.9$ kN/m³ and $k_o = 18.642$ MPa/m²) is elaborated here, concerning four typical yielding states (see Tables 8.13 and 8.14), shear modulus, and distribution of stress σ_r across the pile diameter.

1. The z_*/l (tip-yield) is obtained as 0.5007 using Equation 8.18, from which relevant responses are calculated and are tabulated in

Table 8.12 Parameters for the model piles (Gibson *k*/Constant *k*)

D_r	*0.25*	*0.50*	*0.75*
A_r (kN/m³)	244.9	340.	739.
k_o^c / [k]	18.64 / [3.88]	48.2 / [12.05]	81.43 / [16.96]
Predicted G_L (MPa)	0.31 / [0.105]	0.801 / [0.327]	1.353 / [0.461]
Measured G_L (MPa)[a,b]	0.385[a] / [0.193][b]	0.631 / [0.316]	0.94 / [0.47]
Angle at $\bar{z}_*$ (deg.)	3.94 / [14.0]	2.11 / [10.6]	2.72 / [14.4]

Source: Guo, W. D., *Can Geotech J* 45, 5, 2008.

[a] Via multiplying the values of 3.78, 6.19, and 9.22 with the diameter *d* (0.102m), or
[b] Via multiplying by 0.5*d*.
[c] k_o in MPa/m², [k] in MPa/m.

Numerator: Gibson *k*
Denominator: Constant *k*

Table 8.13 Response of model piles (any *Dr*, Gibson *k*/[Constant *k*])

Items	z_o/l	$u_g k_o/A_r$ $[u_g k/A_r l]$	$H/(A_r d l^2)$	$\omega k_o l/A_r$ $[\omega k/A_r]$	z_m/l	$M_m/(A_r d l^3)$
Pre-tip yield	0.30	1.695 [0.5517]	0.0647 [0.0494]	−2.318 [−0.8391]	0.364 [0.3151]	0.031 [0.0225]
	0.50	3.005 [1.691]	0.0838 [0.0773]	−4.005 [−2.382]	0.409 [0.3931]	0.0434 [0.0392]
	0.535[a] [0.5885[b]]	3.426 [2.86]	0.087 [0.0885]	−4.530 [−3.86]	0.4171 [0.4208]	0.0453 [0.0465]
YRP	$z_r/l = 0.774$, $z_m/l = 0.445$, $H/(A_r d l^2) = 0.099$, and $M_m/(A_r d l^3) = 0.0536$.					

Source: Guo, W. D., *Can Geotech J*, 45, 5, 2008.

[a] Post-tip yield
[b] Tip-yield state

Table 8.14 k Profiles on predictions (Gibson *k*/[constant *k*])

z_o/l	u_g *(mm)*	H *(kN)*	ω *(degree)*	z_m*(m)*	M_m *(kNm)*
0.30	22.3 [21.3]	0.605 [0.462]	−0.050(2.85°) [−0.053(3.03°)]	0.223 [0.193]	0.181 [0.129]
0.50	39.5 [65.3]	0.783 [0.723]	−0.087(4.93°) [−0.15(8.61°)]	0.251 [0.241]	0.248 [0.224]
0.535[a] [0.5885[b]]	45.0 [110.4]	0.814 [0.8285]	−0.097(5.57°) [−0.244(13.95°)]	0.255 [0.257]	0.261 [0.266]
YRP	∝	0.926	π/2 (90°)	0.272	0.307

Source: Guo, W. D., *Can Geotech J*, 45, 5, 2008.

[a] Post-tip yield
[b] Tip yield state

Table 8.13. (a) $H/(A_r dl^2) = 0.0837$ from Equation 8.1; and $H = 0.784$ kN ($= 0.0837 \times 244.9 \times 0.102\text{m} \times 0.612^2$ kN/m); (b) $u_g k_o/A_r = 3.005$ using Equation 8.2, and $u_g = 39.5$ mm ($= 3.005 \times 244.9/18642$ m); (c) $z_m/l = 0.4093$ using Equation 8.20, as z_m ($= 0.251$ m) $< z_o$ ($= 0.306$ m); and (iv) in view of $z_m < z_o$, $M_m/(A_r dl^3) = 0.0434$ [$= (2/3 \times 0.4093 + 0.245) \times 0.0837$] using Equation 8.22 and $M_m = 0.248$ kNm. The tip-yield state is plotted in Figure 8.17a through c. The moment is of similar order to that recorded in similar piles tested under soil movement (Guo and Ghee 2005).

2. Two z_o/l of 0.3 and 0.5 for pre-tip yield state are considered: At $z_o/l = 0.3$ ($< z_r/l$), the z_m/l is computed as 0.364 using Equation 8.21, which offers $H = 0.605$ kN, $u_g = 22.3$ mm, $z_m = 0.223$ m, and $M_m = 0.18$ kNm using the right M_m for $z_m > z_o$ in Table 8.1.

As z_o/l increases to 0.5, H increases to 0.783 kN, M_m to 0.248 kNm, and z_m to 0.251 m. The two points, each given by a pair of M_m and z_m, are plotted in Figure 8.17e.

3. The value of C_y (at tip-yield state) is calculated as 0.3326 [= 244.9/(0.0395 × 18,642)] with u_g = 39.5 mm. Assigning $C = 0.2919$ (< C_y), for instance, the z_r/l is calculated first, and followed by these responses:
   i. The z_r/l is evaluated in steps (see Table 8.15), using D_1 = 0.6968 and D_0 = 2.52049 from Equation 8.12, and A_1 = 3.9127×10^{-6}, and A_0 = 0.627 from Equation 8.11, and finally $z_r/l = 0.756$ (= $0.627^{1/3} + 3.9127^{1/3} \times 0.01 - 0.6968/6$) using Equation 8.10.
   ii. The H is calculated as 0.814 kN, as $H/(A_r dl^2) = 0.087$ using Equation 8.13, u_g = 45 mm [= 244.9/(18642 × 0.29193)], as per Equation 8.14, $z_o/l = 0.535$ [= (1 – 0.29193) × 0.7555], z_m = 0.255 m with z_m/l = 0.4171 [= $(2 \times 0.087)^{0.5}$] per Equation 8.20, and M_m = 0.261 kNm with $M_m/(A_r dl^3)$ = 0.0453 from Equation 8.22.
4. At YRP state, the z_r/l is calculated as 0.774 (= $0.68^{1/3} + 0.017 - 0.735/6$) with $C = 0$ and the estimated values of D_1, D_o, A_o, and A_1 in Table 8.15. The z_m/l, $H/(A_r dl^2)$, and $M_m/(A_r dl^3)$ are calculated as 0.445 (< z_o/l), 0.099, and 0.0536, respectively (see Table 8.13) using Equations 8.20, 8.23, and 8.24. Upon YRP state, it follows H = 0.926 kN, z_m = 0.272 m, and M_m = 0.307 kNm.
5. The bending moment $M(z)$ and the shear force $H(z)$ with depth were determined using the expressions in Table 8.6 (Guo 2008). With the ratios of z_o/l = 0.3 and 0.5, the predicted profiles are plotted in Figure 8.17e and f, and agree with the M_{max} and z_{max} predicted before (Figure 8.17a through c). For instance, the $H(z)$ profile at $z_o = 0.5l$ was gained using ω = –0.087, u_g = 39.5 mm (see Table 8.14), and Equations 8.33 and 8.34:

$$H(z) = 0.5A_r d\left(z_o^2 - z^2\right) + k_o d\omega\left(l^3 - z_o^3\right)/3 + k_o du_g\left(l^2 - z_o^2\right)/2 \quad (z \le z_o) \tag{8.33}$$

$$H(z) = k_o d\omega\left(l^3 - z^3\right)/3 + k_o du_g\left(l^2 - z^2\right)/2 \quad (z > z_o) \tag{8.34}$$

Table 8.15 Calculation of z_r/l for Post-Tip and/or YRP States

Gibson k	*Equation 8.12*			*Equation 8.11*		*Equation 8.10*
(for any D_r)	C	D_o	D_1	A_o	A_1	z_r/l
Post-tip yield	0.2919	2.5205	0.6968	0.627	3.912×10^{-6}	0.756
YRP	0	2.735	0.735	0.680	4.969×10^{-6}	0.774
Constant *k* (for any D_r)				Equation 8.11g		Equation 8.10g
	C			A_o	A_1	z_r/l
YRP	0			5.4405	3.975×10^{-5}	0.774

Source: Guo, W. D., *Can Geotech J*, 45, 5, 2008.

Local shear force $H(z)$ ~ displacement $u(z)$ relationships were predicted (Guo 2008), using expressions in Table 8.6, for the normalized depths z/l of 0, 0.3, 0.5, 0.62, and 0.9. They are plotted in Figure 8.17d.

6. The shear modulus G_L (at the pile-tip level) is equal to 0.310 MPa (= $2\tilde{G}$), as the $\tilde{G}$ is estimated as 0.1549 MPa (= $0.5k_o dl/3.757$), in light of k_o = 18.642 MPa/m^2 in Table 8.12, l = 0.612m, and d = 0.102m, with Equation 3.62, Chapter 3, this book.
7. The measured σ_r on the pile surface occurred at a local displacement u of 21.3 mm, or a ground displacement u_g, of 57 mm (z_o = 0.361 m and z_r = 0.470 m, post-tip yield).

The discussion for Equation 3.62 (Chapter 3, this book) indicates γ_b = 0.178 and $K_1(\gamma_b)/K_0(\gamma_b)$ = 2.898. The $\tilde{G}$ = 0.1549 MPa and u = 21.3 mm, thus allow the maximum σ_r (with $r = r_o$ = 0.051 m and θ_p = 0) to be obtained using Equation 3.48 (Chapter 3, this book), with $\sigma_r = 2 \times 154.9 \times 0.0213 \times 0.178/0.051 \times 2.898$. The stress σ_r across the diameter is predicted as $\sigma_r = 66.85\cos\theta_p$ (compared to $\tau_{r\theta} = -33.425\sin\theta_p$) in light of Equation 3.48. It compares well with the measured data (Prasad and Chari 1999), as shown in Figure 3.25 (Chapter 3, this book).

The maximum σ_r was cross-examined using $p_u = A_r zd$. The predicted total net pressure of 52.5 kPa (Guo 2008) compares well with 67.6 kPa (= 244.9 × 0.276) from $A_r z$, in view of the difference between the measured and predicted force (see Figure 8.17a).

Other features noted are: (1) the moment M_m locates below the slip depth ($z_m > z_o = 0.3l$ = 0.1836 m) under H = 0.605 kN, and moves into upper plastic zone ($z_m < z_o = 0.5l$ = 0.306 m) at H = 0.783 kN (see Figure 8.17); (2) during the test, the pile rotates about a depth (z_r) largely around $0.62l$, at which a negligible displacement of u is evident (see Figure 8.17d), though an opposite direction of force below the depth is observed, as with some field tests; (3) the nondimensional responses [e.g., $H/(A_r dl^2)$] are independent of the parameters A_r and k_o and are directly applicable to other tests.

Example 8.15 Analysis using constant *k*

The solutions for a constant k and Gibson p_u [in Table 8.1 (right column) and Table 8.4] were matched with each measured pile-head and groundline displacement (H ~ u_g) curve, as indicated by the *dashed lines* in Figure 8.17a. The k thus deduced (using the same A_r as that for Gibson k), and the shear modulus G (= G_L) and the angle at tip-yield are furnished in Table 8.12. The associated curves of M_{max}~H and M_{max}~z_m are also plotted in Figure 8.17b and c, respectively. This analysis indicates:

- The shear moduli deduced are 0.105 MPa (D_r = 0.25), 0.327 MPa (0.5), and 0.461 MPa (0.75), respectively, exhibiting –45.6%, 3.5%, and –1.9% difference from the revised measured values of 0.193 MPa, 0.316 MPa, and 0.470 MPa, respectively.

- The tip-yield (thus pile-head force H_o; see Figure 8.17a), occurs at a displacement far greater than $0.2d$ (= 20.4 mm) and at a rotation angle ~5 times those inferred using a Gibson k (i.e., 10~15 degrees, see Table 8.12).

In parallel to the Gibson k, the calculation for $D_r = 0.25$ was elaborated with $A_r = 244.9$ kN/m^3 and $k = 3.88$ MN/m^3 (Guo 2008) and was focused on the difference from the Gibson k analysis. The results encompass H, u_g, z_m, and M_m at the tip-yield state, at z_o/l of 0.3 and 0.5, and at post-tip yield state; profiles of bending moment, $M(z)$, and shear force, $H(z)$, at the slip depths of $z_o = 0.3l$ and $0.5l$ (see Figure 8.17e and f), and local shear force-displacement relationships at five different depths (see Figure 8.17d). The conclusions are that the two points given by the pairs of M_m and z_m, from Equation 8.20g and the right M_m for $z_m > z_o$ in Table 8.1 agree well with the respective $M(z)$ profiles. Upon the rotation point yield, an identical response to that for a Gibson k (see Table 8.15) was obtained. The G was estimated as 0.105 MPa (= 3.88 × 0.102/3.757 MPa) using Equation 3.62 and a head-displacement u_g of 92 mm (prior to tip yield) was needed to mobilize the radial pressure σ_r of 66.85 MPa at a local displacement u of 31.3 mm, with identical distribution of the σ_r to that for Gibson k. In brief, the results presented in Figure 8.17d, e, and f for constant k largely support the comments on Gibson k about the M_m, z_m, z_r, and the nondimensional responses. The underestimation of the measured force at u_g of 92 mm (see Figure 8.17a) indicates the impact of stress hardening. The actual k should be bracketed by the uniform k and Gibson k.

Example 8.16 Effect of k profiles

The impact of the k profiles is evident on the predicted $H \sim u_g$ curve; whereas it is noticeable on the predicted M_m only at initial stage (see Figure 8.17). The latter is anticipated, as beyond the initial low load levels, the M_m is governed by the same value of A_r and the same Equations 8.20 and 8.22. The A_r deduced is within ± ~15% difference of $\gamma_s' K_p^2$. The k deduced (constant) is within ± ~3.5% accuracy against the revised measured k, except for the pile in $D_r = 0.25$, and those deduced from Gibson k, which are explained herein.

Assuming Gibson k, the local displacement u of 21.3 mm for inducing the measured σ_r involves 64% [= $(u - u^*)/u$ = (21.3 − 13)/13] plastic component, as $u > u^*$ (= 13 mm = 244.9/18640 m). Stipulating a constant k, the local u of 31.3 mm required encompasses 68% [= (31.3 − 18.6)/18.6] plastic component, as $u > u^*$ (= 18.6 mm = 244.9 × 0.276/3880 m). The actual local limit u^* may be higher than 13.0~18.6 mm adopted, owing to the requirement of a higher A_r (thus p_u) than 244.9 kPa/m adopted to capture the stress hardening exhibited (see Figure 8.17a) beyond a groundline displacement u_g of 57 mm. With stress $\sigma_r \propto ku$ (i.e., Equation 3.48, Chapter 3, this book) and good agreement between measured and predicted σ_r, the k may supposedly be underestimated by 64~68% to compensate overestimation of the "elastic" displacement u (no additional stress induced for the plastic component). The hardening effect reduces slightly the percentage of plastic components. Thus the actual underestimation of the measured modulus (Table 8.12, $D_r = 0.25$) was 19.5~45.6%, and the predicted G_L was 0.105~0.31 MPa.

In contrast to $D_r = 0.25$, overestimation of the (Gibson) k is noted for $D_r = 0.5$~0.75, although the displacement of 0.2d and the angle (slope) for the capacity H_o are close to those used in practice. Real k profile should be bracketed by the constant and Gibson profiles.

In brief, the current solutions cater for net lateral resistance along the shaft only, and neglect longitudinal resistance along the shaft, transverse shear resistance on the tip, and the impact of stress hardening (as observed in the pile in $D_r = 0.25$). They are conservative for very short, stub piers that exhibit apparent shear resistance (Vallabhan and Alikhanlou 1982). The linear p_u profile is not normally expected along flexible piles, in which the $p_u \propto z^{1.7}$ (Guo 2006), and a uniform p_u may be seen around a pile in a multilayered sand. The modulus varies with the pile diameter.

8.6 SUMMARY

Elastic-plastic solutions were developed for modeling nonlinear response of laterally loaded rigid piles in the context of the load transfer approach, including the expressions for pre-tip and post-tip yield states. Simple expressions for determining the depths z_o, z_1, and z_r were developed for constructing on-pile force profiles and calculating lateral capacity H_o, maximum moment M_m, and its depth z_m. The solutions are provided for typical ratio of e/l (e.g., $e = 0$ and ∞). The solutions and expressions are underpinned by two parameters, k (via G) and A_r (via s_u, or ϕ and γ_s'). They are consistent with available FE analysis and relevant measured data. They were complied into a spreadsheet program (GASLSPICS) and used to investigate a well-documented case. Comments are made regarding estimation of lateral capacity. Salient features are noted as follows:

1. The on-pile force *(p)* profile is a result of mobilization of slip along the p_u profile (*LFP*). It may be constructed for any states (e.g., pre-tip yield, tip-yield, post-tip yield, and rotation-point yield states), as the p_u profile is unique, independent of the loading level.
2. Nonlinear, nondimensional response (e.g., load, displacement, rotation, and maximum bending moment) is readily estimated using the solutions by specifying the slip depth z_o (pre-tip yield) and rotation depth z_r (via a special parameter C, post-tip yield). Dimensional responses are readily gained for a given pair of k and A_r (or k and

$N_g s_u$). Conversely, the two parameters can be legitimately deduced using two measured nonlinear responses (or even one nonlinear curve) as stress distributions along depth and around pile diameter are integrated into the solutions.

3. For the investigated model piles in sand, the deduced A_r is generally within ±~15% of $\gamma_s' K_p^2$ and the shear moduli is only ~±3.5% discrepancy from the measured data, although ~46% underestimation of the modulus is noted for the stress hardening case, owing to the plasticity displacement for the rigid pile. With piles in clay, N_g ranges from 2.5 to 11.9, depending on pile movement modes as discussed elsewhere by the author.
4. Maximum bending moment raises 1.3 times as the tip-yield state moves to the YRP state and 2.1~2.2 times as the e increases from 0 to $3l$ at either state (N. B. $M_{max} \approx M_o$ given e/l >3) with a Gibson p_u. The raise becomes 1.3~1.5 times and 2.5~2.9 times, respectively, assuming a constant p_u. The eccentricity has 1.5~2 times higher impact on the M_{max} for constant p_u compared to Gibson p_u.

The impact of the p_u profiles on the response is highlighted by the constant and Gibson p_u. The longitudinal resistance along the shaft and transverse shear resistance on the base (or tip) are neglected. As such, the current solutions are conservative for short, stub piles. Rotation point alters from footings to rigid piles under the combined lateral-moment loading, depending to a large extent on boundary constraints (e.g., pile-head and base restraints), and depth of stiff layers. The maximum bending moment changes remarkably, as such its evaluation is critical to assess structural bending failure. Further research is warranted to clarify mechanism of fixity at different depths on pile response. The current solutions can accommodate the increase in resistance owing to dilation by modifying A_r, while not able to capture the effect of stress-hardening.

Chapter 9

Laterally loaded free-head piles

9.1 INTRODUCTION

Piles are often subjected to lateral load exerted by soil, wave, wind, and/or traffic forces (Poulos and Davis 1980). The response has been extensively studied using the p-y curves model (Reese 1958) in the framework of load transfer model, in which the pile–soil interaction is simulated using a series of independent (uncoupled) springs distributed along the pile shaft (Matlock and Reese 1960). Assuming nonlinear p-y curves at any depths, the solution of the response of the piles has been obtained using numerical approaches such as finite difference method (e.g., COM624; Reese 1977). The solutions, however, often offer different predictions against 3-dimensional (3-D) continuum-based finite element analysis (FEA) (Yang and Jeremić 2002). Assuming a displacement mode for soil around the pile, Guo and Lee (2001) developed a new coupled load transfer model to capture the interaction among the springs. The model also offers explicit expressions for modulus of subgrade reaction, k, and a fictitious tension, N_p (see Chapter 7, this book). The coupled model compares well with the finite element analysis, but it is confined to elastic state.

The maximum, limiting force on a lateral pile at any depth is the sum of the passive soil resistance acting on the face of the pile in the direction of soil movement, and sliding resistance on the side of the pile, less any force due to active earth pressure on the rear face of the pile. The net limiting force per unit length p_u with depth (referred to as LFP) mobilized is invariable. A wealth of studies on constructing the LFP (or p_u) has been made to date, notably by using force equilibrium on a passive soil wedge (Matlock 1970; Reese et al. 1975), upper-bound method of plasticity on a conical soil wedge (Murff and Hamilton 1993), and a "strain wedge" mode of soil failure (Ashour and Norris 2000). All these are underpinned by stipulating a gradual development of a wedge around a pile near ground level and lateral flow below the wedge. For instance, Reese et al. (1974, 1975) developed the popular p_u (thus the LFP) for piles in sand using force equilibrium on a wedge. However, the accuracy of the p_u is not warranted for large proportion of piles (Guo 2012), especially for large-diameter piles. Perhaps the scope of the wedge should be associated with rotation center for a rigid pile at one extreme (see Chapter 8, this book) and would not occur for some batter (or capped) piles at the other extreme. Guo (2006) developed a nonlinear expression of LFP that are independent of soil failure modes.

An ideal elastic-plastic load transfer (p-y) curve at any depth is assumed, with a gradient of the k deduced from the coupled model (Chapter 3, this book) and a limiting p_u. The transition from the initial elastic to the ultimate plastic state actually exhibits strong nonlinearity, such as those proposed for soft clay, stiff clay, and sand (Matlock 1970; Reese et al. 1974; Reese et al. 1975). These forms of p-y curves are proven very useful in terms of instrumented piles embedded in uniform soils (Reese et al. 1975) but are insufficiently accurate for a large proportion of piles. In addition, to synthesize the curves, a number of parameters are required to be properly determined (Reese et al. 1981), which is only warranted for large projects. In contrast, a simplified elastic-plastic p-y curve is sufficiently accurate (Poulos and Hull 1989) and normally expeditious.

A rigorous closed-form expression should be developed, as it can be used to validate numerical solutions and develop a new boundary element. Elastic, perfectly plastic solutions (typically for free-head piles) have been proposed for either a uniform or a linearly increase-limiting force profile (Scott 1981; Alem and Gherbi 2000) using the (uncoupled) Winkler model. The solutions were nevertheless not rigorously linked to properties of a continuum medium such as shear modulus. The two LFPs have limited practical use, as the LFPs reported to date is generally non-uniform with depth even along piles embedded in a homogeneous soil (Broms 1964a). Guo (2006) developed a new elastic-plastic solution to the model response of a pile (e.g., load-deflection and load-maximum bending moment relationships), which is consistent with continuum-based analysis. The solution also allows input parameters to be back-estimated using measured responses of a pile regardless of mode of soil failure.

This chapter presents simple design expressions to capture nonlinear response of laterally loaded, free-head piles embedded in nonhomogeneous medium. First, new closed-form solutions are established for piles underpinned by the two parameters k and p_u. Second, the solutions are verified using the 3-D FE analysis for a pile in two stratified soils. Third, guidelines are established for determining the input parameters k and p_u in light of back-estimation against 52 pile tests in sand/clay, 6 piles under cyclic loading, and a few typical free-head pile groups. The selection of input parameters is discussed at length, which reveals the insufficiency of some prevalent limiting force profiles.

9.2 SOLUTIONS FOR PILE–SOIL SYSTEM

The problem addressed here is a laterally loaded pile embedded in a non-homogenous elastoplastic medium. No constraint is applied at the pile-head and along the effective pile length, except for the soil resistance. The free length (eccentricity) measured from the point of applied load, H, to the ground surface is written as e. The pile–soil interaction is simulated by the load transfer model shown previously in Figure 3.27 (Chapter 3, this book). The uncoupled model indicated by the p_u is utilized to represent the plastic interaction and the coupled load transfer model indicated by the k and N_p to portray the elastic pile–soil interaction, respectively. The two interactions occur respectively in regions above and below the "slip depth," x_p. The following hypotheses are adopted (see Chapter 3, this book):

1. Each spring is described by an idealized elastic-plastic p-y curve (y being written as w in this chapter).
2. In elastic state, equivalent, homogenous, and isotropic elastic properties (modulus and Poisson's ratio) are used to estimate the k and the N_p.
3. In plastic state, the interaction among the springs is ignored by taking the N_p as zero.
4. Pile–soil relative slip occurs down to a depth where the displacement w_p is just equal to p_u/k and net resistance per unit length p_u is fully mobilized.
5. The slip (or yield) can only occur from ground level and progress downwards.

All five assumptions are adopted to establish the closed-form solutions presented here. Influence of deviation from these assumptions is assessed and commented upon in the later sections.

In reality, each spring has a limiting force per unit length p_u at a depth x [FL^{-1}] (Chapter 3, this book). If less than the limiting value p_u, the on-pile force (per unit length), p, at any depth is proportional to the local displacement, w, and to the modulus of subgrade reaction, k [FL^{-2}]:

$$p = kw \tag{9.1}$$

$$p_u = A_L(x + \alpha_o)^n \tag{9.2}$$

where k, A_L, and α_o are given by Equation 3.50, 3.68, and other expressions in Chapter 3, this book. The A_L and α_o are related to N_g and N_{co} by $N_g = A_L/(\tilde{s}_u d^{1-n})$ or $N_g = A_L/(\gamma_s' d^{2-n})$ and $N_{co} = N_g(\alpha_o/d)^n$, as used later on.

9.2.1 Elastic-plastic solutions

Nonlinear response of the lateral pile is governed by two separate differential equations for the upper plastic (denoted by the subscript A), and the lower elastic zones (by the subscript B), respectively (Chapter 3, this book). Within the plastic zone ($x \le x_p$), the uncoupled model offers

$$E_p I_p w_A^{IV} = -A_L (x+\alpha_o)^n \ (0 \le x \le x_p) \tag{9.3a}$$

where w_A = deflection of the pile at depth x that is measured from ground level; w_A^{IV} = fourth derivative of w_A with respect to depth x; I_p = moment of inertia of an equivalent solid cylinder pile. Below the depth x_p (elastic zone, $x_p \le x \le L$), the coupled model furnishs the governing equation of (Hetenyi 1946; Guo and Lee 2001)

$$E_p I_p w_B^{IV} - N_p w_B'' + k w_B = 0 \ (0 \le z \le L\text{-}x_p) \tag{9.3b}$$

where w_B = deflection of the pile at depth z ($= x - x_p$) that is measured from the slip depth; w_B^{IV}, w_B''= fourth and second derivatives of w_B with respect to depth z. Equation 9.3b reduces to that based on Winkler model using $N_p = 0$. By invoking the deflection and slope (rotation) compatibility restrictions at $x = x_p$ ($z = 0$) for the infinitely long pile, Equations 9.3a and 9.3b were resolved (Guo 2006), as elaborated next.

9.2.1.1 Highlights for elastic-plastic response profiles

Integrating Equation 9.3a for plastic state yields expressions for shear force, $Q_A(x)$, bending moment, $M_A(x)$, rotation, $w_A'(x)$ [i.e., $\theta_A(x)$], and deflection, $w_A(x)$ of the pile at depth x, as functions of unknown parameters C_3 and C_4, which are then determined from the conditions for a free-head pile (at $x = 0$). They are provided in Table 9.1. Some salient steps/features are explained next by rewriting the critical response in nondimensional form, such as

$$w_A(\bar{x}) = \frac{4A_L}{k\lambda^n}\left\{-\bar{F}(4,\bar{x}) + \bar{F}(4,0) + \bar{F}(3,0)\bar{x} + [\bar{F}(2,0) + \bar{M}_o]\frac{\bar{x}^2}{2} + [\bar{F}(1,0) + \bar{H}]\frac{\bar{x}^3}{6}\right\} + C_3\bar{x} + C_4 \tag{9.4}$$

$$\bar{F}(m,x) = \frac{(\bar{x}+\bar{\alpha}_o)^{n+m}}{(n+m)\ldots(n+1)} \ (m = 1\text{–}4) \tag{9.5}$$

Table 9.1 Expressions for response profiles of a free-head pile

Responses in plastic zone ($x \le x_p$)

$$w_A(x)=\frac{A_L}{E_pI_p}\left\{-F(4,x)+F(4,0)+F(3,0)x+\left[F(2,0)+\frac{M_o}{A_L}\right]\frac{x^2}{2}+\left[F(1,0)+\frac{H}{A_L}\right]\frac{x^3}{6}\right\}+\theta_g x+w_g \qquad (9T1)$$

$$\theta_A(x)=\frac{A_L}{E_pI_p}\left\{-F(3,x)+F(3,0)+\left[F(2,0)+\frac{M_o}{A_L}\right]x+\left[F(1,0)+\frac{H}{A_L}\right]\frac{x^2}{2}\right\}+\theta_g \qquad (9T2)$$

$$-M_A(x)=\left\{-F(2,x)+F(2,0)+\left[F(1,0)+\frac{H}{A_L}\right]x\right\}A_L+M_o \qquad (9T3)$$

$$-Q_A(x)=A_L\left[-F(1,x)+F(1,0)+\frac{H}{A_L}\right] \qquad (9T4)$$

In elastic zone with subscript "B" ($x > x_p$), they are given by the following:

$$w_B(z)=\frac{E_pI_p}{ke^{\alpha z}}\left\{[2\alpha w_p'''+(3\alpha^2-\beta^2)w_p'']\cos\beta z+\left[\frac{\alpha^2-\beta^2}{\beta}w_p'''+\frac{\alpha}{\beta}(\alpha^2-3\beta^2)w_p''\right]\sin\beta z\right\} \qquad (9T5)$$

$$-w_B'(z)=\frac{e^{-\alpha z}}{2\lambda^2}\left\{(w_p'''+2\alpha w_p'')\cos\beta z+\left[\frac{\alpha w_p'''+(\alpha^2-\beta^2)w_p''}{\beta}\right]\sin\beta z\right\} \qquad (9T6)$$

$$-M_B(z)=E_pI_pw_B''(z)=E_pI_pe^{-\alpha z}\left[w_p''\cos\beta z+\left(\frac{w_p'''+\alpha w_p''}{\beta}\right)\sin\beta z\right] \qquad (9T7)$$

$$-Q_B(z)=E_pI_pw_B'''(z)=E_pI_pe^{-\alpha z}\left[w_p'''\cos\beta z-\left(\frac{\alpha w_p'''+2\lambda^2w_p''}{\beta}\right)\sin\beta z\right] \qquad (9T8)$$

where w_p'' and w_p''' are values of 2nd and 3rd derivatives of $w(x)$ with respect to $z\ (= x - x_p)$.

$$F(m,x)=\frac{(x+\alpha_o)^{n+m}}{(n+m)\ldots(n+1)}\quad (m = 1\text{–}4) \qquad (9T9)$$

$$w_p'''=\frac{A_L}{E_pI_p}\left[-F(1,x_p)+F(1,0)+\frac{H}{A_L}\right] \qquad (9T10)$$

$$w_p''=\frac{A_L}{E_pI_p}\left[-F(2,x_p)+F(2,0)+F(1,0)x_p+\frac{Hx_p}{A_L}+\frac{M_o}{A_L}\right] \qquad (9T11)$$

In particular, at $n_2 = 1$ (stable layer, Chapter 12, this book), the following simple expressions are noted:

$$w_{p2}''=\frac{A_{L2}}{E_pI_p\lambda_2^3}\frac{x_{p2}^2}{6}\frac{2x_{p2}+3}{1+x_{p2}}$$

$$w_{p2}'''=\frac{A_{L2}}{E_pI_p\lambda_2^2}\frac{-x_{p2}}{6}\frac{2x_{p2}^2-3}{1+x_{p2}}.$$

where C_3, C_4 = unknown parameters; $\bar{H}$ (= $H\lambda^{n+1}/A_L$), normalized load, and H = a lateral load applied at a distance of e above mudline; λ = the reciprocal of characteristic length, $\lambda = \sqrt[4]{k/(4E_pI_p)}$; $\bar{M}_o$(= $He\lambda^{n+2}/A_L$), normalized bending moment, with $M_o = He$, bending moment at ground level; $\bar{x}$ (= $x\lambda$) normalized depth from ground level; and $\bar{\alpha}_o = \alpha_o\lambda$. The free-head conditions are:

$$-Q_A(0) = E_pI_pw'''_A(0) = H, \quad -M_A(0) = E_pI_pw''_A(0) = He \tag{9.6}$$

In particular, at the transition (slip) depth $x = x_p$, we have $w_A(x = x_p) = w_p$, $w'_A(x = x_p) = w'_p$, $w''_A(x = x_p) = w''_p$, $w'''_A(x = x_p) = w'''_p$, and $w'^{IV}_A(x = x_p) = w^{IV}_p$. The w''_p and w'''_p are provided in Table 9.1.

Equation 9.3b for the elastic state may be solved as ($N_p < 2\sqrt{kE_pI_p}$),

$$w_B(z) = e^{-\alpha z}(C_5\cos\beta z + C_6\sin\beta z) \tag{9.7}$$

where C_5, C_6 = constants and α, β are given by the following:

$$\alpha = \sqrt{\sqrt{k/(4E_pI_p)} + N_p/(4E_pI_p)} \qquad \beta = \sqrt{\sqrt{k/(4E_pI_p)} - N_p/(4E_pI_p)} \tag{9.8}$$

Equation 9.7 offers $w^{IV}_B(z)$, $w'''_B(z)$, $w''_B(z)$, and $w'_B(z)$ in Table 9.1.

The constants C_i (i = 3~6) are determined using the compatibility conditions at the slip depth $x = x_p$ (z = 0) from elastic to plastic state.

Conditions of $w'''_B(z=0) = w'''_p$ and $w''_p(z=0) = w''_p$ may be written as two expressions of the unknown constants C_5 and C_6, which were resolved to yield

$$C_5 = \frac{E_pI_p}{k}[2\alpha w'''_p + (3\alpha^2 - \beta^2)w''_p] \tag{9.9}$$

$$C_6 = \frac{E_pI_p}{k\beta}[(\alpha^2 - \beta^2)w'''_p + \alpha(\alpha^2 - 3\beta^2)w''_p] \tag{9.10}$$

Conditions of $w'_A(x = x_p) = w'_B(z = 0)$ and $w_A(x = x_p) = w_B(z = 0)$ allow C_3 and C_4 to be determined, respectively, as

$$C_3 = \frac{4A_L\lambda^{1-n}}{k}\left[\bar{F}(3,\bar{x}_p) - \bar{F}(2,0)\bar{x}_p - (\bar{F}(1,0) + \bar{H})\frac{\bar{x}_p^2}{2} - \bar{H}\bar{e}\bar{x}_p\right] - \alpha C_5 + \beta C_6 \tag{9.11}$$

$$C_4 = \frac{4A_L}{k\lambda^n}\left[\bar{F}(4,\bar{x}_p) - (\bar{F}(1,0) + \bar{H})\frac{\bar{x}_p^3}{6} - (\bar{F}(2,0) + \bar{H}\bar{e})\frac{\bar{x}_p^2}{2}\right] - C_3\frac{\bar{x}_p}{\lambda} + C_5 \tag{9.12}$$

where $\bar{e} = e\lambda$ and $\bar{x}_p = x_p\lambda$. Substituting Equations 9.9 and 9.10 into Equation 9.11, a normalized C_3 is derived as

$$\frac{C_3 k\lambda^{n-1}}{A_L} = 4[\bar{F}(3,\bar{x}_p) - \bar{x}_p\bar{F}(2,0)] + 4\alpha_N[\bar{F}(2,\bar{x}_p) - \bar{F}(2,0)] + 2\bar{F}(1,\bar{x}_p)$$
$$-2(1+2\alpha_N\bar{x}_p + \bar{x}_p^2)\bar{F}(1,0) - 2[\bar{x}_p^2 + 1 + 2\alpha_N\bar{x}_p + 2(\alpha_N + \bar{x}_p)\bar{e}]\bar{H} \tag{9.13}$$

where $\alpha_N = \alpha/\lambda$. In light of Equations 9.9 and 9.13, the normalized form of C_4 is obtained

$$\frac{C_4 k\lambda^{n}}{A_L} = 4[\bar{F}(4,\bar{x}_p) - \bar{x}_p\bar{F}(3,\bar{x}_p)] + 4(1-\alpha_N^2)[\bar{F}(2,\bar{x}_p) - \bar{F}(2,0)] + 2\bar{x}_p^2\bar{F}(2,0)$$
$$-2\bar{x}_p(1-2\alpha_N^2)\bar{F}(1,\bar{x}_p) + (1+2\alpha_N\bar{x}_p)\bar{F}(0,\bar{x}_p) + (4\bar{x}_p^2/3 - 2)\bar{x}_p\bar{F}(1,0)$$
$$+[4\bar{x}_p^3/3 - 2\bar{x}_p + (2\bar{x}_p^2 - 4 + 4\alpha_N^2)\bar{e}]\bar{H}. \tag{9.14}$$

At ground level, with Equation 9.4, the rotation w_g' (θ_g) and ground-level deflection w_g are expressed, respectively, in normalized form of $\bar{w}_g'$ and $\bar{w}_g$ as

$$\bar{w}_g' = \frac{w_g' k\lambda^{n-1}}{A_L} = -4\bar{F}(3,0) + \frac{C_3 k\lambda^{n-1}}{A_L} \quad \text{and} \tag{9.15}$$

$$\bar{w}_g = \frac{w_g k\lambda^{n}}{A_L} = -4\bar{F}(4,0) + \frac{C_4 k\lambda^{n}}{A_L} \tag{9.16}$$

Conditions of $w_B'(z=0) = w_p'$ and $w_B(z=0) = w_p$ allow the normalized rotation and deflection at the slip depth to be written, respectively, as

$$\frac{w_p' k\lambda^{n}}{A_L} = -\alpha\frac{C_5 k\lambda^{n}}{A_L} + \beta\frac{C_6 k\lambda^{n}}{A_L} \quad \text{and} \tag{9.17}$$

$$\frac{w_p k\lambda^{n}}{A_L} = \frac{C_5 k\lambda^{n}}{A_L} \tag{9.18}$$

Conditions of $w_B^{IV}(z=0) = w_p^{IV}$, $w_B'''(z=0) = w_p'''$, and $w_B''(z=0) = w_p''$ render the following relationship at the slip depth to be established:

$$0.5w_p^{IV} + \alpha w_p''' + \lambda^2 w_p'' = 0 \tag{9.19}$$

Responses of the pile along its length are predicted separately using elastic ($x > x_p$) and plastic solutions, with Equations 9T1, 9T2, 9T3, and 9T4 (see Table 9.1 for these equations) for deflection, rotation, moment, and shear force, respectively in plastic zone; otherwise, Equations 9T5 through 9T8 in Table 9.1 should be employed, as derived from Equations 9.7, 9.9, and 9.10.

The solutions allow response of the pile at any depth to be predicted readily. In particular, three key responses were recast in dimensionless forms and are discussed next.

9.2.1.2 Critical pile response

1. Lateral load: In terms of Equation 9.3a, normalized w_p''', and w_p'', the normalized load $\bar{H}$ of Equation 9T12 (see Table 9.2) is deduced from Equation 9.19 and presented in the explicit form. A slip depth under a given load may be computed, which is then used to calculate other pile response.
2. Groundline deflection: The normalized pile deflection at ground level is obtained as Equation 9T13 using Equations 9.16 and 9.14 and $\bar{H}$. At a relative small eccentricity, e, pile-head deflection w_t may be approximately taken as $w_g + e \times w_g'$, where w_g' = mudline rotation angle (in radian) obtained using Equation 9.15.
3. Maximum bending moment: The maximum bending moment, M_{max}, occurs at a depth x_{max} (or z_{max}) at which shear force is equal to zero. The depth could locate in plastic or elastic zone, depending on $\psi(\bar{x}_p)$, which is deduced from $Q_B(z_{max}) = 0$ (see Table 9.1):

$$\psi(\bar{x}_p) = \beta_N / (\alpha_N + 2\lambda\, w_p''/w_p''') \tag{9.20}$$

$$z_{max} = \tan^{-1}(\psi(\bar{x}_p))/\beta \tag{9.21}$$

Where $\beta_N = \beta/\lambda$, and the depth z_{max} is given by Equation 9.21. More specifically, if $\psi(\bar{x}_p) > 0$, then $z_{max} > 0$. The depth of maximum moment M_{max} is equal to $x_p + z_{max}$. The value of M_{max} may be estimated using Equation 9T15 by replacing z with z_{max}, $M_{max} = M_B(z_{max})$. However, if $\psi(\bar{x}_p) < 0$, then $z_{max} < 0$ (normally expected at a relatively high value of $\bar{x}_p$). The M_{max} locates at depth x_{max}, with $\bar{x}_{max}$ $(= x_{max}\lambda)$ of

$$x_{max}\lambda = [(\alpha_o\lambda)^{n+1} + (n+1)\bar{H}]^{1/(n+1)} - \alpha_o\lambda \tag{9.22}$$

Table 9.2 $\bar{H}$, $\bar{w}_g$, and $\bar{M}(\bar{x})$ of a free-head pile

(a) Normalized pile-head load, $\bar{H}$

$$\bar{H} = -\frac{\bar{F}(1,0)(\bar{x}_p + \alpha_N)}{\bar{x}_p + \alpha_N + \bar{e}} + \frac{\bar{F}(2,\bar{x}_p) - \bar{F}(2,0) + \alpha_N \bar{F}(1,\bar{x}_p) + 0.5\bar{F}(0,\bar{x}_p)}{\bar{x}_p + \alpha_N + \bar{e}} \quad (9T12)$$

Given $\bar{x}_p = 0$, the minimum head load to initiate slip is obtained.

(b) Normalized mudline deflection, $\bar{w}_g$

$$\begin{aligned}\bar{w}_g = {} & 4[\bar{F}(4,\bar{x}_p) - \bar{x}_p \bar{F}(3,\bar{x}_p) - \bar{F}(4,0)] + [4(1-\alpha_N^2) + C_m]\bar{F}(2,\bar{x}_p) \\ & + [(-2 + 4\alpha_N^2)\bar{x}_p + \alpha_N C_m]\bar{F}(1,\bar{x}_p) + [1 + 2\alpha_N \bar{x}_p + C_m/2]\bar{F}(0,\bar{x}_p) \\ & + \{2\bar{x}_p^2 - [4(1-\alpha_N^2) + C_m]\}\bar{F}(2,0) + [4\bar{x}_p^3/3 - 2\bar{x}_p - (\bar{x}_p + \alpha_N)C_m]\bar{F}(1,0)\end{aligned} \quad (9T13)$$

and $C_m = [\frac{4}{3}\bar{x}_p^3 - 2\bar{x}_p + (2\bar{x}_p^2 + 4\alpha_N^2 - 4)\bar{e}]\frac{1}{\bar{x}_p + \alpha_N + \bar{e}}$.

(c) Normalized bending moment, $\bar{M}(\bar{x})$ ($x < x_p$) offers the following:

$$-\frac{M_{max}}{A_L} = \frac{1}{n+2}\left[\alpha_o^{n+1} + (n+1)\frac{H}{A_L}\right]^{(n+2)/(n+1)} - \left(\frac{\alpha_o^{n+2}}{n+2} + \frac{\alpha_o H}{A_L}\right) + \frac{He}{A_L} \quad (9T14)$$

$$-\bar{M}_B(z) = e^{-\alpha z}[C_1(\bar{x}_p)\cos\beta z + C_2(\bar{x}_p)\sin\beta z] \quad (9T15)$$

and $C_1(\bar{x}_p) = E_p I_p \lambda^{n+2} w_p'' / A_L$, $C_2(\bar{x}_p) = E_p I_p \lambda^{n+2}(w_p''' + \alpha w_p'')/(\beta A_L)$

where $M_B(z)$ = bending moment at depth z that is derived as

$$z_{max} = \frac{1}{\beta}\tan^{-1}(\beta w_p''' / [\alpha w_p''' + 2\lambda^2 w_p'']) \quad (9T16)$$

$$x_{max}\lambda = [(\alpha_o\lambda)^{n+1} + (n+1)\bar{H}]^{1/(n+1)} - \alpha_o\lambda \quad (9T17)$$

Source: Guo, W. D., *Computers and Geotechnics*, 33, 1, 2006.

Note: The constants C_i are determined using the compatibility conditions of $Q(\bar{x})$, $M(\bar{x})$, $w'(\bar{x})$, and $w(\bar{x})$ at the normalized slip depth, $\bar{x}_p$ [$\bar{x}=\bar{x}_p$ or $\bar{z} = 0$]. Elastic solutions validated for $N_g < 2(kE_pI_p)^{0.5}$ are ensured by L being greater than the sum of L_c and the maximum x_p. Pile-head rotation angle is given by Equation 9.15.

The M_{max} may be calculated using Equation 9T14, derived from Equation 9.7 by replacing x with x_{max}. The M_{max} of Equation 9T14 is not written in dimensional form to facilitate practical prediction. However, normalized $M_{max}\lambda^{n+2}/A_L$ will be used later to provide a consistent presentation from elastic through to plastic state.

In summary, response of the laterally loaded piles is presented in explicit expressions of the slip depth, at which the maximum p_u normally occurs. The expressions are valid for an infinitely long ($L > L_c$ + max x_p, max x_p = slip depth x_p under maximum imposed load H) pile embedded in a soil of a constant modulus (k) with depth. The solutions are based on the p_u profile rather than mode of soil failure. The impact of the failure mode is catered for by parameters and/or free-head or fixed-head (Guo 2009) solutions.

9.2.2 Some extreme cases

The normalized slip depth under lateral loads may be estimated using Equation 9T12 that associates with the LFP (via A_L, α_o, n), and the pile–soil relative stiffness (via λ). The minimum load, H_e, required to initiate the slip at mudline ($\bar{x}_p = 0$) is given by

$$H_e\lambda^{n+1}/A_L = (\alpha_o\lambda)^n/[2(\alpha_N + \bar{e})] \quad (9.23)$$

Under the conditions of $\alpha_o = 0$, and $\alpha_N = 1$ (i.e., $\beta_N = 1$), the current solutions may be simplified (for instance, Equation 9T12 may be replaced with Equation 9.24)

$$\frac{H\lambda^{n+1}}{A_L} = \frac{0.5\bar{x}_p^n[(n+1)(n+2) + 2\bar{x}_p(2+n+\bar{x}_p)]}{(\bar{x}_p + 1 + \bar{e})(n+1)(n+2)} \quad (9.24)$$

$$\frac{w_g k\lambda^n}{A_L} = \frac{2}{3}\bar{x}_p^{n+3}\frac{2\bar{x}_p^2 + (2n+10)\bar{x}_p + n^2 + 9n + 20}{(\bar{x}_p + 1 + \bar{e})(n+2)(n+4)} + \frac{(2\bar{x}_p^2 + 2\bar{x}_p + 1)\bar{x}_p^n}{\bar{x}_p + 1 + \bar{e}} + \frac{2\bar{x}_p^4 + (n+4)(\bar{x}_p + 1)[2\bar{x}_p^2 + (n+1)(\bar{x}_p + 1)]}{(\bar{x}_p + 1 + \bar{e})(n+1)(n+4)}\bar{e}\bar{x}_p^n \quad (9.25)$$

$$\psi(\bar{x}_p) = \frac{-2\bar{x}_p(2+n)\bar{e} - 2(1+n)\bar{x}_p^2 + (n+2)(n+1)}{2(n+2)(\bar{x}_p + 1 + n)\bar{e} + [2\bar{x}_p^2 + 2(2+n)\bar{x}_p + n + 2](n+1)} \quad (9.26)$$

With the same conditions, the constants $C_1(\bar{x}_p)$, and $C_2(\bar{x}_p)$ for bending moment in Equation 9T15 may be replaced with

$$C_1(\bar{x}_p) = \frac{0.5\bar{x}_p^n}{\bar{x}_p + 1 + \bar{e}}\left(\frac{1+n+2\bar{x}_p}{1+n}\bar{e} + \frac{2+n+2\bar{x}_p}{2+n}\bar{x}_p\right) \quad C_2(\bar{x}_p) = 0.5\bar{x}_p^n \quad (9.27)$$

Equation 9.26 offers a critical normalized slip depth, $\bar{x}_p$ (rewritten as $\bar{x}'$), at $\psi(\bar{x}') = 0$:

$$\bar{x}' = \frac{-0.5(2+n)}{1+n}\bar{e} + 0.5\sqrt{[\frac{2+n}{1+n}\bar{e}]^2 + 2(n+2)} \quad (9.28)$$

The condition of $\bar{x}_p > \bar{x}'$ is equivalent to $\psi(\bar{x}_p) < 0$ and renders the maximum bending moment to occur above the slip depth. Equation 9.24 may also be rewritten as

$$(\bar{x}_p + 1)H\lambda^{1+n}/A_L + M_o\lambda^{2+n}/A_L = \bar{x}_p^n\left[0.5 + \frac{\bar{x}_p(2+n+\bar{x}_p)}{(1+n)(2+n)}\right] \quad (9.29)$$

With $\psi(\bar{x}_p) \geq 0$, a normalized ratio of the maximum moment $M(z_m)$ over the load H ($\neq 0$) can be obtained

$$-M(z_m)\lambda/H = e^{-\lambda z_m}[D_1(\bar{x}_p)\cos(\lambda z_m) + D_2(\bar{x}_p)\sin(\lambda z_m)] \tag{9.30}$$

where

$$D_1(\bar{x}_p) = C_1(\bar{x}_p)/\bar{H} \text{ and } D_2(\bar{x}_p) = C_2(\bar{x}_p)/\bar{H} \tag{9.31}$$

Equations 9.27 through 9.31 can be used to deduce some available expressions.

1. Imposing $n = e = 0$, $\alpha_N = 1$, and $\alpha_o = 0$, Equations 9T12 and 9.25 reduce to Equations 9.32 and 9.33, respectively

   $$H\lambda/A_L = (1 + \bar{x}_p)/2 \tag{9.32}$$

   $$w_g k / A_L = \frac{1}{6}[(\bar{x}_p + 1)^4 + 2(\bar{x}_p + 1) + 3] \tag{9.33}$$

   Equations 9.32 and 9.33 are essentially identical to those brought forwarded previously (Rajani and Morgenstern 1993) using Winkler model ($N_p = 0$) for a pipeline that is embedded in a homogenous soil and has a constant ($n = 0$) limiting force (resistance) along its length.
2. $M_{max} = H^2/(2A_L)$ and $x_{max} = H/A_L$ ($n = 0$), while $M_{max} = \sqrt{8H^3/9A_L}$ and $x_{max} = \sqrt{2H/A_L}$ ($n = 1$), with M_{max} being obtained from Equation 9T14 and x_{max} from Equation 9.22 by setting $\alpha_o = 0$ and $e = 0$, which were put forwarded previously (Ito et al. 1981).
3. Introducing $\bar{x}_p = 0$ (elastic state) and at $e = 0$, Equation 9T11 offers $w_p'' = 0$. Furthermore, using $N_p = 0$ in Equations 9T15 and 9T16, z_{max} and M_{max} obtained, respectively, are virtually identical to the results obtained using Winkler model (Rajani and Morgenstern 1993).

Example 9.1 Elastic-plastic solutions for $n = 0$ and $n = 1.0$

Elastic-plastic solutions were developed for a homogenous limiting force profile (LFP), or a linear increasing LFP profile between pile and soil. Such profiles are seldom observed in practice on flexible piles. However, they are useful as extreme cases of the current solutions for piles in nonhomogeneous LFP, as illustrated below. To allow such comparisons, nondimensional parameters of the following for $n = 0$ were introduced (Hsiung 2003):

$$y_p = 2H\lambda/A_L, \bar{H}_c/\bar{N}_c = 0.5y_p, \text{ and } y_m = M_o/M_c \tag{9.34}$$

where $M_c = 2E_pI_p\lambda^2 w_p$. As $w_p = p_u/k$, the M_c can be rewritten as $M_c = A_L/(2\lambda^2)$.

9.1.1 Slip Depth under a Given Load

Setting $n = 0$ in Equation 9.24, the normalized slip depth $\bar{x}_p$ under the load H is derived as

$$\bar{x}_p = -1 + H\lambda/A_L + \sqrt{(H\lambda/A_L)^2 + 2M_o\lambda^2/A_L} \tag{9.35}$$

Equation 9.35 is consistent with the previous expression (Alem and Gherbi 2000). Taking $n = 1$ and $e = 0$ in Equation 9.24, a nonlinear equation for $\bar{x}_p$ results from which the $\bar{x}_p$ is derived as

$$\bar{x}_p = (8H\lambda^2/A_L)^{1/2}\cos(\varpi) - 1 \tag{9.36}$$

where $\cos(\varpi)$ is gained from H using $\cos(3\varpi) = (32H\lambda^2/A_L)^{-3/2}$. Equation 9.36 is applicable to any load levels. Equation 9.36 can be rewritten in another form (Motta 1994).

9.1.2 Piles in Homogenous Soil ($n = 0$) due to Lateral Load H

Equation 9.33 can be converted to the following expression (Rajani and Morgenstern 1993):

$$w_g k/A_L = \frac{1}{2} + \frac{2}{3}\bar{H} + \frac{8}{3}\bar{H}^4 \text{ or } w_g k/A_L = \frac{1}{2} + \frac{2}{3}\frac{\bar{H}_c}{\bar{N}_c} + \frac{8}{3}\left(\frac{\bar{H}_c}{\bar{N}_c}\right)^4 \tag{9.37}$$

Setting $n = 0$ in Equation 9.28, the critical value, $\bar{x}'$ (rewritten as $\bar{x}'_o$) may be expressed as

$$\bar{x}'_o = -\bar{e} + (1 + \bar{e}^2)^{1/2} \tag{9.38}$$

A characteristic load, $\bar{H}^*$, was previously defined as (Alem and Gherbi 2000):

$$\bar{H}^* = \bar{H}(\bar{H} + 2\bar{e}) - 1 \tag{9.39}$$

In light of Equation 9.35, the condition of $\bar{H}^* < 0$ is equivalent to that of $\bar{x}_p < \bar{x}'_o$. "$\psi(\bar{x}_p) \ge 0$" is equivalent to $\bar{x}_p \le \bar{x}'_o$, requiring that the applied force, H^*, be equal to the total limiting (probably not shear) force developed in the plastic zone, $H^* = A_L \times x'_p$. Together with Equation 9.35, this equilibrium offers a unit value of $\bar{x}'_o$, identical to that obtained from Equation 9.38 at $e = 0$.

Under $\psi(\bar{x}_p) \ge 0$ or $\bar{x}_p \le \bar{x}'$, setting $n = e = 0$ in Equation 9.26, the $\psi(\bar{x}_p)$ is simplified as

$$\psi(\bar{x}_p) = \frac{1 - \bar{x}_p}{1 + \bar{x}_p} \tag{9.40}$$

Equation 9.40 in conjunction with Equation 9.21 allow $\tan(\lambda z_m)$ to be determined, which in turn gives

$$\cos(\lambda z_m) = (1+\bar{x}_p)\Big/\sqrt{2(1+\bar{x}_p^2)} \tag{9.41}$$

Equation 9.41 allows the bending moment of Equation 9T15 at $n = 0$ to be rewritten as

$$-M(z_m) = 0.5\frac{A_L}{\lambda^2}e^{-\lambda z_m}\sqrt{0.5(1+\bar{x}_p^2)} \tag{9.42}$$

Equation 9.42 may be written as

$$-M(z_m) = 0.25\frac{A_L}{\lambda^2}e^{-\lambda z_m}\sqrt{2(y_p^2 - 4y_p + 4)} \tag{9.43}$$

Using Equation 9.24 and setting $n = 0$ in Equation 9.31, the D_1 and D_2 are simplified as

$$D_1(\bar{x}_g) = 1 - 0.5\bar{N}_c/\bar{H}_c,\ D_2(\bar{x}_g) = 0.5\bar{N}_c/\bar{H}_c \tag{9.44}$$

The D_1 and D_2 allow Equation 9.30 to be rewritten in the form given previously (Rajani and Morgenstern 1993).

The maximum bending moment may occur at a depth x_m offered by Equation 9.22, and locate within the plastic zone, thus

$$x_m = H/A_L \tag{9.45}$$

The moment furnished by Equation 9T14 can be rewritten as

$$-(M_m + He)/A_L = 0.5(H/A_L)^2 \tag{9.46}$$

At $e = 0$, Equations 9.45 and 9.46 are identical to those proposed before (Hsiung 2003).

9.1.3 Piles in Homogenous Soil ($n = 0$) due to Moment Loading Only ($e = \infty$)

Substituting $H = n = 0$ into Equation 9.29, we obtain

$$M_o\lambda^2/A_L = 0.5(\bar{x}_p + 1)^2 \tag{9.47}$$

Substituting M_c with $A_L/(2\lambda^2)$, Equation 9.47 allows the y_m (= M_o/M_c) to be established

$$y_m = (\bar{x}_p + 1)^2 \tag{9.48}$$

From Equation 9.25, at $e = \infty$, the normalized displacement may be rewritten as

$$w_g k / A_L = 0.5 + 0.5(1+\bar{x}_p)^4 \tag{9.49}$$

Therefore,

$$w_g k / A_L = 0.5 + 0.5 y_m^2 \tag{9.50}$$

Equation 9.50 is consistent with the previous expression (Hsiung 2003).

9.1.4 Piles in Gibson Soil (n = 1.0) due to Lateral Load H Only (e = 0)

Considering $z = x - x_p$, Equation 9T15 can be expanded in terms of the normalized slip depth $\bar{x}_p$, thus

$$-M(z) = 2E_p I_p \lambda^2 e^{-\lambda x}(C_5(\bar{x}_p)\cos(\lambda x) + C_6(\bar{x}_p)\sin(\lambda x)) \tag{9.51}$$

Substituting $2E_pI_p\lambda^2$ with $k/(2\lambda^2)$, the coefficients C_5 and C_6 can be written as

$$C_5(\bar{x}_p) = e^{\lambda x_p}\frac{2A_L}{k\lambda}(C_1(\bar{x}_p)\cos(\lambda x_p) - C_2(\bar{x}_p)\sin(\lambda x_p)) \text{ and} \tag{9.52}$$

$$C_6(\bar{x}_p) = e^{\lambda x_p}\frac{2A_L}{k\lambda}(C_1(\bar{x}_p)\cos(\lambda x_p) + C_2(\bar{x}_p)\sin(\lambda x_p)) \tag{9.53}$$

Provided that n = 1 and e = 0, the $C_1(\bar{x})$ and $C_2(\bar{x})$ given by Equation 9.27 can be simplified as

$$C_1(\bar{x}_p) = \frac{\bar{x}_p^2}{6}\frac{2\bar{x}_p + 3}{1+\bar{x}_p} = \frac{H\lambda^2 \bar{x}_p}{A_L} - \frac{\bar{x}_p^3}{6} \quad C_2(\bar{x}_p) = 0.5\bar{x}_p \tag{9.54}$$

Thus

$$\frac{2A_L}{k\lambda}C_1(\bar{x}_p) = \frac{1}{2E_pI_p\lambda^2}\left(Hx_p - \frac{A_L x_p^3}{6}\right) \tag{9.55}$$

$$\frac{2A_L}{k\lambda}C_2(\bar{x}_p) = \frac{A_L x_p}{k} \tag{9.56}$$

With Equations 9.55 and 9.56, Equation 9.51 can be transformed into the expressions derived earlier (Motta 1994), though a typing error is noted in the latter.

9.1.5 Piles in Gibson Soil (n = 1.0) due to Moment Loading Only

Setting n = 1 and $e = \infty$ in Equation 9.25, the normalized pile-head displacement is

$$w_g k\lambda / A_L = (\bar{x}_p^4 + 5\bar{x}_p^3 + 10\bar{x}_p^2 + 10\bar{x}_p + 5)\bar{x}_p / 5 \tag{9.57}$$

Similarly, from Equation 9.29, the normalized bending moment is rewritten as

$$M_o\lambda^3/A_L = \bar{x}_p(\bar{x}_p^2 + 3\bar{x}_p + 3)/6 \tag{9.58}$$

Example 9.2 A lateral pile with uniform p_u profile

For the example pile (Hsiung 2003), let's estimate the deformation and maximum bending moment for the following two cases: In Case A, the free-head pile is subjected to a lateral load H of 400 kN and a moment M_o of 1,000 kN-m. In Case B, the free-head pile is subjected to a lateral load H of 280 kN, together with a moment M_o of 700 kN-m.

The following parameters were known: $k = 40$ MPa, $\lambda = 0.4351$/m, $w_p = 0.01$ m, $A_L = w_pk = 0.01 \times 40$ MPa $= 400$ kPa, and $e = 1{,}000/400 = 2.5$ (Case a) or $e = 700/280 = 2.5$ (Case b). First, from Equation 9.24, the minimum load to initiate a slip is found to be 220.17 kN [$= 0.5A_L/(\lambda(1 + e\lambda)$]. The given loads ($H$) for both cases are greater than 220 kN, therefore, the elastic-plastic solution presented herein is applicable.

Case A: Equation 9.29 may be transformed into the following form:

$$(\bar{x}_p + 1)^2 - 2\frac{H\lambda}{A_L}(\bar{x}_p + 1) - 2\frac{M_o\lambda^2}{A_L} = 0$$

Input $H = 400$ kN, $M_o = 1{,}000$ kN-m and other parameters, it follows: $\bar{x}_p = 0.5009$. In Equation 9.25, because $\bar{e} = 1.088$ ($= 2.5 \times 0.4351$), the normalized displacement is obtained as

$$w_gk/A_L = \frac{1.0875((1.5009)^4 + 1)}{2(1.5009 + 1.0875)} + \frac{1.5009((1.5009)^4 + 2\times 1.5009 + 3)}{6(1.5009 + 1.0875)} = 2.3465$$

Therefore, the displacement w_g is about 23.5 mm.

Using Equation 9.26, the $\psi(\bar{x}_p)$ is estimated to be −0.0618. With Equation 9.38, the $\bar{x}'$ is found as 0.1926. Either $\psi(\bar{x}_p) < 0$ or $\bar{x}_p > \bar{x}'$ means that the maximum bending moment should occur in plastic zone. The depth x_m is 1 m as determined from Equation 9.22. The moment is 1,200 kN-m as estimated below using Equation 9T14:

$$-M_m = 0.5(H/A_L)^2 \times A_L + He = 0.5 \times 400 + 400 \times 2.5 = 1{,}200 \text{ kN-m}.$$

Case B: With $H = 280$ kN and $M_o = 700$ kN-m, it follows: $\bar{x}_p = 0.17368$; thus w_g is 13.24 mm.

From Equation 9.26, $\psi(\bar{x}_p) = 0.223$, and using Equation 9.38, the $\bar{x}'$ maintains at 0.1926. The condition of $\psi(\bar{x}_p) > 0$ or $\bar{x}_p < \bar{x}'$ implies that the maximum bending moment should occur at a depth of $z_m + x_p$ below the ground level (x_p = 0.1737/0.4351 = 0.399 m). From Equation 9.21, the depth z_m is found to be 0.504 m (= arc*tan*(0.223)/0.4351). Thus, λz_m is equal to 0.21495. The coefficient $C_1(\bar{x}_p)$ is given by Equation 9.27:

$$C_1(\bar{x}) = 0.5\frac{(0.17368+1)\times 0.17368+1.0878\times(1+2\times 0.17368)}{1+0.17368+1.0878}$$
$$= 0.36911$$

Thus, the moment can be estimated by Equations 9T15 and 9.27:

$$-M(z_m) = \frac{400}{0.4351^2}e^{-0.1495}(0.36911\cos(0.1495\times 180/\pi)$$
$$+0.5\sin(0.1495\times 180/\pi))$$

Thereby, $M(z_m)$ is 795.88 kN-m that occurs at a depth of 0.903 m below the ground level.

Finally, the statistical relationship between load and displacement (Hsiung and Chen 1997) compares well with the analytical solution only when their normalized displacement is less than 8, otherwise it underestimates the displacement. Similar discussion can be directed toward fixed-head piles. In summary, even for piles in a homogenous soil, the closed-form solution can be presented in various, complicated forms.

9.2.3 Numerical calculation and back-estimation of LFP

Although they may appear complicated, the current solutions can be readily estimated using modern mathematical packages. They have been implemented into a spreadsheet operating in EXCEL™ called GASLFP. At $\alpha_o = 0$ and $\alpha_N = 1$, simplified Equations 9.24, 9.25, 9.26, and 9.27 are also provided.

Using Equation 9T12, the slip depth x_p for a lateral H may be obtained iteratively using a purposely designed macro (e.g., in GASLFP). Conversely, slip depth x_p may also be assigned to estimate the load H. Either way, normalized slip depth $\bar{x}_p$ is then calculated with the λ, which allows the ground-level deflection, w_g, and the $\psi(\bar{x}_p)$ to be calculated. The latter in turn permits calculation of the maximum bending moment, M_{max}, and its depth, x_{max}. The calculation is repeated for a series of H or assumed x_p, each offers a set of load H, deflection w_g, bending moment M_{max}, and its depth x_{max}, thus the H-w_g, M_{max}-w_g, and x_{max}-M_{max} curves are determined. These calculations are compiled into GASLFP and are used to gain numerical values of the current solutions presented subsequently (except where specified). For comparison, the predictions using simplified expressions are sometimes provided as well.

With available measured data, the three input parameters N_g, N_{co}, and n (or A_L, α_o, and n) may be deduced by matching the predicted w_g, M_{max}, and x_{max} (using GASLFP) across all load levels with the corresponding measured data, respectively. They may also be deduced using other expressions shown previously, for instance, Equation 9.15 for rotation, the solutions for a shear force profile, or displacement profile (see Chapter 12, this book). The values obtained should capture the overall pile–soil interaction rather than some detailed limiting force profile (in the case of a layered soil). The back-estimated parameters will be unique, so long as three measured curves are available (Chapter 3, this book, option 6). However, if only one measured (normally w_g) response is available, N_g may be back-estimated by taking N_{co} as 0 ~ 4 and n as 0.7 and 1.7 for clay and sand, respectively. Should two measured (say w_g and M_{max}) responses be known, both N_g and n may be back-estimated by an assumed value of N_{co}. Measured responses encompass the integral effect of all intrinsic factors on piles in a particular site, which is encapsulated into the back-figured LFP. The LFP obtained for each pile may be compared, allowing a gradual update of nonlinear design.

Example 9.3 Validation against FEA for piles in layered soil

Yang and Jeremić (2002) conducted a 3-D finite element analysis (FEA) of a laterally loaded pile. The square aluminium pile of 0.429 m in width, 13.28 m in length, and with a flexural stiffness E_pI_p of 188.581 MN-m^2, was installed to a depth of 11.28m in a deposit of clay-sand-clay profile and sand-clay-sand profile, respectively. The pile was subjected to lateral loads at 2 m above ground level. The clay-sand-clay profile refers to a uniform clay layer that has a uniform interlayer of sand over a depth of (4~8)d (or 1.72~3.44 m, d = width of the pile). Conversely, the clay layer was sandwiched between two sand layers (i.e., the sand-clay-sand profile). In either profile, the clay has an undrained shear strength s_u of 21.7 kPa, Young's modulus of 11 MPa, Poisson's ratio ν_s of 0.45, and a unit weight γ_s' of 13.7 kN/m^3. The medium-dense sand has an internal friction angle ϕ of 37.1°, shear modulus of 8.96 MPa at the level of pile base, $\nu_s = 0.35$, and $\gamma_s' = 14.5$ kN/m^3.

The FE analysis of the pile in either soil profile provided the following results: (1) p-y curves at depths up to 2.68 m; (2) pile-head load (H) and mudline displacement (w_g) relationship; (3) w_g versus maximum bending moment (M_{max}) curve; and (4) profiles of bending moment under ten selected load levels. With the p-y curves, the limiting p_u at each depth was approximately evaluated, thus the variations of p_u with depth (i.e., LFP) was obtained (Chapter 3, this book, option 4) and is shown in Figure 9.1a$_1$ and a$_2$ as *FEA*. Using the bending moment profiles, the depths of maximum bending moment (x_{max}) were estimated and are shown in Figure 9.1c$_1$ and c$_2$.

First, closed-form solutions were obtained for the clay-sand-clay profile. An average shear modulus of the soil $\tilde{G}$ (bar "~"denotes average) was calculated as 759.5 kPa (= $35s_u$; Poulos and Davis 1980). The k and N_p were estimated as 2.01 MPa, and 892.9 kN using Equations 3.50 and 3.51 (Chapter 3, this book), respectively. In light of Chapter 3, option 5 for constructing LFP for a stratified soil, Reese LFP *(C)* and LFP*(S)* (see Table 3.9, Chapter 3, this book) were obtained using properties of the clay and sand, respectively. The p_u of the clay layer over a depth x of (2~4)d was then increased in average by 30 percent due to the underlying stiff sand layer, while the p_u of the sand at z = (4~6)d was reduced by 20 percent due to the overlying weak clay layer (Yang and Jeremić 2005). The overall p_u from the upper to the low layer is fitted visually by $n = 0.8$ (see Figure 9.1). As the maximum x_p (max x_p, gained later) will never reach the bottom clay layer, the bottom layer is not considered in establishing the LFP. The LFP for the pile in the stratified soil is described by $n = 0.8$, $N_g = 6$, and $\alpha_o = 0$.

The CF predictions using the LFP are plotted in Figure 9.1 a_1 through c_1, which compares well with the *FEA* results, in terms of displacement w_g and moment M_{max} (Figure 9.1b_1). The depth of the M_{max}, x_{max} is, however, overestimated by up to 20 percent (Figure 9.1c_1), which is resulted from using an equivalent homogeneous medium (Matlock and Reese 1960). A stratified profile reduces the depth x_{max}. The max x_p (under the maximum pile-head load) reaches 4.63d, which locates at a distance of 3.37d (= 8d – 4.63d) above the bottom layer. The current p_u over x = (1.5~8)d slightly exceeds that obtained using the previous instruction (Georgiadis 1983) for a layered soil (not shown here), and it is quite compatible with the overall trend of FEA result within the max x_p. Interestingly, replacing the current Guo LFP with the Reese LFP*(C)*, the current solutions still offer good predictions to 30 percent of the maximum load, where the response is dominated by the upper layer.

Next, the current solutions were conducted for the piles in the sand-clay-sand profile. Shear modulus G was equal to 1.206 MPa by averaging the sand modulus at mid-depth (= 0.5 × 8.96 MPa) and the clay modulus (= 0.759 MPa), and thus k = 3.33 MPa and N_p = 5.731MN. Again, both Reese LFP*(S)* (also Broms LFP) and Reese LFP*(C)* were ascertained. Importantly, the effect of the bottom sand layer on LFP is considered in this case, as the interface (located at $x = 8d$) is less than 2d from the max x_p of 6.75d obtained subsequently. The p_u was maintained similar to Broms LFP over x = (0~4)d and reduced approximately by 20% over x = (6~8)d. For instance, $p_u/(\gamma_s' d^2)$ at $x = 8d$ was found to be 70.0. An overall fit to the p_u of the top layer and the point (70, 8) offers the current Guo LFP described by $n = 0.8$, $N_g = 16.32$ [$= 14.5 \times \tan^4(45° + 37.1°/2)$], and $\alpha_o = 0$. The current predictions are presented in Figure 9.1a_2 through c_2. They agree well with the FE analysis in terms of the mudline displacement w_g and the maximum bending moment M_{max} (Figure 9.1b_2). Only the depth of the M_{max} was overestimated (Figure 9.1c_2) at $H < 270$ kN, for the same reason noted for the clay-sand-clay profile. The max x_p reaches 6.75d at $H = 400$ kN. The Guo LFP is close to the overall trend of that obtained from FEA within the max x_p.

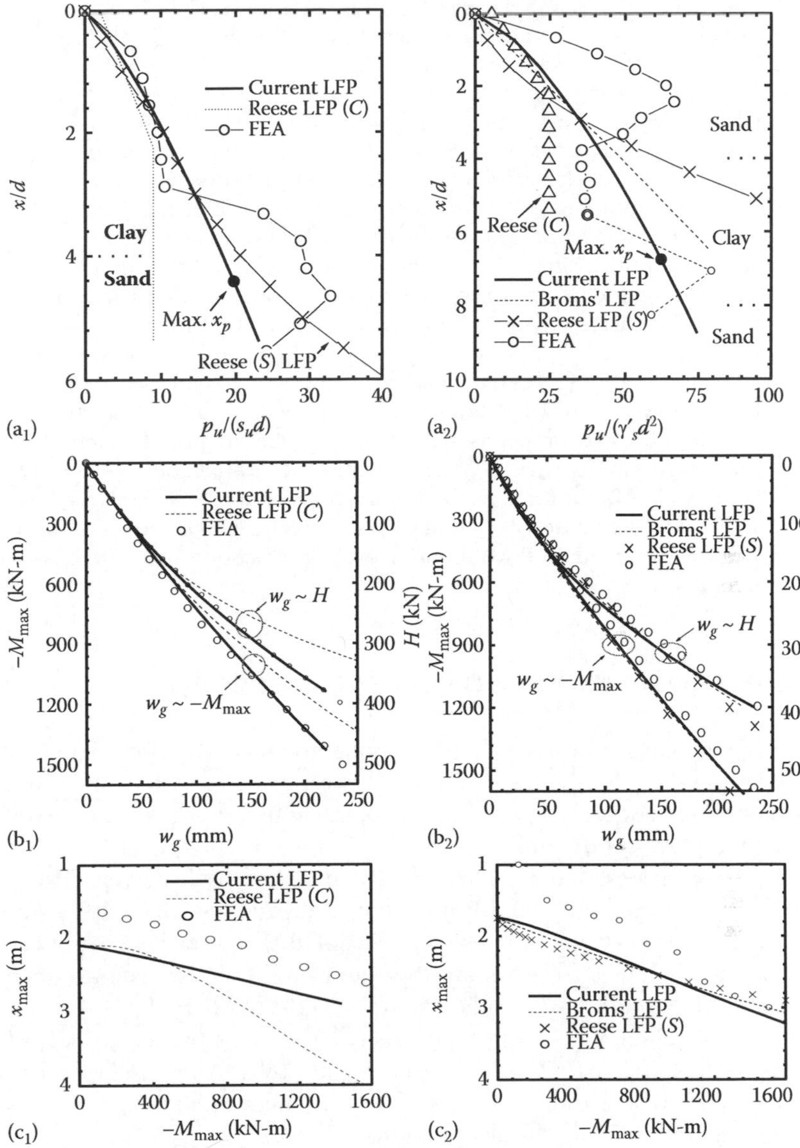

Figure 9.1 Comparison between the current predictions and FEA results (Yang and Jeremić 2002) for a pile in three layers. (a_i) p_u. (b_i) H-w_g and w_g-M_{max}. (c_i) Depth of M_{max} (i = 1, clay-sand-clay layers, and i = 2, sand-clay-sand layers). (After Guo, W. D., *Computers and Geotechnics* 33, 1, 2006.)

With the pile in sand-clay-sand profile, at the depth of $4d$ (of clay-sand interface), the w_p is calculated as 25.2 mm (= 9 × 21.7 × 0.429/3.33) using the clay strength; or as 39.7mm (– 85.71 × (4 × 0.429)$^{0.8}$/3330) using the sand properties and A_L = 85.71 kN/m$^{1.8}$. The pile–soil relative slip occurred in the clay before it did in the overlying sand. This sequence is opposite to the assumption of slip progressing downwards. This incomparability does not affect much the overall good predictions, which indicate that response of the pile is dominated by the overall trend of the LFP within the max x_p and the sufficient accuracy of the LFP for the stratified soil; and the limited impact of using the sole x_p to represent a transition zone that should form for a nonlinear p-y curve (Yang and Jeremić 2002). Initially, a "deep" layer exceeding a depth of $8d$ may be excluded in gaining a LFP. Later, the sum of $2d$ and the max x_p should be less than the depth (of $8d$), otherwise the "deep" layer should be considered in constructing the LFP. These conclusions are further corroborated by the facts that *Broms LFP* or *Reese LFP(S)* also offer excellent predictions up to a load of 400 kN, but for a gradual underestimation of the displacement w_g and the moment M_{max} at higher load levels. A 10% reduction in the gradient of the LFP would offer better predictions against the FEA results.

9.3 SLIP DEPTH VERSUS NONLINEAR RESPONSE

Nonlinear responses of a laterally loaded pile are generally dominated by the LFP within the maximum slip depth. The responses of $\bar{H}$, $\bar{w}_g$, $\bar{M}_{max}$, and $\bar{x}_{max}$ are obtained concerning $n = 0.5 \sim 2.0$, $\alpha_o\lambda < 0.3$, and $\bar{x}_p < 2.0$, which are related to practical design.

Figure 9.2 indicates that (1) $\bar{H}$ normally increases with decrease in n, and/or increase in $\alpha_o\lambda$ ($\bar{x}_p$ < 2.0) and $\bar{H}$ < 2; (2) $\bar{w}_g$ = ~20$\bar{H}$ (< 50); for instance, at $\bar{x}_p$ = 2.0 (n = 0.5, $\alpha_o\lambda$ = 0), it is noted that $\bar{H}$ = 1.38 and $\bar{w}_g$ = 21.1, respectively; (3) $\bar{w}_g$ follows an opposite trend to $\bar{H}$ in response to n and $\alpha_o\lambda$; (4) $\bar{M}_{max}$ and $\bar{w}_g$ are similar in shape to the load $\bar{H}$; (5) $\bar{M}_{max}$ < 5.0 and $\bar{x}_{max}$ < 3.0; (6) at $\bar{x}_p$ = 2 and n = 1, as $\alpha_o\lambda$ increases from 0 to 0.2, $\bar{H}$ increases by 20% (from 1.47 to 1.77) and $\bar{M}_{max}$ by 13% (from 1.68 to 1.90); and (7) $\bar{w}_g$ and $\bar{M}_{max}$ are more susceptible to $\bar{x}_p$ than $\bar{H}$ and $\bar{x}_{max}$ are, as demonstrated in Table 9.6, discussed later.

9.4 CALCULATIONS FOR TYPICAL PILES

9.4.1 Input parameters and use of GASLFP

The closed-form solutions or GASLFP (Guo 2006) are for an infinitely long pile (i.e., pile length > a critical length, L_{cr} in elastic zone), otherwise

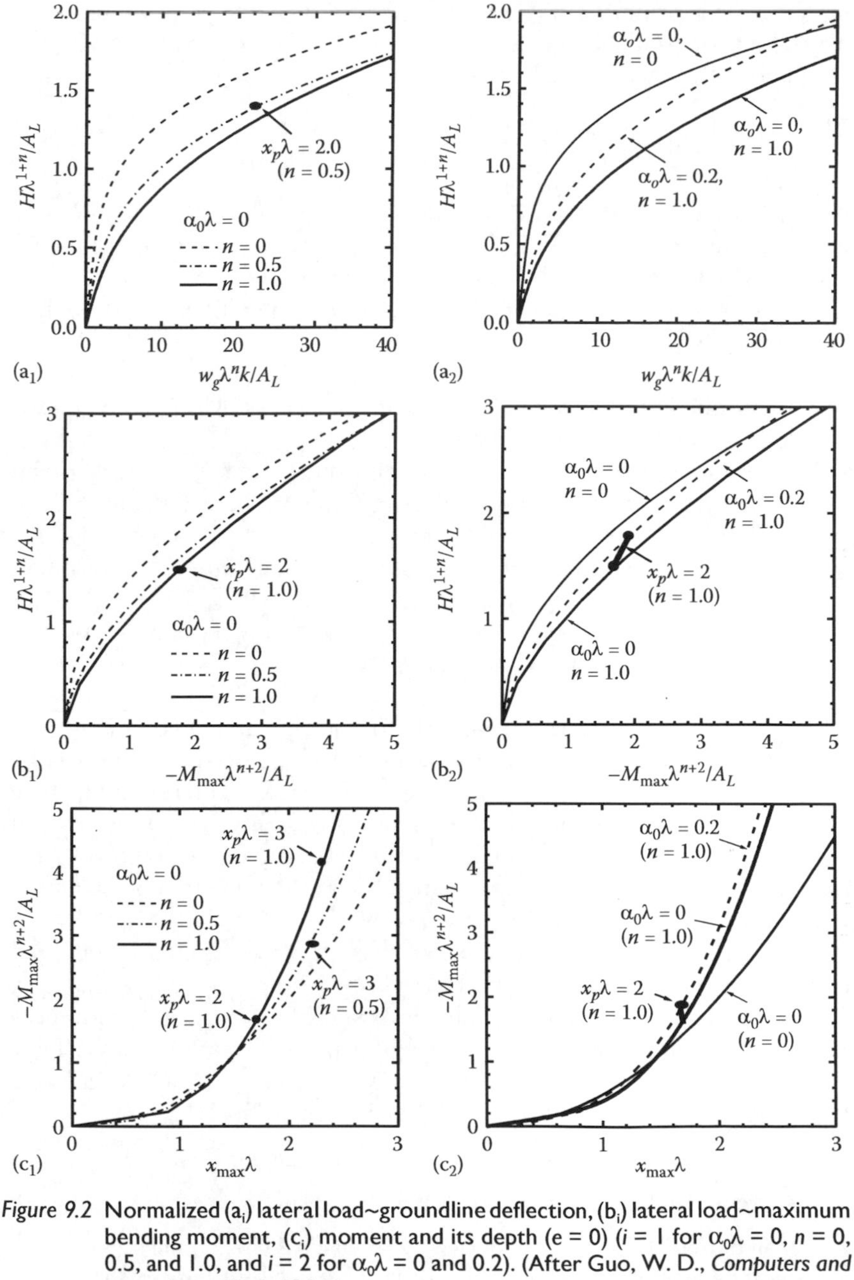

Figure 9.2 Normalized (a_i) lateral load~groundline deflection, (b_i) lateral load~maximum bending moment, (c_i) moment and its depth (e = 0) (i = 1 for $\alpha_0\lambda = 0$, n = 0, 0.5, and 1.0, and i = 2 for $\alpha_0\lambda$ = 0 and 0.2). (After Guo, W. D., *Computers and Geotechnics* 33, 1, 2006.)

the solutions for rigid piles (Guo 2008) should be employed. A number of expressions are available for determining the L_{cr}. To be consistent with the current model, Equation 9.59 is used (Guo and Lee 2001):

$$L_{cr} = 1.05d(E_p / G)^{0.25} \tag{9.59}$$

With a pile–soil relative stiffness E_p/G between 10^2 and 10^5 (commonly seen), L_{cr} is equal to (3.3~18.7)d. Over the length L_{cr}, using an average G, it is recursive to determine L_{cr} using Equation 9.59. The procedure for conducting a prediction for piles in clay (e.g., using GASLFP; Guo 2006) may be summarized as

1. Input pile dimensions (d, L, t), flexural stiffness of E_pI_p (and equivalent E_p), loading eccentricity, e.
2. Calculate shear modulus G using an average s_u ($\tilde{s}_u$) over the critical length L_{cr} estimated iteratively using Equation 9.59 (assuming L_{cr} = 12d initially, and a Poisson's ratio, ν_s).
3. Compute the k and the N_p by substituting the G into Equations 3.50 and 3.51, respectively (Chapter 3, this book).
4. Estimate $\tilde{s}_u$ over a depth of 8d for w_g = 0.2d (or a depth of 5d for w_g= 0.1d) and the parameters α_o, n, and N_g.

Likewise, the procedure is applicable to piles in sand. The modulus G, however, is averaged initially over a depth of 14d (guessed L_{cr}) and may be correlated with SPT (Chapter 1, this book). The process can be readily done (e.g. in GASLFP). The input parameters such as α_o, n, and N_g are discussed amply in later sections (e.g. n = 0.7, α_o = 0.05 ~ 0.2 m, and N_g = 0.6 ~ 3.2 for piles in clay), which may be used herein. They may be deduced by matching prediction with measured data.

The studies on four typical piles using the current solutions are presented next, which are installed in a two-layered silt, a sand-silt layer, a stiff clay, and a uniform clay, respectively. The best predictions or match with measured data are highlighted in bold, solid lines in the pertinent figures. They yield parameters N_g, N_{co}, and n (thus the LFPs) and are provided in Table 9.3. These predictions are elaborated individually in this section.

Example 9.4 Piles in two-layered silt and sand-silt layer

Kishida and Nakai (1977) reported two individual tests on single piles A and C driven into a two-layered silt and a sand-silt layer, respectively. Each pile was instrumented to measure the bending strain down the depth.

Pile A was 17.5 m in length, 0.61 m in diameter, and had a flexural stiffness E_pI_p of 298.2 MN-m². The pile was driven into a two-layered

Table 9.3 Parameters for the "bold lines" predictions in Figures 9.3 and 9.4

Figs.	$\tilde{s}_u$ (kPa) or (N)[a]	α_c or (G/N)[b]	E_pI_p (MN-m^2)	e (m)	L (m)	r_o (m)	N_g/N_{co}	$\bar{N}_c/\alpha_o\lambda$	n
9.3	15	121	298.2	0.1	17.4	0.305	4.53/3.28	5.8/.08	0.5
9.3	(12)[a]	(640)[b]	169.26	0.2	23.3	0.305	8.3/4.67	____	1.0
9.4	153	545.1	493.7	0.31	14.9	0.32	0.85/0.0	1.67/0	1.5

Source: Guo, W. D., *Computers and Geotechnics*, 33, 1, 2006.

[a] values of SPT blow count.

[b] *G/N* (shear modulus over N) in kPa.

silt, with a uniform undrained shear strength (s_u) of 15 kPa to a depth of 4.8 m, and 22.5 kPa below the depth. Assuming $G = 121s_u$ (s_u = 15 kPa) (Kishida and Nakai 1977), L_c/r_o was estimated as 26.06, and the pile was classified as infinitely long.

The shear modulus G of 1.82 MPa offers E_p/G^* of 18,588. The factor γ_b, modulus of subgrade reaction, k, and normalized fictitious tension, $N_p/(2E_pI_p)$, were obtained as 0.0835, 5.378 MPa, and 0.0169 using Equations 3.54, 3.50, and 3.51 (Chapter 3, this book), respectively. The stiffness factors α and β were calculated using Equation 9.8 and the definition of the λ. All of these values for the elastic state are summarized in Table 9.4. The p_u for the upper layer (s_u = 15 kPa) was determined using *Reese LFP(C)*. Likewise, for the lower layer ($x \geq 7.87d$ = 4.8 m), the p_u was taken as a constant of $9d \times 22.5$ kN/m [with $p_u/(s_ud) = 13.5$, taking s_u (upper layer) = 15 kPa as the normalizer]. Using Chapter 3, this book, option 5, the overall LFP for the two-layered soil is found close to the *Reese LFP(C)* near the ground level and passes through point (13.5, 7.87), as indicated by "$n = 0.5$" in Figure 9.3a$_1$. The LFP is thus expressed by $n = 0.5$, $A_L = 53.03$ kPa/m$^{0.5}$, and α_o = 0.32 m in Equation 9.2 (or $N_g = 4.53$, $N_{co} = 3.28$, and $n = 0.5$). The higher strength s_u of 22.5 kPa of the lower layer renders 0~50% (an average of 25%) increase in the p_u over $x = (4\sim7.87)d$, which resembles the effect of an interlayer of sand on its overlying clay deposit (Murff and Hamilton 1993) (Yang and Jeremić 2005).

The set of parameters offer excellent predictions of the pile response, as shown in Figure 9.3. Critical responses obtained are shown in Table 9.4, including minimum lateral load, H_e, to initiate the slip at ground level; lateral load, H^{**}, for the slip depth touching the second soil layer, as shown in Figure 9.3; maximum imposed lateral load, H_{max}; and slip depth, x_p, under the H_{max}.

Those critical loads (H_e, H^{**}, and H_{max}) and the slip depth are useful to examine the depth of influence of each soil layer. Pile–soil relative slip occurred along the pile A at a low load, H_e of 54.6 kN, and developed onto the second layer (i.e., x_p = 4.8 m) at a rather high load, H^{**} of 376.9 kN. Influence of the second layer on the pile A is well captured through the LFP. This study is referred to Case I. Further analysis indicates that as n increases from 0.5 to 0.7, N_g drops from 4.53 to 3.2,

Table 9.4 Parameters for the "bold lines" (Examples 9.4 and 9.5)

Piles	E_p/G^*	γ_b	k (MPa)	$N_p/(2E_pI_p)$	α (m^{-1})	β (m^{-1})	λ (m^{-1})
A	18,587.7	.0856	5.378	.0169	.2750	.2422	.2591
C	2,739.2	.1382	26.400	.0674	.4808	.4047	.4444

	Input parameters			*Predictions*			
Plastic	N_g *and* ϕ	A_L (kN/m^{n+1})	α_o (m)	H_e (kN)	H^{**} (kN)	H_{max} (kN)	$x_p(\bar{x}_p)$ at H_{max}
A	$N_g = 4.53$	53.03	.320	54.57	376.9	393.	8.36*d*(1.32)
C	$\phi' = 28°$	83.58	.563	48.90	244.4	440.	4.72*d*(1.28)

Source: Guo, W. D., *Computers and Geotechnics*, 33, 1, 2006.

Note: G = 1.82 and 7.68 MPa for piles A and C, respectively. If $N_p = 0$, then $\alpha = \beta = \lambda$.

$H_e = H$ at $x_p = 0$; $H_e^{**} = H$ at the slip depth x_p shown in Figure 9.3. $\gamma_s' = 16.5$ (kN/m³).

α_o reduces from 0.31 to 0.1, and G/s_u increases from 109 to 131 in order to gain good match with measured data. The LFP is well fitted using Hensen's expression with $c = 6$ and $\phi = 16°$.

9.4.1 Hand calculation (Case II)

The simplified Equations 9.23 through 9.28 (referred to as "S. Equations") were used to predict the response of the pile A (Case II), as indicated in Table 9.5. The LFP (assuming $\alpha_o = 0$) reduces to the triangular dots shown in Figure 9.3a₁. The calculation was readily undertaken in a spreadsheet. It starts with a specific x_p followed by $\bar{x}_p$, $\bar{H}$, $\bar{w}_g$, and $\bar{M}_{max}$ as shown in Table 9.6. This calculation is illustrated for two typical $x_p = 1$ m, and $x_p = 5$ m (for $e = 0$ m). With $\lambda = 0.2591$/m (Table 9.4), $\bar{x}_p$ (at $x_p = 1$ m) is equal to be 0.2591. Using $n = 0.5$ (Table 9.3) and $e = 0$, $\bar{x}'$ is computed as 1.118 from Equation 9.28, thus Equations 9.26, 9.24, and 9.25 offer the following:

$$\psi(\bar{x}_p) = \frac{-2\times1.5\times0.2591^2 + 2.5\times1.5}{[2\times0.2591^2 + 2\times2.5\times0.2591 + 2.5]\times1.5} = 0.6020$$

$$\frac{H\lambda^{1.5}}{A_L} = \frac{0.5\times0.2591^{0.5}[1.5\times2.5 + 2\times0.2591\times(2.5+0.2591)]}{1.2591\times1.5\times2.5} = 0.2792$$

and

$$\frac{w_g k\lambda^{0.5}}{A_L} = \frac{2}{3}0.2591^{3.5}\frac{2\times0.2591^2 + (2\times0.5+10)\times0.2591 + 0.5^2 + 9\times0.5 + 20}{1.2591\times2.5\times4.5}$$
$$+ \frac{(2\times0.2591^2 + 2\times0.2591 + 1)\times0.2591^{0.5}}{1.2591} = 0.6796$$

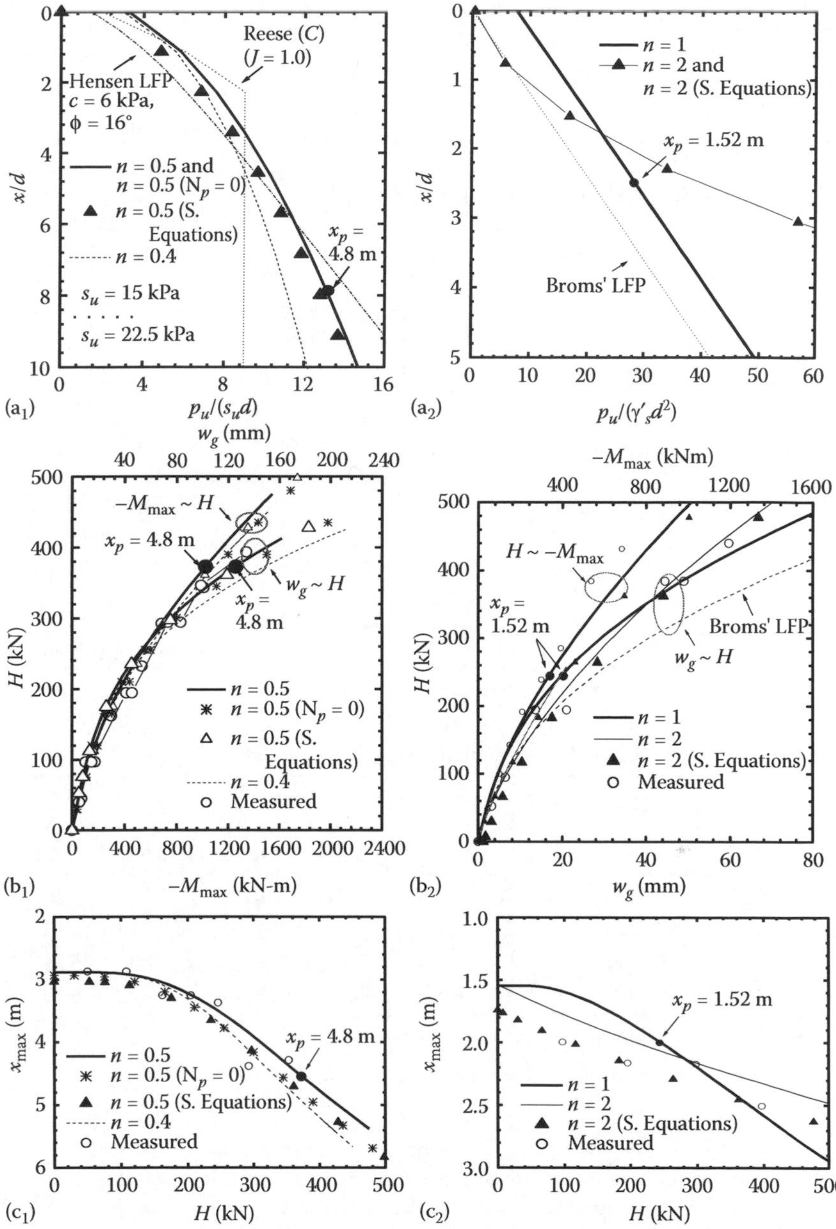

Figure 9.3 Calculated and measured (Kishida and Nakai 1977) response of piles A and C. (a_i) p_u. (b_i) H-w_g and H-M_{max}. (c_i) Depth of M_{max} (i = 1, 2 for pile A and pile C). (After Guo, W. D., *Computers and Geotechnics* 33, 1, 2006.)

Table 9.5 Sensitivity of current solutions to $k(N_p)$, LFP, and e

Cases	*Limiting force profiles*	*Remarks*	*References*
		Pile A	*Figure 9.3*
I	N_g, N_{co}, and n provided in Table 9.3	Using Guo LFP	$n = 0.5$
II	$N_g = 4.53$, $N_{co} = 0$, and $n = 0.5$	Using $e = N_p = 0$	$n = 0.5$ (S. Equations)
III	N_g, N_{co}, and n provided in Table 9.3	$N_p = 0$	$n = 0.5$ ($N_p = 0$)
IV	$N_g = 4.79$, $N_{co} = 3.03$, and $n = 0.4$	Same total force as that using Reese *LFP(C)* over the max x_p	$n = 0.4$
		Pile C	*Figure 9.3*
I	N_g, N_{co}, and n provided in Table 9.3	Using Guo *LFP*	$n = 1$
II	$N_g = 3K_p$, $N_{co} = 0$, and $n = 1.0$	Using *Broms LFP*	*Broms LFP*
III	$N_g = 8.3$, $N_{co} = 0$, and $n = 2$	$e = 0.31$m, $G = 7.68$ MPa	$n = 2$
VI		With $e = N_p = 0$.	$n = 2$ (S. Equations)

Source: Guo, W. D., *Computers and Geotechnics,* 33, 1, 2006.

Table 9.6 Response of pile A using simplified expressions and e = 0 (Figure 9.3)

x_p *(m)*	$\bar{x}_p$	$\frac{H\lambda^{n+1}}{A_L}$	$\frac{w_g k\lambda^n}{A_L}$	$-\frac{M_{max}\lambda^{n+2}}{A_L}$	H *(kN)*	w_g *(mm)*	$-M_{max}$ *(kN-m)*	x_{max} *(m)*
1	0.2591	0.2792	0.6797	0.1079	112.3	13.2	167.4	3.09
3	0.7774	0.5851	2.3435	0.3218	235.2	45.4	501.1	3.65
5	1.2957[a]	0.8982	6.1526	0.6574	361.1	119.2	1020.0	4.71
8	2.0730[a]	1.4187	20.327	1.3716	570.4	393.8	2185.1	6.38

Source: Guo, W. D., *Computers and Geotechnics*, 33, 1, 2006.

[a] $\bar{x}_p > 1.118$ and M_{max} in the upper plastic zone (i.e., $x_{max} < x_p$).

As $\bar{x}_p < \bar{x}'$ [$\psi(\bar{x}_p) > 0$], the maximum bending moment occurs below the slip depth. Substituting $\psi(\bar{x}_p) = 0.602$, and $\beta = \lambda$ into Equation 9.21, z_{max} is calculated as 2.091 m, $tan(\lambda z_{max}) = 0.602$, and $cos(\lambda z_{max}) = 0.8567$. Also Equation 9.27 offers

$$C_1(\bar{x}_p) = \frac{0.5\times(2.5+2\times0.2591)\times0.2591^{1.5}}{2.5\times1.2591} = 0.06323$$

$$C_2(\bar{x}_p) = 0.5\times0.2591^{0.5} = 0.25451$$

The values of C_1 and C_2 allow the normalized moment to be estimated using Equation 9T15 as

$$-M_B(z_{max})\lambda^{n+2}/A_L = e^{-0.54188}(0.06323+0.25451\times 0.602)$$
$$\times 0.8567 = 0.10785$$

Assigning x_p = 5 m (e = 0), $\bar{x}_p$ is equal to 1.2957, and, thus, $H\lambda^{1.5}$/ A_L = 0.8982 and $w_g k\lambda^{0.5}/A_L$ = 6.1526. As $\bar{x}_p > \bar{x}'$, the depth of x_{max} is calculated directly from Equation 9.22 as

$$x_{max} = (1.5\times 0.8982)^{1/1.5}/0.2591 = 4.71(m)$$

The moment M_o (=$H \times e$) is zero. Thus, the normalized maximum moment is estimated from Equation 9T14 as

$$-M_{max}\lambda^{n+2}/A_L = (1.5\times 0.8982)^{2.5/1.5}/2.5 = 0.6574$$

With N_p = 0 and α_0 = 0, the predicted responses are indicated by "(S. Equations)" in Figure 9.3. They are close to those bold lines obtained earlier ($\alpha_0 \neq 0$, $N_p \neq 0$).

9.4.2 Impact of other parameters (Cases III and IV)

To examine the effect of the parameters on the predictions against Cases I and II, the following investigations (Table 9.5) are made:

Case III: Taking N_p as 0, the responses obtained are shown in Figure 9.3 as "n = 0.5 (N_p = 0)," which are slightly softer (higher w_g) than, indicating limited impact of the elastic coupled interaction.

Case IV: A new LFP of n = 0.4 (with N_g = 4.79, N_{co} = 3.03, and n = 0.4) is utilized, which offers a similar total resistance on the pile to that from the *Reese LFP(C)* profile to the maximum x_p of 8.36d. This LFP offers very good predictions against the measured data, indicating limited impact of the n value on the prediction. The n value may be gauged visually for layered soil.

Example 9.5 Pile C in sand-silt layer

Pile C with L = 23.3 m, d = 0.61 m, and E_pI_p = 169.26 MN-m^2, was driven through a sand layer (in depth z = 0~15.4 m) and subsequently into an underlying silt layer with a s_u of 55 kPa. The blow count of SPT, N of the sand layer was found as: 12 (z = 0~11.0 m), 8 (11.0~13.8 m), and 16 (13.8~15.4 m), respectively. The effective unit weight γ_s' was 16.5 kN/m^3. The shear modulus, G, and angle of friction of the sand, ϕ', were estimated using the blow count by (Kishida and Nakai 1977) G = 640N kPa, and $\phi' = \sqrt{8(N-4)} + 20$, respectively. They were estimated to be 7.68 MPa and 28°, respectively, with N = 12. This offers L_c/r_o of 15.8, and the pile is classified as infinitely long. With the effective pile length L_c located within the top layer, the problem becomes a pile in a single layer.

As with pile A (Example 9.4), the parameters for elastic state were estimated and are shown in Table 9.4. An apparent cohesion was reported in the wet sand near the ground level around the driven pile, which is considered using N_{co} = 4.67 (of similar magnitude to $\tilde{N}_c$). As N_g = 8.31 (Equation 3.66, Chapter 3, this book) and n = 1.0, A_L was computed as 83.58 kPa/m and α_o as 0.563 m. The LFP is then plotted as n = 1 in Figure 9.3a$_2$. The G and LFP offer close predictions of the pile responses to the measured data (Figure 9.3b$_2$ and c$_2$). The responses at a typical slip depth of 2.5d (= 1.52 m) are also highlighted. The critical values are tabulated in Table 9.4, including H_e = 48.9 kN and x_p = 2.88 m upon H_{max} = 440.1 kN.

In the study, the n = 1 and N_{co} > 0 are divergent from n = 1.3 ~ 1.7 and N_{co} = 0 normally employed for sand, which are examined (Table 9.5) as follows:

- Case II: Ignoring the apparent cohesion, the n = 1.0 LFP then reduces to the *Broms* LFP (Figure 9.3a$_2$), which incurs overestimation of displacement (Figure 9.3b$_2$) against measured data. The *Reese* LFP*(S)* (with depth corrections) happens to be nearly identical to the *Broms*' one, which results in overestimation of w_g, as noted previously (Kishida and Nakai 1977) and maximum bending moment, M_{max}, and depth of the M_{max} (not shown).
- Case III: Use of α_o (N_{co}) = 0 and n = 2 leads to a lower limiting force than that derived from the n = 1 LFP (see Figure 9.3a$_2$) above a depth of 1.8d and vice versa. This slightly overestimates the deflection, Figure 9.3b$_2$, and bending moment, Figure 9.3c$_2$, up to a load level of 380 kN, to which the total limiting force in the slip zone reaches that for the n = 1 case, Figure 9.3a$_2$.
- Case IV: By changing e to zero in Case III, the predictions using the simple expressions are slightly higher than those obtained earlier.

Example 9.6 A pipe pile tested in stiff clay

Reese et al. (1975) reported a test on a steel pipe pile in stiff clay near Manor. The pile was 14.9 m in length, 0.641 m in diameter, and had a moment of inertia, I_p, of 2.335 × 10^{-3} m^4 (thus E_pI_p = 493.7 MN-m^2). The undrained shear strength s_u increases linearly from 25 kPa at the ground level (z = 0) to 333 kPa at z = 4.11 m. The submerged unit weight γ_s' was 10.2 kN/m^3. The test pit was excavated to 1 m below the ground surface. The average undrained shear strength over a depth of 5d and 10d was about 153.0 and 243 kPa, respectively. Lateral loads were applied at 0.305 m above the ground line. The instrumented pile offered measured responses of H-w_g, H-M_{max}, and M_{max}-x_{max}. The α_c (=G/s_u) was back-estimated as 545.1 by substituting k of 331.3 MPa (Reese et al. 1975) into Equation 3.50 (Chapter 3, this book). With L_c/r_o = 10.8, the pile was infinitely long. The p_u profile was estimated by using Equation 3.67 (Chapter 3, this book) with J = 0.92 and modified using the depth factor (Reese et al. 1975). The LFP thus obtained was fitted using N_g = 0.961, N_{co} = 0.352, and n = 1.0. It is plotted in Figure 9.4a$_1$

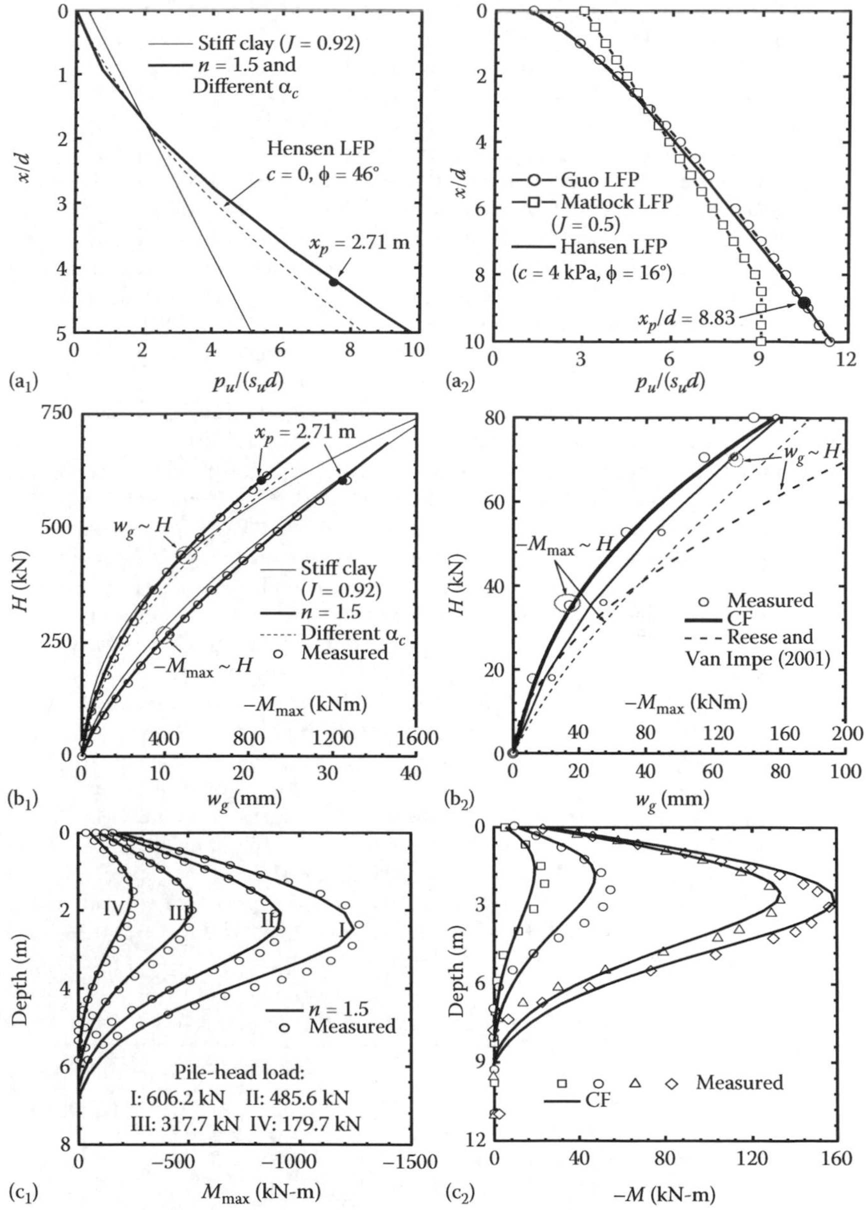

Figure 9.4 Calculated and measured response of: (a_i) p_u, (b_i) H-w_g, and H-M_{max}, and (c_i) bending moment profile [i = 1, 2 for Manor test (Reese et al. 1975), and Sabine (Matlock 1970)]. (After Guo, W. D., *Computers and Geotechnics* 33, 1, 2006; Guo, W. D. and B. T. Zhu, *Australian Geomechanics* 40, 3, 2005a.)

as "Stiff clay (J = 0.92)" in which an average s_u of 153 kPa was used to gain the $p_u/(ds_u)$. This LFP renders good predictions of the pile responses to a load level of 450 kN against the measured data (Figure 9.4b_1 and c_1).

In view of the similarity in the strength profile between stiff clay and sand, an "n = 1.5" was adopted to conduct a new back-estimation (with excellent match with measured data), which offered $N_g = 0.854$, and $N_{co} = 0$ (n = 1.5). This verifies the use of "n = 1.5 ~ 1.7" for a linear increasing strength profile. The max x_p of 2.71 m implies the rational of the back-figured LFP. Bending moment profiles were computed for lateral loads of 179.7, 317.7, 485.6, and 606.2 kN, respectively, using Equations 9T3 and 9T7. They exhibit, as depicted in Figure 9.4c_1, an excellent agreement with the measured profiles, despite overestimating the depth of influence by ~1 m (compared to the measured ones).

Assuming a constant k in the current solutions, the limiting deflection w_p should increase with depth at a power n of 1.5 from zero at mudline, compared to a conservative, linearly increasing k (=150x, MPa) (adopted in the COM624P) (Reese et al. 1975). The average k over the maximum slip depth of 2.71 m is 203.3 MPa (α_c = 348.4). Using this k value, the predicted H-w_g is shown in Figure 9.4b_1 as "Different α_c." Only slight overestimation of mudline deflection is noted in comparison with those obtained from the "n = 1.5" case utilizing k = 331.3 MPa (see Table 9.7). Thus, the effect of k on the predictions is generally not obvious. The pronounced overestimation using characteristic load method (based on COM624P) for this case may thus be attributed to the LFP (Duncan et al. 1994).

The sensitivity of this study to the parameters was examined against the two measured relationships of H-M_{max} and H-w_g shown in Figure 9.4. With $\nu_s = 0.3$ and $\alpha_0 = 0.1$ m, values of n = 1.7 and $N_g = 0.6$ were deduced by matching GASLFP with the measured relationships. The value of G of 76.5 MPa was taken as 500 times the average shear strength s_u within a depth of 5d. Increases in the n from 1.5 to 1.7 and the α_0 from 0 to 0.1 lead to a drop of N_g from 0.854 to 0.6. A good prediction by Reese et al. (1975) is also noted.

Table 9.7 Effect of elastic parameter α_c on the Manor test (Figure 9.4)

	Input parameters				*Calculated elastic parameters*		
Cases	n	$\alpha_o(m)$	$A_L(kN/m^{n+1})$	α_c	E_p/G^*	k *(MPa)*	$N_p/(2E_pI_p)$
Stiff clay[a]	1.0	0.234[b]	147.0	545.1	551.20	331.28	0.1683
n = 1.5[c]	1.5	0	163.3	545.1	551.16	331.31	0.1683
Different α_c	1.5	0	163.3	348.4	862.44	203.30	0.1234

Source: Guo, W. D., *Computers and Geotechnics, 33,* 1, 2006.

[a] LFP for stiff clay (J = 0.92)
[b] Corresponding H_e is 24.5 kN
[c] Maximum x_p = 4.23d and $\bar{x}_p$ = 1.73

For all cases: $\gamma_s' = 10.2$ kN/m³

At the maximum load H of 596 kN, the calculated M_{max} was 1242.3 kN-m (less than the moment of initial yield of 1,757 kN-m) and the maximum x_p was 2.71 m (4.24d). Within the maximum x_p, the LFP is remarkably different from the Matlock LFP obtained using J = 0.25, but it can be fitted well by using Hansen LFP along with two hypothetical values of c = 0 kPa and ϕ = 46°. The similar shape of the current Guo LFP to that for sand was attributed to the similar increase in the soil strength with depth (Guo and Zhu 2004). The critical length L_{cr} was found to be 3.56 m (≈5.5d, close to the assumed 5d used for estimating the G). Thus the pile was "infinitely long" and the use of GASLFP was correct.

Example 9.7 A pile in uniform clay (CS1)

Matlock (1970) presented static and cyclic, lateral loading tests, respectively, on two steel pipe piles driven into soft clay at Sabine, Texas. Each pile was 12.81 m in length, 324 mm in diameter, 12.7 mm in wall thickness, and had a bending stiffness of 31.28 MN-m^2 (i.e., E_p = 5.79 × 10^4 MPa). The cyclic loading test can be modeled well by the method described later in this chapter. The static test is analysed here. The pile (extreme fibers) would start to yield at a bending moment of 231 kN-m and form a fully plastic hinge at 304 kN-m (Reese and Van Impe 2001). As the critical bending moment is higher than the value of M_{max} (under maximum load) calculated later, and crack would not occur. The bending stiffness is taken as a constant in the current solutions (see Chapter 10, this book). The Sabine Pass clay was slightly overconsolidated marine deposit with s_u = 14.4 kPa, and a submerged unit weight γ_s' of 5.5 kN/m^3. Lateral loads were applied at 0.305 m above the ground line. The measured H-w_g curve and bending moment profiles for four typical loading levels are plotted in Figure 9.4b$_2$ and c$_2$.

The shear modulus G was 1.29 MPa (=90s_u), ν_s = 0.3, and k/G = 2.81. As E_p = 5.79 × 10^4 MPa, L_{cr} is equal to 4.96 m (15.3d) (i.e., an "infinitely long" pile). The best match between the predicted and the measured three responses was obtained (see Figure 9.4), which offered n = 1.7, α_o = 0.15 m, and N_g = 2.2. The corresponding Guo LFP is depicted in Figure 9.4a$_2$, along with the Matlock LFP (J = 0.5) (see Table 3.9, Chapter 3, this book). The Guo LFP is well fitted by the Hansen LFP (with a fictitious cohesion c of 4 kPa and an angle of friction ϕ of 16°). The predicted w_g, M_{max} and the moment distribution (using GASLFP) are presented in Figure 9.4b$_2$ and c$_2$, together with the measured data and Reese and Van Impe's (2001) prediction. The CF prediction (GASLFP) is rather close to the measured data compared to that by Reese and Van Impe (2001). The maximum load H of 80 kN would induce a M_{max} of 158.9 kN-m (less than the bending moment of yielding of 231 kN-m, no crack occurred as expected, see Chapter 10, this book) and a maximum slip depth x_p of 2.86 m (8.83d). Within this max x_p, the average limiting force was close to that from the Matlock LFP.

9.5 COMMENTS ON USE OF CURRENT SOLUTIONS

The current solutions (in form of GASLFP) were used to predict response of 32 infinitely long single piles tested in clay and 20 piles in sand due to lateral loads. Typical piles in clay showed that $N_g = 0.3 \sim 4.79$ (clay); $N_{co} = 0 \sim 4.67$ [or $\alpha_o\lambda < 0.8$ (all cases), and $\alpha_o\lambda < 0.3$ (full-scale piles)]; (c) $n = 0.5 \sim 2.0$ with $n = 0.5 \sim 0.7$ for a uniform strength profile, $n = 1.3 \sim 1.7$ for a linear increase strength profile (similar to sand); $\bar{x}_p = 0.5 \sim 1.69$ at maximum loads [or $x_p = (4 \sim 8.4)d$]; and $\alpha_c = 50 \sim 340$ (clay) and 556 (stiff clay).

These magnitudes are consistent with other suggestions for piles in clay. For example, (a) $N_g = 2 \sim 4$ for $n = 0$ (Viggiani 1981); (b) $N_{co} = 2$ for a smooth shaft (Fleming et al. 1992), 3.57 for a rough shaft (Mayne and Kulhawy 1991), and 0.0 for a pile in sand; (c) "$n > 1$" from theoretical solution (Brinch Hansen 1961), and the upper bound solutions for layered soil profiles (Murff and Hamilton 1993); and the values of n tentatively deduced against the reported p_u profiles (Briaud et al. 1983); and finally, (d) $\alpha_c = 80 \sim 140$ (Poulos and Davis 1980); $210 \sim 280$ (Poulos and Hull 1989); $175 \sim 360$ (D'Appolonia and Lambe 1971), $330 \sim 550$ (Budhu and Davies 1988), and the α_c values summarized previously (Kulhawy and Mayne 1990).

Each combination of n, N_g, and N_{co} produces a specific LFP. The existing *Matlock LFP*, *Reese LFP(C)* and *LFP(S)* may work well for relevant piles. They need to incorporate the impact of a layered soil profile (Figure 9.3), an apparent cohesion around a driven pile in sand (Figure 9.3) and so forth. By stipulating an equivalent, homogeneous modulus (elastic state), and the generic LFP (plastic state), the current solutions are sufficiently accurate for analyzing overall response of lateral piles in layered soil, regardless of a specific distribution profile of limiting force. The analysis indicates that $k = (2.7\text{~}3.92)G$ with $\tilde{k} = 3.04G$; $G = (25\text{~}340)s_u$ with $\tilde{G} = 92.3s_u$; and $n = 0.7$, $\alpha_0 = 0.05\text{~}0.2$ m ($\tilde{\alpha}_o = 0.11$ m), and $N_g = 0.6\text{~}4.79$ (1.6) may be used for an equivalent, uniform strength profile.

Study so far on 20 piles installed in sand shows that $k = (2.4\text{~}3.7)G$ with $\tilde{k} = 3.2G$; $G = (0.25\text{~}0.62)N$ (MPa) with $\tilde{G} = 0.5N$ (MPa); and $n = 1.7$, $\alpha_0 = 0$, and $N_g = (0.4\text{~}2.5)\ K_p^2$ for an equivalent uniform sand.

9.5.1 32 Piles in clay

Table 9.8 summarized the properties of the 32 piles in clay, which encompass (1) a bending stiffness E_pI_p of 0.195~13,390 MN-m^2 (largely 14~50 MN-m^2); (2) a diameter d of 90~2,000 mm; and (3) a loading eccentricity of 0.06~10 m. Table 9.9 provides the subsoil properties, such as (1) undrained shear strength s_u of 14.4~243 kPa; (2) a shear modulus G of (25~315)s_u (with $\tilde{G} = 92.3s_u$) (Guo 2006); and (3) the values of k, α_0, and N_g deduced using GASLFP and $n = 0.7$ (but 1.7 for CS2).

Table 9.8 Summary of piles in clay investigated

Cases	*Pile type*	L *(m)*	d *(mm)*	E_p *(10^4MPa)*	E_pI_p *(MN-m^2)*	e*(m)*	*References*	
CS1	Steel pipe pile	12.81	324	5.78	31.28	0.305	Matlock (1970)	Sabine
CS2	Steel pipe pile	15.2	641	5.94	493.7	0.305	Reese et al. (1975)	Single pile
CS3	Steel pipe pile	12.81	324	5.78	31.28	0.0635	Matlock (1970)	Austin
CS4	Open-ended pipe pile	5.537	114.3	7.43	0.623	0.813	Gill (1969)	P1
CS5		6.223	218.9	4.83	5.452	0.813		P2
CS6		5.08	323.8	5.79	31.279	0.813		P3
CS7		8.128	406.4	3.62	48.497	0.813		P4
CS8		5.537	114.3	7.43	0.623	0.813		P5
CS9		6.223	218.9	4.83	5.452	0.813		P6
CS10		5.08	323.8	5.79	31.279	0.813		P7
CS11		8.128	406.4	3.62	48.497	0.813		P8
CS12	Cased concrete pile	35.05	254	5.34	10.905	0.305	Cappozoli (1968)	St. Gabriel
CS13	Steel pipe pile	3.43	160	5.69	1.83	1.17	Wu et al. (1999)	P1
CS14		3.43	90	6.05	0.195	0.28		P3
CS15		5.71	200	7.16	5.62	0.29		P13
CS16		14.0	500	3.19	97.9	0.72		P17
CS17	Cased concrete pile	15.2	244	8.10	14.1	0.38	Long et al. (2004)	P1
CS18				8.10	14.1	0.267		P2a
CS19				8.10	14.1	0.343		P3
CS20				8.10	14.1	0.356		P4

continued

Table 9.8 (Continued) Summary of piles in clay investigated

Cases	*Pile type*	L *(m)*	d *(mm)*	E_p *(10^4MPa)*	E_pI_p *(MN-m^2)*	e *(m)*	*References*	
CS21				8.10	14.1	0.356		P2b
CS22				8.10	14.1	0.356		P6
CS23				8.10	14.1	0.381		P10
CS24				8.10	14.1	0.254		P11
CS25				8.10	14.1	0.406		PT
CS26	Closed-ended pipe pile	13.115	273	5.15	14.04	0.305	Brown et al. (1987)	Single pile
CS27	Steel pipe pile	16.5	406	3.85	514.0	1.0	Price and Wardle (1981)	Single pile
CS28	Closed-ended pipe pile	5.16	305	1.62	686.8	0.201	Committee on Piles Subjected to Earthquake (1965)	Japan
CS29	Steel pipe pile	22.39	1500	1.39	3459.0	10.0	Nakai and Kishida (1982)	Pile B
CS30	Steel pipe pile	40.0	2000	1.71	1.339×10^4	6.77		Pile C
CS31	Steel pipe pile	40.4	1574	1.59	4779.4	0.5	Kishida and Nakai (1977)	Pile B
CS32	Steel pipe pile	17.5	609.6	4.32	292.5	0.1		Pile A
Range of the values		3.43~40.4	90~2,000	1.39~8.10	0.195~13,390	0.06~10		
Average		13.88	421.4	5.65	334.91	0.997		

Source: Guo, W. D. and B. T. Zhu, *Australian Geomechanics*, 40, 3, 2005a.

Table 9.9 Soil properties for 32 piles in clay

Items	*Soil properties*	*Input parameters*						*Derived c and* ϕ			
Case	Soil type	s_u (kPa)	G (MPa)	α_0 (m)	N_g	k/G[a]	G/s_u	c^b (kPa)	ϕ^b	λ	$e\lambda$
CS1	Slightly OC Sabine soft clay	14.4	1.29	0.15	2.2	2.81	90	4	16°	0.4126	0.1258
CS2	OC stiff clay*	153 (243)	76.5	0.1	0.6	3.92	500 (315)	0	46°	0.6242	0.1904
CS3	Austin soft clay	38.3	1.30	0.1	1.5	2.81	34	6	22°	0.4134	0.0262
CS4	Submerged silt clay	37.5	0.94	0.1	0.8	2.70	25	5	18°	1.0046	0.8167
CS5		34.3	1.10	0.1	1.5	3.10	32	6	22°	0.6288	0.5112
CS6		34.3	1.72	0.1	0.7	2.87	50	3	14°	0.4457	0.3623
CS7		34.3	1.20	0.1	1.25	2.89	35	4	18°	0.3657	0.2973
CS8	Dry silt clay	82.5	2.64	0.1	1	2.91	32	11	23°	1.3251	1.0773
CS9[c]		66.5	2.33	0.06	2.5	2.97	35	10	32°	0.7505	0.6102
CS10		64.5	1.93	0.06	0.8	2.90	30	4.5	21°	0.4599	0.3739
CS11		64.5	1.93	0.1	1.2	3.00	30	6	24°	0.4156	0.3379
CS12[d]	Soft to medium clay	28.7	3.16	0.1	2.2	3.02	110	7	22°	0.6839	0.2086
CS13[d]	Shanghai soft clay	40	2.4	0.2	2.2	2.95	60	11	25°	0.9917	1.1603
CS14			3.2	0.06	2.0	3.0	80	10	24°	1.8730	0.5244
CS15			1.6	0.1	1.6	2.81	40	6	25°	0.6687	0.1939
CS16[c]			4.0	0.15	2.0	3.20	100	6	23°	0.4252	0.3061
CS17[c]	Medium stiff clay	43.5	2.61	0.2	2.1	2.89	60	8	27°	0.6047	0.2298

continued

Table 9.9 (Continued) Soil properties for 32 piles in clay

Items	*Soil properties*			*Input parameters*				*Derived c and φ*			
CS18			1.52	0.1	1.2	2.78	35	4	24°	0.5232	0.1397
CS19			2.18	0.1	2	2.85	50	7.5	26°	0.5761	0.1976
CS20			4.35	0.1	1.7	3.0	100	5	27°	0.6936	0.2469
CS21*			0.87	0.0	0.15	2.67	20	0.5	9°	0.4505	0.1604
CS22			3.48	0.1	1.9	2.95	80	7.3	26.0°	0.6532	0.2325
CS23[d]			4.35	0.2	2.1	3.0	100	11	24°	0.6936	0.2642
CS24			4.35	0.1	1.5	3.0	100	5	25.5°	0.6936	0.1762
CS25			2.61	0.1	1.4	2.89	60	5	24.5°	0.6047	0.2455
CS26	Stiff clay	73.4	22.02	0.1	1.2	3.55	300	7	26.5°	1.0862	0.3313
CS27[d]	OC stiff London clay	44.1	11.0	0.1	2.1	3.43	250	6	27°	0.3681	0.3681
CS28	Submerged soft silt clay	27.3	2.73	0.05	1.2	3.28	100	3	19°	0.2389	0.0480
CS29	Soft clay	26.0	1.82	0.1	0.8	3.21	70	2	25.0°	0.1433	1.4335
CS30	Soft clay	41.6	2.91	0.1	0.8	3.28	70	2	6.5°	0.1155	0.7822
CS31	Soft clay	15.6	4.68	0.1	1.4	3.39	300	2	5.5°	0.1697	0.0849
CS32[c]	Silt	16.6	2.17	0.2	3.2	2.98	131	6	16.0°	0.2727	0.0273
Ranges of the values		14.4~243	0.94~76.5	0~0.2	0.15~3.2	2.67~3.92	25~315	0~11	0~11		
Average value		41.75	5.653	0.107	1.52	3.03	81.5	5.7	5.7		

Source: Guo, W. D., J. of *Geotechnical and Geoenvironmental Engineering*, 2012. With permission of ASCE.

[a] Guo and Lee 2001

[b] Values of *c* and φ deduced by matching Hensen's equation (Chapter 3, this book) with the deduced Guo LFP, rather than measured data. $n = 0.7$ for all piles but for 1.7 for CS2

[c] Matlock LFP being good

[d] p_u from Matlock LFP being lower than the deduced one

Salient features noted are as follows (Guo and Zhu 2005): (1) $\tilde{G}$ may be calculated over a depth L_{cr} = (5.5~17.6)d with $\tilde{L}_{cr}$= 12.5d. (2) At a displacement w_g (x = 0, ground level) of 0.2d: (a) max x_p = (4.1~10.1)d with $\tilde{x}_p$ = 7.2d, with 47% of the $\tilde{x}_p$ values ≤ 6.5d (see Figure 9.5c); (b) x_{max} (depth of M_{max}) = (4.3~8.1)d with $\tilde{x}_{max}$ = 6.5d; and (c) θ_g (the slope at x = 0) = (2.12~4.22%) with $\tilde{\theta}_g$ = 2.91%. Thereby, the A_L and $\tilde{s}_u$ may be estimated over a depth of 5d (or 10d) concerning w_g of 0.1d (or 0.2d) or θ_g = 1.5% (3%), respectively.

The deduced parameters offer the λ and A_L, which render the deduced Guo LFP, and the normalization of the measured pile deflection w_g (or w_t) and bending moment M_{max} for each pile. The LFP for each pile is plotted in Figure 9.5 to the max depth of x_p mobilized under the maximum test load. A comparison between the deduced p_u profile and the calculated LFPs using pertinent methods (see Chapter 3, this book) for each of the 32 piles (with a normalized eccentricity eλ of 0~1.43) was made, which demonstrates that Matlock's method (J = 0.25) well predicts the average p_u for 8 piles (CSs 9, 14, 16~17, 19, 22, 24, and 32) (see Figure 9.5b), and the p_u for 5 piles (CSs 12, 13, 20, 23, and 27) using J = 0.5~0.75 [thus closer to *Reese LFP*(C)], but significantly overestimates the p_u for 19 piles (CSs 1~8, 10~11, 15, 18, 21, 25~26, and 28~31, Figure 9.5c) despite using a small J = 0.25, especially for the large diameter piles (CSs 29~31 with d ≥1.5 m). The API code or Matlock LFP underestimates the deflection but overestimates the bending moments of nineteen piles. If CS2 (stiff clay), and the large diameter piles are removed from the data base, the Matlock LFP with J = 0.25 seems to offer a reasonable average p_u (see Figure 9.5) for all piles with average values of $\tilde{d}$ = 0.28 m and $\tilde{s}_u$ = 43.3 kPa. However, this would not warrant a sufficiently accurate prediction of individual pile response. Hensen's method consistently predicts the p_u of each pile well, with the "stipulated" cohesion c and angle of internal friction ϕ in Table 9.9.

The normalized $\bar{w}_g$ or $\bar{w}_t$, and $\bar{M}_{max}$ of the measured data are plotted in Figure 9.6 and Figure 9.7 respectively, together with the simple solutions of Equations 9.24, 9.25, 9T15, and 9T14 for eλ = 0~1.43, α_o = 0, and N_p = 0. These figures indicate negligible impact of simple solutions using α_o = 0 and the remarkable impact of loading eccentricity. In particular, the deduced Young's modulus E is plotted against USC q_u (=2s_u, in MPa) previously in Chapter 3, this book, together with those deduced from six rock-socketed shafts. The data may be correlated with E = (60~400)q_u, which strikingly resembles that proposed for vertically loaded piles (Rowe and Armitage 1987).

9.5.2 20 Piles in sand

Table 9.10 shows the 20 piles in sand, which have an E_pI_p = 8.6 × 10^{-5}~527.4 MN-m^2 (largely 20~70 MN-m^2), d = 18.2~812.8 mm, and were loaded at

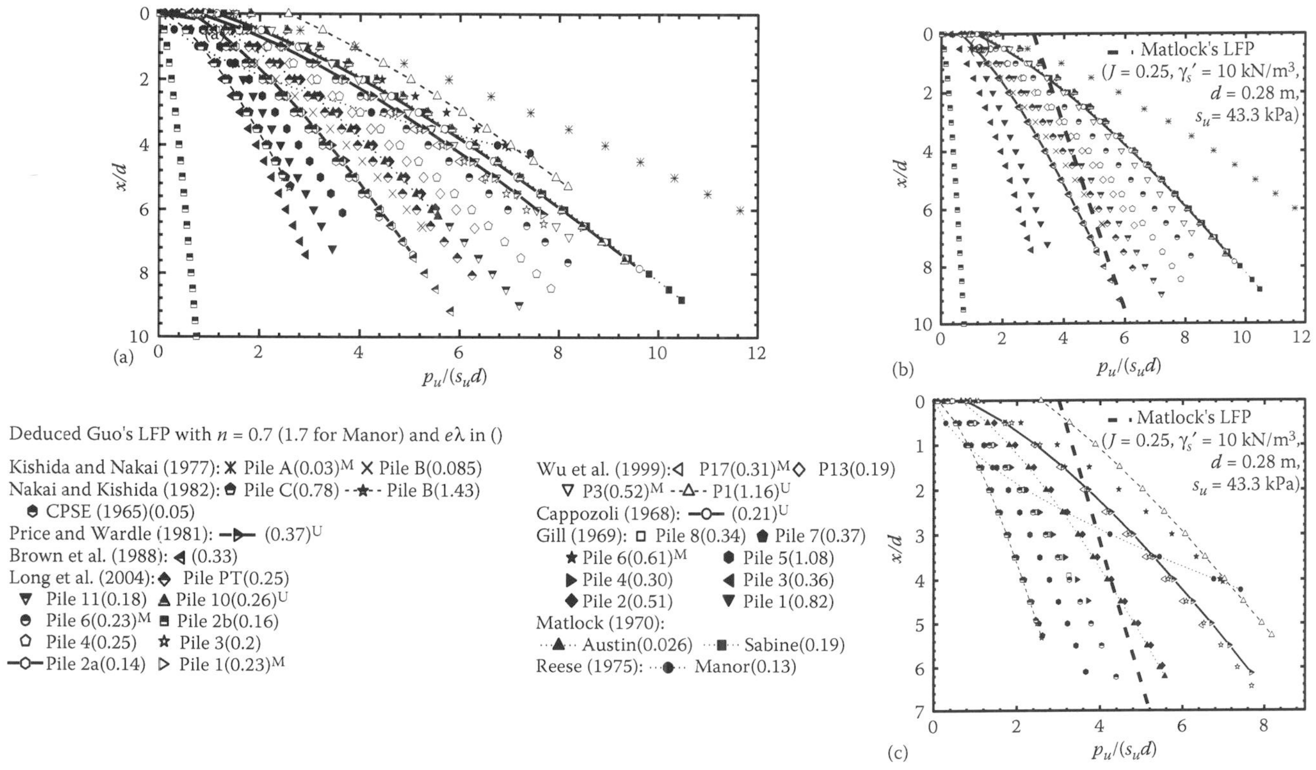

Figure 9.5 LFPs for piles in clay. (a) Thirty-two piles with max x_p/d <10. (b) Seventeen piles with 6.5 < max x_p/d < 10. (c) Fifteen piles with max $x_p/d \leq 6.5$. Note that Matlock's LFP is good for piles with superscript "M," but it underestimates the p_u with "U" and overestimates the p_u for the remaining 22 piles. (After Guo, W. D., *J Geotech Geoenviron Engrg*, ASCE, 2012.)

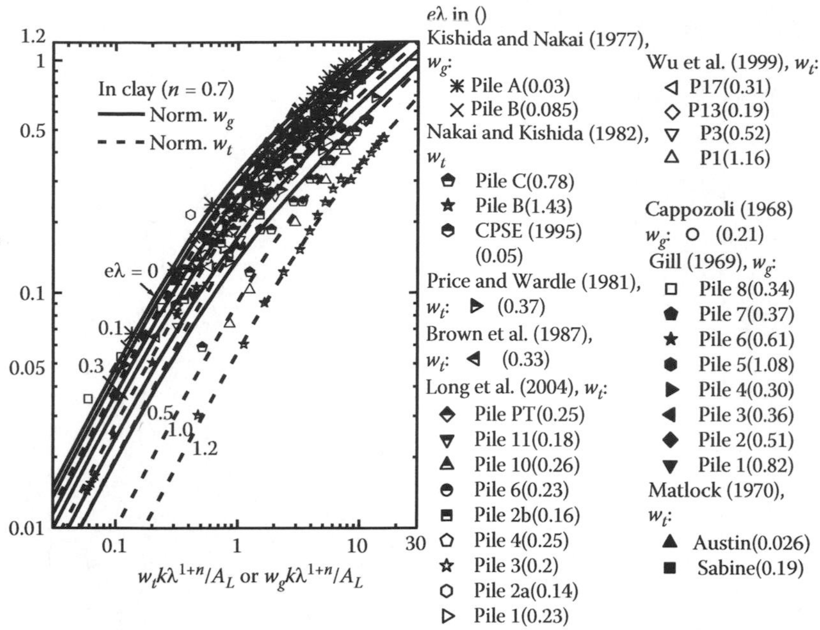

Figure 9.6 Normalized load, deflection (clay): measured versus predicted (n = 0.7). (After Guo, W. D., *J Geotech Geoenviron Engrg*, ASCE, 2012.)

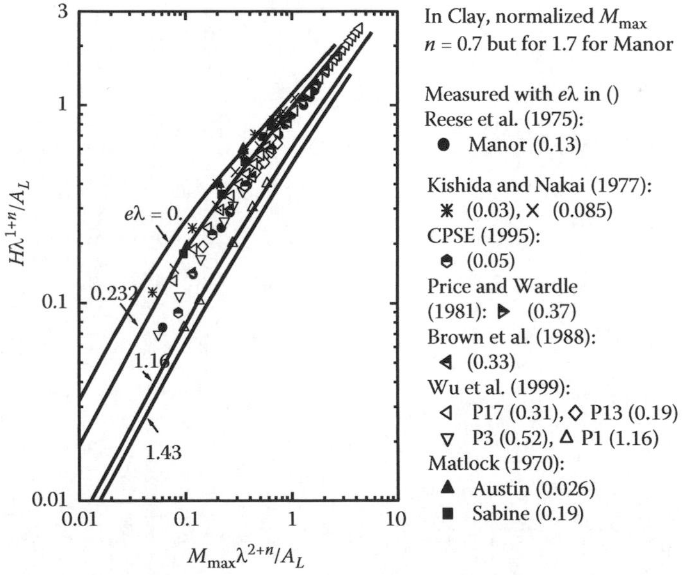

Figure 9.7 Normalized load, maximum M_{max} (clay): measured versus predicted (n = 0.7). (After Guo, W. D., *J Geotech Geoenviron Engrg*, ASCE, 2012.)

Table 9.10 20 Piles and sands

Cases	*Pile type*	L *(m)*	d/t *(mm)*	E_p*(GPa)*	E_pI_p *(MNm²)*	e *(m)*[a]	γ_s (γ_s') *(kN/m³)*	*SPT* N	G *(MPa)*		*Reference*
PS1	Open-ended pipe pile	21.05	610/9.53	24	163	0.305	(10.4)	18	10.5	Mustang	Cox et al. (1974)
PS2	Open-ended pipe pile	13.12	273/9.27	50.3	13.708	0.305	15.4	35	16.5	Single pile	Brown et al. (1988)
PS3	Driven pipe pile	23.5	609.6	24.5	166.04	0.2	16.5	12	3.23	Pile C	Kishida & Nakai (1977)
PS4		44	812.8	24.6	527.4	0.6	17.0 [c]	9	4.15	Pile D	
PS5	Pipe pile	1.8	60.5/2.3	55.1	3.626×10^{-2}	0.12	18.0 [c]	27	5	Pile A	Nakai & Kishida (1982)
PS6	Driven open-ended pipe pile	11.5	324/9.5	52.9	28.6	0.69	(10.3)	10	6.15	Single pile	Rollins et al. (2005)
PS7	Open-ended pipe pile	16.07	406.4/7.9	51.5	69.016	0	(9.87)	12	9.54	Pile 16	Alizadeh & Davisson (1970)
PS8[b]	Reinforced pipe pile	16.1	480.3/7.9	26.7	69.877	0.031	(10.9)	12	11.8	Pile 2	
PS9[b]	Reinforced H pile (14BP73)	12.2	444.8[e]	32.1	61.698	0.031	(9.87)	12	16.6	Pile 6	
PS10[b]	Reinforced pipe pile	16.2	480.3/7.9	26.7	69.877	0.107	(9.87)	12	16.6	Pile 10	
PS11[b]	Reinforced H pile	12.96	434.5[e]	40.7	71.225	0.153	(9.87)		6.65	Pile 13A	

PS12	Driven aluminum pipe pile	0.5	18.2/0.75	16	8.6×10^{-5}	0.01	16.22		0.45	Model tests	Gandhi & Selvam (1997)
PS13[c]						0.01			0.45		
PS14[c]	Bored pipe pile					0.047[d]			0.3		
PS15	Open-ended pipe pile	5.537	120.6	59.9	0.623	0.813	19.63	29.2	18	Pile 9	Gill (1969)
PS16		7.315	218.9	55.2	6.227	0.813	19.63	26.9	16.6	Pile 10	
PS17		9.296	323.9	39.6	21.408	0.813	19.63	24.9	11.5	Pile 11	
PS18		9.296	406.4	36.2	48.497	0.813	19.63	24.9	11.5	Pile 12	
PS19	Open-ended pipe pile	11.1	430	43	72.1	2.2	15.18		6	Centrifuge tests	McVay et al. (1995)
PS20							14.51		3.2		
Ranges of the values		0.5~44		16~59.9	8.6×10^{-5}~527.4	0.01~2.2	9.87~19.63	9 ~ 35	0.3 ~ 18.0		
Average value		12.2	346/–	36.7	73.07	0.513	14.7	18.9	8.74		

Source: Guo, W. D., *J. of Geotechnical and Geoenvironmental Engineering, 2012*. With permission of ASCE.

[a] Free length e was taken as the height of the top of the pile cap.
[b] Using $\phi = 43°$ (Reese and Van Impe 2001) rather than $\phi = 35°$ (Alizadeh and Davisson 1970). An enlarged diameter 480.3 mm for PS8 was used instead of the OD of inner pipe pile. Using $\phi = 35°$, the obtained s_g will be doubled.
[c] Assumed ones.
[d] All piles are free-head, but PSs 13 and 14 are fixed-head. The latter response was checked using a finite difference program.
[e] Equivalent diameter.

an eccentricity e of 0.01~2.2 m (i.e., $e\lambda$ = 0~1.12). The parameters G, N_g, and k for each pile were deduced using an excellent match with one measured curve for eight piles of PSs 9, 14~20 or with three measured curves for the rest piles. The parameters are tabulated in Table 9.11. Typical highlights are as follows:

- G = 0.3~0.45 MPa (model piles), G = 3.2~18.0 MPa (prototype piles) with SPT N = 9~35; and G = (0.25~0.62)N (MPa) with $\tilde{G}$ = 0.50N (MPa) (Guo 2006).
- L_{cr} = (6.7~15.9)d with $\tilde{L}_{cr}$ = 9.5d. At w_g = 0.2d, max x_p = (2.22~6.39)d; $\tilde{x}_p$ = 4.61d and 50% $\tilde{x}_p$ being less than 3d; $\tilde{L}_c + \tilde{x}_p$ = 14d; x_{max} = (3.85 ~ 6.55)d, with $\tilde{x}_{max}$ = 4.67d; and θ_g = (2.85~5.41)% with $\tilde{\theta}_g$ = 4.03%. The $\tilde{G}$ may be calculated over a depth of 14d.
- $\tilde{\gamma}_s$ and $\tilde{\phi}$ may be averaged over a maximum slip depth of 5d to construct the LFP concerning w_g = ~0.2d. Angle of internal friction ϕ is 29.6~43°.

These features resemble those obtained for free-head piles in calcareous sand (Guo and Zhu 2005b). In particular, the features of the deduced G (recast as E using ν = 0.25) and p_u are as follows:

- E = (0.65~1.6)N (MPa) deduced is accord with E = (0.5~1.5)N (MPa) (Kulhawy and Mayne 1990) and E = (1.4~1.8)N (MPa) (Kishida and Nakai 1977).
- The average s_g (=N_g/K_p^2) of 1.27 exceeds s_g of 1.0 (Barton 1982; Zhang et al. 2002) by 27%. The s_g distribution is: (1) s_g = 0.4~1.0 for the bored pile (PS14) or the open-ended pipe piles (PSs 1, 2, 7, 19, and 20); (2) s_g = 1.1~2.5 for the driven pipe piles (PSs 3, 4, 12, 13, and 15~18), including pile PS3 (s_g = 2.5, x_p = 1.7d), PS6 (1.5, 3.6d), PS12 (2.0, 5.7d), PS15 (2.8, 4.3d), PS16 and PS18 (1.5~1.8, 4.3d and 2d); and (3) s_g = 1.6~1.8 (with max x_p = 1.5~2d) for the reinforced pipe piles (PSs 8 and 10) and large sectional-area pile (PS11).
- The Reese LFP(S) or API code method (Reese et al. 1974) well predicts the net resistance p_u of the open-ended pile PS1, reinforced H pile PS9, and bored aluminum (model) pipe pile PS14 (s_g = 0.4~0.55), respectively. Nevertheless, with α_p = A_L of Equation 9.2 over A_L of Reese LFP(S) (see Figure 9.8b and c), it is noted that: (1) α_p = 4 for the five piles of the reinforced pipe pile PS8 and PS10, and the H pile PS11, the driven pile PS12; and the open-ended pipe pile PS15; and (2) α_p = 1.5~3.5 for ten piles. The Guo LFPs along the piles PSs16 and 18 may fluctuate slightly without measured bending moments. In particular, a higher limiting force for seventeen out of twenty piles (with α_p = ~4) is deduced, but it is associated with a shallower max x_p of ~5.7d. Displacement of 85% piles is overestimated and bending moment underestimated using API code method.

Table 9.11 Parameters for 20 piles in sand[a]

Case	*Soil type*	N_g/K_p^2	G/N *(MPa)*	k/G[b]	α_p *ratio*[c]	γ_s/γ_s'	s_r[d]	λ *(1/m)*	ϕ *(degree)*	A_L *(kNm$^{-2.7}$)*	$e\lambda$
PS1	Submerged dense sand	0.55	0.58	3.55	1.0	1.94	0.740	0.4893	39	95.08	0.14924
PS2	Medium dense sand	0.66	0.47	2.98	1.8	1	0.973	1.0107	38.5	127.2	0.30826
PS3		2.5	0.27	3.22	3.0	1	2.98	0.3537	35	484.2	0.07074
PS4	Loose sand	1.1	0.46	3.28	1.7	1	1.329	0.2835	29.6	190.6	0.1701
PS5	Medium dense sand	0.85		3.12	2.0	1	1.255	3.221	38	116.5	0.38652
PS6	Submerged medium dense sand	1.5	0.62	3.18	3.5	1.95	2.036	0.6433	35.3	153.9	0.44388
PS7		0.78	0.35	3.30	1.6	1.993	0.892	0.5808	43	164.4	0
PS8		1.6	0.58	3.56	4.0	1.899	1.631	0.6225	35	322.8	0.0193
PS9		0.49	0.62	3.61	1.0	1.993	0.611	0.7020	35	105.4	0.02176
PS10		1.6	0.62	3.73	4.0	1.993	1.711	0.6833	35	345.2	0.07311
PS11		1.8	0.25	3.27	4.0	1.993	1.93	0.5256	35	387.3	0.08042
PS12	Medium dense sand	2.0		2.86	4.0	1	2.861	7.8206	36.3	148.5	0.07821
PS13		1.0		2.44	2.1	1	1.430	7.5150	36.3	74.3	0.07515
PS14		0.4		2.38	1	1	0.625	6.6929	36.3	29.7	0.31457
PS15	Compacted granular soil	2.8	0.62	3.44	4.0	1	3.52	2.2322	34.0	364.7	1.81478
PS16		1.8	0.62	3.44	2.5	1	2.43	1.2298	33.5	269.5	0.99983
PS17		1.1	0.46	3.43	2.0	1	1.51	0.8238	32.9	175.7	0.66975
PS18		1.5	0.46	3.46	3.0	1	1.97	0.6727	32.9	256.5	0.54691
PS19	Medium dense sand	0.65		3.23	2.0	1	0.895	0.5092	39	148	1.12024

continued

Table 9.11 (Continued) Parameters for 20 piles in sand[a]

Case	*Soil type*	N_g/K_p^2	G/N *(MPa)*	k/G[b]	α_p *ratio*[c]	γ_s/γ_s'	s_r[d]	λ *(1/m)*	ϕ *(degree)*	A_L *(kNm$^{-2.7}$)*	$e\lambda$
PS20	Loose sand	0.8		3.07	1.5	1	0.987	0.4298	34.0	112.7	
Ranges of the values		0.49~2.5	0.25~0.62	2.38~3.73	1.0~4.0		0.611~2.861	0.3537~7.8206	29.6~39	29.7~484.2	
Average value		1.27	0.5	3.23	2.49	1.338	1.618	1.8521	35.7	203.6	

Source: Guo, W. D., *J. of Geotechnical and Geoenvironmental Engineering*, 2012a. With permission of ASCE.

[a] n = 1.7 and α_0 = 0 for all cases.
[b] Guo and Lee 2001.
[c] $\alpha_p = A_L$ for Guo LFP over A_L for Reese LFP(S).
[d] $s_r = (\gamma_s/\gamma_s')k_p/(d^{n_p}K_p^2)$, k_p = 3.5~27, n_p = 0.2~0.25, γ_s and γ_s' = unit weight, and effective unit weight of sand, respectively $p_u/\gamma_s' d^2 = s_r K_p^2 (x/d)^{1.52}$ (Guo and Zhu 2010). The s_r and s_g are slightly different, as is the power 1.52 from n = 1.7 in Equation 9.2.

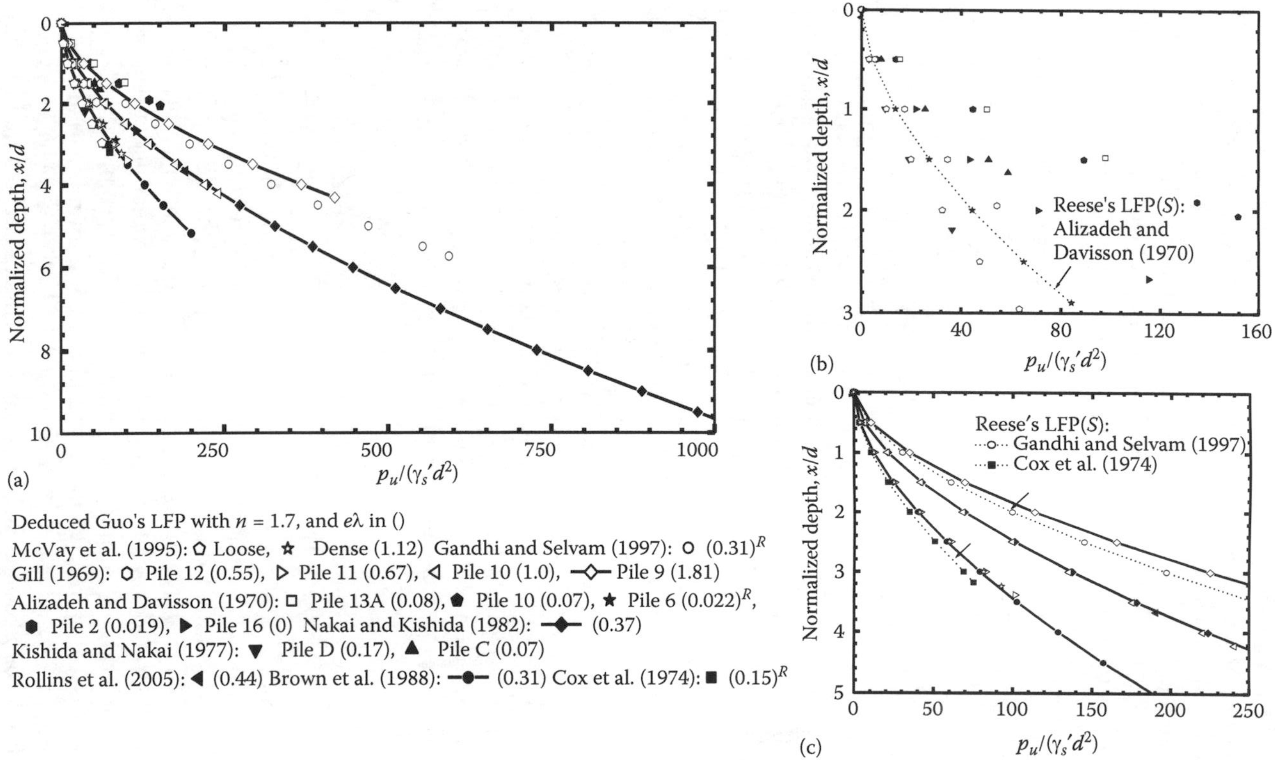

Figure 9.8 LFPs for piles in sand. (a) Eighteen piles with max $x_p/d < 10$. (b) Nine piles with max $x_p/d \leq 3$. (c) Nine piles with max $x_p/d > 3$. (Note that Reese's LFP is good for piles with superscript R, but it underestimates the p_u for the remaining piles.) (After Guo, W. D., *J Geotech Geoenviron Engrg*, ASCE, 2012.)

A fixed-head (translation only) condition would reduce the gradient of the p_u to 25% that on a free-head pile (Guo 2009), as is noted here on pile PSs 12 and 13 compared to PS 14. This implies Reese LFP(S) may incur overestimation of displacement and underestimation of maximum bending moment for a single pile, which is opposite to the likely underestimation of the displacement for single piles in clay using the Matlock LFP (J = 0.25~0.5). The deduced parameters may be used along with the closed-form solutions to design free-head (Guo 2006) and fixed-head piles (Guo 2010).

As for piles in layered soil, Guo (2006) indicates within the max x_p, the p_u for each layer may be still obtained using Equation 9.2 (or Chapter 3, this book), but the calculated value should be increased by ~40% in a weak layer or decreased by ~30% in a stiff layer. The average trend of the p_u for all layers may be adopted for a pile design, with due consideration that a sharp increase in shear strength with depth renders a high n (e.g., n = 1.7 for CS2, and n = 2.0 for a better simulation for PS3).

The max x_p was 6.4d (sand)~7.2d (clay) for all test piles. With d < 0.6 m, response of the majority of the piles was dominated by the upper ~4.5 m soil. With d = 1.5~2.0 m (e.g., CS 29, 30, and 31), the response will be dominated by soil down to a depth of ~10 m, as the max x_p reached 8.0 m (=5.32d), 9.9 m (=4.92d) and 10.4 m (=6.56d), respectively, which is a striking contrast to the max x_p of 1.8 m (=2.19d) for pile PS4 (d = 0.813 m). Thus, the p_u profile is often constructed as if in one layer.

The deduced parameters enable the Guo LFPs to be plotted in Figure 9.8 and the measured data for all piles to be normalized and plotted together in Figures 9.9 and 9.10 for deflection and bending moment, respectively. Except for pile D with the largest E_pI_p of 527.4 MN-m^2, the measured data again compare well with Equations 9.24, 9.25, 9T15, and 9T14. Thus the simple expressions may be used for piles in clay, sand, and/or in multilayered soil.

9.5.3 Justification of assumptions

Finally, the salient features and hypothesis are compared in Table 9.12, concerning the current CF solutions and the numerical program COM624P. Both are capitalized on load transfer model, but only the CF solutions are rigorously linked to soil modulus via Equation 3.50 (Chapter 3, this book). COM624P allows incorporating various forms of nonlinear p-y curves, but the resulting overall pile response is negligibly different from that obtained using the current solutions. COM624P and the CF solutions actually cater for a linearly increasing and a uniform profile of k, respectively. Their predictions should bracket nonhomogeneous k normally encountered in practice. To capture overall pile–soil interaction, only the parameters for the LFP vary with modes of soil failure. The current solutions offer very good

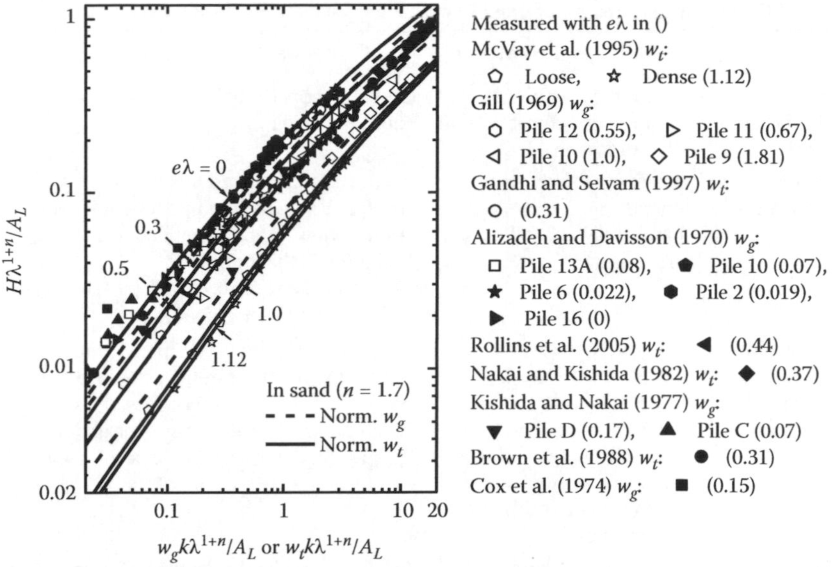

Figure 9.9 Normalized load, deflection (sand): Measured versus predicted (n = 1.7). (After Guo, W. D., *J Geotech Geoenviron Engrg*, ASCE, 2012.)

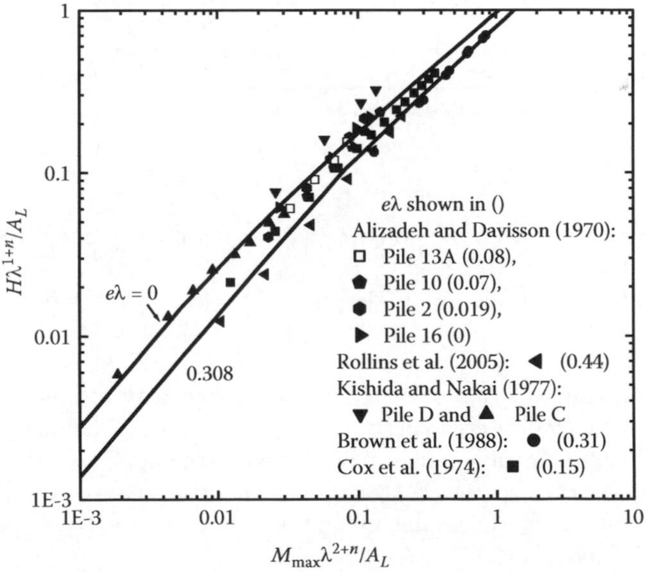

Figure 9.10 Normalized load, maximum M_{max} (sand): Measured versus predicted (n = 1.7). (After Guo, W. D., *J Geotech Geoenviron Engrg*, ASCE, 2012.)

Table 9.12 Salient features of COM624P and the current CF solutions

Item	*COM624P (FHWA 1993)*	*GASLFP (CF solutions)*
Pile–soil interaction model	Uncoupled, incompatible with continuum-based numerical analysis	Coupled, consistent with continuum-based numerical analysis
Subgrade modulus, k	Increase linearly with depth. Empirically related to soil properties.	A constant calculated from an average modulus, G, over the effective pile length of L_c + max x_p and using Equation 3.50, Chapter 3, this book. Theoretically related to soil, pile properties, pile-head, and base conditions.
Limiting force per unit length (LFP)	Many parameters, various expressions, and procedures for different soils. Parameters derived from soil failure modes of wedge type and lateral plastic flow.	Three parameters n, N_c, (or α_o), and N_g, a unified expression of Equation 9.2, and procedure for all kinds of soils. Parameters deduced from overall pile response, regardless of mode of failure.
p-y curve	Consisting of four piecewise curves.	An elastic-perfectly plastic curve, solid line.
Computation	Finite difference method	Explicit expressions of the x_p using spreadsheet program GASLFP or by hand
Advanced use	In form of numerical program; no other specified use.	In form of explicit expressions; capturing overall pile–soil interaction by LFP, and indicating the effective depth by x_p. May be used as a boundary element for advanced numerical simulation.

Source: Guo, W. D., *Computers and Geotechnics* 33, 1, 2006.

to excellent prediction of response of ~70 tested piles in comparison with measured data.

The current solutions are underpinned by five hypotheses mentioned previously.

- In contrast to assumption 1, the *p*-y curve may follow a parabola (Matlock 1970) or a hyperbola (Jimiolkwoski and Garassino 1977). There exists a transition zone in between the upper plastic zone and the lower elastic zone. Fortunately, use of the idealized elastic-plastic *p*-*y*(*w*) curve has negligible impact on the prediction (Poulos and Hull 1989; Castelli et al. 1999).
- Use of an equivalent modulus (assumption 2) may overestimate the x_{max} by ~20%.

- Ignoring coupled impact in plastic zone (assumption 3) is sufficiently accurate, as slip generally occurs under a very low load level, but it extends to a limited depth under a maximum load level.
- Assumptions 4 and 5 are satisfied by the single slip depth (Figure 9.1b) rather than a transition zone, the p_u profile, and an infinite length (with $L > L_c$ in elastic zone). Otherwise, another slip may be initiated from a short, rigid pile base at a rather high load level (Guo 2003a).

A single modulus and p_u profile allow a gradual increase in w_p with depth. This is not always true in a stratified soil, as the w_p of a deeply embedded weak layer may be lower than that of a shallow, stiff layer. However, the use of an overall p_u is sufficiently accurate as against 3-D FEA results. In particular, pile–soil relative slip may never extend to an underlying weak layer if a very stiff upper layer is encountered. The assumption is thus valid. Deriving from the normal and shear stresses, respectively, on the pile–soil interface (Baguelin et al. 1977; Briaud et al. 1985), the resistance in elastic state may be sufficiently accurately evaluated using elastic theory (Guo and Lee 2001), and in the plastic (slip) state by the p_u profile. In rare cases, the nonhomogeneous modulus may markedly affect the pile response, for which the previous numerical results (Davisson and Gill 1963; Pise 1982) may be consulted along with the current predictions.

9.6 RESPONSE OF PILES UNDER CYCLIC LOADING

Case studies to date demonstrate that (a) with idealized *p-y*(*w*) curves, GASLFP well captures the static and cyclic responses of piles in calcareous sands; (b) critical length L_{cr} of lateral piles is (9~16)d, within which soil properties govern the pile response; (c) the LFPs may be described by $n = 1.7$, and $\alpha_0 = 0$ (uncemented sands) or $\alpha_0 > 0$ (cemented sands), along with $N_g = (0.9\sim2.5)K_p^2$ for static loading (using post peak friction angle); (d) under cyclic loading, the N_g reduces to (0.56~0.64) times N_g for static loading; (e) the value of max x_p at maximum test load is (3~8.3)d, within which limiting force reduces with cyclic loading or gapping effect.

9.6.1 Comparison of *p-y*(*w*) curves

A fictitious pile, with $d = 2.08$ m, and $E_p = 3.0 \times 10^4$ MPa, was installed in calcareous, Kingfish sand (Kingfish B platform site, Bass Strait). The sand featured $\gamma_s' = 8.1$ kN/m², $\phi = 31°$, $G = 5.0$ MPa, and a cone tip resistance, q_c, increasing linearly (at a gradient of 400 kPa/m) with depth.

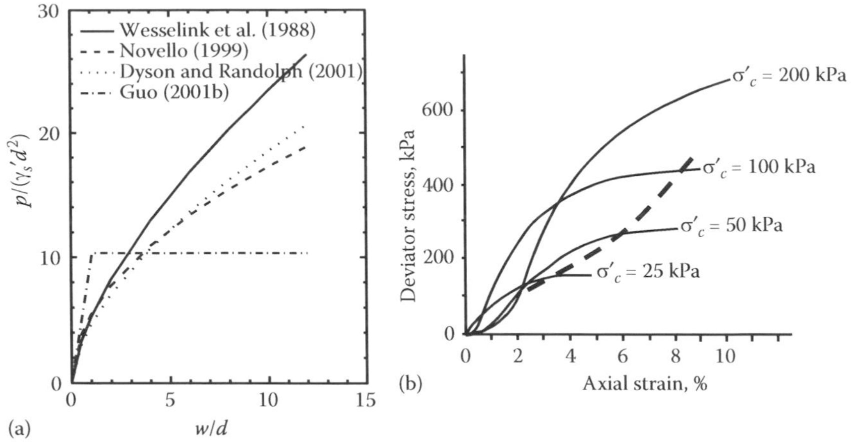

Figure 9.11 (a) *p-y(w)* curves at $x = 2d$ (d = 2.08 m). (b) Stress versus axial strain of Kingfish B sand. (After Hudson, M. J., G. Mostyn, E. A. Wiltsie and A. M. Hyden, *Proc Engrg for Calcareous Sediments*, Balkema, Perth, Australia, 1988; Guo, W. D., and B. T. Zhu, *International Symposium on Frontiers in Offshore Geotechnics*, Taylor & Francis/Balkema, Perth, Australia, 2005b.)

Table 9.13 Static Hardening *p-y* curves for calcareous sand

Author	*p-y model*	*Parameters for Kingfish B*
Wesselink et al. (1988)	$p = Rd(x/x_0)^n (y/d)^m$	x_0 = 1 m, n = 0.7, m = 0.65, R = 650
Novello (1999)	$p = Rd\sigma'^{n}_{v0} q_c^{1-n} (y/d)^m < q_c d$	$R = 2, n = 0.33, m = 0.5, \sigma'_{v0} = \gamma_s x$
Dyson and Randolph (2001)	$p = R\gamma_s d^2 (q_c/\gamma_s d)^n (y/d)^m$	$R = 2.7, n = 0.72, m = 0.6$

Source: Guo, W. D., and B. T. Zhu, *International Symposium on Frontiers in Offshore Geotechnics*, Taylor & Francis/Balkema, Perth, Australia, 1, 2005b.

A sample *p-y* curve at $x = 2d$ is plotted in Figure 9.11a, together with the predictions using the existing three *p-y* curve (hardening) models (see Table 9.13) exhibiting continual increase in resistance with the pile deflection. Assuming $n = 1.7$, $\alpha_0 = 0$, and $N_g = 0.33K_p^2$, the predicted response (using GASLFP) is shown in Figure 9.12.

In contrast, the idealized *p-y* curve by Guo (2001b) somehow resembles the stress-strain relationships of the (near surface) Kingfish B sand shown in Figure 9.11b (Hudson et al. 1988). Such a similarity was examined previously by McClelland and Focht (1958). At each cell pressure, σ'_c, the deviator stress increased with axial strain, and reaches a constant once the strain is beyond the dashed line. The intersection of the dashed line with each stress-strain curve may be viewed as the yield point. At σ'_c of 25~100 kPa, the yield strain is 3%~8% for the uncemented sand. The yield strain is 2%~7% at

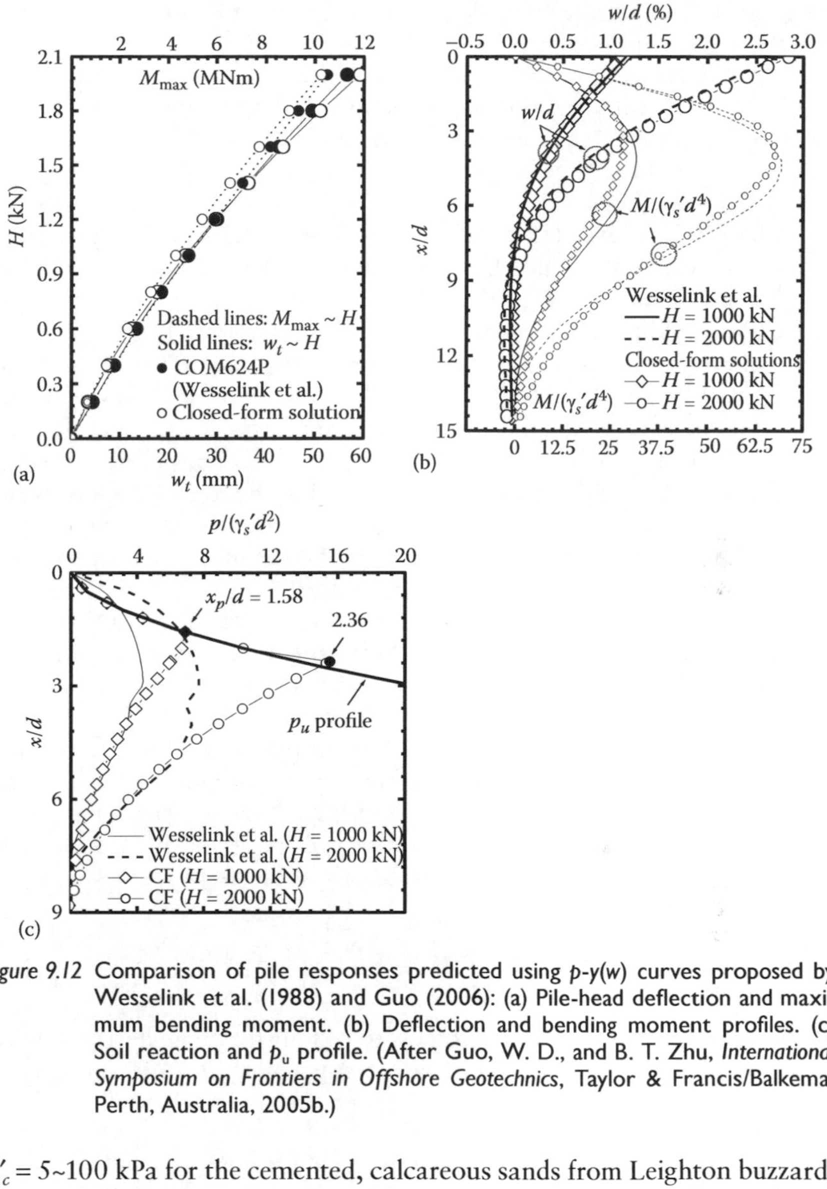

Figure 9.12 Comparison of pile responses predicted using *p-y(w)* curves proposed by Wesselink et al. (1988) and Guo (2006): (a) Pile-head deflection and maximum bending moment. (b) Deflection and bending moment profiles. (c) Soil reaction and p_u profile. (After Guo, W. D., and B. T. Zhu, *International Symposium on Frontiers in Offshore Geotechnics*, Taylor & Francis/Balkema, Perth, Australia, 2005b.)

$\sigma_c' = 5\sim100$ kPa for the cemented, calcareous sands from Leighton buzzard, Dogs Bay, Ballyconneely and Bombay Mix (Golightly and Hyde 1988).

9.6.2 Difference in predicted pile response

Using suitable parameters, the three hardening models produce similar *p-y(w)* curves and pile response. Thus, only the typical *p-y* curve (Wesselink et al. 1988) was used to predict the response of the fictitious pile using program COM624P (FHWA 1993). This is shown in Figure 9.12 together with those predicted using GASLFP. Using $E_p = 3.0 \times 10^4$ MPa and G = 5.0 MPa, the critical pile length L_{cr} (Guo and Lee 2001) was estimated to be 9.24d (=19.22 m). Thus the pile deflection and soil reaction mainly take place within this depth as shown in Figure 9.12c.

Figure 9.12 indicates that:

1. The p-$y(w)$ curves from the ideal p-y curve and working hardening models offer similar pile-head deflection, w_t (Figure 9.12a), maximum bending moment, M_{max}, distributions of normalized deflection, y/d, and normalized bending moment, $M/(\gamma_s' d^4)$ (Figure 9.12b).
2. The p-$y(w)$ models alter slightly the distribution of soil reaction, $p/(\gamma_s' d^2)$ (Figure 9.12c), especially within a depth of (1.5~5)d. Using the elastic-plastic model, the normalized soil resistance increases, following the LFP, from zero at groundline to the maximum value at the slip depth, x_p (=1.58d, and 2.36d at H = 1.0 and 2.0 MN, respectively) (Figure 9.12c). Below the depth, the resistance decreases with depth in the elastic zone. The soil reaction peaked at the slip depth is due to ignorance of the transition zone (Guo 2001b).
3. Depth of maximum bending moment, x_m, at H = 2.0 MN reduces from 9.3 m (4.47d) to 8.06 m (3.88d) using the hardening and idealized models, respectively.

The soil reaction is generally deduced by differentiating twice from discrete measured values of bending moment from instrumented piles. The value is very sensitive to the function adopted to fit the measured moments, and generally results in different values of p from very same results. The current back-estimation is deemed sufficiently accurate.

9.6.3 Static and cyclic response of piles in calcareous sand

Using GASLFP, static and cyclic responses of four different steel pipe piles in calcareous sand (see Table 9.14) are investigated against measured data, and are shown next in Examples 9.8 through 9.10.

Example 9.8 Kingfish B: Onshore and centrifuge tests

9.8.1 Kingfish B: Onshore tests

Two tubular steel piles (piles A and B) are 356 mm OD, 4.8-mm wall thickness, 6.27 m in length, and 24 MN-m^2 in flexural stiffness. They were driven into an onshore test pit that was filled with saturated, uncemented calcareous (Kingfish B) sand (Williams et al. 1988). Both piles were laterally loaded. The bending moments, head rotation, head displacement, displacement of the sand surface, and pore pressure were

Table 9.14 Properties of piles and soil

Cases[b]	*Loading*	L *(m)*	d *(m)*	e *(m)*	EI *(MN-m²)*	E_p *(MPa)*
1 & 2	Static & cyclic	5.9	0.356	0.37	24.0	3.044×10^4
3	Static	32.1	2.137	2.4	79506	7.77×10^4
4	Static	6.0	0.37	0.45	98.0	1.065×10^5
5 & 6	Static & cyclic	5.9	0.0254	0.254	1.035×10^{-3}	5.065×10^4
Cases	γ_s' (kN/m³)	ϕ	G (MPa)	E_p (MPa)	N_g	L_{cr}/d
1/2	8.04	31	2.2	3.044×10^4	0.9/1.4[a]	11.4
3	8.04	31	3.45	7.77×10^4	1.2	12.9
4	6.4	30	2.0	1.065×10^5	1.0	15.8
5/6	8.45	28	3.4	5.065×10^4	2.4/1.5[a]	11.6

Source: Guo, W. D. and B. T. Zhu, *International Symposium on Frontiers in Offshore Geotechnics*, Taylor & Francis/Balkema, Perth, Australia, 1, 2005b.

[a] 0.9/1.4: N_g = 0.9 and 1.4 for static and terminal cyclic loading respectively.
[b] Cases 1–4 are uncemented soil with α_0 = 0(m), while cases 5 and 6 are cemented soil with α_0 = 0.15 (m).

n = 1.7
Cases 1, 2, and 3: Kingfish B Tests A1, B, and centrifuge tests; Case 4: North Rankin test, and Cases 5 and 6: Bombay High tests.

recorded. Pile A was initially "pushed" monotonically (one load-unload loop at each load level) to 106 kN at a free-length of 0.37 m above groundline, and at loading rates of 1 kN/min (virgin loading), and 5 kN/min (unloading and reloading), respectively. The pile was then "pulled" monotonically to failure in the opposite direction. Only the response for the "push" test was modeled here (Case 1). Pile B was subjected to a series of two-way cyclic loadings with 250 sec period in the initial cycle, 60 sec period up to 100 cycles and 250 sec period in the 101st cycle.

9.8.2 Test A1 (static loading)

In Test A, the sand has a void ratio of 1.21, a dry unit weight of 12.4 kN/m³, and a saturated unit weight of 17.85 kN/m³. The friction angle was 31° (from a cone resistance of 1.5 to 3 MPa) (Kulhawy and Mayne 1990). The modulus G was 2.2 MPa ($\nu_s = 0.3$), using a secant Young's modulus of 5.6 MPa (at 50% ultimate deviator stress), in light of the consolidated drained triaxial tests (Hudson et al. 1988). This G and the E_p of 3.044×10^4 MPa [$=24/(\pi \times 0.356^4/64)$] allow L_{cr} to be calculated (Guo and Lee 2001) as 11.4d (=4.05 m < embedment length of 5.9 m).

With uncemented sand, using $\alpha_0 = 0$, and $n = 1.7$, the N_g was deduced as $0.9K_p^2$ by matching the GASLFP prediction with the measured $H\sim w_g$ (at groundline) relationship and bending moment distributions (see Figure 9.13a$_1$ and b$_1$). The maximum bending moment at $H = 106$ kN was overestimated by 5.9% against the measured value, and pile head displacements and bending moments at other load levels

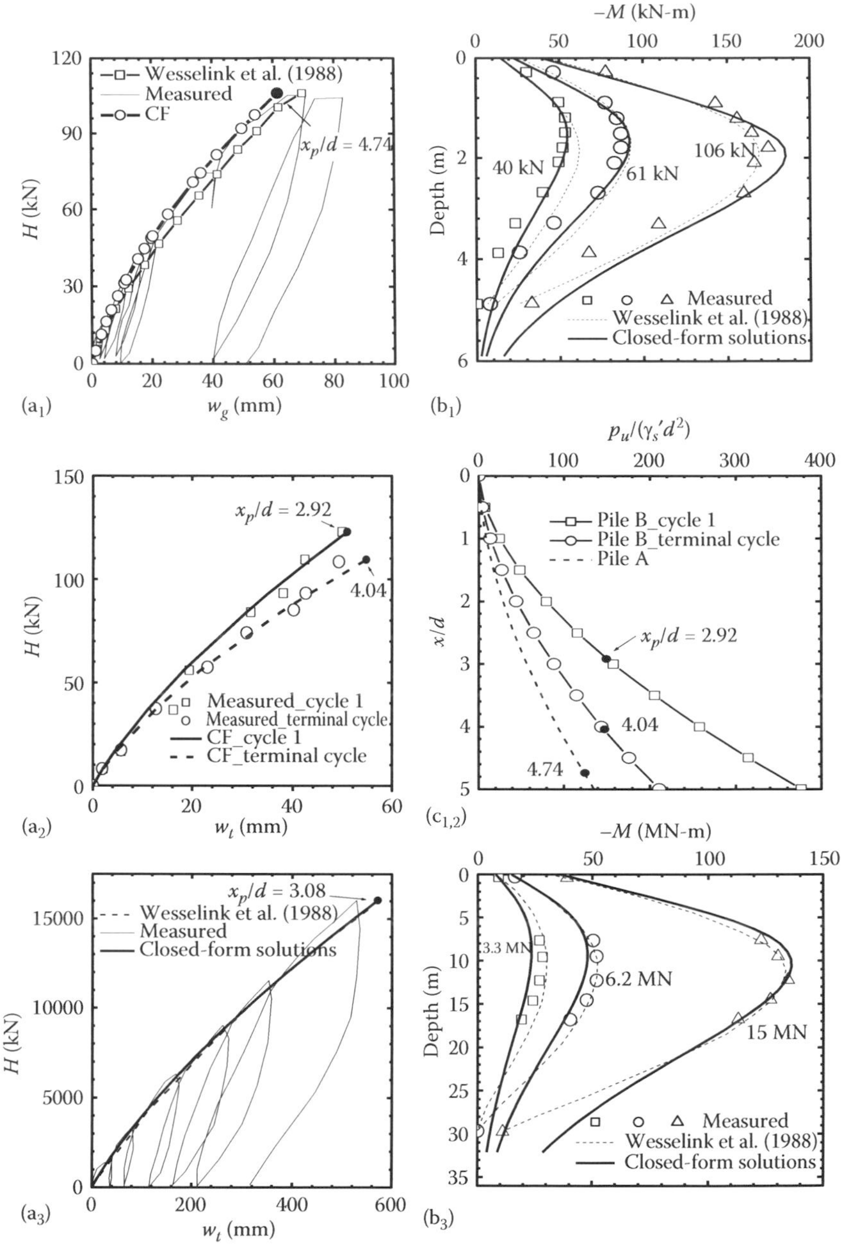

Figure 9.13 Predicted versus measured pile responses for Kingfish B. (a_1) H-w_g. (a_2 and a_3) H-w_t. (b_i) M profile. (c_i) LFPs (i = 1, and 2 onshore test A, and test B, i = 3 centrifuge test). (After Guo, W. D., and B. T. Zhu, *International Symposium on Frontiers in Offshore Geotechnics*, Taylor & Francis/Balkema, Perth, Australia, 2005b.)

were well predicted (Figure 9.13), despite of the remarkable difference in the p-y curves [e.g. Figure 9.11(a)]. The overestimated bending moment near the pile tip indicates a higher modulus G than used here.

A slip depth x_p of 1.689 m (4.74d) is obtained at a maximum H of 106 kN, which induces a pile deflection at ground level of 22.68 mm (i.e., $w_p/d = 6.37\%$), and corresponds to an effective overburden stress of 13.6 kPa (=1.689 × 8.04). This stress (see Figure 9.11b) would have a peak deviator stress at an axial strain at about 2.5% (≈40% w_p/d). The soil within the depth x_p must have yielded, thus use of the idealized p-y curves is more suitable than that of the hardening model proposed by Wesselink et al. (1988), although the latter can also give a good prediction on the pile response.

9.8.3 Test B (cyclic loading)

With those for Test A, N_g for pile B (Case 2) was deduced as $2.5K_p^2$ and $1.4K_p^2$, respectively, for cycle 1 and the terminal cycle against measured pile deflection at 0.11 m height above groundline (see Figure 9.13a$_2$). The increased N_g values are partly contributed to local densification of the soil during cyclic loading (Wesselink et al. 1988). The corresponding LFPs were plotted in Figure 9.13c$_{1,2}$ along with that for Test A. The value of N_g at the terminal cycle is 0.56 times that of cycle 1, showing impact of increased depth of gap, x_p (e.g., 2.73d at cycle 1 to 4.04d at the terminal cycle at H = 110 kN).

9.8.4 Kingfish B: Centrifuge tests

Wesselink et al. (1988) conducted a series of centrifuge tests to mimic the behavior of one typical prototype pile installed in a submerged uncemented Kingfish B calcareous sand (Case 3). The pile was 2.137 m in diameter, 34.5 m in length, and had an E_pI_p of 79506.0 MN-m^2 (E_p = 7.77 × 10^4 MPa). Load was applied at 2.4 m above the sand surface on the onshore pile (Wesselink et al. 1988). The sand had a submerged unit weight of 8.04 kN/m^3 and a post peak friction angle of 31°. The secant moduli at half of ultimate deviator stress, E_{50}, were 8.0, 5.6, 12.3, and 10.0 MPa at σ'_c of 25, 50, 100, and 200 kPa, respectively, or the values of G were 3.1, 2.2, 4.7, 3.8 MPa ($\nu_s = 0.3$). This offers an average G of 3.45 MPa and an L_{cr} of 12.9d (= 27.48 m).

Using GASLFP and a LFP described by n = 1.7, α_0 = 0, and N_g = $1.2K_p^2$, the pile-head displacement and bending moment were predicted and are shown in Figure 9.13a$_3$ and b$_3$. They demonstrate close agreement with the measured H-w_t (at groundline) relationships and bending moments (Figure 9.13b$_3$) and the predictions using complicated p-y curves (Wesselink et al. 1988). Similar to Test A, the bending moment around the pile tip (Figure 9.13b$_3$) was overestimated.

The slip depth x_p at the maximum load H of 16.03 MN was 6.59 m (=3.08 d), at which the pile deflection at ground level was 323.78 mm (i.e., w_p/d = 15.15%); the effective overburden stress was 53.0 kPa (= 6.59 × 8.04); and the peak deviator stress attained at an axial strain of ~6% (0.4w_p/d) (Figure 9.11b). This large-diameter pile has a smaller ratio of x_p/d than that for the small-diameter, onshore test piles, and a 30% higher value of N_g (also noted previously) (Stevens and Audibert 1979). Overall, the onshore and centrifuge tests (static loading) in Kingfish B sand are well modeled using GASLFP and a LFP described by $n = 1.7$, $\alpha_0 = 0$, and $N_g = (0.9\sim1.2)K_p^2$.

Example 9.9 North Rankin B: In situ test

Renfrey et al. (1988) conducted lateral loading test on a free head pile in situ in North Rankin B (Case 4) on the northwest shelf of Western Australia at an eccentricity of 0.45 m above groundline. The pipe pile of ~6.0 m consists of an upper 3.5 m long, 370 mm OD, and 30-mm thick pipe that is threaded with a lower 2.35 m long pipe of 340 mm OD and 12.5 mm thickness. The E_pI_p of the pile was calculated as 98.0 MN-m^2 ($E_p = 1.065 \times 10^5$ MPa). At the site, uncemented to weakly cemented calcareous silt and sand sediments extended to a depth of 113 m below seabed. To a depth of 30 m, the submerged unit weight γ_s' was 6.4 kN/m^3, and ϕ was 30° (Reese et al. 1988).

Using $\alpha_0 = 0$, $N_g = 1.1K_p^2$ (thus LFP shown in Figure 9.14b), and G = 2.0 MPa, the pile displacements at locations of the upper and lower jacks were predicted using GASLFP. As shown in Figure 9.14a, they compare quite well with the measured data. The value of k was calculated as 5.56 MPa (Guo and Lee 2001), and L_{cr} was 5.9 m (=15.95d). The k value agrees well with 5.72 MPa used previously (Reese et al. 1988), but it is higher than 3~4 MPa recommended by Renfrey et al. (1988). At the maximum load H of 185 kN, the slip depth x_p was calculated as 1.97 m (=5.32d) and the pile deflection w_g at groundline was ~67.1 mm (=18.1%d).

Example 9.10 Bombay High: Model tests

Golait and Katti (1988) presented static (Case 5) and cyclic (Case 6) model tests on a pile with L/d = 50. As tabulated in Table 9.14, the stainless steel pipe pile was 25.4 mm OD and had a flexural rigidity of 1.035 kN-m^2 (thus $E_p = 5.065 \times 10^4$ MPa). It was tested in artificially prepared calcareous mix made with 40% beach sand, 56% calcium carbonate, and 2.5% sodium-meta-silicate. The sand was equivalent to Bombay High cemented calcareous sand in respect of strength, plasticity, and stress-strain relationship. It was placed in the 1.4 × 1.0 × 2.0m (height) model tank, compacted to a void ratio of 1.0 ± 0.05 and then saturated (γ_s' = 8.45 kN/m^3). CU tests on the sand (Golait and Katti 1987) offered a secant Young's modulus (at 50% ultimate stress) of 8.8 MPa and ϕ of 28°. The G was obtained as 3.4 MPa, and L_{cr} was 11.6d (0.295 m < L).

Taking n = 1.7, the match between the GASLFP prediction and the measured pile displacements at groundline (Figure 9.15a) render α_0 = 0.15 m, $N_g = 2.4K_p^2$ and $1.54K_p^2$, respectively, for static and cyclic loading, showing the adhesion of the cemented sand

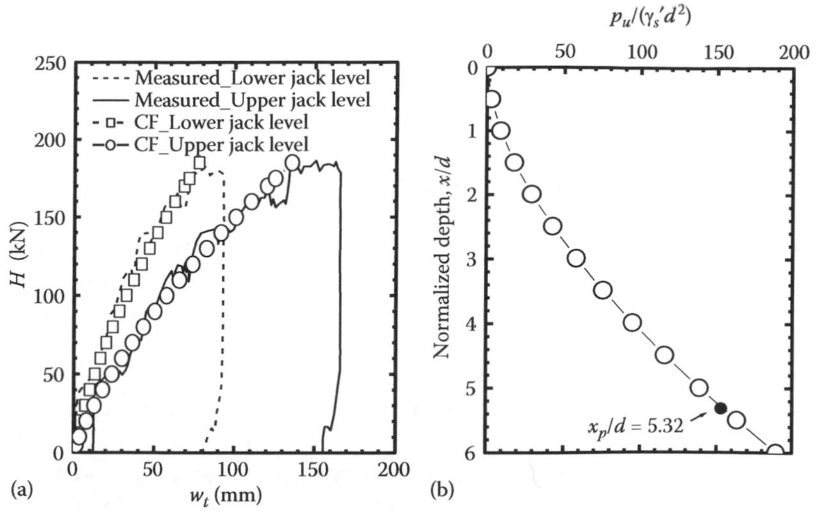

Figure 9.14 Predicted versus measured in situ pile test, North Rankin B. (a) H-w_t. (b) LFP. (After Guo, W. D., and B. T. Zhu, *International Symposium on Frontiers in Offshore Geotechnics*, Taylor & Francis/Balkema, Perth, Australia, 2005b.)

(Guo 2001b) (α_0 >0) and similar values of N_g to those for the Kingfish B onshore tests (Cases 1 and 2). Within the slip depths of 7.68d and 8.32d at the maximum static and cyclic loads, respectively, the N_g for cyclic loading was ~0.64 times that for static loading (see Table 9.14).

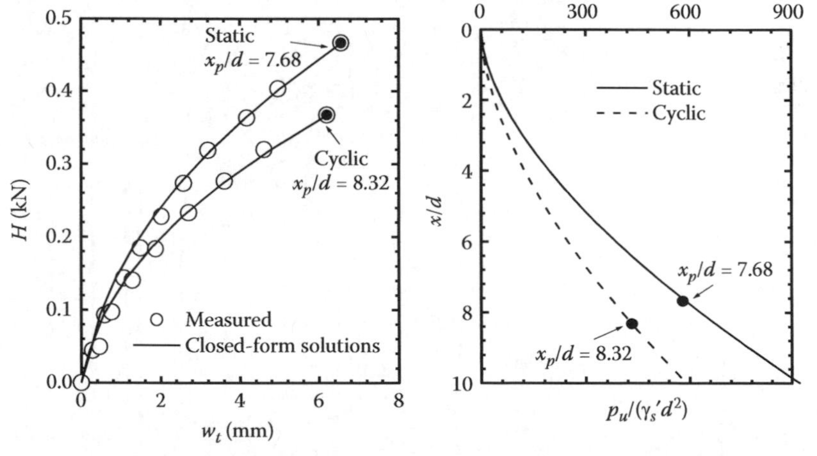

Figure 9.15 Predicted versus measured pile responses for Bombay High model tests. (a) H-w_t. (b) LFPs. (After Guo, W. D. and B. T. Zhu, *International Symposium on Frontiers in Offshore Geotechnics*, Taylor & Francis/Balkema, Perth, Australia, 2005b.)

The good comparisons between the measured and predicted pile responses at cycle 1 and the terminal cycle in Cases 2 and 5 indicate the gapping effect can be well captured by simply reducing the N_g to $0.56\sim0.64N_g$ (static).

9.7 RESPONSE OF FREE-HEAD GROUPS

Piles generally work in groups (see Figure 9.16). To estimate the group response under lateral loading, a widely accepted approach is to use the curve of soil resistance, p versus local pile deflection, y *(p-y curve)* along individual piles. The p attains a limiting force per unit length of $p_m p_u$ (p_m = p-multipliers) (Brown et al. 1988) within a depth normally taken as $8d$. The p_m is used to capture shadowing impact owing to other piles. Elastic-plastic solutions for free-head piles (Section 9.2.1) or fixed-head piles (Chapter 11, this book) are directly used for piles in groups, with due values of the p_m. In other words, irrespective of the head constraints, within elastic state, each spring has a subgrade modulus, kp_m, and limiting force per unit length $p_m p_u$. Note the p_u (via N_g) should be reduced further under cyclic loading. These solutions compare well with finite element analysis (FEA) and measured data, and have some distinct advantages (Guo 2009). In particular, various expressions for estimating p_m have been developed previously (McVay et al. 1998; Rollins et al. 2006a; Rollins et al. 2006b), such as Equation 3.69 (Chapter 3, this book) from thirty tests on pile groups (Guo 2009). Overall, it shows that the existing methods are probably unnecessarily complicated, underpinned by too many input parameters.

Using GASLGROUP, the effect of group interaction, pile stiffness, and loading properties (cyclic/static) on response of free-head pile groups is examined next, against typical in situ tests in clay, whereas that for fixed-head groups is addressed in Chapter 11, this book.

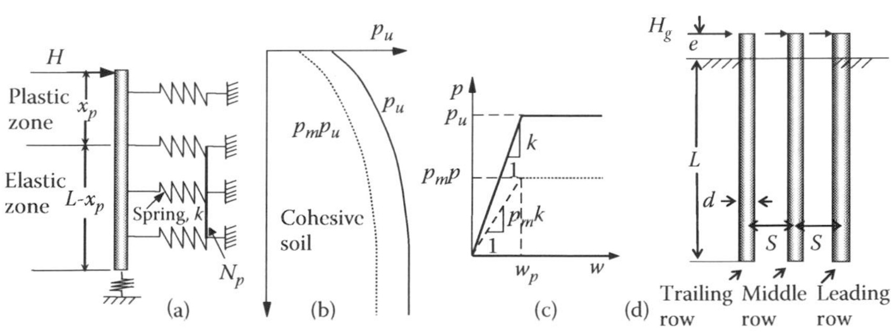

Figure 9.16 Modeling of free-head piles and pile groups. (a) A free-head pile. (b) LFPs. (c) *p-y* curves for a single pile and piles in a group. (d) A free-head group. (After Guo, W. D., *Proc 10th ANZ Conf on Geomechanics*, Brisbane, Australia, 2007.)

9.7.1 Prediction of response of pile groups (GASLGROUP)

Lateral response of a pile group may be estimated using the aforementioned closed-form solutions for a single pile. The difference is the reduction in modulus k and the p_u owing to the shadowing effect on each row of piles in a group. This is well catered for by using the p-multipliers concept (Brown et al. 1988) via the following procedure: For any displacement, the k and p for a single pile in isolated form is multiplied respectively with the p-multiplier to gain those for a pile in a group (Figure 9.16c). Using the closed-form solutions, a slip depth can be obtained for a desired pile-head displacement, which in turn allows a pile-head load for each pile in a row to be estimated. Multiplying the load by number of piles in the row gives the total load on the row. This calculation is replicated for other rows in a group with the corresponding p-multipliers, thus the total load on the group is calculated. Given a series of displacements, a number of the total loads are calculated accordingly, thus a load-displacement curve is obtained. This procedure has been implemented into the program GASLGROUP.

Example 9.11 Pile group response

Rollins et al. (2006a) conducted lateral loading tests on two isolated single piles, and three (3×3, 3×4, and 3×5) groups under free-head condition (via a special loading apparatus). The subsoil profile was simplified (Rollins et al. 2006b) as follows: An overconsolidated stiff clay with s_u of 70 kPa to a depth of 1.34 m, a sand interlayer of 0.31 m (between a depth of 1.34 m and 1.65 m), and overconsolidated stiff clay with s_u of 105 kPa from 1.65 m to 3.02 m. The sand has a relative density D_r of 60% and an angle of friction of 36°. Groundwater was located at a depth 1.07 m. The closed-ended, steel pipe piles were driven to a depth of 11.9 m, with 324 mm OD, 9-mm wall thickness. It had a moment of inertia of 1.43×10^8 mm^4 (owing to irons attached for protecting strain gauges), thus E_pI_p = 28.6 MN-m^2. The center-to-center spacings were 5.65 pile diameters for the 3 × 3 group and 4.4 pile diameters for the 3 × 4 group in longitudinal (loading) direction, where they were 3.3 pile diameters (both cases) in the transverse direction.

Lateral load was exerted at (e=) 0.38 m above ground line for the virgin (single) pile test, and at e = 0.49 m for the 15 cyclic loadings, which furnished the measured responses of the single piles, as presented in Figure 9.17a. Lateral (static) load was applied on the 3 × 3 group also at e = 0.38 m, and the measured results are shown in Figure 9.17(b). The measured bending moment distributions along each pile in the three different rows are depicted in Figure 9.18 (at a displacement w_g of 64 mm). Lateral load test on the 3 × 4 group was undertaken under e = 0.48 m. The measured responses are shown in Figure 9.19a, and bending moment distributions along each pile are presented in Figure 9.20.

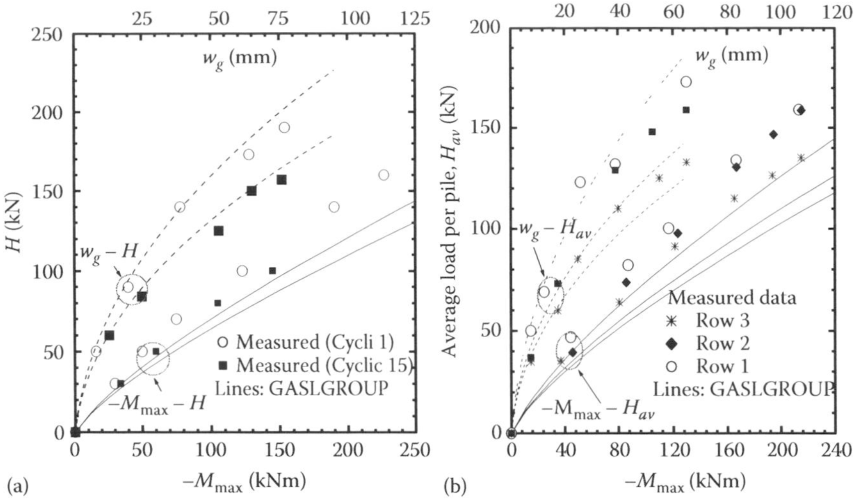

Figure 9.17 Predicted versus measured (Rollins et al. 2006a) response: (a) Single pile. (b) 3×3 group (static, 5.65*d*). (After Guo, W. D., *Proc 10th ANZ Conf on Geomechanics*, Brisbane, Australia, 2007.)

Cyclic loading tests were also conducted on the 3 × 4 group, with measured the response being plotted in Figure 9.19b. The program GROUP (Reese et al. 1996) was used to predict the response of the pile groups using back-figured *p*-multipliers (Rollins et al. 2006b), which are plotted in Figures 9.18 and 9.20.

GASLGROUP predictions were made using n = 1.6, α_0 = 0.05 m, $G = 190s_u$ (s_u = 75 kPa), along with N_g of 0.55 for single pile under static loading. A ground heaving of 75~100 mm was observed within the pile group during driving, which rendered the use of N_g = 0.6 for pile groups, whereas other parameters n and α_o remained unchanged. Values of *p*-multipliers were calculated using Equation 3.69 (Chapter 3, this book) (see Table 9.15), which are smaller than those adopted previously (Rollins et al. 2006b). The predictions using GASLGROUP under free-head condition are plotted in Figures 9.17 through 9.20 for all the single and group piles under static or cyclic loading. In particular, the elastic static response of the single piles is based on k = 48.818 MPa (G = 14.25 MPa, γ_b = 0.1368) and λ = 0.8082/m (α_N = 0.8713 and β_N = 0.7398). The figures indicate the current predictions compare well with the measured load-displacement curves. The maximum bending moment was overestimated at large load levels against measured data. This may be attributed to the use of a large equivalent stiffness E_p [$= E_pI_p/(\pi d^4/64)$] determined from the enlarged cross-section, otherwise the moment can also be well predicted using a low stiffness and slightly higher value of N_g.

The GASLGROUP calculation utilizes the E_p for an equivalent solid circular pile. Without this conversion of E_p, a good solution would also

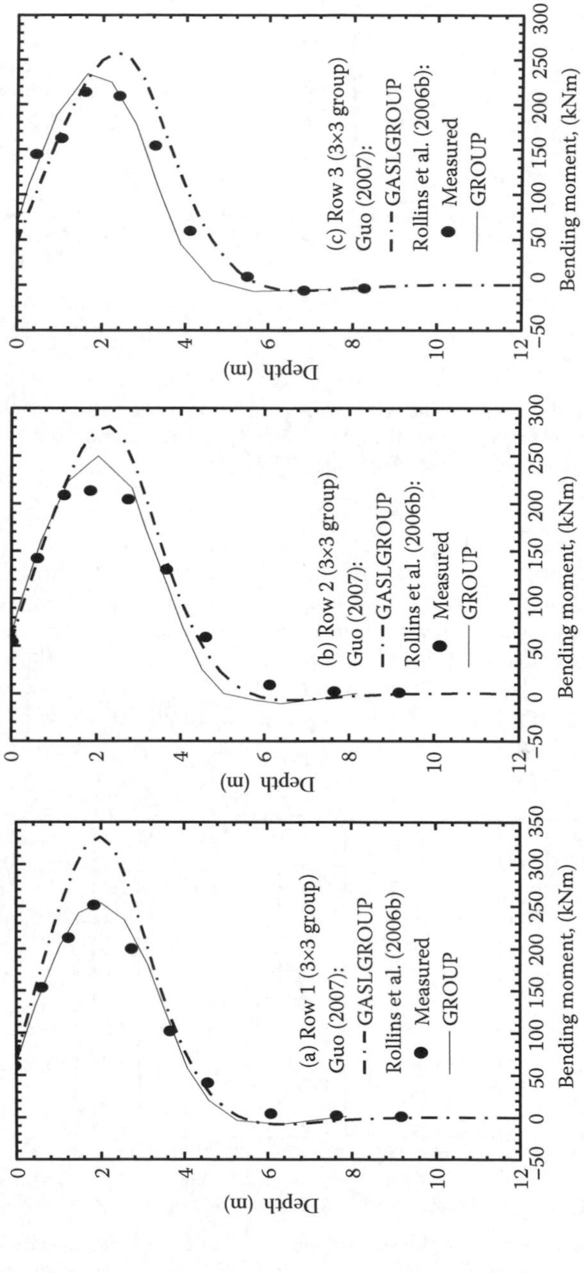

Figure 9.18 Predicted versus measured (Rollins et al. 2006a) bending moment profiles (3×3, w_g = 64 mm). (a) Row 1. (b) Row 2. (c) Row 3. (After Guo, W. D., *Proc 10th ANZ Conf on Geomechanics*, Brisbane, Australia, 2007.)

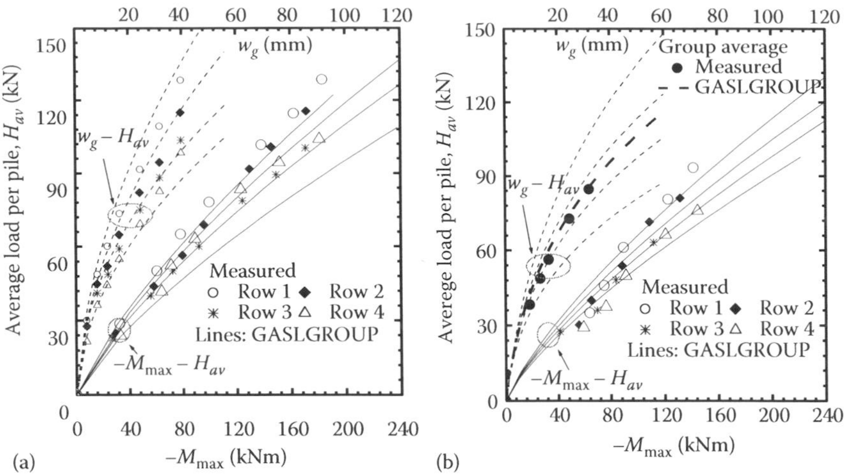

Figure 9.19 Predicted versus measured (Rollins et al. 2006a) response of 3×4 group. (a) Static at 4.4*d*, (b) Cyclic at 5.65*d*. (After Guo, W. D., *Proc 10th ANZ Conf on Geomechanics*, Brisbane, Australia, 2007.)

be gained, but yield untrue values of the parameters G and N_g. This example is for free-head piles in clay. The variation of p-multipliers is observed in comparison with the previous calculations for fixed-head (capped) piles in clay and sand (see later Chapter 11, this book). However, the prediction is affected more by the values of E_p than the p-multipliers. The input values (n = 1.6 and N_g = 0.55~0.6) are quite consistent with previous conclusions (e.g., Example 9.6 for a pile in stiff clay).

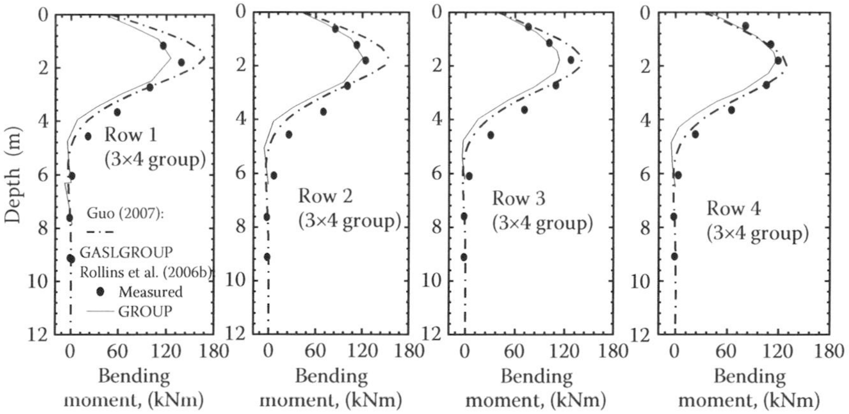

Figure 9.20 Predicted versus measured (Rollins et al. 2006a) bending moment (3×4, w_g = 25 mm). (After Guo, W. D., *Proc 10th ANZ Conf on Geomechanics*, Brisbane, Australia, 2007.)

Table 9.15 Input parameters (static loading)

References	Group	N_g[c]	s/d[b]	p_m *by row*				*n and* α_o
				1st	2nd	3rd	4th	
In situ tests (Rollins et al. 2006b) in clay. s_u = 75 kPa, γ_s = 15.3 kN/m³	3×3	0.60	5.65	$\frac{0.87}{0.95^d}$	$\frac{0.49}{0.88}$	$\frac{0.37}{0.77}$		For all cases: n = 1.6, α_o = 0.05, G = 14.25 MPa[a], N_g = 0.55 for single pile
	3×4	0.60	4.4	$\frac{0.85}{0.90}$	$\frac{0.6}{0.80}$	$\frac{0.45}{0.69}$	$\frac{0.32}{0.73}$	

Source: Guo, W. D., *Proceedings of 10th ANZ Conference on Geomechanics*, Brisbane, Australia, 2007.

[a] Shear modulus G (=190s_u), and p_mG for a pile in a group.
[b] Normalized center to center spacing in loading direction, but s = 3.3d within any a row.
[c] Under cyclic loading, the N_g was multiplied by a ratio of 0.65 to obtain that for single pile and by 0.8 for group piles.
[d] p_m in denominator was adopted by (Rollins et al. 2006b).

9.8 SUMMARY

New elastic-plastic solutions were presented for laterally loaded, infinitely long, free-head piles. They have been calibrated against FEA results for a pile in two different types of stratified soils. The solutions permit nonlinear response of the piles to be readily estimated right up to failure. Presented in explicit expressions of slip depth and LFP, the closed-form solutions may be used as a boundary element to represent pile–soil interaction in the context of analyzing a complicated soil-structure interaction. The analysis of ~70 pile tests to date provides the ranges of input parameters and allows the following remarks:

- The generic expression of p_u is applicable to all types of soils. It can generally accommodate existing LFPs through selecting a suitable set of parameters.
- Nonlinear response of free-head piles is dominated by the LFP and the maximum slip depth. It may be predicted by selecting a series of slip depth x_p, using GASLEP or the simplified expressions provided. Three input parameters, A_L, α_0, and n are sufficient for an accurate prediction of the nonlinear response.
- The pile response is insensitive to the shape of the LFP, under similar total resistance over a maximum slip depth x_p. Available LFPs may be fitted using Equation 9.2 and used in current solutions.
- The LFP may be generated using Equation 9.2 along with n = 0.5~2.0. A low value of n corresponds to a uniform strength profile, and a high one corresponds to a sharply changed strength profile. For a layered soil, the generated LFP may not be able to reflect

a detailed distribution profile of limiting force along a pile but an overall trend.

- The 32 free-head piles tested in cohesive soil are associated with (a) $k = (2.7\sim3.92)G$ with an average of $3.04G$; (b) $G = (25\sim315)s_u$ with an average of $92.3s_u$; (c) $n = 0.7$, $\alpha_0 = 0.05\sim0.2$ m (average of 0.11m), and $N_g = 0.6\sim3.2$ (1.6) for LFP, except for $n = 0.5\sim0.7$ for some atypical decrease shear strength profile, or $n = 1.0\sim2.0$ for a dramatic increase in strength with depth. The 20 free-head piles tested in sand are pertinent to (a) $k = (2.38\sim3.73)G$ with an average of $3.23G$; (b) $G = (0.25\sim0.62)N$ (MPa) with $G = 0.50N$ (MPa); and (c) $n = 1.7$, $\alpha_0 = 0$, and $N_g = (0.5\sim2.5)K_p^2$ with an average of $1.27K_p^2$ for the LFP.
- The 6 single piles under cyclic loading are well modeled using a reduced N_g of $(0.56\sim0.64)N_g$ (static). The N_g reduces to 80% that was used for static loading in simulating piles in groups.
- The LFP should be deduced using current solutions along with measured data to capture overall pile–soil interaction rather than sole soil failure mechanism and to cater for impact of various influence factors.
- Responses of laterally loaded pile groups are largely affected by the equivalent pile stiffness for an enlarged cross-section and insensitive to an accurate determination of p-multipliers. The p-multipliers concept generally works well.

Chapter 10

Structural nonlinearity and response of rock-socket piles

10.1 INTRODUCTION

Behavior of an elastic pile subjected to lateral loads was modeled using *p-y* concept (McClelland and Focht 1958; Matlock 1970; Reese et al. 1974; Reese et al. 1975). The model is characterized by the limiting force per unit length (p_u) mobilized between the pile and soil, especially at high load levels (Randolph et al. 1988; Guo 2006). The profile of p_u (or LFP) along the pile has generally been constructed using empirical or semi-empirical methods (Brinch Hansen 1961; Broms 1964a, 1964b; Matlock 1970; Reese et al. 1974; Reese et al. 1975; Barton 1982; Guo 2006) (see Chapter 3, this book). In light of a generic LFP, Guo (2006, 2009) developed elastic-plastic, closed-form (CF) solutions for laterally loaded free and fixed-head piles. The solutions well-capture the nonlinear response of lateral piles in an effective and efficient manner (Chapter 9, this book). They also enable the LFP (p_u profile) to be deduced against measured pile response. Nonetheless, they are confined to elastic piles with a constant E_pI_p. As noted in Chapter 9, this book, structural nonlinearity of pile body is an important issue at a large deflection (Nakai and Kishida 1982; Reese 1997; Huang et al. 2001; Ng et al. 2001; Zhang 2003), in particular for rock socket piles.

In foundations (for bridge abutments), locks and dams, transmission towers, in retaining walls, or in stabilizing sliding slopes, shafts (drilled piers or bored piles) are often rock socketed to take large lateral forces and overturning moments generated by traffic load, wind, water current, earth pressure, etc. Techniques for laterally loaded piles in soil are routinely employed to examine the response of such shafts (Carter and Kulhawy 1992; Reese 1997; Zhang et al. 2000). Shaft head deflection and slope were generally underestimated (DiGioia and Rojas-Gonzalez 1993), owing to neglect of (1) near-rock surface softening during post-peak deformation (Carter and Kulhawy 1992); (2) nonlinear flexural rigidity of the shafts; and (3) tensile slippage between the shaft and the rock, and owing to uncertainty about loading eccentricity. Those points are critical to rock-socketed piles.

Young's modulus of rock E was empirically correlated with the soil/rock mass classification and/or property indices, such as rock mass rating (RMR or RMR_{89}), uniaxial compressive strength (UCS) of intact rock, geology strength index (GSI), and rock quality designation (RQD) (Serafim and Pereira 1983; Rowe and Armitage 1987; Bieniawski 1989; Sabatini et al. 2002; Liang et al. 2009). These correlations are proved to be useful for vertically loaded shafts and may be further verified for lateral piles.

Reese (1997) proposed a p-$y(w)$ curve-based approach to incorporate nonlinear shaft-rock interaction and nonlinear flexural rigidity. As mentioned before, Guo (2001, 2006) developed the closed-form (CF) solutions for lateral piles (Figure 3.27, Chapter 3, this book) using a series of spring-slider elements along the pile shaft. Each element is described by an elastic-perfectly plastic p-$y(w)$ curve (p = resistance per unit length, and y = local shaft deflection). Under a lateral load H and M_o, limiting resistance per unit length p_u [FL^{-1}] is fully mobilized to the depth x_p (called slip depth) along the LFP. The solutions were proven sufficiently accurate compared to existing numerical methods (Guo 2006). They were thus extended to incorporate structural nonlinearity of piles (Guo and Zhu 2011).

Reese's approach is based on p_u estimated using an empirical curve and a parameter α_r (see Table 3.9, Chapter 3, this book). The p_u was late correlated to rock roughness by a factor g_s ($= p_u/q_u^{0.5}$, and q_u= average uniaxial compressive strength (UCS) of intact rock in MPa except where specified [FL^{-2}]) (Zhang et al. 2000), in light of the Hoek-Brown failure criterion (Hoek and Brown 1995). The latter resembles the resistance per unit area on a vertically loaded shaft in rock or clay (Horvath et al. 1983; Seidel and Collingwood 2001) using g_s = 0.2 (smooth) ~0.8 (rough). Reese's approach is good for pertinent cases, but it is not adequate for other cases (Vu 2006; Liang et al. 2009). As with lateral piles in sand or clay, a profile of limiting force per unit depth (LFP) for piles in rock is described as (see Chapter 3, this book):

$$p_u = A_L(\alpha_o + x)^n \tag{10.1}$$

In particular, the A_L is correlated to a limiting force factor N_g, pile diameter d; and the average q_u [FL^{-2}] for rock by $A_L = N_g q_u^{1/n} d$, as the p_u should resemble the τ_{max}. A ratio g_s ($= p_u/q_u^{1/n}$) of 0.2~0.8 is stipulated, and n = 0.7~2.3 as deduced for piles in clay, sand, and later on in rock. Equation 10.1 well-replicates the existing p_u expressions by Reese and Zhang et al. for rock-socketed shafts.

This chapter studies response of 6 nonlinear piles and 16 rock-socketed shafts to gain values of n and g_s, to assess the impact of loading eccentricity, and to formulate a simple approach to capture impact of structural nonlinearity on response of lateral shafts. The analysis uses closed-form solutions and measured data.

10.2 SOLUTIONS FOR LATERALLY LOADED SHAFTS

The solutions for a free-head pile are directly used here, but with a new p_u profile and a constant k, see Figure 3.27a (Chapter 3, this book) with a lateral load H applied at an eccentricity e above soil/rock surface, creating a moment M_o (= He) about the soil/rock surface. For instance, the normalized pile-head load $\bar{H}(= H\lambda^{n+1}/A_L)$ and the mudline deflection $\bar{w}_g$ $(= w_g k\lambda^n/A_L)$ are given by

$$\bar{H} = \frac{H\lambda^{n+1}}{A_L} = \frac{0.5\bar{x}_p^n[(n+1)(n+2) + 2\bar{x}_p(2+n+\bar{x}_p)]}{(\bar{x}_p + 1 + \bar{e})(n+1)(n+2)} \tag{10.2}$$

$$\begin{aligned}\bar{w}_g = \frac{w_g k\lambda^n}{A_L} = \frac{2}{3}\bar{x}_p^{n+3}\frac{2\bar{x}_p^2 + (2n+10)\bar{x}_p + n^2 + 9n + 20}{(\bar{x}_p + 1 + \bar{e})(n+2)(n+4)} + \frac{(2\bar{x}_p^2 + 2\bar{x}_p + 1)\bar{x}_p^n}{\bar{x}_p + 1 + \bar{e}} \\ + \frac{2\bar{x}_p^4 + (n+4)(\bar{x}_p + 1)[2\bar{x}_p^2 + (n+1)(\bar{x}_p + 1)]}{(\bar{x}_p + 1 + \bar{e})(n+1)(n+4)}\bar{e}\bar{x}_p^n\end{aligned} \tag{10.3}$$

Other expressions are provided in Chapter 9, this book, concerning the normalized maximum bending moment $\bar{M}_{max}$ $(= M_{max}\lambda^{2+n}/A_L)$ and its depth d_{max}; and profiles of deflection, bending moment, and shear force using the normalized depths $\bar{x}$ $(= \lambda x)$ and $\bar{z}$ $(= \bar{x} - \bar{x}_p$, with $\bar{x}_p = \lambda x_p)$, respectively, for plastic and elastic zones. The solutions are characterized by the inverse of characteristic length λ $[= \sqrt[4]{k/(4E_pI_p)}]$, the slip depth x_p, and the LFP. In particular, it is often sufficiently accurate to stipulate $\alpha_o = 0$ (zero resistance at ground level) and $N_p = 0$ (uncoupled soil layers), which result in the simplified expressions. The numerical values of these expressions are obtained later using the program GASLFP (see Figure 10.1), which may also be denoted as CF.

10.2.1 Effect of loading eccentricity on shaft response

Laterally loaded (especially rock-socketed) shafts may be associated with a high loading eccentricity. The response to normalized eccentricity $e\lambda$ of a shaft was gained using the simplified closed-form solutions and is presented (a) in Figure 10.2a$_1$ and a$_2$, against the normalized maximum bending moment $\bar{M}_m$, the normalized applied moment $\bar{M}_o$; (b) in Figure 10.2b$_1$ and b$_2$, against the normalized pile-head load $\bar{H}$; and (c) in Figure 10.3c$_1$, and c$_2$ against the inverse of pile-head stiffness $\bar{w}_t/\bar{H}$. Given n = 1, the same moment and load for rigid piles are also obtained using the rigid pile

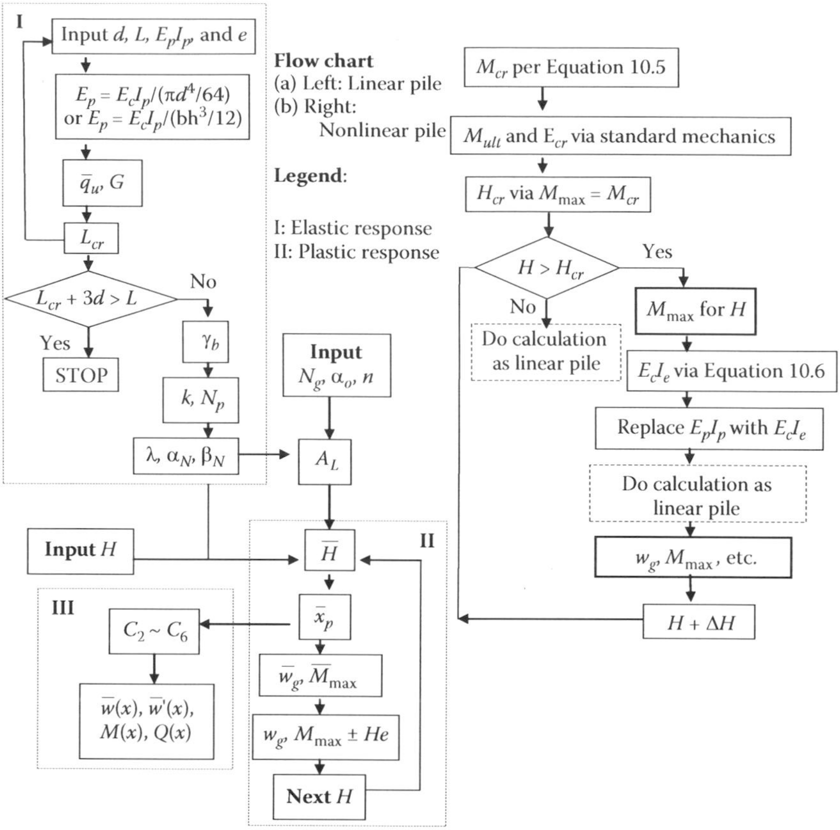

Figure 10.1 Calculation flow chart for the closed-form solutions (e.g., GASLFP) (Guo 2001b; 2006).

solutions (Guo 2008). The normalized response is plotted versus the normalized eccentricity of $\pi e/L$ (= $e\lambda$, as $\lambda = \pi/L$, L = length for a rigid pile) in Figure 10.3a$_1$ and b$_1$, for tip-yield state, and for both tip-yield and rotation-point yield state, respectively. Note the tip-yield state implies the soil just yields at pile-tip level; rotation-point yield state means full mobilization of limiting strength along entire pile length (see Chapter 8, this book). The rigid piles are classified by $\lambda L > 2.45\sim3.12$ using $k/G = 2.4\sim3.9$ (Guo and Lee 2001), and λ is thus taken as π/L.

Figure 10.2a$_1$, b$_1$, and a$_2$ show that (1) the M_o or M_m at $x_p\lambda = 0.172$ (n = 1) ~ 0.463 (n = 2.3) of a flexible shaft (with $L > L_c$) may match the ultimate values of a rigid shaft rotating about its tip or head; (2) only flexible shafts have $x_p\lambda > 0.172$ (n = 1) ~ 0.463 (n = 2.3); (3) a flexible shaft may allow a large slip depth x_p and thus a higher M_m than a rigid short shaft; and (4) a high $e\lambda$ (e/L) > 3 renders $M_m \approx M_o$ (= He), and allows pile-head deflection w_t to be approximated by the following "cantilever beam" solution:

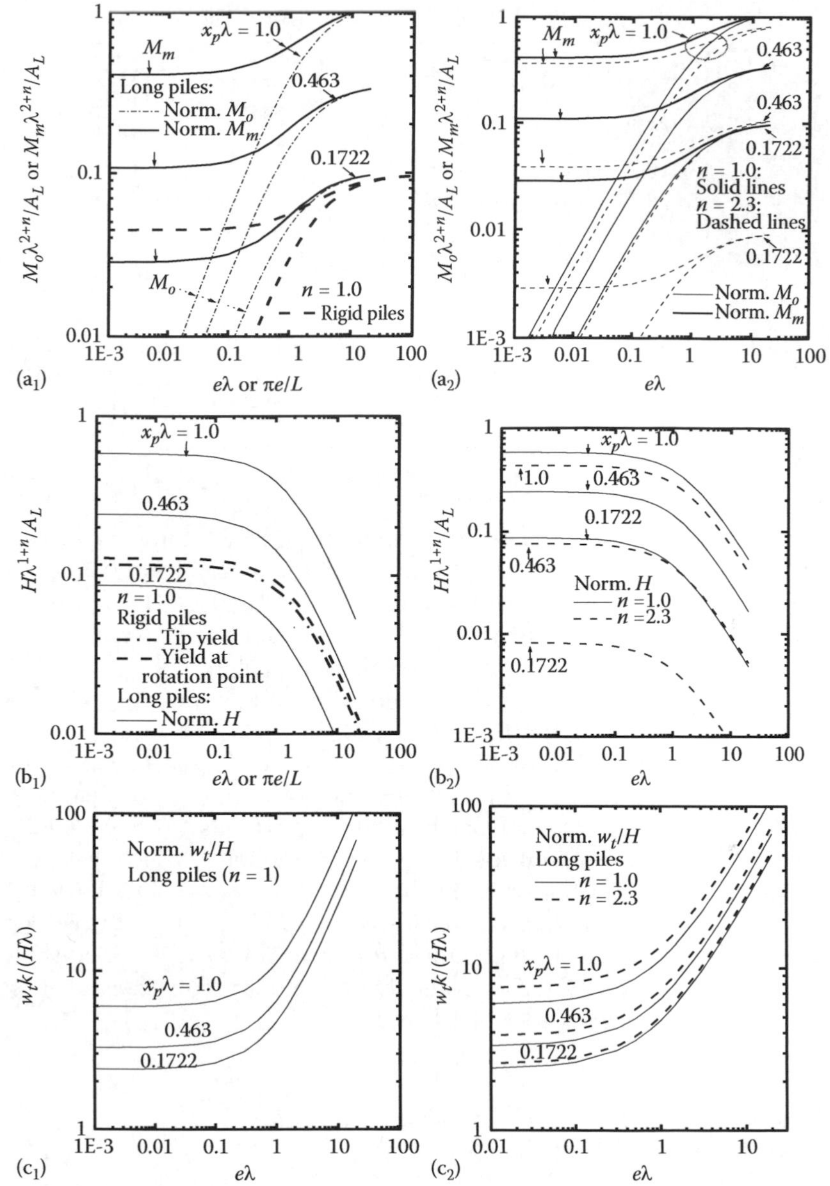

Figure 10.2 Normalized bending moment, load, deflection for rigid and flexible shafts. (a_1–c_1) $n = 1.0$. (a_2–c_2) $n = 1.0$ and 2.3.

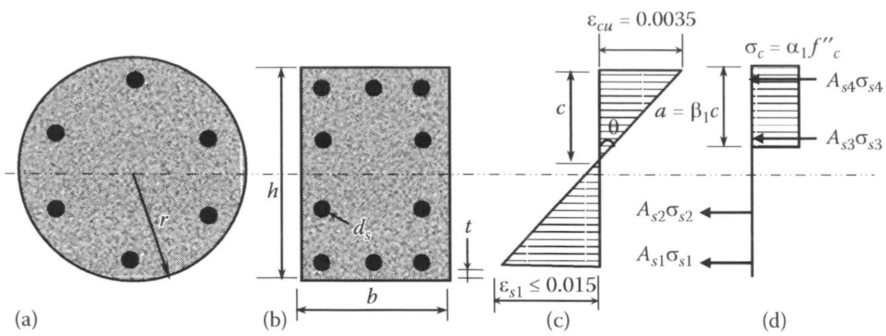

Figure 10.3 Simplified rectangular stress block for ultimate bending moment calculation. (a) Circular section. (b) Rectangular section. (c) Strain. (d) Stress. (After Guo, W. D., and B. T. Zhu, *Australian Geomechanics* 46, 3, 2011.)

$$w_t = w_g + He^3 / (3E_c I_e) + \theta_g e \tag{10.4}$$

where w_g and w_t = pile deflections at groundline and pile-head level, respectively, θ_g = rotation or slope at ground level, $E_c I_e$ = an effective flexural rigidity. The load H and displacement w_t are calculated by $H = \bar{H}A_L / \lambda^{n+1}$ and $w_g = \bar{w}_g A_L / k\lambda^n$ using Equations 10.2 and 10.3, respectively.

As discussed in Chapter 9, this book, the parameters A_L, α_o, n, and k may be readily deduced by matching the CF solution with the observed shaft response spectrum of (1) $H - M_{max}$; (2) $H - w_t$; (3) $H - x_{max}$ (x_{max} = depth of maximum bending moment); and/or (4) $H - \theta_t$ (shaft head slope). Three measured spectrums (or profiles) would suffice unique deduction of n, A_L (α_o or N_{co}), and k for an elastic shaft. In particular, the obtained k from piles in clay and sand offered the Young's modulus, which was plotted against USC q_u in Chapter 1, this book, and shows $E = (60\text{~}400)q_u$ in which q_u is in MPa. The E/q_u ratio featured by USC $q_u < 50$ MPa (weak r ock) is strikingly similar to that gained for vertically loaded drilled shaft (Rowe and Armitage 1987), as explained later on. Further back-estimation is conducted here for rock-socketed shafts.

10.3 NONLINEAR STRUCTURAL BEHAVIOR OF SHAFTS

10.3.1 Cracking moment M_{cr} and effective flexural rigidity $E_c I_e$

Under lateral loading, crack occurs at extreme fibers of a concrete shaft once the tensile stress approaches the modulus of rupture f_r (= $M_{cr} y_t / I_g$). The cracking moment M_{cr} is taken as a maximum bending moment M_{max} in the shaft at which the crack incepts. I_g is moment of inertia of the shaft cross-section about centroidal axis neglecting reinforcement, and y_r is the distance between extreme (tensile) fibers and the centroidal axis. With empirical correlation of $f_r = k_r(f_c')^{0.5}$, the cracking moment M_{cr} is given by (ACI 1993):

$$M_{cr} = k_r\sqrt{f_c'}I_g / y_r \tag{10.5}$$

where f_c' = characteristic compressive strength of concrete (kPa); k_r = 19.7~31.5, a constant for a normal weight concrete beam. The k_r for drilled shafts may be gained using M_{cr} under a critical load H_{cr} beyond which deflection increases drastically (shown later).

An effective flexural rigidity E_cI_e is conservatively taken for any section of the shaft. The Young's modulus of concrete E_c may be correlated by E_c = $151{,}000(f_c')^{0.5}$ (f_c' and E_c in kPa). The effective moment of inertia I_e reduces with increase in $M_{\max}$ and is given by (ACI. 1993):

$$E_cI_e = \left(\frac{M_{cr}}{M_{\max}}\right)^3 E_cI_p + \left[1 - \left(\frac{M_{cr}}{M_{\max}}\right)^3\right] E_cI_{cr} \tag{10.6}$$

where I_{cr} = moment of inertia of cracked section at which a fictitious hinge (attaining ultimate bending moment, M_{ult}) occurs. Note E_cI_e and E_cI_p are equivalent to E_pI_p and EI, respectively, and $E_cI_{cr} = (EI)_{cr}$. The M_{ult} and I_{cr} are obtained by using bending theory (moment-curvature method) (Hsu 1993) and limit state-simplified rectangular stress block method (Whitney 1937; BSI 1985; EC2 1992; ACI 1993). They may be numerically estimated using nonlinear stress-strain relationships for concrete and steel (Reese 1997). The limit-state method in classical mechanics is used herein for drilled shafts, as presented previously (Guo and Zhu 2011).

10.3.2 M_{ult} and I_{cr} for rectangular and circular cross-sections

The ultimate bending moment M_{ult} and I_{cr} may be obtained by bending theory via moment-curvature method involving stress-strain relationships of concrete and reinforcement (Hsu 1993). For instance, Reese (1997) provided the M_{ult} and I_{cr} for closely spaced-crack assumption using a Hognestaad parabolic stress-strain relationship for concrete and an elastic perfectly plastic stress-strain relationship for steel. The cracks may be initiated at different locations, and the concrete stress-strain relationship depends on construction and strength, rate and duration of loading, etc. In particular, a rational flexural theory for reinforced concrete is yet to be developed (Nilson et al. 2004). Pragmatically, limit-state design underpinned by simplified rectangular stress block method (referred to as RSB hereafter) (Whitney 1937) has been widely adopted to calculate the M_{ult} and I_{cr} (BSI 1985; EC2 1992; ACI 1993).

At ultimate state, the pile may fail either by crushing of the concrete in the outmost compression fiber or by tension of the outmost steel rebar. The compression failure may occur once a maximum strain reaches ε_{cu} of 0.0035 (BSI 1985; EC2 1992; Nilson et al. 2004) or 0.003 (ACI 1993). The tension failure has not been defined (BSI 1985; EC2 1992; ACI 1993), perhaps owing to a diverse steel-failure strain (i.e., 10% to 40%) (Lui 1997). In this investigation, the failure is simply defined once the maximum steel strain reaches ε_{su} of 0.015 (Reese 1997), as the definition has limited impact on analyzing lateral piles, which are dominated by compressive failure.

Figure 10.3 shows typical cracked cross-sections for a circular pile (Figure 10.3a) and a rectangular pile (Figure 10.3b) with four rows of rebars. A linear strain distribution is stipulated (Figure 10.3c). The compressive stress in concrete is simplified as a rectangular stress block (Figure 10.3d) characterized by those highlighted in Table 10.1, including the intensity of the stress, σ_c, the depth a of the stress block, and the stress induced in a rebar in the i-th row, σ_{si}.

The stresses must meet the equilibrium of axial force of Equation 10.7 and bending moment of Equation 10.8 in either section:

$$\int_A \sigma \, dA = P_{xc} + P_{xs} = P_x \tag{10.7}$$

$$\int_A \sigma x_1 \, dA = M_c + M_s = M_n \tag{10.8}$$

Table 10.1 Stress block for calculating M_{ult}

Items	*Description*
σ_c	$\sigma_c = \alpha_1 f_c''$ with $\alpha_1 = 0.85$, and $f_c'' = 0.85 f_c'$.
a	$a = \beta_1 c$, where $\beta_1 = 0.85 - 0.05(f_c' - 27.6)/6.9$ and $\beta_1 \geq 0.65$, and c = distance from the outmost compression fiber to neutral axis: (1) $c = 0.0035/\theta$ if concrete fails, otherwise (2) $c = 2r - t - d_s/2 - 0.015/\theta$ concerning the debound (tension) failure. The diameter of "$2r$" (r = radius of a circular pile) is replaced with h for a rectangular pile; θ = curvature at limit state, d_s = diameter of rebars, and t = cover thickness (Figure 10.3b).
σ_{si}	σ_{si} = stress in a rebar in i-th row, $\sigma_{si} = \phi_r f_y$ (yield stress) if $\varepsilon_{si} > \varepsilon_{sy}$ otherwise, $\sigma_{si} = \varepsilon_{si} E_s$. Note the i-th row of rebars are counted from the farthest tensile row towards the compressive side (see Figure 10.3d); $\varepsilon_{sy} = \phi_r f_y / E_s$, E_s = Young's modulus of reinforcement, typically taken as 2×10^8 kPa; and $\phi_r = 0.9$, reinforcement reduction factor for tension and flexure.

Source: Guo, W. D., and B. T. Zhu, *Australian Geomechanics* 46, 3, 2011.

where A = area of cross-section excluding the concrete in tension; σ = normal stress in concrete (σ_c) or rebars (σ_{si}); $P_{xc} = \sigma_c A_c$, axial load taken by the concrete; A_c= area of concrete in compression; $P_{xs} = \Sigma\sigma_{si}A_s$, axial load shared by the rebars; A_{si} = total area of rebars in the ith row; P_x = imposed axial load; x_1 = distance from neutral axis; M_c and M_s = moments with regard to the neutral axis respectively induced by normal stress in the concrete and rebars; and M_n = nominal or calculated ultimate moment. Generally, it is straightforward to obtain the values of P_{xc}, P_{xs}, and M_s. Nevertheless, integration involved in estimating M_c may become tedious, particularly for an irregular cross-section. For a rectangular or a circular cross-section, M_c may be calculated by Equations 10.9 and 10.10, respectively

$$M_c = f_c''ab(2c-a)/2 \tag{10.9}$$

or

$$M_c = 2r^2 f_c''\left\{\left[\frac{a}{3r}\left(2-\frac{a}{r}\right)+\frac{1}{2}\left(1-\frac{c}{r}\right)\left(1-\frac{a}{r}\right)\right]\sqrt{2ra-a^2} + \frac{r-c}{2}\left[\arcsin\left(\frac{r-a}{r}\right)-\frac{\pi}{2}\right]\right\} \tag{10.10}$$

where b = width of a rectangular pile.

The steps for computing M_{ult} and I_{cr} are tabulated in Table 10.2. The nominal ultimate bending moment M_n of $M_c + M_s$ is calculated first by Equation 10.8; it is next reduced to the ultimate bending moment, M_{ult} (= ϕM_n) by a factor of ϕ (see Table 10.2); and finally the cracked flexural rigidity, $E_c I_{cr}$, is obtained by

Table 10.2 Calculation of M_{max} and I_{cr}

Steps	*Actions*
1	Select an initial curvature θ (see Figure 10.3) of $0.0035/r$;
2	Calculate position of neutral axis by evaluating c and $a = \beta_1 c$;
3	Compute the axial forces in concrete P_{xc} and rebars P_{xs};
4	Find squash capacity P_{xu} by: $P_{xu} = f_c'A_c + E_s(f_c'/E_c)A_c$, where A_c, A_s = total areas of concrete and rebars, respectively;
5	If $P_{xc} + P_{xs} - P_{xu} > 10^{-4}P_{xu}$, increase θ (otherwise reduce θ) by a designated increment, say $0.0035/1000r$. Repeat steps 1~4 until the convergence is achieved.
6	Estimate the nominal ultimate bending moment of $M_c + M_s$ by Equation 10.8.
7	Compute M_{ult} using $M_{ult} = \phi(M_c + M_s)$, in which ϕ = a reduction factor, to accommodate any difference between actual and nominal pile dimension, rebar cage off-position, and assumptions and simplifications (Nilson et al. 2004). • $\phi = 0.493 + 83.3\varepsilon_{sl}$ and $0.65 \le \phi \le 0.9$, for laterally tied rebar cage • $\phi = 0.567 + 66.7\varepsilon_{sl}$ and $0.70 \le \phi \le 0.9$, for spirally reinforced rebar cage.
8	Compute the cracked flexural rigidity, $E_c I_{cr}$ by $E_c I_{cr} = M_{ult}/\theta$.

Source: Guo, W. D., and B.T. Zhu, *Australian Geomechanics* 46, 3, 2011.

$$E_c I_{cr} = M_{ult} / \theta \qquad (10.11)$$

The aforementioned calculation procedure has been entered into a simple spreadsheet program that operates in EXCEL™. The program offers consistent results against the ACI and AS3600 methods (not shown), and thus was used.

10.3.3 Procedure for analyzing nonlinear shafts

Equation 10.6 indicates the degeneration of the rigidity E_cI_e from the intact elastic value E_cI_p to the ultimate cracked value of E_cI_{cr} (nonlinear shaft). This is normally marked with an increase in shaft deflection. The nonlinear increase may be resolved into incremental elastic, and thus calculated using the solutions for a linear elastic shaft by replacing the rigidity E_pI_p in the solutions with the variable rigidity E_cI_e for each loading. The $M_{\max}$ and the deflection w_g at each load level were first estimated assuming an elastic shaft (discussed earlier). Subsequently, incremental elastic response (Figure 10.1) is obtained by calculating: (1) M_{cr} using Equation 10.5; (2) M_{ult} and E_cI_{cr} in terms of pile size, rebar strength and layout, and concrete strength (Guo and Zhu 2011); (3) H_{cr} (the cracking load) by taking $M_{\max}$ as M_{cr} (via limiting state method); (4) E_cI_e (effective bending rigidity for $H > H_{cr}$) using Equation 10.6 and values of E_cI_p, E_cI_{cr}, M_{cr} and $M_{\max}$; and (5) $M_{\max}$ and w_g at the H by using the calculated E_cI_e to replace E_pI_p. The process is repeated for a set of H (thus $M_{\max}$) and conducted using the GASLFP as well.

The difference between using E_cI_p and E_cI_e is remarkable in the predicted w_g, but it is negligible in the predicted $M_{\max}$. As outlined before, the parameters n, A_L (α_o or N_{co}), and k may be deduced using the elastic shaft response. The value of k_r (via M_{cr}) has to be deduced from crack-inflicted nonlinear shaft response as is elaborated next.

10.3.4 Modeling structure nonlinearity

As mentioned early, the pile analysis incorporating structural nonlinearity is essentially the same as that for a linear (elastic) pile but with new values of bending rigidity E_pI_p beyond cracking load. In estimating E_pI_p using Equation 10.6, the $M_{\max}$ may be gained using elastic-pile (EI) analysis, but for a pile with an extremely high EI (shown later on), as development of crack generally renders little difference in values of $M_{\max}$. The analysis for an elastic pile was elaborated previously (Guo 2006) and is highlighted in Table 10.3. The new value of E_pI_p for each $M_{\max}$ (or load) may be calculated using Equation 10.6, together with M_{cr} from Equation 10.5, and $(EI)_{cr}$ (via Equation 10.11) from the rectangular block stress method (RBS).

Table 10.3 Procedure for analysis of nonlinear piles

Steps	*Actions*
Elastic pile	
1	Input pile d, L, E_pI_p, I_p, and e [with $E_p = E_pI_p/(\pi d^4/64)$ or $= E_pI_p/(bh^3/12)$]
2	Determine parameters α_o, n, and N_g for LFP and k and N_p (see Chapter 3, this book).
3	Calculate critical length L_c if the value of $L_c < 5d$ for sand (or $< 10d$ for clay), the average shear modulus is reselected to repeat the calculation of 1 and 2.
Nonlinear pile	
1	Calculate the M_{max} and pile deflection w_g at each load level for elastic pile (EI).
2	Determine M_{cr} using Equation 10.5.
3	Calculate the critical load H_{cr} by taking M_{max} in step 2 as the M_{cr}.
4	Compute the M_{ult} and $(EI)_{cr}$ using the rectangular block stress method (RBS), in terms of pile size, rebar strength and its layout, and concrete strength.
5	Calculate bending rigidity E_pI_p for each load level above the H_{cr} by substituting values of EI, $(EI)_{cr}$, M_{cr}, and a series of M_{max} into Equation 10.6.
6	Compute pile deflection using new E_pI_p and program GASLFP for each load.

Source: Guo, W. D., and B. T. Zhu, *Australian Geomechanics* 46, 3, 2011.

The M_{cr} (thus k_r) and E_pI_p may be deduced using GASLFP and measured pile response, as are parameters G and LFP for linear piles. The measured M_{cr} defines the start of deviation of the predicted H–w_g curve (using EI) from measured data. In other words, beyond M_{cr}, the E_pI_p has to be reduced to fit the predicted curve with the measured H–w_g curve until final cracked rigidity $(EI)_{cr}$. The back-figured value of k_r is used to validate Equation 10.5, and the values of E_pI_p for various M_{max} together with M_{cr} and $(EI)_{cr}$ are used to justify Equation 10.6.

The impact of reduced rigidity (cracking) is illustrated in terms of a fictitious pile predicted using GASLFP. The pile with $d = 0.373$ m, $L = 15.2$ m, and $EI = 80.0$ MN-m^2 was installed in sand having $\phi = 35°$, $\gamma_s' = 9.9$ kN/m^3, $G = 11.2$ MPa, and $\nu_s = 0.3$. The LFP was described by $\alpha_0 = 0$, $n = 1.6$, and $N_g = 0.55K_p^2$. Under a head load H of 400 kN, the predicted profiles of deflection y, bending moment M, slope θ, shear force Q, and on-pile force per unit length p are illustrated in Figure 10.4, together with the pile having a reduced stiffness E_pI_p of 8.0 MN-m^2. The figures indicate that the reduction in rigidity leads to significant increase in the pile deflection and slope (all depth), and some increases in local maximum Q and p and in slip depth, x_p; however, there is little alteration in the bending moment profile. The results legitimate the back-estimation of E_pI_p using measured w_g and the predicted M_{max} from elastic-pile in Equation 10.6.

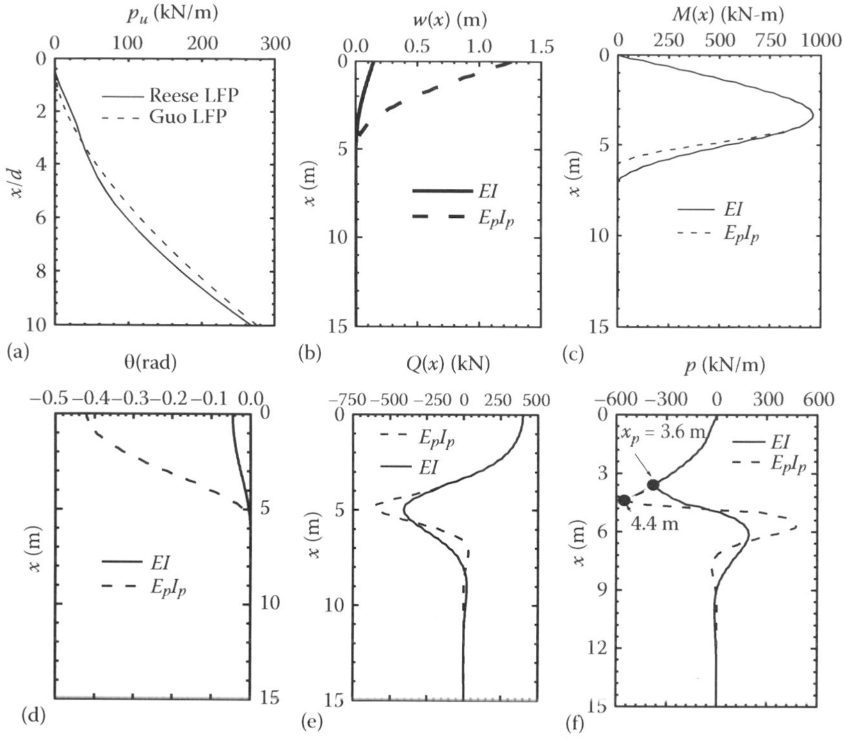

Figure 10.4 Effect of concrete cracking on pile response. (a) LFPs. (b) Deflection ($w(x)$) profile. (c) Bending moment ($M(x)$) profile. (d) Slope (θ) profile. (e) Shear force ($Q(x)$) profile. (f) Soil reaction (p) profile. (After Guo, W. D., and B. T. Zhu, *Australian Geomechanics* 46, 3, 2011.)

10.4 NONLINEAR PILES IN SAND/CLAY

Investigation was conducted into four piles tested in sand (SN1–4) and two in clay (CN1–2) that show structure nonlinearity (see Figure 10.5). The pile properties are tabulated in Tables 10.4 and 10.5, respectively. The sand properties are provided in Table 10.6. In all cases, (1) shear modulus was taken as an average value over 10d; (2) Poisson's ratio was assumed to be 0.3; and (3) γ_s' and ϕ' were taken as the average values over 5d.

10.4.1 Taiwan tests: Cases SN1 and SN2

A bored pile B7 (termed as Case SN1) and a prestressed pile P7 (Case SN2) were instrumented with strain gauges and inclinometers and tested individually under a lateral load applied near the GL (e = 0) (Huang et al. 2001). The soil profile at the site consisted of fine sand (SM) or silt (ML), and occasional silty clay. The ground water was located 1 m below the GL. Over a

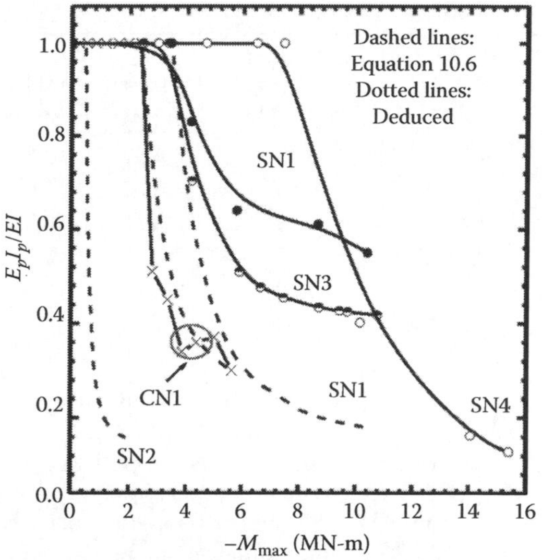

Figure 10.5 Comparison of $E_pI_p/EI \sim M_{max}$ curves for the shafts.

Table 10.4 Summary of information about piles investigated

		Pile details						
Case	*Reference*	d *(m)*	L *(m)*	EI *(GN-m²)*	e *(m)*	f'_c *(MPa)*	f_y *(MPa)*	t *(mm)*
SN1	Huang et al. (2001)	1.5	34.9	6.86	0	27.5	471	50
SN2	Huang et al. (2001)	0.8	34.0	0.79	0	78.5 (20.6)[a]	1226 (471)	30
SN3	Ng et al. (2001)	1.5	28	10	0.75	49	460	
SN4[b]	Zhang (2003)	0.86	51.1	47.67	0	43.4	460	75
CN1	Nakai and Kishida (1977)	1.548	30	16.68	0.5	153.7		
CN2		1.2	9.5	2.54	0.35	27.5		

[a] 78.5 is for the outer prestressed pipe pile and 20.6 for the infill
[b] 0.86 × 2.8(m^2) for a rectangular pile.

depth of 15 m below the GL, the average SPT blow count $\tilde{N}$ was 16.9; the friction angle ϕ' was 32.6° (Teng 1962), and the effective unit weight γ'_s was 10 kN/m³.

Example 10.1 Case SN1

Pile B7 was 34.9 m in length and 1.5 m in diameter. It was reinforced with 52 rebars of $d = 32$ mm and a yield strength f_y of 471 MPa. The pile

Table 10.5 Nonlinear properties of piles investigated

Cases	λ	M_{cr} *from ACI method (MN-m) (1)*	M_{cr} *back-estimated (MN-m) (2)*	k_r *derived from (2)*	M_{ult} *(MN-m)*	$(EI)_{cr}$/EI (%)	L_c/d
SN1	0.19298	1.08~1.73	3.45	62.7	10.50	55.0	8.9
SN2	0.3287	0.28~0.44	0.465	33.0	1.89	14.6	9.8
SN3	0.1748	1.44~2.31	3.34	45.5	10.77	40.0	9.9
SN4	0.242	4.61~7.38	7.42	31.7	10.13	1.9	27.7
CN1	0.3619	2.81~4.50	2.38	16.7	5.57	24.1	13.3
CN2	0.2984	0.2984	0.63	22.3		0.5	7.1

has a concrete cover t of 50 mm thick with a compressive strength f_c' of 27.5 MPa. The EI and I_g were estimated as 6.86GN-m^2 and 0.2485 m^4 respectively, which gives $E_p = 2.76 \times 10^7$ kPa. The cracking moment M_{cr} was estimated as 1.08~1.73 MN-m using Equation 10.5, k_r= 19.7~31.5 and $y_r = d/2 = 0.75$ m. The RSB method predicts a squash compression capacity P_x of 135.2 MN, an ultimate moment M_{ult} of 8.77 MN-m (see Table 10.7), and $(EI)_{cr}$ of 1.021 GN-m^2 [i.e., $(EI)_{cr}/EI = 0.149$].

Response of an elastic pile (with EI) was predicted using the p_u profile described by $n = 1.7$, $\alpha_0 = 0$, and $N_g = 0.9K_p^2 = 10.02$ (see Figure 10.6a), as deduced from the DMT tests (Huang et al., 2001); and G = 10.8 MPa (= 0.64N) (Guo 2006). The predicted load H – deflection w_g curve is shown in Figure 10.6b as EI, together with measured data. The figure indicates a cracking load H of 1.25 kN and an ultimate capacity of 2,943 kN (indicating by a high rate of measured deflection). These leads to a moment M_{cr} of 3.45 MN-m (at 1.25 kN) and k_r = 62.7 (i.e., $k_r \approx$ 2~3 times that suggested by ACI) and a M_{ult} of 10.5 MN-m (at 2943 kN), which also exceeds the calculated M_{ult} (using RSB method) by ~20%, being deduced.

Furthermore, the moment-dependent E_pI_p was deduced by fitting the measured H–w_g curve and is provided in Table 10.8. In particular,

Table 10.6 Sand properties and derived parameters for piles in sand

		Soil properties				*Input parameters (deduced)*		
Case	*Soil type*	γ_s' *(kN/m³)*	ϕ' *(o)*	N	D_r *(%)*	N_g/K_p^2	*G/N (MPa)*	n
SN1	Submerged silty sand	10	32.6	16.9		0.9	0.64	α_0=0 n = 1.7
SN2	Submerged silty sand	10	32.6	16.9		1.0	0.64	
SN3	Submerged sand	11.9	35.3	17.1	44	1.2	0.64	
SN4	Silty sand with occasional gravel	13.3	49	32.5	61	0.55	0.4	

Table 10.7 M_{ult} determined for typical piles in sand using RSB method

Case	β	θ (10^{-3} m^{-1})	c (mm)	ε_{sl} (10^{3})	a (mm)	P_{xc} (kN)	P_{xs} (kN)	M_c (MN-m)	M_s (MN-m)	ϕ	M_n (MN-m)	M_{ult} (MN-m)
SN1	0.85	8.59	408	5.87	347	6143.3	–6143.3	1.25	8.50	0.9	9.75	8.77
SN2	0.65	16.4	214	5.86	139	2865.1	–2865.4	0.38	1.72	0.9	2.10	1.89
SN4	0.74	11.39	307	14.9	226	6092.8	–6083.5	1.18	10.07	0.9	11.25	10.13

Source: Zhu, B.T., unpublished research report, Griffith University, 2004.

$(EI)_{cr}/EI$ was deduced as 0.55 and $(EI)_{cr}$ as 3.773GN-m^2. The $(EI)_{cr}$ is 3.7 times the calculated value of 1.021GN-m^2 using the RSB method.

Using the calculated $(EI)_{cr}$ of 1.021GN-m^2, M_{cr} = 3.45 MN-m, and M_{ult} = 10.5 MN-m, the E_pI_p was estimated using Equation 10.6 for each M_{max} (obtained using EI) and is plotted in Figure 10.5 as a dashed line. This E_pI_p allows w_g, M_{max}, and x_p to be predicted for each specified load, as are tabulated in Table 10.8 (analysis based on Equation 10.6). The predicted $H–w_g$ and $H–M_{max}$ curves are denoted as dash dot lines in Figure 10.6b. Comparison between the predictions using elastic EI and nonlinear E_pI_p indicates that the M_{max} predicted using E_pI_p is ~ 3.2% less than that gained using EI; and the slip depth x_p at H= 2,571 kN increases to 3.08d (using E_pI_p) from 2.49d (using EI) and the deflection w_g exceeds the measured value by 115%. Both conclusions are consistent with the fictitious pile. Nonetheless, using the calculated $(EI)_{cr}/EI$ of 0.1497 would render the deflection to be overestimated by ~2.3 times (compared to the measured 128.3 mm), and an accurate value of $(EI)_{cr}$ is important.

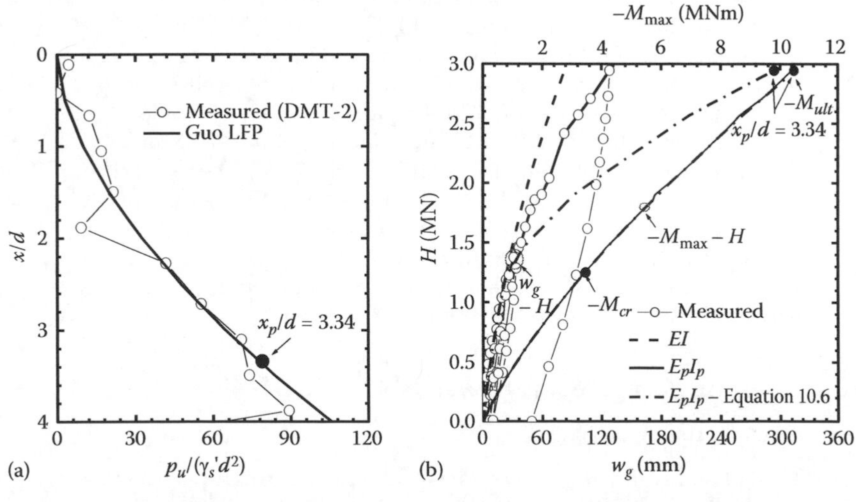

Figure 10.6 Comparison between measured (Huang et al. 2001) and predicted response of Pile B7 (SN1). (a) LFPs. (b) H-w_g and H~M_{max} curves.

Table 10.8 Analysis of pile B7 (SN1) using elastic pile, Equation 10.6, and deduced E_pI_p

	H (kN)	1250	1462	1903	2571	2943
Elastic pile analysis	M_{max} using EI (kN-m)	3447.0	4206.5	5910.0	8773.5	10501
	w_g using EI (mm)	25.9	32.13	46.58	72.4	88.79
	x_p/d	1.65	1.80	2.10	2.49	2.69
Analysis based on Equation 10.6	E_pI_p/EI	1	0.617	0.318	0.200	0.179
	w_g using E_pI_p (mm)	25.9	41.05	92.3	208.9	294.2
	M_{max} using E_pI_p (kN-m)	3447.0	4092.9	5721.9	8618.3	10372.1
	x_p/d	1.65	1.94	2.47	3.08	3.34
Back-analysis against measured deflection	E_pI_p/EI	1	0.83	0.64	0.61	0.55
	w_g using E_pI_p (mm)	25.9	35.2	59.40	97.1	128.3
	M_{max} using E_pI_p (kN-m)	3447.0	4156.8	5802.0	8668.5	10395.9
	x_p/d	1.65	1.86	2.24	2.67	2.92

Example 10.2 Case SN2

The prestressed pipe pile P7 was infilled with reinforced concrete. It was 34.0 m long with 0.8 m O.D. and 0.56 m I.D. and with 19@19 rebars and 38@9 high strength steel wires, and a concrete cover of 30 mm. The strength values are f_y (outer pipe pile) = 1226 MPa, f_y (infilled material) = 471 MPa, f_c' (prestressed concrete) = 78.5 MPa, and f_c' (infilled concrete) = 20.6 MPa, respectively; and an equivalent yield strength of the composite cross-section of 67.74 MPa (= $(E_p/151{,}000)^2$), as EI was 0.79 GN-m^2 and thus $E_p = 3.93 \times 10^7$ kPa. The parameters allow the following to be calculated: M_{cr} = 277.4~443.9 kN-m using k_r = 19.1~31.5, y_r = 0.4 m, and Equation 10.5; M_{ult} = 1.89 MN-m and $(EI)_{cr}/EI$ = 0.146 using the RSB method (see Table 10.7).

The elastic-pile analysis using the LFP in Figure 10.7 (Huang et al. 2001) and G = 10.8 MPa (= 0.64N) (Guo 2006) agrees with the measured deflection w_g until H of 284 kN, at which M_{cr} = 464.7 kN-m (k_r of 33.0). A M_{ult} of 1.82~2.0 MN-m is noted at H = 804–863 kN, which agrees well with 1.89 MN-m gained using the RBS method. With $(EI)_{cr}$ of 0.1152 GN-m^2 (as per RSB method), M_{cr} of 464.7 kN-m, and M_{ult} of 1.89 MN-m (or 1.82~2.0 MN-m), the E_pI_p was estimated using Equation 10.6 for each M_{max} (gained using EI). This E_pI_p allows new w_g and M_{max} to be estimated, which are presented in Figure 10.7b. The deflection w_g was overestimated by ~28%, indicating the accuracy of the estimated $(EI)_{cr}$ and Equation 10.6.

10.4.2 Hong Kong tests: Cases SN3 and SN4

Example 10.3 Case SN3

The lateral loading test was conducted on a bored single pile in Hong Kong (Ng et al. 2001) at a site that consisted of very soft fill, followed by sandy estuarine deposit and clayey alluvium. The ground water was

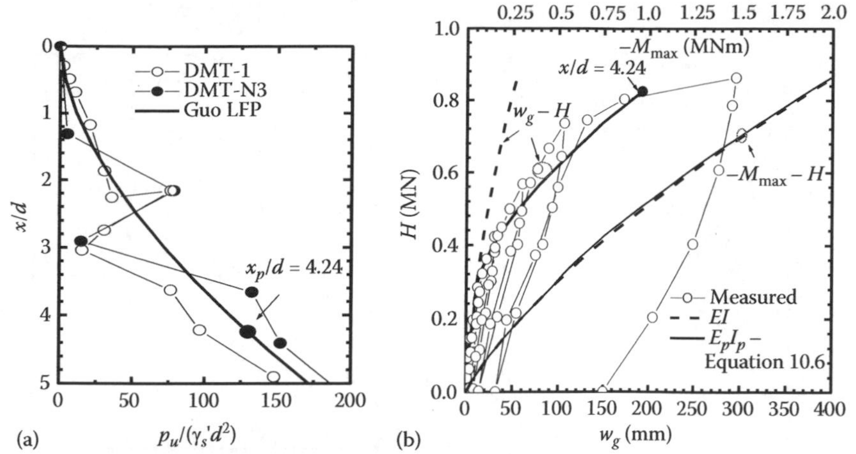

Figure 10.7 Comparison between measured (Huang et al. 2001) and predicted response of pile P7 (SN2). (a) LFPs. (b) H-w_g and H~M_{max} curves.

located at 1.0 m below the GL. The sandy soil extended into a depth of 15 m, which has $\tilde{N} = 17.1$, $\phi' = 35.3°$ with a relative density of 50% (Teng 1962), and $\gamma_s' = 11.0$ kN/m³. Lateral loads were applied at the middle level of the 1.5 m (thickness) pile-cap.

The pile was 28 m in length and 1.5 m in diameter and had the following properties: $f_y = 460$ MPa, $f_c' = 49$ MPa, $E_c = 32.3$ GPa, EI = 10 GN-m², $(EI)_{cr} = 4$ GN-m², $I_g = 0.2485$ m⁴, and $E_p = 4.02 \times 10^7$ kPa. The M_{cr} was estimated as 1.44~2.31 MN-m using $k_r = 19.7$~31.5, y_r = 0.75 m, and Equation 10.5. The M_{ult} was not estimated without the reinforcement detail.

The elastic pile analysis, using the LFP in Figure 10.8a and G = 10.9 MPa, agrees with the measured w_g up to the cracking load H of 1.1 MN, at which M_{max} = 3.34 MN-m (i.e., k_r = 45.5), which again is about twice that of the ACI's suggestion. Note to the maximum slip depth x_p of 2.1d, the limiting force per unit length (see Figure 10.8a) exceeds other predictions (Broms 1964a; Reese et al. 1974). The E_pI_p was deduced using the measured H–w_g data and is denoted as "E_pI_p" in Figure 10.8b. The associated H~M_{max} and E_pI_p/EI~M_{max} curves are plotted in Figures 10.8b and 10.5, respectively. In particular, it is noted that $(EI)_{cr}/EI = 0.4$, and at a load of 2.955 MN, M_{max} = 10.77 MN-m and x_p = 2.1d. Equation 10.6 was not checked without the M_{ult}.

Example 10.4 Case SN4

Lateral load tests on two big barrettes (DB1 and DB2) were carried out in Hong Kong (Zhang 2003) in a reclaimed land to a depth of 20 m. Load was applied along the height direction and at near the GL (e = 0) on DB1. The subsoil consisted either of sandy silty clay or loose to

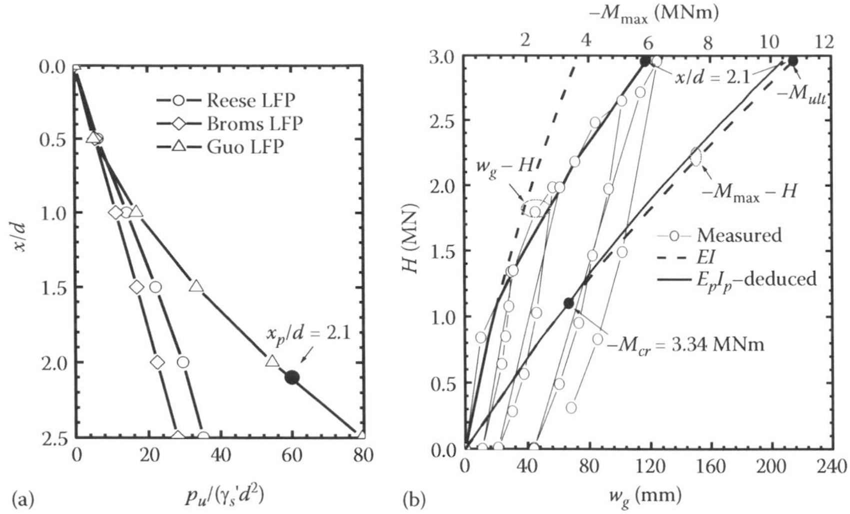

Figure 10.8 Comparison between measured (Ng et al. 2001) and predicted response of Hong Kong pile (SN3). (a) LFPs. (b) H-w_g and H~M_{max} curves.

medium dense sand, with cobbles in between the depths of 10.6~18.9 m below the GL. The ground water was located at a depth of 2.5 m. The soil properties within the top 15 m are: $\tilde{N}$ = 32.5, ϕ' = 49°, and γ_s' = 13.3 kN/m³.

Only the barrette DB1 (i.e., SN4) was modeled. It had a length of 51.1 m and a rectangular cross-section of 2.8 m (height) × 0.86 m. Other properties are: f_y = 460 MPa, f'_c = 43.4 MPa (= f_{cu}/1.22; Beckett and Alexandrou 1997), a cubic concrete compressive strength f_{cu} of 53 MPa, E_c = 3.03 × 10⁷ kPa, and t = 75–100 mm, together with I_g = 1.5732 m⁴, EI = 47.67 GN-m², and E_p = 1.78 × 10⁹ kN-m². The M_{cr} was estimated as 4.61~7.38 MN-m using k_r = 19.7~31.5 and y_r = 1.4 m; and M_{ult} as 10.13 MN-m and $(EI)_{cr}$ as 0.89 GN-m² for the upper 15 m (Table 10.7), using the RSB method.

The elastic-pile analysis employs N_g = 28.15 (= 0.55K_p^2), α_o = 0, n = 1.7, and G = 13.0 MPa (= 0.4N). The associated LFP to a depth of 1.5d, as shown in Figure 10.9a, lies in between Reese's LFP and Broms' LFP; and its average over a depth of ~2.8d is close to that from Reese's LFP. The predicted deflection w_g, as shown in Figure 10.9b, agrees well with the measured data until the cracking load of 2.26 MN, at which the M_{cr} was measured as 7.42 MN-m (k_r = 31.7).

Using M_{cr} of 7.42 MN-m and $(EI)_{cr}$ of 0.89 GN-m², the E_pI_p, as shown in Figure 10.5, was deduced by matching the predicted with the measured deflection w_g and the associated H~M_{max} curve in Figure 10.9b. These figures show that the M_{ult} of 10.13 MN-m occurs at $H \approx 3.61 MN$ (associated with a sharp increase in the measured w_g); at the maximum load H of 4.33 MN, 27.8% increase in the x_p and 26.9% increase in the M_{max} (owing to excessively high pile–soil

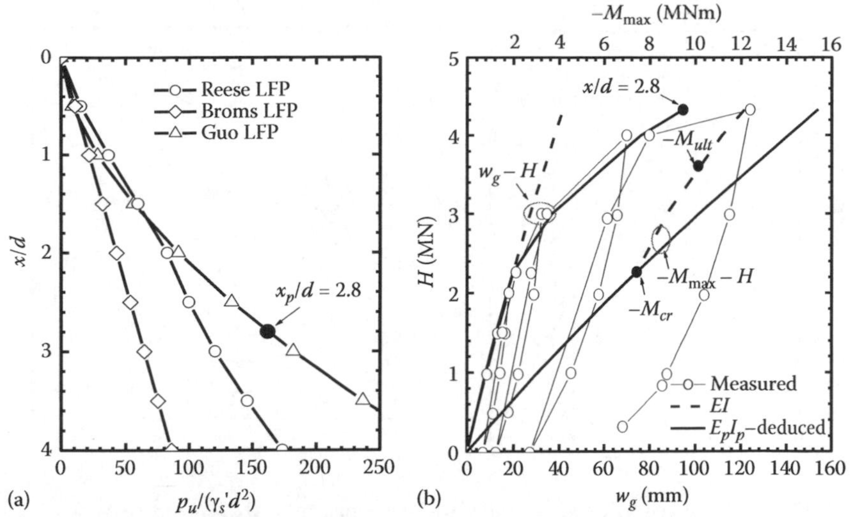

Figure 10.9 Comparison between measured (Zhang 2003) and predicted response of DBI pile (SN4). (a) LFPs. (b) H-w_g and H~M_{max} curves.

relative stiffness E_p/G^* (of 1.1×10^5), which is not generally expected. In brief, the increased M_{max} from nonlinear analysis should be used in Equation 10.6.

10.4.3 Japan tests: CN1 and CN2

Example 10.5 Case CN1

Pile D was tested at a site with an undrained shear strength $s_u = 35 + 0.75x$ (kPa, x in m < 20 m) and an average $\tilde{s}_u$ of 43 kPa over 15.48 m (= 10d). The lateral load was applied at 0.5 m above the GL. The pile had $L = 30$ m, $d = 1.548$ m, $EI = 16.68$ GN-m^2, $E_p = 5.92 \times 10^7$ kPa, and $f_c' = 153.7$ *MPa* [$\approx (E_p/151{,}000)^2$]. The M_{cr} was estimated as 2.81~4.50 MN-m (using $k_r = 19.7$~31.5, $y_r = 0.774$ m and Equation 10.5), which exceeds 1.33 MN-m (Nakai and Kishida 1982) by 2.0~3.4 times. The ultimate bending moment M_{ult} was not estimated without the reinforcement information.

The elastic pile reaction was simulated using $n = 0.7$, $\alpha_o = 0.1$ and $N_g = 2.0$ (see LFP in Figure 10.10a), $G = 5.62$ MPa (= $130.8s_u$) (Kishida and Nakai 1977), and $k = 3.13G$. In particular, Figure 10.10a demonstrates that an LFP within 5d is close to Hansen's LFP gained using c = 8 kPa, $\phi' = 10°$, but it is far below the Matlock's LFP. The predicted pile deflection compares well with the observed one until the cracking load H of 690 kN (see Figure 10.10b), with M_{cr} (M_{max} at $H = 690$ kN) of 2.38 MN-m ($k_r = 16.7$).

The E_pI_p was deduced using the measured deflection w_g (see Figure 10.10b), which in turn offers the bending moment M_{max} in the figure.

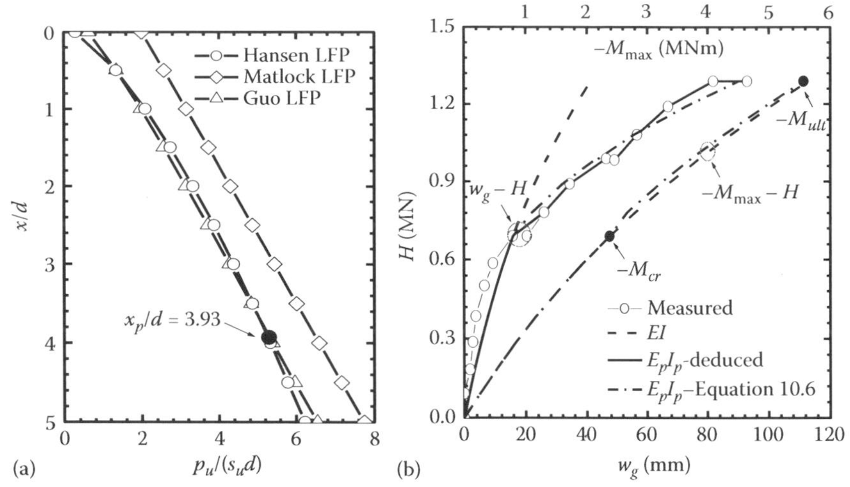

Figure 10.10 Comparison between calculated and measured (Nakai and Kishida 1982) response of Pile D (CN1). (a) LFPs. (b) H-w_g and H~M_{max} curves.

For instance, an H of 1289 kN incurs E_pI_p = 5.0 GN-m^2 (= 0.3EI), w_g = 90.1 mm, M_{max} = 5.57 MN-m, and x_p = 6.09 m (= 3.93d). The cracking renders a 118.2% increase in w_g, 1.24% reduction in M_{max}, and 29.3% increase in x_p, which resemble those noted for piles in sand.

Using M_{max} and E_pI_p at H = 1289 kN, the $(EI)_{cr}$ was calculated to be 4.01 GN-m^2 using Equation 10.6 and $(EI)_{cr}/EI$ = 0.241. Substituting $(EI)_{cr}$ and M_{max} into Equation 10.6, E_pI_p (thus deflection w_g) was calculated. This deflection, designated as E_pI_p – Equation 10.6, compares well with measured one (see Figure 10.10b). Equation 10.6 is validated for this case.

Example 10.6 Case CN2

Test Pile E was conducted at a clay site where the SPT values (depth) were unfolded as N = 0 (0~2 m) and N = 4 ~6 (2~10 m). The s_u was 163.5 kPa at a depth of 3 m below the GL. Load was applied 0.35 m above the GL. The pile properties are: L = 9.5 m, d = 1.2 m, EI = 2.54 GN-m^2, and $f_c' \approx$ 27.5 MPa [= $(E_c/151{,}000)^2$]. The EI was estimated using E_c (E_p) = 2.5 x 10^7 kPa. The M_{cr} was estimated as 554.2~886.7 kN-m (using k_r = 19.7~31.5, and y_r = 0.6 m), which is slightly higher than 527.6 kN-m by Nakai and Kishida (1977), implying the rationale of using the current EI. The ultimate bending moment M_{ult} is not determined without the reinforcement detail.

The behavior of the elastic-pile was modeled using the LFP in Figure 10.11a, G = 21.38 MPa (= 130.8S_u) (Nakai and Kishida 1982), and k = 3.77G. The predicted pile deflection agrees well with the observed data up to H of 470 kN (see Figure 10.11b). The measured M_{cr} (M_{max} at H_{cr} = 470 kN) was computed as 628.4 kN-m and k_r as 22.3.

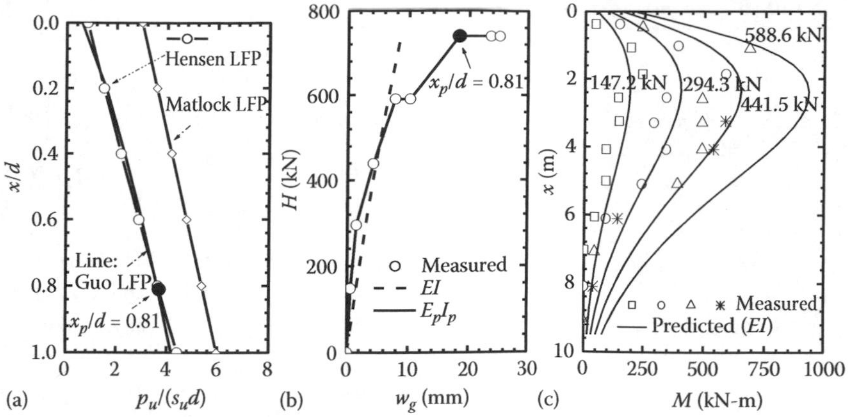

Figure 10.11 Comparison between calculated and measured (Nakai and Kishida 1982) response of pile E. (a) LFPs. (b) H-w_g curves. (c) M profiles for H = 147.2 kN, 294.3 kN, 441.5 kN, and 588.6kN.

Table 10.9 Nonlinear response of pile *E* (CN2)

		Using elastic-pile EI			*Using cracked* E_pI_p *(back-estimated)*				
H *(kN)*	*Measured* w_g *(mm)*	M_{max} *(kN-m)*	w_g *(mm)*	x_p/d	E_pI_p/EI	M_{max} *(kN-m)*	w_g *(mm)*	x_p/d	$(EI)_{cr}/EI$
588.6	7.8	808.2	6.54	0.32	0.60	748.0	7.81	0.40	0.35
735.8	18.3	1046.5	8.54	0.45	0.18	892.5	18.25	0.81	0.005

The E_pI_p was deduced using the measured deflection w_g(see Figure 10.11b). This in particular offers $E_pI_p/EI = 0.6$ and 0.18, respectively, for H = 588.6 kN and 735.8 kN. As tabulated in Table 10.9, the $(EI)_{cr}/EI$ ratio is thus deduced as 0.35 and 0.005, respectively, in light of Equation 10.6. The drastic drop in $(EI)_{cr}$ implies pile failure prior to 735.8 kN. This is perhaps true in terms of the erratic scatter of measured moment at H= 588.6 kN and no record of the moment for H= 735.8 kN (Nakai and Kishida 1982). The predicted bending moment profiles compare well with those measured as shown in Figure 10.11c.

10.5 ROCK-SOCKETED SHAFTS

Sixteen rock-socketed drilled shafts (Reese 1997; Zhang et al. 2000; Nixon 2002; Yang et al. 2005; Cho et al. 2007) are studied. The properties of rock masses around each shaft are tabulated in Table 10.10. A shaft rock relative stiffness ratio E_p/G of 1 and 10^3 would have a L_c = (1~6)d, in which deflection of laterally loaded rock-socketed piles mainly occurs. The stress relief and weathering and pile construction may render

Table 10.10 Rock mass properties (All cases)

Case	Reference	Rock type	Reported properties: γ_m[a] (kN/m³)	q_u (MPa)	RQD (%)	GSI[b,e]
RS1	Reese (1997)	Limestone	23	3.45	0[d]	29
RS2		Sandstone	23	2.77 (5.7)[c]	45	47
RS3	Zhang et al. (2000)	Sandy shale	23	3.26	55	52
RS4		Sandstone	23	5.75	45	48
RS5	Nixon (2002)	Claystone-siltstone	25	12.2	84.6	49
RS6			25	11.3	95	82
RS7		Siltstone-sandstone	15	25.0	44	38
RS8			—	12.2		59
RS9	Yang et al. (2005)	Shale-limestone	10.5	39.1	11	41~61
RS10		Shale-siltstone	16.4	26.2	38	28~45
RS11	Cho et al. (2007)	Crystalline granite	27.01	62.6	59	40
RS12			26.7	57.6	13	25
RS13		Meta-argillite	24.6	27.7	<25	
RS14			24.6	27.7	<25	
RS15		Gneiss	26.67	59.0	30	30
RS16			27.01	61.6	27	30

[a] $\gamma_m = 23$ kN/m³, if not provided.
[b] Ignoring the adjustment of discontinuity orientation.
[c] Average (representative) value.
[d] Assumed as zero by (Reese 1997).
[e] Determined using RMR_{89} (Bieniawski 1989).

a lower value of E than those around tunnel support structures or deep excavations. The real value of G (thus E), along with the parameters A_L, k, and n are thus deduced by matching measured with predicted pile response for each shaft using the GASLFP and are classified into three groups: (a) $n = 0.7$ for long shafts with large diameters; (b) $n = 1.5$ for short~long shafts; and (c) $n = 2.3$ for short shafts, characterized by rigid rotation about tip and $L < 0.7L_c$ (see Tables 10.11 and 10.12). The parameters allow the measured H and w_t to be normalized and plotted in Figures 10.12 and 10.13, in which the CF solutions for $n = 0.7$ and 1.7 and for $n = 2.3$ are also provided, respectively. Typical calculations are elaborated next concerning the tests in Islamorada (Nyman 1980), San Francisco (Reese 1997), and Pomeroy (Yang 2006), respectively, as they show deviations from majority of test data (see Figure 10.13b) and structural nonlinearity.

The back-estimation provides the empirical correlations to laterally loaded rock-socketed piles (e.g., the shear modulus of an isotropic rock

Table 10.11 Properties for shafts in rock

		Details of shaft and loading					Derived parameters (linear shafts)				
Case	*Test name*	d (m)	L (m)	E_cI_p (MN-m^2)	e (m)	H_{max} (kN)	L_c (m)	L_c/d	*Max.* x_p(m)[a]	*Max.* $x_{p/}d$	*Max.* $x_p\lambda$
RS1[c]	Islamorada	1.22	13.3	3730	3.51	667	2.58	2.1	0.778	0.64	0.741
RS2[c]	San Francisco	2.25	13.8	35150	1.24		4.97	2.2	1.13[b]	0.50[b]	0.523
RS3[d]	Sandy shale	0.22	3.97	6.687	0.61	115.6	1.03	4.7	0.672	3.05	1.513
RS4[d]	Sandstone	0.22	3.75	6.687	0.82	177.9	0.95	4.3	0.681	3.09	1.624
RS5[d]	140 long	0.762	4.057	915.4	0.33	1512.0	4.39	5.8	0.900	1.18	0.453
RS6[d]	140 short	0.762	3.356	915.4	0.43	1512.0	4.19	5.5	0.455	0.597	0.241
RS7[d]	185 long	0.762	4.21	915.4	0.70	1334.5	4.55	6.0	0.784	1.10	0.345
RS8[e]	185 short	0.762	2.789	915.4	0.68	1334.5	5.52	7.25	1.107	1.45	0.434
RS9[c]	Dayton	1.83	4.89	14993.2	0.	5000	4.41	2.41	0.016	0.01	0.009
RS10[c]	Pomeroy	2.44	17.31	49362.0	17.1	1230	8.26	3.39	0.092	0.038	0.038
RS11[d]	Wilson short	0.762	4.89	868.63	0.3	1680	5.82	6.72	1.079	1.42	0.459
RS12[d]	Wilson long	0.762	5.71	868.63	0.3	1680	5.94	7.50	1.141	1.50	0.431
RS13[e]	Nash short	0.762	3.35	868.63	0.3	534	7.62	10.0	0.682	0.896	0.189
RS14[e]	Nash long	0.762	4.57	868.63	0.3	979	6.967	9.14	0.934	1.225	0.285
RS15[d]	Caldwell long	0.762	4.89	868.63	0.3	1334	5.335	7.00	0.834	1.094	0.339
RS16[e]	Caldwell short	0.762	4.00	868.63	0.3	1334	8.73	11.46	0.816	1.071	0.195

[a] Slip depth at the maximum testing load H_{max}
[b] Slip depth at the ultimate load $H = 8{,}620$ kN (not the H_{max})
[c] $n = 0.7$
[d] $n = 1.5$
[e] $n = 2.3$

Table 10.12 Backfigured E, LFPs, and characteristic length: Shafts in rock

	E (GPa)		Guo's LFP (Back-figured)[a]						
Case	Measured	Back-figured	N_g	α_o/d	A_L (MN/m^{1+n})[b]	N_{co} (%)	λ (1/m)	g_s[c]	$e\lambda$
RS1	7.24	5.175	0.491	0	3.519	0	0.953	0.6	3.3450
RS2		3.56	0.997	0	9.614	0	0.463	0.8	0.5741
RS3		0.41	2.727	0	1.319	0	2.251	0.6	1.3731
RS4		0.50	3.182	0.05	2.247	3.2	2.385	0.7	1.9557
RS5	0.32[d]	0.153	0.525	0	2.120	0	0.503	0.4	0.1660
RS6	0.20[d]	0.185	1.119	0.22	4.294	0	0.5291	0.38	0.2275
RS7	0.39[d]	0.188	0.373	0	2.431	0	0.4399	0.2	0.3079
RS8	0.44	0.061	0.498	0.0	1.127	0	0.3921	0.38	0.2666
RS9	0.17	2.44	0.328	0.0	112.9	0	0.5486	0.6	0.0000
RS10		0.655	0.082	0.0	21.241	0	0.2821	0.2	4.8239
RS11		0.047	0.328	0.0	3.940	0	0.3704	0.25	0.1111
RS12		0.043	0.262	0.0	2.982	0	0.3776	0.2	0.1133
RS13		0.016	0.262	0.0	0.847	0	0.2764	0.2	0.0829
RS14		0.023	0.262	0.0	0.847	0	0.3047	0.2	0.0914
RS15		0.052	0.459	0.0	5.304	0	0.3801	0.35	0.1140
RS16		0.009	0.262	0.0	1.20	0	0.2387	0.2	0.0716

[a] Averaged or equivalent LFP using Reese's LFP refers to Zhu, B.T (2004), Griffith University report.
[b] $n = 0.7$ and E = (60~400)q_u (Cases 1, 2, 9, and 10), long shafts with large diameters; $n = 1.5$ and E = 4,800q_u for long shafts (Cases 3 and 4), or near long shafts (Cases 5, 6, 7, 11, 12, and 15), as is noted in measured data of Cases 11 and 12; $n = 2.3$ and E = (1,000~1,500)q_u (Cases 8, 13, 14, and 16), short shafts, characterized by rigid rotation about tip.
[c] $g_s = A_L/(q_u)^{1/n}$; $p_u = g_s q_u^{1/n}$.
[d] Average value of upper three layers below rock line.

mass and unconfined strength in Chapter 3, this book). For the cases RS1–7, the GSI was calculated and provided in Table 10.13, and the calculated moduli are provided in Table 10.14.

Example 10.7 Nonlinear shaft of Islamorada test (Case RS1)

Nyman (1980) reported a loading test on a bored shaft in the Islamorada Florida Keys. The shaft was retained by a steel casing in the sand layer overlying the rock to neutralize the sand resistance. The shaft with $d = 1.22$ m, $L = 13.3$ m was laterally loaded at $e = 3.51$ m above the rock surface (Table 10.11). The shaft properties are: $I_p = 0.109$ m^4 and $E_p = 3.43 \times 10^7$ kPa (i.e., $E_cI_p = 3.73$ GN-m^2, $E_c \approx E_p$); $f_c' = 51.6$ MPa; and $M_{cr} = 798{\sim}1276$ kN-m (using $k_r = 19.7{\sim}31.5$, $y_r = d/2 = 0.61$ m in Equation 10.5). The rock around the shaft was a brittle, vuggy, coral limestone characterized by (Reese 1997) (see Table 10.10): (1) RQD = 0; (2) $q_u = 3.45$ MPa; (3) RMR$_{89}$ = 34 (Bieniawski 1989); (4) GSI = 29; and (5) $E = 7.24$ GPa (i.e., $G = 2.9$ GPa).

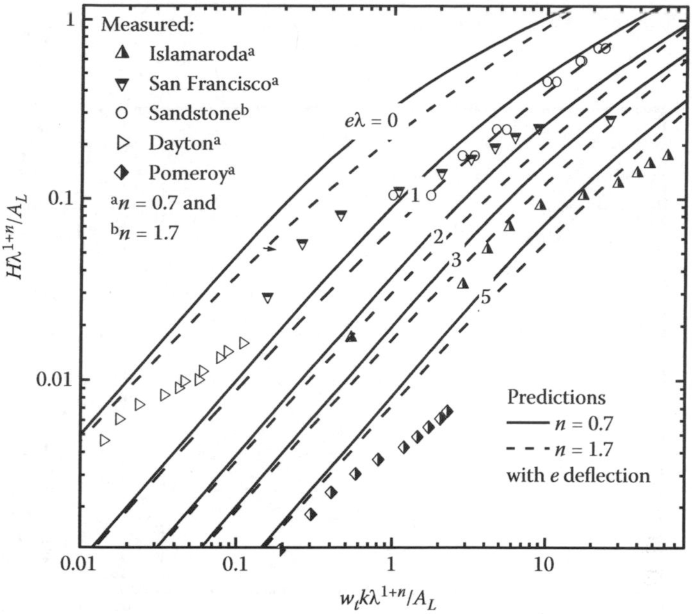

Figure 10.12 Normalized load, deflection: measured versus predicted (n = 0.7 and 1.7).

With $n = 0.7$, $A_L = 3.519$ MPa, $N_g = 0.491$ (obtained using $g_s = 0.6$, see Table 10.12), and k = 12.296 GPa (or G = 5.2 GPa), the H~w_t and H~M_{max} curves were predicted and are plotted in Figure 10.14 as "Constant E_cI_p." The normalized limiting force $p_u/(q_ud)$ is shown in Figure 10.14c. With L_c (= 2.38 m) < L, the deduced G is valid and close to the reported G. The predicted moment M_{max} compares well with the measured one over the entire load regime, see Figure 10.14b. However, ignoring the structural nonlinearity, the predicted H~w_t curve diverges from the measured one (see Figure 10.14a) beyond the cracking load H of 300 kN (Reese 1997).

The nonlinear interaction is captured via the steps outlined before: (1) At the cracking H of 300 kN, the M_{cr} (= M_{max}) was estimated as 1.161 MN-m (thus k_r = 28.7 from Equation 10.5); (2) without reinforcement details, the E_cI_e was "back-estimated" against the measured data (see Figure 10.14a), which shows E_cI_{cr} = 0.71GN-m^2; A M_{max} was estimated as 2.674 MN-m at the maximum imposed H of 667 kN; (3) $H_{cr} \approx$ 300 kN; and (4) values of E_cI_{cr}, M_{cr}, and M_{max} allow the I_{cr} to be estimated as 0.0129m^4 in light of Equation 10.6 and E_cI_{cr}/E_cI_p = 0.118. The calculation was repeated to gain H~w_t and M_{max}~E_cI_e curves. The curves with Δ dots agree well with Equation 10.6 (i.e., the ACI method) shown in Figure 10.14d and with the measured deflection w_t shown in Figure 10.14a. The accuracy in the estimated w_t is similar to that of M_{cr} (thus E_cI_e). At H = 667 kN, the use of a M_{cr} of 0.798 MN-m (31% less than 1.161 MN-m) would render the w_t to be

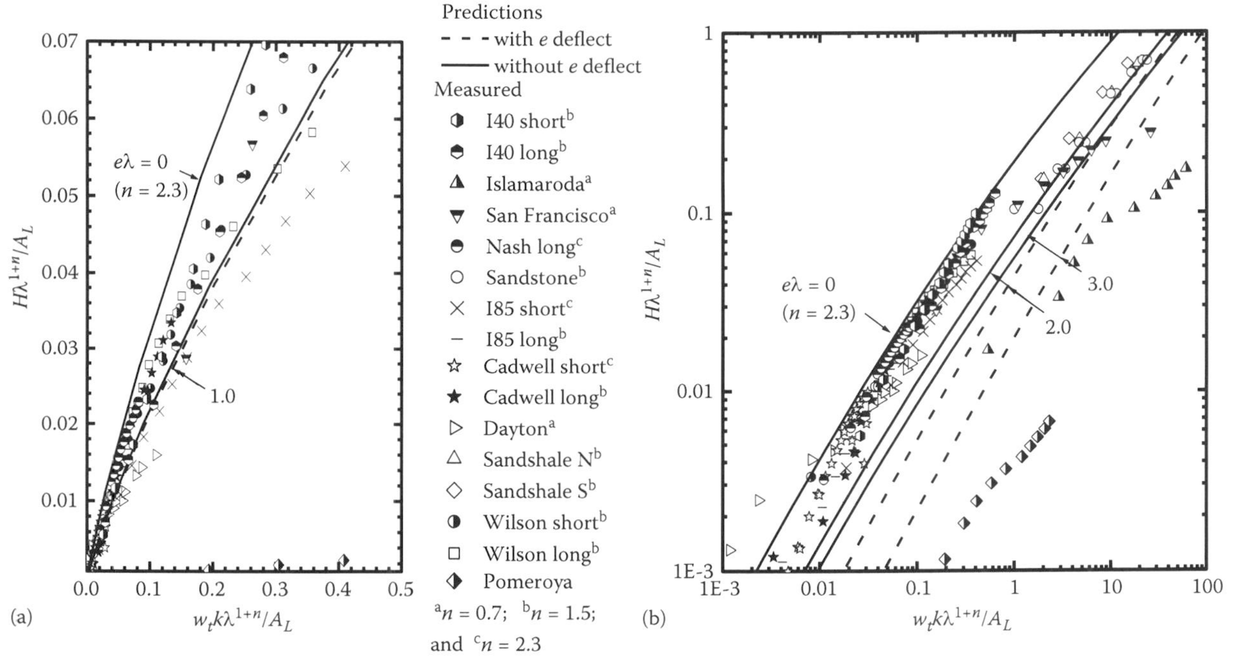

Figure 10.13 Normalized load, deflection: measured versus predicted (n = 2.3). (a) Normal scale. (b) Logarithmic scale.

Table 10.13 Summary of index properties of rock masses

			Derived rock mass classification indices							
			RMR_{89} (Bieniawski 1989)							
Case	*Reference*	*Rock type*	R_A	R_B	R_C	R_D	R_E	*Adj*	*Total*	*GSI*
RS1	Reese et al.	Limestone	1	3	5	10	15	0	34	29
RS2	(1997)	Sandstone	1	8	8	20	15	0	52	47
RS3	Zhang et al. (2000)	Sandy shale	1	13	8	20	15	0	57	52
RS4		Sandstone	2	8	8	20	15	0	53	48
RS5	Nixon	Claystone-	2	17	10	10	15	0	54	49
RS6	(2002)	siltstone	2	20	30	25	10	0	87	82
RS7		Siltstone	4	8	10	6	15	0	43	38

Source: After Zhu, B. T., unpublished research report, Griffith University, 2004.

overestimated by 28.5%; whereas, a M_{cr} of 1.276 MN-m (10% higher) would underestimate the w_t by 9.4%.

The high $e\lambda$ of 3.345 (> 3.0, see Table 10.12) allows the deflection w_t to be readily calculated using elastic shaft stiffness E_pI_p (Guo 2006) for $H < H_{cr}$ or otherwise using Equation 10.4. For instance, at H = 350 kN, the elastic shaft analysis yields λ = 0.95279/m, x_p = 0.441 m, $w_g\lambda^n k/A_L$ = 1.75634, and $M_{max}\lambda^{2+n}k/A_L$ = 0.32565. This offers w_g = 0.52 mm and M_{max} = 1305.96 kNm. The M_{max} allows E_cI_e to be deduced as 1.8477 GNm^4 using Equation 10.6, and thus M_{cr} = 983.6 kNm (less than 1,161 kNm) and I_{cr} = 0.0129 m^4. This new E_cI_e allows the additional deflection from loading eccentricity to be calculated as 2.73 mm [= 350 × 3.51^3/(3 × 1.8477 × 10^3)], and results in a total deflection w_t of 4.78 mm. The calculation was repeated for five typical loads H of 127.8~667 kN. The

Table 10.14 Summary of mechanical properties of rock masses

	E *(GPa)*								
Case	M1	M2	M3	M4	M5	*Back-figured*	*Measured*	k/G	G/q_u
RS1		3.98	0.40	0.55	0.02	5.2	7.24	6.2	839.4
RS2	4.0	11.22	0.11	1.40	0.03[a]	3.56		5.88	273.7
RS3	14.0	14.96	0.39	2.03	0.03	0.41		4.01	35.48
RS4	6.0	11.89	0.52	2.14	0.05	0.50		4.31	30.18
RS5	24.0	19.95	0.75	2.94	0.16	0.15	0.32[b]	3.74	3.63
RS6	74.0	84.14	0.73	22.04	0.62	0.18	0.20[b]	3.9	6.53
RS7		6.68	1.16	2.69	0.18	0.19	0.39[b]	3.72	1.77

Source: Revised from Zhu, B. T., unpublished research report, Griffith University, 2004.

[a] Assuming open joints.

[b] Average value of upper three layers below rock line.

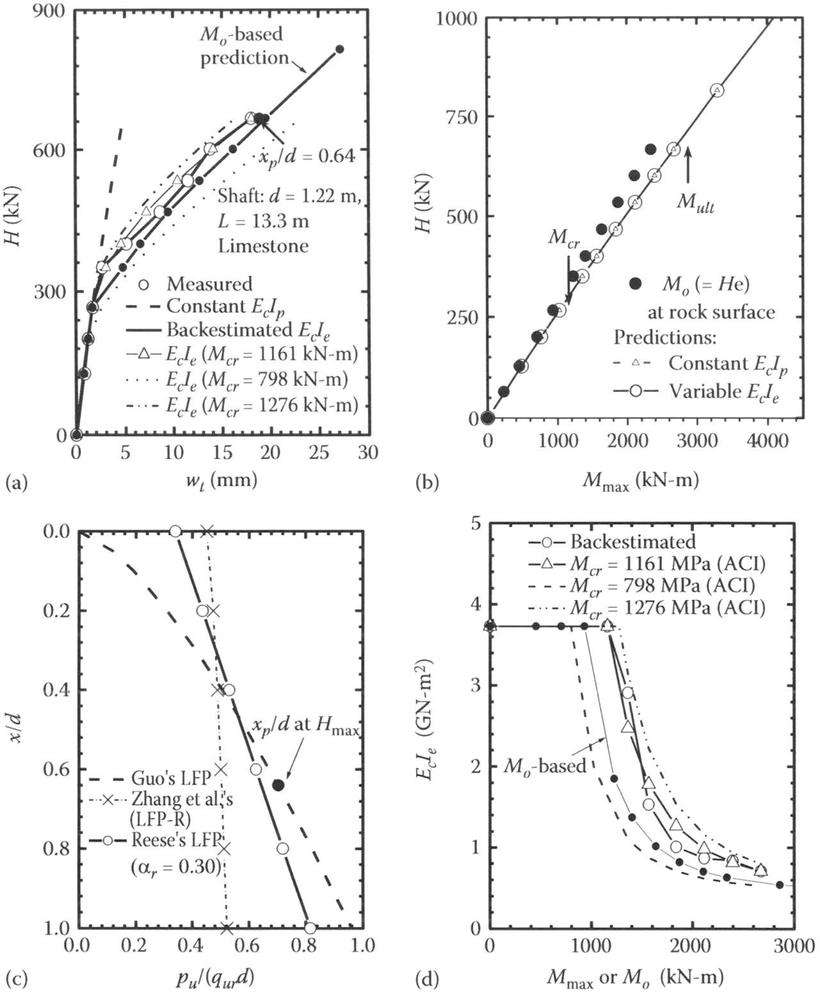

Figure 10.14 Analysis of shaft in limestone (Islamorada test). (a) Measured and computed deflection. (b) Maximum bending moment. (c) Normalized LFPs. (d) E_cI_e.

results are provided in Table 10.15, and are plotted in Figure 10.14 as "M_o based" prediction. They compare well with the GASLFP prediction.

Example 10.8 Nonlinear shafts in San Francisco test (Case RS2)

The California Department of Transportation (Reese 1997) reported the two bored shafts loaded simultaneously near San Francisco at an eccentricity of 1.24 m (see Figure 10.15a). The shafts were 2.25 m in diameter and installed 12.5 m (Shaft A) and 13.8 m (Shaft B) into

Table 10.15 Simple prediction of the nonlinear deflection for high $e\lambda$ (Case RS1)

H (kN)	M_o (kNm)	M_m (kNm)	I_e (m^4)	E_cI_e (GNm^4)	$He^3/(3E_cI_e)$ (mm)	w_t (mm)
127.8	448.7	464.4	0.109	3.73	0.49	1.22
266.0	933.7	983.75	0.109	3.73	1.03	2.56
350.0	1228.5	1305.96	0.05387	1.8477	2.73	4.78
500.0	1755.0	1891.5	0.02638	0.90498	7.97	11.01
667.0	2341.2	2556.88	0.01836	0.62972	15.27	19.56

medium-to-fine grained, well-sorted and thinly bedded (25–75-mm thick) sandstone. The sandstone was very intensely to moderately fractured with bedding joints, joints and fracture zones, and was characterized by: (1) RQD = 0~80% with an average of 45%; (2) q_u = 5.7 MPa (0~0.5 m below the rock surface), 1.86 MPa (0.5–3.9m), 6.45 MPa (3.9–8.8 m), and 16.0 MPa (>8.8 m); (3) RMR_{89} = 52; and (4) GSI = 47.

The shaft B (Case RS2) was reinforced with 40 rebars that has a tensile strength f_c' of 34.5 MPa, a diameter d_s of 43 mm, and a cover thickness t of 0.18 m. Other properties are f_y = 496 MPa; E_cI_p = 35.15 GN-m^2 [from $E_c = 151{,}000(f_c')^{0.5}$ = 28.05 GPa and I_p = 1.253 m^4]; and M_{cr} = 4,074~6,518 kN-m (using Equation 10.5 and k_r = 19.7~31.5).

Taking n = 0.7 and A_L = 9.614 MN/m$^{1.7}$ (see Table 10.12), the N_g was calculated as 0.997 using Equation 10.1 and g_s = 0.8. The normalized limiting force $p_u/(q_ud)$ against depth is plotted in Figure 10.15d as Guo's LFP. The predicted $H–w_t$ ($w_t \approx w_g + \theta_g e$) curve using GASLFP was matched with the measured one to ~2.6 MN (see "CF-constant E_cI_p" in Figure 10.15b), which offered E = 3.56 GPa (k = 12.296 GPa), L_c = 4.97 m (< L, thus validated), and $\tilde{q}_u$ = 2.77 MPa (over the depth of L_c). The measured M_{max} was well predicted as well. At the load H of 2,620 kN, the calculated M_{max} of 4,527 kN-m ($k_r \approx$ 21.8) falls in the predicted M_{cr} range of 4,074~6,518 kN-m.

Structural nonlinearity was incorporated to estimate the w_g and M_{max}. A nominal ultimate bending moment M_n was obtained as 21.57 MN-m using Equation 10.10. This offers a "beam" moment M_{ult} of 19.413 MN-m (= ϕM_n, using a reduction factor ϕ = 0.9 in Equation 10.11) and a "shaft" M_{ult} of 17.472 MN-m (ϕ = 0.81) for drilling shafts. The beam M_{ult} was 6.3% higher than M_{max} of 18,188 kN-m obtained using E_cI_p at H = 9 MN, from which a drastic increase in measured shaft deflection (or flexural failure) was noted. The shaft M_{ult} agrees with 17,740 kN-m obtained previously (Reese 1997). The shaft M_{ult} offers a cracked flexural rigidity E_cI_{cr} of 2.6 GN-m^2 (= 17472/6.72 × 10^{-3}). The E_cI_{cr}/E_cI_p thus was 0.074, and the ultimate load was 8.62 MN. The "Variable E_cI_e" was computed for each M_{max} using Equation 10.5, allowing new values w_t and M_{max} to be estimated. The estimations compare well with measured data across the entire load spectrum (Figure 10.15b and c). At

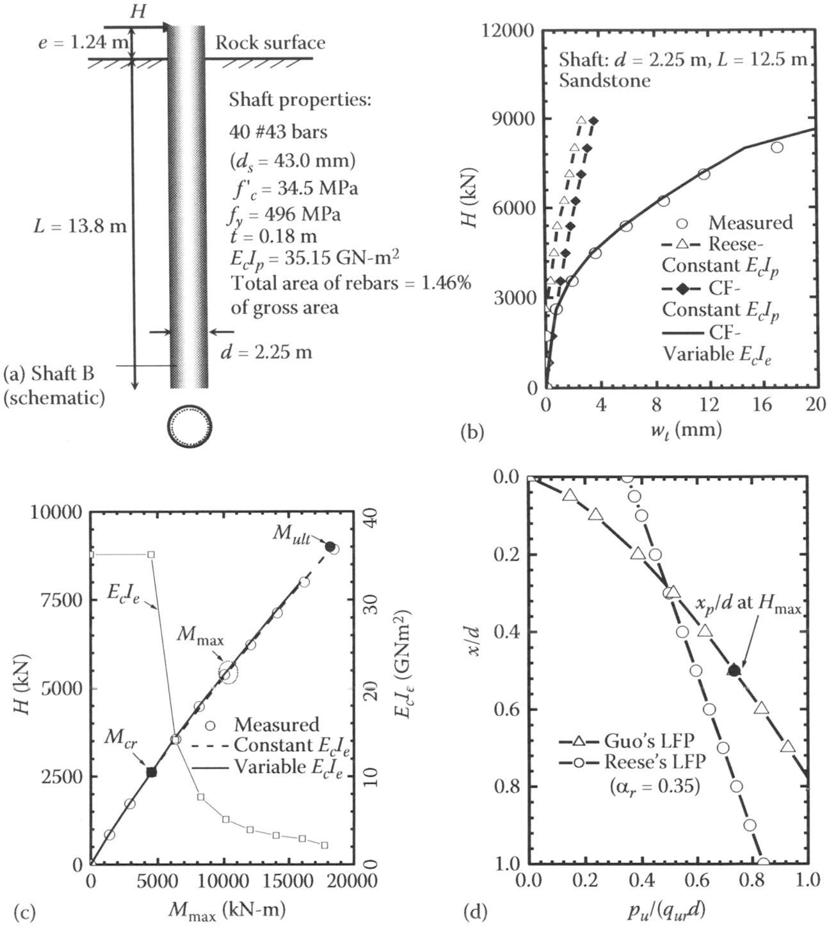

Figure 10.15 Analysis of shaft in sandstone (San Francisco test). (a) Schematic drawing of shaft B. (b) Measured and computed deflection. (c) Maximum bending moment and E_cI_e. (d) Normalized LFPs.

a failure load H of 8.62 MN, it is noted that the deflection w_g raised 4.7 times (from 3.5 to 19.9 mm); the slip depth x_p increased by 47.9% (from 0.887 to 1.312 m), but the M_{max} altered only 2.2%.

Example 10.9 Nonlinear shafts of Pomeroy-Mason test (Case RS10)

Pomeroy-Mason shaft had a high $e\lambda$ value of 4.82 (> 3.0, Table 10.12). The nonlinear shaft response was also well modeled (see Figure 10.16) in the manner to that for Case RS1, including H–w_t and H–M_{max} curves, and bending moment profiles for three typical loads.

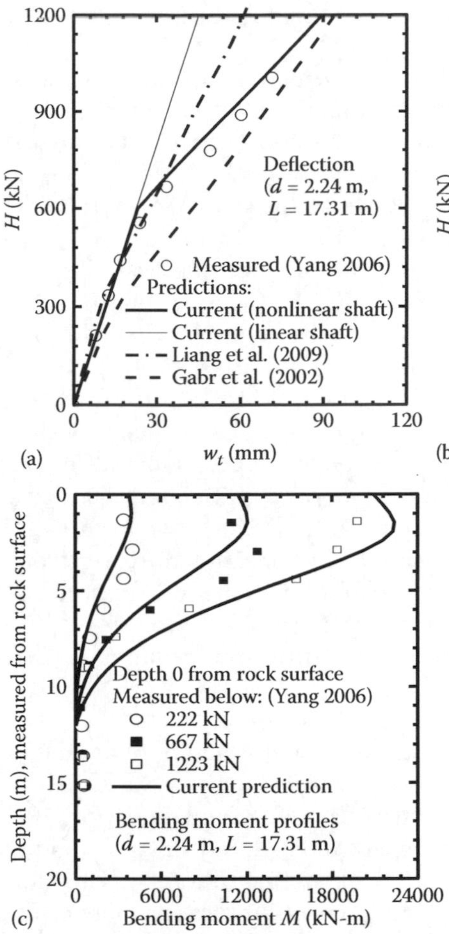

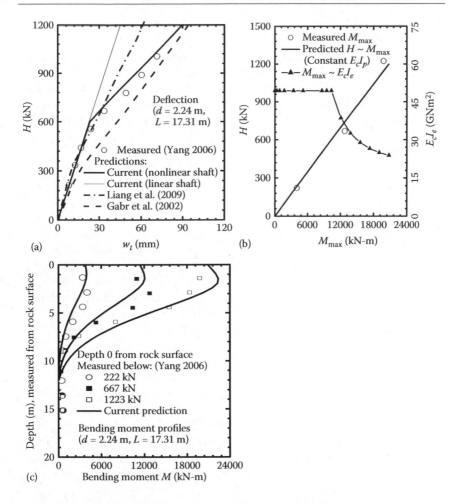

Figure 10.16 Analysis of shaft in a Pomeroy-Mason test. (a) Measured and computed deflection. (b) Bending moment and E_eI_e. (c) Bending moment profiles (n = 0.7).

10.5.1 Comments on nonlinear piles and rock-socketed shafts

The investigation into the 6 nonlinear piles in sand/clay and 16 rock-socketed shafts offers the following conclusions:

- The parameters for elastic piles are quite consistent with the previous findings by Guo (2006): (1) The 4 piles in sand have $k = (2.38{\sim}3.73)G$ and $G = (0.4{\sim}0.64)N$ (MPa); $n = 1.7$, $\alpha_0 = 0$, and $N_g = (0.9{\sim}1.2)K_p^2$ with $N_g = 0.55K_p^2$ for large diameter piles. (2) The 2 piles in cohesive soil have k = (3.13~3.77)G, and G = 130.8s_u; and n = 0.7, α_o = 0.06~0.1 m, and N_g = 2.

- Table 10.11 shows that L_c = (2.1~11.5)d; maximum slip depth x_p = (0.01~3.1)d; $x_p\lambda$ = 0.09~0.741 (n = 0.7); and 0.189~1.624 (n = 1.5~2.3). The shaft I40 short with a maximum $x_p\lambda$ of 0.241 may just attain the upper limit (of rigid shafts) of 0.172 (n = 1) ~0.463 (n = 2.3), beyond which n = 2.3 should be adopted, such as the shaft I85 short, Nash short, Nash long, and Caldwell short.
- Table 10.12 shows that the E currently deduced agrees with that gained from in situ tests (see Table 10.14). In Equation 10.1, n = 0.7~2.3 and α_o = (0~0.22)d may be used to construct p_u with $p_u = g_s q_u^{1/n}$. Long shafts with large diameters (in Cases RS1, 2, 9, and 10) use n = 0.7 and E = (60~400)q_u (E, q_u in MPa); otherwise n = 1.5 and E = 5,000q_u for normal diameters (in eight Cases: RS3–7, 11, 12, and 15). The former follows Reese's suggestion and the latter observes the stiff clay model (Gabr et al. 2002). Short shafts use n = 2.3 and E = 1,000q_u. The deduced g_s (= $p_u/q_u^{1/n}$) and n = 1.7~2.3 for normal diameter shafts are similar to those noted for vertically loaded shafts (n = 2), but for an abnormal reduction in modulus (for short shafts) with increase in UCS from 33 to 63 MPa. The latter may be owing to an increasing strain that requires further investigation.

Pertinent to types of shafts, the currently suggested E and p_u allow the w_t to be well predicted, albeit for the overestimation for I40 shaft. Salient features are that a high $e\lambda$ (> 3) renders $M_o \approx M_{max}$ and Equation 10.4 sufficiently accurate; the cracking moment and reduced flexural rigidity may be computed by using Equations 10.5 and 10.6 and k_r = 21.8~28.7. Nevertheless, without measured bending moment profiles for most of the shafts, the n values may falter slightly.

10.6 CONCLUSION

With piles exhibiting structure nonlinearity, the following are deduced:

- The ultimate bending moment M_{ult}, flexural rigidity of cracked cross section E_pI_p may be evaluated using the method recommended by ACI (1993), and the variation of the E_pI_p with the moment may be based on Equation 10.6. However, this would not always offer a good prediction of measured pile response, such as cases SN1 and SN4.
- Using Equation 10.5 to predict the cracking moment, the k_r should be taken as 16.7~22.3 (clay) and 31.7~62.7(sand). The k_r for sand is 2~3 times that for clay and structural beams.

Study on 16 laterally loaded rock-socketed shafts (4 nonlinear and 12 linear shafts) demonstrates:

- $E \approx (60\sim400)q_u$, except for the atypical short shafts ($L < L_c$)
- $n = 0.7\sim2.3$, $\alpha_o = (0\sim0.22)d$, and $p_u = g_s q_u^{1/n}$, featuring strong scale or roughness effect
- An effective depth to ~10*d* and a maximum slip depth to ~3*d*
- For a layered soil with shear strength increases or decreases dramatically with depth, *n* may be higher or lower than 0.7, respectively. A high *n* of 1.7~2.3 is anticipated for a shape strength increase profile (PS3). Higher than that obtained using existing methods, the deduced gradient of LFP should be reduced for capped piles.

Design of laterally loaded rock-socketed shafts may be based on the closed-form solutions and the provided *E* and p_u. The reduction in flexural rigidity of shafts with crack development obeys that of reinforced concrete beams featured by M_{cr}, M_{ult}, and I_{cr}; and response of nonlinear shaft can be hand-predicted at a high loading eccentricity. Not all cases investigated herein are typical, but the results are quite consistent. In particular, for rigid short shafts, an increasing strain level (thus reduced modulus *E*) is deduced with increasing undrained shear strength p_u, which requires attention and further investigation.*

* Dr. Bitang Zhu assisted the preliminarily calculation of the cases.

Chapter 11

Laterally loaded pile groups

11.1 INTRODUCTION

Laterally loaded piles may entail large deformation in extreme events such as ship impact, earthquake, etc. These piles are often cast into a pile-cap that restrains pile-head rotation but allows horizontal translation. The pile–soil interaction has been preponderantly captured using a series of independent springs-sliders distributed along the shaft linked together with a membrane. In particular, the model allows elastic-plastic closed-form solutions for free-head (FreH), single piles (Guo 2006) to be developed. The solutions are sufficiently accurate for modeling nonlinear pile–soil interaction compared to experimental observation and rigorous numerical approaches (Chapter 9, this book). They have various advantages over early work (Scott 1981), as outlined previously. In particular, the nonlinear response is well captured by the p_u profile and the depth of full mobilization (termed as slip depth, x_p, under a pile-head load H) of the p_u, along with a subgrade modulus (k) that embraces the effect of pile–soil relative stiffness and head and base conditions on the pile response (Chapter 7, this book).

With laterally loaded pile groups, the following facts are noted.

- Fixed-head, elastic solutions generally overestimate maximum bending moment and deflection of capped piles against measured data, and the impact of partially (semi-) fixed-head conditions and loading details (Duncan et al. 2005) needs to be considered.
- Nonlinear response of piles is by and large dominated by the p_u profile (Guo 2006) and the slip depth, x_p. The p_u profile alters significantly with pile-head restraints (Guo 2005). The slip depth is essential to identifying failure mode of individual piles in a group.
- Numerical approaches, such as finite element method (FEM) and finite difference approach (FDA), have difficulty in gaining satisfactory predictions of load distributions among piles in a group (Ooi et al. 2004), but those underpinned by the less rigorous concept of p-multipliers (Brown et al. 1998) offer good predictions of the distributions.

These uncertain points were clarified recently by elastic-plastic closed-form solutions for a laterally loaded, fixed-head (FixH) pile, which cater to semi-fixed-head (free-standing) conditions and incorporate the impact of group interaction and the failure modes. The solutions are underpinned by the input parameters k, p_u, and p-multiplier p_m, and are provided in a spreadsheet program called GASLGROUP (in light of a purposely designed macro operating in EXCEL™). They were substantiated using FEM and/or FEM and FDA results regarding a single pile in sand for a few typical pile groups and were used to predict successfully the nonlinear response of 6 single piles and 24 pile groups in sand and clay.

11.2 OVERALL SOLUTIONS FOR A SINGLE PILE

Figure 3.29 (Chapter 3, this book) shows the springs-sliders-membrane model for a laterally loaded pile embedded in a nonhomogeneous elastoplastic medium. The model of the pile–soil system is underpinned by the same hypotheses as those for free-head piles (Guo 2006) but for the restraining of rotation at head level. Typically, it is noted that governing equations for the pile (see Figure 11.1) are virtually identical to those for free-head piles (Guo 2006) in the upper plastic ($0 \le x \le x_p$) and the lower elastic ($x_p \le x \le L$) zones, as is the methodology for resolving the equations. Compared to FreH piles, the alterations are the slope $w'_A(0) = 0$, and the shear force $-Q_A(0) = H$ (H = pile-head load) to enforce the FixH restraint at $x = 0$.

In light of a constant modulus of subgrade reaction (k) and a limiting force per unit length $p_u = A_L(x + \alpha_o)^n$ (see Chapters 3 and 9, this book), solutions for fixed-head piles are established (Guo 2009). The solutions are provided in Table 11.1 in form of the response profiles using the normalized depths $\bar{x}$ (= λx) and $\bar{z}$ (= $\bar{x} - \bar{x}_p$, with $\bar{x}_p = \lambda x_p$), respectively, for plastic and elastic zones. Note that the schematic profiles of the on-pile force per unit length p (with $p = p_u$ at $x \le x_p$) and the moment $M(x)$ are depicted in Figure 11.2. The λ is the reciprocal of a characteristic length given by $\lambda = \sqrt[4]{k/(4E_pI_p)}$, which controls the attenuation of pile deflection with depth; as well as in Table 11.2 in the form of normalized pile-head load $\bar{H}$ (= $H\lambda^n/A_L$), mudline deflection $\bar{w}_g$ (= $w_g k\lambda^n/A_L$), and bending moment at depth x ($< x_p$) $\bar{M}(\bar{x})$ (= $M(x)\lambda^{2+n}/A_L$).

The normalized $\bar{H}$, $\bar{w}_g$, and $\bar{M}_{max}$ for the single pile are characterized by the p_u profile and the $\bar{x}_p$. Given the λ for the pile–soil system and the p_u for the limiting force profile, the response is characterized by the soil slip depth x_p. The on-pile resistance at mudline and the coupled interaction may be ignored by taking $\alpha_o \approx 0$ and $N_p \approx 0$, respectively, which render the following simplified equations.

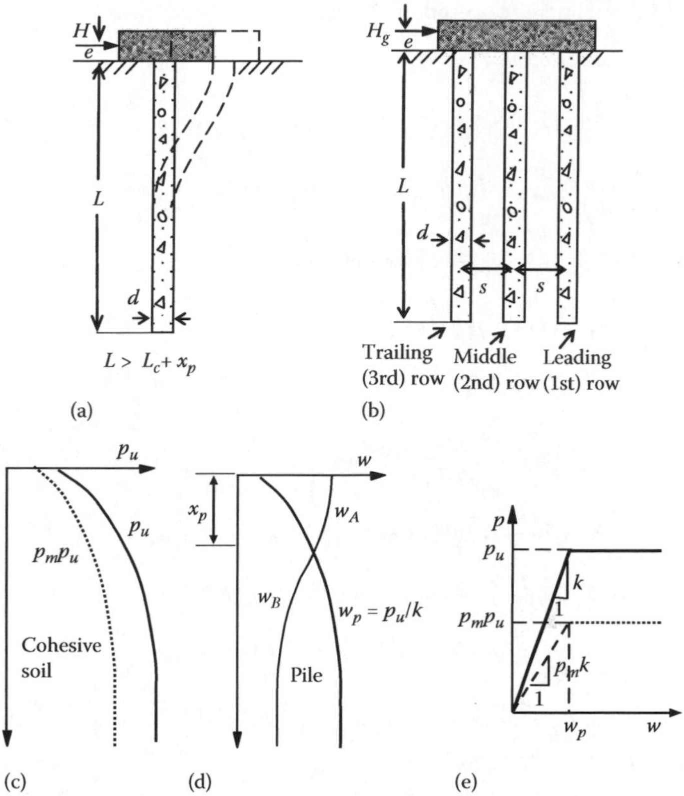

Figure 11.1 Schematic limiting force and deflection profiles. (a) Single pile. (b) Piles in a group. (c) LFP. (d) Pile deflection and w_p profiles. (e) *p-y(w)* curve for a single pile and piles in a group. (After Guo, W. D., *Int J Numer and Anal Meth in Geomech* 33, 7, 2009.)

$$\overline{H} = \overline{x}_p^n \frac{[n^2(1+\overline{x}_p) + n(2\overline{x}_p^2 + 5\overline{x}_p + 4) + 2\overline{x}_p^3 + 6\overline{x}_p^2 + 6\overline{x}_p + 3]}{(1+\overline{x}_p)^2(n+3)(n+1)} \tag{11.1}$$

$$\begin{aligned}\overline{w}_g = \frac{\overline{x}_p^n}{3(n+4)(n+3)} \Bigg\{ & \frac{\overline{x}_p^3 + 3(\overline{x}_p^2 + \overline{x}_p + 1)}{(1+\overline{x}_p)} n^2 \\ & + \frac{2\overline{x}_p^5 + 11\overline{x}_p^4 + 28\overline{x}_p^3 + 21(2\overline{x}_p^2 + 2\overline{x}_p + 1)}{(1+\overline{x}_p)^2} n \\ & + \frac{2[\overline{x}_p^6 + 6\overline{x}_p^5 + 15\overline{x}_p^4 + 24\overline{x}_p^3 + 18(2\overline{x}_p^2 + 2\overline{x}_p + 1)]}{(1+\overline{x}_p)^2} \Bigg\}\end{aligned} \tag{11.2}$$

Table 11.1 Expressions for response profiles of a fixed-head pile

Responses in plastic zone ($\bar{x} \le \bar{x}_p$)

$$w(\bar{x}) = \frac{4A_L}{k\lambda^n}\left[-F(4,\bar{x}) + F(4,0) + F(3,0)\bar{x} + (F(1,0)+\bar{H})\frac{\bar{x}^3}{6} + \bar{H}\bar{e}\frac{\bar{x}^2}{2}\right] + \frac{C_2}{\lambda^2}\frac{\bar{x}^2}{2} + w_g$$

$$w'(\bar{x}) = \frac{4A_L\lambda}{k\lambda^n}\left[-F(3,\bar{x}) + F(3,0) + (F(1,0)+\bar{H})\frac{\bar{x}^2}{2} + \bar{H}\bar{e}\bar{x}\right] + \frac{C_2}{\lambda}\bar{x}$$

$$-M(\bar{x}) = \frac{A_L}{\lambda^{2+n}}[-F(2,\bar{x}) + (F(1,0)+\bar{H})\bar{x} + \bar{H}\bar{e}] + C_2E_pI_p$$

$$-Q(\bar{x}) = \frac{A_L}{\lambda^{1+n}}[-F(1,\bar{x}) + F(1,0) + \bar{H}]$$

where $\bar{e} = \lambda e$; $F(m,\bar{x}) = (\bar{x}+\bar{\alpha}_o)^{n+m}/(n+m)...(n+2)(n+1)$ $(1 \le m \le 4)$; $\bar{\alpha}_o = \lambda\alpha_o$; $F(0,\bar{x}) = (\bar{x}+\bar{\alpha}_o)^n$

$$C_2 = (-\alpha_N C_5 + \beta_N C_6)\frac{\lambda^2}{\bar{x}_p} + \frac{4A_L\lambda^2}{k\lambda^n}\left\{[F(3,\bar{x}_p) - F(3,0)]/\bar{x}_p - [F(1,0)+\bar{H}]\frac{\bar{x}_p}{2} - \bar{H}\bar{e}\right\}$$

$$C_5 = \frac{A_L}{k\lambda^n(\alpha_N + \bar{x}_p)}\{2(1-2\alpha_N^2)[F(3,0) - F(3,\bar{x}_p) + \bar{x}_pF(2,\bar{x}_p)]$$
$$-(1+2\alpha_N\bar{x}_p)F(1,\bar{x}_p) + [1+2\alpha_N\bar{x}_p - (1-2\alpha_N^2)\bar{x}_p^2][F(1,0)+\bar{H}]\}$$

$$C_6 = \frac{A_L}{k\beta_N\lambda^n(\alpha_N + \bar{x}_p)}\{-2\alpha_N(2\alpha_N^2-3)[F(3,0) - F(3,\bar{x}_p) + \bar{x}_pF(2,\bar{x}_p)]$$
$$-[\alpha_N + 2(\alpha_N^2-1)\bar{x}_p]F(1,\bar{x}_p) + [\alpha_N + 2(\alpha_N^2-1)\bar{x}_p + \alpha_N(2\alpha_N^2-3)\bar{x}_p^2][F(1,0)+\bar{H}]\}$$

$$\alpha_N = \sqrt{1+N_p/\sqrt{4E_pI_pk}},\ \beta_N = \sqrt{1-N_p/\sqrt{4E_pI_pk}}$$

Responses in elastic zone ($\bar{x} > \bar{x}_p$, *or* $\bar{z} > 0, \bar{z} = \lambda z = \lambda(x - x_p)$)

$$w(\bar{z}) = e^{-\alpha_N\bar{z}}[C_5\cos(\beta_N\bar{z}) + C_6\sin(\beta_N\bar{z})]$$

$$w'(\bar{z}) = \lambda e^{-\alpha_N\bar{z}}[(-\alpha_NC_5 + \beta_NC_6)\cos(\beta_N\bar{z}) + (-\beta_NC_5 - \alpha_NC_6)\sin(\beta_N\bar{z})]$$

$$-M(\bar{z}) = E_pI_p\lambda^2e^{-\alpha_N\bar{z}}\{[(\alpha_N^2-\beta_N^2)C_5 - 2\alpha_N\beta_NC_6]\cos(\beta_N\bar{z})$$
$$+[2\alpha_N\beta_NC_5 + (\alpha_N^2-\beta_N^2)C_6]\sin(\beta_N\bar{z})\}$$

$$-Q(\bar{z}) = E_pI_p\lambda^3e^{-\alpha_N\bar{z}}\{[-\alpha_N(\alpha_N^2-3\beta_N^2)C_5 + \beta_N(3\alpha_N^2-\beta_N^2)C_6]\cos(\beta_N\bar{z})$$
$$+[-(3\alpha_N^2-\beta_N^2)\beta_NC_5 - \alpha_N(\alpha_N^2-3\beta_N^2)C_6]\sin(\beta_N\bar{z})\}$$

The last four expressions are independent of the head constraint and are identical to those for free-head piles.

$$-\bar{M}_{max} = -0.5\bar{x}_p^n\frac{(\bar{x}_p+1)n^2 + (2\bar{x}_p^2+5\bar{x}_p+5)n + 2(\bar{x}_p^3+3\bar{x}_p^2+3\bar{x}_p+3)}{(\bar{x}_p+1)(n+3)(n+2)} \tag{11.3}$$

Note that the M_{max} is equal to the moment at mudline M_o for a FixH pile (see Figure 11.2). The eccentricity e (Figure 11.1) vanishes from the expressions

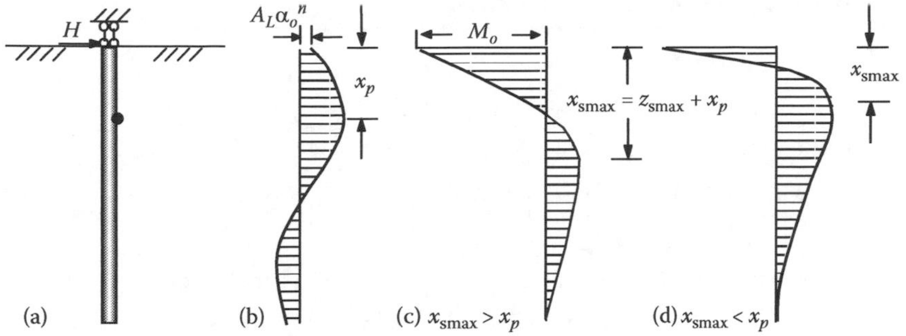

Figure 11.2 Schematic profiles of on-pile force and bending moment. (a) Fixed-head pile. (b) Profile of force per unit length. (c) Depth of second largest moment x_{smax} ($>x_p$). (d) Depth of x_{smax} ($<x_p$).

of $\bar{H}$, $\bar{w}_g$, and $\bar{M}_{max}$, which needs to be considered via a treatment H2 shown later. These solutions have similar features to FreH solutions (Guo and Lee 2001; Guo 2006) (see Chapters 3 and 9, this book) and are directly used to predict nonlinear response of piles and pile groups.

11.3 NONLINEAR RESPONSE OF SINGLE PILES AND PILE GROUPS

A pragmatic approach for estimating response of pile groups is depicted herein in terms of the benchmark solutions for a single pile and the *p*-multipliers p_m for pile-pile interaction.

11.3.1 Single piles

The current solutions outlined in Tables 11.1 and 11.2 allow nonlinear responses of a pile to be predicted. They are compiled into the spreadsheet program GASLGROUP, with the calculation flow chart being depicted in Figure 11.3. Given a mudline deflection w_g (thus $\bar{w}_g$), a slip depth x_p is iteratively obtained using the $\bar{w}_g$ expression (see Table 11.2) or Equation 11.2, which allows other responses (e.g., using Equation 11.1 for the load H) to be evaluated, including the profiles of deflection, rotation, bending moment, and shear force (see Table 11.1) for elastic and plastic zones, respectively. The subsequent solutions were generally computed using GASLGROUP, except where specified. They, however, may be referred to as "CF(FixH)," "CF(FreH)" for FixH and FreH piles, respectively, and as "current CF" for either head restraint.

The impact of ground level resistance α_o and the distribution of limiting force profile n is presented in Figure 11.4a$_1$ and a$_2$ for displacement and Figure 11.4b$_1$ and b$_2$ for maximum bending moment. Given n = 0.7 for clay, α_o = 0, the fixed-head solution is plotted in Figure 11.5a together with

Table 11.2 $\bar{H}$, $\bar{w}_g$, and $\bar{M}(\bar{x})$ of a fixed-head pile

(a) Normalized pile-head load, $\bar{H}$

$$\bar{H} = -\frac{F(0,\bar{x}_p)(\bar{x}_p+\alpha_N)}{1-2\alpha_N^2-2\alpha_N\bar{x}_p-\bar{x}_p^2} - \frac{[1-2\alpha_N^2-2\alpha_N\bar{x}_p][F(1,0)-F(1,\bar{x}_p)]}{1-2\alpha_N^2-2\alpha_N\bar{x}_p-\bar{x}_p^2}$$
$$-\frac{2[F(3,0)-F(3,\bar{x}_p)+\bar{x}_pF(2,\bar{x}_p)-0.5F(1,0)\bar{x}_p^2]}{1-2\alpha_N^2-2\alpha_N\bar{x}_p-\bar{x}_p^2}$$

The $\bar{H}$ is deduced from the following relationship obtained for the depth x_p ($z = 0$):
$(\alpha_N+\bar{x}_p)w_p^{IV} - (1-2\alpha_N^2-2\alpha_N\bar{x}_p)\lambda w_p''' - 2\lambda^2(\lambda w_p' - \bar{x}_p w_p'') = 0$

where w_p', w_p'', w_p''', and w_p^{IV} are values of 1st, 2nd, 3rd, and 4th derivatives of $w(x)$ with respect to depth z. Given $\bar{x}_p = 0$, the minimum head-load to initiate slip is obtained.

(b) Normalized mudline deflection, $\bar{w}_g$

$$\bar{w}_g = 4[F(4,\bar{x}_p)-F(4,0)] - 2\bar{x}_p[F(3,\bar{x}_p)+F(3,0)]$$
$$+\frac{2(1-2\alpha_N^2-\alpha_N\bar{x}_p)}{\alpha_N+\bar{x}_p}[F(3,0)-F(3,\bar{x}_p)+\bar{x}_pF(2,\bar{x}_p)]$$
$$-\frac{1+2\alpha_N\bar{x}_p+\bar{x}_p^2}{\alpha_N+\bar{x}_p}F(1,\bar{x}_p) + \left(\frac{\bar{x}_p^3}{3}+\frac{1+2\alpha_N\bar{x}_p+2\alpha_N^2\bar{x}_p^2+\alpha_N\bar{x}_p^3}{\alpha_N+\bar{x}_p}\right)[F(1,0)+\bar{H}]$$

Note: w_g is deduced from $w(x)$ (see Table 11.1) (independent of e), along with C_5 and C_2.

(c) Normalized bending moment, $\bar{M}(\bar{x})$ ($x < x_p$). The expression is deduced using expressions of C_j (j = 2, 5, and 6) given in Table 11.1:

$$-\bar{M}(\bar{x}) = \frac{1}{\alpha_N+\bar{x}_p}[F(3,\bar{x}_p)-F(3,0)+\alpha_NF(2,\bar{x}_p)+0.5F(1,\bar{x}_p)]+F(2,\bar{x})$$
$$+\left[\bar{x}-\frac{2\alpha_N\bar{x}_p+\bar{x}_p^2+1}{2(\alpha_N+\bar{x}_p)}\right][F(1,0)+\bar{H}]$$

Maximum bending moment M_{max} is obtained by substituting $\bar{x} = 0$.

Note: The constants C_j are determined using the compatibility conditions of $Q(\bar{x})$, $M(\bar{x})$, $w'(\bar{x})$, and $w(\bar{x})$ at the normalized slip depth, $\bar{x}_p$ [$\bar{x} = \bar{x}_p$ or $\bar{z} = 0$]. Elastic solutions validated for $N_p < 2(kE_pI_p)^{0.5}$ are ensured by L being greater than the sum of L_c and the maximum x_p.

the free-head solutions (see Chapter 9, this book) for displacement, as the letter does compare well with measured data of 31 piles in clay (see Figure 11.5b and Chapter 9, this book). In Figure 11.5b, the numbers 1 through 32 indicate the pile CS number in Table 9.9. The same solutions for bending moment are plotted in Figure 11.6a and b, respectively. Given $n = 1.7$ for sand, $\alpha_o = 0$, the fixed-head solution is plotted in Figure 11.7a together with the free-head solutions for displacement. They are compared with measured data of 18 piles in sand in Figure 11.7b, in which the numbers 1 through 20 indicate pile PS number in Table 9.10 (Chapter 9, this book). The same solutions for bending moment are plotted in Figure 11.8. These curves may be used for pertinent design. The values of M_{max} may occur at either elastic

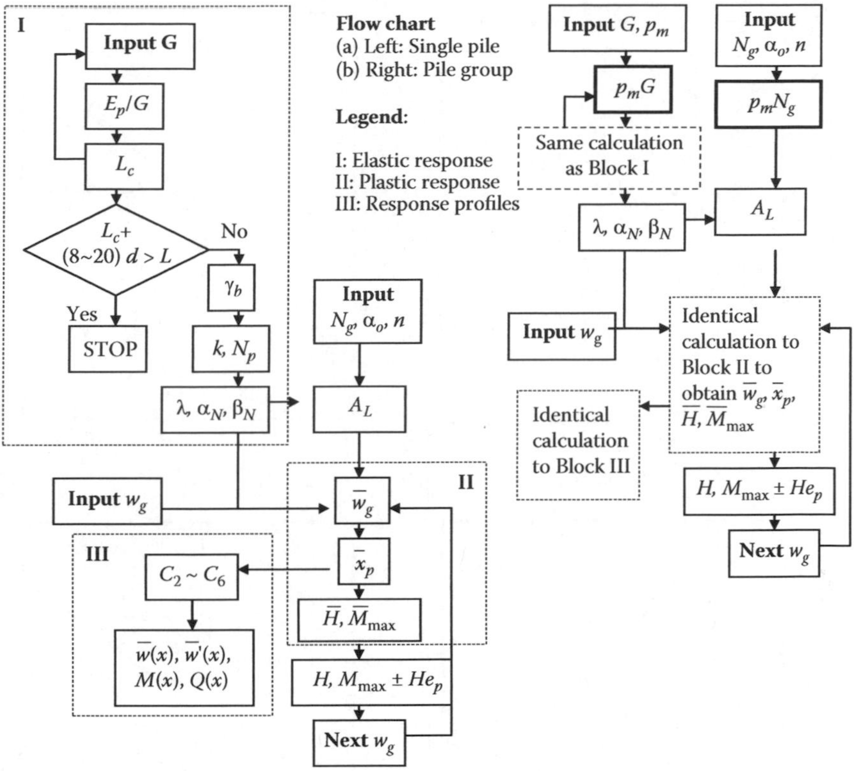

Figure 11.3 Calculation flow chart for current solutions (e.g., GASLGROUP).

or plastic zones. To facilitate practical predictions, both values are provided for any load levels in Figure 11.6 for clay and Figure 11.8 for sand.

11.3.2 Group piles

Among a laterally loaded pile group, soil resistance along piles in a trailing row (see Figure 11.1b) is reduced owing to the presence and actions of the piles ahead of the row (Brown et al. 1998). The gradient of a p-y curve is taken as kp_m (p_m = a p-multiplier, ≤ 1, see Chapter 3, this book) and the limiting force per unit length as $p_u p_m$. This generates a squashed p-y curve (dashed line in Figure 11.1e) for piles in the same row. In particular, the shape of LFP (p_u profile) is also allowed to be altered via a new "n" (G1-2, Table 3.10, Chapter 3, this book). Typical steps are highlighted in Figure 11.3. Each pile or row in a group is analyzed as if it were a single pile for a specified mudline deflection w_g, as the w_g may be stipulated identical for any piles in the group unless specified. The modeling employs the single pile solutions but with a modulus of kp_m and a limiting force per unit length

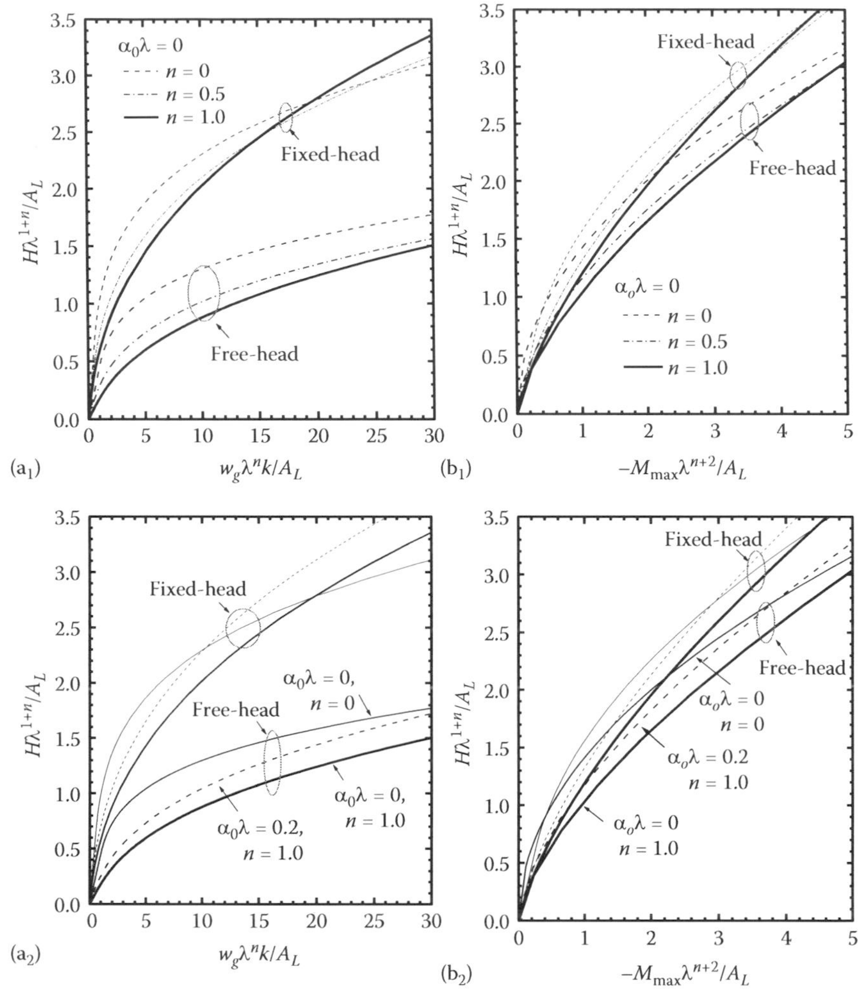

Figure 11.4 Normalized pile-head load versus (a_1, a_2) displacement, (b_1, b_2) versus maximum bending moment.

of $p_u p_m$. The p_m may be obtained from Equation 3.69 in Chapter 3, this book. The calculation is repeated for each row using the associated p_m. The H offers the total load H_g on the group for the prescribed w_g. These steps are repeated for a series of desired w_g, which allows load-deflection curves of each pile (H-w_g), each row, and the group (H_g-w_g) to be generated. Likewise, other responses are evaluated, such as the moment using $\bar{M}_{max}$. This calculation is incorporated into GASLGROUP.

Pile-head load may be exerted at a free length e_p and deflection w_t be measured at a free length e_w above mudline on a free-standing pile-cap. A FixH restraint for each capped pile in a group is not warranted, particularly at a

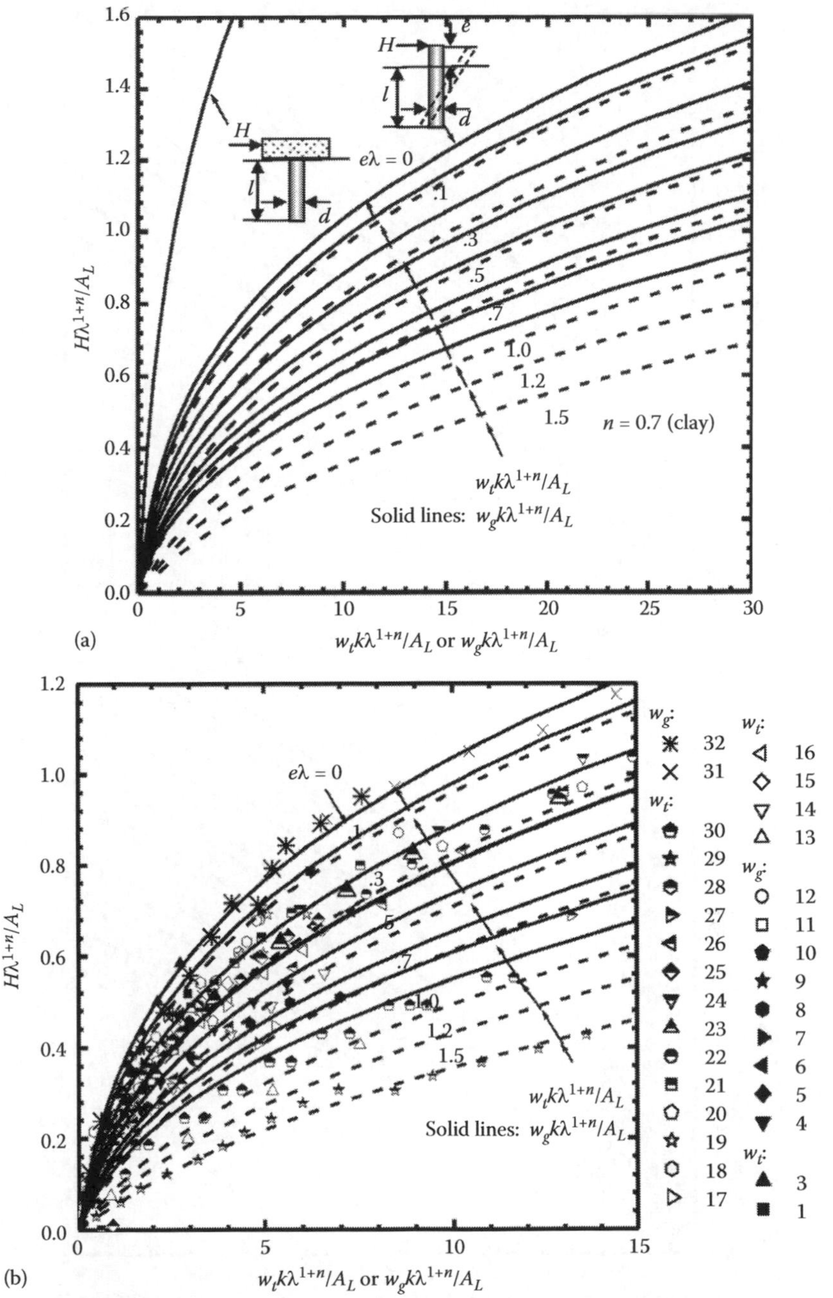

Figure 11.5 Normalized load, deflection (clay): measured versus predicted (n = 0.7) with (a) fixed-head solution, or (b) measured data.

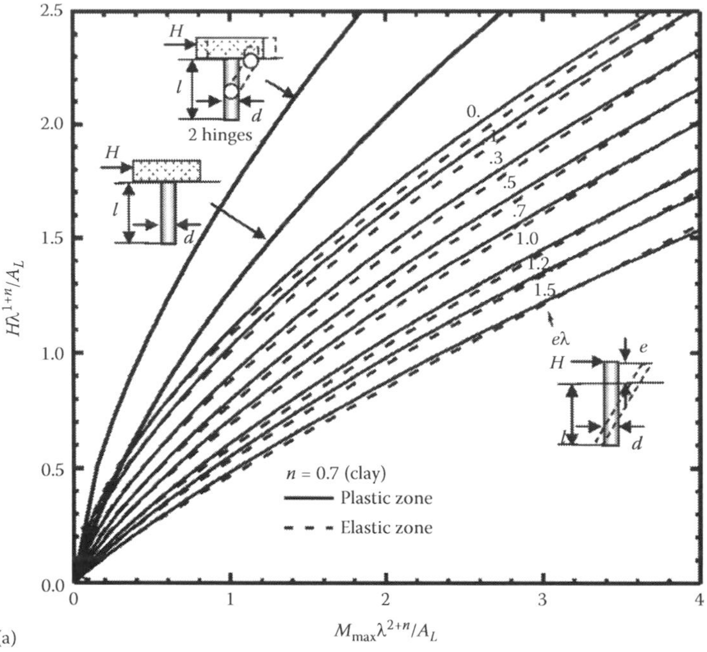

(a)

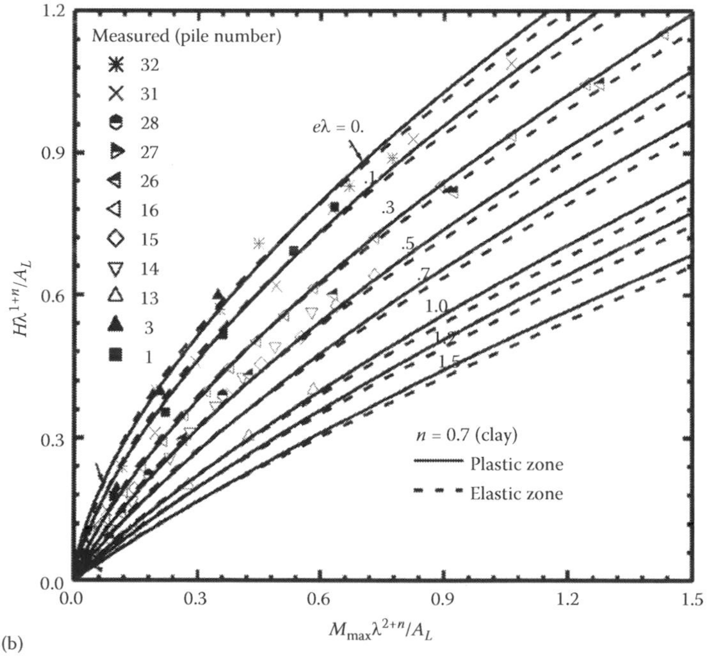

(b)

Figure 11.6 Normalized load, maximum M_{max} (clay): measured versus predicted (n = 0.7) with (a) fixed-head solution, or (b) measured data.

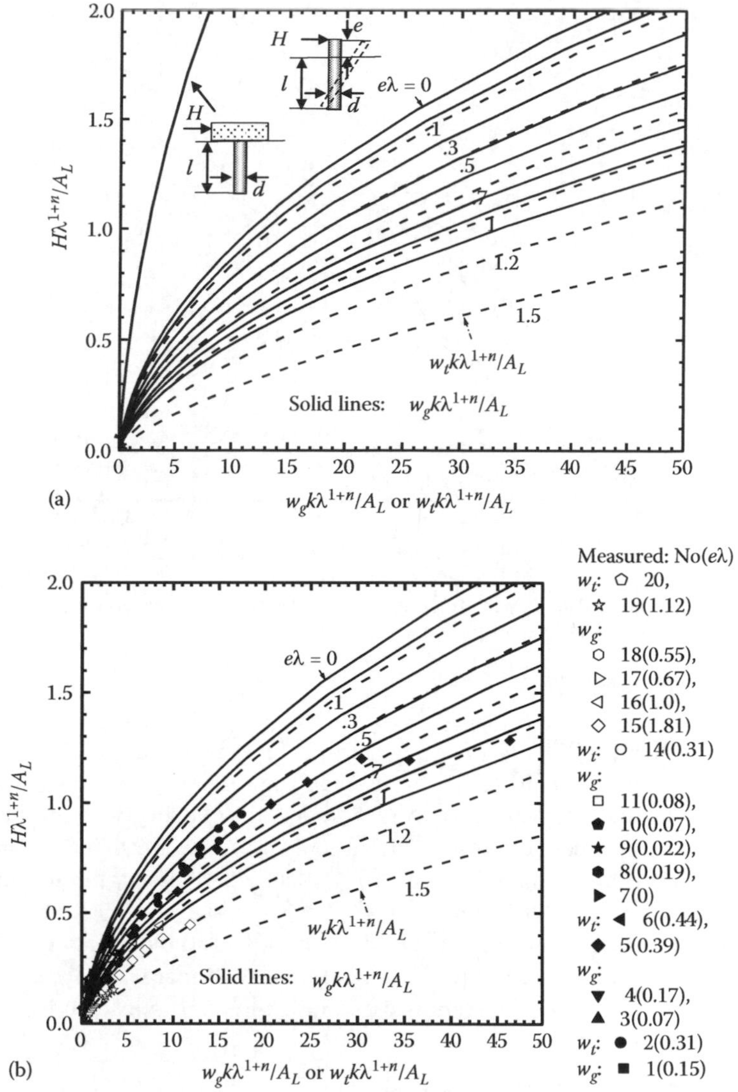

Figure 11.7 Normalized load, deflection (sand): measured versus predicted (n = 1.7) with (a) fixed-head solution, or (b) measured data.

large deflection (thus referred to as semi-FixH pile) (Ooi et al. 2004) (observation *H1*). The current solutions should be revised using an approximate treatment *H2* for Type A or B head restraint (see Figure 11.9). It is given as treatments *H2a*, *H2b*, and *H2c*, respectively, for assessing M_0 (Matlock et al. 1980), moment profile for $e_p \neq 0$, and the difference between the w_t and w_g (Prakash and Sharma 1989; Zhang et al. 1999).

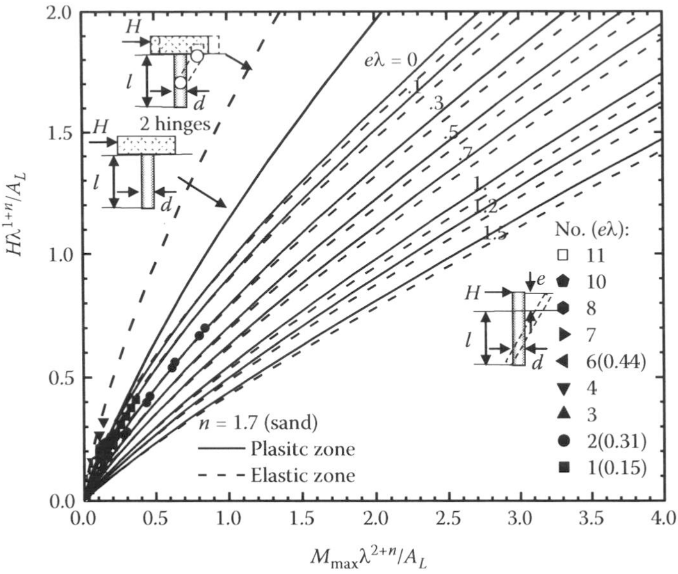

Figure 11.8 Normalized load, maximum M_{max} (sand): measured versus predicted (n = 1.7).

11.4 EXAMPLES

The closed-form (CF) solutions (e.g., Table 11.1) were validated for elastic state (Guo and Lee 2001) and free-head case. They were shown herein against FEM analysis and measured data (Wakai et al. 1999) regarding single and group piles (Examples 11.1 and 11.2) to complement the previous corroborations using full-scale and model tests (Ruesta and Townsend 1997; Brown et al. 1998; Rollins et al. 1998; Zhang et al. 1999; Rollins et al. 2005). Calculations for typical offshore piles and model piles are also provided in Examples 11.3 and 11.4, respectively. Equations 11.1 through 11.3 are used to predict the response of single piles and pile groups.

Example 11.1 Model piles in sand

Wakai et al. (1999) conducted tests on two model single piles and two nine-pile groups in a tank of 2.5m × 2.0m × 1.7m (height). Each aluminum pipe pile was 1.45 m in length (L), 50 mm in outside diameter (d_o), 1.5 mm in wall thickness (t), and with a flexural stiffness E_pI_p of 4.612 kNm^2. The piles were installed into a dense sand that had an angle of internal friction ϕ of 42°, unit weight γ_s of 15.3 kN/m^3 (Poisson's ratio ν_s = 0.4) (see Tables 11.3 and 11.4). Under FreH and capped-head restraints, respectively, lateral load (H) was imposed at 50 mm above mudline (i.e., e_p = 50 mm) on the single piles, at which deflection (w_t) was measured ($e_w = e_p$ = 50 mm). The measured H–w_t relationships are

Illustrations	Descriptions of H2 treatment
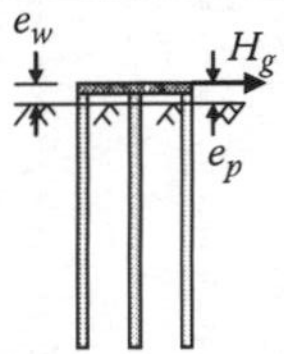 Example 11.1 ($e_w = e_p = 50$ mm) Example 11.4 ($e_w/e_p = 35/85$ mm)	Type (A) H2a: Maximum bending moment • With a lateral load $\boldsymbol{H_{av}}$ ($=H_g$/Number of piles in a group) at a free-length $\boldsymbol{e_p}$ on a capped pile, M_{max} is predicted using CF(FixH) solutions. Value of $H_{av}e_p$ is then subtracted from the M_{max} to gain the moment M_o at mudline (Matlock et al. 1980), as conducted in Examples 11.1 through 11.4. H2b: Bending moment profile
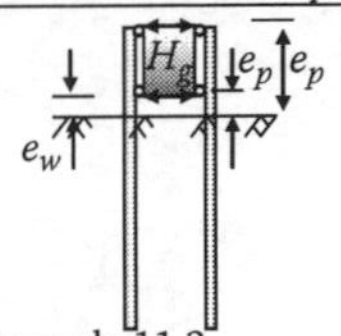 Example 11.3 ($e_w = 0.305$ m $e_p = 0.305$ and 1.83 m)	• Bending moment profile is estimated using CF(FixH) solutions for $e_p = 0$. It is then translated, through reducing depth by an amount of the e_p, to generate the moment profile for $e_p \neq 0$. This approximation (Matlock et al. 1980) is equally applicable to Type (B). H2c: Deflection profile over free length and its impact • Pile deflection w_t at an eccentricity of e_w is the sum of deflection due to free-length, $e_w\theta_w$ (θ_w = rotation angle at mudline, in radian) (Matlock et al. 1980), and the mudline deflection w_g, i.e., $w_t = w_g + e_w\theta_w$. The $e_w\theta_w$ is negligible for a completely fixed-head pile; but it increases with e_w and number and/or flexbility of piles (Examples 11.3 and 11.4).
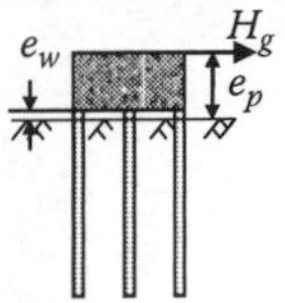 Example 11.2 ($e_w = 0.3$ m. $e_p = 1.3$ m)	Type (B) • Given the typical loading cap shown in the left-head diagram (Case II), an additional bending moment of the cap, $\boldsymbol{He_p}$ (H = load in a pile) needs to be added to the estimated M_{max} (FixH) to gain the true maximum bending moment in each pile. Afterwards, it should be analyzed as a Type (A) head condition.

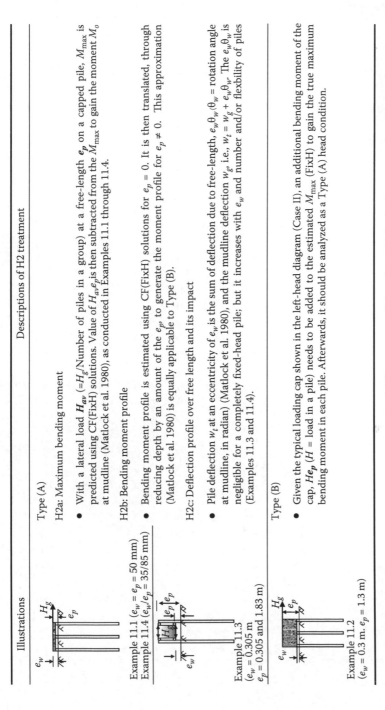

Figure 11.9 Three typical pile-head constraints investigated and H2 treatment. (After Guo, W. D., *Int J Numer and Anal Meth in Geomech* 33, 7, 2009.)

Table 11.3 Parameters for Examples 11.1, 11.2, 11.3, 11.5, and 11.6

References	G *(MPa)*	*Group*	s/d[a]	p_m *by row* *1st*	*2nd*	*3rd*	*4th*	n
Example 11.1: Model test (Wakai et al. 1999) Sand: ϕ = 42.0°, γ_s = 15.3 kN/m³	2.35	3 × 3	2.5	0.80	0.40	0.30		0.58
		3 × 3[b]	2.5	0.80	0.40	0.30		1.15[b]
Example 11.2: In situ test (Wakai et al. 1999): ϕ = 34.4° γ_s' = 12.7 kN/m³	10.2	3 × 3	2.5	0.80	0.40	0.30		0.85 or 1.20
Example 11.3: FLAC3D modeling Soft silty clay, s_u = 5 + 1.25*x* (kPa), *x* in m	1.125	1 × 3	3	0.9	0.75	0.6		0.70
		2 × 3	3	0.9	0.55	0.5		
		3 × 3	3	0.55	0.40	0.25		
		4 × 4	3	0.43	0.30	0.15	0.3	
Example 11.4: Model tests (Gandhi and Selvam 1997) Sand: ϕ = 36.3° γ_s = 14.6~17.3 kN/m³	0.3	1 × 2	4	0.80	0.80			1.35
			8	0.90	0.90			
			12	1.00	1.00			
	3.0	2 × 2	4	0.80	0.80			1.10
			6	1.00	0.80			
			8	1.00	1.00			
	0.8	3 × 2	4	0.80	0.30			
			6	0.80	0.35			
			8	0.80	0.60			1.15
	3.0	1 × 3	4	0.80	0.40	0.30		
	0.3		8	0.90	0.70	0.60		1.20
			12	1.0	0.80	0.80		
	0.8[c]		4	0.80	0.40	0.30		1.25
	3.0[c]	2 × 3	6	0.80	0.50	0.40		
	0.8	3 × 3	4	0.80	0.40	0.30		1.15
			6	0.80	0.45	0.35		
			8	0.80	0.50	0.40		
	0.4	2 × 1	3	0.80				1.35
	0.3	3 × 1	3	0.80				1.20
Centrifuge tests (McVay et al. 1998) Sand: ϕ = 37.1° γ_s' = 14.5 kN/m³ (Guo 2010)	0.33	3 × 3	3	0.80	0.40	.28		0.50

Source: Guo, W. D., *Int J Numer and Anal Meth in Geomech*, 33, 7, 2009.

[a] Normalized center to center spacing in loading direction, but s = 3*d* within any row.

[b] Free-head group, Shear modulus *G* should be multiplied by p_m to gain modulus for a pile in a group.

Table 11.4 Pile properties α_o and N_g in current analysis

Examples	*11.1*	*11.2*	*11.3*	*11.4*
Pile length/diameter (*m*)	1.45/0.05	14.5/0.319	13.4/0.084	0.75/0.0182
Flexural stiffness (MN-m^2)	4.61×10^{-3}	5.623	2.326	86×10^{-6}
α_o for LFP (*m*)	0	0	Table 11.7	0
N_g for LFP	K^2_p	$2.4K^2_p$	1.0 (FixH), 4.0 (FreH)	$(1\text{-}2)K^2_p$ [a]

Source: Guo, W. D., *Int J Numer and Anal Meth in Geomech*, 33, 7, 2009.

[a] 1.0 for single FixH piles and group piles, and 2.0 for FreH single piles.

plotted in Figure 11.10a. The measured moment profiles are plotted in Figure 11.10b for $w_t = w_g = 0.5$, 2.0, and 5.0 mm.

The pile groups had a center-to-center spacing, *s*, of 2.5 pile diameters. Deflections (at an $e_w = 50$ mm) were measured under the total loads of H_g exerted on the group (with $e_p = e_w$ see Figure 11.9) for both free-head and fixed-head (capped) groups, respectively. The average lateral load per pile H_{av} ($H_{av} = H_g/9$) is plotted against the measured w_t in Figure 11.11 for each test. The measured moments (at cap level) for leading and back rows are given in Table 11.5 for $w_g = 5$ mm.

Wakai et al. (1999) also conducted three-dimensional finite element analysis (FEM^{3D}). The predicted H–w_t relationships are plotted in Figure 11.10a, the moment profiles in Figure 11.10b (for $w_t = w_g = 0.5$, 2.0, and 5.0 mm), and H_{av}-w_t relationships in Figure 11.11.

Equations 11.1 through 11.3 (or GASLGROUP) were used to predict the responses of the two single piles and the two pile groups using G = 2.35 MPa (from reported Young's modulus), $n = 1.15$ (or $n = 0.575$ for the FixH group, *G2*, Table 3.10), $N_g = 25.45$ ($= K^2_p$), and $\alpha_o = 0$ m. The

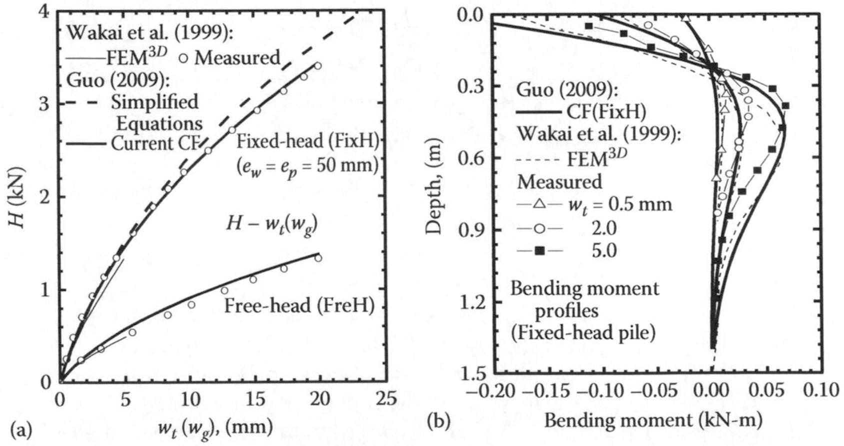

Figure 11.10 The current solutions compared to the model tests and FEM^{3D} results (Wakai et al. 1999) (Example 11.1). (a) Load. (b) Bending moment. (After Guo, W. D., *Int J Numer and Anal Meth in Geomech* 33, 7, 2009.)

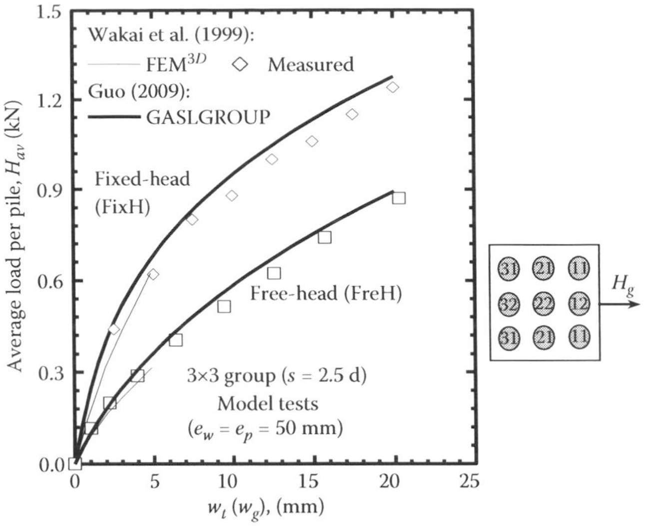

Figure 11.11 The current solutions compared to the group pile tests, and FEM^{3D} results (Wakai et al. 1999) (Example 11.1). (After Guo, W. D., *Int J Numer and Anal Meth in Geomech* 33, 7, 2009.)

Table 11.5 A capped 3 × 3 group in sand at w_g = 5 mm (Type A, Example 11.1)

Row	Measured (kNm)	FEM^{3D} (kNm)	n	H_{av} (kN)	$H_{av}e_p$(kNm)	M_o (kNm)
			Current predictions (GASLGROUP)			
Leading[a]	−0.148	−0.239	0.58	0.959	−0.048	−0.25
			1.15	1.253	−0.063	−0.292
Back[a]	−0.115	−0.182	0.58	0.489	−0.025	−0.158
			1.15	0.464	−0.023	−0.182

Source: Guo, W. D., *Int J Numer and Anal Meth in Geomech*, 33, 7, 2009.

[a] "Leading"' refers to side pile 11, and "Back" to middle pile 32 (see Figure 11.11).

impact of the cap fixity condition on the LFP was neglected as justified subsequently.

11.1.1 Single piles

The predicted H-w_g responses for the FreH pile (Guo 2006) and the capped pile are depicted in Figure 11.10a. They all well replicate the FEM^{3D} analysis and the measured H-w_t curves (implying $w_t \approx w_g$ for the FixH pile). In particular, ignoring the coupled impact (taking $N_p = 0$), Equations 11.1 through 11.3 slightly overestimated the head stiffness (load over displacement) of the FixH pile, as is noted later in other cases. Given G = 2.35 MPa, the calculated k of 1.50 MPa (Equation 3.50, Chapter 3, this book) for the FreH pile exceeds that for the FixH pile by 18.2%, as is noted previously (Syngros 2004). GASLGROUP and the FEM^{3D} predict similar bending moment profiles along the capped pile (see Figure 11.10b). Both, however, overestimate the measured moment M_o at ground level, exhibiting semi-fixed-head features as also seen by a 1.3° rotation (at the e_w level for a similar cap group) at a pile-cap deflection w_t of 20 mm, which renders w_g = 4 mm at w_t = 5 mm; a revised depth of $x - e_p$ (e_p = 50 mm) (*H2*) for the predicted moment profiles (not shown herein); and a total resistance T_R of 2.153 kN (less than H of 3.4 kN) (*G4*), in terms of Equation 3.68, Chapter 3, this book.

11.1.2 Group piles

The responses of the two groups were predicted using p_m = 0.8, 0.4, and 0.3 (either head constraints, *G5*) for the leading (LR), middle (MR), and trailing rows (TR) (Brown et al. 1998), and the aforementioned G, N_g, α_o, and n (see Table 11.3), but for n = 0.58 (FixH, *G2*). Typical load H_{av} and moment M_{max} predicted are tabulated in Table 11.5 for w_g = 5 mm.

The predicted load (H_{av})–deflection (w_g) curves agree with the measured H_{av}-w_t relationships, and the FEM^{3D} predictions (Wakai et al. 1999) (see Figure 11.11) for either group. The predicted maximum moment M_o (= M_{max} + $H_{av}e_p$) of 0.25–0.292 kNm (leading row) and 0.158–0.182 kNm (trailing row) is consistent with the FEM^{3D} prediction (Table 11.5) of 0.239 kNm and 0.182 kNm, respectively. Both nevertheless overestimate the respectively measured 0.148 kNm (leading) and 0.115 kNm (trailing), pinpointing again the effect of w_t–w_g. In the capped group, at w_g = 20 mm, x_p was gained as 8.5*d* (H_{av} = 0.99 kN), 6.1*d* (0.64 kN) and 6.2*d* (0.53 kN), for piles in the leading, middle, and trailing rows, respectively, under the load H_{av}. Values of the x_p are slightly less than 8.6*d* obtained for the single pile at the same w_g.

Overall, the semi-FixH restraint reduces the moment and increases the difference of w_t–w_g. The current solutions for FixH piles agree with the FEM^{3D} analysis (Wakai et al. 1999).

Example 11.2 In situ piles in "silty sand": Cap effect

Wakai et al. (1999) reported in situ tests on a steel pipe pile under FreH conditions, and nine-pile group (a spacing s = 2.5d_o, d_o = outside diameter) under a restrained head. The pile had L = 14.5 m, d_o = 0.319 m, and t = 6.9 mm, and flexural stiffness of E_pI_p = 5.623 MNm^2. The "silt sand" (assumed) to a depth of 2.7 m is featured by c' (cohesion) = 0–16.7 kPa, ϕ' = 34.4°, and γ'_s = 12.7 kN/m^3. Lateral load H was applied, and deflection w_t was measured at the same eccentricity (e_p = e_w) of 0.5 m on the single pile. The load H_g was applied on the group (see Figure 11.9) at e_p = 1.3 m and deflection w_t measured at e_w = 0.3 m. The measured curves of H-w_t (single pile) and H_g-w_t (the pile group) are plotted in Figure 11.12a and b, respectively. The measured bending moments M_o for the piles in leading and trailing rows are provided in

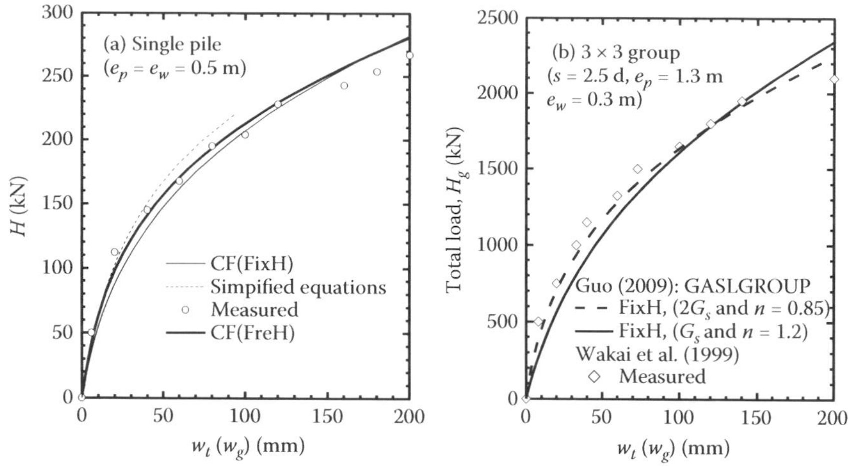

Figure 11.12 Predicted $H_g(H)$-w_g versus measured $H_g(H)$-w_t response of (a) the single pile and (b) pile group (Wakai et al. 1999) (Example 11.2). (After Guo, W. D., *Int J Numer and Anal Meth in Geomech* 33, 7, 2009.)

Table 11.6 A 3 × 3 pile group in sand (Type B, Example 11.2)

Load, H_g (kN)	*392*			*1570*		
Row number	*Leading*	*Middle*	*Trailing*	*Leading*	*Middle*	*Trailing*
Predicted, H_{av}(kN) (1)	62.04	37.86	30.78	240.53	154.51	128.5
Measured[a], H_{av}(kN) (2)	48.81	41.2	40.68	271.86	150.88	112.92
[(1)−(2)]/(2),% (3)	27.1	−8.1	−24.3	−11.5	2.4	13.4
M due to e_p = 1.3 m (4)	80.65	49.22	40.0	312.69	200.86	167.05
GASLGROUP, M_{max} (5)	46.93	33.33	28.87	297.51	220.57	199.69
Predicted, (4)+(5), M_o(kNm) (6)	127.56	82.55	68.88	610.19	421.43	361.74
Measured, M_o(kNm) (7)	112.48	84.36	84.35	674.84	492.07	393.74
[(6)−(7)]/(7),% (8)	−13.4	−2.1	−18.3	−9.6	−14.36	−8.1

Source: Guo, W. D., *Int J Numer and Anal Meth in Geomech*, 33, 7, 2009.

[a] Calculated from reported load ratios of 1, 0.843, and 0.833 on leading, middle, and trailing rows at H_g = 392 kN, and the ratios of 1, 0.555, and 0.37 at H_g = 1570 kN, respectively.

Table 11.6 for two load levels. They were predicted using the FixH and FreH solutions.

11.2.1 Single piles

Assuming FixH condition, the H-w_g curve was predicted using G = 10.2 MPa, $n = 0.5$, $\alpha_o = 0$ m, and $N_g^{FixH} = 31.3$ [$= 2.4K_p^2$] (*G2*), together with the predictions of Equations 11.1 and 11.2. The G is increased by 1.8 times and N_g by 2.4 times ($s_g = 2.4$) compared to "initial" values to cater for pile driving action (Wakai et al. 1999). Assuming FreH condition (Wakai et al. 1999), the H-w_g curve was also predicted by employing G = 48.9 MPa and N_g^{FreH} = 125.17 (i.e., four times these used for FixH piles as per *G1*).

The three predictions closely trace the measured H-w_t data (see Figure 11.12a), but for the slight overestimation of pile-head stiffness using $N_p = 0$. The $n = 0.5$ vindicates a typical feature of piles in clay (Guo 2006). A weaker than fully fixed condition is inferred, as a less than four times increase in the N_g and G would avoid the stiffer response of the FreH (see Figure 11.12a) than fixed-head cases. The x_p (FixH) was gained as 7.6*d* (w_g = 100 mm) and 10*d* (w_g = 200 mm).

11.2.2 Group piles

The group response was predicted using the p_m specified in Table 11.3, α_o = 0 m, and N_g^{FixH} = 31.3, along with either $n = 1.2$ and $G = 10.2$ MPa [i.e., "sand" (*G2*)]; or (2) $n = 0.85$ and $G = 20.44$ MPa (i.e., clay, *G1*). Assuming fully fixed-head conditions, the two predicted total load (H_g)-deflection (w_g) curves follow closely with the measured H_g versus w_t relationship, as is evident in Figure 11.12b, and in particular the 'n = 0.85' prediction. The predicted load H_{av} and moment M_{max} for the "$n = 0.85$" are provided in Table 11.6 for two typical levels of H_g of 392 kN and 1570 kN. The thick pile-cap generates a moment of $1.3H_{av}$ kNm (H_{av} in kN), thereby $M_o = M_{max} + 1.3H_{av}$. The M_o obtained for each row at the two load levels is ~18.3% less than the measured values, indicating the validity of the H_2 treatment (Type B). The slip depths at w_g = 200 mm were deduced as 9.06*d* (LR), 10.2*d* (MR), and 10.7*d*(TR), respectively using the "clay ($n = 0.85$)" analysis, which are ~45% more than 6.6*d*, 7.2*d*, and 7.4*d* obtained from the "sand (n = 1.2)" analysis (compared to 41% increase in the n values from 0.85 to 1.2). They are slightly higher (clay) or less (sand) than 10*d* obtained for the single pile.

Example 11.3 In situ full-scale tests on piles in clay

Matlock et al. (1980) performed lateral loading tests on a single pile, and two circular groups with 5 piles and 10 piles, respectively (see Figure 11.13a). The tubular steel piles had properties of L = 13.4 m, d_o = 168 mm, t = 7.1 mm, and E_pI_p = 2.326 MN-m^2. They were driven 11.6 m into a uniform soft clay that had a s_u of 20 kPa. The center-to-center spacing between adjacent piles was 3.4 pile diameters in the 5-pile group and 1.8 pile diameters in the 10-pile group. Deflections of the pile and each group during the tests were enforced at two support levels with e_p = 0.305 and 1.83 m above mudline to simulate FixH restraints (see Figure 11.9). Measured relationships of H-w_t (with e_w = 0.305 m) and w_t-M_{max} (with e_m = 0.305 m) are plotted in Figure 11.14a and b, respectively, for the single pile. Measured H_{av}-w_t and w_t-M_{max} curves for a pile in the 5-pile and the 10-pile groups are plotted in Figure 11.14c and d. These measured responses were simulated using the current solutions, in light of the parameters given in Table 11.7.

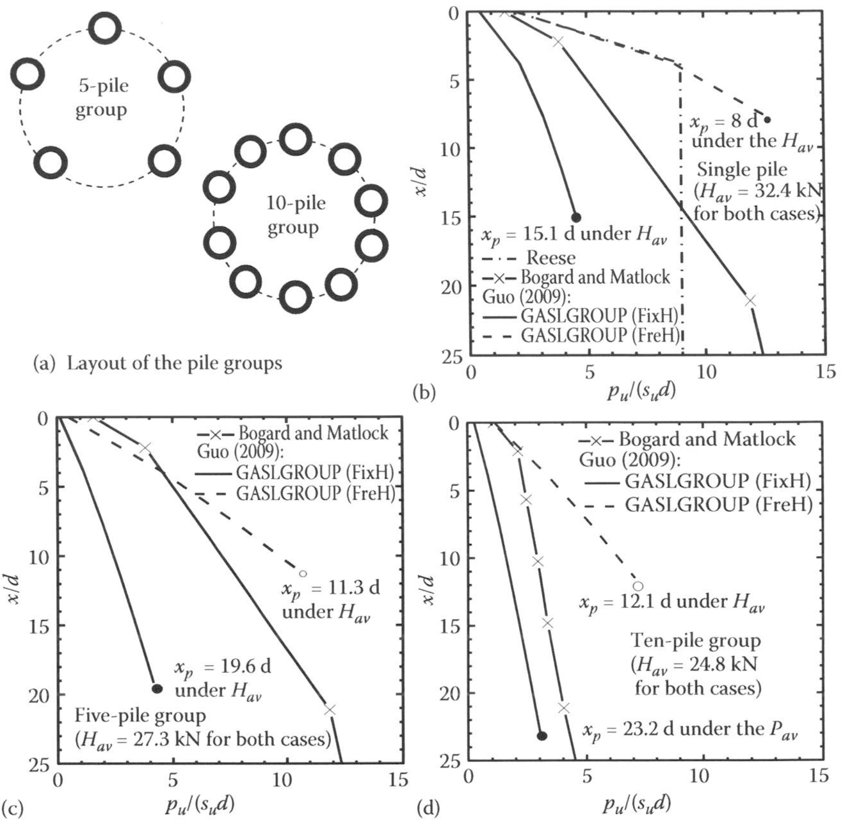

Figure 11.13 (a) Group layout. (b) P_u for single pile. (c) P_u for a pile in 5-pile group. (d) P_u for a pile in 10-pile group. (Matlock et al. 1980) (Example 11.3). (After Guo, W. D., *Int J Numer and Anal Meth in Geomech* 33, 7, 2009.)

11.3.1 Single piles

The single-pile predictions employed (see Table 11.7) $G = 100s_u$ and $N_g^{FreH} = 4.0$ in the FreH solutions (Guo 2006); or utilized $G = 50s_u$, and $N_g^{FixH} = 1.0$ in the FixH solutions (*G1* and *G2*), in addition to $n = 0.55$ and $\alpha_o = 0.05$ m (*G2*) for either prediction. The low values, especially N_g, reflect a reduction in limiting force per unit length caused by the restraint on cap rotation. The corresponding LFPs were plotted in Figure 11.13b as "FreH" and "FixH" to the maximum slip depths determined subsequently.

As plotted in Figure 11.14a, the predicted and measured H-w_g curves show slight discrepancy to ~75% maximum imposed load. Overestimation of FixH pile-head stiffness using $N_p = 0$, in this case, is likely offset by reduction in the stiffness using $\alpha_o = 0$ (the actual $\alpha_o \neq 0$) and $w_t \approx w_g$. The pile-head must be semi-fixed as outlined in the following:

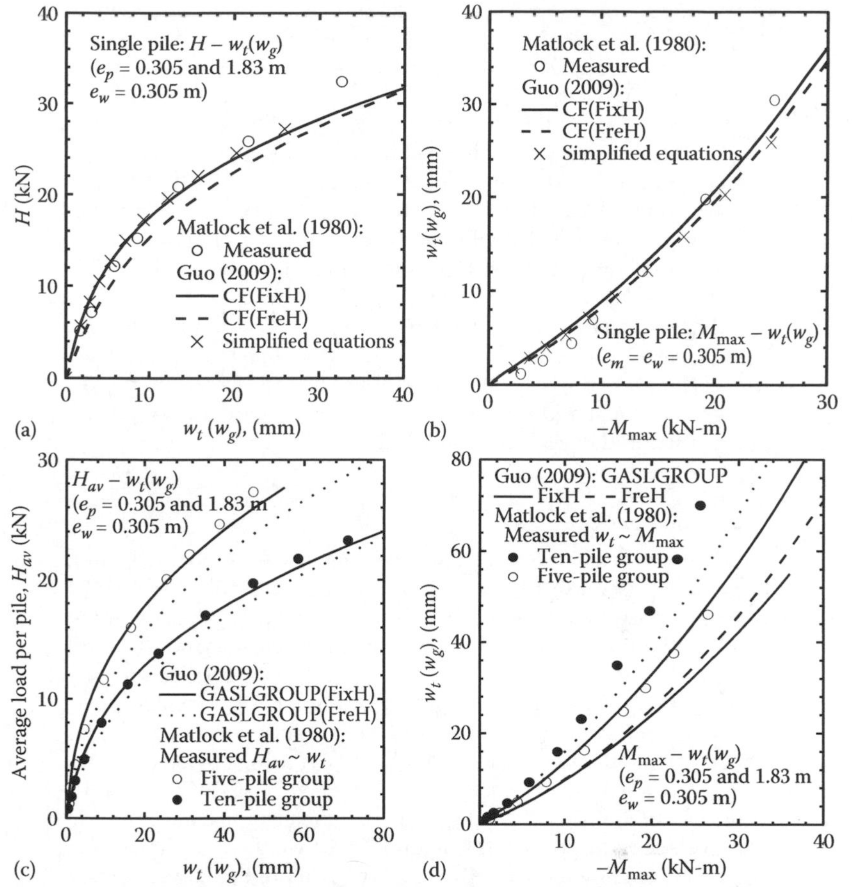

Figure 11.14 Predicted versus measured (Matlock et al. 1980) (Example 11.3). (a) and (c) Load and deflection. (b) and (d) Maximum bending moment. (After Guo, W. D., *Int J Numer and Anal Meth in Geomech* 33, 7, 2009.)

Table 11.7 In situ tests on piles and pile groups in clay (Example 11.3)

	Input parameters					*Calculated parameters*
Items	G/s_u	n[a]	α_o *(m)*[a]	N_g	p_m[a]	x_p/d @ H_{max}
Single pile	50[b]/100[c]	0.55	0.05	1.0[b]/4.[c]	1.0	15.1[b]/8.0[c]
5-pile group	75/100	0.85	0.05	1.0/4.	0.333	19.6/11.3
10-pile group	17/33	0.85	0.25	1.0/4.	0.20	23.2/12.1

Source: Guo, W. D., *Int J Numer and Anal Meth in Geomech*, 33, 7, 2009.

[a] Single values for both FixH and FreH piles.
[b] Values for FixH.
[c] Values for FreH.

1. Owing to the upper level loading [Type A, *H2b*], a deduction of 0.305*H* kNm (*H* in kN, and e_p = 0.305 m) from the computed M_{max} (using the FixH solutions) offers close agreement between the three predicted w_g-M_{max} curves (either head conditions), see Figure 11.14b and the measured w_t-M_{max} relationship.
2. The pile-cap must be semi-FixH once $x_p/d > 4$, to ensure the LFP move from the "dash line" of FixH (see Figure 11.13b) towards the "Bogard and Matlock" curve and to generate a realistic ratio $p_u/(s_u d)$, which should not exceed 9.14–11.94 (Guo 2006) (Randolph and Houlsby 1984), see Figure 11.13b.
3. The total resistances T_R^{FixH} and T_R^{FreH} were 36.35 kN and 24.42 kN, respectively (i.e., $T_R^{FreH} < H_{max} < T_R^{FixH}$, *G3*). Under the load H_{max} of 32.4 kN, limiting force is fully mobilized to a depth x_p of 8*d* (FreH) to 15.1*d* (FixH) (see Table 11.7, and *G4*, Table 3.10). An additional 4 kN (= 36.35-32.4 kN) resistance was mobilized to restrain the head rotation.

11.3.2 Group piles

To study the two pile groups, identical *n* and p_m were adopted using either solutions (see Table 11.7). Typical features are as follows: (1) 70% higher *n* (compared to the single pile) is used to capture impact of more "sharp" strength variation with depth from group disturbance (*G2* and *G5*); (2) a very low shear modulus of $17s_u$ (FixH) or $33s_u$ (FreH) for the 10-pile group, which are 0.34 (= p_m) or 0.33 times that *G* for single piles, and are different from p_m = 0.2 (Matlock et al. 1980) (*G5*). This may be a result of a combination of the unusually small spacing of 1.8*d* (*G1*) and the special pile (circular) layout, as normal values of $G = 75s_u$ (FixH), or = $100s_u$ (FreH) for the five-pile group are noted at a normal $s = 3.4d$.

The average load per pile (H_{av})–mudline deflections (w_g) curve for either group is presented in Figure 11.14c. The predicted M_{max} [after deducting the amount of $H_{av}e_p$ (H2a) for the FixH groups, see Figure 11.9] is plotted in Figures 11.14d and 11.15b. The respectively measured H_{av}-w_t curves (implying $w_t \approx w_g$) are well predicted, but the measured values of M_{max} were markedly overestimated. The semi-fixed head groups reduce the M_{max} more than the amount of $H_{av}e_p$, in response to the increased number of piles and loading levels, and the impact of w_t-w_g (H2c). The LFPs obtained for each group are plotted in Figure 11.13c and d, respectively, exhibiting similar features to those of the single pile. The max x_p along each pile increases by 18% (from 19.6*d* to 23.2*d*) from the 5-pile to the 10-pile groups, although the average load H_{av} reduces by 9% (from 27.3 to 24.8 kN).

Example 11.4 Model piles in sand: Small values of $e_w = e_p$

Gandhi and Selvam (1997) conducted laboratory tests on 2 single piles and 19 pile groups. Each pile was 0.75 m in length, 18.2 mm in outside diameter, 0.75 mm in wall thickness, and had a flexural stiffness E_pI_p of 0.086 kN-m^2. The aluminium pipe piles were driven to a depth of 0.5 m

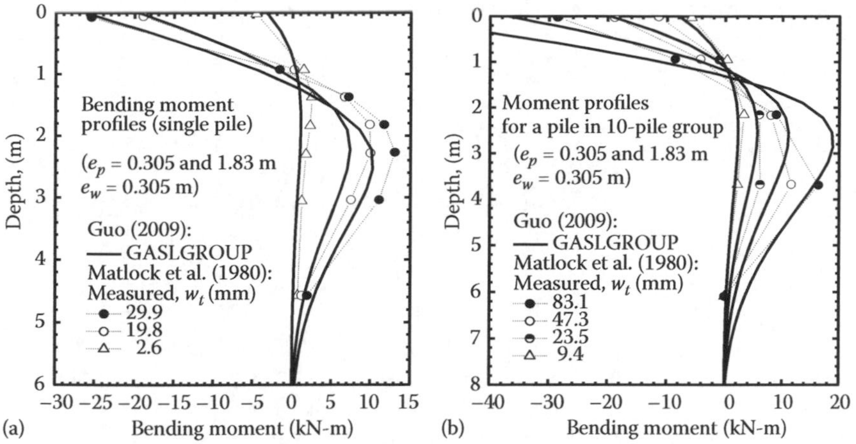

Figure 11.15 **Predicted versus measured (Matlock et al. 1980) (Example 11.3) moment profiles of (a) single pile, and (b) a pile in 10-pile groups (Example 11.3). (After Guo, W. D., *Int J Numer and Anal Meth in Geomech* 33, 7, 2009.)**

below the sand surface. The sand was clean and dry, and fine to medium size, with a frictional angle of 36.3° (a relative density of 60%) and a unit weight of 14.6~17.3 kN/m³. One constrained-head, single pile was laterally loaded at an eccentricity of 0.047 m (= $e_p = e_w$). Another single pile was loaded under FreH restraint. The groups had a normalized spacing, *s*/*d*, of 4~12 (loading direction) and *s*/*d* of 3 (within each row), respectively. They were laterally loaded at e_p = 35 mm with displacement measured at $e_w \approx$ 85 mm (gained from the test setup). The load *H* versus deflection w_t curves for the single FreH and capped piles are measured, along with the two values of M_{max} for the single piles. The moment profiles along the piles in 2 × 3 group were also recorded at w_t = 10 mm.

Guo (2005) provided a good prediction for the two single piles compared to measured data, in light of GASLGROUP, and examined the sensitivity of the solutions to the input parameters. Guo (2009) further conducted prediction of all the 19 groups using N_g = 15.23, N_{co} = 0, *n* = 1.1~1.35, G, and the *p*-multipliers (see Table 11.3). In particular, soil shear modulus *G* for 13 out of the 19 groups increased by 2~10 times compared to 0.3 MPa for the single pile, to cater for impact of sand densification among the piles [e.g., in 1 × 3 group at *s* = 4*d*, and in 2 × 3 and 3 × 3 groups at *s* = (4~8)*d*]. The predicted total load (H_g)~displacement (w_g) curves (assuming FixH) agree well with measured data.

The prediction of nonlinear response can be readily conducted using Equations 11.1~11.3 with typical parameters provided in Table 11.8. Here comes an example for the capped single (see Table 11.9) and the 2 × 3 group piles (*s* = 4*d*) at a mudline displacement w_g of 10 mm (assuming *e* = 0). As shown in Table 11.10, the calculation was conducted in sequence from p_m and *k* to λ until $-M_o$ was obtained, as is illustrated below for a pile in leading row (see Figure 11.16 with *R* = p_m for individual pile).

Table 11.8 Parameters for single piles in sand (Example 11.4)

	Input data	*Calculated elastic parameters (as per expressions in Table 11.1)*					
Items	G *(kPa)*	γ_b	k (kPa)	$N_p/(2E_pI_p)$	α_N	β_N	λ *(1/m)*
FixH	300	0.03805	711.7	26.71	1.1373	0.8425	6.7443
FreH		0.06872	831.1	11.62	1.0575	0.9390	7.0109

	Input data			*Calculated (plastic) parameters*			
Items	n	α_o *(m)*	s_g	N_g	A_L *(kPa/m^n)*	x_p/d @ H *(kN)*	T_R *(kN)*
FixH	1.7	0	0.45	6.86	33.41	12.55 @ 0.332	1.005
FreH[b]		0	2.0	30.45	148.50	6.47 @ 0.255	0.171
FixH[a,b]	1.35	0	1.0	15.23	18.26	12.6 @ 0.332	1.060
FreH[a]		0.05	1.0	15.23	18.26	10.85 @ 0.255	0.165

Source: Guo, W. D., *Int J Numer and Anal Meth in Geomech,* 33, 7, 2009.

$\phi' = 36.3°$, $\gamma_s = 16.22$ (kN/m^3), $K_p^2 = 15.23$, $N_g = s_gK_p^2$.

[a] Guo 2005.

[b] Preferred analysis.

Table 11.9 Response of capped single pile (e = α_o = 0, and N_p = 0)

x_p *(m)*	$\bar{x}_p$	$\bar{H}$	$\bar{w}_g$	$-\bar{M}_{max}$	H *(kN)*	w_g *(mm)*	$-M_o$ *(kN-m)*[a]
	Below: n = 1.35, $N_g = K_p^2$, α_o = 0, G = 0.3 MPa (Guo 2005)						
0.10	0.6744	0.5048	0.8043	0.3488	0.104	1.6	0.0058
0.20	1.3489	1.3442	3.7829	1.2026	0.277	7.4	0.0237
0.25	1.6861	1.8969	7.1205	1.9000	0.391	13.9	0.0396
0.30	2.0233	2.5463	12.63	2.828	0.524	24.6	0.0617
	Below: n = 1.7, $N_g = 0.45K_p^2$, α_o = 0, G = 0.3 MPa (Guo 2009)[b]						
0.10	0.6744	0.4217	0.6992	0.2989	0.081	1.3	0.0047
0.20	1.3489	1.3816	4.1418	1.2797	0.267	7.6	0.0241
0.25	1.6861	2.079	8.3607	2.1599	0.401	15.3	0.043
0.30	2.0233	2.94	15.675	3.319	0.568	28.7	0.0704

Source: After Guo, W. D., *Proceedings GeoFlorida 2010 Conference*, GSP 199, ASCE, West Palm Beach, Florida, 2010.

[a] M_o is the difference between M_{max} obtained from Equation 11.3, and the value of *He* (e = 0.047m).

[b] Elastic properties are shown previously (Guo 2009).

1. As G = 0.8MPa and p_m = 0.8, G for the pile in the row is 0.64 MPa (= $0.8p_m$).
2. With L/r_o = 54.95, $E_p = 1.598 \times 10^7$kN-m^2, and $G^* = (1 + 0.75 \times 0.4)G$, it follows E_p/G^* = 21020.8, and γ_b = 0.046 [= $0.65(21020.8)^{-0.25}(54.95)^{-0.04}$].
3. $K_1(\gamma_b)$ = 21.659 and $K_0(\gamma_b)$ = 3.1975 enable k = 1592 kPa (Guo and Lee 2001), thus λ = 8.248/m.

Table 11.10 Capped single pile and 2 × 3 group at w_g = 10 mm (s = 4d)

	p_m/k (kPa)	λ(1/m)	A_L(kN/m)	$\bar{w}_g$	$\bar{x}_p$	$\bar{H}$	$-\bar{M}_{max}$	x_p (m)	H_{av} (kN)	$-M_o$ (kN-m)
Single pile[c]	1.0/712	6.744	18.268	5.125	1.468[a]/1.506[b]	1.494/1.591	1.400/1.502	0.218/0.223	0.308/0.328	0.0283/0.0305
Leading row[c]	0.8/1592	8.248	9.79	22.73	2.368/2.456	3.183/3.349	3.920/4.143	0.287/0.298	0.27/0.284	0.0309/0.0327
Middle row	0.4/762	6.861	4.90	17.29	2.191/2.678	2.797/2.941	3.273/3.458	0.319/0.331	0.178/0.189	0.0243/0.0258
Trailing row	0.3/562	6.357	3.67	15.44	2.121/2.193	2.65/2.789	3.036/3.208	0.336/0.345	0.152/0.159	0.022/0.0233

Source: After Guo, W. D., *Proceedings GeoFlorida 2010 Conference*, GSP 199, ASCE, West Palm Beach, Florida, 2010.

n = 1.35, $N_g = K_p^2$, α_o = 0, G = 0.3 MPa (single piles); whereas n = 1.25, $G = 0.8p_m$ MPa, $N_g = p_m K_p^2$ (group piles).

[a] Numerator, $N_p \neq 0$ using GASLGROUP.

[b] Denominator, assuming N_p = 0.

[c] e_p = 0.047 m and 0.035 m for single and group piles, respectively.

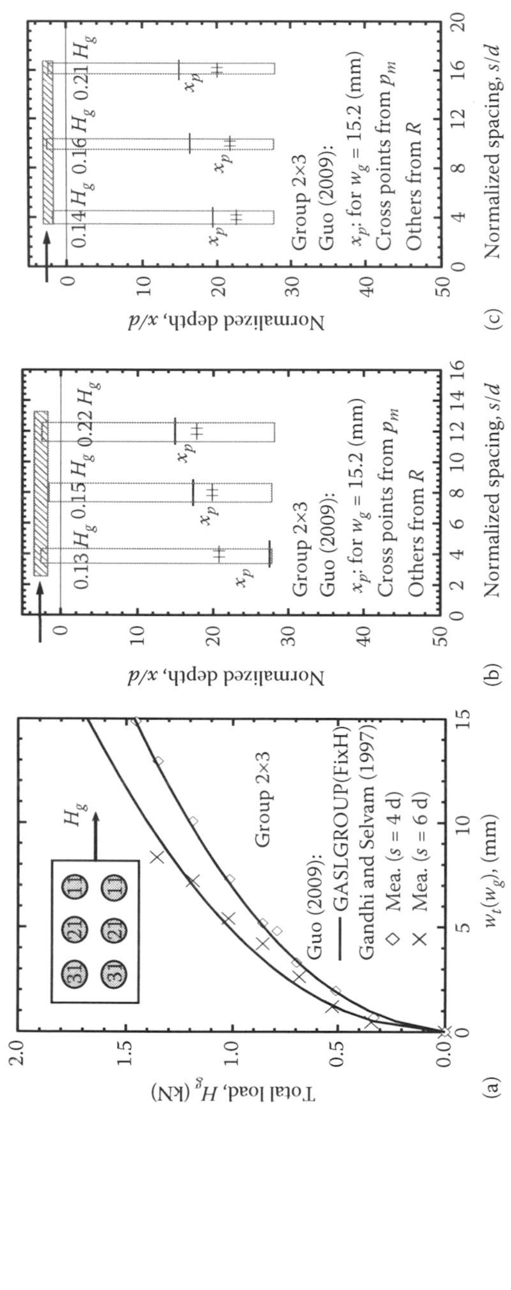

Figure 11.16 Predicted H_g-w_g versus measured (Gandhi and Selvam 1997) H_g-w_t response of various groups (Example 11.4, three rows). (a) Load – displacement. (b) x_p for s = 4*d*. (c) x_p for s = 6*d*. (After Guo, W. D., *Int J Numer and Anal Meth in Geomech* 33, 7, 2009.)

4. $n = 1.25$, $d = 0.0182$ m, $\gamma_s' = 16.22$ kN/m^3, $N_g = 15.23$, thus $A_L = 9.79$ kN/m$^{2.35}$.
5. As $\bar{w}_g = 22.73$, $\bar{x}_p = 2.456$ using Equation 11.2, and $\bar{H} = 3.349$ (via Equation 11.1) and $\bar{M}_{max} = 4.413$ (Equation 11.3), respectively.
6. H (rewritten as H_{av}) = 0.284 kN (= 3.349 × 0.70/8.248), and $M_{max} = 0.043$ kN-m (= 4.143 × 0.70/8.248^2). The M_o is revised as 0.033 kN-m after deducting He_p (Guo 2009).

Steps 1 through 6 were repeated for the piles in middle and trailing rows using $p_m = 0.4$ and 0.3, respectively. The calculated results are given in Table 11.10.

The total load H_g was calculated as 1.264 kN [= (0.284 + 0.189 + 0.159) × 2] at the displacement of $w_g = 10$ mm, which compares well with measured data (see Figure 11.16a). The predicted moment M_o at mudline has considered the impact of loading eccentricity ($e_p = 35$ mm). As shown in Table 11.11, the measured M_o of 0.0225 kNm was overestimated by 45.3% (predicted $M_o = 0.0327$ kNm) for the pile in leading row. The measured values were overestimated by 78% and 79% for those in middle and trailing rows, respectively.

The impact of w_t-w_g is evident between deflections at pile-head and mudline levels. The $w_t = 10$ mm at $e_w \approx 85$ mm leads to a w_g of ~7.0 mm (rotation angle at e_w, $\theta_w = 0.03$). The angle θ_w altering from 0 at pile-head level to 0.03 at e_w renders a value of $w_g = 7.6$ mm, at which new $H_{av} = 0.2497$ kN, and $M_o = -0.0268$ kNm are predicted for piles in the leading row (Table 11.11). This M_o still exceeds the measured value by 19%, and similarly calculated M_o still exceeds the measured value by 47% for piles in trailing row. The differences, although large, seem to have comparable accuracy to experimental results, as explained previously (Guo 2010).

11.5 CONCLUSIONS

Elastic-plastic solutions are developed for laterally loaded, infinitely long, fixed-head piles. They are employed to simulate nonlinear response of capped pile groups by adopting a new treatment H2 and *p*- (or *R*) multiplier via a spreadsheet program GASLGROUP. The impact of free-length, head-fixity, and group interaction is quantified, along with failure modes. The results are synthesized into the guidelines (*G1–5*, Chapter 3, this book). It should be noted that:

- Nonlinear pile responses are characterized by the slip depth x_p and should be predicted. Existing *p-y* curve based approach generally overestimates the pile-head stiffness.

Table 11.11 Calculated and measured response (2 × 3 Group) at w_g = 10 and 7.6 mm

	Measured		*Calculated for w_t = 10 mm*				*Calculated for w_t = 7.6 mm*		
	$-M_o$ (kN-m) (1)	Assum. H_{av}(kN) (2)	H_{av} (kN)/ $-M_{max}$ (kN-m)	$-M_o$[a] (kN-m) (3)	$\frac{(3)-(1)}{(1)}$ (%)	θ @ e_w (radian)	H_{av} (kN)	$-M_o$[d] (kN-m) (4)	$\frac{(4)-(1)}{(1)}$ (%)
Leading row	0.0225[b] 0.0181[c]	0.278[b] 0.278[c]	0.284/0.0427	0.0327	45.3 80.7	0.030	0.2497	0.0268	19.1 48.1
Middle row	0.0145 0.011	0.241 0.217	0.189/0.0324	0.0258	77.9 1.34	0.0238	0.1656	0.0212	46.2 92.7
Trailing row	0.013 0.012	0.18 0.0895	0.159/0.0289	0.0233	79.2 94.2	0.0216	0.1395	0.0191	46.9 59.2

Source: After Guo, W. D., *Proceedings GeoFlorida 2010 Conference*, GSP 199, ASCE, West Palm Beach, Florida, 2010.

G = 0.8MPa, e_p = 0.035 m, e_w = 0.085 m, and N_p = 0.

[a] Calculated as $M_{max} + H_{av}e$.

[b] Spacing s = 6*d*.

[c] Spacing s = 4*d*.

[d] Calculated using a displacement of w_g = 7.6 mm at pile-cap level.

- With the few parameters k, p_u, and p_m, the current calculations are readily conducted and show similar accuracy to complex numerical approaches for layered nonhomogeneous soil. They can capture well the impact of head restraint by the treatment *H2*. Conversely, the parameters are readily deduced by matching the current solutions with measured pile response.

Chapter 12

Design of passive piles

12.1 INTRODUCTION

Piles may be subjected to soil movement, as encountered in stabilizing a sliding slope (Viggiani 1981) (see Figure 12.1a) and supporting bridge abutments (Springman 1989; Stewart et al. 1994). Laterally loaded piles may be subjected to passive forces once subjected to the driving action of adjacent piles (Henke 2009) or nearby excavation activity (Chen and Poulos 1997; Leung et al. 2000; Choy et al. 2007). Predicting the response of these "passive" piles by and large has resorted to numerical solutions (Byrne et al. 1984; Chen and Poulos 1997; Chen and Martin 2002; Mostafa and Naggar 2006). The solutions are useful but would not warrant consistent predictions (Poulos 1995; Chow 1996; Potts 2003) in view of using a uniform limiting-force profile that is not observed along test piles (Matlock 1970; Yang and Jeremic 2002; Guo 2006).

Elastic solutions were proposed to simulate slope stabilizing piles subjected to a uniform soil movement (see Figure 12.1a) (Fukuoka 1977) and to model piles under an inverse triangular profile of moving soil (Cai and Ugai 2003). The solutions compare well with measured pile response of six in situ piles, albeit using measured sliding thrust and gradient of soil movement with depth for each pile. The prediction seems to be highly sensitive to the gradient and is not related to magnitude of the soil movement (see Figure 12.1). The drawbacks are noted in almost all simple solutions (Ito and Matsui 1975; De Beer and Carpentier 1977; Viggiani 1981; Chmoulian and Rendel 2004). Recent study is aimed at linking the pile response to soil movement.

The inadequacy of existing methods such as the *p*-*y*–based analysis (Chen et al. 2002; Smethurst and Powrie 2007; Frank and Pouget 2008) may be owing to the difficulty in gaining a right p_u profile. Nonlinear response of about 70 laterally loaded piles is well captured using elastic-plastic solutions underpinned by an increasing p_u with a power of 0.7~1.7 to depth and its depth of mobilization. The solutions require only two soil parameters for gaining the p_u profile and the modulus k, and exhibit a few advantages over numerical approaches (Guo 2006, 2008). These benefits prompt the

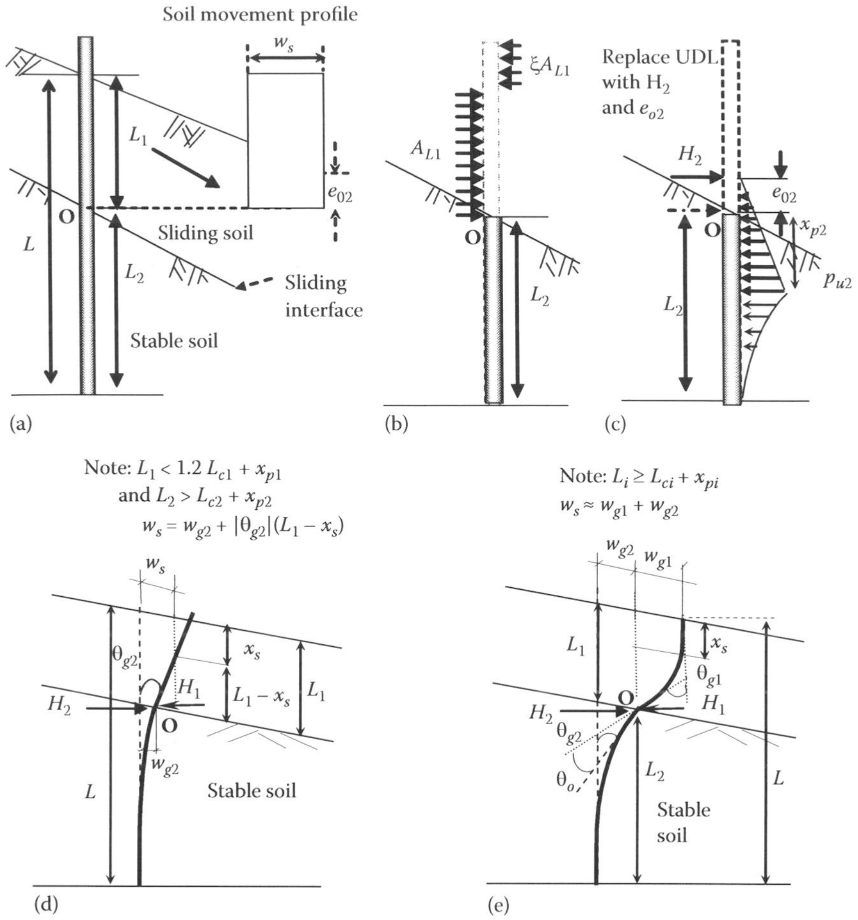

Figure 12.1 An equivalent load model for a passive pile: (a) the problem, (b) the imaginary pile, (c) equivalent load H_2 and eccentricity e_{o2}, (d) normal sliding, and (e) deep sliding (Guo in press).

development of similar solutions to capture nonlinear interaction of passive piles (Guo 2012), as available analytical solutions (Fukuoka 1977; Viggiani 1981; Cai and Ugai 2003; Dobry et al. 2003; Brandenberg et al. 2005) are not applicable to passive piles.

12.1.1 Flexible piles

Figure 12.1a shows a passive pile in an unstable slope. The pile has an embedded length L_i in i^{th} layer and is subjected to a lateral, uniform soil movement, w_s. Note subscript $i = 1$ and 2 denote the sliding and stable layer, respectively. The impact of the movement w_s on the pile is encapsulated into an equivalent load (thrust) H_2. By incorporating boundary conditions, behavior of the pile is modeled using the solution for an active pile under the load H_2 at an eccentricity e_{o2} (see Figure 12.1c), for the stable layer, or under H_1 (= $-H_2$) for the sliding layer (Figure 12.1d and e). The use of the concentrated force H_i at sliding depth is sufficiently accurate (Fukuoka 1977) to model the pile response. The thrust H_2 and the eccentricity e_{o2} (= $-e_{o1}$) above the point O at sliding level (Figure 12.1c) causes a dragging moment M_{o2} (= H_2e_{o2}) above the point, and M_{o1} = M_{o2}. Note the H_2 is the horizontal component of net sliding force (thrust) along the oblique sliding interface. A high eccentricity e_{o2} (thus M_{o2}) renders a low shear force and deflection (Matlock et al. 1980; Guo 2009) at the point O, which need to be assessed.

Design of passive piles generally requires determination of the maximum shear force H_2 or H_1 in each pile (Poulos 1995; Guo and Qin 2010) and the dragging moment M_{o2}, which resemble active piles and are dominated by the limiting force per unit length p_{ui} and the depth of its mobilization x_{pi} between the pile and the soil. However, the use of the p_{ui} profile for an active pile (Chen et al. 2002) significantly overestimates the resistance on piles adjacent to excavation (Leung et al. 2000).

12.1.2 Rigid piles

The slope stabilizing piles (see Figure 12.1) are classified as rigid, once the pile–soil relative stiffness, $E_p/\tilde{G}$, exceeds $0.4(l/d)^4$, with 20% longer critical length than a laterally loaded free-head pile (Guo 2006; Guo 2008). (Note that E_p = Young's modulus of an equivalent solid pile; d = an outside diameter of a cylindrical pile; and $\tilde{G}$ = average shear modulus over the embedment l). Figure 12.2a shows that the pile rotates rigidly to an angle ω and a mudline deflection u_g under a sliding movement w_s, which is equal to the pile deflection u at the depth x_s.

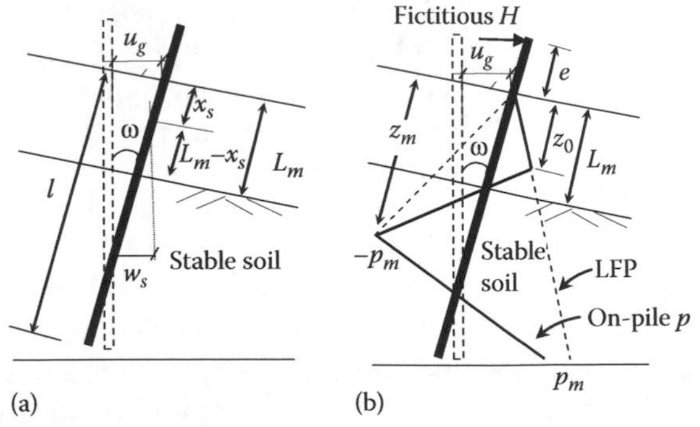

Figure 12.2 A passive pile modeled as a fictitious active pile. (a) Deflection profile. (b) H and on-pile force profile.

12.1.3 Modes of interaction

Four pile–soil interaction modes have been revealed to date, for which analytical solutions are established.

- Plastic flow mode: Soil flows around a stationary pile, and the induced ultimate pressure on pile surface is estimated using plasticity theory (Ito and Matsui 1975; De Beer and Carpentier 1977).
- Rigid pile mode: A rigid pile rotates with sliding clay. The associated maximum shear forces and bending moments are obtained by assuming uniform resistances along the pile in sliding and stable clay layers (Viggiani 1981; Chmoulian 2004; Smethurst and Powrie 2007; Frank and Pouget 2008).
- Normal and deep sliding modes: A pile rotates rigidly only in the sliding layer with an infinite length in stable layer and is termed as normal sliding mode, see Figure 12.1d; or with infinite lengths in either layer, a pile may deform flexibly and move with sliding soil and is referred to as deep sliding mode (Figure 12.1e). Elastic solutions are available for gaining profiles of bending moment, deflection and shear force, in light of a measured sliding force H (= H_1, Figure 12.1c), and a measured differential angle θ_o at the point O between the slope angles θ_{g1} and θ_{g2} of the pile in sliding and stable layers, respectively (see Figure 12.1d and e). Note the angle θ_o is equal to $-\theta_{g2}-\theta_{g1}$, with $\theta_{g1} > 0$ and $\theta_{g2} < 0$. It is negligibly small for a uniform soil movement (Fukuoka 1977) induced in a deep sliding mode; or the angle θ_o is taken as the gradient of a linear soil movement with depth (Cai and Ugai 2003) concerning the normal sliding mode. Guo (2003b) linked the H_2 to soil movement w_s, from which elastic-plastic solutions for a laterally loaded pile (Figure 12.1c) (see Chapter 9, this book) are used to model passive piles without dragging (i.e., zero bending moment at the sliding level).

The aforementioned solutions were generally validated using pertinent instrumented pile response, but the coupled impact of soil movement and any nonzero dragging moment (dragging case) on the piles remains to be captured.

This chapter provides methods for estimating response of piles under "normal sliding mode" and "rigid pile mode", which consist of (a) a correlation between sliding force (H) and soil movement (w_s) for the normal and deep sliding modes, respectively; (b) a practical P-EP solution to capture impact of dragging, and rigid rotation, with a plastic (P) pile–soil interaction in sliding layer, and elastic-plastic (EP) interaction in stable layer; and (c) A new E-E (coupled) solution to model deep sliding case, in which E refers to "elastic" pile–soil interaction in sliding and stable layers, respectively, and "coupled" interaction among different layers is incorporated (Guo in press). The chapter also explores a simple and pragmatic approach for estimating nonlinear response of a rigid, passive pile, regardless of soil movement profiles.

The solutions are entered into programs operating in Mathcad™ (www.PTC.com) and EXCEL™. They are compared with boundary element analysis (BEA) and used to study eight instrumented piles to unlock any salient features of the passive piles. The P-EP solution is employed to develop design charts, and their use is illustrated via a typical example. The simple approach for rigid piles is developed from measured limiting-force profile and is used to predict response of model piles and an instrumented pile.

12.2 MECHANISM FOR PASSIVE PILE–SOIL INTERACTION

12.2.1 Load transfer model

An EP solution was developed for an infinitely long, active pile (Guo 2006), see Figure 12.1c, in the context of the load transfer model (Chapter 3, this book). The pile–soil interaction is captured using a series of springs distributed along the shaft. Each spring is governed by an ideal elastic-plastic p_i-y_i (w_i) curve, with a gradient (i.e., modulus of subgrade reaction) k_i, and a net ultimate lateral resistance per unit pile length p_{ui} [p_i and $y_i(w_i)$ = local net force per unit length and pile deflection, respectively]. The modulus and w_i are rewritten as kd and u for a rigid pile. The coupled effect among the springs is captured by a fictitious tension membrane (N_{pi}) in the elastic zone, and it is neglected in the plastic zone.

The EP solution is dominated by the critical parameters k_i, N_{pi}, and p_{ui}, which are calculated from shear modulus G_i and the gradient A_{Li} of net resistance (Chapter 3, this book). Values of k_i/G_i and $4N_{pi}/(\pi d^2 G_i)$ vary with pile slenderness ratio L_i/d, loading eccentricity e_{oi}, and pile–soil relative stiffness E_p/G_i (L_i = pile length in ith layer, and G_i = soil shear modulus). N_p is taken as zero for rigid piles. The resistance is neglected at sliding interface. The net limiting force on a unit length p_{ui} with depth is simplified as (Guo 2003b, 2006)

$$p_{ui} = A_{Li} x_i^{n_i} \tag{12.1}$$

where x_i = depth measured from point O; n_i = power to the equivalent depth of x_i; and A_{Li} = gradient of the p_{ui} profile with depth (see Chapter 3, Table 3.8, this book, for each layer). The net mobilized p_i attains the p_{ui} within the thickness of the plastic zone x_{pi} from the point O, otherwise, beyond the x_{pi}, it is proportional to pile deflection w_i.

$$p_i = k_i w_i \tag{12.2}$$

where k_i = constant (see Figure 3.27 and Equation 3.50, Chapter 3, this book). Especially, the p_{ui} for layered soils must satisfy Equation 12.1 within the maximum x_{pi} (induced by maximum load H_2) and $p_{ui} \leq 11.9d(\max s_{ui})$ (Note max s_{ui} = maximum s_{ui}) (Randolph and Houlsby 1984).

As with laterally loaded piles, the load transfer model is equally valid to rigid passive piles but for the following special features.

1. The soil may be taken as one layer and the p_u modeled with $n = 1.0$, $A_L = (0.4\text{~}2.5)\gamma_s' K_p^2 d$ and $(0.4\text{~}1.0)\gamma_s' K_p^2 d$ for active and passive piles (Guo 2011), respectively; $K_p = \tan^2(45° + \phi'/2)$, coefficient of passive earth pressure; ϕ' = an effective frictional angle of soil; and γ_s' = an effective unit weight of the soil (dry weight above the water table, buoyant weight below). The p_u may generally be independent of pile properties under lateral loading, though it alters with soil movement profiles.
2. The net on-pile force per unit length p [FL^{-1}] in elastic zone is given by

$$p = kdu \text{ (Elastic state)} \tag{12.3}$$

where u = pile–soil relative displacement. The gradient k [FL^{-3}] may be written as $k_o z^m$ [k_o, FL^{-m-3}] and referred to as constant k ($m = 0$) and Gibson k ($m = 1$) hereafter. The k_o depends on loading eccentricity e with $e = 0$ (pure lateral loading) and $e = \infty$ (pure moment), respectively. The modulus of subgrade reaction kd is obtained from $\tilde{G}$ using the expression shown in Equation 3.62 (Chapter 3, this book). Given $l/r_o = 2\text{~}10$, it follows that $k_o(0.5l)^m d = (5.12\text{~}13.1)\,\tilde{G}$ ($e = \infty$) or $k_o(0.5l)^m d = (4.02\text{~}8.92)\,\tilde{G}$ ($e = 0$). Typically, for a model pile having $l = 0.7$ m and $d = 0.05$ m, the $k_o(0.5l)d$ is equal to $(2.2\text{~}2.85)\,\tilde{G}$.

12.2.2 Development of on-pile force p profile

With respect to a flexible pile, the p_u profile in sliding layer consists of a restrianing zone over depth $0\text{~}x_s$ and a thrust zone in depth $x_s\text{~}L_1$. The p profile is the same as that along an active pile, except for the p_{u2} increasing linearly in the plastic zone over the depths $0\text{~}x_{p2}$ below sliding interface.

Model tests (see Chapter 13, this book) provide the on-pile force per unit length (p) profile on a rigid passive pile. The p profile to a depth z_m of maximum bending moment (see Figure 12.3) resembles that on the entire length of an active pile (Figure 3.28, Chapter 3, this book) in two aspects:

1. A linear limiting force profile p_u is observed, regardless of the test types.
2. The on-pile force per unit length p, mobilized along the positive p_u line to the slip depth z_o from ground level, follows Equation 12.1.

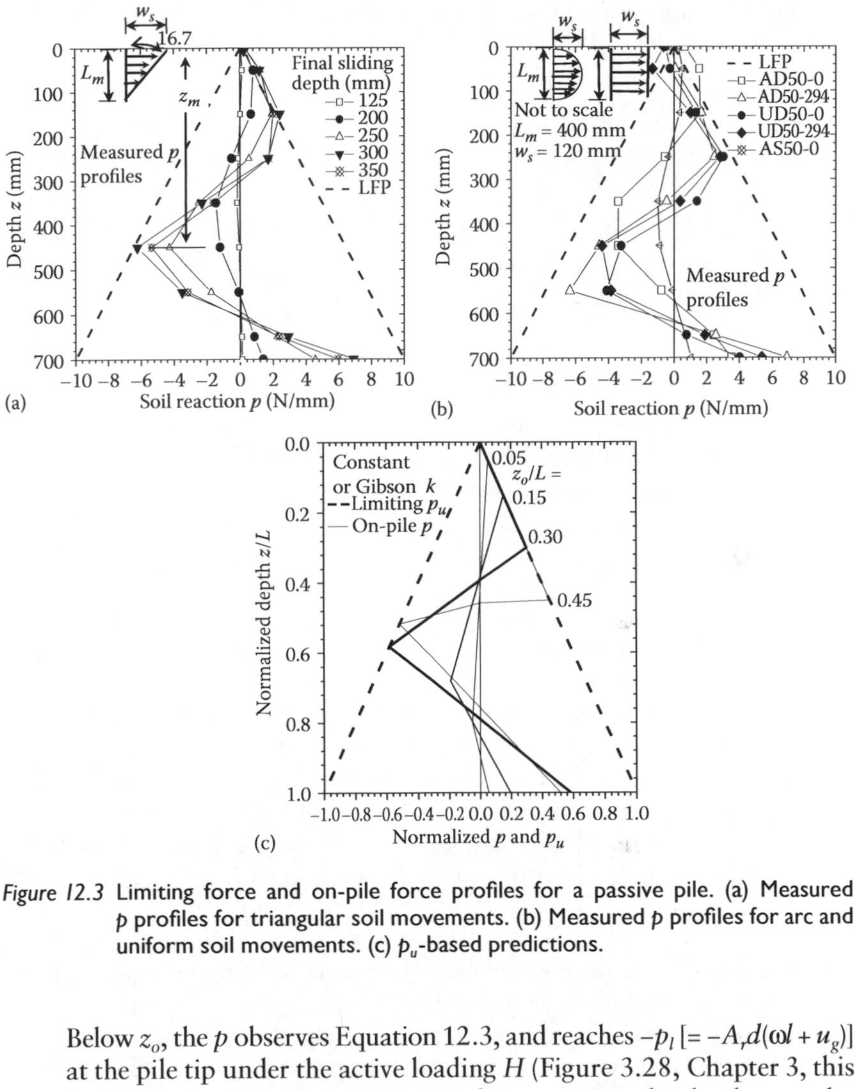

Figure 12.3 **Limiting force and on-pile force profiles for a passive pile. (a) Measured p profiles for triangular soil movements. (b) Measured p profiles for arc and uniform soil movements. (c) p_u-based predictions.**

Below z_o, the p observes Equation 12.3, and reaches $-p_l$ [$= -A_r d(\omega l + u_g)$] at the pile tip under the active loading H (Figure 3.28, Chapter 3, this book); or the p reaches $-p_m$ [$= -A_r d(\omega z_m + u_g)$] at the depth z_m under the passive loading (Figure12.3a). The force per unit length at depth z, $p(z)$ in the depth $0 \sim z_m$ is given by two expressions of

$$p(z) = A_r dz (z = 0 \sim z_o);\ p(z) = kd(\omega z + u_g)(z = z_o \sim z_m) \tag{12.4}$$

Neverthertheless, the p profiles under passive loading (see Figure 12.3a and b) all have an additional portion below the depth of maximum bending moment z_m ($\approx$ 0.45~0.55 m in the figure for a model pile with l = 0.7m) compared to the profile along an active pile (see Figure 3.28). The p in the portion varies approximately linearly from $-p_m$ (at $z = z_m$) to an equal but positive magnitude at the pile-base level. The $p(z)$ over the depth z_m to l is thus approximated by:

$$p(z) = kd(\omega z_m + u_g)\left[1 - \frac{2(z - z_m)}{l - z_m}\right] \quad (z = z_m \sim l) \tag{12.5}$$

These features were observed in model tests (see Chapter 13, this book) under a few typical soil movement profiles encountered in practice and were used to develop solutions shown next.

12.2.3 Deformation features

The geometric relation between the uniform soil movement w_s and the rotation angle θ_{gi}, and lateral deflection w_{gi} of the pile at the point O (see Figure 12.1d and e) is as follows

$$w_s = w_{g1} + w_{g2} + \theta_o(L_1 - x_s) \tag{12.6}$$

where x_s = thickness of the resistance zone in which the pile deflection is higher than the soil movement w_s. Equation 12.6 is intended for $w_{g2} + |\theta_{g2}|L_1 > w_s > w_{g2}$ (i.e., $x_s \leq L_1 \leq 1.2L_{c1} + x_{p1}$). Otherwise, some component of the w_s (>pile movement) would flow around the pile ("plastic flow mode"). As shown in Figure 12.1d, a rigid rotational movement $\theta_o(L_1 - x_s)$ appears over the thickness of $L_1 - x_s$, but it becomes negligible ($\theta_o \approx 0$) for deep sliding case. Equation 12.6 is used to develop P-EP and EP-EP solutions.

Deformation characteristics of rigid, passive piles resemble those for a rigid active pile under a lateral load H (Figure 3.28, Chapter 3, this book) in that:

- The pile deflects to u (= $\omega z + u_g$) by mainly rotating about a depth z_r (= $-u_g/\omega$), at which $u = 0$ (see Figure 12.2b).
- The p reaches the p_u to a depth z_o (called slip depth), below which the p at any depth increases with the local displacement, u. This gives rise to a p profile resembling the solid line shown in Figure 3.28a$_1$.
- The p stays as p_u once the deflection u exceeds the limting value u^* [= A_r/k_o (Gibson k) or = $A_r z/k$ (constant k)]. In particular, once the tip deflection u touches the u^* (or $u^* l/z_o$), or the p_l touches $A_r ld$, the pile is at tip-yield state. These features are shown in Figure 3.28a.

The difference is that under passive loading, an additional displacement component is induced owing to rotation about pile-head level caused by the (sliding) maximum bending moment. For instance, the interaction among pile–soil–shear box (see Chapter 13, this book) encompasses "rigid movement" between pile(s) and shear apparatus (termed herein as "overall interaction," see Figure 12.4a). The interaction between pile(s) and surrounding soil (referred to as "local interaction") resembles laterally loaded piles. The overall interaction is associated with a rather low shear modulus compared to that for local interaction, as discussed in Example 12.2.

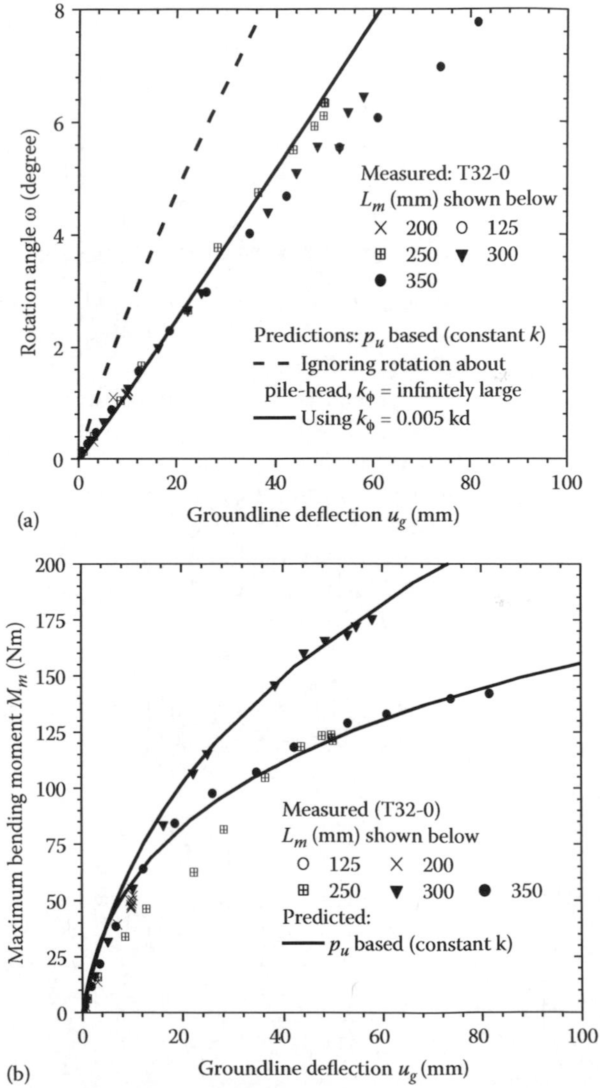

Figure 12.4 Comparison between the current p_u-based solutions and the measured data (Guo and Qin 2010): (a) u_g~ω. (b) u_g~M_m.

12.3 ELASTIC-PLASTIC (EP) SOLUTIONS

The EP solution capitalized on a constant k_i and the aforementioned p_{ui} is indeed sufficiently accurate for modeling active piles (Murff and Hamilton 1993; Guo 2006) compared to numerical approaches (Yang and Jeremic 2002; Guo 2009). It is also useful to calibrating pertinent numerical results (Guo 2010). It is thus used here to model passive piles in stable layer. The values of the parameters A_{Li}, n_i, and G_i for active piles (see Table 3.8, Chapter 3, this book) are generally valid for passive piles (Guo and Ghee 2004; Guo and Qin 2010), but for (1) a much lower p_{ui} on piles adjacent to excavation (Leung et al. 2000; Chen et al. 2002); and (2) an increased critical length of $1.2L_{c1} + x_{p1}$ in the sliding layer (in which L_{c1} = the critical length of an active pile in sliding layer), owing to dragging eccentricity e_{o2}. The thrust H on a passive pile is determined next for "normal" sliding.

12.3.1 Normal sliding (upper rigid–lower flexible)

A passive pile may rotate rigidly about the point O (i.e., $w_{g1} \approx 0$, $\theta_{g1} \approx 0$, and $\theta_o = -\theta_{g2}$) in the sliding layer but behaves as infinitely long in the stable layer. This deformation feature is termed "normal sliding mode" (see Figure 12.1d) and occurs once $L_1 \le 1.2L_{c1} + x_{p1}$. The total movement of the pile is equal to w_{g2} at sliding level (i.e., $x_s = L_1$) and $w_{g2} + |\theta_{g2}|L_1$ at ground level ($x_s = 0$), respectively.

Assuming a plastic pile–soil interaction in sliding layer, the sliding force per unit length p_{u1} is stipulated as $A_{L1}\xi$ in the resistance zone of $x = 0\text{~}x_s$ and as A_{L1} in the thrust zone of $x = x_s\text{~}L_1$, respectively (see Figure 12.1b and d). The factor ξ is used to capture the combined impact of pile-head constraints and soil resistance, etc. Integrating the p_{u1} over the sliding depth offers

$$H_1 = A_{L1}[L_1 - (1+\xi)x_s] \tag{12.7}$$

Equation 12.6 offers $L_1 - x_s = [w_s - (w_{g1} + w_{g2})]/\theta_o$, which allows H_1 of Equation 12.7 to be recast into

$$H_1 = (1+\xi)A_{L1}\frac{w_s - w_{g2}}{|\theta_{g2}|} - \xi A_{L1}L_1 \tag{12.8}$$

where w_{g2} and θ_{g2} may be determined using the normalized $\overline{w}_{g2}$ and $\overline{\theta}_{g2}$ from elastic-plastic solutions of an infinitely long pile in stable layer (Guo 2006).

$$\overline{w}_{g2} = \frac{w_{g2}k_2\lambda_2^{n_2}}{A_{L2}} = \frac{2}{3}\overline{x}_{p2}^{n_2+3}\frac{2\overline{x}_{p2}^2 + (2n_2+10)\overline{x}_{p2} + n_2^2 + 9n_2 + 20}{(\overline{x}_{pi}+1)(n_2+2)(n_2+4)} + \frac{(2\overline{x}_{p2}^2 + 2\overline{x}_{p2} + 1)\overline{x}_{p2}^{n_2}}{\overline{x}_{p2}+1} \tag{12.9}$$

$$\bar{\theta}_{g2} = \frac{\theta_{g2}k_2\lambda_2^{n_2-1}}{A_{L2}} = \frac{-2(\bar{x}_{p2}+1)[\bar{x}_{p2}+1]}{\bar{x}_{p2}+1}\left[\frac{\bar{x}_{p2}^2}{(n_2+1)(n_2+2)} + \frac{\bar{x}_{p2}}{(n_2+1)} + \frac{1}{2}\right]\bar{x}_{p2}^{n_2} + \frac{4\bar{x}_{p2}^2 + 4(n_2+3)\bar{x}_{p2} + 2(n_2+2)(n_2+3)}{(n_2+1)(n_2+2)(n_2+3)}\bar{x}_{p2}^{n_2+1} \tag{12.10}$$

where $\lambda_2 = (k_2/4E_pI_p)^{0.25}$, the reciprocal of characteristic length; I_p = moment of inertia of an equivalent solid pile; and $\bar{x}_{p2} = \lambda_2 x_{p2}$, normalized slip depth.

12.3.2 Plastic (sliding layer)–elastic-plastic (stable layer) (P-EP) solution

A plastic interaction between a passive pile and surrounding soil is normally anticipated in sliding layer, concerning the normal sliding mode. The stable portion is associated with elastic-plastic interaction. The plastic (P) solution (e.g., Equation 12.7) for the pile portion in the sliding layer is resolved together with an EP solution for the pile portion in the stable layer, using the interface conditions (at the point O) of

$$\theta_{g1} + \theta_o = -\theta_{g2},\ M_{o1} = M_{o2},\ H_1 = H, \text{ and } -H_2 = H \tag{12.11}$$

where M_{oi} = dragging moment at sliding level. This solution is referred to as P-EP solution. With the pressure distribution in sliding layer, the dragging moment M_{o1} is obtained as

$$M_{o1} = 0.5A_{L1}[(x_s - 2L_1)x_s\xi + (L_1 - x_s)^2] \tag{12.12}$$

In sliding layer, the shear force Q and bending moment M are determined separately concerning the resistance and thrust zone. In the resistance zone ($x = 0{\sim}x_s$), the force Q and moment M at depth x (from ground level) are given by

$$Q_{A1}(x) = \xi A_{L1}x \quad M_{A1}(x) = 0.5\xi A_{L1}x^2 \tag{12.13a, b}$$

In the thrust zone ($x = x_s{\sim}L_1$), they are given by

$$Q_{A1}(x) = H_1 + A_{L1}(L_1 - x) \tag{12.14a}$$

$$M_{A1}(x) = -\frac{A_{L1}}{2}x^2 + (H_1 + A_{L1}L_1)x - \frac{(H_1 + A_{L1}L_1)^2}{2(1+\xi)A_{L1}} \tag{12.14b}$$

Equation 12.13 has considered the conditions of zero bending moment and shear force at ground level. By rewriting x_s from Equation 12.7, the dragging moment M_{o1} from Equation 12.12 is recast into

$$M_{o1} = \frac{H_1^2}{2A_{L1}}\left[1 - \frac{\xi}{1+\xi}\left(1 + \frac{A_{L1}L_1}{H_1}\right)^2\right] \tag{12.15}$$

The moment M_{o1} is now converted to H_2e_{o2} (see Figure 12.1c), assuming a linearly or a uniformly distributed p_{u2} over the "dragging" zone from depth $L_1 - e_{o2}$ to L_1 (Matlock et al. 1980; Guo 2009). A real value of e_{o2} may be bracketed by the two values of e_{o2} obtained using Equation 12.16 (linear p_{u2}) and Equation 12.17 (uniform p_{u2})

$$e_{o2} = \sqrt[3]{\frac{3M_{o1}}{A_{L2}} + \sqrt{\left(\frac{-3M_{o1}}{A_{L2}}\right)^2 + \left(\frac{-2H_1}{A_{L2}}\right)^3}} + \sqrt[3]{\frac{3M_{o1}}{A_{L2}} - \sqrt{\left(\frac{-3M_{o1}}{A_{L2}}\right)^2 + \left(\frac{-2H_1}{A_{L2}}\right)^3}} \tag{12.16}$$

$$e_{o2} = [H_1 - \sqrt{H_1^2 - 2A_{L2}M_{o1}}\,]/A_{L2} \tag{12.17}$$

Unsure about the p_{u2} profile over the zone, the two values of e_{o2} are estimated using Equations 12.16 and 12.17 for all cases (shown later). They, however, offer only slightly different and often negligible steps in the moment profile at the depth of $L_1 - e_{o2}$. Thereby, only these response profiles estimated using Equation 12.16 (linear p_{u2}) will be presented later. The maximum bending moment in the pile within the sliding layer is given by

$$M_{max1} = \frac{\xi(H_1 + A_{L1}L_1)^2}{2(1+\xi)^2 A_{L1}} \tag{12.18}$$

Dragging does not occur, if $M_{o1} = 0$ or $\xi = \xi_{min}$ in Equation 12.15 with

$$\xi_{min} = 1/\left[\left(1 + \frac{A_{L1}L_1}{H_1}\right)^2 - 1\right] \tag{12.19}$$

On the other hand, the M_{max2} in the low layer may move to sliding level and become the dragging moment M_{o1}. An expression for M_{max2} can be derived, but it is unnecessarily complicated and perhaps of not much practical use. Instead, a maximum value ξ of ξ_{max} is simply taken as $1/\xi_{min}$. Generally speaking, it follows $\xi = \xi_{min}$ for flexible piles, $\xi = \xi_{max}$ for rigid piles, and $\xi_{min} \le \xi \le \xi_{max}$ for upper rigid (in sliding layer) and lower (in stable layer) flexible piles.

The thrust H_2 is correlated to stable layer properties by the normalized $\bar{H}_2$

$$\bar{H}_2 = \frac{H_2\lambda_2^{n_2+1}}{A_{L2}} = \frac{0.5\bar{x}_{p2}^{n_2}[(n_2+1)(n_2+2) + 2\bar{x}_{p2}(2 + n_2 + \bar{x}_{p2})]}{(\bar{x}_{p2}+1)(n_2+1)(n_2+2)} \tag{12.20}$$

The steps for the prediction using P-EP solution are as follows:

1. Obtain parameters A_{Li}, L_1, k_2, and λ_2 ($n_1 = 0$, and $n_2 = 1.0$).
2. Stipulate $\xi = 0\sim0.8$ for flexible piles or $\xi = 3$–6 for rigid piles.
3. Determine normalized slip depth $\bar{x}_{p2}$ using H_1 (from Equation 12.8) = H_2 (from Equation 12.20) for each soil movement w_s.
4. Calculate H_2, w_{g2}, and θ_{g2} using Equations 12.20, 12.9, and 12.10, respectively for the given set of A_{Li} and n_i and the $\bar{x}_{p2}$ gained.
5. Calculate M_{o1}, $e_{o2,}$ ξ_{min}, and ξ_{max} using Equations 12.15, 12.16, 12.19, and $1/\xi_{min}$, respectively.
6. Determine distribution profiles of the displacement, rotation, moment, and shear force in stable layer using Equations 12.13 and 12.14 and the EP solutions provided in Table 9.1 (Chapter 9, this book).
7. Ensure a smooth transition of the obtained moment profile over the dragging zone, otherwise repeating steps 3 through 7 for a new stipulated value of ξ ($\xi_{min} \le \xi \le \xi_{max}$). The prediction is readily done using Mathcad.

Using the EP solutions, the depth x_2 should be replaced with $x_2 - L_1 + e_{o2}$ (sliding layer), as the depth x_i is measured from the depth L_1 (see Figure 12.1); and the rigid rotation angle θ_o (= $-\theta_{g2}$) and the loading zone $L_1 - x_s$ [= $(w_s - w_{g2})/|\theta_{g2}|$] are readily calculated.

12.3.3 EP solutions for stable layer

The elastic-plastic solutions of Table 9.1 (Chapter 9, this book) are used to gain the response profiles of the pile portion in stable layer such as $Q_{A2}(x_2)$, $M_{A2}(x_2)$, $w_{A2}(x_2)$ and $\theta_{A2}(x_2)$ at depth x_2 ($\le x_{p2}$) and $Q_{B2}(x_2)$, $M_{B2}(x_2)$, $w_{B2}(x_2)$ and $\theta_{B2}(x_2)$ at depth x_2 ($> x_{p2}$). The maximum bending moment M_{max2} and its depth x_{max2} depend on the normalized depth $\bar{x}_{max2}$ (= $x_{max2}\lambda_2$).

$$\bar{x}_{max2} = a\tan\left\{1/\left\{1+2\left[\frac{-\bar{x}_{p2}^{n_2+2}}{(n_2+1)(n_2+2)} + (\bar{x}_{p2}+\bar{e}_{o2})\bar{H}_2\right] / \left(\frac{-\bar{x}_{p2}^{n_2+1}}{n_2+1} + \bar{H}_2\right)\right\}\right\} \tag{12.21}$$

At $n_2 = 1$, simple Equation 12.22 and Equation 12.13 or 12.14 are noted for calculating x_{max2} and M_{max2}, respectively:

$$x_{max2} = \frac{-1}{\lambda_2}\tan^{-1}\left[\frac{2x_{p2}^2 - 3}{2x_{p2}^2 + 3 + 6x_{p2}}\right] \tag{12.22}$$

If $\bar{x}_{max2} < 0$, the M_{max2} occurs at the depth x_{max2} in plastic zone and is given by

$$-M_{max2} = \frac{A_{L2}}{\lambda_2^{n_2+2}}\left\{\frac{1}{n_2+2}[(n_2+1)\bar{H}_2]^{(n_2+2)/(n_2+1)} + \bar{H}_2\bar{e}_{o2}\right\} \quad (x_{max2} < 0) \tag{12.23}$$

Otherwise, if $x_{max2} \geq 0$, the M_{max2} occurs in elastic zone (see Chapter 9, this book) and is given by

$$-M_{max2} = \frac{A_{L2}}{\lambda_2^{n_2+2}} e^{-\bar{x}_{max2}} \left\{ \left[\frac{-\bar{x}_{p2}^{n_2+2}}{(n_2+1)(n_2+2)} + (\bar{x}_{p2} + \bar{e}_{o2})\bar{H}_2 \right] \cos(\bar{x}_{max2}) + \left[\frac{-\bar{x}_{p2}^{n_2+1}}{n_2+1} + \bar{H}_2 + \frac{-\bar{x}_{p2}^{n_2+2}}{(n_2+1)(n_2+2)} + (\bar{x}_{p2} + \bar{e}_{o2})\bar{H}_2 \right] \sin(\bar{x}_{max2}) \right\} \quad (x_{max2} \geq 0) \tag{12.24}$$

The solution neglects the coupling impact among soil layers and any friction on sliding interface. It stipulates a uniform soil movement profile (Figure 12.1d), although modifying Equation 12.6 will render other profiles to be accommodated.

The impact of dragging on piles is captured through the eccentricity e_{o2} and the parameter ξ. Conversely, three values of A_{Li} (or H_2), θ_o (or θ_{g2}), and ξ may be deduced by matching predicted with measured profiles of bending moment, shear force, and pile deflection (for a particular w_s) using the solutions. The e_{o2} may then be back-estimated from measured M_{o1} and H_1 at sliding level using Equation 12.16. They may be compared with the E-E prediction to examine the effect of plasticity (ξ and x_{p2}), dragging (e_{o2}), and the nonhomogeneous p_{u2} against the P-EP solution (underpinned by the p_{u2} and a constant k_2). The use of this solution is elaborated on in Example 12.5.

Example 12.1 Nonhomogeneous p_{u2} versus k (numerical)-based solutions

The current P-EP solutions are underpinned by a nonhomogeneous p_{u2} and a uniform k_2, which are different from some numerical solutions based on a nonhomogeneous k_2 and a uniform p_{u2} (Poulos 1995; Chow 1996). The impact of this difference on a pile in normal sliding Case 12.1 is examined next.

Esu and D'Elia (1974) reported a reinforced concrete pile installed in a sliding clay slope (Case 12.1). The pile has an outside diameter d of 0.79 m, a length L of 30 m, and a bending stiffness E_pI_p of 360 MNm^2. The measured "ultimate" shear forces, bending moments, and deflections along the pile are plotted in Figure 12.5. Chen and Poulos (1997) conducted boundary element analysis (BEA) on the pile using $E_i = 0.533x$ (MPa, x depth from ground level, and $n_2 = 1.0$), $s_{ui} = 40$ kPa, $p_{u1} = 3ds_{u1}$, $p_{u2} = 8ds_{u2}$, and $w_s = 110$ mm uniform from ground level to a sliding depth of 7.5 m. The predicted bending moments, deflections, and shear forces are plotted in Figure 12.5 along with the measured data.

The following parameters were determined: $A_{L1} = 94.8$ kN/m (= 3 × 40 × 0.79 kPa, $N_{g1} = 3.0$), $A_{L2} = 52$ kPa ($N_{g2} = 1.3$), $\xi = 0.5$, $k_1 = 2.5$ MPa, and $k_2 = 7$ MPa, together with $n_1 = 0$, $n_2 = 1.0$. The low

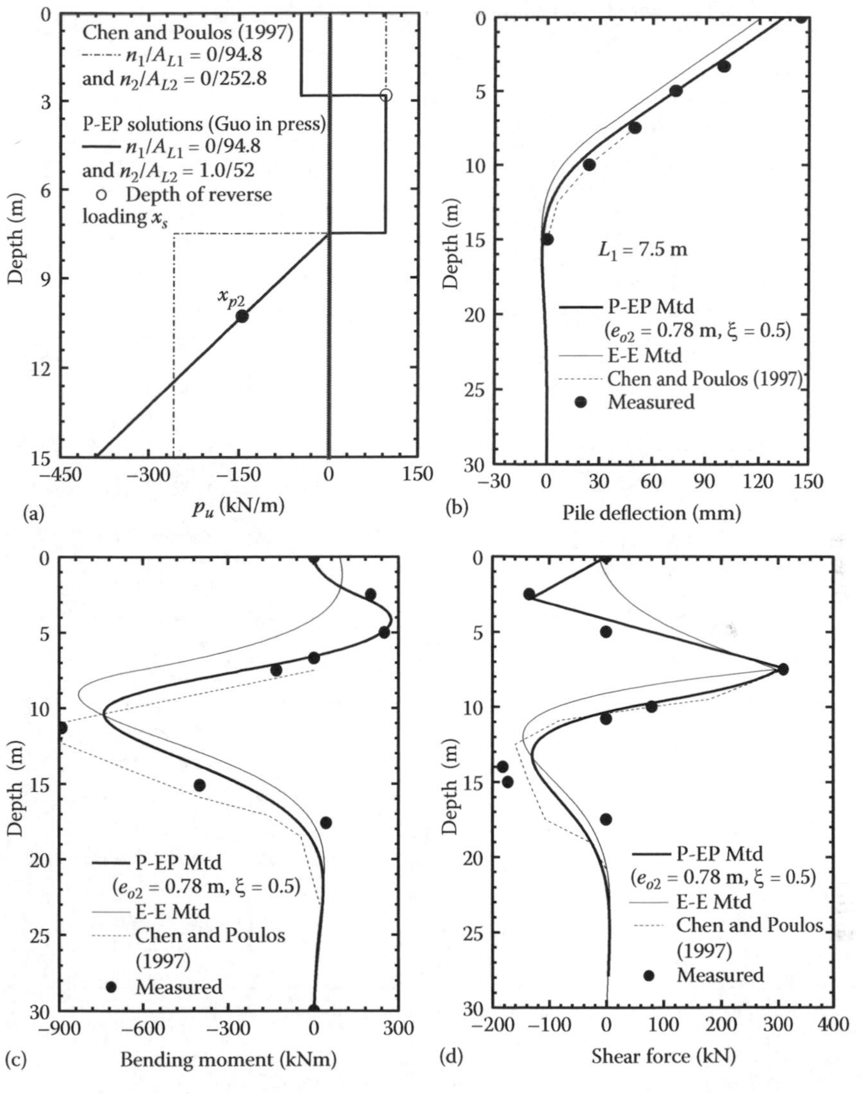

Figure 12.5 Predicted (using $\theta_o = -\theta_{g2}$) versus measured (Esu and D'Elia 1974) responses (Case 12.1). (a) p_u profile. (b) Pile deflection. (c) Bending moment. (d) Shear force. (After Guo, W. D., *Int. J. of Geomechanics*, in press)

p_{u2} (compared to p_{u1}) is based on that for embankment piles (Stewart et al. 1994), as is used again in Cases 12.7 and 12.8. The k_i is equal to $3.65G_i$ and $G_i = 54.8\ s_{ui}$, for which the k_i/G_i is estimated using k_{1i} = 1.5 (in Equation 3.54, for e_{oi}/d = 2~3), γ_i = 0.164, and E_p/G_i = 8,589. The critical length L_{ci} is estimated as 8.1 m (>7.7 m obtained using k_i/G_i = 3, see Table 12.1). With the parameters, the P-EP (via GASMove) predictions and E-E solutions were obtained and are plotted in Figure

Table 12.1 Input properties and parameters for E-E solutions

Piles			Soil	Sliding parameters				References	
d / t (mm)	E_p (GPa)	L_1 / L_2 (m)	N_i / s_{u1}/s_{u2} [a]	L_{c1} / L_{c2} (m)[b]	k_1 / k_2 (MPa)	θ_0 ($\times 10^{-3}$)	H (kN)	Cases	
790	20	7.5	–	7.7	8.0	13	296	(Esu and D'Elia 1974)	12.1
395		22.5	40 / 40	7.7	8.0				
318.5	210	11.2	7.9 / 12.6	6.3	5.0	26[c]	150	Hataosi-2 (Cai and Ugai 2003)	12.2
6.9		12.8	–	5.6	8.0				
318.5	210	8.0	7.9 / 23.6	6.3	5.0	4[c]	70	Hataosi-3 (Cai and Ugai 2003)	12.3
6.9		9.0	–	4.9	15.0				
318.5	210	6.5	7.9 / 12.6	6.3	5.0	−8[c]	300	Kamimoku-4 (Cai and Ugai 2003)	12.4
6.9		7.5	–	5.6	8.0				
318.5	210	4.0	7.9 / 12.6	6.3	5.0	40[c]	250	Kamimoku-6 (Cai and Ugai 2003)	12.5
6.9		6.0	–	5.6	8.0				
300	20	7.3	$s_{u1} = 30$[d]	3.2	6.0	25[c]	40	Katamachi-B (Cai and Ugai 2003)	12.6
60		5.7	$N_2 = 16.7$	2.8	10.0				
1200	20	9.5	–	12.7	15	9.0	500	Carrubba (Carrubba et al. 1989)	12.7
600		13.0	30 / 30	12.7	15				
630	28.45	2.5		23.1	14.4	0.5	60	Leung (2.5 m)	12.8
315		10.0		23.1	28.8				
630	28.45	2.5		23.1	14.4	0.5	85	Leung (3.5 m)	12.8
315		10.0		23.1	28.8				
630	28.45	2.5		23.1	14.4	0.5	100	Leung (4.5 m)	12.8
315		10.0		23.1	28.8				

Source: Guo, W. D., *Int. J. of Geomechanics*, in press. With permission of ASCE.

a s_{ui} (in kPa).

b $G_i = k_i/3$ (k_i in Table 12.6) for estimating L_{ci} using Equation 3.53, Chapter 3, this book.

c Calculated using measured pile deflection profiles (Cai and Ugai 2003).

d Sliding layer s_{u1} = 30 kPa.

12.5. They agree well with the measured data and the BEA. Note with $L_1 < L_{c1}$, the pile deflection is equal to the sum of the deflection w_{g2} and the rigid rotation $\theta_o(L_1\text{-}x)$.

Guo (2003b), in light of the P-EP solution and assuming $\xi = 0$ and no resistance to the depth x_s, predicted the profiles for the stable layer using three pairs of parameters A_{L2}/n_2 (= 120/0, 51/0.5, and 25/1.0) and a uniform subgrade modulus k_2. Each prediction agrees well with the BEA and with the measured profiles of deflection, rotation, and bending moment, respectively. However, the shear force profile is only well replicated using $n_2 = 1.0$, with which the BEA solutions were obtained. The use of "$n_2 = 1.0$" is legitimate as it yields the lowest thrust (otherwise $H_2 = 458.6$ at $n_2 = 0.5$). A typical calculation is elaborated in Example 12.5.

12.4 P_U-BASED SOLUTIONS (RIGID PILES)

Using the p profile of Equations 12.4 and 12.5, see Figure 12.2b, solutions for the passive rigid piles were newly deduced here and are provided in Table 12.2. Independent of overall relative pile-shear box movement during a shear test (see Chapter 13, this book), the solutions are, as tabulated in the table, (1) intended to capture local pile–soil interaction for either a constant k or a Gibson k; (2) underpinned by the aforementioned load transfer model and the observed on-pile force profile p between depth z_m and l of Equation 12.5; and (3) confined to pre-tip yield state. The solutions, referred to as p_u-based solutions hereafter, are adequate, as the tip-yield is unlikely to be reached under a dominantly dragging movement. For instance, with a uniform k, the normalized $\bar{z}_m$ is determined, using Equation 12T5, as

$$\bar{z}_m = 1/(2 - \bar{z}_o) \tag{12.25}$$

where $\bar{z}_m = z_m/l$ and $\bar{z}_o = z_o/l$. Normalized mudline displacement, rotation angle, and maximum bending moment are given, respectively, by

$$\frac{u_g k}{A_r l} = \frac{\bar{z}_m^2 \bar{z}_o}{[\bar{z}_m - \bar{z}_o]^2}, \tag{12.26}$$

$$\frac{\omega k}{A_r} = \bar{z}_o \frac{-2\bar{z}_m + \bar{z}_o}{[-\bar{z}_m + \bar{z}_o]^2}, \text{ and} \tag{12.27}$$

Table 12.2 p-Based solutions (local interaction)

Gibson k (m = 1)	*Constant* k (m = 0)	*Equation*
$u = \omega z + u_g$ and $z_r/l = -u_g/\omega l$		12T1
$\frac{u_g k_o}{A_r} = \frac{2\bar{z}_m^3 + \bar{z}_o^3}{[\bar{z}_o + 2\bar{z}_m][-\bar{z}_m + \bar{z}_o]^2}$	$\frac{u_g k}{A_r l} = \frac{\bar{z}_m^2 \bar{z}_o}{[\bar{z}_m - \bar{z}_o]^2}$	12T2
$\frac{\omega k_o l}{A_r} = \frac{-3\bar{z}_m^2}{(2\bar{z}_m + \bar{z}_o)[-\bar{z}_m + \bar{z}_o]^2}$	$\frac{\omega k}{A_r} = \bar{z}_o \frac{-2\bar{z}_m + \bar{z}_o}{[-\bar{z}_m + \bar{z}_o]^2}$	12T3
$\frac{M_m}{A_r d l^3} = \frac{\bar{z}_m^2}{12(\bar{z}_o + 2\bar{z}_m)}(3\bar{z}_o^2 + 2\bar{z}_o\bar{z}_m + \bar{z}_m^2)$	$\frac{M_m}{A_r d l^3} = \frac{1}{6}\bar{z}_o \bar{z}_m^2$	12T4
$\bar{z}_m^4 + (\bar{z}_o - 4)\bar{z}_m^3 + [\bar{z}_o^2 - 4\bar{z}_o + 2]\bar{z}_m^2$ $+[3\bar{z}_o^3 + 2\bar{z}_o - 4\bar{z}_o^2]\bar{z}_m + 2\bar{z}_o^2 = 0$	$\bar{z}_m = 1/(2 - \bar{z}_o)$	12T5

Note: Bar "—" denotes depths normalized with pile embedment length l.

$$\frac{M_m}{A_r dl^3} = \frac{1}{6}\bar{z}_o \bar{z}_m^2 \tag{12.28}$$

The z_m is first obtained using Equation 12.25 for a specified z_o, which renders the values of $u_g k/A_r l$, $\omega k/A_r$, and $M_m/(A_r dl^3)$ using Equations 12.26, 12.27, and 12.28, respectively.

Equations 12.4 and 12.5 allow the on-pile force profile p to be constructed by linking the adjacent points of (0, 0), $(z_o A_r d, z_o)$, $(0, z_r)$, $(-p_m, z_m)$, and (p_m, l), in which $p_m = kd(\omega z_m + u_g)$, and $z_r = -u_g/\omega$ (note $\omega < 0$). As an example, the p profiles for z_o/l of 0.05, 0.15, 0.3 and 0.45 were obtained, and the p and z normalized by $z_o A_r l$ and l, respectively. These predicted profiles, as shown in Figure 12.3c, are independent of k profiles and generally agree with the measured ones depicted in Figure 12.3a and b.

The p_u-based solutions permit the normalized moment $M_m/(A_r dl^3)$, depth z_m/l, and mudline deflection $u_g k_o/(A_r l^{1-m})$ to be predicted for each z_o/l. The predicted moments and displacements for a series of z_o/l are plotted in Figure 12.6a and b. The figures show the impact of the k profiles on the predicted M_m and u_g. Especially, the solutions for laterally loaded piles (termed as H-l solutions, see Chapter 8, this book) are also provided. The normalized force was obtained using H-z_m solutions deduced using the similarity between active and passive piles. The latter is not explained herein, but it is very similar to that estimated using the shear force profile based on the p_u profile.

As with flexible piles, the response profiles of shear force and bending moment are derived for a linear distributed p_u profile for rigid piles. They consist of three typical zones:

1. In the slip zone of depth z ($\le z_o$), the shear force $Q(z)$ and bending moment $M(z)$ are given by

$$Q(z) = 0.5A_r dz^2 \quad M(z) = A_r dz^3/6 \tag{12.29a, b}$$

2. Between the depths z_o and z_m (i.e., $z = z_o \sim z_m$), the $Q(z)$ and $M(z)$ are expressed as

$$Q(z) = 0.5A_r dz_o^2 + kd[(z - z_o)u_g + 0.5(z^2 - z_o^2)\omega] \tag{12.30a}$$

$$M(z) = kd(z - z_o)^2\left[\frac{u_g}{2} + \frac{(z + 2z_o)\omega}{6}\right] + \frac{A_r z_o^2(3z - 2z_o)}{6} \tag{12.30b}$$

3. Below the depth z_m (i.e., $z = z_m \sim l$), the $Q(z)$ and $M(z)$ are described by

$$Q(z) = A_r dz_o z_m \frac{(z - z_m)}{(z_o - z_m)}\frac{(l - z)}{(l - z_m)} \tag{12.31a}$$

$$M(z) = \frac{A_r dz_o z_m}{6}\frac{(lz_m z_o - z_o z_m^2 + 2lz_m^2 - 6lz_m z + 3(z_m + l)z^2 - 2z^3)}{(z_o - z_m)(l - z_m)} \tag{12.31b}$$

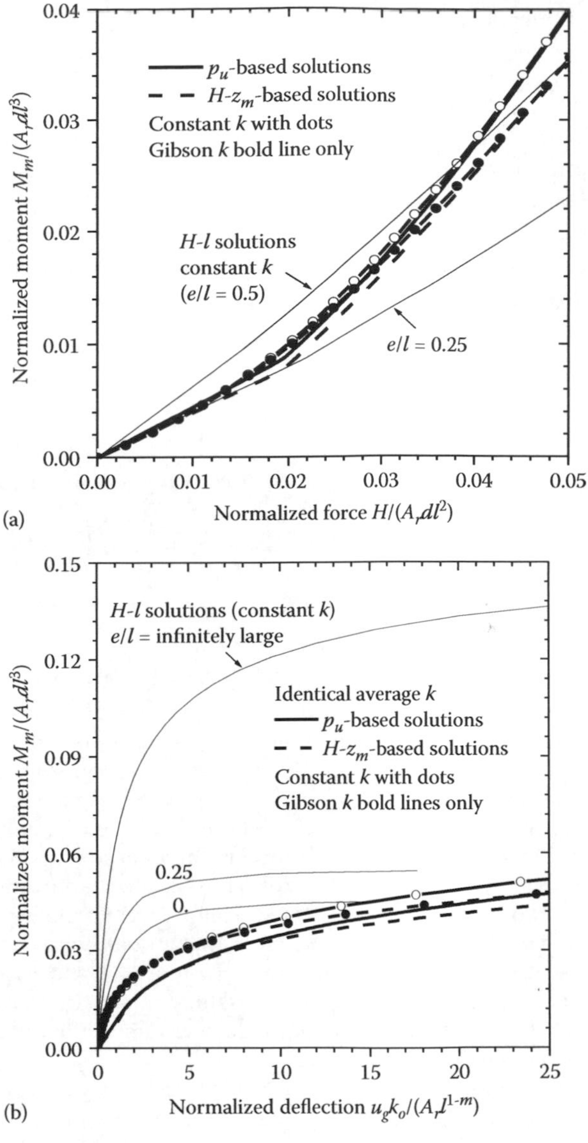

Figure 12.6 Normalized response based on p_u and fictitious force. (a) Normalized force and moment. (b) Normalized deflection and moment.

The absolute pile displacement at any depth z, $u(z)$ is described by

$$u(z) = u_g + \omega z + \frac{M_m}{k_\varphi} z \ (z = 0 \sim l) \tag{12.32a}$$

where M_m = the maximum bending moment at the depth z_m; and k_ϕ = pile rotational (constraining) stiffness about the head (e.g., the k_ϕ for the model tests in Chapter 13, this book, is about $0.005kd$ in kNm/rad; i.e., kd (in kPa) × m^3/rad). The total pile slope (rotation) is equal to the sum of ω and M_m/k_ϕ (compared to θ_o in Equation 12.6). The deflection u consists of relative movement $w(z)$ about a depth in sliding layer of

$$w(z) = u_g + \omega z \ (z = 0 \sim z_m) \tag{12.32b}$$

and relative rotation of $M_m z/k_\phi$ about the pile-head, which is originated from the moment M_m in depth z_m and does not contribute to force and moment equilibrium of the p_u-based solutions. The p_u solutions are not directly related to soil movement but through the H-l based solutions, as shown next.

Example 12.2 Analysis of a typical pile–arc profile

Shear tests of a pile installed in the shear box (in Chapter 13, this book) were conducted using an arc-shaped loading block to induce soil movement to a depth of 200 mm at a distance of 500 mm from a model pile 700 mm in length. The values of M_m, ω, and u_g were measured (during the shear frame movement w_s) and are depicted in Figure 12.7. The evolution of M_m with the frame movement w_s reflects the overall shear box-pile interaction, whereas the relationships of u_g~ω, and u_g~M_m (during the movement w_s) indicate the local pile–soil interaction within the box. These two aspects of the interaction require different moduli of k_od (overall) and kd (local) to be modeled, but a similar or identical p_u profile.

Input parameters A_r (overall or local), k_od (overall), and kd (local) were deduced using H~l and p_u-based solutions and the measured response of the model piles at various w_s to assess impact of soil movement profiles on passive piles.

The parameters A_r and k for the test AS50-0 were deduced against the measured values of M_m, ω, and u_g in Figure 12.7. Initial values of A_r, k_od, and kd were calculated as follows: (1) with γ_s' = 16.27 kN/m^3 and ϕ' = 38°, the A_r was calculated as 134.7 kPa/m (= $0.46\gamma_s'K_p^2$) for either aspect; (2) assuming $\tilde{G}$ = 6.52 kPa (Chen and Poulos 1997), and with $k_od(0.5l) = 2.85\tilde{G}$, the k_od (overall) was calculated as 53.2 kPa/m; and (3) the G (local) was taken as 493 kPa (Hansen 1961; Guo 2006; Guo 2008) in view of a low shear strain (similar to active piles). The associated kd and k (local) were calculated as 1,407.4 kPa and 28,148 kPa/m, respectively, with $kd = 2.85G$ and d = 50 mm. These values of A_r, k_od, and kd actually offer best match with measured response as elaborated next.

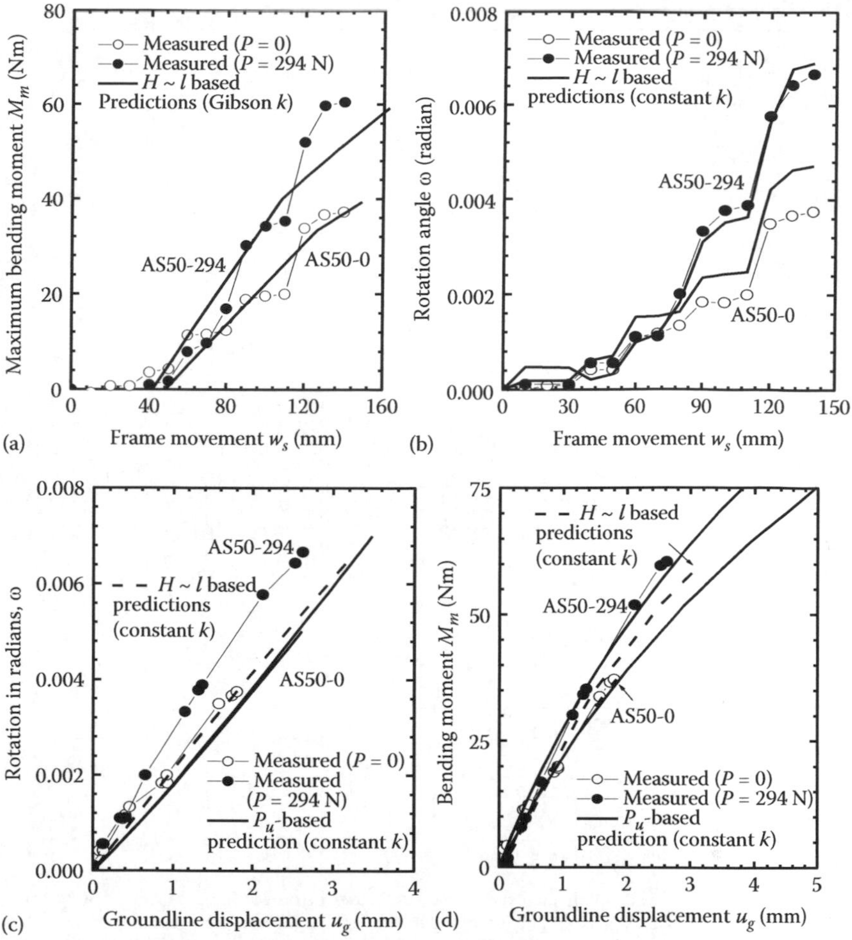

Figure 12.7 Current predictions versus measured data (AS50-0 and AS50-294). (a) M_m during overall sliding. (b) Pile rotation angle ω. (c) Mudline deflection u_g. (d) M_m for local interaction.

12.2.1 *H-l*–based solutions

H-l–based solutions refer to those developed in Chapter 8, this book, for a laterally loaded rigid pile. The solutions for Gibson k and constant k may be used to deduce parameters for overall and local interaction, respectively.

1. For overall sliding, the A_r and kd were deduced by matching the Gibson k (H-l) solutions with the measured w_s~M_m curve. For instance, at z_o = 0 m and e = 0 m, the solutions of Equations 12T10, and 12T11 in Table 12.3 offer $H(A_r dl^2) = 0.0556$ and $u_g k_o / A_r = 1$, respectively. The condition of $z_m > z_o$ renders $z_m/l = 0.4215$ and $M_m/$

Table 12.3 Solutions for an active rigid pile (H-l–based solutions)

Gibson k (m = 1)	*Equation*
$u = \omega z + u_g$ and $z_r / l = -u_g / (\omega l)$	12T1
$k_o\left(\frac{l}{2}\right)^m d = \frac{3\pi\tilde{G}}{2}\left\{2\gamma_b\frac{K_1(\gamma_b)}{K_o(\gamma_b)} - \gamma_b^2\left[\left(\frac{K_1(\gamma_b)}{K_o(\gamma_b)}\right)^2 - 1\right]\right\}$	12T6
$\frac{M_m}{A_r dl^3} = \left(\frac{2\bar{z}_m}{3} + \bar{e}\right)\frac{H}{A_r dl^2}$ and $z_m = \sqrt{2H/(A_r d)}$ $(z_m \le z_o)$	12T7
$\frac{M_m}{A_r dl^3} = \left(\frac{3 - 2\bar{z}_m^3\bar{z}_o + 2\bar{z}_m\bar{z}_o^3 - 4\bar{z}_o + \bar{z}_m^4}{(1-\bar{z}_o)^2(3+2\bar{z}_o+\bar{z}_o^2)}\bar{e} + \bar{z}_m\right)\frac{H}{A_r dl^2}$ $(z_m > z_o)$ $+\frac{1}{6}\left\{\bar{z}_o^2(2\bar{z}_o - 3\bar{z}_m) - (\bar{z}_o - \bar{z}_m)^2\left(\bar{z}_m + 2\frac{3\bar{z}_o(1-\bar{z}_o) + (\bar{z}_o^5 - \bar{z}_m^2)}{(1-\bar{z}_o)^2(3+2\bar{z}_o+\bar{z}_o^2)}\right)\right\}$	12T8
$\bar{z}_m = \frac{\sqrt{3[\bar{z}_o^3(\bar{z}_o+2\bar{e})+16\bar{e}+11][3\bar{z}_o^3(\bar{z}_o+2\bar{e})+1]}}{8(2+3\bar{e})} + \frac{3\bar{z}_o^3(\bar{z}_o+2\bar{e})+1}{8(2+3\bar{e})}$ $(z_m > z_o)$	12T9
$\frac{H}{A_r dl^2} = \frac{1}{6}\frac{1+2\bar{z}_o+3\bar{z}_o^2}{(2+\bar{z}_o)(2\bar{e}+\bar{z}_o)+3}$	12T10
$\frac{u_g k_o}{A_r} = \frac{3+2[2+\bar{z}_o^3]\bar{e}+\bar{z}_o^4}{[(2+\bar{z}_o)(2\bar{e}+\bar{z}_o)+3](1-\bar{z}_o)^2}$	12T11
$\omega = \frac{A_r}{k_o l}\frac{-2(2+3\bar{e})}{[(2+\bar{z}_o)(2\bar{e}+\bar{z}_o)+3](1-\bar{z}_o)^2}$	12T12

Source: Guo, W. D., *Can Geotech J*, 45, 5, 2008.

Note: $u, u_g, \omega, z, z_o, z_r, e$, and l are defined in Figure 8.1 (Chapter 8, this book); $\gamma_b = k_1 r_o / l$; r_o = an outside radius of a cylindrical pile; k_1 increases hyperbolically from 2.14 to 3.8 as e increases from 0 to ∞; $k = k_o$ for constant k (m = 0); $K_i(\gamma_b)$ = modified Bessel function of second kind of i-th order; $\tilde{G}$ = average shear modulus G[FL^{-2}] over pile embedment. The expresssions are used directly along with e = 0 for *H-l*–based solutions.

$(A_r dl^3)$ = 0.0144 in light of Equations 12T9 and 12T8, respectively. Therefore, with a measured M_m of 33.36 Nm, the A_r was deduced as 134.7 kPa/m, and further with a measured u_g^{all} of 126.6 mm, the $k_0 d$ was deduced as 53.2 kPa/m. Note the corresponding fictitious H of 183.3N and z_m of 0.295 m are generally not equal to the maximum shear force and the depth of M_m in the pile, which should be based on local interaction (see later p_u-based solutions).

2. For local interaction, the *kd* was deduced by fitting the constant *k*–based *H-l* solutions (see Table 8.1, Chapter 8, this book) with the measured u_g~ω and u_g~M_m curves. The same $H/(A_r dl^2)$ of 0.0556 (from overall sliding) now corresponds to a new z_o/l of 0.25, as determined using Equation 8.1g (Chapter 8, this book).

Accordingly, the $u_g k/(lA_r)$ and $\omega k/A_r$ were estimated as 0.396 and 0.5933, respectively, using Equations 8.2g and 8.3g (Chapter 8, this book). With l = 0.7m, A_r = 134.7kPa/m and a measured u_g of 1.3 mm, the k(local) was deduced as 28,148 kPa (see Table 12.4). This k allows a ω of 0.0028 to be predicted, which compares well with the measured data.

3. Calculations for various w_s were performed. The measured and predicted pile deflection u_g was 1.3~1.6 mm from the local interaction; as such, the w_s was equal to 127.9~128.2 mm ($\approx u_g^{all}$ + 1.3~1.6 mm). Nevertheless, it is sufficiently accurate to take $u_g^{all} \approx w_s$, as presented in Table 12.4 under "Overall Sliding (Gibson k)".

The back-estimation verified the calculated values of A_r, $\tilde{G}$, and G. The values of A_r = 134.7kPa/m, $\tilde{G}$(overall) = 6.52 kPa, and G (local) = 493 kPa were thus employed to predict M_m, u_g, and ω for other z_o/l (or w_s). The predicted w_s~M_m curve (overall) and the w_s~ω, u_g~ω, and u_g~M_m curves (local) are provided in Figure 12.7a, b, c, and d, respectively, which agree with the respective measured data. The A_r, $\tilde{G}(k_o d)$, and G (kd) are further justified against the entire nonlinear response. Typical results are tabulated in Table 12.4 for measured w_s of 120, 130, and 140 mm. In particular, Figure 12.7a shows the elastic interaction from (w_i, 0) to (126.6, 33.36) (i.e., M_m = 33.36 Nm at w_s = 126.6 mm) and negligible pile response for $w_s \leq w_i$ = 40 mm (Guo and Qin 2010). Note that if Equation 12T7 for $z_m \leq z_o$ were incorrectly used, the ratio $H/(A_r dl^2)$ and the fictitious force H would be deduced as 0.062 and 205.2 N, respectively. Figure 12.7b demonstrates a good capture of the step raises in the rotation as well, using the force H (up to 218.9 N) via the ratio of z_o/l.

Using a fictitious load H, the solutions for a lateral pile may be used to analyze the rigid, passive pile, see Figure 12.2b. This is underpinned by two principles: (1) the value and location of H must be able to replicate the featured on-pile force distribution (Figure 12.3), and the rigid pile deflection; and (2) the resulting solutions must compare well with the newly established p_u-based solutions (next section). They are satisfied, respectively, by (1) substituting l with z_m, in view of negligible total net resistance over the depth of z_m~l; and (2) by taking e/l as $z_o/(3z_m)$ that offers best agreement with the p_u-based solutions. It should be stressed that the fictitious H locates at a distance 0~l/3 above ground surface rather than at the centroid of the sliding force to balance the additional bending moment M_m induced at depth z_m. The H~z_m solutions are thus formulated from the H~l solutions. The solutions are plotted in Figure 12.6a and b.

12.2.2 p_u-based solutions

The A_r, k and k_ϕ were deduced using the p_u-based solutions against the measured u_g~ω and u_g~M_m curves (see Figure 12.7c and d), with a z_o/l of 0.058. The ratios of $u_g k/(lA_r)$, $M_m/(A_r dl^3)$, and $\omega k/A_r$ were obtained as 0.074, 0.00256, and 0.270, respectively, by using p_u-based, right Equations 12T2, 12T4, and 12T3 in Table 12.2, respectively. Using the A_r and k deduced above, we have u_g = 1.49 mm, M_m = 32.0 Nm, and

Table 12.4 Calculated versus measured responses (H-I–based solutions)

Model pile test AS50-0										
Measured data				*Overall sliding (Gibson k)*[a]			*Pile response (constant k)*[b]			
w_s (mm)	M_m (Nm)	u_g (mm)	$-\omega$	H (N)	z_o/l	w_s (mm)	H (N)	z_o/l	u_g (mm)	$-\omega$
120	33.77	1.58	.0035	183.3[c]	0.0[c]	126.6[c]	183.3[d]	0.25[d]	1.30[d]	*.0028*[d]
130	36.64	1.74	.0037	216.7	0.119	150.6	216.7	0.302	1.81	*.0038*
140	37.20	1.80	.0038	218.9	0.1264	178.9	218.9	0.306	1.84	*.0038*
Embankment pile										
Measured data				*Overall sliding (Gibson k)*[a]			*Pile response (constant k)*[b]			
w_s (mm)	M_m(kNm)	u_g (mm)	ω	H (kN)	z_o/l	w_s (mm)	H (kN)	z_o/l	u_g (mm)	ω
5.5	70	6	.0007	28	0.169	5.5	28[e]	0.328	4.3	*.0007*
26	170	37	.0049	50.5	0.662	25.0	50.5[f]	0.684	36.3	*.0046*

[a,b] e = 0 m, and values of *k* are listed in Table 12.5.
[c] Pile–soil relative slip just initiated with w_s = 126.6 mm.
[d] $z_o < z_m$, otherwise $z_o \geq z_m$ for H = 216.7~218.9*N*.
[e] The predicted Q_m and M_m (p_u-based) were 26.16 kN (in L_m layer)~31.32 kN (in stable layer) and 87.3 kNm, respectively.
[f] The predicted Q_m and M_m (p_u-based) were 66.17~74.20 kN and 187.77 kNm, respectively.

ω = −0.00781. The difference between this calculated ω and the measured ω of −0.0031 allows the k_ϕ to be obtained as 6.28 kNm = [32.0 × 0.001/(−0.0031 + 0.00781), i.e., $k_\phi = 0.005kd$].

In light of p_u-based solutions, the response profiles of test AS50-0 were predicted for the two typical values of z_o/l of 0.0171 (w_s = 80 mm) and 0.058 (120 mm), which compare well with the measured profiles of force per unit length, shear force, and bending moment as shown in Figure 12.8a through d. Figure 12.8b and c indicate a maximum shear force Q_m of −39.2 kN and 39.2 kN (w_s = 60 mm), and −140.7 kN and

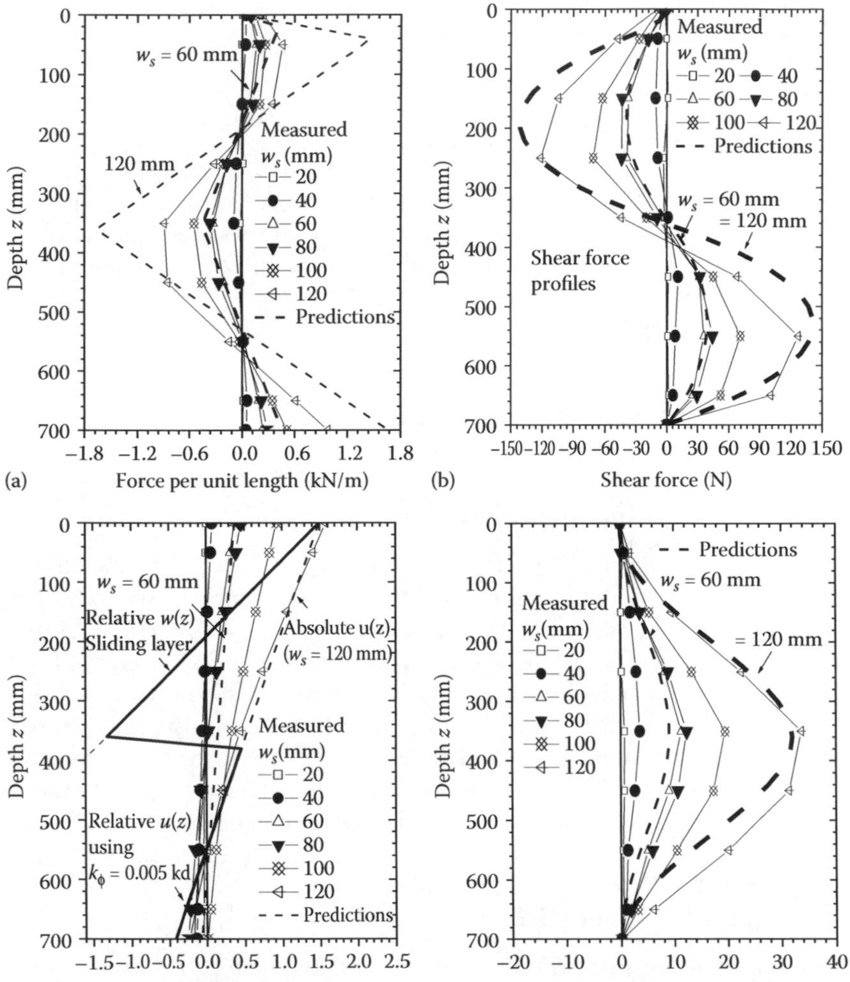

Figure 12.8 Current prediction (p_u-based solutions) versus measured data (AS50-0) at typical "soil" movements. (a) p profiles. (b) Shear force profiles. (c) Pile deflection profiles. (d) Bending moment profiles.

Table 12.5 p_u-Based solutions (AS50-0)

z_o/l	0.01	0.0171	0.035	*0.058*	0.07	0.15	0.25
z_m/l	0.503	0.504	0.509	*0.515*	0.518	0.541	0.571
u_g (mm)	0.21	0.37	0.82	*1.49*	1.9	5.82	16.0
M_m (Nm)	5.3	9.1	18.9	*32.0*	39.2	91.3	170

Note: e = 0 m, and values of *k* are listed; A_r = 729 kPa/m and *k* = 25.2 MPa/m. The values for AS50-294 are identical to those presented here for AS50-0, except for a slight increase in u_g.

141.1 kN (120 mm), respectively, which are slightly higher than the measured (absolute) values of 38.7 and 36.1 kN, and 121.3 and 127.1 kN, respectively. A sufficient accuracy of the k_ϕ in gaining deflection profiles is also observed for the test TD32-0 at z_o/l = 0.243 (w_s = 80 mm) and 0.358 (= 120 mm) (not shown here).

Using A_r = 729 kPa/m (= $2.5\gamma_s'K_p^2$, reflecting sand densification), and kd = 1,260 kPa (k = 25.2 MPa/m), the z_m/l, u_g, and M_m were predicted for seven typical values of z_o/l and are provided in Table 12.5. The u_g~ω and u_g~M_m curves are plotted in Figure 12.7. They agree well with the measured ones, justifying the deduced parameters A_r, kd, and k_ϕ. As anticipated, the A_r of 134.7 kPa/m in the *H-l*–based solutions (local) increases to 729 kPa/m in the p_u-based solutions, in spite of a similar *G* of 442~493 kPa for either solution. Similar study on 14 model tests will be presented elsewhere.

12.5 E-E, EP-EP SOLUTIONS (DEEP SLIDING–FLEXIBLE PILES)

12.5.1 EP-EP solutions (deep sliding)

Contrary to the normal sliding, the condition of $L_1 > 1.2L_{c1} + x_{p1}$ warrants a deep (soil-carrying piles) sliding mode to occur (see Figure 12.1e). Equation 12.6 may be used by neglecting rigid movement [i.e., $(L_1 - x_s)\theta_o \approx 0$] and loading eccentricity ($e_{o2} \approx 0$). The mode is characterized by $H_1(x_{p1}) = |-H_2(x_{p2})|$ and $w_s = w_{g1}(x_{p1}) + w_{g2}(x_{p2})$, which may be resolved to gain the slip depth x_{p1} and x_{p2} (thus H_2, w_{gi}, and θ_{gi}, etc.) for the w_s, thus the pile response. The force H_i and deflection w_{gi} are estimated using elastic-plastic (EP) solutions for sliding and stable layer [thus called EP (sliding layer)-EP (stable layer) solutions]. The input parameters resemble those for an active pile discussed in Chapter 9, this book.

12.5.2 Elastic (sliding layer)–elastic (stable layer) (E-E) solution

Assuming an elastic pile–soil interaction simulated by load transfer model, the deflection w_{Bi} at depth z_i for an infinitely long pile in i^{th} layer with a constant k_i (see Figure 12.1e) (Guo 2006) is resolved as

$$w_{Bi}(z_i) = e^{-\alpha_i z_i}(C_{5i}\cos\beta_i z_i + C_{6i}\sin\beta_i z_i) \tag{12.33}$$

where

$$\alpha_i = \sqrt{\lambda_i^2 + N_{pi}/(4E_pI_p)} \quad \beta_i = \sqrt{\lambda_i^2 - N_{pi}/(4E_pI_p)} \tag{12.34}$$

The four equations of Equation 12.11 for sliding interface were expanded and resolved to yield the four constants C_{51}, C_{52}, C_{61}, and C_{62}, which are condensed into C_{5i} and C_{6i}:

$$C_{5i} = (-1)^i \frac{H}{4} \frac{2\alpha_i\alpha_j + 2\alpha_i^2 - \lambda_i^2 + \lambda_j^2}{\lambda_i^2(\alpha_i\lambda_j^2 + \alpha_j\lambda_i^2)E_pI_p} + \frac{\theta_o}{2} \frac{(2\alpha_i^2 - \lambda_i^2)\lambda_j^2}{\lambda_i^2(\alpha_i\lambda_j^2 + \alpha_j\lambda_i^2)} \tag{12.35}$$

$$C_{6i} = (-1)^i \frac{H}{4} \frac{\alpha_i(2\alpha_i^2 - 3\lambda_i^2 + \lambda_j^2) + 2\alpha_j(\alpha_i^2 - \lambda_i^2)}{\beta_i\lambda_i^2(\alpha_i\lambda_j^2 + \alpha_j\lambda_i^2)E_pI_p} + \frac{\theta_o}{2} \frac{(2\alpha_i^2 - 3\lambda_i^2)\alpha_i\lambda_j^2}{\beta_i\lambda_i^2(\alpha_i\lambda_j^2 + \alpha_j\lambda_i^2)} \tag{12.36}$$

where i = 1, j = 2 or i = 2, j = 1, respectively. Equation 12.33 implies the similarity of the profiles of shear force $Q_{Bi}(z_i)$, bending moment $M_{Bi}(z_i)$, and slope $\theta_{Bi}(z_i)$ in either layer to those of active piles (e.g., Guo 2006). This solution incorporates the coupled effect (with $N_{pi} \neq 0$) among different soil layers and is termed the "E-E (coupled)" solution. It reduces to the (uncoupled) E-E solution (Cai and Ugai 2003) assuming $N_{pi} = 0$ [thus $\lambda_i = \alpha_i = \beta_i$ from Equation 12.34]. Unfortunately, the input values of H_2 and θ_o for the E-E solutions are not directly related to soil movement w_s.

Example 12.3 Derivation of C_{5i} and C_{6i}

The rotation angle (slope), bending moment, and shear force are gained from the first, second, and third derivatives of Equation 12.33, respectively, as shown in Table 7.1 (Chapter 7, this book). The pile responses at the sliding interface (z_i = 0) satisfy the following conditions:

1. Slope condition of $-\theta_{g1} + \theta_o = -\theta_{g2}$. With $\theta_{gi} = -\theta_i(z_i)$, and $\theta_{gi} = -\alpha_i C_{5i} + \beta_i C_{6i}$, it offers

$$-(\alpha_1 C_{51} - \beta_1 C_{61}) + \theta_o = \alpha_2 C_{52} - \beta_2 C_{62} \tag{12.37}$$

2. Moment equilibrium of $M_{o1} = M_{o2}$. With $-w''_{Bi}(z_i) = M_{oi}/E_pI_p = (\alpha_i^2 - \beta_i^2)C_{5i} - 2\alpha_i\beta_i C_{6i}$, the equality becomes:

$$\alpha_1^2 C_{51} - 2\alpha_1\beta_1 C_{61} - \beta_1^2 C_{51} = \alpha_2^2 C_{52} - 2\alpha_2\beta_2 C_{62} - \beta_2^2 C_{52} \tag{12.38}$$

3. Force equilibrium. The equations of $-E_pI_p w'''_{B1}(0) = H_1 = H$, and $-E_pI_p w'''_{B2}(0) = H_2 = -H$ are written as

$$(\alpha_1^3 C_{51} - 3\alpha_1^2\beta_1 C_{61} - 3\alpha_1\beta_1^2 C_{61} + \beta_1^3 C_{61})E_pI_p = H \tag{12.39}$$

$$(\alpha_2^3 C_{52} - 3\alpha_2^2\beta_2 C_{62} - 3\alpha_2\beta_2^2 C_{62} + \beta_2^3 C_{62}) E_p I_p = H_2 = -H \tag{12.40}$$

Equations 12.37, 12.38, 12.39, and 12.40 were resolved together to obtain the four factors, which were then combined into C_{5i} and C_{6i} of Equations 12.35 and 12.36. The depth of maximum bending moment $z_{\max i}$ occurs at $Q_{Bi}(z_{\max i}) = 0$ and is given by

$$z_{\max i} = \frac{1}{\beta_i}\tan^{-1}\left[\frac{-\alpha_i(\alpha_i^2 - 3\beta_i^2)C_{5i} + (3\alpha_i^2 - \beta_i^2)\beta_i C_{6i}}{(3\alpha_i^2 - \beta_i^2)\beta_i C_{5i} + \alpha_i(\alpha_i^2 - 3\beta_i^2)C_{6i}}\right] \tag{12.41}$$

Using the Winkler model (assuming $N_{pi} = 0$, uncoupled), the C_{5i}, C_{6i}, and $z_{\max i}$ reduce to previous solutions, as shown in Example 12.4.

Example 12.4 C_{5i} and C_{6i} for E-E uncoupled solutions

Using the conventional Winkler model ($N_{pi} = 0$, $\alpha_i = \beta_i = \lambda_i$), the C_{5i} and C_{6i} of Equations 12.35 and 12.36 then reduce to

$$C_{5i} = (-1)^i \frac{H}{4}\frac{\lambda_i + \lambda_j}{\lambda_i^3 \lambda_j E_p I_p} + \frac{\theta_o}{2}\frac{\lambda_j}{\lambda_i(\lambda_i + \lambda_j)} \tag{12.42}$$

$$C_{6i} = (-1)^i \frac{H}{4}\frac{\lambda_j - \lambda_i}{\lambda_i^3 \lambda_j E_p I_p} - \frac{\theta_o}{2}\frac{\lambda_j}{\lambda_i(\lambda_i + \lambda_j)} \tag{12.43}$$

The expressions of C_{5i} and C_{6i} are essentially identical to those deduced previously (Cai and Ugai 2003). Accordingly, based on the elastic equations in Table 7.1 (see Chapter 7, this book), the rotation, bending moment, and shear force at depth z are respectively given by

$$\theta_{Bi}(z_i) = w'_{Bi}(z_i) = \lambda_i e^{-\lambda_i z_i}[(-C_{5i} + C_{6i})\cos\lambda_i z_i + (-C_{5i} - C_{6i})\sin\lambda_i z_i] \tag{12.44}$$

$$-M_{Bi}(z_i)/E_p I_p = w''_{Bi}(z_i) = 2\lambda_i^{\ 2} e^{-\lambda_i z_i}\{-C_{6i}\cos\lambda_i z_i + C_{5i}\sin\lambda_i z_i\} \tag{12.45}$$

$$\begin{aligned} -Q_{Bi}(z_i)/E_p I_p &= w'''_{Bi}(z_i) \\ &= 2\lambda_i^3 e^{-\lambda_i z_i}\{[C_{5i} + C_{6i}]\cos\lambda_i z_i - [C_{5i} - C_{6i}]\sin\lambda_i z_i\} \end{aligned} \tag{12.46}$$

The z_i is measured from sliding interface. Finally, the depth of $M_{\max i}$ is simplified to

$$z_{\max i} = \frac{1}{\lambda_i}\tan^{-1}\left[\frac{C_{5i} + C_{6i}}{C_{5i} - C_{6i}}\right] \tag{12.47}$$

In summary, the E-E (coupled) and P-EP solutions are underpinned by the compatible conditions of Equation 12.11 and $L_2 > L_{c2} + x_{p2}$; otherwise,

p_u-based solutions are applicable. The solutions were all entered into a program called GASMove operating in the mathematical software Mathcad™. The GASMove was used to gain numeric values presented subsequently.

12.6 DESIGN CHARTS

Slope stabilizing piles predominately exhibit the normal sliding mode. The piles may be designed initially by using the P-EP solution (assuming n_1 = 0, n_2 = 1.0, e_{oi} = 0, and ξ = 0 for nondragging case). Given a defined relative layer stiffness $(A_{L2}/A_{L1})/\lambda_2^{n2}$, for each normalized soil movement $\bar{w}_s$(= $w_s k_2 \lambda_2^{n_2} / A_{L2}$), Equations 12.8 and 12.20 are resolved to find $\bar{x}_{p2}$, and thereafter the displacement ratio w_s/w_{g2} via Equation 12.9, the normalized load $\bar{H}_2$ via Equation 12.20, moment $\bar{M}_{max2}$ [$=M_{max2}\lambda_2^{n_2+2}/A_{L2}$ via Equation 12.23 or 12.24], and slope $\bar{\theta}_{g2}$ via Equation 12.10. The calculation is repeated for a series of $\bar{w}_s$ to gain the nonlinear response for each of the six layer stiffnesses of 1, 2, 4, 6, 8, or 10. They are plotted in Figure 12.9a, b, c, and d as the curves of $\bar{w}_s$~w_s/w_{g2}, $\bar{H}_2$~w_s/w_{g2}, $-\bar{\theta}_{g2}$~w_s/w_{g2}, and $\bar{M}_{max2}$~w_s/w_{g2}, respectively. These figures may overestimate the w_{g2} and H_2 at sliding level without the dragging impact (e_{o2} = 0), but they are on safe side, and may be employed to calculate the pile response for a known w_s. The use of n_1 = 0 has limited impact on the overall prediction, but it allows the shape of measured moment profiles to be well modeled.

Example 12.5 A design example using P-EP solution

Response of the pile in Example 12.1 is predicted using the design charts for ξ = 0. With E_pI_p = 360 MNm^2, k_i = 8MPa, and w_s = 110 mm, it follows λ_i = 0.273 [= $(8/4 \times 360)^{0.25}$], $(A_{L2}/A_{L1}) \times \lambda_2^{(n1-n2)}$ = 2.01, and $\bar{w}_s$ = 4.62 (= 0.11 × 8,000 × 0.273/52).

The layer stiffness and $\bar{w}_s$ offer w_s/w_{g2} = 1.836 (see Figure 12.9a), and thus $\bar{H}_2$ = 0.49 (Figure 12.9b), $\bar{\theta}_{g2}$= −2.136 (Figure 12.9c), and $\bar{M}_{max2}$ = 0.32 (Figure 12.9d). Consequently, at w_s of 110 mm, the pile responses are H = 342 kN, w_{g2} = 59.9 mm, M_{max2} = 816.8 kNm, and θ_{g2} = −0.0139, which are ~15% larger than 316.15 kN, 52.1 mm, 739.0 kNm, and θ_{g2} = −0.012, obtained using ξ = 0.5 and GASMove, and on the safe side. The accuracy of ξ is not critical, which is noted later in all other cases.

The response profiles are obtained using P-EP solutions (Chapter 8, this book), with θ_o = 0.012, θ_{g2} = −0.012, H_1 = 316.15 kN, A_{L1} = 94.8 kPa (thrust), and ξ = 0.5 (resistance). The depth x_s is calculated as 2.78 m using Equation 12.7, and the M_{o1} and the e_{o2} as 253.07 kNm and 0.93 m using Equations 12.15 and 12.16, respectively. The bending moment and shear force profiles in sliding layer are obtained using Equations 12.13 and 12.14. The pile movement is taken as w_{g2} + $\theta_{g2}(L_1\text{-}x)$, as is

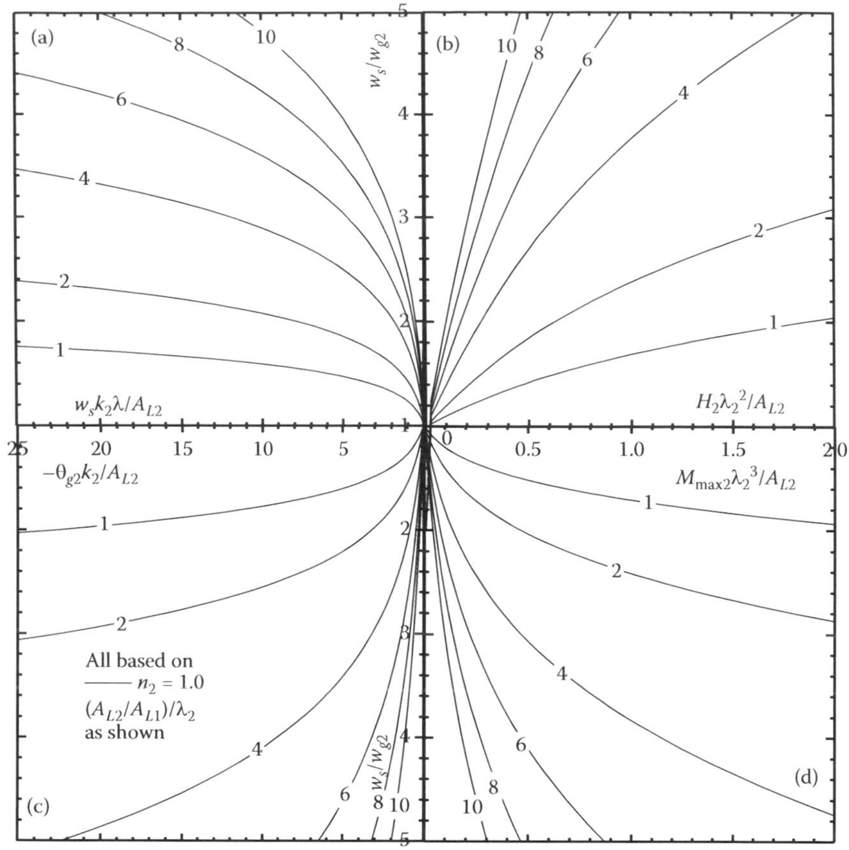

Figure 12.9 Normalized pile response in stable layer owing to soil movement w_s (n_2 = 1, n_1 = 0, e_{o2} = 0, ξ = 0). (a) Soil movement. (b) Load. (c) Slope. (d) Maximum moment. (After Guo, W. D., *Int. J. of Geomechanics*, in press.)

adopted in the GASMove prediction (see Figure 12.5b). With n_2 = 1 and A_{L2} = 52 kPa, the slip x_{p2} was estimated as 2.96 m using Equation 12.20. Pertinent expressions in Table 9.1 (Chapter 9, this book) may be estimated to give $F(2,0) = F(1,0) = 0$, $w''_{p2} = 1.976 \times 10^{-3}$, $w'''_{p2} = 2.443 \times 10^{-4}$, and thereby

$$w_{B2}(z_2) = e^{-\lambda_2 z_2}(0.01926 \cos\lambda_2 z_2 - 0.01325 \sin\lambda_2 z_2) \qquad (12.48)$$

$$-M_{B2}(z_2) = E_p I_p e^{-\lambda_2 z_2}[1.976\times10^{-3} \cos\lambda_2 z_2 + 2.871\times10^{-3} \sin\lambda_2 z_2] \qquad (12.49)$$

$$-M_{A2}(x_2) = \left[-\frac{x_2^{n_2+2}}{(n_2+2)(n_2+1)} A_{L2} + H_2 x_2\right] + M_{o2} \qquad (12.50)$$

These equations offer $w_{B2}(0) = w_{g2} = 52.1$ mm at $z_2 = 0$ of the stable layer; and $M_{max2} = -739.04$ kNm that occurred at 3.631 m (= x_{p2} of 2.963 + z_{max2} of 0.669). The bending moment (also shear force) profile obtained is subsequently shifted upwards by replacing x_2 with $x_2 - L_1 + e_{o2}$ (stable layer, $z_2 = x_2 - x_{p2}$). This results in smooth bending moment and shear force profiles (across the sliding interface).

12.7 CASE STUDY

The E-E and P-EP solutions were utilized to study seven instrumented piles (i.e., Cases 12.2 through 12.8). The pile and soil properties are provided in Table 12.1, including the outside diameter d, wall thickness t, Young's modulus E_p, thicknesses of sliding layer L_1/stable layer L_2, and SPT blow counts N_i and/or undrained shear strength s_{ui}. All tests provide the profiles of bending moment and deflection, but only Cases 12.6 and 12.7 (also Case 12.1 studied earlier) furnish the shear force profiles. They are plotted in Figures 12.10 through 12.12. The critical lengths L_{c1}/L_{c2} were calculated using $k_i = 3G_i$. The angle θ_o and the thrust H were deduced using the E-E solution and the measured response. They are tabulated in Table 12.1 as well. The input parameters A_{Li}, k_i, w_s, and ξ ($n_1 = 0$, and $n_2 = 1$) for the P-EP solution are given in Table 12.6, together with the calculated values of e_{o2}, x_s, x_{p2}, H_2, θ_o, w_{g2}, λ_2, ξ_{min}, and ξ_{max}. An example of a calculation using p_u-based solutions and H-l solutions is presented in Figure 12.13 for an equivalent rigid pile. Each case is briefly described next.

Example 12.6 Sliding depth, long/short piles, and θ_o (Cases 12.2 through 12.5)

Steel pipe piles (termed as Cases 12.2 through 12.5) used in stabilizing Hataosi and Kamimoku Landslide all have $d = 318.5$ mm, $t = 6.9$ mm, $E_P = 210$ GPa, and respective measured values of H_2 and θ_o (Cai and Ugai 2003). Using $k_i = 0.64N_i$ and $A_{Li} = (3.4\text{\textasciitilde}8.5)N_i$ kN/m (see Table 12.6), the P-EP predictions were made as is shown next.

The Hataosi Landslide adopts two rows of the piles (at a center-center spacing of 12.5d) installed to 24 m deep, to stabilize the active slide that occurred at $L_1 = 11.2$ m (Case 12.2) and to 17 m (= L) at another location with the slide at $L_1 = 8$m (Case 12.3). The predicted values of $H_2 = 139.1$ kN and $\theta_o = 0.029$ (Case 12.2) (at a uniform w_s of 153 mm) agree well with the measured 150 kN and 0.026, respectively. The predicted H_2 of 69.6 kN and θ_o of 0.0085 (at $w_s = 27$ mm, Case 12.3) again compare well with the measured 70 kN and 0.004, respectively.

The Kamimoku Landslide utilizes piles with column spacing of 4 m and row spacing of 2 m. They were installed to 14 m deep to arrest the slide at $L_1 = 6.5$m at one location of Kamimoku-4 (Case 12.4).

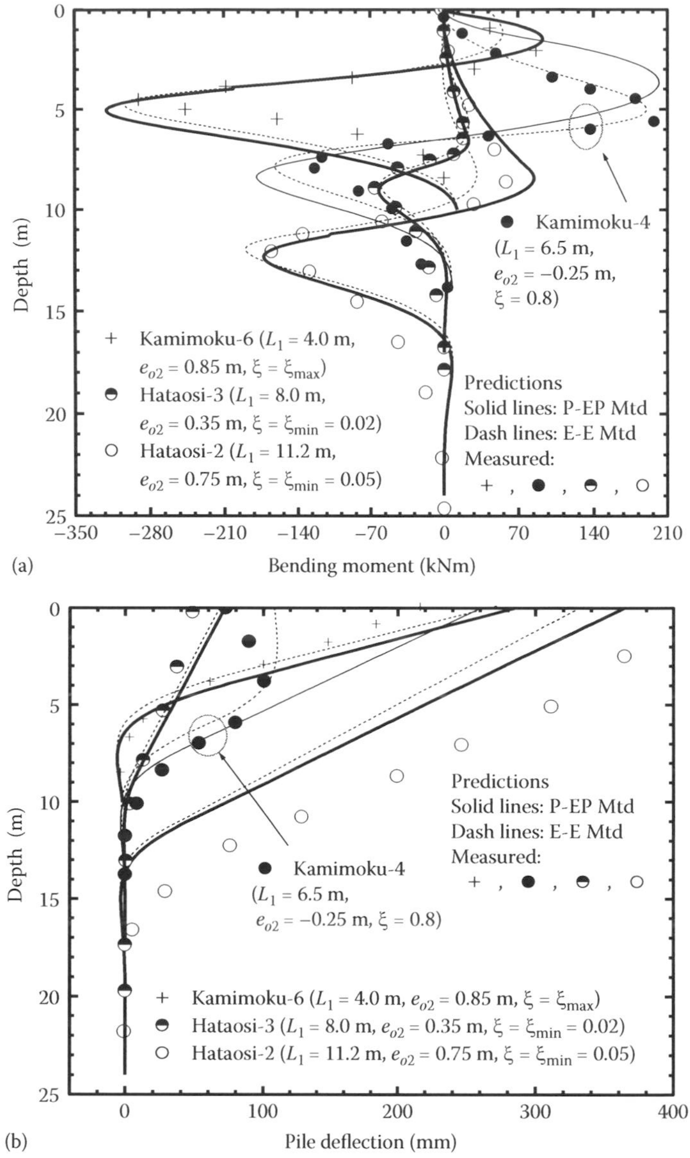

Figure 12.10 Predicted versus measured pile responses: (a) moment and (b) deflection for Cases 12.2, 12.3, 12.4, and 12.5. (After Guo, W. D., *Int. J. of Geomechanics*, in press.)

The predicted H_2 of 123.3 kN at w_s = 150 mm agrees with 139.1 kN noted in the similar Case 12.2, otherwise the H_2 is "overestimated" as 300 kN using an "abnormal" negative θ_o (Cai and Ugai 2003). At another location of Kamimoku-6 (Case 12.5), the piles were installed to a depth of 10 m to arrest the sliding at L_1 = 4 m. The predicted H_2

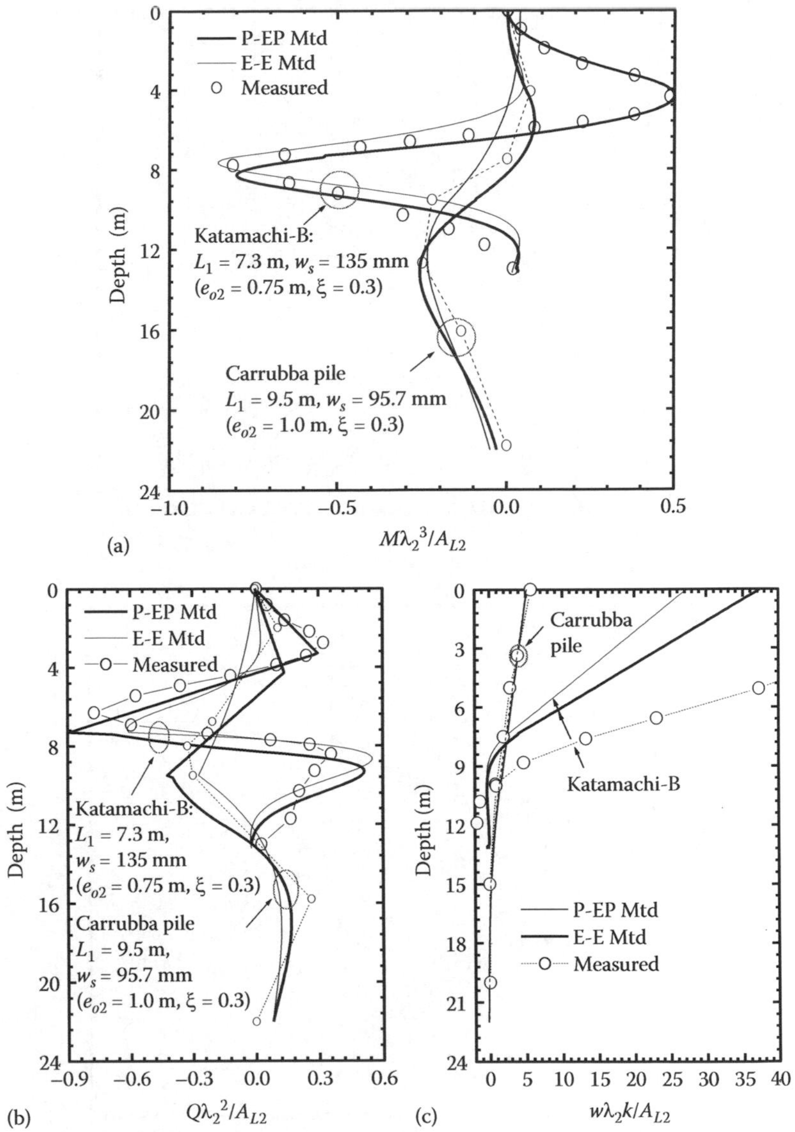

Figure 12.11 Predicted (using $\theta_o = -\theta_{g2}$) versus measured normalized responses (Cases 12.6 and 12.7). (a) Bending moment (b) Shear force. (c) Deflection. (After Guo, W. D., *Int. J. of Geomechanics*, in press.)

of 253.7 kN and θ_o of 0.054 at w_s = 320 mm match well with the measured 250 kN and 0.04, respectively, despite the pile being classified as short in both layers.

The predicted bending moment and deflection profiles for each case are plotted in Figure 12.10a and b, respectively. Note rotation

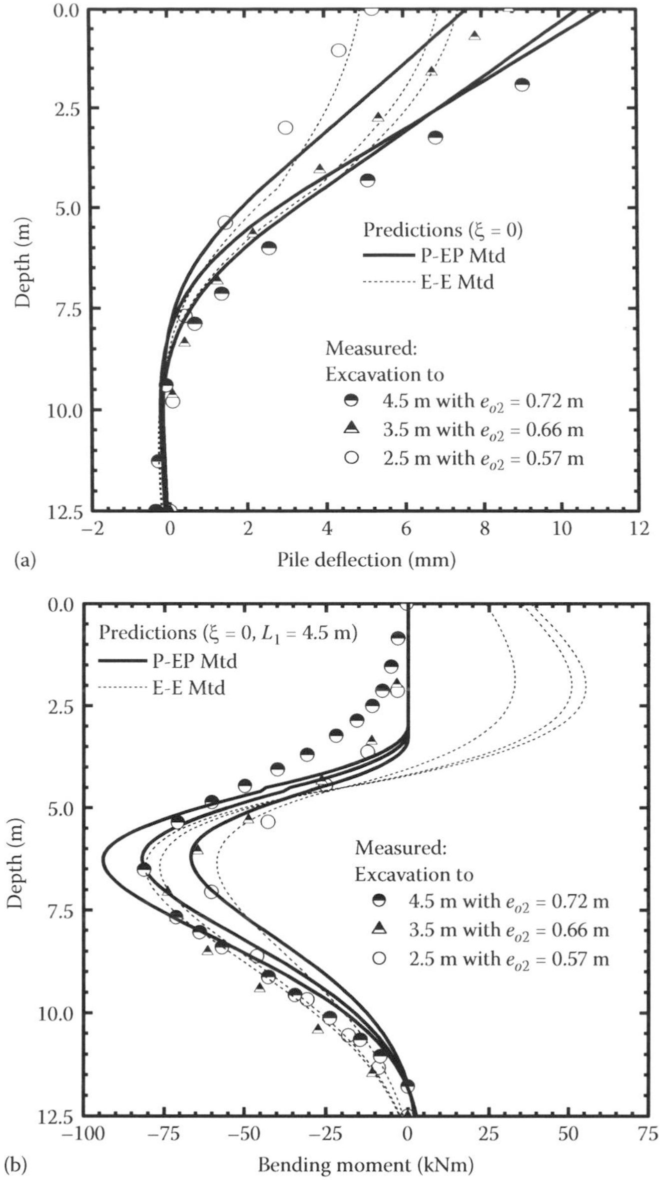

Figure 12.12 Predicted versus measured pile response during excavation (Leung et al. 2000) (Case 12.8): (a) deflection and (b) bending moment at 2.5–4.5m. (After Guo, W. D., *Int. J. of Geomechanics*, in press.)

Table 12.6 P-EP Solutions for Cases 12.1 through 12.9 ($H = H_1 = -H_2$, $n_1/n_2 = 0/1.0$)

Input Data					*Output*			*Meas*	*Case*
A_{L1} / A_{L2} (kN/m^{ni+1})	k_1 / k_2 (MPa)	w_s *(mm)* / ξ	e_{o2}[a] *(m)*	x_s / x_{p2} *(m)*	*H(kN)*[b] / $\theta_o(\times 10^{-3})$	w_{g2}*(mm)*[b] / λ_2	ξ_{min} / ξ_{max}	M_{max2}*(kNm)* / H_2*(kN)*	
94.8	2.5[c]	110	0.78	2.78	316.2	52.1	0.105	903	12.1
52.0	8.0	0.50	0.93	2.96	12.3	0.273	9.556	310	
50.2	5.0	153	0.75	8.03	139.1	61.0	0.05	165.2	12.2
80.4	8.0	0.05	0.77	1.73	29.0	0.584	24.45	--	
50.2	5.0	27.0	0.35	6.45	69.6	14.1	0.022	65.7	12.3
80.4	15.0	0.02	0.42	1.07	8.47	0.683	44.85	--	
50.2	5.0	150	−0.25	2.25	123.3	49.0	0.081	123.2	12.4
80.4	8.0	0.8	−0.24	1.56	24.0	0.584	12.31	--	
100.3	5.0	320	0.85	0.22	253.7	116	0.176	290.3	12.5[d]
140.5	8.0	5.67	0.91	1.79	54.4	0.584	5.67	--	
19.8	6.0	135	0.75	3.32	59.1	42	0.09	69.0	12.6
39.6	10.0	0.3	0.91	1.82	23.4	0.775	10.88	51.0	
198	15	95.7	1.0	4.28	778.7	49.2	0.09	2200	12.7
79.2	15	0.3	0.87	3.05	8.9	0.207	10.67	1460	
37.8	28.8	4.5	0.57	3.30	45.2	3.1	0.05	60.0	12.8
18.9	28.8	0.0	0.60	1.83	1.15	0.425	21.64	55.0	
		6.0	0.66	3.30	53.1	4.1	0.06	75.0	12.8
		0.00	0.69	2.06	1.44	0.425	17.18	70.0	
		7.3	0.72	2.99	57.2	4.8	0.07	82.0	12.8
		0.0	0.76	2.23	1.68	0.425	14.79	83.9	

Source: Guo W. D., *Int J of Geomechanics*, in press. With permission of ASCE.

[a] e_{o2}, denominator and numerator calculated using Equations 12.16 and 12.17, respectively.
[b] $\theta_o = -\theta_{g1} - \theta_{g2}$, with $\theta_o = -\theta_{g2}$ for rigid rotation when $L_1 < 1.2L_{c1} + x_{p1}$.
[c] L_{c1}, L_{c2} = 10.35 and 8.0 m.
[d] Use of positive θ_o rather than −0.008 (see Table 12.1); short piles, elastic analysis is only approximate.

is excluded in the deflection profile for the deep sliding pile of the Kamimoku-4. Impact of sliding depth on the responses is observed from the figures.

Example 12.7 Diameter and plastic hinge (Cases 12.6 and 12.7)

During the Katamachai Landslide (Case 12.6), reinforced piles (with d = 300 mm, t = 60 mm, and L = 10 m) failed at L_1 = 7.3 m. The soil has s_{u1} = 30 kPa and N_2 (SPT) = 16.7, which give $k_1 = 200s_{u1}$, $k_2 = 0.6N_2$,

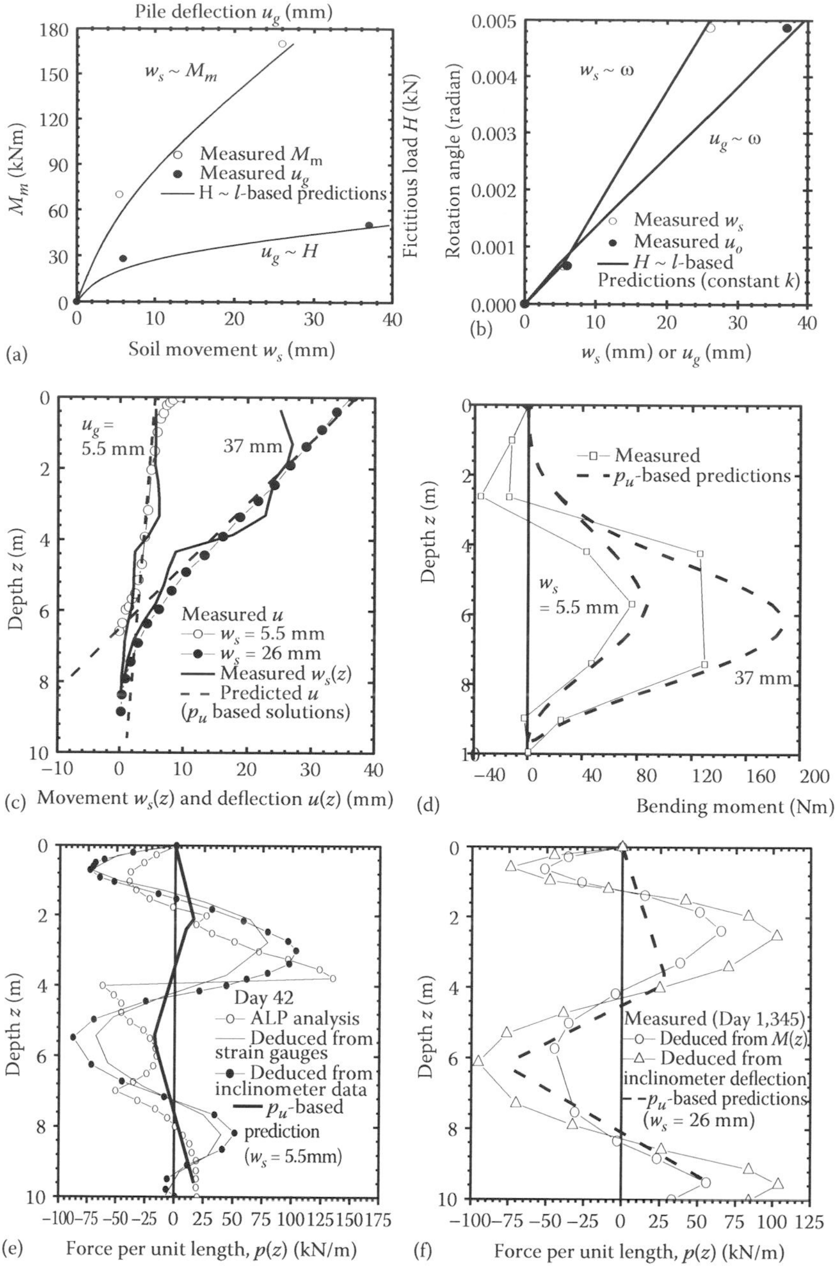

Figure 12.13 Current predictions versus measured data (Smethurst and Powrie 2007). (a) w_s~M_m and u_g~H. (b) w_s~ω and u_g~ω. (c) Soil movement w_s and pile deflection u(z). (d) Bending moment profile. (e) p profiles (day 42). (f) p profiles (day 1,345). (After Guo, W. D., *Int. J. of Geomechanics*, in press.)

$N_{g1} = 2.2$, and $A_{L2} = 2.37N_2$ kN/m^2. The P-EP solution thus predicts $H_2 = 59.1$ kN and $\theta_o = 0.0234$ at $w_s = 135$ mm, which are in close proximity to the measured 51 kN and 0.025, respectively, in spite of underestimating the pile deflection (see Figure 12.11). (Note some portion of rigid movement may exist in the measured pile deflection.) The predicted profiles of bending moment (M), shear force (Q), and pile deflection (w) are provided in nondimensional form in Figure 12.11 to compare with a large diameter (Case 12.7).

Carrubba et al. (1989) reported tests on a large diameter, reinforced concrete pile ($d = 1.2$ m and $L = 22$ m, Case 12.7) utilized to stabilize a slope with an undrained shear strength s_{ui} of 30 kPa (similar to Case 12.6). The pile had a plastic hinge at a depth of 12.5 m after a slide occurred at $L_1 = 9.5$ m and with a 2 m transition layer (see Figure 12.11). The P-EP prediction was made using (Chen and Poulos 1997). (1) $k_i \approx 15$ MPa ($= 500s_{ui}$); (2) $E_pI_p = 2{,}035.8$ MNm^2; (3) $A_{L1} =$ 198 kN/m ($N_{g1} = 5.5$) and $A_{L2} = 79.6$ kN/m^2 ($N_{g2} = 2.65$); and (4) a uniform w_s of 95 mm to a depth of 7.5 m. The pile observes normal sliding mode (see Table 12.6), and the P-EP prediction compares well with the measured response including the depth of the plastic hinge (at the M_{max}).

Figure 12.11 indicates the large diameter pile was subjected to much lower normalized values of bending moment, shear force, and deflection, which is more effective than the normal diameter pile.

Example 12.8 Piles: Retaining wall excavation (Case 12.8)

Leung et al. (2000) conducted centrifuge tests on a model single pile located 3 m behind a retaining wall at 50 g, to simulate a bored pile, 0.63 m in diameter, 12.5 m in length, and 220 MNm^2 in flexural stiffness E_pI_p. The pile was installed in a sand that has a unit weight of 15.78 kN/m^3, a relative density of 90%, an angle of internal friction of 43°, and an average E of 27 MPa (with Poisson's ratio $\nu_s =$ 0.4, $E = 6x$ MPa) over a sliding depth of 4.5 m (at which sand moves significantly).

The input parameters were gained as $A_{L1} = 37.8$ kN/m, $A_{L2} = 18.9$ kN/m^2, $\xi = 0$, and $L_1 = 4.5$ m (see Table 12.6). The P-EP predictions were thus made for $w_s = 4.5$, 6.0, or 7.3 mm to simulate the impact of excavation to a depth of 2.5, 3.5, or 4.5 m, respectively, on the pile. Values of the free-field w_s were determined using the measured w_s (e.g., a linear reduction from 14 mm at the surface to zero at a depth of 7.5 m, as recorded at 3 m from the wall following the excavation to 4.5 m).

The predicted and measured bending moment and deflection profiles agree well with each other for each excavation depth (see Figure 12.12). The slip depth x_{p2} reached 1.8~2.23 m, which indicates a limited impact of the p_{ui} (compared to the modulus k_i) on the prediction. The pile is rigid in either layer with $L_i < L_{ci}$ (= 23.1 m), which explains the divergence between the predictions and the measured data.

Example 12.9 Rigid pile subjected to trapezoid movement

Smethurst and Powrie (2007) observed the response of bored concrete piles used to stabilize a railway embankment. The piles were instrumented, with 10 m in length, 0.6 m in diameter, and a flexural rigidity E_pI_p of 10^5 kNm^2 (including the impact of cracking). They were installed at a center-to-center spacing of 2.4 m in a subsoil described as rockfill (depth 0~2 m with $\phi' = 35°$), embankment fill (depth 2~3.5m with $\phi' = 25°$), and weathered Weald clay (depth 3.5~4.5 m with $\phi' = 25°$); followed by intact Weald clay that extended well below the pile tip (depth 4.5~10m, with $\phi' = 30°$).

The subsoil around the pile moved laterally 7 mm on day 42, and 25 mm on day 1,345, uniform to a depth of 3.5 m, and then approximately linearly reduced to zero at around 7.5 m. This trapezoid w_s profile induces the following response of Pile C as measured in situ: by day 42, pile-head movement u_g^{all} ($\approx w_s$), maximum moment M_m (at a depth z_m of 5.6 m), and head rotation angle ω were 6–8 mm, 81.8 kNm, and 0.67×10^{-4}, respectively; and by day 1,345, it follows u_g^{all} = 35–38 mm, M_m = 170.7 kNm at z_m = 5.8 m; and $\omega = 48.7 \times 10^{-4}$. M_m was 70 kNm at w_s = 5.5 mm.

Maximum thrust (sliding force) H may be gained from slope analysis and then used to estimate the A_r by stipulating a linearly distributed net thrust over a sliding depth, or a uniform thrust on piles in clay (Viggiani 1981). A laterally loaded, flexible pile or an embankment pile may be analyzed as rigid (Sastry and Meyerhof 1994; Stewart et al. 1994), with an equivalent length $L_e = 2.1r_o(E_p/G)^{0.25}$ ($\leq l$). Assuming a Gibson p_u, the force satisfies H > $A_rd^2z_o^2/2$ [Figure 12.1c, i.e., $A_r < 2H/(d^2z_o^2)$]. Instead of using slope analysis, the H (see Table 12.4) was deduced here as 28 kN using the measured M_m of 70 kNm at w_s = 5.5 mm and e = 0 in Equation 12T7 (see Table 12.3). This H along with d = 0.6 m and z_o = 3.5 m allow a net pressure A_r of less than 12.7 kPa [$= 2 \times 28/(0.6^2 \times 3.5^2)$] (or A_rd = 7.62 kPa/m) to be obtained, although large pressures may be deduced on the front or the back pile surface.

More accurately, using the H-l–based solutions, the measured M_m at w_s = 5.5 mm allows the A_r for a Gibson p_u profile to be iteratively deduced as 6.61 kPa/m, and the k_o (overall, Gibson k) as 1.552 MPa, and further with a measured u_g of 4.3 mm, the k (local, constant k) was deduced as 9.31 MPa. These resemble the analysis on the AS50-0 test as:

1. Overall sliding: The H of 28 kN offers a ratio $H/(A_rdl^2)$ of 0.0705. The z_o/l was obtained as 0.169 using Equation 12T10, and u_gk_o/A_r = 1.291 using Equation 12T11. With the measured $u_g^{all} \approx w_s$ (= 5.5 mm), the k_o was deduced as 1.552 MPa (= $1.291A_r/u_g$), which offers k_od = 930.9 kPa/m and $\tilde{G}$ = 1.711 MPa (see Table 12.4).
2. Local interaction: The $H/(A_rdl^2)$ of 0.0705 (at w_s = 5.5 mm) corresponds to a new ratio z_o/l of 0.328 as gained using the "right" Equation 8.1g in Table 8.1 (Chapter 8, this book) (constant k and Gibson p_u). As a result, the $u_gk/(A_rl)$ and ω were calculated

as 0.62477 and 6.42×10^{-4}, respectively, using Equations 8.2g and 8.3g. With an equivalent E_p of 1.572×10^7 kPa and G of 1.71 MPa, the equivalent rigid length in either layer was calculated as 6.17 m [$= 2.1 \times 0.3 \times (1.572 \times 10^7/1{,}710)^{0.25}$]. The total rigid pile length l was 9.67 m (i.e., 3.5 m in sliding layer and 6.17 m in stable layer). The measured u_g of 4.3 mm and $u_g = 0.62477lA_r/k$, together with $l = 9.67$ m, allow the k(local) to be deduced as 9.31 MPa/m (kd = 5.586 MPa and G = 2.052 MPa). Note the use of the smaller $\tilde{G}$= 1.171 MPa for gaining the equivalent length is justified.

Slope stability analysis (Smethurst and Powrie 2007) indicates that a concentrated force of 60 kN is required to arrest the failing embankment by a factor of safety of 1.3. With the deduced A_r and k_o (or k), a typical maximum shear force Q_m (= H) of 50.5 kN (< 60 kN) would induce a bending moment M_m of 170.7 kNm at w_s = 26 mm (H-l solutions, overall sliding). The corresponding z_o/l (local, Table 12.4) is 0.684, thus u_g = 36.3 mm and ω = 0.0046. Likewise, a set of forces H (< 60 kN) were selected to gain M_m, w_s, u_g, and ω, which yielded the nonlinear w_s~M_m and u_g~H curves depicted in Figure 12.13a and the w_s~ω and u_g~ω curves in Figure 12.13b. The curves agree well with the measured data on day 42 and day 1,345 (see also Table 12.4), respectively.

The p_u-based solutions were obtained in the same manner as that for test AS50-0. To determine p_u and k_ϕ, a similar pile subjected to lateral spreading (Dobry et al. 2003) is compared here. The pile with l = 8 m and d = 0.6 m was embedded in "sand" with ϕ' = 34.5°, and c = 5.1 kPa. It was well modeled using a uniform p_u = 10.45 kN/m, and k_ϕ = 5738 kNm/rad via simple expressions. Thereby, the p_u based solutions for the embankment pile was obtained by assuming A_r = 12.7 kPa (i.e., 1.22 times 10.45 kPa) and k_ϕ = 7447.2 kNm/rad (1.298 times 5,738 kNm/rad). For w_s = 5.5 mm and 26 mm, the predicted profiles of deflection, bending moment, and on-pile force per unit length (at day 42 and day 1,345) are plotted in Figure 12.13c, d, e, and f, respectively. They agree well with the measured data, but for the net on-pile force p (Figure 12.13e). At the low w_s of 5.5 mm, the p is sensitive to pile flexibility and constantly underestimated against the measured data, whereas at the large w_s of 26 mm, it becomes nearly independent of the flexibility, and it is thus well predicted (see Figure 12.13f). The predicted maximum M_m and shear force Q_m are 87.3 kNm and 26.16~31.32 kN (w_s = 5.5 mm) and 187.77 kNm and 66.17~74.20 kN (w_s = 26 mm), respectively, which yield a ratio of $M_m/(Q_m l)$ of 0.29~0.35 and 0.26~0.29, individually. They agree well with 0.15~0.39 deduced from eight in situ piles (Guo and Qin 2010).

The embankment pile has an equivalent slenderness ratio l/d of 16 (= 9.67/0.6) and was subjected to a sliding depth ratio L_m/l of 0.362~0.776 (= 3.5~7.5/9.67), which are comparable to the UD tests on d_{50} model piles (see Chapter 13, this book) with l/d = 14 and L_m/l = 0.571. The current $G/\tilde{G}$ of 1.752 (= 2.052/1.71) and $M_m/(Hl)$ of 0.262~0.345 are also comparable to 1.13~1.81 (= 37~59/32.6) and 0.357 of the model piles, respectively. The similarity in the response profiles between the current model piles and the in situ tested piles (Dobry et al. 2003; Smethurst and Powrie 2007; White et al. 2008) is anticipated and observed. The current approach should be useful to relevant design.

12.7.1 Summary of example study

Response of eight passive piles (flexible) was studied. The critical values of M_{max2}, H_2, λ_2, k_2, n_2 (see Table 12.6), and θ_{g2} (not shown here, but can be deduced from Table 12.6) were obtained for each case. They offer the pairs of $\bar{M}_{max2}$ versus $\bar{H}_2$ and $\bar{M}_{max2}$ versus $\bar{\phi}_{g2}$ as plotted in Figure 12.14a and b, respectively, which compare well with the P-EP predictions.

Tables 12.1 and 12.6 indicate slightly different thrust H and slope θ_o between the E-E and P-EP solution for all piles in normal sliding mode. The main features of the instrumented piles (Cases 12.1 through 12.6) are as follows:

1. Response of Case 12.4 pile exhibits deep sliding mode. It can be well predicted using the E-E solution (the only case with $M_{max1} > M_{max2}$).
2. All piles but Case 12.4 exhibit normal sliding mode, which shows:
   - rigid rotation in sliding layer (in Cases 12.2 and 12.6, see Tables 12.1 and 12.6), which legitimizes the new critical length of $1.2L_{c1}+x_{p1}$
   - the sufficient accuracy of the P-EP solutions (n_2 = 1.0) for predicting the moments and deflections (Figures 12.10~12.12) by using $\theta_o = -\theta_{g2}$ (see Table 12.6, but for the deflection in Case 12.6) and the feature of $M_{max1} < M_{max2}$
   - the variation of ξ with $\xi = \xi_{min}$ if $\xi_{max} - \xi_{min} > 20$ (flexible piles), $\xi = \xi_{max}$ if $\xi_{max} - \xi_{min} < 7$ ("rigid" piles in both layer), and $\xi = \xi_{min}$ ~ξ_{max} if $7 \le \xi \le 20$.

The values of $A_{L,i}$ and k_i (or G_i) may be determined in the same manner as those for active piles. The p_{ui} and k_i resulted may differ from those adopted in other suggestions (Stewart et al. 1994; Chen et al. 2002; Cai and Ugai 2003), but their impact is limited as long as they yield a similar total resistance over the slip depths as with laterally loaded piles.

12.8 CONCLUSION

New E-E and P-EP solutions are established for passive piles and presented in closed-form expressions. The P-EP solutions are complied into a program called GASMove and presented in nondimensional charts. Nonlinear

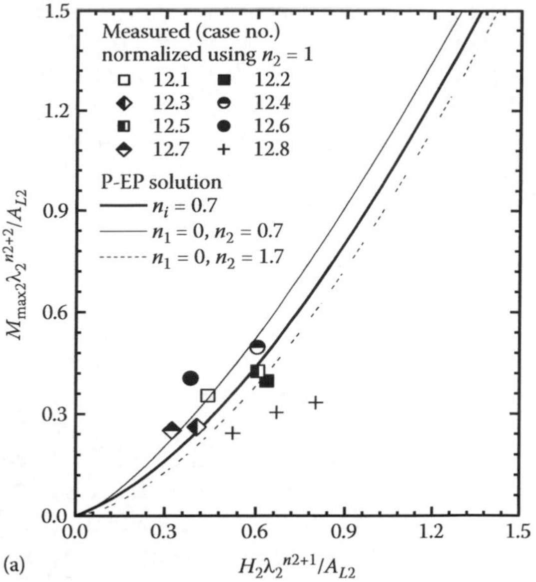

(a)

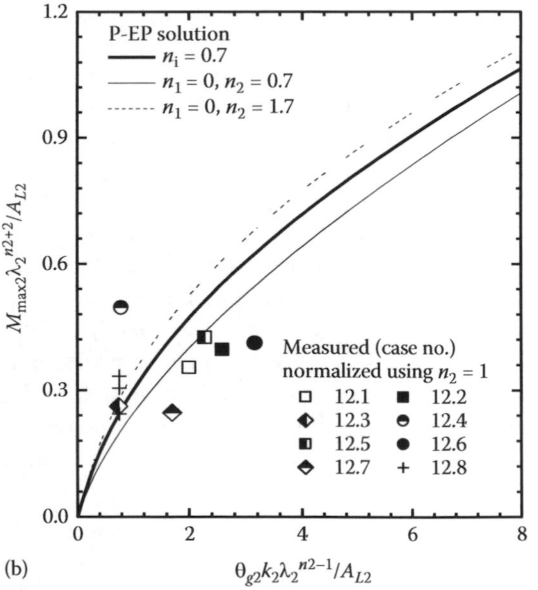

(b)

Figure 12.14 Predicted versus measured pile responses for all cases. (a) Normalized load versus moment. (b) Normalized rotation angle versus moment. (After Guo, W. D., *Int. J. of Geomechanics*, in press. With permission of ASCE.)

response of passive rigid piles exhibits overall sliding and local interaction. It shows new components of on-pile force profile, free-head and constraining rotations compared to active piles. New p_u-based solutions were developed to capture local interaction, and the *H-l* solutions are used to correlate the bending moment empirically with soil movement (overall interaction).

The solutions are used to study eight instrumented piles, which show the following:

1. The proposed H and w_s relationship of Equation 12.8 works well, although underpinned by a uniform soil movement profile. The obtained angle θ_o and shear force H are generally slightly different between the P-EP solution and the E-E solution, respectively.
2. The P-EP solution well captures the pile response under normal sliding mode (with $L_1 < L_{c1}$ and $L_2 > L_{c2}$); and the E-E solution works well for "deep sliding" mode, in light of input soil parameters G_i (for k_i), ξ, w_s, L_1, and A_{Li} (for p_{ui}). Similar predictions may be gained from different sets of p_u and k profiles, as with laterally loaded piles.
3. The solutions are readily evaluated using professional math programs (e.g., Mathcad™). The design charts allow nonlinear response to be hand-calculated.

The *H-l* and p_u-based solutions well capture measured nonlinear response (M_m, ω, and u_g) of 17 model piles and an in situ instrumented pile using the gradient A_r, moduli of subgrade reaction k_o (overall) and k (local), and rotational stiffness k_ϕ, irrespective of the arc, triangular, or uniform profiles of soil movement. Input values of A_r and k_o from overall sliding agree with other numerical studies on passive piles; whereas those of A_r, k, and k_ϕ for local interaction resemble those for laterally loaded piles. The impact of soil movement profiles on passive piles may be modeled by varying A_r and k. The current solutions are applicable to equivalent rigid piles in any soils exhibiting a linear p_u profile. They are sufficiently accurate, as indicated by the examples, for back analysis and design of slope stabilizing piles. Finally, to improve our understanding about passive piles, pile tests should provide both bending moment and shear force profiles for each magnitude of soil movement w_s.

Chapter 13

Physical modeling on passive piles

13.1 INTRODUCTION

Centrifuge tests (Stewart et al. 1994; Bransby and Springman 1997; Leung et al. 2000) and laboratory model tests (Poulos et al. 1995; Pan et al. 2002; Guo and Ghee 2004) were conducted to model responses of piles subjected to soil movement. The results are useful. The correlation between maximum bending moment (M_{max}) and lateral thrust (i.e., shear force, T_{max}) in a pile is needed in design (Poulos 1995), but it was not provided in majority of the model tests.

Theoretical correlation between M_{max} and T_{max} was developed using limit equilibrium solutions for piles in a two-layered cohesive soil (Viggiani 1981; Chmoulian 2004). It is also estimated using *p-y* curve–based methods in practice. However, the former is intended for clay, whereas the *p-y* method significantly overestimated pile deflection and bending moment (Frank and Pouget 2008). The correlation is also established using elastic (Fukuoka 1977; Cai and Ugai 2003) and elastic-plastic solutions (Guo 2009). The solutions nevertheless cannot capture the effect of soil movement profiles coupled with an axial load on pile response. The measured correlation from in situ slope stabilizing piles (Esu and D'Elia 1974; Fukuoka 1977; Carrubba et al. 1989; Kalteziotis et al. 1993; Smethurst and Powrie 2007; Frank and Pouget 2008) is very useful and valuable. However, only small-scale experiments can bring about valuable insight into pile–soil interaction mechanisms in an efficient and cost-effective manner (Abdoun et al. 2003).

Model tests were conducted under rotational soil movement (Poulos et al. 1995; Chen et al. 1997), translational-rotational movement (Ellis and Springman 2001), and translational movement (Guo and Qin 2010). As reviewed later, the 1*g* (*g* = gravity) rotational tests provide useful results of ground level deflection y_o, maximum bending moment M_{max}, and pile rotation angle ω for single piles and group piles. The translational-rotational centrifuge tests offer ultimate response for full-height piled bridge embankment under two typical shallow and deep sliding depths. The translational tests provide the impact of pile diameter, soil movement profiles (Guo and

Qin 2005), sliding depth, and axial load on response of single piles in sand and reveal 3~5 times increase in maximum bending moment from rotational to translational soil movement (Guo and Qin 2010).

This chapter presents 19 model tests on single piles (14 under an inverse triangular loading block, with 2 diameters, and 5 under uniform) and 11 tests on free-standing two-pile groups (5 under uniform and 6 under triangular loading block). They were conducted under translational movement for two axial load levels, a few typical pile center-center spacings, and sliding depths. The tests were studied to be establish: (1) relationships between maximum bending moment (M_{maxi}) and thrust (sliding force T_{maxi}) for single piles, and two-pile groups in sliding ($i = 1$) and stable ($i = 2$) layers for each frame (soil) movement w_f; (2) a restraining bending moment of pile cap (M_o) that renders capped piles to be analyzed as free-head single piles; and (3) a model of response of passive piles with effective soil movement (w_e).

The model tests provide profiles of bending moment, shear force, and deflection along the pile for each typical soil (frame) movement and, accordingly, maximum thrust (T_{max}) and maximum moment (M_{max}). The T_{max}~M_{max} plot allows the restraining moment M_o to be obtained. Theoretical correlations between M_{max} and T_{max} and between T_{max} and w_e are established and compared with 22 model pile tests. Pile-pile interaction is assessed using the subgrade modulus of the soil k and the M_{max} for piles in groups against pertinent tests on single piles and numerical results. The simple solutions (Guo and Qin 2010) were substantiated using tests under translational or rotational modes of soil movement. They are used to predict measured response of eight in situ test piles and one centrifuge test pile.

13.2 APPARATUS AND TEST PROCEDURES

13.2.1 Salient features of shear tests

The current model tests are described previously (Guo and Qin 2010) concerning single piles. Focusing on two-pile groups, some salient features include:

1. The apparatus (see Figure 13.1) mainly consists of a shear box and a loading system. The latter encompasses a lateral jack on the aluminum frames and vertical weights on the pile-cap. The jack pushes the rectangular (uniform) or triangular block to generate the translational soil movement profiles, and the weights give the desired vertical loading. The pile-cap is equipped with two LVDTs. Each of two typical piles (see Figure 13.1b) was instrumented with ten pairs of strain gauges along its shaft. Readings from the strain gauges and the LVDTs during a test are recorded via a data acquisition system, from which measured responses are deduced (see Section 13.2.4).

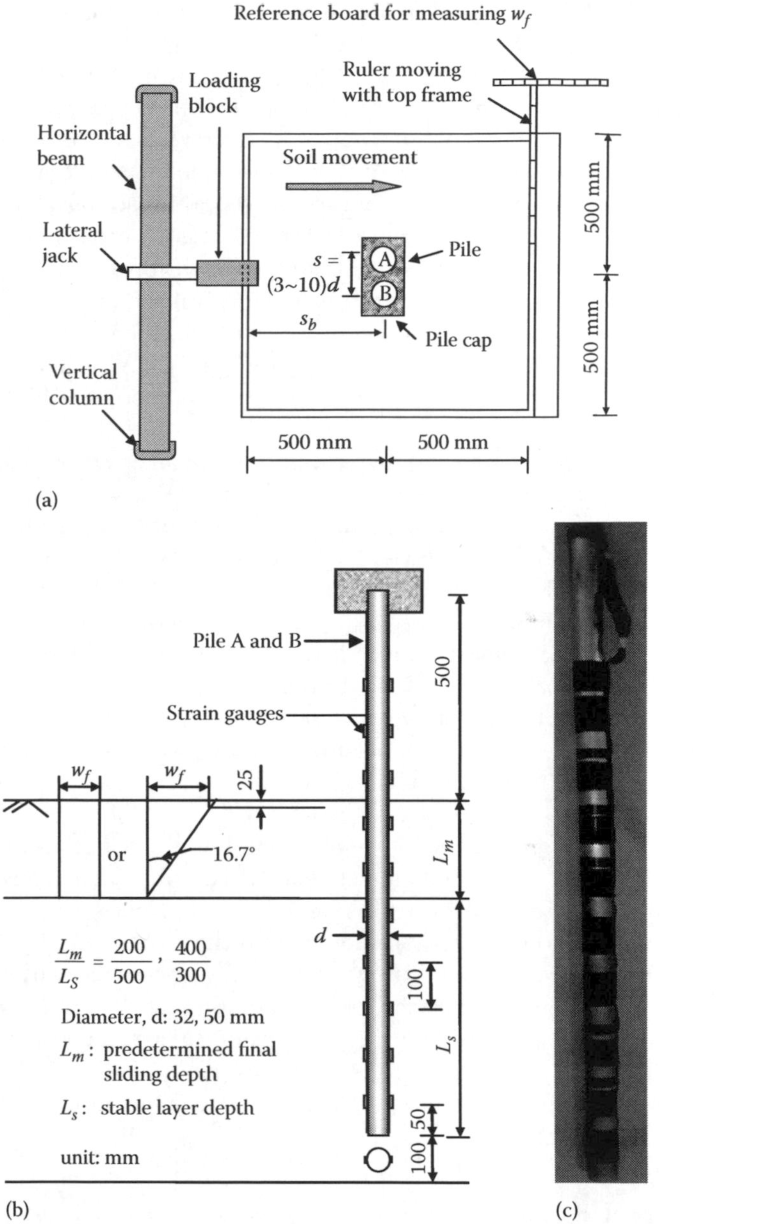

Figure 13.1 Schematic of shear apparatus for modeling pile groups. (a) Plan view. (b) Testing setup. (c) An instrumented model pile.

2. The shear box has internal dimensions of 1 × 1 m, and 0.8 m in height. Its upper portion is movable, consisting of a desired number of 25-mm-thick square laminar frames and is underlined by a fixed timber box 400 mm in height. The upper frames are forced to move by a triangular (with an angle of 16.7° from vertical line) or a rectangular loading block (see Figure 13.1b) to generate soil movement to a depth of L_m (≤ 400 mm). The movement profiles actually vary across the shear box and with depth at the loading location. For instance, the triangular block induces a triangular soil movement profile to a depth of $3.33w_f$ (w_f = frame movement) until the pre-selected depth L_m (e.g., a sliding depth L_m of 200 mm is reached at w_f = 60 mm). Further increase in w_f causes a trapezoidal soil movement profile. The movement rate is controlled by a hydraulic pump and a flow control valve.
3. The tests use oven-dried medium-grained quartz, Queensland sand, with an effective grain size D_{10} = 0.12 mm, D_{50} = 0.32 mm, a uniform coefficient C_u = 2.29, and a coefficient of curvature C_c = 1.15. The model sand ground was prepared using a sand rainer at a falling height of 600 mm, which has a dry density of 16.27 kN/m^3 (or a relative density index of 0.89), and an internal frictional angle of 38°.
4. The aluminum pipe piles tested were 1,200 mm in length and with two sizes of either (a) 32 mm in outer diameter (*d*), 1.5 mm in wall thickness (*t*), and a bending stiffness of 1.28 × 10^6 $kNmm^2$ (referred to as d32 piles), or (b) d = 50 mm, *t* = 2 mm and a stiffness of 5.89 × 10^6 $kNmm^2$ (referred to as d50 piles).

A pile cap was fabricated from a solid aluminum block of 50 mm thick; into which two *d* ± 0.5 mm diameter holes were drilled at a center to center spacing of 3*d*, 5*d*, 7*d*, or 10*d*. Two d32 or d50 piles were first socketed into a cap and then jacked together into the model sand in the shear box (see Figure 13.1a). The tests were generally conducted without axial load, but for some tests using the uniform loading block. An axial load of 294 N per pile was applied by using weights on the pile cap, or exerted at 500 mm above the sand surface for single piles (Figure 13.1b). The load was ~10% the maximum driven resistance of 3~4 *kN* (gained using the jack-in pressure) for a single pile.

13.2.2 Test program

A series of tests were conducted using the triangular or uniform loading block. Each test is denoted by two letters and two numbers, such as "TS32-0 and TD32-294": (1) the triangular loading block is signified as *T* (and U for uniform block); (2) the "S" and "D" refer to pre-selected sliding depths (L_m) of 200 mm and 350 mm, respectively (compared to SD = sliding depth used later); (3) the "32" indicates 32 mm in diameter; and (4) the "0" or "294" denotes without (0) or with an axial load of 294 N.

1. With the T block, 14 typical tests on single piles are presented. They are summarized in Table 13.1, which encompass three types of "Pile location," "Standard," and "Varying sliding depth" tests. The "Pile location" tests were carried out to investigate the effect of relative distance between the loading block and pile location. The "Standard" tests were performed to determine pile response to the two pre-selected final sliding depths of 200 mm and 350 mm; and the "Varying sliding depth" tests were done to gain bending moment raises owing to additional movement beyond the triangular profile.
2. Using the T block, five tests on groups of two piles in a row (see Table 13.2) were conducted. They consist of tests T2, T4, and T6 on the capped piles located at a loading distance s_b of 340, 500, or 660 mm from the initial position of loading block and with s/d = 3 (s = pile center-center spacing, d = 32 mm, outer diameter of the pile); and tests T8 and T10 on capped piles (centered at s_b = 500 mm) with s/d = 5 and 7. Two single pile tests of TS32-0 and TS32-294 at a s_b = 500 mm (Guo and Qin 2010) are used as reference.
3. Using the U block, the test program (see Table 13.3) includes six tests on two-pile in row groups (a) test U3 with s/d = 3 and without load, and test U4 with s/d = 3 and with a load of 596 N (some loading problem), respectively; (b) tests U5 and U6 without load, and at s/d = 5 and 10, respectively; and (c) tests U9 and U10 with s/d = 3 without load or with a load of 596 N, but at a new sliding depth of 400 mm, respectively. Two single d32 pile tests of US32-0 and US32-294 at s_b = 500 mm are used as reference, along with three single d50 piles of US50(250), 50(500), and 50(750) tested without load and located respectively at a distance s_b of 250, 500, or 750 mm from the initial (left side) boundary.

The loading frames were enforced to translate incrementally by 10 mm to a total top-frame movement w_f of 140 mm. It generally attained a final sliding depth (SD, L_m) of 200 mm (L_m/L = 0.286) for tests T2 through T10 (i.e., tests T2, T4, T6, T8, and T10) and tests U3 through U6, although tests U9 and U10 were conducted to a sliding depth L_m of 400 mm (L_m/L = 0.57). The tests are generally on d32 piles but for three single d50 piles. Excluding one problematic test U4 (explained later), 11 tests on capped pile groups were discussed here against 7 single piles.

13.2.3 Test procedure

To conduct a test, first the sample model ground was prepared in a way described previously to a depth of 800 mm. Second, the instrumented pile

Table 13.1 Summary of 14 typical pile tests (under triangular movements)

Test description	*Frame movement at ground surface* w_f *(mm)*	*Maximum bending moment* M_{max} *(kNmm)*		*Depth of* M_{max}, d_{max} *(mm)*	*Max shear force,* T_{max} *(N)*		*Pile deflection at groundline,* y_o *(mm)*	L_m/L_s *(mm)*	*Remarks*
		Tension side	*Compression side*		*Stable layer*	*Sliding layer*			
TS32-0 (s_b = 340)[a]	60/80	63.8/81.0		400	266.9/327.7	266.6/325.8	11.5/14.8	200/500	Pile location
TS32-0 (s_b = 660)	60/80	30.0/40.0		400	114.9/150.3	120.4/153.7	7.8/10.8	200/500	
TS32-0 (L_m = 200)[a]	60/70	39.3/49.7	−34.2/−45.0	370	147.2/183.8	159.8/201.1	7.1/10.3	200/500	Standard Tests (s_b = 500)
TS32-294	60/90	29.8/78.6	−26.8/−76.5	375	108.5/295.5	98.0/279.9	5.4/13.1	200/500	
TS50-0	60/80	45.8/89.2	−37.9/−80.2	380	191.9/363.9	180.3/355.7	2.9/7.2	200/500	
TS50-294	60/80	58.5/115.6	−59.2/−120.0	400	229.6/445.5	241.4/467.5	3.5/7.3	200/500	
TD32-0	120	119.5	−112.1	450	495.9	414.8	58.7	350/350	
TD32-294	120	124.6	−117.5	465	532.4	463.5	73.8	350/350	
TD50-0	120	93.2	−84.4	450	393.8	353.1	58.9	350/350	
TD50-294	120	143.0	−135.1	450	577.6	453.4	67.5	350/350	
T32-0 (L_m = 125)	40/60	5.2/5.7		325	18.9/18.2	22.8/22.5	0.5/0.6	125/575	Varying sliding depth (s_b = 500)
T32-0 (L_m = 250)	80/120	62.6/123.5		450	258.1/509.4	233.9/457.3	22.4/47.7	250/450	
T32-0 (L_m = 300)	100/150	115.3/175.0		450	450.6/675.2	399.4/619.6	25.1/54.8	300/400	
T32-0 (L_m = 350)	120/150	118.1/140.0		475	471.7/557.3	406.7/535.3	42.2/73.9	350/350	

Source: Guo, W. D. and H.Y. Qin, *Can Geotech J*, 47, 2, 2010.

[a] s_b and L_m in mm. Except for L_m/L_s, the "/" separates two measurements at the two specified values of w_f.

Table 13.2 Summary of test results on single piles and two piles in rows, triangular movement

Test description	Maximum bending moment M_{max} (kNmm)	Depth of M_{max}, d_{max} (mm)	Maximum shear force T_{max} (N)		Pile deflection at groundline y_0 (mm)	y_0 =	w_i(mm)/k (kPa)	Remarks
			Stable layer	Sliding layer				
TS32-0	49.7	370	183.8	201.1	10.3		37.0/34	Single pile
TS32-294	77.55		295.5	279.9	13.1		37.0/34	Single pile
T2	41.4	450	160.7	162.5	5.5	$0.3(w_f - 50)$	45.0/40	s_b = 340 (mm), s_v/d = 3
T4	31.0	450	124.4	120.3	6.3	$0.3(w_f - 50)$	45.0/30	s_b = 500 (mm), s_v/d = 3
T6	15.3	450	58.1	65.5	3.1	$0.12(w_f - 35)$	35.0/30	s_b = 660 (mm), s_v/d = 3
T8	50.0	350	198.2	205.5	11.8	$0.32(w_f - 35)$	30.0/35	s_b = 500 (mm), s_v/d = 5
T10	54.6	350	224.0	210.6	8.6	$0.23(w_f - 45)$	40.0/31	s_b = 500 (mm), s_v/d = 7

Note: Frame movement, w_f = 70 mm (T2–T8 tests), and 90 mm (T10), and L_m/L_s = 200 mm/500.

Table 13.3 Summary of test results on single piles and two piles in rows (uniform movement)

Test description	Maximum bending moment M_{max} (kNmm)	Depth of M_{max}, d_{max} (mm)	Maximum shear force T_{max} (N)		Pile deflection at groundline y_0 (mm)	y_o =	w_i(mm)/k (kPa)	Remarks
			Stable layer	Sliding layer				
US32-0	25.31~35.81[a]	460	92.43~133.39	91.47~131.28	6.67~9.2	$0.25w_f$	100	Single pile
US32-294	39.21~44.9	460	146.9~172.8	125.4~151.0	10.1~11.8	$0.8w_f$	60	Single pile
US50(250)	56.7~94.9	260	428.9~588	292.7~433	86.8~123			Single pile, s_b = 250 mm
US50(500)	57.88~67.7	360	294.6~327.2	216~233.19	2.47~2.52			Single pile, s_b = 500 mm
US50(750)	40.65~35.38	460	172.3~161.6	155~150.7	5.52~5.31			Single pile, s_b = 750 mm
U3 (32-0)	26.84~23.84	360	108.7~94.3	106.53~86.0	12.29	$0.5w_f$	/30	$s_v/d = 3$
U4(32-294)	5.31~6.20	360	16.84~22.06	31.59~28.57	12.32		/30	$s_v/d = 3$
U5 (32-0)	25.66~55.78	360	96.83~193.9	82.6~183.8	4.7~10.24			$s_v/d = 5$
U6 (32-0)	13.57~19.77	360	41.4~53.2	53.0~72.2	3.91~5.92			$s_v/d = 10$
U9(32-0)	26.41~24.18	460	118.8~93.0	193.~108.8	46.60	$0.81w_f$	/30	$s/d = 3$
U10 (32-294)	33.63~21.2	460	152.73~97.27	116.9~131.1	31.90	$0.71(w_f-10)$	/30	$s_v/d = 3$

[a] at frame movement, w_f = 60~140mm, w_i = 0~10mm and L_m/L_s = 200 mm/500 mm(US) for all tests, but U9 and U10 with 400mm/300mm.

was jacked in continuously to a depth of 700 mm below the surface, while the (driving) resistance was monitored. Third, an axial load was applied on the pile-head using a number of weights (to simulate a free-head pile condition) that were secured (by a sling) at 500 mm above the soil surface. Fourth, the lateral force was applied via the T (or U) block on the movable frames to enforce translational soil movement towards the pile. Finally, the sand was emptied from the shear box after each test. During the passive loading, the gauge readings, LVDT readings, and the lateral force on the frames were taken at every 10 mm translation of the top steel frame to a total frame movement w_f of ~150 mm. Trial tests prove the repeatability and consistency of results presented here.

13.2.4 Determining pile response

The data recorded from the 10~20 strain gauges and the two LVDTs during each test were converted to the measured pile response via a purposely designed spreadsheet program (Qin 2010). In particular, the first- and second-order numerical integration (trapezoidal rule) of a bending moment profile along a pile offers rotation (inclination) and deflection profiles. Single and double numerical differentiations (finite difference method) of the moment profiles furnish profiles of shear force and soil reaction. The program thus offers five "measured" profiles of bending moment, shear force, soil reaction, rotation, and deflection for each typical frame movement w_f. The profiles provide the maximum shear force T_{max} (i.e., thrust), the bending moment M_{max}, depth of the moment d_{max}, the pile-rotation angle ω, and the pile deflection y_o at model ground level. Typical values are furnished for single piles in Table 13.1 and for two-pile groups in Table 13.2 for tests T2 through T10 and in Table 13.3 for tests U3 through U6, U9, and U10.

13.2.5 Impact of loading distance on test results

The current test apparatus allows non-uniform mobilization of soil movement across the shear box. The associate impact is examined by the "pile location" tests via the distance s_b (see Figure 13.1) and in terms of the measured maximum bending moments and maximum shear forces. The d32 pile was installed at a loading distance s_b (see Figure 13.1a) of 340 mm, 500 mm (center of the box), or 660 mm from the loading jack side. The loading block was driven at the pre-specified final sliding depth of 200 mm. The measured data indicate a reduction in M_{max} of ~32 Nm (at w_f = 70~80 mm) as the pile was relocated from s_b = 340 mm to 500 mm (center), and a further reduction of ~10 Nm from the center to s_b of 660 mm. The reductions are 20~70% compared to a total maximum moment of 45~50 Nm for the pile tested at the center of the box.

The normalized moments and forces at s_b = 340, 500, and 660 mm (T block) (Qin 2010) are plotted in Figure 13.2, together with those at

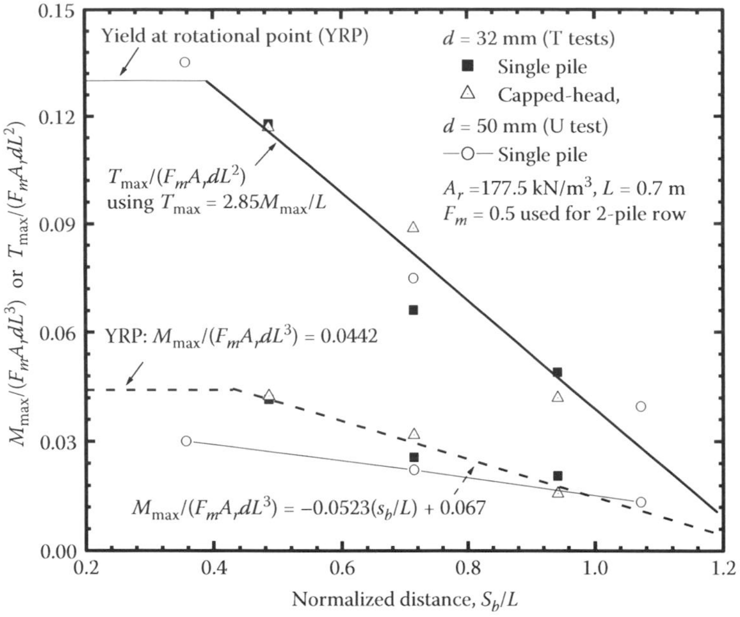

Figure 13.2 Impact of distance of piles from loading side s_b on normalized M_{max} and T_{max}.

s_b = 250, 500, and 750 mm (U block), in which the parameters used are the gradient of the limiting force A_r (see Chapter 8, this book), pile embedment depth L, and the pile-pile interaction factor F_m. The moment decreases proportionally with the distance s_b and may be described by $M_m/(F_mA_rdL^3)$ = $0.067 - 0.0523(s_b/L)$ (T block) or = $0.0386 - 0.0234(s_b/L)$ (U block), which is subjected to $0 \le M_m/(F_mA_rdL^3) \le 0.0442$ and $F_m = 0.5$. The values of F_m and A_r are explained in Section 13.4.4. The shear force variation for all tests may be described by $T_{ult}/(F_mA_rdL^2) = 0.188 - 0.149(s_b/L)$ that is bracketed by $0 \le T_{ult}/(F_mA_rdL^2) \le 0.13$ (T_{ult} = ultimate T_{max}). The upper limits of 0.0442 and 0.13 are gained using solutions for lateral piles at "yield at rotation point" (Chapters 3 or 8, this book). A normalized distance of $s_b/L > 1.29$ (T block) or 1.65 (U block) would render the moment and thrust negligibly small. The results presented for two-pile groups (d = 32 mm) are generally centered at s_b = 500 mm and SD = 200 mm, except where specified.

13.3 TEST RESULTS

13.3.1 Driving resistance and lateral force on frames

The total jack-in forces were monitored during the installation of six single piles and three two-pile groups using a mechanical pressure gauge attached to the hydraulic pump (Ghee 2009). As shown in Figure 13.3, the driving force on the single piles increases with the penetration. At the final penetration of 700 mm, the average total forces of the same diameter piles reach 5.4 kN (d = 50 mm piles) and 3.8 kN (d = 32 mm piles), respectively, with a variation of ~ ±20%. Note the axial load of 294 N on the pile head (used later) is 7~9% of this final jacking resistance. These results reflect possible variations in model ground properties, as the jack-in procedure was consistent. The associated average shaft friction was estimated as 54 kPa (d = 32 mm) and 49.1 kPa (d = 50mm), respectively, if the end-resistances was neglected on the open-ended piles. The average forces per pile obtained for the two-pile rows at s/d = 3, 5, and 10 are plotted in Figure 13.3 as well. They are quite consistent with the single piles, indicating consistency of the model sand ground for all group tests as well.

Total lateral forces on the shearing frames were recorded via the lateral jack during the tests at each 10 mm incremental frame movement (w_f). They are plotted in Figure 13.4 for the six tests (T block) on single piles and eight tests on two-pile groups [T block (Qin 2010) and U block (Ghee 2009)]. The figure demonstrates the force in general increases proportionally with the frame movement until it attains a constant. This offers a shear modulus G of 15~20 kPa (see Example 13.1). Figure 13.5 provides evolution of maximum shear forces T_{max} with the total lateral forces exerted on the shear box. The ultimate (maximum) shear resistance offered by the pile is ~0.6 kN, which accounts for ~10% of the total applied forces of 5~8 kN on

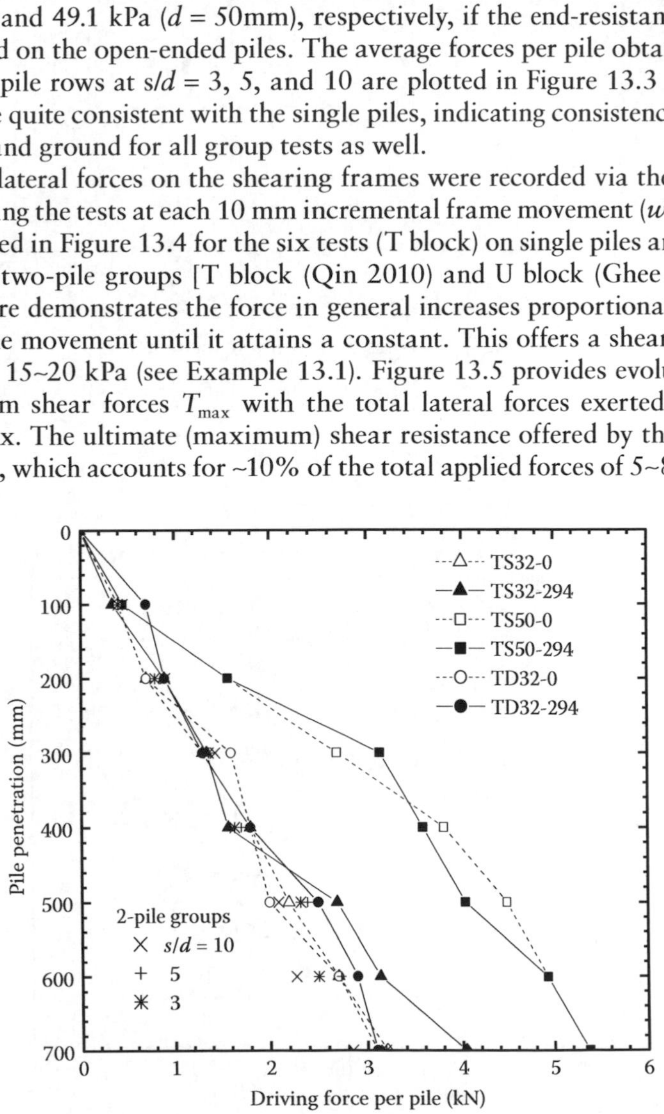

Figure 13.3 Jack-in resistance measured during pile installation.

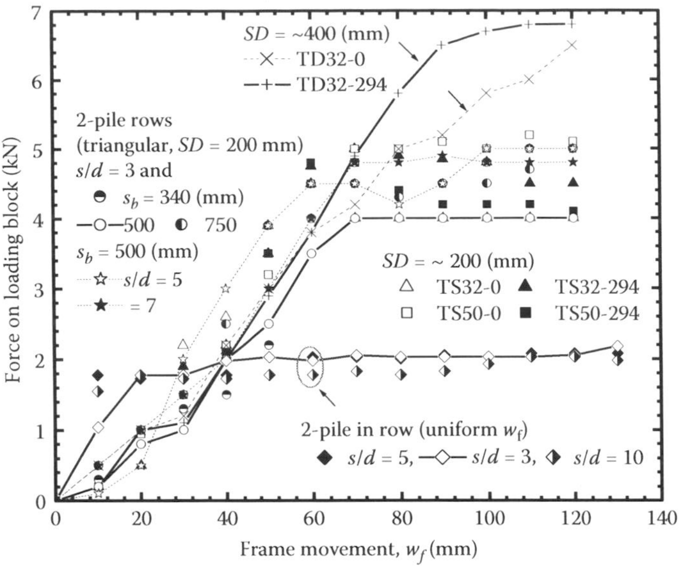

Figure 13.4 Total applied force on frames against frame movements.

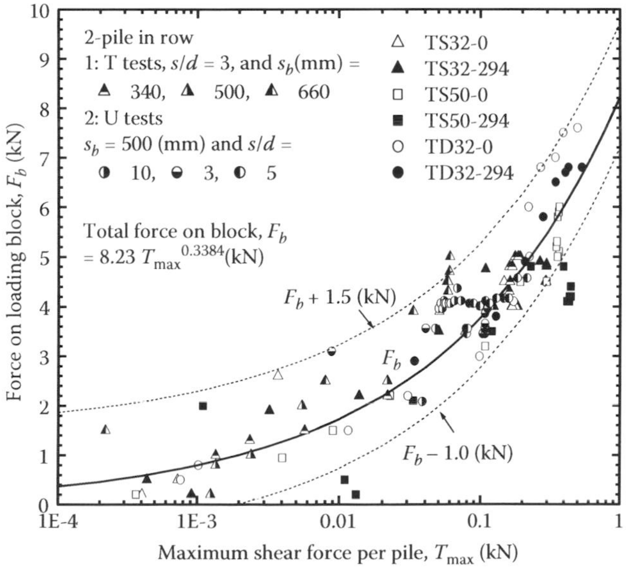

Figure 13.5 Variation of maximum shear force versus lateral load on loading block.

the shearing frames. The shear stress and modulus thus may be ~10% less for the tests without the pile. The average overburden stress σ_v at a sliding depth of 200 mm is about 1.63 kPa (= 16.3 × 0.1). At this low stress level, sand dilatancy is evident during the tests, and appears as “heaves” (Guo and Qin 2010).

Example 13.1 Determining shear modulus

During the TS tests, the maximum shearing stress τ is estimated as 4.5~5.0 kPa (= 4.5~5.0 kN on loading block/shear area of 1.0 m^2). The maximum shear strain γ is evaluated as 0.25~0.3 (= w_f/L_m, with w_f = 50~60 mm and L_m = 200 mm), assuming the shear force is transferred across the sliding depth L_m of 200 mm.

Figure 13.4 also indicates that the total lateral force (on frames) attained maximum either around a w_f of ~60 mm (TS series) or 90~120 mm (TD series) and dropped slightly afterwards. The latter indicates residue strength after the dilating process, which is evident by the gradual formation of heaves mentioned earlier. As shown later, the pile response, however, attained maximum at a higher w_f of either 70~90 mm (TS series) or 120 mm (TD series), indicating ~30 mm movement loss in the initial w_f (denoted as w_i later) (i.e., $w_i \approx 30$ mm in transferring the applied force to the pile). As for the two-pile groups, Figures 13.4 and 13.5 demonstrate the maximum force per pile is close to single piles, which is 2~2.5 kN at a w_f of 60 mm (T2~T10, T block at various s_b or s/d = 3–7) or 1.8~2 kN at 20 mm (U block at s/d = 3–10) for all piles at a final SD = 200 mm; and the G is 13~16 kPa, which is also comparable to 15~20 kPa for single piles (Guo and Qin 2010) using the T block, but it is ~50% of 20~35 kPa (with τ = 3.5 kPa at γ = 0.1) using the U block.

As seen in Figure 13.5, the induced forces of T_{max} in the single d_{32} piles are 168 N (without P) ~295 N (with P = 294 N) in the TS tests. The total resistance T_{max} of two-piles is 246 N (without P). Overall, a two-pile group offers ~25% higher resistance than that offered by the single d_{32} pile. A factor of safety (FS) owing to two-piles, defined as resistance over the sliding force, is thus close to 1.37 (= 1.25 × 1.1), considering the ~10% increase in resistance of a single pile to sliding over the “no pile” case. The two piles increase the resistance to sliding by 37%.

Example 13.2 Determining T_{max}

The induced shear force in each pile T_{max} was overestimated significantly (not shown here) using the plasticity theory (Ito and Matsui 1975). In contrast, this force was underestimated by using T_{max} = $0.2916\delta F\gamma_s L_m^2$ (Ellis et al. 2010). The current tests have a unit weight γ_s of 16.27 *kN/m*^{3} and L_m = 0.2 m. Using δF = 0.37 (T block), the T_{max} was estimated as 70.2 N, which is 25~50% of the measured 168~295 N (for s_b = 500 mm, in Figure 13.5). With δF = 0.2 for the U block tests, the T_{max} was estimated as 37.6 N, which is only 17~52% the measured T_{max} of 72~216 N for the two-pile groups at s/d = 3~10.

A sliding depth ratio R_L is defined herein as the ratio of thickness of moving soil (L_m) over the pile embedment (i.e., $L = L_m + L_s$). For instance, in the TS test series, a w_f of up to 60 mm (or up to a sliding depth L_m of 200 mm) corresponds to a triangular profile, afterwards, the w_f induces a trapezoid soil movement profile (with a constant $R_L = 0.29$).

13.3.2 Response of M_{max}, y_o, ω versus w_i (w_f)

The pile deflection at sand surface y_o and the rotation angle ω are measured during the tests. Using the T block, the y_o~w_f and y_o~ω curves are depicted in Figure 13.6a and using the U block in Figure 13.6b. These curves manifest the following features:

- An initial movement w_f (i.e., the w_i mentioned earlier) of 37~40 mm (triangular) or of 0~10 mm (uniform) causes negligible pile response; and an effective frame movement w_e (= $w_f - w_i$) is used later on. This w_i alters with the loading distance s_b and the pile center-center spacing (Figure 13.6a), irrespective of single piles or two-pile groups. The w_i indicates the extent and impact of the evolution of strain wedges carried by the loading block, as is vindicated by the few "sand heaves" mentioned earlier.
- The single piles and two-pile rows observe, to a large extent, the theoretical law of $\omega = 1.5y_o/L$ (pure lateral loading) or $\omega = 2y_o/L$ (pure moment loading) (Guo and Lee 2001) (see Figure 13.6a and b), which are salient features of a laterally loaded, free-head, floating base pile in a homogeneous soil.
- The ratio of y_o/w_e varies from 0.12~0.32 in the T block tests (see Figure 13.6a) to 0.25~0.8 (single) or 0.5 (s/d = 3~5) in the U block tests (see Figure 13.6b).

The overall interaction between shear box and pile of y_o~w_f and local interaction between pile and soil of y_o~ω are evident. These features provide evidence for the similarity in deflection between active and passive piles. Next the response of maximum bending moment M_{max} is discussed, whereas the measured T_{max} is provided in form of $0.357LT_{max}$ to examine the accuracy of $M_{max} = 0.357LT_{max}$ (discussed later).

First, as an example, the profiles of ultimate bending moment and shear force for single piles are presented in Figure 13.7a$_1$ and b$_1$ together with those for d50 piles. They were deduced from the two tests on the d32 piles without axial load (TS32-0) and with a load of 294 N (TS32-294) and at w_f = 70~90 mm. Second, the evolution of the maximum M_{max} and shear force T_{max} (T block) with the movement w_f is illustrated in Figure 13.8a,

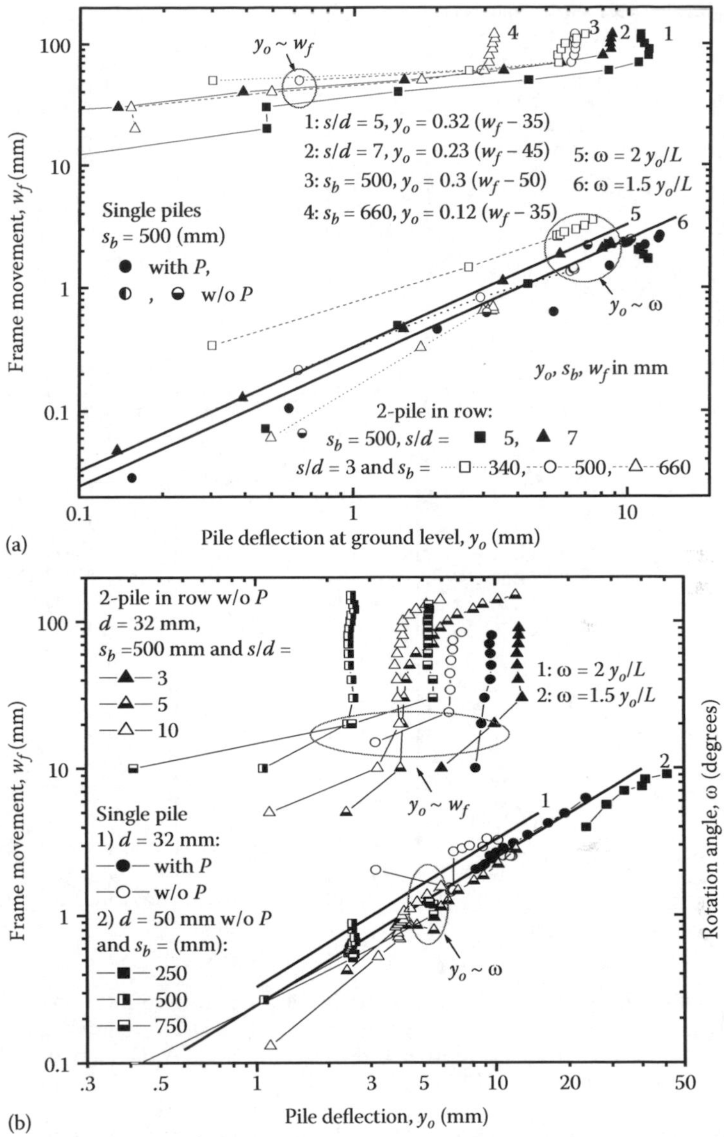

Figure 13.6 **Development of pile deflection y_o and rotation ω with w_f. (a) Triangular block. (b) Uniform block.**

together with those for d_{50} piles. The figures indicate a w_i of ~37 mm; a linear increase in M_{max} for all tests at w_f = 37~80 mm (R_L = 0.17~0.29); and a nearly constant M_{max} for a w_f beyond 80~90 mm (which is associated with a trapezoid soil movement at L_m = 200 mm). As for the tests using the U block, the evolution of M_{maxi} and T_{maxi} (i = 1 and 2 for sliding layer and

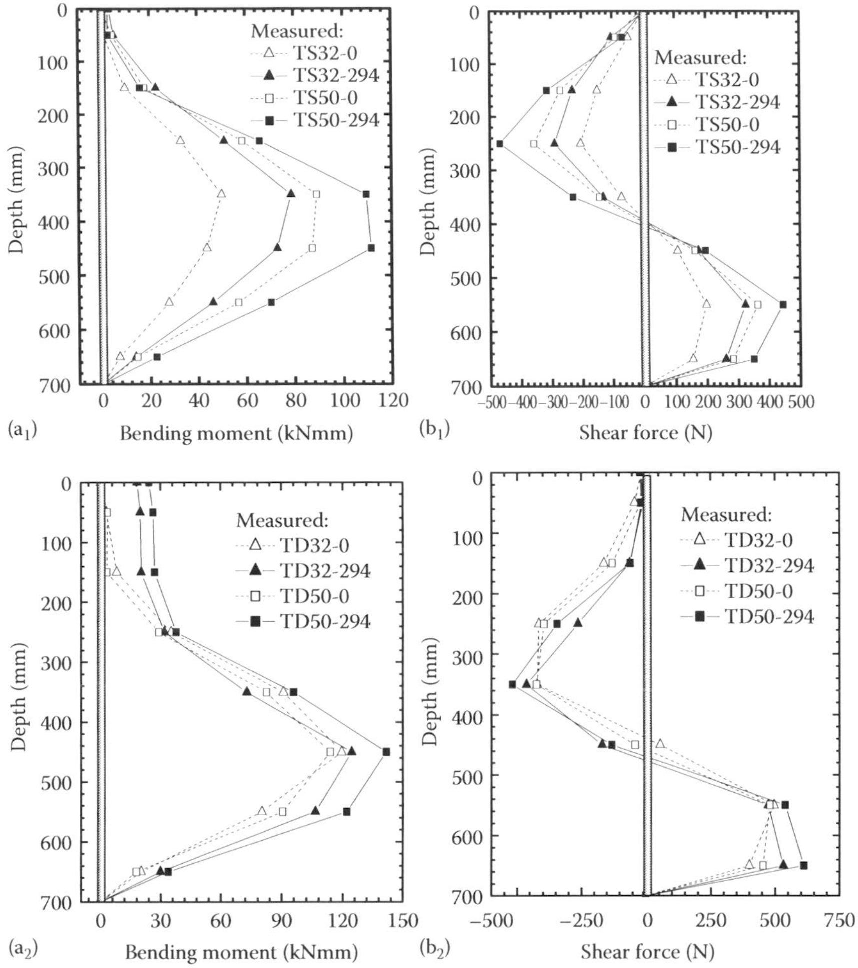

Figure 13.7 Maximum response profiles of single piles under triangular movement. (a_i) Bending moment. (b_i) Shear force (i = 1, final sliding depth = 200 mm, and i = 2, final sliding depth 350 mm). (After Guo, W. D., and H. Y. Qin, *Can Geotech J* 47, 22, 2010.)

stable layer) in a single pile as the frame advances is shown in Figure 13.8b. They exhibit similar features to T block tests. Overall, the correlation of $M_{max} = 0.357LT_{max}$ seems to be sufficiently accurate for any w_f of either T or U block tests.

As for two-pile groups, under the T block, the development of maximum moment M_{max2} and shear force T_{max2} with the effective w_e are depicted in Figure 13.9a (w_i = 35~42 mm). It indicates the impact of the loading distance s_b (see Section 13.2.5) and a higher M_{max2} for a larger spacing s/d = 5~7 than s/d = 3. Table 13.2 shows M_{max} = 49.7 kNmm (at w_f = 70 mm or

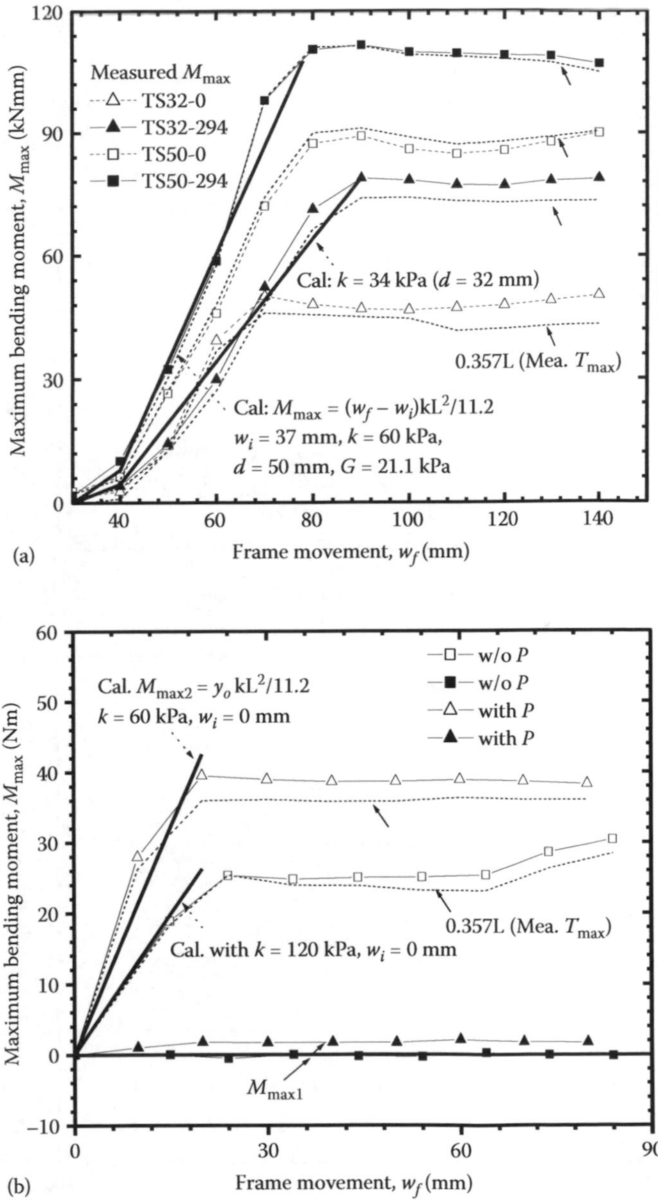

Figure 13.8 Evolution of maximum response of single piles (final sliding depth = 200 mm). (a) Triangular movement. (b) Uniform movement.

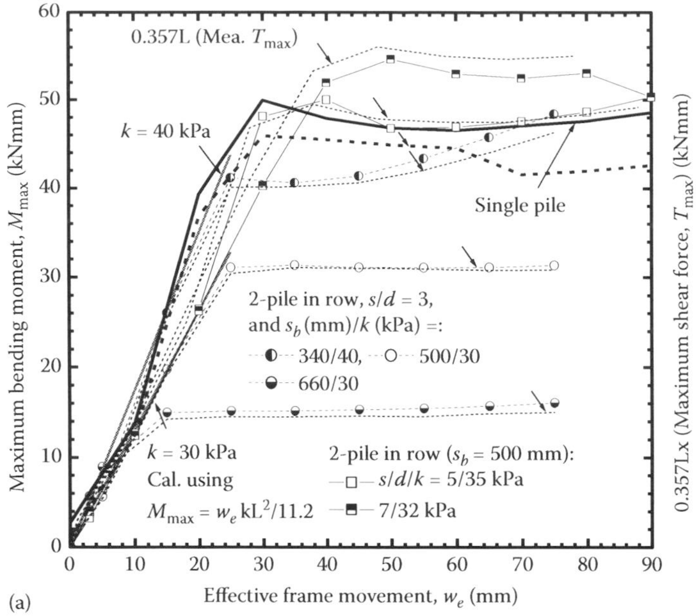

(a)

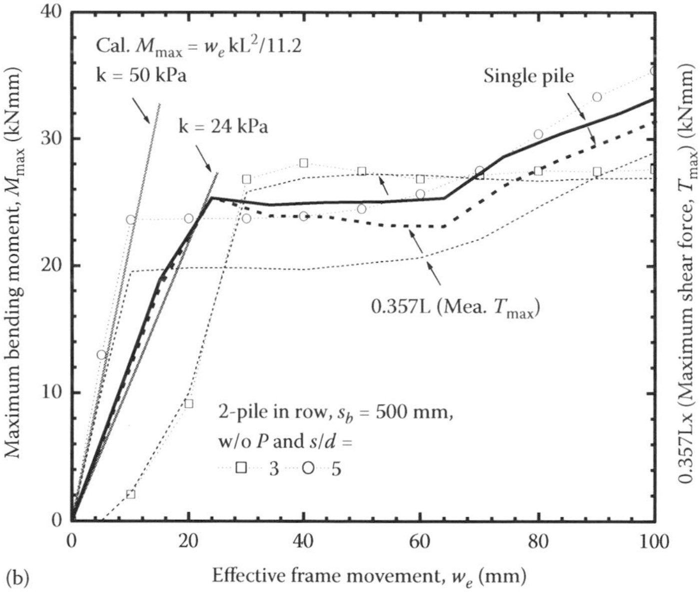

(b)

Figure 13.9 Evolution of M_{max} and T_{max} with w_e for two piles in a row (w/o axial load). (a) Triangular movement. (b) Uniform movement.

$w_e \approx 30$ mm) for a single pile; and that only 62% the M_{max} (i.e., 31.0 Nm, or $F_m = 0.56$) is mobilized along the piles in the 2-pile rows (Figure 13.9a) at $w_e = 15$~40 mm. Using the U block, the evolution in pile A (see Figure 13.1) is plotted in Figure 13.9b for the capped 2- piles with s/d = 3 and 5, respectively. Note the "abnormal" reduction in M_{max} with increase in s/d resembles previous report by Chen and Poulos (1997), and may be attributed to difficult testing conditions. These results would not affect the conclusions drawn herein. These figures indicate a w_e of 20~30 mm (w_i = 0~8 mm) is required to mobilize the ultimate M_{max} for a single pile and 2-piles in a row. The M_{max} for the single d_{32} pile is ~25.3 kNmm (w/o load) and ~39.2 kNmm (with load) at w_f = 70 mm, which are ~50% of 49.7 Nm and 77.6 Nm, respectively under T block tests. About 70% ($F_m = 0.7$) the M_{max} of the single pile (~ 35.0 kNmm) is mobilized in the piles in a row, with values of M_{max} = 23~24 kNmm at s/d = 3 and w_f = 140 mm (without axial load).

Now let's examine the impact of a higher sliding depth on the M_{max}. At a large sliding depth of 350 mm, the moment and shear force profiles for the d32 piles and the d50 piles (Guo and Qin 2006) at the maximum state (w_f = 120 mm) are depicted in Figure 13.7a_2 and b_2, respectively (T block). The evolution of the maximum moments and shear forces with the advance of the T block and the frames is illustrated in Figure 13.10. In comparison with aforementioned results for SD = 200 mm, these figures show (1) a reduced w_i with increase in diameter from w_i = 30 mm (d = 50 mm) to 37 mm (d = 32 mm); (2) a nearly constant bending moment down to a depth of 200 mm owing to axial load (Figure 13.7a_2), otherwise a similar moment distribution to that from TS tests (Figure 13.7a_1); (3) a slight increase in M_{max} to ~143 kNmm (see Figure 13.7a_2) that occurs at a depth d_{max} of 0.465 m owing to the load; (4) a consistently close match between 0.357LT_{max} and M_{max}; and (5) higher values of T_{max} and M_{max} may attain if w_f exceeds 140 mm, showing moment raises, as is further explored next.

13.3.3 M_{max} raises (T block)

The evolution of M_{max} is now re-plotted against the normalized sliding depth R_L in Figure 13.11 to highlight the moment raises at either $R_L = 0.29$ or 0.5 caused by uniform movement beyond the triangular movement. The raises at R_L = 0.179, 0.357, 0.429, and 0.5 were also determined from four more tests on the d32 piles (without axial load) to the pre-selected final sliding depths of 125(R_L = 0.179), 250(0.357), 300(0.429), and 350 mm (0.5), respectively. The values of M_{max} obtained were 5.2, 62.6, 115.3, and 118.1 Nm upon initiating the trapezoid profile, and attained 5.7, 123.5, 175.0, and 140.0 (not yet to limit) Nm at the maximum w_f, respectively. These values are plotted in Figure 13.11 against the R_L, together with the TS32-0 test. The raises are 0.5, 60.9, 59.3, and 21.9(?) kNmm, respectively. The trapezoidal movement doubled the values of M_{max} at a R_L = 0.357,

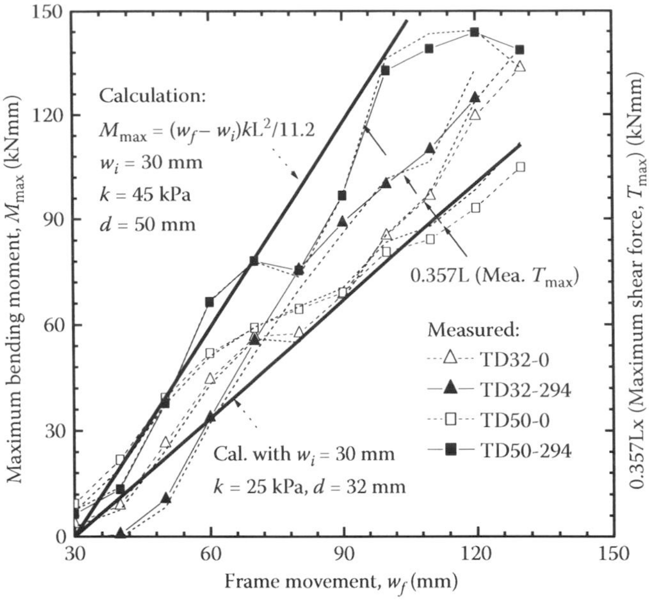

Figure 13.10 Evolution of maximum response of piles (final sliding depth = 350 mm). (After Guo, W. D., and H. Y. Qin, *Can Geotech J* 47, 22, 2010.)

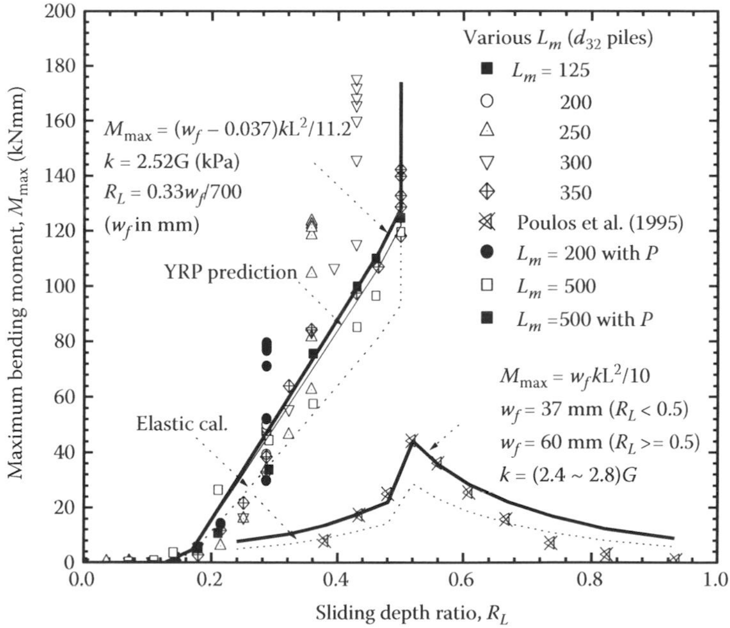

Figure 13.11 Variation of maximum bending moment versus sliding depth ratio. (Revised from Guo, W. D., and H. Y. Qin, *Can Geotech J* 47, 22, 2010.)

although it has negligible impact at $R_L = 0.179$. The M_{max} along with d_{max}, T_{max}, and y_o are also provided in Table 13.1. Note that the M_{max} and T_{max} from T32-0 ($L_m = 350$ mm at $w_f = 120$ mm) are 1.2% and ~5% less than those from TD32-0, showing the repeatability and accuracy of the current tests.

13.4 SIMPLE SOLUTIONS

As it is mentioned before, a correlation between the M_{max} and T_{max} is evident, which is explored theoretically next.

13.4.1 Theoretical relationship between M_{max} and T_{max}

Chapter 12, this book, demonstrates the use of analytical solutions for laterally loaded (active) piles to study passive piles by taking the lateral load H as the maximum sliding force, T_{max}, induced in a pile. This is approximately corroborated by the ratio of deflection over rotation (y_o/ω) mentioned earlier. Next, the use is substantiated from the measured relationships between M_{max} and T_{max} and between the effective w_e and T_{max}.

Elastic solutions for a free-head, floating-base, laterally loaded, rigid pile offer (Scott 1981)

$$M_{max} = (0.148\sim0.26)HL \text{ and } d_{max} = (0.33\sim0.42)L \tag{13.1}$$

where H = lateral load applied at pile-head level and d_{max} = depth of maximum bending moment. The coefficients of 0.148/0.33 are deduced for a uniform k and 0.26/0.42 for a Gibson k. Furthermore, elastic-plastic solutions provide at any stress state (Guo 2008)

$$M_{max} = \left(\frac{2}{3}d_{max} + e\right)H \tag{13.2}$$

where e = the real or fictitious free-length of the lateral load above the ground surface. The ratios of M_{max}/HL are equal to (1) 0.183 with a constant k and p_u at tip-yield state; (2) 0.208 using either k at yield at rotation point (YRP); (3) 0.319 adopting Gibson k and p_u at tip-yield state; (4) 0.322 (based on constant k and Gibson p_u at tip-yield state); and (5) 0.34 assuming a Gibson p_u and either k at YRP. In brief, elastic-plastic solutions offer a value of M_{max}/HL = 0.148~0.34. Moreover, model tests on passive piles show $d_{max} = (0.5\sim0.6)L$ and $M_{max} = (0.33\sim0.4)LT_{max}$ for the passive piles (see Figure 13.7a$_1$ and a$_2$); and $M_{max} = 0.357T_{max}L$ (see Figures 13.8 through 13.10). In other words, the ratio of 0.357 is slightly higher than

0.34 for YRP state, perhaps owing to dragging moment. By taking H = T_{max}, Equation 13.1 may be modified for passive piles as

$$M_{max} = (0.148\sim0.4)T_{max}L \tag{13.3}$$

13.4.2 Measured M_{max} and T_{max} and restraining moment M_{oi}

Table 13.3 provides the critical values of maximum bending moment M_{max2}, shear force T_{maxi}, and maximum pile deflection y_o, w_i and k at w_f = 60~140 mm. The measured M_{max2}~T_{max2} curves are plotted, respectively, for T block tests in Figure 13.12a and for U block tests in Figure 13.12b. This reveals that an interceptor M_{oi} at T_{maxi} = 0 of the M_{maxi}~T_{maxi} line for stable and sliding layer may emerge, compared to M_{oi} = 0 for the single piles (e.g., Figure 13.12a and b)

$$M_{max\,i} = T_{max\,i}L/\alpha_{si} + M_{oi} \tag{13.4}$$

The M_{oi} is termed as restraining bending moment, as it captures the impact of the cap rigidity, P-δ effect, and sliding depth. The values of M_{oi} for typical pile groups were determined and are tabulated in Table 13.4. For instance, at SD = 200 mm, the M_{o1} of U3 tests is −3 Nm [= ~ 6.7%, the M_{max2} (Pile A or B)], and M_{o2} is 0.

The linear relationships between M_{maxi} and T_{maxi} are evident and are characterized by (1) α_{s1} = −(6.75~30), showing impact of cap fixity (see Tables 13.4 and 13.5); (2) α_{s2} = 2.8 (T block) and 2.76~2.83 (U block) for stable layer; and (3) nonzero moments M_{o1} and M_{o2} for U block, see Figure 13.12b, but for M_{o2} = 0 for the T block (Figure 13.12a). The difference in ratios of M_{maxi}/T_{maxi} is negligible between single piles and piles in the two-pile groups.

The M_{max} measured for piles in groups at w_f = 60~70 mm (see Tables 13.2 and 13.3 for d32 piles without axial load) is 13.6~33.6 Nm (uniform, *s*/*d* = 3~5) and 15.3~54.6 Nm (triangular, *s*/*d* = 3~7), which are largely a fraction of 31.1~71.6 Nm (Chen et al. 1997), despite of a similar scale. These differences are well captured using simple calculations, as shown later in Examples 13.6 and 13.7.

The aforementioned Tests U3–U6 and T2–T10 were generally conducted to a final sliding depth (SD) of 200 mm on two piles capped in a row (Guo and Ghee 2010). An increase in SD to 400 mm leads to: (1) reverse bending (from negative to positive M_{max1}) and 20% reduction in the positive M_{max2}; (2) near-constant bending moment and shear force to a depth of 200 mm (at w_f = 110~140 mm); (3) a high ratio of y_o/w_e of 0.81 (w_i = 0) during the test without (w/o) axial load (see Table 13.3) (~1.62 times y_o/w_e = 0.5 for SD = 200 mm), or a ratio of 0.7 (w_i = 10 mm) with the axial load; (4) shifting of depth of M_{max2} towards sand surface and generation of restraining moment

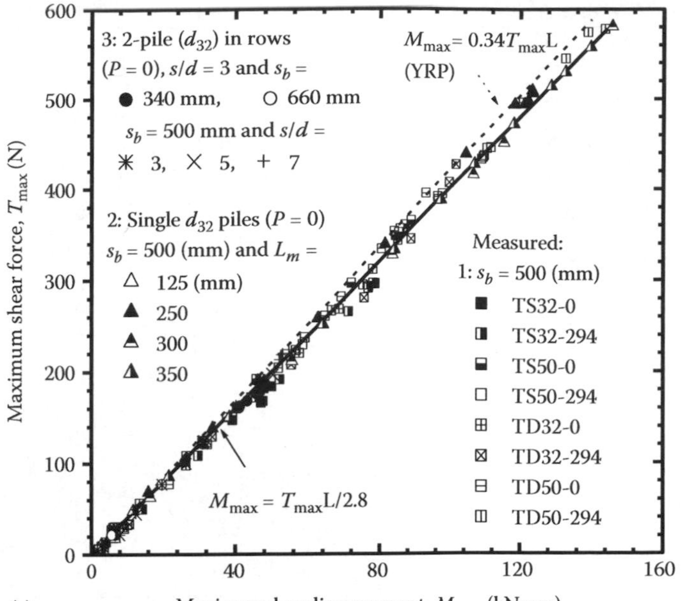

(a)

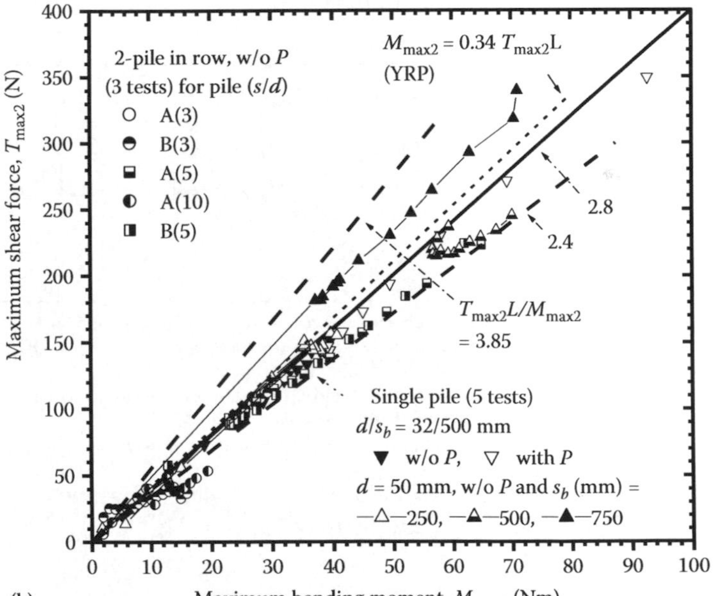

(b)

Figure 13.12 Maximum shear force versus maximum bending moment. (a) Triangular movement. (b) Uniform movement.

Table 13.4 Prediction of model pile groups and response (Uniform Soil Movement)

Test no.	Group	Pile	$\frac{T_{max2}L}{M_{max2}-M_{o2}}$ L_2 *layer*	$\frac{T_{max1}L}{M_{max1}-M_{o1}}$ L_1 *layer*	M_{oi} *(Nm)* L_1 *layer*	M_{oi} *(Nm)* L_2 *layer*
			Without axial load			
US32-0	Single		2.8	−28.0[a]	0	0
U3	1×2	A	2.83	−6.75	−3.0	0
	In row	B	2.76	−6.75	0	0
U9	1×2	A	2.8	−17.0	0	0
	(SD[c] = 400 mm)	B	2.8	−(17~28)	0	0
			With axial load			
U2	Single		2.8	−10.8	0	0
U4[b]	1×2	A	–	–	–	–
	In row	B	–	–	–	–
U10	1×2	A	2.8	−6.75	−7	0
	(SD[c] = 400 mm)	B	2.8	−6.75	−14	0

[a] This should be −∝ if the loading eccentricity was zero.
[b] Results are not presented owing to unfit pile and cap connection.
[c] Sliding depth (SD = L_m) is 200 mm except where specified.

Table 13.5 y_o, M_{oi}, and M_{maxi} relationships for any w_f

Items		$\frac{y_o kL}{T_{maxi}}$ [a]	$T_{maxi}L/(M_{maxi}-M_{oi})$ [b] *Piles in line*	$T_{maxi}L/(M_{maxi}-M_{oi})$ [b] *Pile in row*	$\frac{k_{single}}{k_{group}}$ [a]
Tests U3–U10	Sliding layer	3	4.5~6.75	−(6.75~28)[c]	0.64~1.08 (0.84)
	Stable layer	4	2.8		
Tests T2–T20	Sliding layer	1	10~30	—	0.4~1.4
	Stable layer	4	2.8		

[a] k_{single} = (2.4~3)*G*, modulus of subgrade reaction for a single isolated pile obtained using Equation 3.62, which is the *kd* defined previously (Guo 2008); $k = k_{single}$ = 45~60 kPa for d = 50 mm, and k = 28.8 ~ 38.4 kPa for d = 32 mm.
[b] Theoretically = 3.85 (a linear *k* with depth) and 6.75 (a uniform *k*) for free-head piles in elastic medium. M_{oi} relies on cap stiffness, fixity, etc.
[c] It depends on loading eccentricity e, which is ∝ for e = 0 and free-head pile.

(M_{o1} = −7 Nm) under an axial load of 294 N per pile; (5) on-pile force profiles shown in Figure 13.13; and finally (6) a ratio of $M_{max2}/T_{max2}L = 1/2.8$, concerning SD = 200~400 mm.

13.4.3 Equivalent elastic solutions for passive piles

The effective frame movement w_e (= $w_f - w_i$) causes the groundline deflection y_o (local interaction) during the passive loading process of overall

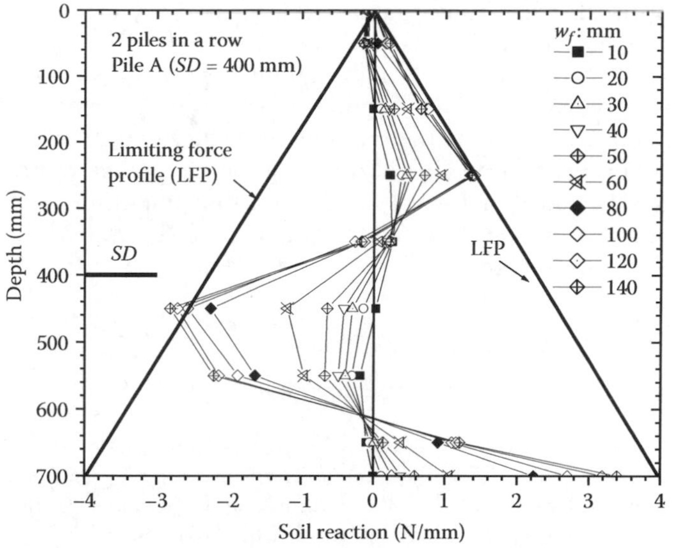

Figure 13.13 Soil reaction deduced from uniform tests (SD = 400 mm).

pile–soil movement. The correlation of $y_o \approx \alpha w_e$ may be established. The y_o may be related to sliding thrust T_{max} using elastic theory for a lateral pile in a homogeneous soil (see Chapter 7, this book) by

$$y_o = 4T_{max}/(kL) \tag{13.5}$$

where k = (2.4~3)G and is approximately proportional to pile diameter d (Guo 2008). Alternatively, with Equation 13.3, we have

$$T_{max} = \alpha(w_f - w_i)kL/4 \tag{13.6}$$

$$M_{max} = \alpha(w_f - w_i)kL^2/(10\text{~}27) \tag{13.7}$$

During deep sliding tests, pile–soil relative rigid movement may be incorporated into the w_i and modulus of subgrade reaction k. Thus, α may be taken as unity. The initial frame movement w_i depends on the position s_b, pile diameter, and loading manner: w_i = 0.03~0.037 m (the current translational tests), and w_i = 0.0 for the rotational tests reported by Poulos et al. (1995). A denominator of 15.38~27 corresponds to elastic interaction for Gibson~constant k, a value of 11.2 reflects plastic interaction as deduced from tests, and 11.765 for YRP state. Later on, all "elastic calculation" is based on a denominator of 15.38 (Gibson k) unless specified. The length L is defined as the smallest values of L_i (pile embedment in ith layer, i = 1, 2 for sliding and stable layer, respectively) and L_{ci}. Without a clear sliding layer, the length L for the model piles was taken as the pile embedment length. The L_{ci} is given by Equation 3.53 in Chapter 3, this book, using Young's modulus of an equivalent solid cylinder pile E_p and average shear modulus of soil over the depth of ith layer $\tilde{G}$. It must be stressed that the modulus k deduced from overall sand–pile-shear box interaction should not be adopted to predict groundline y_o and those provided in Table 13.1, as it encompasses impact of "rigid" movement between shear box and test pile(s). The T_{max} and M_{max} must be capped by ultimate state.

Example 13.3 Determining T_{max} from y_o

Two local-interaction cases are as follows:

1. The deflection at groundline y_o was measured as 46 mm for TD32-0 under w_f = 110 mm. Given the measured T_{max} = 0.4 kN, k = 50 kPa, and L = 0.7 m, Equation 13.5 yielded a similar deflection y_o of 45.7 mm to the measured one.
2. The y_o at w_f = 110 mm was calculated as 63.5 mm for TD32-294, in light of the measured T_{max} = 0.5 kN, k = 45 kPa, and L = 0.7 m, which also compares well with the measured y_o of 62.5 mm. The k is 45~50 kPa for the local interaction.

13.4.4 Group interaction factors F_m, F_k, and p_m

Interaction among piles in a group is captured by a group bending factor, F_m, and a group deflection factor F_k defined, respectively, as $F_m = M_{mg}/M_{ms}$ (Chen and Poulos 1997), and $F_k = k_{group}/k_{single}$ (Qin 2010) under an identical effective soil movement. Note that the subscripts g and s in M_{mg} and M_{ms} indicate the values of M_{max} induced in a pile in a group and in the standard single pile under same vertical loading and lateral soil movement (Table 13.2), respectively. The factors F_m and F_k exhibit similar trends with the distance s_b (two piles), and the normalized pile spacing (s/d), respectively. The values of F_m were determined for laboratory test results and/or numerical simulations under various head/base constraints (Chen 1994; Chen et al. 1997; Pan et al. 2002; Jeong et al. 2003; Qin 2010). The linear relationships between M_{maxi} and T_{maxi} (Chen and Poulos 1997) and between T_{maxi} and k render resemblance between values of F_m and F_k for the same tests, but for the impact of M_{oi}. Moreover, the F_m, and F_k are compared with the p-multiplier (p_m) for laterally loaded piles (Brown et al. 1988; Rollins et al. 2006; Guo 2009) calculated using $p_m = 1 - a(12 - s/d)^b$ in which $a = 0.02 + 0.25\,Ln(\mathrm{m})$; $b = 0.97(\mathrm{m})^{-0.82}$ (Guo 2009). The calculated p_m with m = 1 and 1.2 (based on curve fitting) largely brackets the aforementioned values of F_m and F_k, indicating $p_m \approx F_m \approx F_k$. This group interaction is illustrated in the soil reaction profiles in Figures 13.13 and 13.14.

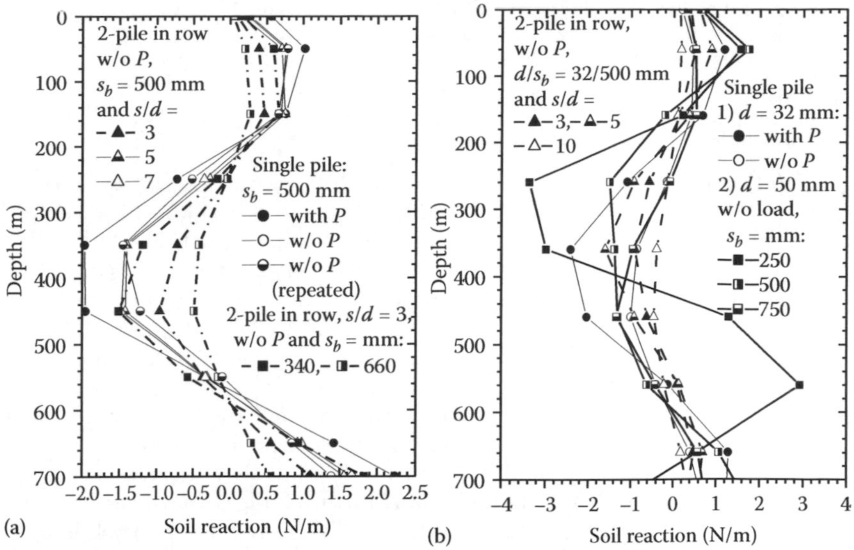

Figure 13.14 Soil reaction. (a) Triangular block. (b) Uniform block.

13.4.5 Soil movement profile versus bending moments

Soil movement profiles alter the evolution of maximum bending moment and shear force. For instance, under the U block, the "ultimate" bending moments induced in free-head single piles are 15~91% those under the T block, depending on *s/d*, s_b and loading level (see Tables 13.2 and 13.3). In other words, under same conditions, a uniform soil movement has a 1.1~6.6 times higher factor of safety than a triangular profile, which grossly agrees with the numerical findings (Jeong et al. 2003).

Example 13.4 Limiting y_o^* and p_u profiles

The T_{maxi} and M_{maxi} should be capped by the limiting force per unit length p_u, or by the upper elastic limiting deflection y_o^* (= p_u/k. N.B. the asterisk denotes upper limit herein). Study on 20 laterally loaded piles in sand indicates $p_u = A_r dz$ at depth z with $A_r = s_g \gamma_s K_p^2$, s_g = 0.4~2.5 and an average s_g of 1.29 (Chapter 9, this book). The current tests have γ_s = 16.27kN/m^3, ϕ = 38°, d = 32 mm, and an "average" condition initially with s_g = 1.29. As discussed previously, the jacked-in operation of the piles causes the resistance increase by 37%. This renders the s_g revised as 1.77 (= 1.29 × 1.37). The A_r is thus estimated as 177.545 kN/m^3 with $K_p = \tan^2(45 + 38°/2)$, and the p_u at the pile tip (z = 0.7 m) is 3.98/Nmm. This offers the bold lines of limiting force profiles of $p_u = \pm$ 177.545dz in Figure 13.13, which serve well as the upper bounds of the measured data. The y_o^* was calculated as 6~12 mm under a sliding depth of ~200 mm (see Figure 13.6). Otherwise, y_o^* at L_m = 400 mm was indeterminate (see Table 13.3), as the deep sliding results in continual increase in y_o with w_f. The y_o^* allows the ultimate M_{max2}^* to be calculated as $y_o^*kL^2/11.2 + M_{o2}$, as deduced from Equations 13.4 and 13.7.

Example 13.5 Limiting T_{maxi} and M_{maxi}

With respect to actively loaded piles, Guo (2008) shows $M_{max}/(A_r dL^3) \le 0.036$ and $T_{ult}/(A_r dL^2) \le 0.113$ (linear increase k with depth, Gibson k) prior to the pile tip yield (i.e., reaching ultimate stress). The LFP on a passive pile (Figure 13.13) resembles that on an active pile, as do the profiles of on-pile force per unit length within a depth to maximum bending moment (e.g., 0.46 m for L = 0.7 m). With the normalized M_{max} of 0.036, the limiting values of M_{max} upon tip-yield state is estimated as 70.1 Nm (= $0.036 \times 177.545 \times 0.7^3$). This may be the ultimate state for the test without load (Guo and Ghee 2010). At the extreme case of pure moment loading and impossible state of yield along entire pile length (i.e., YRP), the ultimate M_{max}, (rewritten as M_{ult}) satisfies $M_{ult}/(A_r\, dL^3) \le 0.075$. During the tests with load and deep sliding depth, a passive pile is likely to be dragged forward and may be associated with a medium stress that has a normalized M_{max} of ~0.0555 (in between the tip-yield and YRP). The M_{max} is thus equal to 108.1 Nm (= $0.0555 \times 177.545 \times 0.7^3$), which agrees well with the measured value (Guo and Ghee 2010). Likewise, the T_{max} is predicted. As tests (see Figure 13.6) generally exhibit behavior of free-head, floating base pile under pure lateral loading, limits of “yield at rotational point” were obtained using solutions of active piles and are shown in Figure 13.2.

13.4.6 Prediction of T_{maxi} and M_{maxi}

13.4.6.1 Soil movement profile versus bending moments

The evolution of M_{maxi} and T_{maxi} for typical piles with w_f was modeled regarding tests on all the groups using Equations 13.5 through 13.7. The input values of M_{oi}, k, and y_o are provided in Tables 13.2, 13.3, and 13.4. They were synthesized and are provided in Table 13.5. The following features are noted:

- In stable layer, all piles observe $y_o kL/T_{max2} = 4$, $T_{max2}L/(M_{max2} - M_{o2}) =$ 2.4~2.8, and $y_o kL^2/(M_{max2} - M_{o2}) = 11.2$ and $M_{o2} = 0$, which are close to for free-head piles (Guo and Qin 2010).
- The moment M_{max1} ($\approx T_{max1}L/4.5$~30) for all piles is less than M_{max2} ($\approx T_{max2}L/2.8$), although $T_{max1} \approx T_{max2}$, regardless of the cap fixity condition, etc. This is opposite of laterally loaded pile groups for which the M_{max} normally occurs at pile-cap level.
- Comparison shows k_{group}/k_{single} = 0.4~1.4 and y_o/w_e = 0.25~0.4 (w_i = 37~40 mm).

These are common features between T and U block tests. Using the U block, Tables 13.4 and 13.5 indicate the following special features:

- In sliding layer, piles in a row (without load) observe $T_{max1}L/(M_{max1}-M_{o1}) = -(6.75\sim28)$, and with a positive moment M_{max1}.
- The modulus reduces and $k_{group}/k_{single} = 0.64\sim1.08$ (an average of 0.84), as deduced by a good comparison between predicted and measured deflection y_o. The ratio y_o/w_e increases from 0.5 (SD = 200 mm) to 0.7~0.8 (SD = 400 mm).

The T_{maxi} and M_{maxi} were calculated and are plotted together with the measured data in Figures 13.8 and 13.9 (note that w_e is taken as y_o in Figure 13.8). They resemble those gained from U block tests but for $y_o kL/M_{max1} = 1$. Equations 13.1, 13.5, and 13.6 were thus validated against the current model tests.

13.4.7 Examples of calculations of M_{max}

In the current T block tests, the translational movement of the loading block results in increasing sliding depth to a pre-selected final SD. The previous model pile tests on free-head single piles and two, three, or four piles in a row (Poulos et al. 1995) were carried out, in contrast, by rotating loading block (rotational loading) about a constant sliding depth. The current T tests were generally associated with an effective soil movement of 30~70 mm (see Table 13.6), which are rather close to the movement between 37 mm ($R_L < 0.5$) and 60 mm ($R_L > 0.5$) in the rotational tests (Poulos et al. 1995). Nevertheless, Figure 13.11 displays a 3~5 times disparity in the measured moment M_{max} between the current (with L = 700 mm and d = 32 mm) and the previous tests (L = 675 mm and d = 25 mm). This difference was examined using Equations 13.6 and 13.7 and is elaborated in Examples 13.6 for translational movement and 13.7 for rotational moment, respectively.

Example 13.6 Translational movement

13.6.1 TS, TD Tests, and *k*

The measured $M_{max}\sim w_f$ and $T_{max}\sim w_f$ curves (see Figures 13.8 and 13.10) were modeled regarding the pre-specified final sliding depths of 200 mm (TS tests) and 350 mm (TD tests), respectively. With k/G = 2.841 (d = 50 mm) and 2.516 (d = 32 mm) (Guo 2008), and G = 15~21 kPa (deduced previously), the k is obtained as 45~60 kPa (d = 50 mm). With d = 32 mm, the k reduces to 25~34 kPa, as a result of the reduction to 28.8~38.4 kPa (from change in d), and decrease in ratio of k/G by 0.8856 (= 2.516/2.841) times. Given w_i = 30 mm (TD50-294) or 37 mm (TS50-294), the moment is calculated using $M_{max} = (w_f - w_i)$

Table 13.6 "Translating" pile tests TD32-0 and T32-0

Movement		*Calculated*[a]		*Measured* M_{max}*(kNmm)*	
w_f *(mm)*	R_L	T_{max} *(kN)*	M_{max} *(kNm)*	*T32-0*	*TD32-0*
30	0.1270	0	0	0.86	3.66
40	0.1693	0.018	4.62	2.78	8.80
50	0.2116	0.080	20.04	11.68	26.37
60	0.2540	0.142	35.45	21.69	44.42
70	0.2963	0.203	50.86	38.47	56.56
80	0.3386	0.265	66.27	63.96	57.50
90	0.3809	0.327	81.68	84.23	68.40
100	0.4233	0.388	97.09	97.51	85.17
110	0.4656	0.450	112.50	106.98	96.56
120	0.6398	0.512	127.92	118.12	119.50
150[b]	*0.7997*	*0.697*	*174.15*	*139.78*	

Source: Guo, W. D., and H.Y. Qin, *Can Geotech J*, 47, 2, 2010.

[a] $w_i = 37$ mm, $\gamma_b = 0.048$, $k/G = 2.516$, $L = 0.7$ m, $d = 32$ mm, G = 14.0 kPa
[b] Trapezoid movement profile.

$kL^2/11.2$. The results agree well with the measured data, as indicated in the figures for d32 piles. The values of k are good for the $d = 32$ and 50 mm. Equation 13.4 is sufficiently accurate for the shallow sliding case as well.

13.6.2 Translational loading with variable sliding depths (Constant L)

The measured M_{max} of the piles TD32-0 and T32-0 ($L_m = 350$, Table 13.1) tested to a final sliding depth of 350 mm is presented in Table 13.6. Using $M_{max} = (w_f - 0.037)kL^2/11.2$ and $k = 34$ kPa, the M_{max} was estimated for a series of w_f (or R_L) (see Table 13.6) and is plotted in Figure 13.12a. The R_L was based on actual observation during the tests, which may be slightly different from theoretical $R_L = 0.33w_f/L$. The calculated M_{max} is also plotted in Figure 13.12b. The moment raise at $R_L = 0.5$ was estimated using an additional movement of 30 mm (beyond the w_f of 120 mm) to show the capped value. Table 13.6 shows the calculated value agrees with the two sets of measured M_{max}, in view of using a single w_i of 37 mm for either test.

13.6.3 A capped pile

As an example, the M_{maxi} and T_{maxi} for a capped pile with $y_o = 0.5w_e$ ($s/d = 3$) were calculated by three steps: (1) $k/G = 2.1$~2.37 ($L = 0.7$ m and $d = 32$ mm) using Equation 3.62, Chapter 3, this book;

(2) G = 15~21 kPa, and k = 30~50 kPa, respectively (Guo and Qin 2010); and (3) T_{maxi} calculated using Equation 13.6, with $y_o = 0.5w_e$ and $w_i = 0$ (see Table 13.3) or the p_u.

For instance, the ultimate sliding force per unit length p_u was estimated as 1.11 kN/m at a depth of 0.2 m, which gives a total sliding resistance T_{max}^* of 111.0 N (= $0.5p_u \times 0.2$). On the other hand, with $y_o^* = 22.2$ mm (= $0.5w_f^*$, $w_f^* = 44.3$ mm measured), the T_{max}^* was estimated as 116.6 N (= 22.2 × 0.001 × 30 × 0.7/4), using k = 30 kPa (see Table 13.3). The two estimations agree with each other. The M_{max2} was calculated as $T_{max2}\,L/2.8$ (with $M_{o2} = 0$) with a limited M_{max2}^* of 70.1 Nm. The estimated T_{max2} and M_{max2} agree well with the measured values (Guo and Ghee 2010).

As sliding depth increases to 400 mm, the same parameters offer T_{max}^* = 0.444 kN (= 0.5 × 2.22 × 0.4), and M_{maxi} = 111.0 Nm (= $0.7T_{max}^*/2.8$), respectively. The latter agrees with the measured value of 118.12 Nm (Guo and Ghee 2010). Likewise, the M_{maxi} and T_{maxi} were calculated for all other groups, which also agree with measured data (see pertinent figures).

Example 13.7 Rotational movement (Chen et al. 1997)

13.7.1 Rotational loading about a fixed sliding depth (single piles)

The M_{max} was obtained in model pile tests by loading with rotation about a fixed sliding depth (thus a typical R_L) (Poulos et al. 1995). The results for a series of R_L were depicted in Figure 13.11 and are tabulated in Table 13.7. This measured M_{max} is simulated via the following steps:

- The ratio of k/G was obtained as 2.39~2.79 (Guo and Lee 2001), varying with d/L. The shear modulus G (in kPa) was stipulated as $10z$ ($z = L_s+L_m$ in m).
- With $w_i = 0$ (as observed), the T_{max} was estimated using Equation 13.6 for w_f = 37 mm ($R_L < 0.5$) or w_f = 60 mm ($R_L > 0.5$), respectively The M_{max} was calculated as $w_f kL^2/10$ (Equation 13.7).

The test piles are of lengths 375~675 mm, and the G was 3.75~6.75 kPa. The values of the M_{max} calculated for the 10 model single piles are provided in Table 13.7. They are plotted against the ratio R_L in Figure 13.11, which serve well as an upper bound of all the measured data.

13.7.2 Rotational loading about a fixed sliding depth (piles in groups)

Each group was subjected to a triangular soil movement with a fixed sliding depth of 350 mm (Chen et al. 1997). Under a frame movement w_e^* = 60 mm (w_i = 0), it was noted that (1) $y_o^* \approx 0.5w_e^*$, with measured pile displacement y_o^* of 24~30 mm of the groups, resembling the current two piles in a row (see Table 13.1 and Figure 13.6); (2) an angle of rotation ω of 3.5~4.5°; and (3) a maximum moment M_{max} of largely

Table 13.7 Calculation for "rotating" tests

Input data				*Calculated*				*Measured*[a]
Embedded length L *(mm)*	w_f *(mm)*	G *(kPa)*	R_L	*Factor* γ_b *(= 1.05d/L)*	k/G	T_{max} *(kN)*	M_{max} *(Nm)*	M_{max} *(Nm)*
525	37	5.25	0.38	0.05000	2.54	0.0648	*13.62*	8.0
575	37	5.75	0.43	0.04565	2.48	0.0760	*17.47*	17.4
625	37	6.25	0.48	0.04200	2.43	0.0879	*21.97*	25.0
675	60	6.75	0.52	0.03889	2.39	0.1631	44.03	44.2
625	60	6.25	0.56	0.04200	2.43	0.1425	35.63	36.1
575	60	5.75	0.61	0.04565	2.48	0.1232	28.33	25.5
525	60	5.25	0.67	0.05000	2.54	0.1051	22.08	15.8
475	60	4.75	0.74	0.05526	2.61	0.0884	16.79	7.1
425	60	4.25	0.82	0.06176	2.69	0.0729	12.39	3.0
375	60	3.75	0.93	0.07000	2.79	0.0588	8.82	0.8

Source: Guo, W. D., and H.Y. Qin, *Can Geotech J*, 47, 2, 2010.

[a] Poulos et al. 1995

32.9~44.1 Nm (between 31.0 and 71.6 Nm) induced in two, three, and four piles in a row.

The study on the single pile tests shows k (single) = 16.13 kPa (k/G = 2.39, G = 6.75 kPa). With k_{group} = (0.64~1.0)k_{single}, the k_{group} is calculated as 10.32~16.13 kPa. With $M_{max} = y_o kL^2/11.2$, the M_{max} at y_o = 24 mm is calculated as 40.3~63.01Nm (= 10.32~16.13 × 0.024 × 0.675²/2.8). This estimation compares well with the measured values of 40~55 Nm. The angle was estimated as 3.1~3.82° [= 1.5 × (24~30)/675/π × 180°] using free-head solution of $\omega = 1.5y_o/L$. The estimation agrees with the measured values of 3.5~4.5°. The free-standing pile groups under the rotational soil movement thus also exhibit features of free-head, floating-base piles.

Overall, Equations 13.5 and 13.6 offer good estimations of M_{max} (thus T_{max}) for all the current model piles (e.g., Table 13.6) and the previous tests (e.g., Table 13.7). With regard to single piles, the 3~5 times difference in the M_{max} is owing to the impact of the pile dimensions (via L and k/G), the subgrade modulus k, the effective movement w_e, and the loading manner (w_i). As for piles in groups, the high magnitude y_o^* in Chen's tests principally renders a higher M_{max} (of 31.0~71.6 Nm) than the current values (15.3~54.6 Nm, triangular, see Table 13.2).

13.4.8 Calibration against in situ test piles

The simple correlations of Equation 13.4 are validated using measured response of eight in situ test piles (see Chapter 12, this book) and one centrifuge test pile subjected to soil movement. The pile and soil properties are tabulated in Table 13.8, along with the measured values of the maximum bending moment M_{max}. The shear force T_{max}, however, was measured for

Table 13.8 $M_{max}/(T_{max}L)$ calculated for field tests

Piles			*Soil*		*Measured*			
D/t[a] (mm)	E_p[a] *(GPa)*	L_1/L_2[a] *(m)*	k_1/k_2[a] *(MPa)*	Lc_1/Lc_2[a] *(m)*	M_{max}[a,b] *(kNm)*	T_{max}[a] *(kN)*	$\frac{M_{max}}{T_{max}L}$[b]	*References*
790/395	20	7.5/22.5	$\frac{8.0}{8.0}$	19.5/19.5	903	310	0.388	Esu and D'Elia (1974)
318.5/6.9	210	11.2/12.8	$\frac{5.0}{8.0}$	6.3/5.6	165.2	144–150[c]	0.18–0.20	Hataosi-2[d]
318.5/6.9	210	8.0/9.0	$\frac{5.0}{15.0}$	6.3/4.9	65.7	70–71.2[c]	0.15–0.19	Hataosi-3[d]
318.5/6.9	210	6.5/7.5	$\frac{5.0}{8.0}$	6.3/5.6	197.2	143–300[c]	0.10–0.25	Kamimoku-4[d]
318.5/6.9	210	4.0/6.0	$\frac{5.0}{8.0}$	6.3/5.6	290.3	231–250[c]	0.18–0.22	Kamimoku-6[d]
300/60	20	7.3/5.7	$\frac{6.0}{10.0}$	3.2/2.8	69.5	40–56.2[c]	0.38–0.62	Katamachi-B[d]
1200/600	20	9.5/13.0	$\frac{15}{15}$	12.7/12.7	2,250	600	0.395	Carrubba et al. (1989)
630/315	28.45	2.5/10.0	$\frac{14.4}{28.8}$	23.1/23.1	60.2	56–60[c]	0.40–0.43	Leung et al. (2000) (2.5 m)
630/315	28.45	3.5/9.0	$\frac{14.4}{28.8}$	23.1/23.1	73.8	65–85[c]	0.25-0.32	Leung et al. (2000) (3.5 m)
630/315	28.45	4.5/8.0	$\frac{14.4}{28.8}$	23.1/23.1	81.2	72-100[c]	0.18–0.25	Leung et al. (2000) (4.5 m)

continued

Table 13.8 (Continued) $M_{max}/(T_{max}L)$ calculated for field tests

Piles			*Soil*		*Measured*			
D/t[a] (mm)	E_p[a] *(GPa)*	L_1/L_2[a] *(m)*	k_1/k_2[a] *(MPa)*	Lc_1/Lc_2[a] *(m)*	M_{max}[a,b] *(kNm)*	T_{max}[a] *(kN)*	$\frac{M_{max}}{T_{max}L}$[b]	*References*
915/19	1070[e]	6.8/4.2	10.0	11.9/11.9	901.9[f]	532.6[f]	0.249[f]	5*Nov*, 86[g]
			10.0		312.5	221.9	0.207	4*Nov*, 86
					1102.3[f]	642.3[f]	0.252[f]	11*Nov*, 88[g]
					536.6	309.5	0.255	10*Nov*, 88
					1473.5[f]	796.4[f]	0.272[f]	1*Oct*, 92[g]
					544.3	378.5	0.211	30*Sept*, 92
					1434.2[f]	834.7[f]	0.253[f]	6*July*, 95[g]
					756.1	487.0	0.228	5*July*, 95

Source: Guo, W. D., and H.Y. Qin, *Can Geotech* J, 47, 2, 2010.

[a] Chapter 12, this book; *D* = outside diameter; *t* = wall thickness; E_p = Young's modulus of pile; L_1/L_2 = thickness of sliding/stable layer; k_1/k_2 = subgrade modulus of sliding/stable layer; L_{c1}/L_{c2} = equivalent length for rigid pile in sliding/stable layer. In the estimation, G_i was simply taken as $k_i/3$.

[b] M_{max} = measured maximum bending moment; *L* = the smallest value of L_i and L_{ci}.

[c] estimated using elastic and elastic-plastic solutions against measured bending moment and pile deflection and soil movement profiles.

[d] Cai and Ugai (2003).

[e] E_pI_p = flexural stiffness.

[f] All for sliding layers.

[g] Frank and Pouget (2008).

three of the nine piles. The T_{max} for the other six piles was thus taken as that deduced using elastic and elastic-plastic theory (Cai and Ugai 2003; Guo 2009). Modulus of subgrade reaction k_i, and equivalent length of rigid pile L_{ci} were calculated previously (Guo 2009). The length L for each pile was taken as the smallest values of L_i and L_{ci}. This allows the ratio $M_{max}/(T_{max}L)$ for each case to be evaluated. The results are tabulated in Table 13.8 and are plotted in Figure 13.15. The ratios all fall into the range of the elastic solutions capitalized on constant k to the plastic solution of Equation 13.3. The slightly higher ratio for the exceptional Katamachi-B pile is anticipated (Guo 2009). It may argue that the four piles with a ratio of 0.26–0.4 exhibit elastic-plastic pile–soil interaction and with an eccentricity greater than 0 (or dragging moments).

The ratio $M_{max}/(T_{max}L)$ is independent of loading level for either the model tests or the field test. Figure 13.12 shows that the ratio $M_{max}/(T_{max}L)$ from model tests stays almost invariably at 0.357, regardless of loading level. Figure 13.16 indicates the ratio for the in situ pile in sliding layer (Frank and Pouget 2008) stays around 0.25 for the 16 year test duration, although a low ratio is noted for stable layer (not a concern for practical design).

Determination of T_{max} depends on pile-head or base constraints (Guo and Lee 2001): $y_o = T_{max}/(kL)$ or $y_o = (1\sim4)T_{max}/(kL)$ for fully fixed head or semi-fixed head, long piles, respectively.

Example 13.8 Determining T_{max} (free-head piles)

The in situ pile (Frank and Pouget 2008) at pre-pull-back situation is evaluated using the free-head solution. The k obtained was 15 MPa

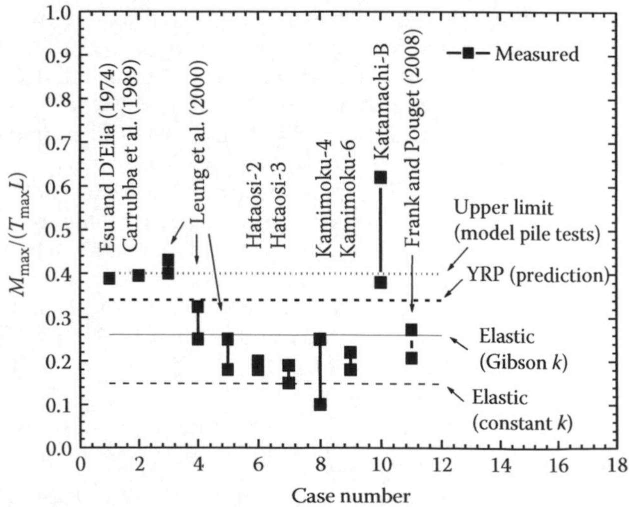

Figure 13.15 Calculated versus measured ratios of $M_{max}/(T_{max}L)$. (Revised from Guo, W. D., and H. Y. Qin, *Can Geotech J* 47, 22, 2010.)

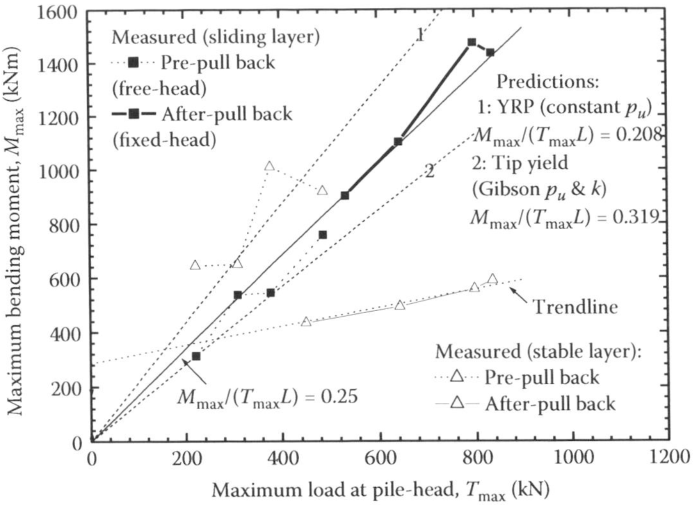

Figure 13.16 Measured M_{max} and T_{max} (Frank and Pouget 2008) versus predicted values. (Revised from Guo, W. D., and H. Y. Qin, *Can Geotech J* 47, 22, 2010.)

(= $150s_u$, undrained shear strength, s_u = 100kPa). At a groundline deflection y_o of 32 mm (recorded on 05/07/1995), the T_{max} was estimated as 816 kN (= $y_okL/4$). This T_{max} agrees well with the measured load of 845 kN. Note the measured pile deflection increases approximately linearly from groundline to a depth of 6.8~8.0m, exhibiting "rigid" characteristics.

Example 13.9 Determining T_{max} (fixed-head piles)

The deflection and bending moment are calculated for the two-row piles used to stabilize a sliding slope (Kalteziotis et al. 1993). The steel piles had a length of 12 m, an external diameter of 1.03 m, a wall thickness t of 18 mm, and a flexural stiffness E_pI_p of 1,540 MNm^2. Given $k = k_1$ = 15 MPa (Chen and Poulos 1997) and an equivalent rigid pile length $L = L_1$ = 4 m (sliding depth), the T_{max} was calculated as 45 kN (= $y_okL/4$) at y_o = 0.003 m. This T_{max} compares well with the measured 40~45 kN. The T_{max} gives a uniform on-pile force per unit length of 10–11.25 kN/m. The moment is thus estimated as 80~90 kNm [= 0.5 × (10~11.25) × 4^2] about the sliding depth, and as 180~202.5 kNm about the depth 6 m. The average moment agrees well with the measured 150 kNm, considering that the depth of sliding may be 4–6 m (Chow 1996; Chen and Poulos 1997).

13.5 CONCLUSION

An experimental apparatus was developed to investigate the behavior of vertically loaded piles and two-pile groups in sand undergoing lateral soil movement. A large number of tests have been conducted to date. Presented here are nineteen typical tests on single piles and eleven tests on capped two piles in a row under a triangular or uniform loading block. The results are provided regarding the total force on shear frames, the induced shear force, bending moment, and deflection along the piles. The tests enable simple solutions to be proposed for predicting the pile response.

The following features are noted from the model tests:

- Maximum bending moment increases by 60% (d = 32 mm piles) or by 30% (d = 50 mm piles), and its depth by ~50% upon applying a static load of 7–9% the maximum driving force.
- The ratio of y_o/ω of two-pile rows exhibits features of free-head, floating-base rigid piles.
- Shear force T_{max} (thrust) in each pile may reach ~$0.6\delta F\gamma_s L_m^2$; ultimate M_{max2} (uniform) reaches (0.15~0.9)M_{max2} (triangular); and limiting force per unit length p_u reaches $s_g\gamma_s K_p^2 dz$.
- Ultimate M_{max} and T_{max} are capped by the limits gained using solutions of laterally loaded piles at yield at rotation point, but for the increase (~5% for model piles) in M_{max} due to dragging moment. Response of the piles becomes negligible, once the relative distance s_b (between loading block and pile center) exceeds ~1.7L.
- A constant ratio of pile deflection y_o over soil movement w_e for each single or capped pile is noted, despite of rigid movement in w_e. y_o/w_e is 0.2~0.8, with low values for a low SD. The pile rows manifest a rigid free-head rotation; as such the deflection is related to bending moment by $M_{max2} - M_{o2} = ky_oL^2/11.2$, and $M_{max}/(A_r dL^3) \leq 0.036$~$0.055$ (for SD = 0.29L~.57L).
- The moment M_{maxi} is largely proportional to the thrust T_{maxi} in stable layer and moving layer, with $M_{max2} - M_{o2} = T_{max2}L/2.4$~$2.8$ and $T_{max1}L/(M_{max1} - M_{o1}) = 4$~$30$.
- The reduction in subgrade modulus k for piles in group resembles the p-multiplier for laterally loaded piles, and $k_{group}/k_{single} = 0.4$~$1.4$. The shear modulus around piles is higher under a uniform loading block than a triangular one. The ratio y_okL/T_{max1} changes from 3 (uniform block) to 1 (triangular block).

With respect to the solutions, the following conclusions can be drawn:

- Equation 13.7 may be used to estimate the maximum bending moment M_{max}, for which the sliding thrust T_{max} is calculated using Equation 13.6 and capped by ultimate plastic state (gained using limiting force profile). The estimation should adopt an effective frame movement of $w_f - w_i$, in which the w_i depends on the pile diameter, pile position, and loading manner.

- The subgrade modulus k may be estimated using the theoretical ratio of k/G and the shear modulus G (varying with diameter). The G increases from overall pile–soil–shear box interaction to local pile–soil interaction.
- The proposed solutions offer good prediction of the translational and rotational tests, in which 3~5 times the current M_{max} are noted, despite of similar dimensions; and correct ranges of $M_{max}/(T_{max}L)$ for eight in situ test piles and a centrifuge test pile. The M_{oi} depends on pile-cap relative stiffness, fixity/connection, and loading eccentricity. Under unsymmetrical loading or even unfit connection between piles and pile cap, the thrust and moment relationship are still valid. The M_{max2} (>M_{max1}) may be employed to design passive piles.

The parameter α correlated soil movement with y_o depends on soil movement profiles and location of movement against pile location, which need to be examined using more in situ tests.

ACKNOWLEDGMENT

Dr. Enghow Ghee and Dr. Hongyu Qin, along with other students, conducted the model tests as part of their PhD research work.

References

Abdoun, T., R. Dobry, T. D. O'Rourke and S. H. Goh. 2003. Pile response to lateral spreads: Centrifuge modeling. *J Geotech and Geoenvir Engrg*, ASCE 129(10):869–78.

ACI. 1993. Bridges, substructures, sanitary, and other special structures-structural properties. *ACI Manual of Concrete Practice 1993*, Part 4. Detroit, MI: ACI.

Alem, A. and M. Gherbi. 2000. Graphs for the design of laterally loaded piles in clay. *Proc Int Conf on Geotechnical and Geological Engrg, GeoEng2000*, Melbourne, Australia.

Alizadeh, M. and M. T. Davisson. 1970. Lateral load test piles, Arkansas river project. *J Soil Mech and Found Engrg*, ASCE 96(5):1583–604.

Anagnostopoulos, C. and M. Georgiadis. 1993. Interaction of axial and lateral pile responses. *J Geotech Engrg*, ASCE 119(4):793–8.

API. 1993. RP-2A: Recommended practice for planning, designing and constructing fixed offshore platforms. Washington, DC: American Petroleum Institute.

Ashour, M. and G. Norris. 2000. Modeling lateral soil-pile response based on soil-pile interaction. *J Geotech and Geoenvir Engrg*, ASCE 126(5):420–8.

Aubeny, C. P., S. W. Han and J. D. Murff. 2003a. Inclined load capacity of suction caissons. *Int J Numer and Anal Meth in Geomech* 27(14):1235–54.

Aubeny, C. P., S. W. Han and J. D. Murff. 2003b. Suction caisson capacity in anisotropic, purely cohesive soil. *Int J of Geomechanics* 3(2):225–35.

Baguelin, F., R. Frank and Y. H. Said. 1977. Theoretical study of lateral reaction mechanism of piles. *Geotechnique* 27(3):405–34.

Balfour-Beatty-Construction. 1986. Report on foundation design for overhead catenary system. London, U.K.: Rep. Tuen Muen Light Railway Transit.

Baligh, M. M. 1985. Strain path method. *J Geotech Engrg*, ASCE 111(9):1108–36.

Banerjee, P. K. and T. G. Davies. 1977. Analysis of pile groups embedded in Gibson soil. *Proc 9th Int Conf on Soil Mech and Found Engrg*, Tokyo, 1:381–6.

Banerjee, P. K. and T. G. Davies. 1978. The behavior of axially and laterally loaded single piles embedded in non-homogeneous soils. *Geotechnique* 28(3):309–26.

Barden, L. and M. F. Monckton. 1970. Tests on model pile groups in soft stiff clay. *Geotechnique* 20(1):94–96.

Barron, R. 1948. Consolidation of fine grained soils by drain wells. *Transactions*, ASCE 113:718–42.

Barton, Y. O. 1982. Laterally loaded model piles in sand: Centrifuge tests and finite element analysis. PhD thesis, University of Cambridge.

Basu, D. and R. Salgado. 2008. Analysis of laterally loaded piles with rectangular cross sections embedded in layered soil. *Int J Numer and Anal Meth in Geomech* 32(7):721–44.

Berezantzev, V. G., V. S. Khristoforv and V. N. Golubkov. 1961. Load-bearing capacity and deformation of piled foundations. *Proc 5th Int Conf on Soil Mech and Found Engrg, speciality session 10*, Paris, 2:11–5.

Bergdahl, U. and G. Hult. 1981. Load tests on friction piles in clay. *Proc 10th Int Conf on Soil Mech and Found Engrg*, Stockholm: A. A. Balkema.

Bieniawski, Z. T. 1989. *Engineering rock mass classifications*. New York: John Wiley & Sons.

Biot, M. A. 1937. Bending of an infinite beams on an elastic foundation. Transaction, ASME, *J of App Mech*, ASCE 59:A1–A7.

Biot, M. A. 1941. General theory of three-dimensional consolidation. *J Applied Physics* 12:155–64.

Bjerrum, L. 1973. Problems of soil mechanics and construction on soft clays and structurally unstable soils: collapsible, expansive and others. *Proc 8th Int Conf Soil Mech Found Engrg*, 111–59.

Bjerrum, L., I. J. Johannessen and O. Eide. 1969. Reduction of skin friction on steel piles to rock. *Proc 7th Int Conf on Soil Mech and Found Engrg*, 27–34.

Bolton, M. D. 1986. The strength and dilatancy of sands. *Geotechnique* 36(1):65–78.

Booker, J. R. and H. G. Poulos. 1976. Analysis of creep settlement of pile foundations. *J Geotech Engrg*, ASCE 102(1):1–14.

Booker, J. R. and J. C. Small. 1977. Finite element analysis of primary and secondary consolidation. *Int J of Solids and Structures* 13(2):137–49.

Bowles, J. E. 1997. *Foundation analysis and design*. New York and London: McGraw-Hill Int.

Bozozuk, M., B. H. Fellenius and L. Samson. 1978. Soil disturbance from pile driving in sensitive clay. *Can Geotech J* 15(3):346–61.

Brand, E. W., C. Muktabhant, and A. Taechathummarak. 1972. Load tests on small foundations in soft clay. *Proc. Performance of Earth and Earth-Supported Structure, ASCE Specialty Conference* 1(2):903–8.

Brandenberg, S. J., R. W. Boulanger and B. L. Kutter. 2005. Discussion of single piles in lateral spreads: Field bending moment evaluation by R. Dobry, T. Abdoun, T. D. O'Rourke, and S. H. Goh. *J Geotech and Geoenvir Engrg*, ASCE 131(4):529–34.

Bransby, M. F. and S. M. Springman. 1997. Centrifuge modelling of pile groups adjacent to surcharge loads. *Soils and Foundations* 37(2):39–49.

Briaud, J. L., T. D. Smith and B. Meyer. 1983. Laterally loaded piles and the pressuremeter: Comparison of exsiting methods. *Laterally loaded deep foundations: Analysis and performance*. ASTM, STP835. J. A. Langer, E. T. Mosley and C. D. Thompson, eds. Kansas City, MO.

Briaud, J. L., T. D. Smith and L. M. Tucker. 1985. A pressuremeter method for laterally loaded piles. *Proc 10th Int Conf Soil Mech and Found Engrg*, San Francisco: A. A. Balkema, 3:1353–6.

Briaud, J. L., L. M. Tucker and E. Ng. 1989. Axially loaded 5 pile group and single pile in sand. *Proc 12th Int Conf on Soil Mech and Found Engrg*, 2:1121–4.

Brinch Hansen, J. 1961. The ultimate resistance of rigid piles against transversal forces. Bulletin No. 12. Copenhagen: The Danish Geotechnical Institute.

Brinch Hansen, J. 1968. A theory for skin friction on piles. Bulletin No. 25. Copenhagen: The Danish Geotechnical Institute.

Broms, B. 1964a. The lateral response of piles in cohesionless soils. *J Soil Mech and Found Engrg*, ASCE 90(3):123–56.

Broms, B. 1964b. The lateral response of piles in cohesive soils. *J Soil Mech and Found Engrg*, ASCE 90(2):27–63.

Broms, B. 1966. Methods of calculating the ultimate bearing capacity of piles: A summary. *Sols-Soils* 18–19:21–32.

Broms, B. 1979. Negative skin friction. *Proc 6th Asian Regional Conf on Soil Mech and Found Engrg*, 41–75.

Brown, D. 2002. Effect of construction on axial capacity of drilled foundations in Piedmont soils. *J Geotech and Geoenvir Engrg*, ASCE 128(12):967–73.

Brown, D. A., C. Morrison and L. C. Reese. 1988. Lateral load behaviour of pile group in sand. *J Geotech and Geoenvir Engrg*, ASCE 114(11): 1261–76.

BSI. 1981. Code of practice for "site investigation." BS 5830, BSI standards.

BSI. 1985. Structural use of concrete. BS 8110, London, Parts 1–3.

BSI. 1994. Code of practice for "Earth retaining strctures". BS 8002, BSI standards.

Budhu, M. and T. G. Davies. 1988. Analysis of laterally loaded piles in soft clays. *J Geotech Engrg*, ASCE 114(1):21–39.

Burland, J. B. 1973. Shaft friction of piles in clay—A simple fundamental approach. *Ground Engineering* 6(3):30–42.

Burland, J. B. and M. C. Burbidge. 1985. Settlement of foundations on sand and gravel. *Proc Institution of Civil Engineers* I(78):1325–81.

Butterfield, R. and P. K. Banerjee. 1971. The elastic analysis of compressible piles and pile groups. *Geotechnique* 21(1):43–60.

Butterfield, R. and R. A. Douglas. 1981. Flexibility coefficients for the design of piles and pile groups. CIRIA: Technical Note 108, London, CIRIA.

Buttling, S. and S. A. Robinson. 1987. Bored piles: Design and testing. *Proc Singapore Mass Rapid Transit Conf*, Singapore, 155–75.

Byrne, P. M., D. L. Anderson and W. Janzen. 1984. Response of piles and casings to horizontal free-field soil displacements. *Can Geotech J* 21(1):720–5.

Cai, F. and K. Ugai. 2003. Response of flexible piles under laterally linear movement of the sliding layer in landslides. *Can Geotech J* 40(1):46–53.

Cappozoli, L. J. 1968. Pile test program at St. Gabriel. Louisiana, Louis J. Cappozolli & Associates, internal report.

Carrubba, P., M. Maugeri and E. Motta. 1989. Esperienze in vera grandezza sul comportamento di pali per la stabilizzaaione di un pendio. *Proc XVII Convegno Nazionale di Geotechica, Assn Geotec*, Italiana, 1:81–90.

Carter, J. P. and F. H. Kulhawy. 1992. Analysis of laterally loaded shafts in rock. *J Geotech Engrg*, ASCE 118(6):839–55.

Casagrande, A. and S. Wison. 1951. Effect of rate loading on the strength of clays and shales at constant water content. *Geotechnique* 2(3):251–63.

Castelli, F., E. Maugeri and M. E. Motta. 1999. Discussion on "Design of laterally loaded piles in cohesive soils using p-y curves" by D. Wu, B. B., Broms, and V. Choa, 1998. *Soils and Foundations* 39(6):133–4.

Chan, K. S., P. Karasudhl and S. L. Lee. 1974. Force at a point in the interior of layered elastic half-space. *Int J Solids Struct* 10(11):1179–99.

Chandler, R. J. 1968. The shaft friction of piles in cohesive soil in terms of effective stress. *Civ Engng Publ Wks Rev* 63:48–51.

Chang, M. F. and A. T. C. Goh. 1988. Performance of bored piles in residual soils and weathered rocks. *Proc Int Geotech Sem,* Deep foundation on bored and auger piles. Rotterdam: Balkema, 303–13.

Chen, C.-Y. and G. R. Martin. 2002. Soil-structure interaction for landslide stabilizing piles. *Computers and Geotechnics* 29(5):363–86.

Chen, L. T. 1994. The effect of lateral soil movements on pile foundations. PhD thesis, civil engineering. Syndey, Australia: University of Sydney.

Chen, L. T. and H. G. Poulos. 1997. Piles subjected to lateral soil movements. *J Geotech and Geoenvir Engrg*, ASCE 123(9):802–11.

Chen, L. T., H. G. Poulos and T. S. Hull. 1997. Model tests on pile groups subjected to lateral soil movement. *Soils and Foundations* 37(1):1–12.

Chen, L. T., H. G. Poulos, C. F. Leung, Y. K. Chow and R. F. Shen. 2002. Discussion of "Behavior of pile subject to excavation-induced soil movement". *J Geotech and Geoenvir Engrg*, ASCE 128(3):279–81.

Cheung, Y. K., L. G. Tham and D. J. Guo. 1988. Analysis of pile group by infinite layer method. *Geotechnique* 38(3):415–31.

Chin, F. K. 1970. Estimation of the ultimate load of piles not carried to failure. *Proc 2nd Southeast Conf on Soil Eng*, Singapore:81–90.

Chin J. T. 1996. Back analysis of an instrumented bored piles in Singapore old alluvium. *Proc 12th Southeast Asian Geotechnical Conf*, 441–46.

Chin, J. T., Y. K. Chow and H. G. Poulos. 1990. Numerical analysis of axially loaded vertical piles and pile groups. *Computers and Geotechnics* 9:273–90.

Chinese Hydaulic Engineering Code (SD128-86). 1987. *Manual for soil testing*, 2nd ed. (in Chinese), Beijing: Chinese hydraulic-electric press.

Chmoulian, A. 2004. Briefing: Analysis of piled stabilization of landslides. *Proc Institution Civil Engineers, Geotechnical Engineering* 157(2):55–6.

Cho, K. H., M. A. Gabr, S. Clark and R. H. Borden. 2007. Field p-y curves in weathered rock. *Can Geotech J* 44:753–64.

Chow, Y. K. 1986a. Analysis of vertically loaded pile groups. *Int J Numer and Anal Meth in Geomech* 10(1):59–72.

Chow, Y. K. 1986b. Discrete element analysis of settlement of pile groups. *Computers and Structures* 24(1):157–66.

Chow, Y. K. 1987. Axial and lateral response of pile groups embedded in nonhomogenous soils. *Int J Numer and Anal Meth in Geomech* 11(6):621–38.

Chow, Y. K. 1989. Axially loaded piles and pile groups embedded in a cross-anisotropic soil. *Geotechnique* 39(2):203–11.

Chow, Y. K. 1996. Analysis of piles used for slope stabilization. *Int J Numer and Anal Meth in Geomech* 20(9):635–46.

Chow, Y. K., J. T. Chin and S. L. Lee. 1990. Negative skin friction on pile groups. *Int J Numer and Anal Meth in Geomech* 14:75–91.

Choy, C. K., J. R. Standing and R. J. Mair. 2007. Stability of a loaded pile adjacent to a slurry-supported trench. *Geotechnique* 57(10):807–19.

Christensen, R. W. and T. H. Wu. 1964. Analyses of clay deformation as a rate process. *J Soil Mech and Found Engrg*, ASCE 90(6):125–57.

Christian, J. T. and J. W. Boehmer. 1972. Plane strain consolidation by finite elements. *J Soil Mech and Found Engrg*, ASCE 96(4):1435–57.
Christie, I. F. 1964. A re-appraisal of Mechant's contribution to the theory of consolidation. *Geotechnique* 14(4):309–20.
Clancy, P. and M. F. Randolph. 1993. An approximate analysis procedeure for piled raft foundations. *Int J Numer and Anal Meth in Geomech* 17(12): 849–69.
Clark, J. I. and G. G. Meyerhof. 1972. The behaviour of piles driven clay. An investigation of soil stress and pore water pressure as related to soil properties. *Can Geotech J* 9(4):351–73.
Clayton, C. R. I. 1978. A note on the effects of density on the results of standard penetration tests in chalk. *Geotechnique* 28(1):119–22.
Coduto, D. P. 1994. *Foundation design*. Englewood Cliffs, NJ: Prentice Hall.
Comodromos, E. M. and K. D. Pitilakis. 2005. Response evaluation for horizontally loaded fixed-head pile groups using 3-D non-linear analysis. *Int J Numer and Anal Meth in Geomech* 29(10):597–625.
Cook, R. D. 1995. *Finite element modeling for stress analysis*. New York: John Wiley & Sons, Inc.
Cooke, R. W. 1974. The settlement of friction pile foundations. *Proc Conf on Tall Buildings*, Kuala Lumpur, Malaysia.
Cooke, R. W., D. W. Bryden-Smith, M. N. Gooch and D. F. Sillett. 1981. Some observations of the foundation loading and settlement of a multi-storey building on a pile raft foundation in London Clay. *Proc Int Civil Eng Conf*, Part 1.
Cooke, R. W., G. Price and K. Tarr. 1980. Jacked piles in London clay: Interaction and group behavior under working conditions. *Geotechnique* 30(2): 97–136.
Cornforth, D. H. 1964. Some experiments on the influence of strain conditions on the strength of sand. *Geotechnique* 16:193.
Cowper, G. R. 1966. The shear coefficients in Timoshenkos beam theory. *J Appl Mech* 33:335–40.
Cox, W. R., L. C. Reese and B. R. Grubbs. 1974. Field testing of laterally loaded piles in sand. *Proc 6th Int Offshore Technology Conference*, Houston, Texas, paper OTC 2079, 459–72.
Coyle, H. M. and L. C. Reese. 1966. Load transfer for axially loaded piles in clay. *J Soil Mech and Found Engrg*, ASCE 92(2):1–26.
CPSE. 1965. Lateral bearing capacity and dynamic behavior of pile foundation [in Japanese]. Committee on Piles Subjected to Earthquake (CPSE), Architectural Institute of Japan, 1–69.
Craig, R. F. 1997. *Soil mechanics*. London: Spon Press, an imprint of Taylor & Francis Group.
Cryer, C. W. 1963. A comparison of the three-dimensional consolidation theories of Biot and Terzaghi. *Quart. J Mech and Math* 16:401–12.
Daloglu, A. T. and C. V. G. Vallabhan. 2000. Values of *k* for slab on Winkler foundation. *J Geotech Engrg*, ASCE 126(5):463–71.
D'Appolonia, D. J. and T. W. Lambe. 1971. Performance of four foundations on end-bearing piles. *J Soil Mech and Found Engrg*, ASCE 97(1):77–93.
Das, B. M. 1990. *Principles of foundation engineering*. Boston, MA: PWS-Kent.
Davis, E. H. and H. G. Poulos. 1972. Rate of settlement under two- and three-dimensional conditions. *Geotechnique* 22(1):95–114.

Davis, E. H. and G. P. Raymond. 1965. A non-linear theory of consolidation. *Geotechnique* 15(2):161–73.

Davisson, M. T. 1972. High capacity piles. *Proc Soil Mech Lecture Series on Innovations in Foundation Construction*, ASCE, Illinois section, Chicago, 81–112.

Davisson, M. T. and H. L. Gill. 1963. Laterally-loaded piles in a layered soil system. *J Soil Mech and Found Engrg*, ASCE 89(3):63–94.

De Beer, E. and R. Carpentier. 1977. Discussion on "methods to estimate lateral force acting on stabilising piles," Ito, T., and Matsui, T. 1975. *Soil and Foundations* 17(1):68–82.

Dickin, E. A. and R. B. Nazir. 1999. Moment-carrying capacity of short pile foundations in cohesionless soil. *J Geotech and Geoenvir Engrg*, ASCE 125(1): 1–10.

Dickin, E. A. and M. J. Wei. 1991. Moment-carrying capacity of short piles in sand. *Proc of Centrifuge*, Rotterdam, Netherlands: A. A. Balkema.

DiGioia, A. M. and L. F. Rojas-Gonzalez. 1993. Discussion on analysis of laterally loaded shafts in rock. *J Geotech Engrg*, ASCE 119(12):2014–5.

Dobereiner, L. and M. H. Freitas. 1986. Geotechnical properties of weak sandstones. *Geotechnique* 36(1):79–94.

Dobry, R., T. Abdoun, T. D. O'Rourke and S. H. Goh. 2003. Single piles in lateral spreads: Field bending moment evaluation. *J Geotech and Geoenvir Engrg*, ASCE 129(10):879–89.

Dodds, A. 2005. A numerical study of pile behaviour in large pile groups under lateral loading. California, University of Southern California; PhD dissertation.

Duncan, J. M. and C. Y. Chang. 1970. Non-linear analysis of stress and strain in soils. *J Soil Mech and Found Engrg*, ASCE 96(5):1629–52.

Duncan, J. M., L. T. Evans and P. S. K. Ooi. 1994. Lateral load analysis of single piles and drilled shafts. *J Geotech Engrg*, ASCE 120(5):1018–33.

Duncan, J. M., M. D. Robinette and R. L. Mokwa. 2005. Analysis of laterally loaded pile groups with partial pile head fixity. *Advances in Deep Foundations*, GAS 132, ASCE.

Dyson, G. J. and M. F. Randolph. 2001. Monotonic lateral loading of piles in calcareous sand. *J. of Geotechnical and Geoenvironmental Engineering*, ASCE 127(4):346–52.

EC2. 1992. Design of concrete structures—Part 1–1 General rules and rules for buildings. ENV 1992-1-1. Brussels, Belgium.

Edil, T. B. and I. B. Mochtar. 1988. Creep response of model pile in clay. *J Geotech Engrg*, ASCE 114(11):1245–59.

Eide, O., J. N. Hutchinson and A. Landva. 1961. Short and long term test loading of a friction pile in clay. *Proc 5th Int Conf on Soil Mech and Found Engrg*, Paris, 2:45–53.

Ellis, E. A., I. K. Durrani and D. J. Reddish. 2010. Numerical modelling of discrete pile rows for slope stability and generic guidance for design. *Geotechnique* 60(3):185–95.

Ellis, E. A. and S. M. Springman. 2001. Full-height piled bridge abutments constructed on soft clay. *Geotechnique* 51(1):3–14.

England, M. 1992. Pile settlement behaviour: An accurate model. Application of the stress wave theory to piles. Rotterdam: Balkema, 91–6.

Esu, F. and B. D'Elia. 1974. Interazione terreno-struttura in un palo sollecitato dauna frana tipo colata. *Rivista Italiana di Geotechica* 8(1):27–38.

Fang, H.-Y. 1991. *Foundation engineering handbook*. New York: Van Nostrand Reinhold.

Fear, C. E. and P. K. Robertson. 1995. Estimating the undrained shear strength of sand: a theoretical framework. *Canadian Geotechnical Journal* 32: 859–70.

Feda, J. 1992. Creep of soils and related phenomena: Developments in Geotechnical Engineering 68. Prague: Publishing House of the Czechoslovak Academy of Sciences.

Fellenius, B. H. 1980. The analysis of results from routine pile load tests. *Ground Engineering* 13(6):19–31.

Fellenius, B. H. and B. B. Broms. 1969. Negative skin friction for long piles driven in clay. *Proc 7th Int Conf on Soil Mech and Found Engrg*, Mexico, 93–8.

Fellenius, B. H. and L. Samson. 1976. Testing of drivability of concrete piles and disturbance to sensitive clay. *Can Geotech J* 13(2):139–60.

FHWA. 1993. COM624P-laterally loaded pile analysis program for the microcomputer. No. FHWA-SA-91–048. Washington, DC, 2.

Flaate, K. 1972. Effects of pile driving in clay. *Can Geotech J* 9(1):81–8.

Flaate, K. and P. Selnes. 1977. Side friction of piles in clay. *Proc 9th Int Conf on Soil Mech and Found Engrg*, Tokyo, Japan, 1:517–22.

Fleming, W. G. K., A. J. Weltman, M. F. Randolph and W. K. Elson. 2009. *Piling engineering*, 3rd ed. London and New York: Taylor & Francis.

Foray, P. Y. and J. L. Colliat. 2005. CPT-based design method for steel pipe piles driven in very dense silica sands compared to the Euripides pile test results. Frontiers in offshore geotechnics, ISFOG. Perth: Taylor & Francis, 669–75.

Frank, R. 1974. Etude theorique du comportement des pieux sous charge verticale. introduction de la dilatance. France, University Paris, Doctor-Engineer thesis.

Frank, R. and P. Pouget. 2008. Experimental pile subjected to long duration thrusts owing to a moving slope. *Geotechnique* 58(8):645–58.

Fujita, H. 1976. Study on prediction of load vs. settlement relationship for a single pile. Special report of Hazama-Gumi, Ltd, No.1. (in Japanese.)

Fukuoka, M. 1977. The effects of horizontal loads on piles due to landslides. *Proc 9th Int Conf on Soil Mech and Found Engrg, speciality session 10*, Tokyo, 1:27–42.

Gabr, M. A., R. H. Borden, K. H. Cho, S. C. Clark and J. B. Nixon. 2002. p-y curves for laterally loaded drilled shafts embedded in weathered rock FHWA/NC/2002-08. North Carolina Dept. of Transportation.

Gandhi, S. R. and S. Selvam. 1997. Group effect on driven piles under lateral load. *J Geotech and Geoenvir Engrg*, ASCE 123(8):702–9.

Gavin, K. G., D. Cadogan and P. Casey. 2009. Shaft capacity of continuous flight auger piles in sand. *J Geotech and Geoenvir Engrg*, ASCE 135(6):790–8.

Gazetas, G. and R. Dobry. 1984. Horizontal response of piles in layered soil. *J Geotech and Geoenvir Engrg*, ASCE 110(1):20–40.

Georgiadis, M. 1983. Development of p-y curves for layered soil. *Proc Geotech Practice in Offshore Engineering*, ASCE, 1:536–45.

Geuze, E. C. W. A. and T. K. Tan. 1953. The mechanical behaviour of clays. *Proc 2nd Int Congress on Rheology*.

Gill, H. L. 1969. Soil-pile interaction under lateral loading. Technical Report R-670. California, U. S. Naval Civil Engineering Laboratory.

Glick, G. W. 1948. Influence of soft ground in the design of long piles. *Proc. 2nd Int Conf on Soil Mechanics and Foundation Eng* 4:84–8.

Golait, Y. S. and R. K. Katti. 1987. Stress-strain idealization of Bombay High calcareous soils. *Proc 8th Asian Conf Soil Mech and Found Engrg*, ICC, 2:173–7.

Golait, Y. S. and R. K. Katti. 1988. Some aspects of behavior of piles in calcareous soil media under offshore loading conditions. *Proc Engrg for Calcareous Sediments*, Perth, Australia., Rotterdam: Balkema, 1:199–207.

Golightly, C. R. and A. F. L. Hyde. 1988. Some fundamental properties of carbonate sands. *Proc Engrg for Calcareous Sediments*, Perth, Australia. Rotterdam: Balkema, 1:69–78.

Green, P. and D. Hight. 1976. The instrumentation of Dashwood House. CIRIA(78).

Guo, D. J., L. G. Tham and Y. K. Cheung. 1987. Infinite layer for the analysis of a single pile. *Computers and Geotechnics* 3(4):229–49.

Guo, W. D. 1997. Analytical and numerical solutions for pile foundations. PhD thesis, Department of civil and environmental engineering, Perth, Australia, The University of Western Australia.

Guo, W. D. 2000a. Vertically loaded single piles in Gibson soil. *J Geotech and Geoenvir Engrg*, ASCE 126(2):189–93.

Guo, W. D. 2000b. Visco-elastic load transfer models for axially loaded piles. *Int J Numer and Anal Meth in Geomech* 24(2):135–63.

Guo, W. D. 2000c. Visco-elastic consolidation subsequent to pile installation. *Computers and Geotechnics* 262):113–44.

Guo, W. D. 2001a. Subgrade modulus for laterally loaded piles. *Proc 8th Int Conf on Civil and Structural Engrg Computing*, paper 112, Eisenstadt, nr Vienna, 19–21 September 2001.

Guo, W. D. 2001b. Lateral pile response due to interface yielding. In *Proc 8th Int Conf on Civil and Structural EngrgComputing*. Stirling, UK: Civil-Comp Press, paper 108.

Guo, W. D. 2001c. Pile capacity in a non-homogenous softening soil. *Soils and Foundations* 41(2):111–20.

Guo, W. D. 2003a. Response of laterally loaded rigid piles. *Proc 12th Panamerican Conf on Soil Mech and Geot Engrg*, Cambridge, Massachusetts, Verlag Gluckauf GMBH, Essen, Germany, 2:1957–62.

Guo, W. D. 2003b. A simplified approach for piles due to soil movement. *Proc 12th Panamerican Conf on Soil Mech and Found Engrg* Cambridge, Massachusetts, Verlag Gluckauf GMBH, Essen, Germany, 2:2215–20.

Guo, W. D. 2005. Limiting force profile and laterally loaded pile groups. *Proc 6th Int Conf on Tall Buildings*. Hong Kong, China: World Scientific, 1:280–5.

Guo, W. D. 2006. On limiting force profile, slip depth and lateral pile response. *Computers and Geotechnics* 33(11):47–67.

Guo, W. D. 2007. Nonlinear behaviour of laterally loaded free-head pile groups. *Proc 10th ANZ Conf on Geomechanics*, Brisbane, Australia, 1:466–71.

Guo, W. D. 2008. Laterally loaded rigid piles in cohesionless soil. *Can Geotech J* 45(5):676–97.

Guo, W. D. 2009. Nonlinear behaviour of laterally loaded fixed-head piles and pile groups. *Int J Numer and Anal Meth in Geomech* 33(7):879–914.

Guo, W. D. 2010. Predicting non-linear response of laterally loaded pile groups via simple solutions. *Proc GeoFlorida 2010 Conf*, West Palm Beach, Florida, GSP 199, ASCE:1442–9.

Guo, W. D. 2012. A simple model for nonlinear response of fifty-two laterally loaded piles. *J Geotech Geoenviron Engrg*, ASCE, doi:http://dx.doi.org/10.1061/(ASCE)GT.1943-5606.0000726.

Guo, W. D. In press. p_u based solutions for slope stabilising piles *Int J of Geomechanics*. ASCE, doi:http://dx.doi.org/10.1061/(ASCE)GM.1943-5622.0000201

Guo, W. D. and E. H. Ghee. 2004. Model tests on single piles in sand subjected to lateral soil movement. *Proc 18th Australasian Conf on Mechanics of Structures and Materials*, Perth, Australia, Rotterdam, The Netherlands: A. A. Balkema, 2:997–1003.

Guo, W. D. and E. H. Ghee. 2005. A preliminary investigation into the effect of axial load on piles subjected to lateral soil movement. *Proc 1st Int Symp on Frontiers in Offshore Geotechnics*, Perth, Australia, Taylor & Francis, 1: 865–71.

Guo, W. D. and E. H. Ghee. 2010. Model tests on free-standing passive pile groups in sand. *Proc 7th Int Conf on Physical Modelling in Geotechnics*. ICPMG 2010, Zurich, Switzerland, 2:873–8.

Guo, W. D. and F. H. Lee. 2001. Load transfer approach for laterally loaded piles. *Int J Numer and Anal Meth in Geomech* 25(11):1101–29.

Guo, W. D. and H. Y. Qin. 2005. Vertically loaded single piles in sand subjected to lateral soil movement. *Proc 6th Int Conf on Tall Buildings*. ICTB-VI, Hong Kong, China: World Scientific Publishing Co. Pte. Ltd, 1:327–32.

Guo, W. D. and H. Y. Qin. 2006. Vertically loaded piles in sand subjected to triangular profiles of soil movements. *Proc. 10th Int. Conf. on Piling and Foundations*, Amsterdam, Netherlands 1, paper no. 1371.

Guo, W. D. and H. Y. Qin. 2010. Thrust and bending moment of rigid piles subjected to moving soil. *Can Geotech J* 47(22):180–96.

Guo, W. D., H. Y. Qin and E. H. Ghee. 2006. Effect of soil movement profiles on vertically loaded single piles. *Proc. Int. Conf. on Physical Modelling in Geotechnics*, Hong Kong, China. London, UK: Taylor & Francis Group plc, 2:841–6.

Guo, W. D. and M. F. Randolph. 1996. Torsional piles in non-homogeneous media. *Computers and Geotechnics* 19(4):265–87.

Guo, W. D. and M. F. Randolph. 1997a. Vertically loaded piles in non-homogeneous media. *Int J Numer and Anal Meth in Geomech* 21(8):507–32.

Guo, W. D. and M. F. Randolph. 1997b. Non-linear visco-elastic analysis of piles through spreadsheet program. *Proc 9th Int Conf of Int Association for Computer Methods and Advances in Geomechanics—IACMAG 97*, Wuhan, China, 3:2105–10.

Guo, W. D., and M. F. Randolph. 1998. Rationality of load transfer approach for pile analysis. *Computers and Geotechnics* 23(1–2):85–112.

Guo, W. D. and M. F. Randolph. 1999. An efficient approach for settlement prediction of pile groups. *Geotechnique* 49(2):161–79.

Guo, W. D. and B. T. Zhu. 2004. Laterally loaded fixed-head piles in sand. *Proc 9th ANZ Conf. on Geomechanics*, Auckland, New Zealand, 1:88–94.

Guo, W. D. and B. T. Zhu. 2005a. Limiting force profile for laterally loaded piles in clay. *Australian Geomechanics* 40(3):67–80.

Guo, W. D. and B. T. Zhu. 2005b. Static and cyclic behaviour of laterally loaded piles in calcareous sand. *International Symposium on Frontiers in Offshore Geotechnics*, Perth, Australia. Taylor & Francis/Balkema, 1:373–9.

Guo, W. D. and B. T. Zhu. 2010. Nonlinear response of 20 laterally loaded piles in sand. *Australian Geomechanics* 45(2):67–84.

Guo, W. D. and B. T. Zhu. 2011. Structure nonlinearity and response of laterally loaded piles. *Australian Geomechanics* 46(3):41–52.

Gurtowski, T. M. and M. J. Wu. 1984. Compression load tests on concrete piles in alluvium. *Anal and Design of Pile Foundations*, ASCE:138–53.

Haberfield, C. M. and B. Collingwood. 2006. Rock-socketed pile design and construction: A better way? *Proc Institution of Civil Engineers, Geotechnial Engineering* 159(GE3):207–17.

Hanbo, S. and R. Källström. 1983. A case study of two alternative foundation principles. *Väg-och Vattenbyggaren* 7-8:23–7.

Hara, A., T. Ohta, M. Niva, S. Tanaka and T. Banno. 1974. Shear modulus and shear strength of cohesive soils. *Soils and Foundations* 14(3):1–12.

Henke, S. 2009. Influence of pile installation on adjacent structures *Int J Numer and Anal Meth in Geomech* 34(11):1191–210.

Hetenyi, M. 1946. *Beams on elastic foundations*. Ann Arbor: University of Michigan Press.

Hewitt, P. B. and S. S. Gue. 1996. Foundation design and performance of a twin 39 storey building. *Proc 7th ANZ Conference on Geomechanics*, Adelaide, Australia, 637–42.

Hirayama, H. 1990. Load-settlement analysis for bored piles using hyperbolic transfer function. *Soils and Foundations* 30(11):55–64.

Ho, C. E. and C. G.Tan. 1996. Barrette foundation constructed under polymer slurry support in old alluvium. *Proc 12th Southeast Asian Geotechnical Conf*, Kuala Lumpur, 379–84.

Hoek, E. 2000. *Practical rock engineering*, http://www.rockscience.com/hoek/Practical RockEngineering.asp.

Hoek, E. and E. T. Brown. 1995. The Hoek-Brown criterion – a 1988 update. *Proc 15th Can Rock Mech Symp, University of Toronto*, Toronto, 31–8.

Hoek, E., P. K. Kaiser and W. F. Bawden. 1995. *Support of underground excavations in hard rock*. Rotterdam, Netherlands: Balkema.

Holtz, W. G. and H. J. Gibbs. 1979. Discussion of SPT and relative density in coase sand. *J Geotech Engrg*, ASCE 105(3):439–41.

Horvath, R. G., T. C. Kenney and P. Kozicki. 1983. Methods for improving the performance of drilled piers in weak rock. *Can Geotech J* 20:758–72.

Hsiung, Y.-M. 2003. Theoretical elastic-plastic solutions for laterally loaded piles. *J Geotech and Geoenvir Engrg*, ASCE 129(5):475–80.

Hsiung, Y.-M. and Y. L. Chen. 1997. Simplified method for analyzing laterally loaded single piles in clay. *J Geotech and Geoenvir Engrg*, ASCE 123(11):1018–29.

Hsu, T. C. 1993. *Unified theory of reinforced concrete*. Boca Raton, FL: CRC Press.

Hudson, M. J., G. Mostyn, E. A. Wiltsie and A. M. Hyden. 1988. Properties of near surface Bass Strait soils. *Proc Engrg for Calcareous Sediments*, Perth, Australia. Rotterdam: Balkema, 1:25–34.

Ilyas, T., C. F. Leung, Y. K. Chow and S. S. Budi. 2004. Centrifuge model study of laterally loaded pile groups in clay. *J Geotech and Geoenvir Engrg*, ASCE 130(3):274–83.

Indraratna, B., A. S. Balasubramaniam, P. Phamvan and Y. K. Wong. 1992. Development of negative skin friction on driven piles in soft Bangkok clay. *Can Geotech J* 29(3):393–404.

Ishihara, K. 1993. Liquefaction and flow failure during earthquakes. The 33rd Rankine Lecture. *Géotechnique* 43(3):349–415.

Itasca, F. 1992. *Users manual*. Minneapolis: Itasca Consulting Group.

Ito, T. and T. Matsui. 1975. Methods to estimate lateral force acting on stabilising piles. *Soils and Foundations* 15(4):43–59.

Ito, T., T. Matsui and W. P. Hong. 1981. Extended design method for multi-row stabilizing piles against landslide. *Soils and Foundations* 21(1):21–37.

Jardine, R. J. and F. C. Chow. 1996. *New design methods for offshore piles*. London: Marine Technology Directorate.

Jardine, R. J., D. M. Potts, A. B. Fourie and J. B. Burland. 1986. Studies of the influence of non-linear stress-strain characteristics in soil-structure interaction. *Geotechnique* 36(3):377–96.

Jeong, S., B. Kim, J. Won and J. Lee. 2003. Uncoupled analysis of stabilizing piles in weathered slopes. *Computers and Geotechnics* 30(8):671–82.

Jeong, S., J. Lee and C. J. Lee. 2004. Slip effect at the pile-soil interface on dragload. *Computers and Geotechnics* 31(2):115–26.

Jimiolkwoski, M. and A. Garassino. 1977. Soil modulus for laterally loaded piles. *Proc 9th Int Conf Soil Mech and Found Engrg, speciality session 10*, Tokyo, 43–58.

Johannessen, I. J. and L. Bjerrum. 1965. Measurement of the compression of a steel pile to rock due to settlement of the surrounding clay. *Proc 6th Int Conf on Soil Mech and Found Engrg*, 261–4.

Johnston, I. W. and T. S. K. Lam. 1989. Shear behaviour of regular triangular concrete/rock joints—analysis. *J Geotech Engrg*, ASCE 115(5):711–27.

Jones, R. and J. Xenophontos. 1977. The Vlasov foundation model. *Int J Mech Sci Div*:317–23.

JSCE. 1986. Interpretation of the design code for Japanese railway structures. Earth Pressure Retaining Structures and Foundations, Japanese Society of Civil Engrg, Tokyo, Japan.

Kalteziotis, N., H. Zervogiannis, R. Frank, G. Seve and J.-C. Berche. 1993. Experimental study of landslide stabilization by large diameter piles. *Proc Int Symp on Geotech Engrg of Hard Soils–Soft Rocks*, A. A. Rotterdam: Balkema, 2:1115–24.

Kaniraj, S. R. 1993. A semi-empirical equation for settlement ratio of pile foundations in sand. *Soils and Foundations* 33(2):82–90.

Karthigeyan, S., VVGST, Ramakrishna and K. Rajagopal. 2007. Numerical investigation of the effect of vertical load on the lateral response of piles. *J Geotech and Geoenvir Engrg*, ASCE 133(5):512–21.

Kenney, T. C. 1959. Discussion. *Proceedings*, ASCE 85(SM3):67–79.

Kishida, H. 1967. Ultimate bearing capacity of piles driven into loose sand. *Soils and Foundations* 7(3):20–9.

Kishida, H. and S. Nakai. 1977. Large deflection of a single pile under horizontal load. *Proc 9th Int Conf Soil Mech and Found Engrg, Speciality session 10*, Tokyo, 87–92.

Knappett, J. A. and S. P. G. Madabhushi. 2009. Influence of axial load on lateral pile response in liquefiable soils. Part II: numerical modelling. *Geotechnique* 59(7):583–92.

Koerner, R. M. and A. Partos. 1974. Settlement of building on pile foundation in sand. *J Geotech Engrg*, ASCE 100(3):265–78.

Koizumi, Y. and K. Ito. 1967. Field tests with regard to pile driving and bearing capacity of piled foundations. *Soils and Foundations* 7(3):30–53.

Kolk, H. J. and E. V. D. Velde. 1996. A reliable method to determine friction capacity of piles driven into clays. *Proc Offshore Technology Conf*, OTC 7993, Houston.

Komamura, F. and R. J. Huang. 1974. New rheological model for soil behaviour. *J Geotech Engrg*, ASCE 100(7):807–24.

Konrad, J.-M., and M. Roy. 1987. Bearing capacity of friction piles in marine clay. *Geotechnique* 372:163–75.

Konrad, J.-M., and B. D. Watts. 1995. Undrained shear strength for liquefaction flow failure analysis. *Canadian Geotechnical Journal* 32:783–94.

Kovacs, W. D. and L. A. Salmone. 1982. SPT hammer energy measurement. *J Geotech Engrg*, ASCE 108:599–620.

Kraft, L. M., R. P. Ray and T. Kagawa. 1981. Theoretical t-z Curves. *J Geotech Engrg*, ASCE 107(11):1543–61.

Kulhawy, F. H. 1984. Limiting tip and side resistance: Fact or fallacy? Analysis and design of pile foundations. *Proc ASCE Geotech Eng Symp*, 80–98.

Kulhawy, F. H. and R. E. Goodman. 1980. Design of foundations on discontinuous rock. *Proc Int Conf on Structural Found on Rock*, Sydney, Australia.

Kulhawy, F. H. and C. S. Jackson. 1989. Some observations on undrained side resistance of drilled shafts. Foundation engineering: Current principles and practices. New York: F. H. Kulhawy, ASCE:1011–25.

Kulhawy, F. H. and P. W. Mayne. 1990. Manual on estimating soil properties for foundation design. Electric power research institute, Palo Alto, CA, Report EL-2870.

Kulhawy, F. H. and K. K. Phoon. 1993. Drilled shaft side resistance in clay soil to rock. *Proc Conf on Design and Perform Deep Foundations: Piles and Piers in Soil and Soft Rock*, ASCE:172–83.

Kuwabara, F. 1991. Settlement behaviour of nonlinear soil around single piles subjected to vertical loads. *Soil and Foundations* 31(1):39–46.

Laman, M., G. J. W. King, and E. A. Dickin. 1999. Three-dimensional finite element studies of the moment-carrying capacity of short pier foundations in cohesionless soil. *Computers and Geotechnics* 25:141–55.

Lambe, T. W. and R. V. Whitman. 1979. *Soil mechanics*. New York: John Wiley & Sons.

Lee, C. Y. 1991. Axial response analysis of axially loaded piles and pile groups. *Computers and Geotechnics* 11(4):295–313.

Lee, C. Y. 1993a. Settlement of pile group-practical approach. *J Geotech Engrg*, ASCE 119(9):1449–61.

Lee, C. Y. 1993b. Pile group settlement analysis by hybrid layer approach. *J Geotech Engrg*, ASCE 119(6):984–97.

Lee, E. H. 1955. Stress analysis in viscoelastic bodies. *Quart J Mech and Math* 13:183.

Lee, E. H. 1956. Stress analysis in viscoelastic bodies. *J of Applied Physics* 27(7):665–72.

Lee, E. H., J. M. Radok and W. B. Woodward. 1959. Stress analysis for linear viscoelastic materials. *Transactions, Society of Rheology* 3:41–59.

Lee, J. H. and R. Salgado. 1999. Determination of pile base resistance in sands. *J Geotech and Geoenvir* Engrg, ASCE 125(8):673–83.

Lchane, B. M., C. Gaudin and J. A. Schneider. 2005. Scale effects on tension capacity for rough piles buried in dense sand. *Geotechnique* 55(10):709–19.

Lehane, B. M. and R. J. Jardine. 1994. Shaft capacity of driven piles in sand:a new design approach. *Proc 7th Int Conf. on Behaviour of Offshore Structure*, Boston, 23–36.

Lehane, B. M., J. A. Schneider and X. Xu. 2005. A review of design methods for offshore driven piles in siliceous sand. Frontiers in offshore geotechnics, ISFOG, Perth., Taylor & Francis, 683–9.

Leonardo, Z. 1973. *Foundation Engineering for Difficult Subsoil Condition*. Van Norstrand Reinhold Company.

Leung, C. F., Y. K. Chow and R. F. Shen. 2000. Behaviour of pile subject to excavation-induced soil movement. *J Geotech and Geoenvir Engrg*, ASCE 126(11): 947–54.

Leussink, H., W. Wittke and K. Weseloh. 1966. Unterschiede im Scherverhalten rolliger Erdstoffe und KugelschuÈttungen im Dreiaxial- und Biaxialversuch. VeroÈff. Inst. Bodenmech. Felsmech. TH Frideric Karlsruhe, 21.

Liang, R., K. Yang and J. Nusairat. 2009. p-y criterion for rock mass. *J Geotech and Geoenvir Engrg*, ASCE 135(1):26–36.

Liu, J. L. 1990. Via Chinese technical code for buidling pile foundations. No. JGJ 94–94.

Liu, J. L., Z. L. Yuan and K. P. Zhang. 1985. Cap-pile-soil interaction of bored pile groups. *Proc 11th Int Conf on Soil Mech and Found Engrg*, Rotterdam: Balkema, 1433–6.

Lo, K. Y. 1961. Secondary compression of clays. *J Soil Mech and Found Engrg*, ASCE 87(4):61–87.

Lo, S.-C. R. and K. S. Li. 2003. Influence of a permanent liner on the skin friction of large-diameter bored piles in Hong Kong granitic saprolites. *Can Geotech J* 40:793–805.

Long, J., M. Maniaci, G. Menezes and R. Ball. 2004. Results of lateral load tests on micropiles. *Proc Sessions of the Geosupport Conference: Innovation and Cooperation in the Geo-industry*, Orlando, Florida.

Lui, E. M. 1997. *Structural steel design*. New York: CRC Press.

Mandel, J. 1957. Consolidation des Couches Dargiles. *Proc 4th Int Conf on Soil Mech and Found Engrg*, England, 1:360–7.

Mandolini, A. and C. Viggiani. 1997. Settlement of piled foundations. *Geotechnique* 47(4):791–816.

Mase, G. E. 1970. *Theory and problems of continuum mechanics*. New York: McGraw-Hill.

Matlock, H. 1970. Correlations for design of laterally loaded piles in soft clay. *Proc 2nd Annual Offshore Technol Conf, OTC 1204*, Dallas, Texas, 577–94.

Matlock, H. and L. C. Reese. 1960. Generalized solutions for laterally loaded piles. *J Soil Mech and Found Engrg*, ASCE 86(5):63–91.

Matlock, H., B. Wayne, A. E. Kelly and D. Board. 1980. Field tests of the lateral load behaviour of pile groups in soft clay. *Proc 12th Annual Offshore Technology Conf, OTC 3871*, Houston, Texas, 671–86.

Mayne, P. W. and F. H. Kulhawy. 1991. Load-displacement behaviour of laterally loaded rigid shafts in clay. *Proc 4th Int Conf Piling and Deep Found*, Stresa, Italy. Rotterdam: A. A. Balkema, 1:409–13.

Mayne, P. W., F. H. Kulhawy and C. H. Trautmann. 1995. Laboratory modeling of laterally-loaded drilled shafts in clay. *J Geotech and Geoenvir Engrg*, ASCE 121(12):827–35.

McClelland, B. 1974. Design of deep penetration piles for ocean structures. *J Geotech Engrg*, ASCE 100(7):705–47.

McClelland, B. and J. A. Focht. 1958. Soil modulus for laterally loaded piles. *Transactions*, ASCE, paper no. 2954, 1049–63.

McCorkle, B. L. 1969. Side-bearing pier foundations. *Civil Engineering*, New York. 39:65–6.

McKinley, J. D. 1998. Coupled consolidation of a solid, infinite cylinder using a Terzaghi formulation. *Computers and Geotechnics* 23:193–204.

McLachlan, N. W. 1955. *Bessel functions for engineers*. New York: Oxford University Press.

McVay, M., R. Casper and T. L. Shang. 1995. Lateral response of three-row groups in loose to dense sands at 3D and 5D pile spacing. *J Geotech Engineer*, ASCE 121(5):436–41.

McVay, M., L. Zhang, T. Molinit and P. Lai. 1998. Centrifuge testing of large laterally loaded pile. *J Geotech and Geoenvir Engrg*, ASCE 124(10):1016–26.

Mesri, G. 2001. Undrained shear strength of soft clays from push cone penetration test. *Geotechnique* 51(2):167–8.

Meyerhof, G. G. 1956. Penetration tests and bearing capacity of cohesionless soils. *J Soil Mech and Found Engrg*, ASCE 82(1):1–19.

Meyerhof, G. G. 1957. Discussion on research on determining the density of sands by spoon penetration testing. *Proc 4th Int Conf on Soil Mech and Found Engrg*, London.

Meyerhof, G. G. 1959. Compaction of sands and the bearing capacity of piles. *J Soil Mech and Found Engrg*, ASCE 85(6):1–29.

Meyerhof, G. G. 1976. Bearing capacity and settlement of pile foundation. *J Geotech Engrg*, ASCE 102(3):197–228.

Meyerhof, G. G., S. K. Mathur and A. J. Valsangkar. 1981. Lateral resistance and deflection of rigid wall and piles in layered soils. *Can Geotech J* 18:159–70.

Meyerhof, G. G. and V. V. R. N. Sastray. 1985. Bearing capacity of rigid piles under eccentric and inclined loads. *Can Geotech J* 22(3):267–76.

Meyerhof, G. G. and A. S. Yalcin. 1984. Pile capacity for eccentric inclined load in clay. *Can Geotech J* 21(3):389–96.

Meyerhof, G. G., Yalcin, A. S. and Mathur, S. K. 1983. Ultimate pile capacity for eccentric inclined load. *J Geotech Engrg*, ASCE 109(3):408–23.

Mitchell, J. K. 1964. Shearing resistance of soils as a rate process. *J Soil Mech and Found Engrg*, ASCE 90(1):29–61.

Mitchell, J. K. 1976. *Fundamentals of soil behavior*. New York: John Wiley and Sons.

Mitchell, J. K., R. G. Campanella and A. Singh. 1968. Soil creep as a rate process. *J Soil Mech and Found Engrg*, ASCE 94(1):231–253.

Mitchell, J. K. and Z. V. Solymar. 1984. Time-dependent strength gain in freshly deposited or densified sand. *J Geotech Engrg*, ASCE 110(11):1559–76.

Mokwa, R. L. and J. M. Duncan. 2005. Discussion on Centrifuge model study of laterally loaded pile groups in clay by T. Ilyas, C. F. Leung, Y. K. Chow and S. S. Budi. *J Geotech and Geoenvir Engrg*, ASCE 131(10):1305-1307.

Montgomery, M. W. 1980. Prediction and verification of friction pile behaviour. *Proc ASCE Symposium on Deep Foundations*, Atlanta, GA, 274–87.

Mostafa, Y. E. and M. H. E. Naggar. 2006. Effect of seabed instability on fixed offshore platforms. *Soil Dynamics and Earthquake Engineering* 2612):1127–42.

Motta, E. 1994. Approximate elastic-plastic solution for axially loaded piles. *J Geotech Engrg*, ASCE 120(9):1616–24.

Motta, E. 1995. Discussion on Pipelines and laterally loaded piles in elastoplastic medium by Ranjani and Morgenstern. *J Geotech Engrg*, ASCE 121(1):91–3.

Motta, E. 1997. Discussion on Laboratory modeling of laterally-loaded drilled shafts in clay by Mayne, P. W., Kulhawy, F. H. and Trautmann, C. H. *J Geotech and Geoenvir Engrg*, ASCE 123(5):489–90.

Murayama, S. and T. Shibata. 1961. Rheological properties of clays. *Proc 5th Int Conf on Soil Mech and Found Engrg*, Paris, France.

Murff, J. D. 1975. Response of axially loaded piles. *J Geotech Engrg*, ASCE 101(3):357–60.

Murff, J. D. 1980. Pile capacity in a softening soil. *Int J Numer and Anal Meth in Geomech* 4:185–9.

Murff, J. D. and J. M. Hamilton. 1993. P-Ultimate for undrained analysis of laterally loaded piles. *J Geotech Engrg*, ASCE 119(1):91–107.

Murray, R. T. 1978. Developments in two- and three-dimensional consolidation theory, Applied Science Publishers, U. K.

Nakai, S. and H. Kishida. 1982. Nonlinear analysis of a laterally loaded pile. *Proc 4th Int Conf on Numerical Methods in Geomechanics*, Edmonton:835–42.

Neely, W. J. 1991. Bearing capacity of auger-cast piles in sand. *J Geotech Engrg*, ASCE 117(2):331–45.

Ng, C. W. W., T. L. Y. Yau, J. H. M. Li and W. H. Tang. 2001. Side resistance of large diameter bored piles socketed into decomposed rocks. *J Geotech and Geoenvir Engrg*, ASCE 127(8):642–57.

Ng, C. W. W., L. Zhang and D. C. N. Nip. 2001. Response of laterally loaded large-diameter bored pile groups. *J Geotech and Geoenvir Engrg*, ASCE 127(8):658-69.

Nilson, A. H., D. Darwin and C. W. Dolan. 2004. *Design of concrete structures. 13th edition.* New York: McGraw Hill.

Nixon, J. B. 2002. Verification of the weathered rock model for p-y curves. Civil Engineering. North Carolina, USA, North Carolina State University. M. Sc. thesis.

Nogami, T. and O. ONeill. 1985. Beam on generalized two-parameter foundation. *J of Engrg Mechanics*, ASCE 111(5):664–79.

Novello, E. A. 1999. From static to cyclic p-y data in calcareous sediments. *Proc Engrg for calcareous sediments*, Perth, Australia, Balkema, Rotterdam, 1:17–27.

Nyman, K. J. 1980. Field load tests of instrumented drilled shafts in coral limestone. Civil Engineering. Houston, Texas, The University of Texas at Austin. MS thesis.

Olson, R. E. 1990. Axial load capacity of steel pipe piles in clay. *Proc 22nd Offshore Technology Conf*, OTC 4464, Houston, U.S.A.:4181–92.

O'Neill, M. W. 1983. Group action in offshore piles. *Geotechnical practice in offshore engineering*, ASCE:25–64.

O'Neill, M. W., O. I. Ghazzaly and H. B. Ha. 1977. Analysis of three-dimensional pile groups with non-linear soil response and pile-soil-pile interaction. *Proc 9th Annual OTC*, Houston, Paper OTC 2838:245–56.

O'Neill, M. W., R. A. Hawkins and L. J. Mahar. 1982. Load-transfer mechanisms in piles and pile groups. *J Geotech Engrg*, ASCE 108(12):1605–23.

O'Neill, M. W. and L. C. Reese. 1999. Drilled shaft: Construction procedures and design methods. Publication No. FHWA-IF-99-025. Washington, DC, U. S. Dept. of Transportation, Federal Highways Administration.

Ooi, P. S. K., B. K. F. Chang and S. Wang. 2004. Simplified lateral load analyses of fixed-head piles and pile groups. *J Geotech and Geoenvir Engrg*, ASCE 130(11):1140–51.

Orrje, O. and B. Broms. 1967. Effect of pile driving on soil properties. *J Soil Mech and Found Engrg*, ASCE 93(5):59–73.

Osterman, J. 1959. Notes on the shearing resistance of soft clays. *Acta polytechnica Scandinavica* 263.

Ottaviani, M. 1975. Three-dimensional finite element analysis of vertically loaded pile groups. *Geotechnique* 25(2):159–74.

Pan, J. L., A. T. C. Goh, K. S. Wong and C. I. Teh. 2002. Ultimate soil pressure for piles subjected to lateral soil movements. *J Geotech and Geoenvir Engrg*, ASCE 128(6):530–5.

Parry, R. H. G. and C. W. Swain. 1977a. Effective stress methods of calculating skin friction on driven piles in soft clay. *Ground Engineering* 10(3):24–6.

Peck, R. B. and A. R. S. Bazaraa. 1969. Discussion on settlement of spread footings on sand. *J Soil Mech Fdns Div Am Soc Ciu Engis* 95:905–9.

Peck, R. B., W. E. Hanson and T. H. Thornburn. 1953. *Foundation engineering*. New York: John Wiley & Sons.

Pedley, M. J., R. A. Jewell and G. W. E. Milligan. 1990. A large scale experimental study of soil-reinforced interaction–Part I. *Ground Engrg*, 45–8.

Pells, P. J. N., G. Mostyn and B. F. Walker. 1998. Foundations on sandstone and shale in the Sydney region. *Australian Geomechanics* 33(4):17–29.

Petrasovits, G. and A. Award. 1972. Ultimate lateral resistance of a rigid pile in cohesionless soil. *Proc 5th European Conf on Soil Mech and Found Engrg*, Madrid.

Pise, P. 1982. Laterally loaded piles in a two-layer soil system. *J Geotech Engrg*, ASCE 108(9):1177–81.

Potts, D. M. 2003. Numerical analysis:a virtual dream or practical reality? *Geotechnique* 53(6):535–73.

Potyondy, J. G. 1961. Skin friction between various soils and construction materials. *Geotechnique* 11(4):339–53.

Poulos, H. G. 1968. Analysis of the settlement of pile groups. *Geotechnique* 18(4):449–71.

Poulos, H. G. 1971. Behaviour of laterally loaded piles: II-pile groups. *J Soil Mech and Found Engrg*, ASCE 97(5):733–51.

Poulos, H. G. 1975. Torsional response of piles. *J Geotech Engrg*, ASCE 101(10):107–18.

Poulos, H. G. 1979. Settlement of single piles in nonhomogeneous soil. *J Geotech Engrg*, ASCE 105(5):627–42.

Poulos, H. G. 1981. Cyclic axial response of single pile. *J Geotech Engrg*, ASCE 107(1):41–58.

Poulos, H. G. 1988. Modified calculation of pile-group settlement interaction. *J Geotech Engrg*, ASCE 114(6):697–706.

Poulos, H. G. 1989. Pile behaviour-theory and application. *Geotechnique* 39(3):365–415.

Poulos, H. G. 1993. Settlement of bored pile groups. *Proc BAP II, Ghent*, Balkema, Rotterdam, 103–17.

Poulos, H. G. 1995. Design of reinforcing piles to increase slope stability. *Can Geotech J* 32(5):808–18.

Poulos, H. G., L. T. Chen and T. S. Hull. 1995. Model tests on single piles subjected to lateral soil movement. *Soils and Foundations* 35(4):85–92.

Poulos, H. G. and E. H. Davis. 1980. *Pile foundation analysis and design*. New York: John Wiley & Sons.

Poulos, H. G. and T. S. Hull. 1989. The role of analytical geomechanics in foundation engineering. *Foundation engineering:current principles and practices*, Evanston, Illinois, ASCE, 2:1578–1606.

Poulos, H. G. and N. S. Mattes. 1969. The analysis of downdrag in end-bearing piles. *Proc 7th Int Conf on Soil Mech and Found Engrg*, Mexico, Sociedad Mexicana de Mecanica de Suelos, 2:203–9.

Poulos, H. G. and M. F. Randolph. 1983. Pile group analysis:A study of two methods. *J Geotech Engrg*, ASCE 109(3):355–72.

Powell, J. J. M. and R. S. T. Quarterman. 1988. The interpretation of cone penetration tests in clays, with particular reference to rate effects. *Proc 1st Int Symp on Penetration Tests*:903–9.

Prakash, S. and H. D. Sharma. 1989. *Pile foundations in engineering practice*. New York: John Wiley & Sons, Inc.

Prasad, Y. V. S. N. and T. R. Chari. 1999. Lateral capacity of model rigid piles in cohesionless soils. *Soils and Foundations* 39(2):21–9.

Pressley, J. S. and H. G. Poulos. 1986. Finite element analysis of mechanisms of pile group behaviour. *Int J Numer and Anal Meth in Geomech* 10(2):213–21.

Price, G. and I. F. Wardle. 1981. Lateral load tests on large diameter bored piles. Contractor Report 46. Transport and Road Research Laboratory, Department of Transport, Crowthorne, Berkshire, England.

Qian, J. H., W. B. Zhao, Y. K. Cheung and P. K. K. Lee. 1992. The theory and practice of vacuum preloading. *Computers and Geotechnics* 13(2):103–18.

Qin, H. Y. 2010. Response of pile foundations due to lateral force and soil movements. PhD thesis, School of Engineering. Gold Coast, Griffith University.

Rajani, B. B. and N. R. Morgenstern. 1993. Pipelines and laterally loaded piles in elastoplastic medium. *J Geotech Engrg*, ASCE 119(9):1431–47.

Rajapakse, R. K. N. D. 1990. Response of axially loaded elastic pile in a Gibson soil. *Geotechnique* 40(22):237–49.

Ramalho Ortigão, J. A. and M. F. Randolph. 1983. Creep effects on tension piles for the design of buoyant offshore structures. *Int Sympo on Offshore Engrg*.

Ramaswamy, S. D. 1982. Pressuremeter correlations with standard penetration and cone penetration tests. *Proc 2nd European Symp on Penetration Testing*, Stockholm.

Randolph, M. F. 1981a. Piles subjected to torsion. *J Geotech Engrg*, ASCE 107(8):1095–111.

Randolph, M. F. 1981b. The response of flexible piles to lateral loading. *Geotechnique* 31(2):247–59.

Randolph, M. F. 1983. Design consideration for offshore piles. *Proc Conf on Geot, Practice in Offshore Engrg, ASCE*, Austin, Texas, 1:422–39.

Randolph, M. F. 1988a. The axial capacity of deep foundation in calcareous soil. *Proc Engrg for Calcareous Sediments*, Perth, Australia, Rotterdam: Balkema, 2:837–57.

Randolph, M. F. 1988b. Evaluation of grouted insert pile performance. *Proc Engrg for Calcareous Sediments*, Perth, Australia, Rotterdam: Balkema, 2:617–26.

Randolph, M. F. 1994. Design methods for pile groups and piled rafts. *Proc 13th ICSMFE* New Delhi, India, 5:61–82.

Randolph, M. F. 2003a. RATZ manual version 4.2: Load transfer analysis of axially loaded piles. RATZ Software Manual. Perth, Australia.

Randolph, M. F. 2003b. Science and empiricism in pile foundation design. *Geotechnique* 53(10):847–75.

Randolph, M. F., J. Dolwin and R. D. Beck. 1994. Design of driven piles in sand. *Geotechnique* 44(3):427–48.

Randolph, M. F. and G. T. Houlsby. 1984. The limiting pressure on a circular pile loaded laterally in cohesive soil. *Geotechnique* 34(4):613–23.

Randolph, M. F., R. J. Jewell and H. G. Poulos. 1988. Evaluation of pile lateral load performance. *Proc Engrg for Calcareous Sediments*, Perth, Australia, Balkema, Rotterdam, 1:639–45.

Randolph, M. F. and B. S. Murphy. 1985. Shaft capacity of driven piles in clay. *Proc 17th Offshore Technology Conf*, Houston, Texas.

Randolph, M. F. and C. P. Wroth. 1978. Analysis of deformation of vertically loaded Piles. *J Geotech Engrg*, ASCE 104(12):1465–88.

Randolph, M. F. and C. P. Wroth. 1979a. An analytical solution for the consolidation around a driven pile. *Int J Numer and Anal Meth in Geomech* 3:217–29.

Randolph, M. F. and C. P. Wroth. 1979b. An analysis of the vertically deformation of pile groups. *Geotechnique* 29(4):423–39.

Reese, L. C. 1958. Soil modulus for laterally loaded piles. *Transactions*, ASCE, 123: 1071–4.

Reese, L. C. 1977. Laterally loaded piles: program documentation. *J Geotech Engrg*, ASCE 103(4):287–305.

Reese, L. C. 1997. Analysis of laterally loaded shafts in weak rock. *J Geotech and Geoenvir Engrg*, ASCE 123(11):1010–7.

Reese, L. C., J. D. Allen and J. Q. Hargrove. 1981. Laterally loaded piles in layered soils. *Proc 10th Int Conf Soil Mech and Found Engrg*, Stockholm: A. A. Balkema, 2:819–22.

Reese, L. C., W. R. Cox and F. D. Koop. 1974. Analysis of laterally loaded piles in sand. *Proc 6th Annual Offshore Technology Conf*, OTC 2080, Dallas, Texas, 2:473–83.

Reese, L. C., W. R. Cox and F. D. Koop. 1975. Field testing and analysis of laterally loaded piles in stiff clay. *Proc 7th Annual Offshore Technology Conf*, OTC 2312, Dallas, Texas, 671–90.

Reese, L. C. and M. W. O'Neill. 1989. New design method for drilled shaft from common soil and rock tests. *Proc Foundation Engr Current Principles and Practices*, ASCE:1026–39.

Reese, L. C. and W. F. Van Impe. 2001. Single piles and pile groups under lateral loading. Rotterdam: A. A. Balkema.

Reese, L. C., S. G. Wright, J. M. Roesset, L. H. Hayes, R. Dobry and C. V. G. Vallabhan. 1988. Analysis of piles subjected to lateral loading by storm-generated waves. *Proc Engrg for Calcareous Sediments*, Perth, Australia. Balkema: Rotterdam, 1:647–54.

Rendulic, L. 1936. Porenziffer und porenwasser druck in tonen. Der Bauingenieur 17(51/53):559–64.

Renfrey, G. E., C. A. Waterton and P. V. Goudoever. 1988. Geotechnical data used for the design of the North Rankin A platform foundation. *Proc Engrg for Calcareous Sediments*, Perth, Australia, Rotterdam: Balkema, 1:343–55.

Riggs, C. O. 1983. Reproducible SPT hammer impact force with an automatic free fall SPT hammer system. *Geotech Testing J*, ASTM 6(4):201–9.

Robertson, P. K. and R. G. Campanella. 1983. Interpretation of cone penetration tests. Part I: Sand. *Can Geotech J* 20(4):718–33.

Rollins, K. M., J. D. Lane and T. M. Gerber. 2005. Measured and computed lateral resistance of a pile group in sand. *J Geotech and Geoenvir Engrg*, ASCE 131(1):103–14.

Rollins, K. M., R. J. Olsen, J. J. Egbert, D. H. Jensen, K. G. Olsen and B. H. Garrett. 2006a. Pile spacing effect on lateral pile group behaviour: Load tests. *J Geotech and Geoenvir Engrg*, ASCE 132(10):1262–71.

Rollins, K. M., K. G. Olsen, D. H. Jensen, B. H. Garrett, R. J. Olsen and J. J. Egbert. 2006b. Pile spacing effect on lateral pile group behaviour: Analysis. *J Geotech and Geoenvir Engrg*, ASCE 132(10):1272–83.

Rollins, K. M., K. T. Peterson and T. J. Weaver. 1998. Lateral load behaviour of full-scale pile group in clay. *J Geotech and Geoenvir Engrg*, ASCE 124(6): 468–78.

Roscoe, G. H. 1983. The behaviour of flight auger bored piles in sand. *Proc Conference on Piling and Ground Treatment*, Institute of Civil Engineers, London, UK, 241–50.

Rowe, P. W. 1972. The relevance of soil fabric to site investigation practice. *Geotechnique* 22(2):195–300.

Rowe, R. K. and H. H. Armitage. 1984. The design of piles socketed into weak rock. Research Report GEOT-11-84, The University of Western Ontario.

Rowe, R. K. and H. H. Armitage. 1987. A design method for drilled piers in soft rock. *Can Geotech J* 24(1):126–42.

Ruesta, P. F. and F. C. Townsend. 1997. Evaluation of laterally loaded pile group at Roosevelt Bridge. *J Geotech and Geoenvir Engrg*, ASCE 123(12): 1153–61.

Sabatini, P. J., R. C. Bachus, P. W. Mayne, J. A. Schneider and T. E. Zettler. 2002. Geotechnical Engineering Circular No. 5: Evaluation of Soil and Rock Properties. Rep. No. FHWA-IF-02-034.

Sastry, V. V. R. N. and G. G. Meyerhof. 1994. Behaviour of flexible piles in layered sands under eccentric and inclined loads. *Can Geotech J* 31:513–20.

Sawaguchi, M. 1971. Approximate calculation of negative skin friction of a pile. *Soils and Foundations* 11(3):31–49.

Sayed, S. M. and R. M. Bakeer. 1992. Efficiency formula for pile groups. *J Geotech Engrg*, ASCE 118(2):278–99.

Schanz, T. and P. A. Vermeer. 1996. Angles of friction and dilatancy of sand. *Geotechnique* 46(1):145–51.

Schmertmann, J. H. 1975. Measurement of in-situ shear strength. *Proc ASCE Speciality Conf on In-Situ Measurement of Soil Properties*, Raleigh, 57–138.

Schneider, J. A., X. Xu and B. M. Lehane. 2008. Database assessment of CPT-based design methods for axial capacity of driven piles in siliceous sands. *J Geotech and Geoenvir Engrg,* ASCE 134(9):1227–44.

Scott, R. F. 1981. *Foundation analysis*. Englewood Cliffs, NJ: Prentice Hall.

Seed, H. B. and L. C. Reese. 1955. The action of soft clay along friction piles. *Transactions, ASCE* 122:731–54.

Seed, R. B. and P. DeAlba. 1986. Use of SPT and CPT tests for evaluating the liquefaction resistance of sand. S. P. Clemence, ed., *Proc In Situ '86*. Virginia Tech, Blacksburg, VA. (GSP 6), 281–302, New York: ASCE.

Seed, R. B., and L. F. Harder. 1990. SPT-based analysis of cyclic pore pressure generation and undrained residual strength. *Proc H. Bolton Seed Memorial Symposium* 2:351–76.

Seidel, J. P. and B. Collingwood. 2001. A new socket roughness factor for prediction of rock socket shaft resistance. *Can Geotech J* 38:138–53.

Seidel, J. P. and C. M. Haberfield. 1995. The axial capacity of pile sockets in rocks and hard soils. *Ground Engineering* 28(2):33–8.

Sen, G. S. and D. T. Zhen. 1984. *Soft soil foundation and underground engineering*. Beijing, China: Construction Industry Press House (in Chinese).

Serafim, J. L. and J. P. Pereira. 1983. Considerations of the geomechanics classification of Bieniawski. *Proc Int Symp on Engrg Geology and Underground Construction*, LNEC, Lisbon, Portugal, II.33-II.42.

Shen, R. F. 2008. Negative skin friction on single piles and pile groups. Department of Civil Engineering, Singapore, National University of Singapore. PhD thesis, 326.

Shi, P., S. Bao and H. Yin. 2006. The origin, application, and development of piles in China. Foundation analysis and design, geotechnical special publication No. 153, Shanghai, ASCE, 285–92.

Shibata, T., H. Sekiguchi and H. Yukitomo. 1982. Model tests and analysis of negative friction acting on piles. *Soils and Foundations* 22(2):29–39.

Shioi, Y. and J. Fukui. 1982. Application of N-value to design of foundations in Japan. *Proc 2nd European Symp on Penetration Testing*, Amsterdam: Balkema.

Skempton, A. W. 1951. Cast in-situ bored piles in London clay. *Geotechnique* 3(1):40–51.

Skempton, A. W. 1953. Discussion: Piles and pile foundations, settlement of pile foundations. *Proc 3rd Int Conf Soil Mech*, 3:172.

Skempton, A. W. 1957. Discussion of planning and design of new Hong Kong airport. *Proc Institution of Civil Engineers, 7*:305–7.

Skempton, A. W. 1959. Cast in-situ bored piles in London clay. *Geotechnique* 9(4):153–73.

Skempton, A. W. 1964. Long-term stability of clay slopes. *Geotechnique* 14(2):77–102.

Skempton, A. W. 1986. Standard penetration tests procedures and the effects in sands of overbudern presssure, relative density, particle size, aging, and overconsolidation. *Geotechnique* 36(3):425–47.

Sliwinski, Z. and W. G. K. Fleming. 1974. Practical consideration affecting the performance of diaphragm walls. *Proc Conf on Diaphragm Walls and Anchors*, ICE, London, 1–10.

Smethurst, J. A. and W. Powrie. 2007. Monitoring and analysis of the bending behaviour of discrete piles used to stabilise a railway embankment. *Geotechnique* 57(8):663–77.

Smith, T. D. 1987. Pile horizontal soil modulus values. *J Geotech Engrg*, ASCE 113(9):1040–4.

Soderberg, L. G. 1962. Consolidation theory applied to foundation pile time effects. *Geotechnique* 12:217–25.

Sogge, R. L. 1981. Laterally loaded pile design. *J Geotech Engrg*, ASCE 107(9):1179–99.

Springman, S. M. 1989. Lateral loading on piles due to simulated embankment contstruction. PhD thesis, University of Cambridge.

Stevens, J. B. and J. M. E. Audibert. 1979. Re-examination of p-y curve formulations. *Proc 11th Annu Offshore Technol Conf, OTC 3402*, Dallas, Texas, 397–403.

Stewart, D. P., R. J. Jewell and M. F. Randolph. 1994. Design of piled bridge abutment on soft clay for loading from lateral soil movements. *Geotechnique* 44(2):277–96.

Stroud, M. A. 1974. The standrad penetration tests in insensitive clays and soft rocks. *Proc European Symp on Penetration Testing*, Stockholm.

Sun, K. 1994. Laterally loaded piles in elastic media. *J Geotech Engrg*, ASCE 120(8):1324–44.

Sutherland, H. B. 1974. Granular materials. Review paper, session 1, conference on settlement of structures, Cambridge, UK.

Syngros, K. 2004. Seismic response of piles and pile-supported bridge piers evaluated through case histories. PhD thesis, civil engineering, City University of New York, 418.

Teng, W. 1962. *Foundation design*. Englewood Cliffs, NJ: Prentice Hall.

Terzaghi, K. 1943. *Theoretical soil mechanics*. New York: John Wiley & Sons.

Terzaghi, K. and R. B. Peck. 1967. *Soil mechanics in engineering practice*. New York: John Wiley & Sons.

Terzaghi, K., B. P. Ralph and G. Mesri. 1996. *Soil mechanics in engineering practice*. New York: John Wiley & Sons.

Timoshenko, S. P. and J. N. Goodier. 1970. *Theory of elasticity*. New York: McGraw-Hill.

Tomlinson, M. J. 1970. Some effects of pile driving on skin friction. Behaviour of piles. London, Institution of Civil Engineers, 107–14.

Tomlinson, M. J. 1994. *Pile design and construction practice*. London, UK: E & FN Spon.

Touma, F. T. and L. C. Reese. 1974. Behaviour of bored piles in sand. *J Geotech Engrg*, ASCE 100(7):749–61.

Trenter, N. A. and N. J. Burt. 1981. Steel pipe piles in silty clay soils at Belavan, Indonesia. *Proc 10th Int Conf on Soil Mech and Found Engrg*, Stockholm: A. A. Balkema, 3:873–80.

Trochanis, A. M., J. Bielak and P. Christiano. 1991. Three-dimensional nonlinear study of piles. *J Geotech Engrg*, ASCE 117(3):429–47.

Vallabhan, C. V. and F. Alikhanlou. 1982. Short rigid piers in clays. *J Geotech Engrg*, ASCE 108(10):1255–72.

Vallabhan, C. V. G. and Y. C. Das. 1988. Parametric study of beams on elastic foundations. *J of Engrg Mechanics*, ASCE 114(12):2072–82.

Vallabhan, C. V. G. and Y. C. Das. 1991. Modified Vlasov model for beams on elastic foundations. *J Geotech Engrg*, ASCE 117(6):956–66.
Vallabhan, C. V. G. and G. Mustafa. 1996. A new model for the analysis of settlement of drilled shaft. *Int J Numer and Anal Meth in Geomech* 20(2):143–52.
Valliappan, S., I. K. Lee and P. Boonlualohr. 1974. Settlement analysis of piles in layered soil. *Proc 7th Int Conf of Aust Road Research Board, part* 7, 144–53.
Van der Veen, C. 1953. The bearing capacity of a pile. *Proc 3rd Int Conf Soil Mech Found Engrg*, Zurich, 84–90.
Van den Elzen, L. W. A. 1979. Concrete screw piles, a vibrationless, non-displacement piling method. *Proc Conf Recent Developments in the Design and Construction of Piles*, Institute of Civil Engineers, London, UK, 67–71.
Van Weele, A. F. 1957. A method of separating the bearing capacity of a test pile into skin firction and point resistance. *Proc 4th Int Conf Soil Mech Found Engrg*, London, 76–80.
Vesić, A. S. 1961a. Bending of beams resting on isotropic elastic solid. *J of Engrg Mechanics*, ASCE 87(22):35–53.
Vesić, A. S. 1961b. Beams on elastic subgrade and Winklers hypothesis. *Proc 5th Int Conf on Soil Mech and Found Engrg*, Paris, 1:845–50.
Vesić, A. S. 1967. Ultimate load and settlement of deep foundations in sand. *Proc Symp on Bearing Capacity and Settlement of Foundations*, Duck University, Durham, NC., 53.
Vesić, A. S. 1967. A study of bearing capacity on deep foundation. Final report, Project B-189, Georgia Institute Technology, Atlanta, GA, 231–36.
Vesić, A. S. 1969. Experments with instrumented pile groups in sand. ASTM, STP 444:177–222.
Vesić, A. S. 1980. Predicted behaviour of piles and pile groups at the Houston site. *Proc Symposium Pile Group Prediction*, F.H.W.A., Washington, DC.
Viggiani, C. 1981. Ultimate lateral load on piles used to stabilise landslide. *Proc 10th Int Conf Soil Mech and Found Engrg*, Stockholm, Sweden, 3:555–60.
Vijayvergiya, V. N. and J. A. J. Focht. 1972. A new way to predict capacity of piles in clay. *Proc 4th Offshore Technology Conf*, OTC 1718, Houston.
Vu, T. T. 2006. Laterally loaded rock-socketed drilled shafts. M. Sc. thesis, Civil and architectural engineering, University of Wyoming.
Wakai, A., S. Gose and K. Ugai. 1999. 3-D elasto-plastic finite element analyses of pile foundations subjected to lateral loading. *Soils and Foundations* 39(1): 97–111.
Webb, D. L. and L. W. Moore. 1984. The performance of pressure grouted auger piles in Durban. *Symposium on Piling along the Natal Coast, S.A.*, Institute of Civil Engineers, Durban, South Africa.
Wesselink, B. D., J. D. Murff, M. F. Randolph, I. L. Nunez and A. M. Hyden. 1988. The performance and analysis of lateral load tests on 356 mm dia piles in reconstituted calcareous sand. *Proc Engrg for Calcareous Sediments*, Perth, Australia. Rotterdam: Balkema, 1:271–80.
Whitaker, T. 1957. Experiments with model piles in groups. *Geotechnique* 4:147–67.
Whitaker, T. and R. W. Cooke. 1966. *An investigation of the shaft and base resistance of large bored piles in London clay.* London: Institution of Civil Engineers.
White, D. J., M. J. Thompson, M. T. Suleiman and V. R. Schaefer. 2008. Behaviour of slender piles subject to free-field lateral soil movement. *J Geotech and Geoenvir Engrg*, ASCE 134(4):428–36.

Whitney, C. S. 1937. Design of reinforced concrete members under flexure or combined flexure and direct compression. *J of ACI* 33:483–98.

Williams, A. F. and P. J. N. Pells. 1981. Side resistance rock sockets in sandstone, mudstone and shale. *Can Geotech J* 18:502–13.

Won, J., S.-Y. Ahn, S. Jeong, J. Lee and S.-Y. Jang. 2006. Nonlinear three-dimensional analysis of pile group supported columns considering pile cap flexibility. *Computers and Geotechnics* 33(6–7):355–70.

Wong, J. and M. Singh. 1996. Some engineering properties of weathered Kenny Hill Formation in Kuala Lumpur. *Proc 12th Southeast Asian Geotech Conf*, Kuala Lumpur.

Wride, C. E., E. C. McRoberts and P. K. Robertson. 1999. Reconsideration of case histories for estimating undrained shear strength in sandy soils. *Can Geotech J* 36:907–33.

Wu, D., B. B. Broms and V. Choa. 1999. Design of laterally loaded piles in cohesive soils using p-y curves. *Soils and Foundations* 38(2):17–26.

Wyllie, D. C. 1999. *Foundation on rock*. London and New York: E & FN Spon.

Yamashita, K., M. Kakurai and T. Yamada. 1993. Settlement behaviour of a five-story building on a piled raft foundation. *Proc 2nd Int Geot Sem on Deep Foundations on Bored and Auger Piles*. Ghent, 351–6.

Yan, L. and P. M. Byrne. 1992. Lateral pile response to monotonic pile head loading. *Can Geotech J* 29:955–70.

Yang, K. 2006. Analysis of laterally loaded drilled shafts in rock. Civil Engineering. Akron, OH, University of Akron PhD thesis.

Yang, K., R. Liang and S. Liu. 2005. Analysis and test of rock-socketed drilled shafts under lateral loads. *Proc 40th U.S. Symp on Rock Mechanics: Rock Mechanics for Energy, Mineral and Infrastructure Development in the North Region*. American Rock Mechanics Association, ARMA/USRMS Paper no. 05-803.

Yang, Z. and B. Jeremic. 2002. Numerical analysis of pile behaviour under lateral load in layered elastic-plastic soils. *Int J Numer and Anal Meth in Geomech* 26(14):1385–406.

Yang, Z. and B. Jeremic. 2003. Numerical study of group effect for pile groups in sands. *Int J Numer and Anal Meth in Geomech* 27(15):1255–76.

Yang, Z. and B. Jeremic. 2005. Study of soil layering effects on lateral loading behaviour of piles. *J Geotech and Geoenvir Engrg*, ASCE 131(6):762–70.

Yin, J.-H. 2000. Comparative modeling study of reinforced beam on elastic foundation. *J Geotech and Geoenvir Engrg*, ASCE 126(3):265–71.

Yoshimine, M., P. K. Robertson, and C. E. Wride. 1999. Undrained shear strength of clean sands to trigger flow liquefaction. *Canadian Geotechnical Journal* 36:891–906.

Yun, G. and M. F. Bransby. 2007. The horizontal-moment capacity of embedded foundations in undrained soil. *Can Geotech J* 44(4):409–24.

Zeevaert, L. 1960. Reduction of point bearing capacity of piles because of negative friction. *Proc 1st Pan-American Conf Soil Mech Found Engrg*: 331–8.

Zhang, L. M. 2003. Behavior of laterally loaded large-section barrettes. *J Geotech and Geoenvir Engrg*, ASCE 129(7):639–48.

Zhang, L. M., M. C. McVay and P. Lai. 1999. Numerical analysis of laterally loaded 3×3 to 7×3 pile groups in sands. *J Geotech and Geoenvir Engrg*, ASCE 125(11):936–46.

Zhang, L. Y. 2010. Prediction of end-bearing capacity of rock-socketed shafts considering rock quality designation. (RQD) *Can Geotech J* 47(10):1071–84.

Zhang, L. Y., H. Ernst and H. H. Einstein. 2000. Nonlinear analysis of laterally loaded rock-socketed shafts. *J Geotech and Geoenvir Engrg*, ASCE 126(11), 955–68.

Zhang, L. Y, F. Silva-Tulla and R. Grismala. 2002. Ultimate resistance of laterally loaded piles in cohesionless soils. *Proc. Int. Deep Foundations Congress, 2002;* An int. perspective on theory, design, construction, and performance, Orlando, FL, ASCE, II, 1364–75.

Index

A

Absolute pile displacement, 424
Aging correction factor, 4
Alluvial/diluvial clay, cohesion of, 5
  N value and friction angle/cohesion, 6
Aluminum pipe piles test, 450
API, undrained shear strength, 21
Average load per pile
  current solutions vs. group pile tests, 390
  mudline deflections, 396
  predicted vs. measured, 336, 338, 395
Axial pile deformation, 109
Axial pile response, numerical solutions, 147

B

Back-estimation procedures effect, on pile response, 51
Back-figured shaft load transfer factors, 52
Base node displacement, 50
Bearing capacity factor, 23, 27, 29
Bending moment
  measurement profiles, 455
  predicted vs. measured, 337, 338, 397
Bentonite displacement, tremie pipe, 30
Berezantev's method, 28
Bessel functions, 83, 105, 119, 164, 165, 186, 207
  load transfer factor, 202
  of non-integer order, 105, 109
  shaft load transfer factor, 119
Block failure hypothesis, 41
Bombay High model tests, 332–334
Bored piles, 20
  β variation with depth, 26
  normalized uniaxial compressive strength, 26
  side support system, 30
  soil-shaft interface friction angle, 24
Boundary condition
  on ground level, 37
  at pile heads, 42
  pile response, 208
  radial consolidation of elastic medium, 163
Boundary element analysis (BEA), 49, 50, 125, 130, 180
  elastic–plastic continuum-based interface model, 125
  pile-head stiffness for different pile groups, 180
  of three dimensional solids, 130
Broms' solutions, 256

C

Calcareous sand, 30
  free-head piles, 318
  static and cyclic response of piles, 328–334
    North Rankin B, in situ pile test, 333
  static hardening *p-y* curves, 326
Calibration
  against in situ test piles, 478–482
  fixed-head piles, 482
  free-head piles, 481–482

M_{max}/T_{max}
calculation for field tests, 479–480
vs. predicted values, 482
Cantilever beam solution, 344
Capacity ratio
degree of slip, 137, 139
vs. slip length, 136
Cap capacity, 43
Cap effect, 391
Capped single pile, 398, 399
Cap-pile groups, model tests, 41
Case studies, 435–444
Dashwood house, 196
diameter and plastic hinge, 439, 441
driven piles in overconsolidated clay, 197
jacked piles in London clay, 196
Napoli Holiday Inn and office tower, 197–198
pile-bearing capacity by Seed and Reese, 171–172
predicted vs. measured pile responses, 445
retaining wall excavation, 441
rigid pile subjected to trapezoid movement, 442–444
San Francisco five-pile group, 198–199
sliding depth, 435–436
Stonebridge Park apartment building, 196–197
5-story building, 194–195
19-story building, 192–194
test reported, pile loaded to failure at intervals after driving, 172–175
Centrifuge tests, 447
Chinese hydraulic engineering code (SD 128-86), 5
Chin's method, 33
Clamped base pile (CP), 83, 216
Clamped pile, radial stress distributions, 226
Clay
drilled shafts, 20
normalized load, 315
piles in, 19, 309–310
32 piles, soil properties, 311–312
in situ full-scale tests, 393
soils
frictional angle, 8
group interaction factor K_g, 45
undrained shear strength, 37
Closed-form (CF)
analysis, 112
critical values, 116
solutions, 121, 341
Cohesionless soil, 98
Cohesive soil, 98
COM624P, salient features of, 324
Concentric cylinder model (CCM)
nonlinear load transfer analysis, 64
base stress-strain nonlinearity effect, 65
shaft stress-strain nonlinearity effect, 64–65
stress and displacement fields, 66
nonlinear load transfer model, 59–64
numerical solutions, calibration
base load transfer factor, 51–52
FLAC analysis, 49–50
load transfer approach, accuracy, 56–59
shaft load transfer factor, 52–55
shaft and base models, 47–49
time-dependent case, 65
base pile–soil interaction model, 77
GASPILE program, for vertically loaded piles, 78
nonlinear visco-elastic stress-strain model, 67–69
shaft displacement estimation, *see* Shaft displacement estimation
visco-elastic model, for reconsolidation, 78
Cone bearing resistance, 11
Cone penetration test (CPT), 1, 10
cone resistance, 11
profiles, 61
soil sensitivity, 10–11
SPT *N*-blow count, 11
SPT, relationship, 12
undrained shear strength, 11
Cone resistance
penetration, 12
shaft friction, 30
Constant rate of penetration (CRP) test, 156
Continuum-based finite difference analysis, 110
Creep
analysis, relaxation factors for, 160

displacement, 68, 73
modification factor, 74
parameters, 156
on excess pore pressure, 167
from loading tests, 76
rate of displacement, 76
pile B1, response, 157
piling approach, 147
settlement, 67
tests, curve fitting parameters, 75
Current predictions vs. measured data, 425
Current solutions, flow chart for, 381
Cyclic amplitude, 143
capacity and, 142–145
yield stress level, 144

D

Darcy's law, pressure gradient, 162
Davisson's approach, 33
Deep sliding tests, 471
Deflection profiles, schematic, 377
Deformation modulus E_m
empirical expressions, 18
moisture content vs. strength, 17
DMT tests, 354, 355, 357
Drilled shafts
limit-state method, 347
net unit end-bearing capacity, 29

E

Edils statistical formula, 75
Elastic analysis, 84
Elastic-pile (EI), 350
analysis of pile B7 (SN1), 356
behavior, 360
reaction, 359
soil interaction, 430
Elastic-plastic (EP)
interaction, 408, 415
interface, 135
solutions, 117, 287
vertically loaded single pile, 113
visco-elastic-plastic solutions, 174
Elastic solutions, 116
uniform soil movement, 405
vertically loaded single pile, 113
Elastic springs, characterization, 201
Elastic theory, 204
Empirical formulas, parameters, 170
End-bearing piles, 44
End-bearing resistance, 39
End-bearing stress, 23
Equivalent solid cylinder pile
cross-sectional area, 109
radial attenuation function, 83
Young's modulus, 49, 472
and moment of inertia, 206, 280

F

Fictitious tension, 82, 89, 211
Finite difference approach (FDA), 375
Finite element method (FEM) analyses, 49, 50, 177, 180, 214, 218, 220, 225, 227, 264, 293, 375
Fixed-head, clamped-base piles (FixHCP), 84, 88, 208, 209, 216
elastic analysis, 84
expressions, 209
normalized pile-head deformation, 218
Fixed-head, floating-base piles (FixHFP), 84, 89, 208, 210, 213, 216, 218
expressions, 210
pile-head and pile-base conditions, 84
Fixed-head (FixH) piles, 93, 216, 390, 391
elastic-plastic solutions, 334
expressions for response profiles, 378
Fixed-head solution, 379, 380
FLAC analyses, 49, 50, 55, 57, 112, 119, 121
pile-head stiffness, 52
SA, pile-head stiffness, 120
vs. VM approach, 122
Flexible piles, 138
Floating pile (FP), 83, 216, 222, 223
excluding base, 209
free-head, 203, 216
Fourier series, 82, 203
Frame movement, total applied force on frames, 458
Free-head, clamped-base piles (FreHCP) expressions, 84, 88, 208, 209, 216
Free-head, floating-base piles (FreHFP) expressions, 84, 208, 210

Free-head piles, 213, 285
  in calcareous sand, 318
  elastic-plastic solutions, 334
  modeling, 334
  response, pile–soil relative stiffness, 215
Free-standing pile groups, 40
FreH restraint, 397

G

GASLFP prediction, 329
  and usage, 292, 296–307
GASLGROUP predictions, 336, 382
  and usage, 334
GASLSPICS, 266
GASMove prediction, 434
GASPILE analyses, 65, 153, 158, 169
  CF and nonlinear, 129
  load–settlement relationships predicted, 174
GASPILE predictions, 64
GASPILE program, 64, 78, 114, 151, 154
  for vertically loaded piles, 78
Generalized analytical solutions for vertically loaded pile groups (GASGROUP), 184, 187, 195
Geology strength index (GSI), 15, 342
Gibson *k*, 250, 271; *see also* Nonlinear piles, response
Gibson p_u, 236, 237, 263, 264
  depth of rotation, 239
  normalized applied, 247
  post-tip yield and YRP states, 242
  responses of piles, 239
  solutions for $H(z)$ and $M(z)$, 240
  tip-yield state, 241
Gibson soil, 107, 130, 230, 290
Grain-size correction factor, 4
Groundline deflection, 284, 413
Group interaction factors, 472–473
Group piles, *see* Piled groups
Group settlement factors, 191
Group stiffness factors, 191

H

Hammer efficiency, 2
Head-stiffness, 51, 55, 57, 59, 61, 121, 192
  ratio, 51
Hooke's law, 162
Hybrid analysis, 105
Hyperbolic stress-strain law, 79

I

Intact rock
  frictional angle, 14
  uniaxial compressive strength, 15, 342
Interactions, elastic-plastic, 408, 415
Interface friction angle, soil-shaft, 24
Interface model, elastic-plastic continuum-based, 125

J

Jack-in resistance measurement, during pile installation, 457

K

Kingfish B, predicted vs. measured pile responses, 330
Kingfish sand, 325, 326

L

Lame's constant, 82, 204
Laplace transform, 162
Lateral load-groundline deflection, 297
Lateral load-maximum bending moment, 297
Laterally loaded free-head piles
  coupled load transfer analysis, 96
  current solutions, 308
    assumptions, justification, 322–325
    32 piles in clay, 308–313
    20 piles in sand, 313–322
  cyclic loading, piles response
    calcareous sand, static/cyclic response, 328–334
    predicted pile response, difference, 327–328
    p–$y(w)$ curves, 325–327
  free-head groups, response, 334
    GASLGROUP program, 335–339
  GASLFP use, 296–307
  pile–soil system, solutions, 279
    critical pile response, 284–285

elastic-plastic response profiles, 280–284
extreme cases, studies, 286–292
LFP, back-estimation of, 292–296
slip depth vs. nonlinear response, 296
subgrade modulus, 90
Laterally loaded piles, 284, 285
features of, 95
governing equations, expressions, 207
pile groups
facts, 375
group piles, 381–386
single pile, 376–381
Laterally loaded piles, clay
loading tests of, 12
Laterally loaded rigid piles, 229
comparison with solutions, 266–268
elastic-plastic solutions, 229, 230
features of, 230–232
maximum bending moment, 245–250
nonlinear response, calculation, 250–253
post-tip yield state, solutions for, 238–245
pre-tip yield state, solutions for, 232–236
yield at rotation point (YRP), 245
load-moment loci
capacity H_o, 255–257
at tip-yield and YRP state, 253–254
moment loading locus
elastic, tip-yield, and YRP loci for constant p_u, 261–264
impact of size/pile-tip resistance, 264–265
YRP state, 257–261
Laterally loaded shafts, 343
Lateral piles
capacity, 94
coupled elastic model, 81; *see also* Vlasov's foundation model
load transfer factor, 83–87
nonaxisymmetric displacement, 82–83
rigid piles in sand, 90–93
short and long, 83–87
stress field, 82–83
subgrade modulus, 87–90
elastic-plastic model
limiting force per unit length (LFP), 98–104
rigid pile, features, 94–98
elastic solutions, 201
analyzing response, 201
boundary conditions, 208–209
critical pile length, 212–213
fixed-head piles, 216
free-head piles, 213–216
impact of pile-head and base conditions, 216–218
loaded piles underpinned by k and N_p, 205–208
load transfer factor γ_b, 209–211
maximum bending moment/critical depth, 213
moment-induced pile response, 218
nonaxisymmetric displacement/stress field, 202–205
pile-head rotation, 218
pile response, 202
short/long piles, 213
subgrade modulus/design charts, 218–220
subgrade reaction modulus/fictitious tension, 211–212
validation, 212
nonlinear response, 280
Lateral pile–soil system
potential energy, 205
schematic presentation, 82
Lateral pile tests
Young's modulus vs. undrained shear strength, 15
Layer thickness ratio, 53
Limestone, shaft analysis, 368
Limiting force
profile, 231
schematic presentation, 377
Limiting force per unit length (LFP), 94, 98–104
back-estimation of, 292–296
cohesionless soil, 101
cohesive soil, 101
elastic pile analysis, 357, 358
layered soil profile, 101
linear, 94
Matlock, 307

normalized slip depth, 286
pile responses, 306
piles in clay, 314
piles in rock, 342
piles in sand, 321
predicted vs. measured, 361
sensitivity of current solutions, 302
typical, 100
Linear elastic (LE), 169
Linear visco-elastic (LVE), 170
media, 72
model, 164
Load-deformation profiles, 116
Load displacement, 64
predicted vs. measured, 400
Load distribution, predictions, 129
Loaded pile, vertically
schematic analysis, 48
Load failure, 40
Loading block
shear force vs. lateral load, 458
triangular/uniform, 450
Loading eccentricity, 88
Loading test, 20, 32, 154
analysis, 126
capacity from, 32–33
Load settlement, 122
homogeneous case, 125
nonhomogeneous case, 126–127
Load transfer
analysis, 122
approach, 47, 93
curves, gradient, 47
factor, 59, 84, 169
vs. E_p/G^*, 85
vs. *H*/*L* ratio, 58, 59
vs. Poisson's ratio relationship, 56, 57
vs. relative stiffness, 60, 61
vs. slenderness ratio, 54, 55
functions, 65
stress and displacement fields, 66
models, 82, 106–108, 202, 230, 410
nonhomogeneity expressions, 106
pile tests, 69
parameters
base factor and realistic value, 62
soil-layer thickness effect, 62
pile–soil interaction, 279
Longitudinal stress gradients, 79

M

Manor test, elastic parameter effect, 306
Marine clays, 6
Maximum bending moment, 284
Maximum end-bearing capacity, 35
Maximum pore pressure, 158
Mechant's model, 159
λ Method
α deduced, variation, 22
Meyerhof correlations, 29, 30
Mobilization, 32, 101
shaft capacity, 134
shaft friction, 93
shaft shear stress, 108
Mochtar's statistical formula, 75
Modeling pile groups, shear apparatus, 449
Model pile data
current predictions, 269
parameters for, 270
response of, 271
Model piles
load tests, 20
in sand, 386, 396
test AS50-0, 428
Modulus of subgrade reaction, 85
Mudline displacement, 267

N

Near-pile drained soil modulus, 13
Negative skin friction (NSF)
group pile, schematic of, 38
single pile, schematic of, 36
Nonaxisymmetric displacement, 202, 203
Nonhomogeneity influence, 182
Nonhomogeneous soil
pile, closed-form solutions, 184
pile group, 185
Nonlinear (NL) analysis
elastic analysis, 182
nondimensional shear stress vs. displacement relationship, 63
stress level, 62, 64
Nonlinear elastic (NLE)
load transfer curves, 72
predictions, 156
Nonlinear piles
procedure for analysis, 351

response, 251
soil interaction, 229
Nonlinear shafts
of Islamorada test, 364
of Pomeroy-Mason test, 370–371
in San Francisco test, 368–370
Nonlinear visco-elastic (NLVE), 156, 169
load transfer (*t-z*) model, 70, 71
Normalized deflection, 423
Normalized load, deflection
measured vs. predicted, 383, 384, 385
predicted vs. measured, 365, 366
Normalized load, maximum
measured vs. predicted, 386
Normalized local stress displacement relationships, 72
Normalized modulus, subgrade reaction, 86
Normalized on-pile force profiles, predicted vs. measured, 235
Normalized pile-head rotation angle, 223
pile-head constraints, 224
Normalized response
based on p_u and fictitious force, 423
Normal sliding mode, 408, 414

O

Offshore piles, 22
On-pile force profiles, 231, 249, 258
bending moment, 379
Overburden pressure, effect of, 4
Overconsolidation correction factor, 4

P

Passive pile, design, 405
case study, 435–444
diameter and plastic hinge, 439, 441
predicted vs. measured pile responses, 445
retaining wall excavation, 441
rigid pile subjected to trapezoid movement, 442–444
sliding depth, 435–436
charts, 433–435
E-E, EP-EP solutions, 430–433
elastic-plastic (EP) solutions
normal sliding mode, 414–415
plastic–elastic-plastic (P-EP) solution, 415–417
p_u-based solutions, 421–430
stable layer, 417–421
equivalent load model, 406
as fictitious active pile, 407
flexible piles, 406–407
limiting force and on-pile force profiles, 411
modes of interaction, 408–409
rigid piles, 407
deformation characteristics, 412
soil interaction, mechanism
deformation features, 412–413
load transfer model, 409–410
on-pile force p profile, development, 410–412
solutions for instrumented piles, 446
Passive pile, physical modeling
apparatus/test procedures, 448
loading distance, impact of, 455–456
pile response, determining, 455
shear tests, salient features of, 448–450
test procedure, 450–455
distance impact of piles, from loading side, 456
equivalent elastic solutions, 470–472
group interaction factors F_m, F_k, and p_m, 472–473
in situ test piles, calibration, 478–482
lateral forces on shearing frames, 457
M_{max} calculations, 475–478
$M_{max}/T_{max}/M_{oi}$, 468–470; *see also* T block
M_{max}/T_{max}, prediction
impact of higher sliding depth, 465
soil movement profile vs. bending moments, 474–475
model tests, 447
features, 483
shear force variation, 456
single piles and two piles in rows
triangular movement, 453
uniform movement, 454

soil movement profile vs. bending moments, 473–474
test results
driving resistance/lateral force on frames, 456–460
M_{max} raises, 465–467
M_{max}, y_o, ω vs. w_i, 460–465
T_{max} from y_o, determination, 472
typical pile tests, under triangular movements, 452
Peck's expression, 6
Pier rotation angle, 267
Pile analysis
closed-form solutions, in nonhomogeneous soil, 184
geometry/elastic property, 74
hyperbolic approach of, 148
pile-arc profile, 424
properties of, 329
rigid and short piles, 87
Pile axial capacity, 37
Pile-cap, 38
Pile center-to-center spacing, 224
Pile deflection, 99, 211, 221, 284, 328, 331, 334, 377, 407, 409, 418, 419, 460
development, and rotation, 461
Pile deformation, 109, 131, 231
limiting force, schematic profiles of, 97
Piled groups, 105, 177, 193, 194, 336, 381, 382, 386, 390, 396, 470
boundary element approach
case studies, *see* Case studies
Generalized analytical solutions for vertically loaded pile groups (GASGROUP), 184
interaction factor, response, 186
load transfer response, 184–185
methods of analysis, 187–192
pile group analysis, 187
pile-head stiffness response, 185–186
clay, in situ tests, 395
elastic compression, 180
elastic continuum-based methods, 177
empirical methods, 177, 178
hybrid load transfer approach, 177
imaginary footing method, 178, 179
interaction factor, response, 220–227
lateral response, 335
layered soil, 181
layout, 394
load transfer approaches, 177
load transmitting area, 179
measured vs. predicted, 392
normalized spacing, 42
numerical methods
boundary element/integral approach, 180–181
infinite layer approach, 181–182
interaction factors/superposition principle, 183–184
nonhomogeneity influence, 182–183
nonlinear elastic analysis, 182
p-multiplier p_m, 104
procedures, 177
response, 335
in sand, 392
settlement, 177, 183
shallow footing analogy/imaginary footing method, 177–180
single pile settlement, 178
stiffness, 188, 189, 190
vertical stress in subsoil, 179
Pile displacement, 24, 83, 94, 208
features, 230
Pile failure, 41, 361
schematic modes, 39
Pile-head constraints, impact
on single pile bending moment, 222
on single pile deflection, 221
Pile-head constraints, normalized pile-head rotation angle, 224
Pile-head deformation, 213
Pile-head displacement, 137, 216
due to moment loading, 219
Pile-head force, 32
Pile head load, 33, 113, 125, 128, 382
loading time t_c/T vs. settlement influence factor, 153
normalized response, 252
settlement, 65
settlement curves, 152
vs. settlement relationship, 33
Pile-head response, slip development effect, 126
Pile-head rotation, 218
Pile-head stiffness, 48, 53, 59, 111, 112, 122, 126, 127, 183, 187
settlement influence factor, 127

slip development, effect, 126
vs. *H*/*L* relationship, 124
vs. slenderness ratio relationship, 121, 123
Pile installation
effects, 20
jack-in resistance measurement, 457
Pile loading tests, 30, 32; *see also* Loading test
Pile location, 451, 455
Pile, modeling radial consolidation, 79
Pile properties, in current analysis, 389
Pile response
concrete cracking effect, 352
effect of α_g, 115
pile body displacement, 208
component, 203
Pile rotation, profiles, 208
Piles, cyclic response, 328
Piles, laterally loaded rigid, *see* Laterally loaded rigid piles
Pile shaft, 19, 20, 22, 35
angle of friction, 23
displacement, 183
earth pressure, 23, 24
soil stress-strain relationship, 59
Pile slenderness ratio, 211, 409
Pile–soil adhesion factor, 169
Pile–soil interaction, 90, 108, 409; *see also* Concentric cylinder model (CCM); Torque-rotation transfer model
elastic-plastic solution, 113–118
elastic solution, 109–110
elastic theory, verification of, 110–113
factors, 181, 184
load transfer, 279
Pile soil interface, 61
creep displacement, 68
nonlinear stress level, 61, 70
pore pressure on, 168
shear stress, 70
Pile–soil relative displacement, 61
Pile–soil relative stiffness, 83, 88, 151, 183, 217, 298, 407
factor, 106
ratio, 131
Pile–soil stiffness, 138, 213
Pile–soil system, 107, 279
elastic-plastic solutions, 280
critical pile response, 284–285
elastic-plastic response profiles, 280–284
equilibrium, 206
extreme cases, studies, 286–292
LFP, back-estimation of, 292–296
schematic presentation, 133
Piles penetration, into stiffer clays, 20
Pile stiffness, 31, 133
Pile surface
mobilized shaft shear stress, 108
shear strength, 24
Pile tip
shaft resistance, 24
yields, 244
Pile $T_{max,}$ determination, 459
Plane-strain triaxial tests, 10
Plastic–elastic-plastic (P-EP) solution, 416, 417, 432
Plastic flow mode, 408
Plasticity index, 23
data friction angle, variations, 9
plasticity vs. undrained shear strength ratio, 5
Plasticity sand, Peck's suggestion for, 8
Plastic pile–soil interaction, 414
Plastic response, critical stiffness, 142
Poisson's ratio, 49, 52, 53, 64, 74, 82, 121, 151, 162, 171
impact of, 212
on piles, 49
pile response, 212
soil response, 205
two-parameter model, 212
Pomeroy-Mason test, 370–371
Post-peak deformation, 341
Post-tip yield state, 95, 237
Pre-tip yield, 94, 257

R

Radial attenuation function, 83
Radial deformation, 226
Radial displacement, potential energy, 202
Radial nonhomogeneity, 175
Radial stress gradients, 79
Raft, *see* Pile-cap
Ramp loading, modification factor, 73
Randolph's equation, 185
RATZ predictions, 33, 64
Real tensile force, 211

Rigid disc
two-dimensional, 207
using intact model, 205
Rigid/flexible shafts
normalized bending moment, load, deflection, 345
Rigid movement, pile–soil, 471
Rigid piles, 88, 134, 140, 143, 227, 422
footing base area, 265
H-*l*–based solutions, 426
modes, 408
movement, 413
at pre-tip and tip-yield states, 233–234
in sand, 255
schematic analysis, 230
surface, 91
reduction of, 92
schematic analysis, 92
Rock
effective friction angle, 14
frictional angle, 14
mass, deformation modulus, 35
moduli deduced from shafts, 13
properties for shafts, 363, 364
shaft friction estimation, 34
shear modulus, 15
single pile capacity, 34–35
strength/stiffness of, 13–18
uniaxial compressive strength, 34
Rock masses
classification indices, 367
index properties of, 367
mechanical properties of, 367
properties, 362
Rock mass rating (RMR), 15
geology strength index, 15
Rock quality designation (RQD), 14, 15, 342
estimation of E_m, 18
Rock-socketed drilled shafts, 361
Guo's LFP, 364
limestone, shaft analysis, 368
mass properties, 362
nonlinear piles, comments, 371–372
normalized load, deflection, 365, 366
Pomeroy-Mason test, shaft analysis, 371
sandstone, shaft analysis, 370
shafts, properties for, 363
Rock-socketed shafts, 34, 343
Rock-socket piles, 341
laterally loaded shafts solutions, 343
loading eccentricity effect on shaft response, 343–346
nonlinear structural behavior, of shafts
analyzing procedure, 350
cracking moment M_{cr}, 346–347
modeling structure nonlinearity, 350–352
M_{ult} and I_{cr}, 347–350
Reese's approach, 342
structural nonlinearity, 341
Young's modulus, 342
Rotational soil movement, 447
Rough-surface pile, soil frictional angle, 23
RSB method, 354

S

SA, pile-head stiffness, 120
Sand/clay, nonlinear piles
Hong Kong tests, 356–359
Japan tests, 359–361
Taiwan tests, 352–356
Sand-clay-sand profile, 296
Sands
bearing stress for piles, 27
group interaction factor K_g, 45
modulus *k* rigid piles, 90–93
pile in dense sand, 268
pile penetration, 28
piles, stress approach
base resistance q_b, 27–29
β methods, 23–27
properties, parameters, 354
RSB method, 355
single piles, parameters, 398
20 piles and, 316–317
parameters for, 319–320
Sand-silt layer, 303
Sandstone
uniaxial compression strengths of, 15
Settlement influence factor, 130
Shaft displacement estimation, 48, 69–77
nonlinear creep displacement, 72–74

shaft model vs. mode loading tests, 74–77
visco-elastic shaft model, 69–72
Shaft friction, 118
cone resistance, 30
dilatancy effect on adhesion, 25
friction fatigue, 31
with undrained shear strength, 34
variation, 31
Shaft load
transfer factor, 48, 119
transfer parameter, 48, 57, 119
Shaft resistance, 23, 24, 168
Shaft shear stress, 113
Shaft stress, 24, 25
Shear force vs. lateral load, 458
Shear modulus, 70, 212
determination, 459
Shear stiffness, of pile, 211
Shear strain rate, 68
Shear stress, 70
distribution, 51
Shear tests, salient features, 448
Short piles, 213
estimating parameters, 85
Silt/clay, shaft component, 43
Silty sand, in situ piles, 391
Simple linear analysis (SL), 62
nondimensional shear stress vs. displacement relationship, 63
stress level, 64
Simple solutions, M_{max} and T_{max}
measurement, restraining moment, 468–470
theoretical relationship, 467–468
Single pile capacity
comments, 30–32
empirical methods, 29–30
group capacity, at high load levels, 38–39; *see also* Piled groups
block failure, 41–44
capped pile groups vs. free-standing pile groups, 42–44
in clay, 39
comments, 44
free-standing groups, 41–42
group interaction, 40–41
spacing, 39–40
weak underlying layer, 44–45
from loading tests, 32–33
negative skin friction, 35–38
in rock, 34–35
stress approach in clay, 19
α methods, 19–21
β methods, 23
λ methods, 22
stress approach in sand
base resistance q_b, 27–29
β methods, 23–27
vertically loaded, *see* Vertically loaded piles
Single pile, measured vs. predicted, 392
Single-pile predictions, 394
Single pile response, *see* Free-head pile, response
Skin friction, 34
negative, 35–38
Slenderness ratio
fictitious tension, 89
Sliding depth, 435
Sliding, elastic-plastic (EP) solutions, 430
Sliding parameters, piles/soil, 420
Slip depth, 331
load, 288
Slip length
load ratio development, 140
vs. capacity ratio, 136
Slope stabilizing piles, 407, 433
Soil
creep/visco-elastic response, 147
friction angle, 6, 8, 25
relative density impact, 8, 9
internal friction angle, 11
nonhomogeneity, 105
piles, in homogenous soil, 289
pile–soil relative slip and strain-softening properties, 142
plasticity, liquid limit (LL), 5
Poisson's ratio, 47, 48, 52, 107, 161, 205
properties of, 329
reaction, T block, 473
relative density, penetration resistance, 4
resistance, 102, 213, 381
rigidity, 28
sensitivity, 10
shear modulus, 47, 106, 204
shear stress, 68
strain-softening soil, 105
strain wedge mode, 277
strength, 175
visco-elastic response, 67
Soil movements

current prediction vs. measured data, 429
normalized pile response in stable layer owing to, 434
profile vs. bending moments, 474–475
limiting profiles, 473–474
M_{max}/T_{max}, prediction, 474–475
Soil-shaft interface friction angle, 24
Soil stiffness effect, 11–13, 32
pile-head load vs. settlement relationship, 33
SPT N correction factor, for overburden pressure, 3
SPT tests
key factors on efficiency, 1
vs. moisture content, 16
sand/silty sand, 3
soil shear modulus, 11
soil strength, 8
Standard penetration tests (SPT), 1, 7
blow counts using q_c, 11
blow counts vs. internal friction angle, 7
friction angle vs. N, D_r, and I_p value, 6–8
parameters affecting strength, 8–10
relative density, 3–4
shaft friction, 30
variation, 31
ultimate base pressure, 32
undrained soil strength vs. N value, 5–6
values modification, 1–3
Steel piles, 25
adhesion, 37
Steel pipe piles, 435
shaft resistance, 22
Stiff clay, 20
London clay, 23
pipe pile tested, 304, 306
Strain path method, 158
Strain-softening factor effect, 134, 144
offshore pile, 145
Strain-softening soil capacity
elastic solution, 132–134
load and settlement response, 135–142
n_{max} vs. π_v, 141
plastic solution, 134
Stress β approach, 23
Stress block, 348
Stress level, s_u/N dependence, 6
Stress of q', on weaker layer, 44
Stress-strain behavior, time-dependent, 67

T

T block, 465–467
M_{max} calculations, 475
capped pile, 476–477
rotational movement, 477–478
translational loading with variable sliding depths, 476
translational movement, 475–476
soil reaction, 473
Tensile fibers, 347
Three-dimensional finite element analysis (FEA^{3D}), 267, 293, 294, 389, 390
current predictions, comparison, 295
hyperbolic stress-strain model, 266
trailing row, 391
Tip-yield states, 94, 257, 258, 260
critical response, 245
piles response, 243
Torque-rotation transfer model, 78
nonhomogeneous soil profile, 79
nonlinear stress-strain response, 79–80
shaft torque-rotation response, 80–81
Torsional loading, on pile, 78
Torsional vs. axial loading, 80
Translational-rotational centrifuge tests, 447
Translational-rotational movement, 447
Triangular movement
M_{max} and T_{max} evolution, 464, 465
single piles, evolution, 463
single piles, maximum response profiles of, 462
T_{50}/T_{90}, variations of, 168
Two typical pile-head constraints investigated
and H2 treatment, 387

U

Uncoupled model, *see* Winkler model
Undrained shear strength, 11, 21
API, 21
normalized strength, 21
Undrained shear strength vs. N_{60}, 7
Uniform block, soil reaction, 473
Uniform movement
M_{max} and T_{max} evolution, 464
single piles, evolution, 463
single piles, maximum response profiles, 462
soil reaction deduction, from uniform tests, 471

V

Van der Veen's approach, 33
Variational method (VM) analysis, 120
Varying sliding depth tests, 451
Vertically loaded piles, 105
single piles
capacity/cyclic amplitude, 142–145
comparison with solutions, 119–122
load settlement, *see* Load settlement
load transfer models, *see* Load transfer, models
paramatric study, 119–122
pile-head stiffness and settlement ratio, 119
pile-head stiffness vs. slenderness ratio relationship, 121, 123
settlement influence factor, 127–131
soil profile effect, 122
strain-softening soil capacity, 132–142
time-dependent load transfer model applications, 152–156
pile response, loading rate effect, 152
time-dependent response
visco-elastic consolidation, *see* Visco-elastic consolidation
visco-elastic load transfer behavior, *see* Visco-elastic load transfer behavior
Vertical soil nonhomogeneity, 182
Vesić's method, 30
Visco-elastic analysis, 171, 173
Visco-elastic behavior, 156
Visco-elastic consolidation, 158
behavior, 169–171
case study, 171–175
current predictions, comments, 175
logarithmic variation of u_o, 165–168
reconsolidations, governing equation for, 159
boundary conditions, 163
diffusion equation, 163
soil skeleton, volumetric stress-strain relation of, 160–162
solution, 163–165
visco-elastic stress-strain model, 159–160
volume strain rate/pore water flow, 162–163
shaft capacity, 168–169
Visco-elastic effect, 166
Visco-elastic load transfer behavior, 147
closed-form solutions, 149–151
time-dependent load transfer model, 148–149
validation, 151–152
Visco-elastic medium, 162, 163, 165
Visco-elastic shaft, 148
model, 69
Visco-elastic soil, shaft displacement, 148
Visco-elastic solutions, 159
Vlasov's foundation model, 81, 202
VM (variational method) analysis, 120
Voigt element, 160

W

Windows Excel, 166, 266, 292, 350, 376, 409
Winkler model, 87, 227, 280, 287, 432

Y

Yield at rotation point (YRP), 96, 231, 242, 259
critical response for tip-yield, 249
Gibson k, 272
M_m response, 248

normalized load, 261, 262
normalized profiles, 254
pile-head displaces, 245
plasticity, tip-yield, 263
predicted vs. measured, 235
rigid pile, 264
Young's modulus, 64, 108, 110, 125, 154, 186, 293, 313, 332, 346, 389
equivalent solid cylinder pile, 49
equivalent solid pile, 407
of rock, 342
vs. undrained shear strength, 15

Z

Zeevaert's method, 37